Sustainable Energy Supply in Asia

SUNSTAINABLE ENERGY SUPPLY IN ASIA

(Proceedings of the International Conference, ASIA ENERGY VISION 2020, organised by The Indian Member Committee – World Energy Council, under The Institution of Engineers (India), during November 15-17, 1996 at New Delhi)

VOLUME ONE

Edited by
PRADEEP CHATURVEDI

World Energy Council
CONSEIL MONDIAL DE L'ENERGIE

For
Indian Member Committee-World Energy Council
The Institution of Engineers (India)

Published by
CONCEPT PUBLISHING COMPANY, NEW DELHI 110059

First-Published 1997

ISBN 81–7022 – 631–7 (Vol. I)
ISBN 81–7022 – 633–3 (Set)

Printed and Published by
Ashok Kumar Mittal
Concept Publishing Company
A/15-16, Commercial Block, Mohan Garden,
New Delhi - 110059 (INDIA)

Phone : 5554042, 5504042
Fax : 091-11-5598898

FOREWORD

The Indian Member Committee of the World Energy Council, IMC-WEC organised an International Conference on Energy - 'Asia Energy Vision 2020' with the theme Sustainable Energy Supply in Asia in New Delhi during November 15-17, 1996. Experts from the Asia region as well as other parts of the world took part in the conference.

This conference provided an opportunity for interaction amongst policy framers, industrialists and researchers to meet on a platform and discuss and synthesise on issues crucial for energy sector development in Asia.

The Papers were of a very high quality. All participants thought that the same should be published and made available to wider readership. The Institution of Engineers (India) is fulfilling its obligation by publishing the proceedings in two volumes.

I would like to compliment M/s Concept Publishing Company for publication of the Two Volumes.

I am sure the proceedings will be found valuable.

N.C. VAISH

Chairman
Indian Member Committee of the World Energy Council
and
President
The Institution of Engineers (India)

November 23, 1996

INTRODUCTION

The Indian Member Committee of the World Energy Council organised 'Asia Energy Vision 2020', an International Conferene on Energy, with the focus on Sustainable Energy Supply in the South Asia Region. This was a major initiative preparatory to the 17th Congress of the World Energy Council planned for Houston, USA, in September, 1998. The significance of the Conference can be realised from the fact that it was inaugurated by the Hon'ble Prime Minister of India, Mr. H.D. Deve Gowda. Its National Advisory Committee was chaired by the Secretary to the Government of India, Ministry of Power, and had representatives from all ministries of the government, and major energy industries from the private and public sectors had supported the Conference. The Ministry of Power played the key role as the sponsoring agency.

The interest aroused in the delegates and the participants from India and other parts of the world has helped in getting papers on policy, international funding, regional cooperation and technical matters. The experts hailed this Conference as having technical inputs of a very high order.

The Technical Committee conducted the technical sessions in the following four divisions:

Division 1: Sustainable Energy Supply Options
Division 2: Energy Sector Developments in South Asia
Division 3: Energy Conservation and Energy Efficiency
Division 4: Funding Mechanism and Sources

The Conference also had three important Discussion Panels as follows:

Panel I : Role of Nuclear Power in attaining Sustainable Energy Supply in Asia.
Panel II : Policy Options for Meeting Long Term Energy Needs in Asia.
Panel III : Strategies for Energy Sector Financing in Asia.

In keeping with the tradition laid down by the World Energy Council, at the 16th Congress at Tokyo, India Energy Seminar was organised to focus on energy requirements in India and the possibility of its fulfillment through international funding and technology transfer. In addition, Overview Presentations on the global, regional and national levels were also organised.

All these papers have been accordingly segregated in different technical divisions and as per their policy relevance. Some of the policy papers were presented by the distinguished experts at the Conference and not submitted in advance so that they are able to update the information to the latest stage. Whereas the First Volume consisting of all the papers received in advance was published earlier and circulated to the participants at the Conference to support the discussions, papers received at the time of the Conference have been published as Volume II of the Proceedings. Volume II of the Proceedings also consists the New Delhi Declaration consisting of Conclusions and Recommendations. These two volumes are complimentary to one another, giving deep insight into the present status, internationally, regionally and nationally.

Wishing everyone a useful information input,

PRADEEP CHATURVEDI
Chairman
Technical Committee
Asia Energy Vision 2020

New Delhi
November 23, 1996

CONTENTS

DIVISION - I
SUSTAINABLE ENERGY SUPPLY OPTIONS

DIVISION II

ENERGY SECTOR DEVELOPMENT IN SOUTH ASIA

DIVISION III
ENERGY CONSERVATION AND ENERGY EFFICIENCY

DIVISION IV
FUNDING MECHANISM AND SOURCES

DIVISION V
INDIA ENERGY SEMINAR

DIVISION - I

SUSTAINABLE ENERGY SUPPLY OPTIONS

ENERGY SCENARIO PRESENT STATUS AND FUTURE TRENDS

1

Kotur S. Narasimhan
Director, Central Fuel Research Institute, Dhanbad 828 108, INDIA

ABSTRACT

Energy is an essential need for human existence besides national development. For many years in the past, per capita energy consumption was in use as an index to measure the health of the economic development of a nation. Recently, the cost of energy as a share of GNP is considered a better norm. Therefore, this is more widely used in measuring the energy efficiency of a nation. This will help in adopting efficient practices to reduce specific energy consumption while achieving the planned development. Recently, because of the environment consciousness, specific emissions of pollutants are considered. Pollutants include particularly carbon dioxide besides oxides of sulphur and nitrogen. Thus, while projecting the demand for energy and planning for a supply pattern, two factors are to be kept in view. One is the cost of energy as a share of GNP and the other is the specific rate of emissions, particularly that of carbon dioxide. This eventually, may become a global mandate with many nations bound to subscribe to the Rio treaty in a bid to preserve our biosphere.

1. Consumption Pattern

Industrially well advanced countries depend almost entirely on the commercial energy sources. For them, planning for an energy balance and to predict its impact on environment is easy. On the other hand, developing countries like India, largely depend on noncommercial energy sources. They are essentially firewood, besides agricultural and livestock waste. When these sources play a significant role, similar assessment cannot be equally accurate for want of adequate data. Forest produce, mainly the firewood, is the main stay of energy for half of the world's population. This accounts for 15 per cent of the total global energy consumption. The figure is 80 per cent for sub-Sahara Africa. Similarly, 80 per cent of the large population, of rural India, depend for their existence on sources like firewood, animal and agricultural wastes. These energy sources apparently do not cost the consumers anything. Adequate statistical information is available on the production of various primary commercial energy sources and the sectoral consumption of the derived forms of secondary energy in India (TEDDY)[1]. However, information on the quantum of noncommercial energy sources consumed in real terms is extremely scanty. In reality, only two recorded references are available on this subject. In the Eighth Five-Year Plan document[2] a mention has been made that the total energy supplies, both commercial and noncommercial forms together, increased from 82.7 million tonnes of oil equivalent (mtoe) in 1950-51 to about 291 mtoe in 1990-91. In this, the share of the noncommercial energy sources declined from 74 per cent in 1950-51 to 41 per cent in 1990-91. Though not indentical, similar information given in TEDDY says that the noncommercial energy consumption has increased from about 143 mtoe in 1972-73 to 161 mtoe in 1982-83. Both these statements corroborate each other on the trend of decrease in the share of noncommercial energy sources. The available dual data however, do not corroborate each other to accurately quantity the rate at which the share of noncommercial energy has declined. While the share of noncommercial sources has declined over a period the absolute quantity of their consumption has steadily increased in absolute terms. This particularly is a matter of great concern as it implies increased deforestation. Further, if such are the facts, the computed total per capita energy consumption over the years in the country is on the decline. Such a situation is highly unlikely, since there are other indicators for the standard of living having perceptibly gone up in the country over these years.

On the contrary, World Bank projected information[3] on the quantum of fuel wood removed in the immediate past is available. It is assumed that all this is used as an energy source. Further, most of the agricultural residues are also consumed as fuel in the rural area. On the assumption that all such available sources are consumed together with the commercial energy sources, the computed per capita energy consumption gives a totally different picture with respect to the level and the trend. The actual position should one imagine is closer to this scenario as this explains all the observations made earlier.

2. Demand Projection

Based on this simplistic model of the computed trend in the per capita energy consumption, an attempt is made to project the energy needs for the future till the year A.D. 2020. The objective of this exercise is to project the minimum energy needs in the future and assess how such a need can be satisfied keeping in mind both the energy costs and the environmental aspects. Such a consideration therefore leads to two extreme possibilities. In the first scenario, the consumption of the noncommercial energy sources also grows at the same rate as in the past and the balance is met from the commercial energy sources. As a second extreme, the growth of the commercial energy is forced towards achieving a zero fuel wood consumption by A.D. 2020.

Energy needs are worked out for these two extremes, at the assumed current constant per capita energy consumption accounting only for the increase in the population. As such there is a wide variation in the commercial energy requirement depending on how the demands are met. Thus, to maintain the same standard of life from now on, measured by the level of the per capita energy consumption, the limits of demand on total commercial energy requirements can vary significantly. Therefore, if the desired dependence on the firewood need to be restricted to stave off deforestation, which is an environmental compulsion, the nation must work towards a target of meeting the demand through commercial energy growth. Options open to meet such a contingency are not many.

The share of the commercial energy consumption in different forms and the likely growth in such a demand on the resources will however be more if the per capita consumption is to grow as well. The preferred energy sources like hydro power or nuclear are not only limited but even now does not bear the major burden of the supply. One therefore wishes that much of the needs are supplemented by the renewable sources of energy. These are also eco-friendly. In such an endeavour the full potential of these sources to the extent practicable must be realised. Subsequent analyses have been carried out on this premise.

Bio-fuels, improved cook stoves, wind farms, small hydro power plants and solar energy are alternate renewable clean sources being promoted. However, bio gas plants and improved stoves to the extent adopted will only improve the efficiency of using the biomass and other solid fuels already consumed. The size of the usage of these alternatives now is given in Table 1 along with the estimated potential.

TABLE 1 : **Potential of Renewable Energy Sources**

Item	*Unit*	*In use*	*Available potential (MW)*
Bio gas plant	no	two million	17,000
Small hydro	MW	120	10,000
Wind Farms	MW	350	4,000
Co-generation	MW		3,500
Solar Cooker	no	34 million	25
Improved stoves	no	19 million	—
Solar Thermal Plant	m^2	2 million	—

Full realisation of the potential of these alternatives will mean around 1,800 MW power or 10 million tonnes coal or 4.5 million tonne oil equivalents. This is the year 1997 would represent less than 1 per cent of the estimated total energy or 1.9 per cent of the commercial energy. In this context, it is worth summing up the report of the World Energy Conference[4] saying that by the year A.D. 2020 the uses of renewable non-conventional energy sources globally can be only 10 per cent of the commercial resources. To achieve such a target, substantial industrial and infrastructure growth and capital availability is necessary. Full realisation of wind, small hydro and co-generation would provide an equivalent of 17,500 MW additional available capacity. However, the availability factor for these sources is dismally low. Now, say for diesel and wind power it is less than 5 per cent. In view of all these the share of these sources can at best meet 2% of the total demand or 4% of the commercial energy by the year A.D. 2020. Assuming that the growth of hydro and nuclear power is maintained at the same rate as in the past, the installed capacity for them will grow to around 42,500 MW. Of this total, 38,000 MW can be derived from hydro sources and 4,500 from nuclear fuels. This means a realisation of 40 per cent of the potential for hydro and 60 per cent of the neclear. Thus, the brunt of the requirement has to be met from fossil fuels.

The coal reserves of the country at 450 million tonnes production is estimated to last for 125 years. Against this, oil at 40 million tonnes and natural gas 22 billion cubic metres may last only for 20 and 32 years respectively. At this juncture, it is ironical that the estimated balance of the reserves, of oil and natural gas with time is on the decline. Thus, the major energy supply at least during the next 30 years will be heavily dependent on coal.

The current sectoral dependence on the secondary energy forms for different end uses and the share on the primary sources indicate that nearly 80 per cent of the primary energy sources and 70 per cent of energy in its end use form are solids. Therefore, the quality of solid fuels, cost of processing and transport and the end use efficiency all plays a vital role. The losses that occur during transformation from the primary form to the secondary form. This is around 14 per cent of the total energy consumed. Seventy per cent of the end energy uses, most of which are noncommercial fuels, is in solid form. Liquid fuels, gaseous fuels and electricity represent 11.0, 1.0 and 3.5 percent. For 100 units of end use energy consumption, 15 units are lost between source and end use. Much of this is accounted during conversion to power and its subsequent transmission and distribution. Demand side management of the power supply, thus, form another important aspect of technology development towards energy efficiency.

3. Environment

As the world is increasingly becoming conscious of the environment, all the present day technological development efforts relate to containing pollution. The mandate on the energy sector is to contain the oxides of sulphur and nitrogen emitted and reduce the specific carbon dioxide emission per unit power generated. By reducing the share of the non commercial energy use, it is possible to achieve a reduction of 5 per cent in carbon dioxide, 10 per cent in oxides of sulphur and 15 per cent in oxides of nitrogen emissions. Further reduction in these emissions is possible by resorting to improved technologies. This will be discussed subsequently. Other major pollutant of the energy sector is ash from coal. A hundred units of coal used produce 40 units of waste. Containment of such massive wastes of roughly 200 million tonnes in the years to come becomes a major consideration for energy management.

4. Energy Transport Systems

While the energy sources are uneven in distribution geographically across the country, the end use points are more diffused. Thus, efficient energy utilisation would need economic energy transport systems. Established transport system for energy is rail, road and pipeline. Use of pipelines is limited to the transportation of natural gas besides to some extent for petroleum and its products. While rail bears the transport need of the primary energy, transmission lines and roads carry most of the secondary forms. Secondary energy as power is transmitted by transmission lines. Most of the losses in energy are in the power sector, mostly during conversion and secondly during transmission and distribution. Besides, land based solid fuel (the major source) transport system is the cost controlling factor as would be discussed subsequently. There is sufficient scope to improve the energy efficiency of transport systems particularly in the power sector. Sixty percent of installed power is coal based in the country. The coal used for power generation contains up to 45 per cent non-energy material. Thus, 45 per cent of energy spent on transportation is unproductive. Similarly, short distance road transport costs around Re.1 per tonne-km (subjected to frequent and rapid increase) 45 per cent of this is unproductive. On the other hand, the in situ ash in most coal seams in India averages at 35 per cent. If large scale mining of these seams is not practical, deshaling can ensure the quality. The inadequate infrastructure is the bottleneck for much needed expansion of the needed power sector. A reduction in ash by 10 per cent would ensure meeting the coal demand required two years therefore without any additional infrastructure creation. Further, at 25 mm sizes required at grinding mills in power stations, beneficiation can reduce ash to 30 per cent without practically any loss in the heat value. Thus, the existing infrastructure can effectively transport 36 per cent additional energy. Development or arterial high voltage direct current (HVDC) on the national power grid is the other step that can reduce cost of energy transport.

5. Economics

As mentioned earlier, the cost of energy the nation pays as part of GNP is one way to measure its energy efficiency. No accurate assessment of this at any given time is available. Nor how it is changing with time is known. While the cost of coal and oil produced in the country and imported are available, it is difficult to assign a value to the firewood and other noncommercial energy sources used. An attempt therefore is made to compute a cost for the total energy in proportion to the value of coal and oil consumed based on heat value of the noncommercial fuels.

Such a computation, indicates that the trend of energy cost, as percentage of GNP is very high, for the commercial sources alone and the trend is also on the increase. This is essentially because of excessive dependence on import of liquid fuels, and high cost of indigenous coal production for a lower quality material. As of today the unit energy cost (excluding duties) of indigenous coal at coastal and many inland place is equivalent to imported coals. The latter are much cleaner.

6. Vision 2020

The projection on the demand for the minimum total energy for the estimated population is, to the best of imagination, accurate. This minimum need are considered based on the current level of per capita energy consumption. The current level of per capita consumption is also not for above the required minimum for sustenance and the pupulation growth rate is realistic. Left uncontrolled, in the absence of any legislation based on a well founded national energy policy the actual scenario will be one nearer to the projection based on noncommercial energy growth. Also it would be a stupendous task for the country to work towards a zero firewood consumption, because it requires a huge capital and a massive infrastructure. Thus, it is necessary to assess to what extent the additional energy demand is met from commercial sources alone and how much of it can come from the indigenous resources.

Before these points are deliberated, it is worthwhile to understand the global situation in terms of the trends, the compulsions and the limitations on the availability of energy. Besides, the status of the technology and the developmental needs are also to be assessed. The global energy situation as foreseeen in the past is[4] :

— Wood, the only significant renewable energy source that is the main source of 50 per cent of world's population, also accounts 15 per cent of total energy consumed globally;
— Hydro power generation grows at 6 per cent compared to 5 per cent of nuclear power and the former will increase in absolute terms compared to nuclear;
— 75 per cent to 80 per cent of the commercial energy needs are met from fossil fuels; at the same time 60 per cent of the total energy is also from these sources;
— Renewable sources, other than wood and hydro, are insignificant in terms of its share in the total energy balance;
— With the increasing shortage of oil which would become dearer with time, the same needs to be restricted to essential applications for which no substitute is possible, like for transport, petrochemicals etc;
— Regional oil shortage would occur because of rising energy prices; and
— Fossil fuels cause local pollution, and loss of habitat and global warning.

In the light of these facts, figures and projections, to foresee a vision for the year 2020 A.D. towards an energy efficient and an environmental friendly nation it is essential and necessary to evolve a national policy and set target for:

— Contain the share of the firewood use towards zero consumption;
— Increase the share of the renewable energy sources and biomass to near potential and increase the efficiency of their end use;
— Restrict the use of natural gas and petroleum prducts to essentials like transport, fertilizers, chemicals and alike;
— Improve the efficiency and the economics of use of indigenous sources; and
— Practice a technology driven demand side management.

These aspects need be discussed further with respect to each primary energy source. Major need for energy is in the domestic sector to provide energy to prepare food. Second in magnitude is industry followed by transport and agriculture sectors.

At this stage one may exercise caution in understanding the energy needs that are very conservative compared to other forecasts. Growth based projection definitely assumes higher energy needs than the minimum considered. However, technical technological problems will remain similar but appear in a greater magnitude.

NOTES

1. Teri Energy Data Directory & Year Book (1995-96) Tata Energy Research Institue (1995).
2. Eighth Five Year Plan (1992-97) Vol. II, Government of India, Planning Commission, New Delhi.
3. World Year Books (1989 to 1993), Published by World Bank, Washington D.C.
4 World Energy : Looking Ahead to 2020, Report of the Conservation Commission of World Energy Conference, IPC Science & Technology Press (1978).

RENEWABLE ENERGY SOURCES AND THE ROLE OF SOLAR ENERGY CENTRE

2

VED MITRA AND ASHVINI KUMAR
Solar Energy Centre
Ministry of Non-conventional Energy Sources
Government of India
New Delhi (INDIA)

ABSTRACT

Energy is vital input for economy and social development in every society. Presently, the global primary energy demands are met largely from oil, natural gas, coal, nuclear energy and huydro-electric. It is, however, believed that the supplies from existing energy sources would not be adequate for a long time, if the earth's population keeps on increasing at its present rate, and all the people achieve standard of economic development. The present global concern regarding deterioration of environmental conditions is also providing added impetus to the development of energy alternatives, in particular renewable sources of energy.

In India, the Ministry of Non-conventional Energy Sources is implementing wide ranging programmes in different aspects of renewable energy sources. The Ministry has also established Solar Energy Centre as an institute to serve an effective interface between the Government, R&D institutes, industry and user organizations for development and utilization of solar energy in the country. This paper discusses present role of the Centre for balanced development and utilization of solar energy including the technical facilities available and its present activities. The paper also outlines the proposed plans for changing the role of Centre to cover entire renewable energy sector by converting it into a National Institute of Renewable Energy.

INTRODUCTION

Energy is a vital input for economy and social development in our society. Presently, the global primary energy needs are met largely from oil, natural gas, coal, nuclear energy and hydro-electric. In some of the developing countries, various other fuels, such as, wood and animal waste are also used for meeting the energy requirements. It however, appears that there will not be adequate supplies from the existing energy sources if the earth's population keeps on increasing at its present rates, and all the people achieve the standard of economic development. In the post era of United Nations Conference on Environment and Development in Rio de Janeiro the concern for deterioration of global environmental conditions has emerged the main driving force towards the development of energy alternatives, in particular renewable sources of energy and improvements in the efficiency of energy use.

In India, the Ministry of Non-Conventional Energy Sources is implementing wide ranging programmes at different aspects of renewable energy sources. In fact, the renewable energy programme in India is one of the largest programmes in the world. A significant R&D base, large manufacturing capabilities, and a country-wide network for distribution and after-sale services is already developed. The Solar Energy Centre has been established by the Ministry as an organization to serve an effective interface between the Government, R&D Institutions, Industry and User Organizations for development and promotion of the solar energy use in the country.

POTENTIAL OF RENEWABLE ENERGY SOURCES IN INDIA

India is blessed with abundant supply of renewable energy resources. The estimated potential of renewable energy sources in the country is given in Table.1[1]. It is clear that the flow of renewable energy to the earth's land surface is substantially greater than the total energy requirement at the present rate of consumption. Even a small fraction of this resources is sufficient enough to provide an alternative for meeting the energy requirements.

TABLE.1 : Estimated Potential of Renewable Energy Sources in India

SOURCE/ TECHNOLOGY	ESTIMATED POTENTIAL
Biogas Plants	12 Million
Biomass	17,000 MW
Improved Chulha	120 Million
Solar Energy	5×10^{15} Kwh/year
Small Hydro	10.000 MW
Wind Energy	20.000MW
Ocean Power	50.000MW
Sea Wave Power	20.000MW
Tidal Power	9.000

SOLAR ENERGY CENTRE

The Solar Energy Centre established during the Sixth Five Year Plan functions as a technical division of the Ministry of Non-Conventional Energy Sources. The Centre began functioning from its present campus at Gwal Pahari since 1992. The principal areas of work in this Centre cover resource assessment, testing of solar thermal and solar photovoltaic systems and components, development of standards, solar passive buildings, design and consultancy services, data base relating to solar energy technologies and international cooperation.

FACILITIES

Buildings

Three main sets of buildings have been constructed in the Centre: Workshop, Guest House and the Administrative and Technical Blocks. The entire arachitectural design of the buildings (and the campus) has been done by the School of Planning and Architecture, New Delhi. The buildings incorporate solar passive features with a view to reducing energy consumption on lighting, cooling and heating, and provide a pleasant environment for work.

Workshop

The Centre has fully equipped workshop for the fabrication of mechanical items to support the development and testing work.

Solar Thermal Test Facility

The Centre has, for several years, been carrying out testing of solar collectors, solar cookers and other devices in outdoor test area. The outdoor test bed is provided with water and power connections. A large area sun simulator and indoor-outdor test rig have also been established with assistance from the Government of the Federal Republic of Germany. This facility enables the testing of full size solar collectors as per National and International test standards. Associated with this are the instruments for calibration of pyranometers and temperature sensors. Overall, the facility is the most modern and sophisticated, and is the only one of its kind in South and South East Asia.

The Centre has developed capabilities in respect of the following:

— Tests on solar cookers and solar collectors as per Indian (BIS) standards; tests on solar collectors as per international standards also.
— Thermal efficiency test under outdoor conditions as well as indoor conditions with the help of sun simulators.
— Incident angle modifier and time constant measurement tests.
— Calibration of pyranometers.
— Calibration of resistance temperature detectors and thermocouples.
— Technical support to the Solar Thermal industry for product improvement

Photovoltaic Test Facility

The Photovoltaic Test facility was established during the year 1992-93. This was taken up as a project with USAID assistance in collaboration with Sandia National Laboratories of U.S.A. Under this project equipments required for performance testing of solar cells, PV modules and systems were acquired. In regard to the qualification testing of PV modules, some equipments were procured under this project and the remaining equipments were designed and developed in India. Now, SEC is equipped to conduct all the qualification tests except the mechanical tests, which are under development. Facilities are also established for testing solar PV lighting system, PV water pumping systems and small capacity inverters and charge controllers. The major PV test facilities presently available are :

- Sun Simulators for performance testing,
- Environmental Chambers for temperature and humidity testing,
- Hot Spot activity measurement test set up,
- Hi-pot test set up,
- UV exposure system,
- Outdoor exposure and NOCT test set up,
- Field testing equipment,
- PV lighting system test facility,
- PV water pumping test facility,
- 4 KW PV power plant for testing inverters and charge controllers.

Presently, the test facility is offering services to PV industries and users in the country.

Keeping in view the demand and the requirement of manufacturers situated in remote corners of the country, additional test centres have been set up at Calcutta, Bangalore and Trivendrum. These are expected to be fully operational during 1996-97.

Library and Information Centre

The Centre has a small library which stocks essential publications relating to solar energy and receives a few important periodicals. A dedicated satellite-based communication link has been established with the National Informataion Centre. It is possible to access international data networks through this link. An E-mail facility is also operational. A small data base on solar energy equipment manufacturers and State level organizations has been created.

Instruments

In addition to the above facilities, the Centre has some other equipments like spectrophotometer, weather-o-meter, magnetron sputtering system and a weather data recorder for carrying out different measurement required in research and development for solar energy technologies.

Design Facilities

The Centre has acquired software and capabilities for the design of a variety of solar thermal systems.

Training Facilities

The Centre has facilities for conducting training programmes and seminars in the form of an auditorium, a seminar room and audio visual equipment.

MAJOR ACTIVITIES AND PROGRAMMES

During the Eighth Plan period, the staff of the Centre have worked closely with the solar energy programme of MNES and have contributed significantly to the development, utilization and dissemination of this technology. The major activities and programmes are as follows:

Solar Thermal Testing and Standards

The facility for indoor/outdoor testing of solar thermal collectors and systems is serving the Indian industry, research and scientific institutions and State nodal agencies/departments etc. by way of testing of solar flat plate collectors, solar cookers and other systems as per the national standards. Technical support is provided to the solar thermal industry for carrying out improvement in their products through testing. Memorandum of understanding have been signed with IIT, Delhi and National Physical Laboratory, Delhi to provide framework for developing joint programmes for carrying out R&D, testing and training in the area of solar enrgy utilization.

Solar Thermal Power Generation

A 50 KW solar thermal power plant was commissioned as R&D-cum-demonstration facility in 1989. Its operation was held up because of non-availability of the critically required spares and deterioration of some of the important components. The plant has recently been revived at a derated capacity in February 1996 after carrying out repairs and using only a part of the plant. Steps have been initiated for its modernization based on latest collectors LS-3 system. Development of indigenous components like receiver tubes, collectors and electronic cards have also been undertaken to achieve the rated capacity of the plant.

Solar Photovoltaic Testing and Standards

This group undertakes routine testing of solar cells, PV modules and lighting systems. Over 100 solar cells, about 500 modules, 300 lanterns and other systems were tested during the period 1994-96. A Spectral Response measurement systems has been set up at the Centre for calibration and providing primary/secondary reference cells. A comprehensive water pumping test facility was established for a thorough testing of PV water pumps.

Systems Development and Resource Assessment

Under this activity, technical specifications of different solar thermal systems are prepared and

booklets are brought out to disseminate information. A joint R&D project was undertaken with the Devi Ahilya Vishwavidyalaya, Indore to develop designs of the demonstration modules for solar hot water systems for training purposes.
A daylight measurement station has been established at Solar Energy Centre as part of the Daylight Measurement Programme in India, coordinated by Central Building Research Institute, Roorkee.

Information Dissemination

An E-mail service is now available at the Solar Energy Centre. A small database has been createad on renewable energy programmes. The Centre has brought out several publications on different aspects of solar energy technologies and their utilization. Recently the publication of a quarterly news letter namely the "New Energy News" and bi-monthly newsletter, namely, "Solar Energy Information Services" have also been started at the Centre.

The Centre has also taken up several actions to encourage the study of renewable energy in shcools and colleges.

International Cooperation

The Solar Energy Centre is a focal point for cooperation with G-15 countries in the field of solar energy. Under this activity, Solar Energy Centre organises training programmes and information dissemination activities for G-15 countries. Under a programme for exchange of data on the performance of solar collectors produced in G-15 countries, a collector has already been received by the Centre from Egypt. A solar collector is also being sent from India to Egypt. SEC also has been implementing a memorandum of understanding with the National Renewable Energy Laboratory of the USA. Exchange of samples, data, testing procedures, standards and also training programmes are the part of this MOU. Other areas of activity include solar radiation analysis and resource assessment, solar thermal testing etc.

An Indo-German project of the establishment of test facilities for solar collectors has been successfully implemented. Second phase of the Indo-German project is also expected soon which would have activities pertaining to the development of test facilities for solar thermal systems, interaction of industry in two countries and development of more efficient designs of solar systems for producing hot water and process heat.

FUTURE PLANS

Keeping in view the importance and long-term relevance of solar energy for the country. It was felt that the Centre should be built into a strong institution that will provide the technical support to the national renewable energy programme. Specifically, it is proposed that the Solar Energy Centre should be converted into an autonomous institution and named as the National Institute of Renewable Energy (NIRE). It is envisaged to become the foremost Centre for research, development, testing, training and maintenance of up-to-date data on renewable energy. As proposed, the institution will continue to be fully supported financially by the Ministry. The set up of the proposed institute is given in Annexure-I.

The proposed activities of the Institute are as follows:

(i) Technology Assessment
(ii) Technical back-up to MNES and Industry and IREDA
(iii) In-house R&D in specific areas
(iv) Assist industry for product upgradation and new product development
(v) Setting up of satellite centres in different parts of the country for specific NRSE areas
(vi) Testing and centification of devices and systems
(vii) Evaluation of renewable energy equipment, systems, power plants, etc.
(viii) Joint technical programmes with other national institutions and testing centres
(ix) Consultancy and advisory services for manufactures and users
(x) Study of the economics of renewable energy technologies and their environmental impact
(xi) Resource survey and assesment of the potential of renewable energy sources in the country
(xii) Compilation and dissemination of information concerning renewable energy technologies, patents and products as well as data on resources
(xiii) Advanced level training and organization of seminars and workshops for NRSE industry and users
(xiv) Cooperation with scientific institutions in other countries and implementation of programmes assigned by government under bilateral and multilateral agreements
(xv) Assistance in curriculum development in renewable energy subjects.

CONCLUSION

It is obvious that the Solar Energy Centre is playing an important role, specially in the area of testing and standardization of solar thermal and solar photovoltaic products and components. The Centre is providing technical support to the other test centres, solar industry and various other R&D institutions for carrying out activities pertaining to the development of solar energy systems. Recently, the Centre has developed strong linkages with the premier R&D and academic institutions in the country (viz. NPL, IITs, IARI, IACS, Calcutta) to have a comprehensive focus on the common technological issues. In case, the proposal for expanding the scope of activities to cover whole renewable energy area is accepted, the role of the Centre for contributing to the national effort in this direction will further be strengthened.

CLEANER POWER GENERATION BY RENEWABLE ENERGY SOURCES

3

Dr. D.K. Sharma, C.S.O. and **Dr. D.P. Kothari**
Professor & Head, Centre for Energy Studies
Indian Institute of Technology,
Hauz Khas, New Delhi-110 016 (India)

ABSTRACT

Presently more than 70% of power generation is based on coal combustion in India. Since the demand for power is increasing, therefore, the pressure on increasing the production of coal is increasing. The quality of Indian coal is not good, therefore, the use of coal leads to increase in the generation of fly ash. However, there is solar, wind, hydropower, ocean energy, biomass and wastes which can be utilised for cleaner power generation. It has been estimated that around 10,000 MW of power can be generated from the municipal solid wastes, potential of other renewable energy sources has also been assessed. A beginning in the use of renewable energy sources has been made. Future may see a cleaner power generation from renewable energly sources.

1. Introduction

World reserves of oil and natural gas are limited. Oil and natural gas reserves are not expected to last for more than 60-70 years at the present rate of consumption. However, the consumption of oil and natural gas is continuously increasing with increase in the population of the world, with increase in the standard of living of masses and due to the rapid industrialisation in the developing countries. The oil and natural gas reserves in India are also not expected to last longer. Most of the power generation in India is based on coal. Coal reserves of the world are expected to last for more than 300 years. The use of coal in power generation is expected to increase with increase in the demand for power. However, the coal is associated with inorganic mineral matter which is analysed as ash. Therefore, the direct combustion of coal leads to the generation of fly ash. The use of fossil fuels also results in an increase in the green house effect. There is a need to supplement the use of fossil fuels with other sources of energy. Fortunately, there are several renewable energy sources which can be used to generate cleaner power. The use of some of these energy sources has been technically tested but it is not yet economically established though attempts are being made. The aim of this article is to discuss the near term and future possibilities of generating power by using these renewable sources of energy. These would supplement the generation of power by conventional energy sources such as coal, diesel, fuel oil, natural gas, hydro and nuclear etc.

2. Energy from Wastes

The four big mega cities of India generate about 16,600 tonnes of garbage every day (Pramod, 1996). Disposal of this waste is a problem. This production of around 10,000 MW of electricity from this municipal solid wastes has been estimated. With increase in the standard of living of masses the amount of waste generated per person is expected to increase. This may pose the problem of disposal of more wastes and there may arise a scarcity of land for the disposal of this waste. Several technology options are available for the treatment of municipal solid wastes and these include: (a) Incineration, (b) Sanitary landfilling, (c) Anaerobic digestion, (d) Pyrolysis, (e) Gasification, (f) Pelletisation or Briquetting. These technologies may be used to utilize municipal solid wastes for obtaining energy which may be used for power generation.

3. Potential of Solar Energy

India is blessed with abundant solar energy which is equivalent to 500 trillion KwH per annum (Naidu, 1996). Renewable energy technologies are being promoted in India for the last 10 years. Research and Development work in the area of harnessing solar energy is being extended. Demonstration projects are being set up and the physical progress in different sectors has been reported recently (Naidu, 1996). Solar energy is a clean source of energy. Large amount of solar energy is available and it has great potential for utilization. This is a renewable source of energy. However, this is a usually low density energy and requires a larger space. Insolation varies seasonably and is not available in the night. Economic viability has yet to be tested on larger scales. However, the solar energy has a potential as a cleaner and renewable energy source for rural, desert and remote areas. There is a need to extend further research and development work on finding out efficient and cost effective technologies for harnessing solar energy. It has been reported that globally around 300 MW photovoltaic and 350 MW solar thermal capacity has been generated (Naidu, 1996). In India also both photovoltaic and solar thermal routes are being considered. The detailed progress of solar energy projects in India has been reported recently (Naidu, 1996). Research work in the area of tapping solar energy by photovoltaic, solar concentration, solar distillation, solar building architecture, solar refrigeration and solar ponds is being extended at the centre for Energy Studies at I.I.T. Delhi

4. Potential of Biomass Energy

Plenty of biomass is also available in India. Biomass is also a renewable source of energy. It has been reported that 14 per cent of the world's energy supply today is based on biomass (Naidu, 1996). Direct combustion of biomass may generate smoke and is not efficient. Several routes are available for generation of cleaner power from biomass wastes such as bagasse, rice husk, rice straw, jute sticks etc. Hydrolysis of agroresidues generates fermentable sugars which can be converted to ethanol. Biomass wastes may be fired in fluidised bed combustors to generate power. Briquetting of biomass directly or after charring (pyrolysis) may also afford cleaner boiler fuels for power generation. Gasification of biomass wastes to produce producer gas is another option for obtaining cleaner fuels from biomass. Research and development work on the production of hydrogen by using cyanobacterium may be extended. Research work on the utilization of biomass wastes is being extended at the Centre for Energy Studies, I.I.T. Delhi.

5. Hydel Potential

About 16 per cent of the hydel potential of the country has been exploited. Hydropower is also a cleaner source of energy. In India, there is also a great potential (> 10,000 MW) for setting up small hydro plants (Naidu, 1996).

6. Energy from Seas and Oceans

Seas and oceans have also great potential to generate power by three different routes namely (1) Ocean Thermal Energy Conversion, (2) Sea or Ocean Wave Energy, (3) Tidal energy. World's largest plant of 600 MW is being set up at Kulasekarapatnam in Tamil Nadu (Naidu, 1996). It has been reported that Indian ocean has a large potential of 50,000 MW and this may afford an opportunity to exploit this energy. In addition, there are hydrocarbon hydrates in ocean bed which may be exploited to obtian gaseous fuels. Attempts may be made to tap tidal and wave power and some efforts are being made in this direction. The details of the same have been reported recently (Naidu, 1996). Aquatic biomass may also be utilized and research work in this direction may be required.

7. Potential of Wind Energy

The global wind power potential is two times the present power production in the world (Naidu, 1996). The wind power is being exploited in North America, Europe, India and other parts of the world. India is among the top countries of the world in exploiting wind energy. Efforts are being made to exploit this renewable source of energy.

8. Option for Cogeneration

There is also a need for recycling or reusing wastes e.g. bagasse, especially for cogeneration. Wastes from industries such as sugar, paper, distilleries and food industries may be tested for the organic contents which are biodegradable by using anaerobic digestion to generate biogas. Biogas obtained may be used for generating power. Studies may also be made to utilize organic industrial wastes for their management and for their utilization to generate power if possible.

9. Conclusions

Research and development work in the area of exploiting renewable energy sources should be extended. Renewable energies afford cleaner power generation. While coal based power generation may have to be increased, however, with the depletion of oil and natural gas reserves, the additional power may have to be generated by exploiting renewable sources of energy. Demand for generation of power and especially for the generation of cleaner power will increase. Therefore, research and development work to make the generation of power from renewable energy sources economically attractive may be required.

Almost eighty percent of the population in India lives in rural areas, therefore, most of the power plants based on renewable energy sources may be established in rural area. Centralised power stations may be set up where cattle dung, agroresidues, rural wastes and other wastes may be collected. Continuous biogas digesters, biomass gasifiers and briquetting units may be set up. Canals and river streams may be used for power generation by mini and microhydels.

Solar concentrators and photovoltaic cells may be designed to generate power for catering to the need of 15-25 villages. Coastal areas where the transportation of coal may be costlier because of high ash contents in coal and due to the larger distances from the coal mines, may make use of wind energy and ocean energy. similarly, wind energy, small hydro and solar energy may be used in hilly areas for power generation.

Research work in the areas of cleaner power generation from coal (Sachdev, 1993), utilization of petrocrops (Sharma, 1994), (laticiferous, resinous and oil seed bearing plants) for obtaining petroleum, acidic and enzymatic hydrolysis of agroresidues (Singh, 1984), hydrogen generation from biomass wastes and cyanobacterium, and fuel cells etc. may be extended.

Gasification and combustion reactivity of Indian coals may be studied for selecting suitable processes of their utilization. Integrated gasification combined cycle power generation may also afford relatively cleaner power generation. This may also help in selecting suitable processes for coal utilization. Renewable energy sources along with the use of washed coals are expected to play important role in power generation in future.

Energy conservation measures and minimum dispatch of power and optimal mix of power may also help (Nagrath, 1994) as power saved is power generated. An optimal mix of power generated by using different sources may ultimately help in meeting the demand of power in India in the twenty-first century.

REFERENCES

Nagrath, I.J. and Kothari, D.P., Power System Engineering, Tata Mc-Graw-Hill, New Delhi,1994.

Naidu, B.S.K., *Natural Energy Review*, pp. 6, Jan. 1996.

Pramod, P., *Natural Energy Review*, pp. 4, Jan. 1996.

Sachdev, R.K., Coal Usage in India - An Approach Towards Cleaner Coal, *Proc. Regional Workshop on Environmentallly Sound Clean Coal Technologies*, Bangkok, 1993.

Sharma, D.K., Tiwari, M. and Behera, B.K., Biores. Tech., 49, 1, 1994.

Singh, A., Das, K. and Sharma, D.K., Ind. Eng. Chem. Prod. Res. Dev., 23, 257, 1984.

DEVELOPMENT OF WIND ENERGY AS A SUSTAINABLE ENERGY OPTION

4

R. Venkatesh, Research & Development Division
Crompton Greaves Ltd, India

ABSTRACT

In the present scenario of increasing shortages in the energy sector, depleting natural resources, ever increasing pollution and resulting eco-degradation and increasing stress on sustainable development programmes, renewables offer a sustainable energy option. Of the various renewable energy options wind, hydro and solar offer the highest potential.

Wind energy has matured to a level of development where it is ready to become a generally accepted utility generation technology. Given the advanced state of wind turbine technology, the availability of wind resource, modularity of wind electric generators and an expected increase in demand for energy sources not damaging to the environment, it is sensible to expect that wind energy will become a significant component of energy supplies in the near future.

Though there had been a tremendous development in the wind generator technology in the last 15 years there are still a lot of limitations such as poor energy capture, Reactive power demand, Operation of IG at non optimum slips, Higher cut-in wind speed, motoring action and import of power and poor grid penetration which act as barriers for the growth of wind power as a utility generation technology and as a sustainable energy option. All these limitations reduce the energy output of a WEG and increases the payback period. By overcoming these limitations not only the performance be improved but the payback period can be reduced thus making wind electric generators more attractive technically and commercially.

This paper examines the limitations of present day wind electric generators and remedial measures to overcome these limitations. The paper also describes a new "Asynchronous link for interfacing wind electric generators to power grid for enhanced energy capture and better VAR management". The proposed scheme under development eliminates all the deficiencies with the present day WEG's. This asynchronous link is based on a new and patented technology. The advantages of the new system are Enhanced energy capture, Wider operating range, Power generation at Unity Power Factor and the capability of generating both real and reactive power and improved power quality in terms of lesser harmonics.

1. Introduction

In the present scenario of increasing demand-supply gap in the energy sector, depleting natural resources, ever increasing pollution and resulting eco-degradation and increasing stress on sustainable development programmes, renewables offer a sustainable energy option. Of the various renewable resources Wind, Solar and Biomass offer the highest potential. Of the various renewable energy options wind, hydro and solar offer the highest potential. This project deals mainly with the wind energy and wind based hybrid systems. The estimated wind potential in India is 20,000 MW spread mainly over 7 States (Tamil Nadu, Gujarat, Andhra Pradesh, Karnataka, Kerala Maharashtra and Madhya Pradesh.

Wind energy has matured to a level of development where it is ready to become a generally accepted utility generation technology. Given the advanced state of wind turbine technology, the availability wind resource, modularity of wind electric generators (WEG) and an expected increase in demand for energy sources not damaging to the environment, it is sensible to expect that wind energy will become a significant component of energy supplies in the near future.

Although wind resource is unlimited and renewable, unfortunately this is not true for the land which is used for setting up wind farms. In the years to come the prime resource for generation of wind power will not be the wind but potential sites. With only limited sites suitable for power generation available, it makes better sense to develop technologies which would increase the efficiency of the wind electric generators, along with array optimization techniques to maximize the energy density (energy output per unit area of land utilized).Though there had been a tremendous development in the wind generator technology in the last 15 years there are still a lot of limitations which act as barriers for the growth of wind power as a utility generation technology and as a sustainable energy option. Limitations such as poor energy capture, Reactive power demand, Operation of induction generator at non optimum slips, Higher cut-in wind speed, motoring action and import of power and poor grid penetration which act as barriers for the growth of wind power as a utility generation technology and as a sustainable energy option.

This paper examines the limitations of present day wind electric generators and remedial measures to overcome these limitations. Development of an asynchronous link for interfacing wind electric generators to power grid is the main theme. This asynchronous link is based on a new and patented technology and will overcome most of the limitations of the present day generators.

2. Barriers to the Development of Wind Energy

The barriers to the growth of wind energy or for that matter any renewable resource could be classified broadly as technological (owing to limitations in technology) and institutional (owing to promotional policies, Socio-economic factors, political factors etc. which are mainly non technological). This papers deals only with those barriers arising out of technological reasons. The data discussed in this section pertain to wind electric generators employing induction generators for converting mechanical energy into electrical energy, which is the case with about 90 % of the present day generators.

The major Technological limitations of present day WEG's are :

2.1 Poor energy capture: The energy capture of present day WEG's is very poor and the typical utilization factor is well below 30 %. This makes the wind power generation less attractive and increases the pay back period. This leads to reduced private investments in this source of power generation.

The reasons for this poor energy capture can be best illustrated by the basic principle of operation of wind electric generators.

The kinetic energy per unit volume of wind is given by

$$KE = 1/2 \, .p.u^2$$ where p is the density of air and u is the velocity of air.

As wind passes over the blades of WEG it imparts a part of its kinetic energy to the blades causing it to rotate. This rotational (mechanical) energy is converted into electrical energy using electric generators. The energy imparted per second by the air on a surface area of A m^2 normal to the wind assuming the wind is brought to rest, is

$$Pw = 1/2 \; pu^3 A \text{ watts} = 1/8 \, \pi \, p \, d^2 u^3 \text{ watts},$$

where A is the area swept by the blades for a practical turbine of rotor diameter d.

It is easy to see that no device can extract all the energy from the wind or the air would have to be brought completely to rest. This would prevent air from passing through the system and would reduce extraction efficiency. In fact, slowing the air too little or too much gives a low efficiency. This defines the efficiency of the turbine blades, C_p. So the amount of power that can be extracted from wind P_t is given by

$P_t = Pw .C_p$, where C_p is the power coefficient or the aerodynamic efficiency.

Applying various fluid mechanics theories, it can be shown that C_p reaches a maximum value of 0.593*(59.3%),called the Betz limit and this defines the maximum energy that can be extracted from wind.

Not only C_p has a low value of 59.3%, it also varies with the tip speed ratio (ratio of angular velocity of blade to wind velocity) and has an 'inverted V' characteristics, having a maximum value of C_p for a single value of tip speed ratio. as shown in Fig. 1.

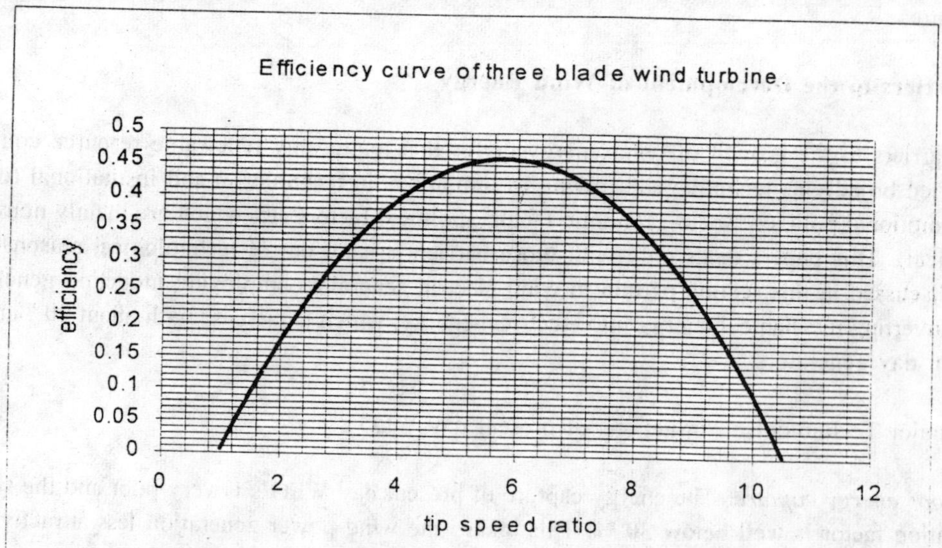

FIG 1. *Variation of C_p (efficiency of turbine) with the tip speed ratio.*

Since for a directly grid connected WEG, the shaft speed has to be a constant (as the grid frequency is fixed), the maximum value of C_p occurs at a single wind velocity. This leads to overall reduction in energy capture over the entire operating (wind velocity) range.

2.2 Reactive power demand : Apart from operation at poor efficiencies, a major problem with the present day WEG is the reactive power demand imposed on the grid. This sometimes leads to stopping of generation, even when the wind is favourable, due reactive loading on the grid. The reactive power consumed by the conventional grid connected WEG is very high and typically a 10MW wind farm would require 6 MVAR reactive power at peak operating point. Typical reactive power consumption in a 250 KW WEG is shown in Table 1. The table shows the reactive requirement, wind velocity profile and KVAR hr consumption as a function of wind velocity. This sometimes forces the WEG to be shutdown when the grid is not able to support the reactive demand. Provision of fixed/switched capacitors helps the situation to some extent. Fixed capacitors do not provide the optimum reactive power and switched capacitors have the reliability problem of capacitors and associated switches. The capacitors provided for reactive support to the generator are kept well below the required level for technical considerations (to prevent the self excitation of the induction generator). So the generator needs reactive support from the grid even if capacitors are provided. Table 1 shows the VAR and KVAR hr consumption of a typical 250 KW generator with different values of capacitors connected to the generator.

The problem of reactive power consumption is amplified due to the synergism in the consumption of reactive powers by wind farms and agricultural pump sets connected to the same local grid. This reactive demand at present is met by central capacitor banks and leads to increased line losses and poor voltage profile.

2.3 Operation of induction generator at non optimum slips: Operation of the induction generator connected to a fixed frequency grid results in variable slip operation for various power outputs. This leads to operation at non optimum slips resulting in reduced induction generator efficiency.

2.4 Higher cut-in wind speed: Present WEG are started as motors (at cut in speeds) and run upto or above synchronous speed before generation can start. The synchronous speed is fixed for a given grid frequency and is high at 750/1000 rpm. This means a longer starting time and higher cut in speed.

2.5 Motoring action & import of power: Most of the present day WEG are run up as induction motors to synchronous speed before generation can start. Running up to a fixed (and higher) speed (for a fixed grid frequency) results in a long motoring action and the associated import of power. The problem is acute during low wind periods when the generator has to be started and stopped many number of times in a day.

2.6 Poor grid penetration: Though with the present advances in meteorology and with the large amount of data base generated it is possible to predict winds fairly accurately, the grid penetration is low for the following two reasons. In spite of being predictable there is still some uncertainty about wind and the velocity of wind (and the power generated) is not steady but varies widely with time. This makes wind energy unsuitable for base load operations. Also wind being uncontrollable wind power is not dispatchable. All this lead to a poor grid penetration. The problem is compounded by weak grids which are used to evacuate power generated by wind electric generators.

2.7 Unstable grid frequency: As most of the present day WEG's are fixed speed machines their output (efficiency) depends on the rotor speed (grid frequency) for a given wind speed. The

TABLE 1 : Typical reactive power consumption in a 250 Kw WEG

w vel	hrs	VAR 27	k VAR HR	C1VAR	C2VAR	C3VAR	Clk VARHR	C2kVARHR	C3kVARHR
2	858	0	0	0	0	0	0	0	0
3	630	0	0	0	0	0	0	0	0
4	823	44	36.212	19	7.5	4	15.637	6.1725	3.292
5	925	56	51.8	31	18.5	16	28.675	17.1125	14.8
6	1095	67	73.365	42	29.5	27	45.99	32.3025	29.565
7	1064	81	86.184	6	0	0	6.384	0	0
8	886	96	85.056	21	8.5	0	18.606	7.531	0
9	683	111	75.813	36	23.5	0	24.588	16.0505	0
10	562	127	71.374	52	39.5	7	29.224	22.199	3.934
11	254	146	37.084	71	58.5	26	18.034	14.859	6.604
12	438	168	73.584	93	80.5	48	40.734	35.259	21.024
13	298	170	50.66	95	82.5	50	28.31	24.585	14.9
14	159	170	27.03	95	82.5	50	15.105	13.1175	7.95
15	57	170	9.69	95	82.5	50	5.415	4.7025	2.85
16	10	170	1.7	95	82.5	50	0.95	0.825	0.5
17	10	170	1.7	95	82.5	50	0.95	0.825	0.5
18	0	170	0	95	82.5	50	0	0	0
19	0	170	0	95	82.5	50	0	0	0
20	0	170	0	95	82.5	50	0	0	0
21	0	170	0	95	82.5	50	0	0	0
	8752		681.252				278.602	195.541	105.919

CAP1= 25/75KVAR
CAP2=37.5/87.5 KVAR
CAP3= 40/120 KVAR

Typical Reactive Power Consumption in a 250 kW WEG

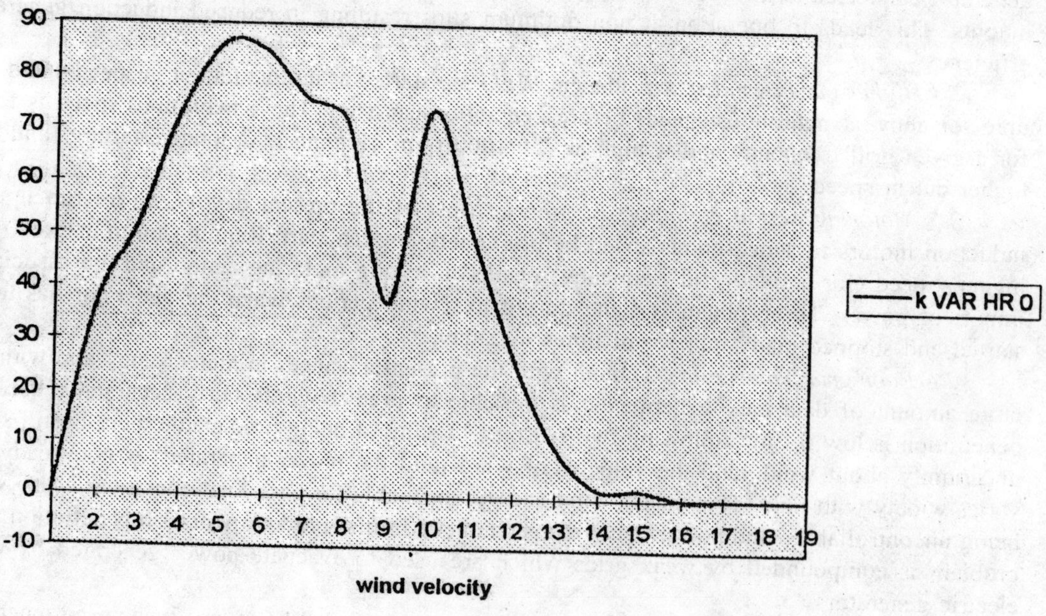

manufacturers of WEG try to maximize the efficiency for a given operating point (depending on the wind velocity profile) for a grid frequency of 50 Hz. In practical cases the frequency is well below the specified frequency for most of the time. This leads to reduced operating efficiency (less than the values certified by power curves) and lower power outputs.

To sum up : All the above mentioned limitations reduce the energy output of a WEG and increases the payback period. By overcoming these limitations not only the performance be improved but the payback period can be reduced thus making wind electric generators more attractive technically and commercially.

3. Remedial Measures for Overcoming Barriers

The remedial measures for overcoming the above mentioned technological barriers are discussed in this section.

3.1 Poor energy capture

One way of overcoming this problem is to have a dual speed operation. This is achieved by having two sets of poles and a pole changing mechanism. By this the generator has two synchronous speeds. This means that the tip speed ratio can be optimized at two wind speeds. This increases the energy capture as the WEG will have Maximum C_p at two operating points, instead of one point as in the case of single fixed speed machines. The limitation with this type of arrangement is that the pole changing mechanisms are difficult to manufacture and maintain. Also the energy capture is not increased appreciably and the increase is site specific.

The best solution would be to have a variable speed operation. By this the speed of the rotor is controllable and continuously variable. This ensures an optimum tip speed ratio and hence a maximum C_p over a wider (entire range-limited only by the rated capacity of the machine) operating range. By having an asynchronous link, it is possible to operate the WEG at different frequencies (and hence variable shaft speeds) and connect it to a grid of fixed frequency. This enables operation at constant tip speed ratio and C_p max over a wide wind velocity range. (refer Table 2). The use of asynchronous link not only results in more energy capture but also wider operating range as seen from Fig.2. This means an improved utilization factor. By using an asynchronous link the terminal voltage and frequency (Rotor speed) can be controlled. By controlling (increasing) the terminal voltage more power than the name plate rating can be generated (provided wind is there) without overloading (stator current) the generator. This effectively increases the operating range.

3.2 Non-optimum Slip Operation

By having a variable speed operation (Asynchronous link) it is possible to operate at optimum slips throughout the entire operating range . This reduces the stator copper losses and improves the generator efficiency.

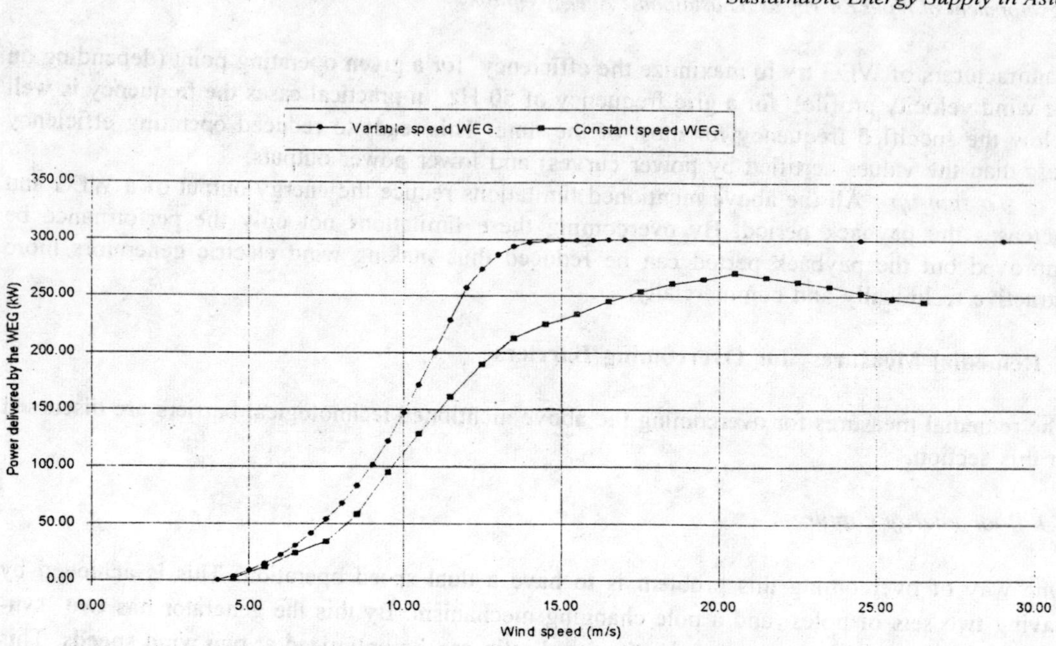

FIG 2 : *Power curve for constant speed and variable speed WEG.*

TABLE 2 : Operation and Analysis of the Developed Variable Speed WEG

Power curve for	Cromption Greaves.					
Rating	250	kW	Core loss at 50Hz	-	4	kW
Diameter	27	m	Frequency exponent		1.45	
Air density	1.2	kg/m3	Machine constant		0.013758	
Cut in speed	4	m/s	Mechanical loss		2	kW
Cut out speed	29	m/s				
Cost per unit		lakhs				
Scaling Factor		kW				

Re calculated for 2 blade turbine.

Wind speed m/s.	Output kW	Pw kW	efficiency	Cp	Pm kW	Pe kW	EFF Cu
4	8.576	21.99	0.390049	0.545185	11.98651	11.73	97.86
4.5	13.420	31.30	0.428695	0.545233	17.06821	16.79	98.37
5	19.504	42.94	0.454203	0.545135	23.409	23.1	98.68
5.5	26.978	57.16	0.472012	0.545163	31.15898	30.81	98.88
6	35.961	74.20	0.484623	0.545131	40.45046	40.05	99.01
6.5	46.579	94.34	0.493721	0.545011	51.4179	50.95	99.09
7	59.026	117.83	0.500935	0.545033	64.22231	63.67	99.14
7.5	73.403	144.93	0.50648	0.544944	78.97762	78.33	99.18
8	89.834	175.89	0.510741	0.544868	95.83627	95.06	99.19
8.5	108.509	210.97	0.514328	0.544856	114.9496	114.03	99.2
9	129.539	250.44	0.517253	0.544912	136.4654	135.36	99.19
9.5	153.054	294.54	0.519644	0.54491	160.4961	159.18	99.18
10	179.204	343.53	0.521649	0.544962	187.2126	185.64	99.16
10.5	208.118	397.68	0.523326	0.544992	216.7339	214.87	99.14
11	239.931	457.24	0.524735	0.545046	249.218	247	99.11
11.5	274.753	522.47	0.525873	0.545009	284.7512	282.16	99.09
12	312.375	593.63	0.526216	0.544491	323.2237	320.25	99.08

3.3 Higher Cut-in Speed

Cut-in speeds can be reduced by reducing the inertia of the rotating mass or by having a variable speed operation (Asynchronous link). Reducing the inertia calls for special designs and materials, as the reduction in inertia should not affect the rigidity and other mechanical parameters adversely. By introducing an asynchronous link and having a variable speed operation this can be accomplished. The generator (and the rotating mass) needs to be runup only to a fraction of its rated speed before generation starts. This makes it possible to extract power from the wind at much lower wind speeds and increases the operating range. This effectively increases the power output.

3.4 Import of Power

Import of power can be reduced by reducing the inertia of the rotating mass or by having a variable speed operation (Asynchronous link). Reducing the inertia calls for special designs and materials, as the reduction in inertia should not affect the rigidity and other mechanical parameters adversely. By introducing an asynchronous link and having a variable speed operation this can be accomplished. The generator (and the rotating mass) needs to be runup only to a fraction of its rated speed before generation starts. This reduces the net power imported from the grid for start up operations.

3.5 Reactive Demand

The reactive power requirement of an WEG is one of the major limiting factors today. This can be overcome by using DC generators or Alternators (Synchronous Generators) in the place of Induction generators. Use of alternators also gives better controllability and scope to generate reactive power, if needed by the grid. But the use of DC generators or alternators has not gained popularity owing to their high cost and maintenance problems. This problem could also be overcome by using an asynchronous link for interconnecting the WEG to the power grid. The asynchronous link being developed and described in this paper, is capable of operating at any power factor as required (UPF, Lead or Lag). This is illustrated in. Figures 3 & 4.

3.6 Poor Grid Penetration

The main reason for poor grid penetration is due to the factors that wind is not (100 %) predictable and is not controllable (i.e. Wind energy is not dispatchable). These two factors make wind energy less reliable. This can be overcome by having integrated (Hybrid) energy systems. By combining two or more energy sources the energy output can be made more predictable and controllable. This combination can be made up of two or more renewable energy sources or one or more renewable sources in combination with a conventional energy source or a storage element. This combination of a RES with a conventional source (100 % Controllable) makes it more reliable and to some extent dispatchable. Some of the examples of integrated energy systems are :

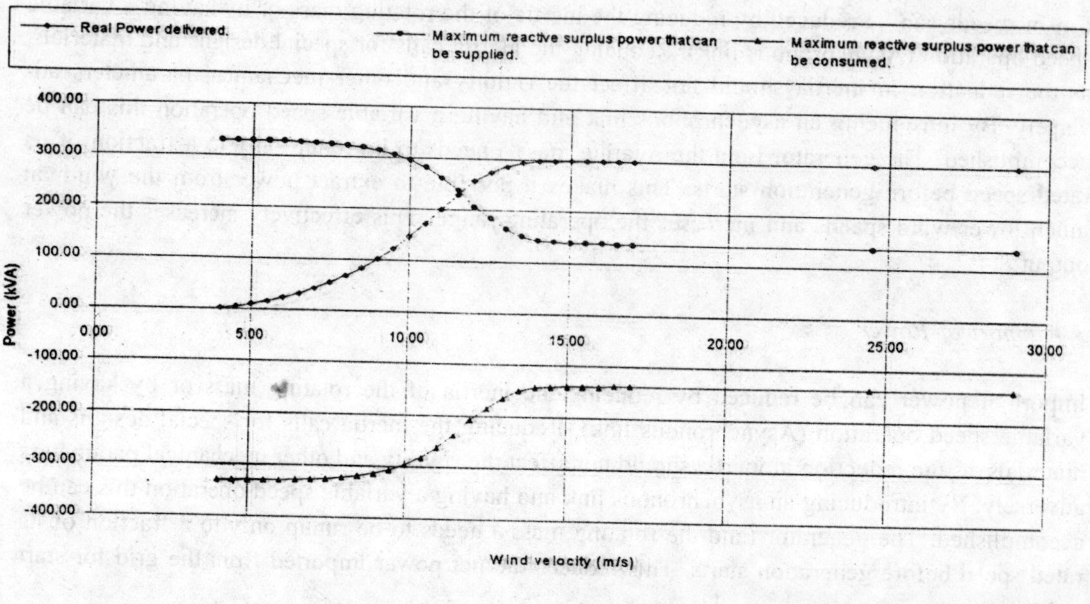

FIG. 3 : *Real Power and reactive power capacities for the proposed scheme.*

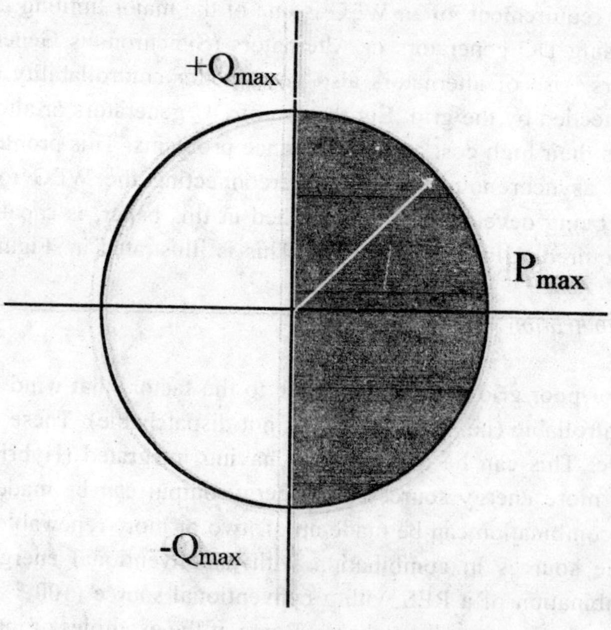

FIG. 4 : Operating circle for new scheme.

— Wind - Diesel
— Wind - Solar - Diesel
— Wind - Battery
— Wind - Pumped Hydro
— Wind - Biomass

In combining such diverse energy sources having a DC (Asynchronous) link helps in proper integration and operation of the system.

Renewable energy source based hybrid systems could also be effectively used as distributed utility. The use of distributed utility helps reduce the capital cost associated with the development of T & D infrastructure and can serve as an effective means to promote rural electrification. Distributed utility also helps reduce line losses by meeting the local demands by locally generated power. These distributed utilities could be based on any of the locally available renewable sources like wind, solar, biomass etc. Distributed utility also helps improve the system reliability and improves the system voltage and frequency profile.

Renewable energy source based hybrid systems could also be effectively used as remote autonomous power stations to supply power to remote villages. This brings power to many villages, without having to wait for the development of T&D infrastructure or generating capacities.

2.7 Unstable Grid Frequency

This problem can be well tackled by having an asynchronous link between the grid and the generator. This link makes the operation of the generator insensitive to grid frequency and voltage conditions.

To sum up from the above discussion it is clear that having an asynchronous link for interfacing WEG to power Grid enables a variable speed operation and eliminates most of the problems associated with present day wind electric generators. This improves the power quality and the payback period, thus making wind power more attractive.

4. The Development

A new "Asynchronous link for interfacing wind electric generators to power grid for enhanced energy capture and better VAR management" is being developed by Crompton Greaves Ltd. The proposed scheme under development eliminates all the above mentioned deficiencies with the present day WEG's. What is new with the proposed scheme is the technology used for the asynchronous link. The Power Conversion System (PCS) technology is a patented technology which is currently being developed by CG jointly with DC Transformation Inc. USA. is a novel concept which has wide applications in the area of power systems. The technology can be used for transformation of power between AC and DC with voltage step up or step down. Some of the applications of this newly developed technology include High Power DC sources, HVDC, Static Var Compensators and Drives.

The salient feature of the technology is that it does not use magnetic transformers for transforming power from one voltage level to another. The technology combines the principle of resonant charging with the principle of voltage multiplication and represents the State of The Art in power electronics and controls. The circuit operates at frequencies of 1KHz. and above.

The asynchronous link developed by CG using PCS technology offers added benefits apart from higher energy capture.

The power curve and operating circle for the new scheme are shown in Fig.3 and 4. It is evident that the system can supply the reactive power needed by the WEG and can also supply reactive (lead or lag) power to the grid. This way the new system can be used not only for enhancing the energy capture, but also for VAR management. The system has a very good dynamic response and typically it would take less than 5 ms. To change the operating point from one extreme (supplying VAR to grid) to the other (consuming VAR from the grid). The operational characteristics and analysis of the proposed variable speed WEG is illustrated in Table 2.

The advantages of the newly developed system are

(i) Enhanced energy capture (typically 10% - 18% depending on site conditions and WEG characteristics)

(ii) Wider operating range and more power output from the generator without overloading the generator. As the power is generated at higher voltages (more than the rated voltage) at higher wind velocities, the cut out velocity can be extended without thermally overloading the generator.

(iii) Eliminates transformer for WEG to grid connection, power factor correction capacitors and associated capacitor switches.

(iv) Power is being generated at Unity Power Factor.(unlike constant speed WEG which operate at 0.6-0.9 pf lag and demands reactive VAR support from the grid.)This eliminates a major disadvantage of the present day systems, that of stopping of generation (even when the wind is favourable) due to reactive loading on the grid.

(v) It is capable of generating both real and reactive power and can be used for MVAR management to support the grid to have improved voltage stability and transmission efficiency.

(vi) Unlike conventional asynchronous links, the new system operates at higher frequencies and hence requires smaller filters. This means improved power quality in terms of lesser harmonics.

(vii) The system is modular and it is easy to upgrade both power and voltage levels (as might be required when the machine size increases or the grid voltage level increases)

(viii) It is compact and efficient.

(ix) Operation at lower speeds during lower wind velocities reduces the noise level and the turbulence created. This improves the overall array efficiency at lower wind velocities.

(x) The developed technology is an integrative and control technology that can be used in conjunction with many renewable energy sources and can play the role of an enabling technology in formation of hybrid energy systems.

5. Conclusion

The paper clearly brings out the limitations with the present day WEG's which act as barriers for the growth of wind energy as a sustainable option. The remedial measures for overcoming these barriers are discussed in detail. The paper also describes a new development, which is a control and integrative technology and could eliminate all the technological limitations associated with present day WEG's. The proposed solution would make wind power generation more attractive by reducing the payback period at the same time offering a superior and diverse performance.

REFERENCES

Wind Energy for the Eighties—British Wind Energy Association, Published by Peter Peregrinus Ltd., 1982.

Wind Energy Conversion Using a Self Excited Induction Generator, G Raina & O P Malik, IEEE Transactions on Power Apparatus and Systems, Vol. PAS-102, No.12, pp. 3933-3936, Dec. 1983

Development of an asynchronous Link for Interfacing Wind Electric Generators to Power Grid for Enhanced energy Capture and Improved VAR Management, R Venkatesh et al, GRIDSAFE, International Conference on Power System Quality and Reliability.

Compact and Efficient Transformerless Power Conversion system,
Rudolf Limpaecher, Topsfield, MA,
U.S.A. Patent Number:5,270,913, Dec.14, 1993
U.S.A. Patent Number:5,357,419, Oct. 1994
International Patent Number: WO93/20610, Oct.14, 1993.

ENVIRONMENTAL CONCERNS AND PROSPECTS OF HYDRO POWER DEVELOPMENT IN INDIA

5

R. S. GOEL
ME (WRD), MSc (Hydrology, Ireland), M. Phill (Social Science),
MDPA, FIE, FIWRS, FIWWA, MIIPA, MICA,
National Council Member,
The Institution of Engineers (India)
Executive Member, Indian Water Resources Society &
Director, Environmental Impacts Assessment
Central Water Commission, New Delhi.

R. N. SRIVASTAVA
B. Tech. (Hons.), FIE, MIWRS, MIIPA
Member (Thermal), Central Electricity Authority &
Ex-Officio Addl. Secy. to Govt. of India and
Former Director-General,
National Power Traning Institute, New Delhi.

ABSTRACT

The hydro power potential in India has been assessed as 84000 MW at 60% load factor. The share of hydel power in the total installed capacity in the country has been declining, Plan after Plan. In the 1962-63, hydro projects had a 50% share in the total installed capacity. This has declined to 26% by 1994-95. Such a dismal share of hydro thermal mix is adversely affecting the optimal utilisation of natural and financial resources. This article reviews the main bottlenecks in hydro development in India while analysing the environmental impacts of river valley projects. The framework for environmental clearance and monitoring of river valley projects in India is also included. The article aims at critical review of the present status and environmental concerns for hydro power development in India; while making concrete suggestions for accelerating the hydro generation.

1. INTRODUCTION

In has not been possible to achieve the capacity addition targets of hydro projects in India due to a variety of reasons. The achievement in hydro capacity additions in the Fifth Plan (1974-79) was over 90% of the target, it declined to 60% in the Sixth Plan. The achievement in the Seventh Plan was also hardly 69%. The likely achievement during the Eighth Plan is expected to be merely 30% of the target of 9282 MW. Hydro thermal mix would come down to an all time low of 25.75% level, which is bound to adversely affect the optimum utilisation of thermal power plants.

The total installed capacity of hydro power plants at the end of Seventh Plan (March 90)

The perceptions in the paper may not belong to the authors' organisations.

was around 18,300 MW—29% of the total capacity. Oly 881 MW of hydro capacity was added during the two annual plans 1990-92. Hydro schemes aggregating to 9,282 MW, which were in the various stages of execution were targeted for commissioning in the Eighth Plan It is unfortunate that only 1,620 MW of hydro capacity could be commissioned so far in the first four years of the Plan. It is anticipated that only about 3,000 MW of hydro capacity (30% of the targetted capacity) may actually materialise in the Plan period; with over 6,000 MW slipping over to the Ninth Plan.

The report on "Perspective on National Power Development upto 2006-07" prepared by Central Electricity Authority (CEA) in September 1991 had recommended that the share of the hydro power in the overall mix of generation on All India basis, needs to be improved to at least 34.66% by the end of 2006-07 (Tenth Plan) by adding hydro capacity of the order of 52,000 MW over 15 years period from 1992-2007. With an anticipated capacity addition of only about 3,000 MW in the Eighth Plan period (1992-97), hydro capacity additions of about 50,000 MW would need to be planned during the Ninth and Tenth Plan periods (about 20,000 MW in the Ninth and about 30,000 in the Tenth Plan). However, keeping in view the average total capacity of 3500-4000 MW getting actually added annually in the previous two plan periods, capacity additions of hydro plants along, averaging to 4000 MW-6000 MW annually during the Ninth and Tenth Plan periods would seem to be not only a formidable task but almost impossible to achieve.

2. HURDLES IN HYDRO POWER DEVELOPMENT

The efforts for the hydro electric development, during fifties and sixties had a meaningful support from surveys conducted from time to time of the hydro electric resources of the country. The first systematic survey was conducted in early fifties, which was based on evolving specific schemes in the river basins; which were considered technically feasible and economically viable. For diverse reasons this discrete methodology has been chosen in preference to the general practice of estimating the exploitable potential from an assessment of theoretical potential. Acording to this survey, which identified 260 possible schemes in different rivers, corresponding capacity of about 42 million KW at 60% load factor. Certain river basins in the Himalayan river systems, specifically in the Upper reaches, were left out due to topographical reasons. The assessment had been in the recent survey conducted by the CEA taking into account the technologies and information in regard to the economic viablility. The economics of power generation also has, in the recent past, undergone a marked change with the energy becoming more and more scarce and costly. Hence CEA undertook an exhaustive and comprehensive assessment exercise taking into account, the latest technological development and availability of much larger and reliable data etc. As a result of this study, the country is now assessed to possess hydro resources of the order of 85,000 MW at 60% load factor equivalent to an annual energy generation of 450 Twh.

It is well recognised that hydro power constitutes the cheapest source of power generation in India even after the recent enormous increases in the cost of the equipment and the implementation of civil works. The economic advantages of hydro power has been enhanced in the recent years with the steep increases in the energy costs from fossil fuel and the rapidly approaching limits to the exploitable resources of such fuels. While the importance of hydro electric development for economic and reliable supply of power is recognised, the dismal share of hydro electric plants in the overall installed generating capacity has been due to the following main factors:

— Uneven distribution of hydro electric resources, among the political divisions of the country.
— Concentration of hydro resources in the States, having neither demand for power nor capability for large hydro development.
— Environmental constraints in terms of deforestation, submergence, rehabilitation and changes in river regimes etc.
— Inter-State and inter-regional disputes.
— Inadequate emphasis during investigation and initial stages of planning.
— Geological uncertainties leading to slowing down of the civil works.
— Socio economic impediments due to far flungness of the sites from the urban centres.
— Attitudinal impediments in terms of choices for shorter gestation schemes (thermal).
— Financial constraints.

3. ENVIRONMENTAL CONCERNS AND HYDRO POWER DEVELOPMENT

The controversy concerning the large versus small dams, fear of earthquakes, silting fo reservoirs, submergence of forests, extinction of "rare" flora and fauna, submergence of ancient cultural heritage, the prospect of outbreak of diseases like malaria and dislocation of population are often cited as reasons for abandoning the large hydro projects. These aspects have been briefly discussed in the following pragraphs. Attempts are made to remove some of the misconceptions, built up over the years, by a torrent of biased media coverage. Basic facts on comparative evaporation, submergence of land, seismic effects and cost-benefit aspects of large-varsus-small dams, water pollution and health impacts; based on significant rigorous studies, carried out by reputed agencies and experts in their respective fields; have been reported here for the purpose of getting the record straight.

3.1 Seismic Impacts

In discussing this much-debated issue, one must not forget the basic conclusion drawn at the International Symposium on Earthquake and dams in China, May 1987. It was pointed out that the dams designed by modern techniques and built according to the latest specifications have considerable reserve strength to withstand severe earthquakes. Because of the application of such advanced techniques in both the design and execution, large dams have larger earthquake resistance capability. On the other hand, the small or medium-size hydraulic structures constructed without the state-of-art techniques remain susceptible to much smaller seismic intensity. Nurek dam of Russia—305 metres high earthen dam sited in highly seismic area and is the highest one of its kind in the world—has already withstood seismic intensity of over 7 on the Richter scale and has remained structurally as sound as at the time of its construction. There is very little proof that can be gathered on the statement that impounding of water in large reservoirs induces seismicity. The fact remains, and this has been adequately documented, that even when such seismic shocks occur, such shocks are much less to the natural tectonic intensity which the dams are designed to withstand. Such large dams as Tarbela, Bhakra, Ramganga and Pandoh are located in the Himalayan region and there is little to prove that these dams have caused RIS or any increase in seismic activity in the area following their construction.

As demonstrated elaborately by Dr. Jai Krishna, President International Society of Earthquake Engg. (ISEE), tall dams in seismic environments are found much more resistant to earthquakes and they are significantly economical as opposed to the dams of lesser height in same areas. Director of Wadia Institute of Himalayan Geology and a noted geologist, Prof. Vikram C. Thakur, has concluded from his studies that the construction of large dams in the outer Himalayan zone does not pose seismic problems and serious flaws exist in the seismic-gap theory. Prof Thakur points out that the basic geological data, now available for all major parts of the Himalayas, shows that large structures can be safely planned and constructed through delineation of active faults. One must also note that present design of dams undergo a detailed and extensive process of field investigations, analytical and laboratory studies, testing of construction materials. Such processes are undertaken in association with an established and wide network of seismological observatories. At every step, adequate care is taken to maintain the desired quality control. Procedures are followed meticulously in the areas of field investigtion, design and construction by qualified personnel and reputed independent technical advisory committee of experts.

3.2 Effects on Wildlife, Monuments and Forests

Irresponsible media coverages concerning the "destruction of flora and fauna" by the large water projects have helped to create misconceptions in public minds. Very few amongst us are aware that the most-frequented tourist and scenic places in India such as the Brindaban and Shalimar gardens, Ukai and Periyar wildlife resorts, Dhyaneshwar and Ramganga Udyans, Kalindi-Kunj, Matatila Lake etc. are diret by-products of various river valley projects. rare species of birds flock to these sanctuaries which have further helped to diversify wildlife significantly. This is most visible in and around Rihand, Matatila and Ramganga dams where the availability of watr all-year round has flourished and diversified the bird species.

Historical and cultural monuments have been successfully relocated from various dam sites giving these a greater lease of life. Such efforts met with great success in Nagarjunsagar, Bargi and Srisailam projects in India and the relocation of Abu Sibel temple in Egypt is wellknown internationally. In the same vein, Jyotirling temple has been preserved in the planning of the Omkareshwar dam and expenses incurred in improving the aproach roads and bridges to the aforesaid temple are a necessary and integral part of the project.

In discussing the fast-dwindling forest land of our country, what is conveniently forgotten are the harmful effects of overcultivation of land to enhance agricultural production, destruction of pasture land through endless grazing of cattle and the shifting jhoom cultivation which still cotinue in some parts of north-east India. Areas submerged by dams are not the real culprits and those above factors along with the illegal felling of trees by rapacious contractors are the fundamental causes behind the denudation of forests.

One other point worth making and that is the loss of forests due to felling of trees to meet the fuelwood requirement of the poor, particularly during the harsh winter days in the mountains. A small amount of hydropower, if made available to the hill people, would reduce such losses significantly. Of the reported 3% loss of the forest area due to river valley projects (Forest Survey of India Report 1991), much of it occur in areas which are often described as forests but do not have the required forest cover and therefore, cannot be described as forest land. People are knowingly confused by not clarifying the terms such as "forest cover" and "forest land".

3.3 Medium and Small River Valley Projects

It has been established that shallow storage causes proportionately greater loss of land arca due to submergence. Shallower storage also means greater evaporation as per studies conducted in many river valleys by reputed scientific agencies. In addition, it is difficult to find large number of alternative sites for medium projects even if such an alternative is preferred over deeper storage. Moreover, small and medim projects have to be constructed generally in the upper reaches of hills, which cause substantial loss of valuable forests. Construction of large dams, on the other hand, in the foothills involves submergence of large areas of cultivated lands per unit of storage. This shortcoming, however, is compensated many ways in multi-purpose projects such as through use of canals *et al.*

For this reason small and medium hydel projects are not only costlier than the large projects but they also submerge larger land areas for the equal amount of storage and in addition, they have increased evaporation losses. Steep gradients in the river beds, large rolling boulders and sedimentation problems further limit the efficacy of small and medium hydel projects. This became particularly evident in case of Ichhari and Maneri Bhali dams (60 metres and 39 metres height respectively), both being run by the river projects. These were filled up to crest by sedimentation during construction itself as planned.

It is evident therefore that large water storage projects are surely better alternatives wherever the parameters such as the volume of water flow, geological and topographical considerations and regional requirements can be satisfied and it can be established that such projects would integrate the environmental, socio-economic and engineering aspects as well. It is also known from historical records that the dynamic nature of environmental effects, which seem adverse to the environment at the time of construction, generally tend to stabilise and become less unfavourable. This has become evident from case studies made for the Aswan, Ramganga Indira Nehar Project, BSL, Bhakra and Hirakud, among other projects.

3.4 Dislodgement of Population

The physical, emotional and financial sufferings of displaced persons caused by the construction of river valley projects cannot be overstated and should be a prime concern for all. Present day trend is to provide liberal compensation, adequate amenities, land and employment to the displaced population. This has been exhibited in the Sardar Sarovar Project, Tehri Dam and such large water projects.

Moreover, attractive provisions made in the Draft National Rehabilitation Policy for WRD projects (approved by the National Water Board) should be implemented *in toto* and the mistakes made in the dam projects of yesteryears must not be allowed to be repeated. Environmental, social and voluntary agencies should make efforts to come up with models of rehabilitation pattern and also cooperate in the healthy execution of viable and highly beneficial multipurpose projects. These agencies must keep in mind the thirst, hunger and employment needs of millions of our countrymen who should be best served through these large water management projects. Successful execution of river valley projects would remove the perpetual economic, educational and primary health-related backwardness prevalent among the tribals, hill people and those residing in the underdeveloped regions of the country. The issue of rehabilitating the displaced people and providing them with long-term meaningful employment out of large water projects is crucial for setting up industrial facilities and power generation stations, building houses for the mass and other vital activities. A planned aproach must be adopted to rehabilitacte displaced people, no matter what sectoral activities caused such displacements rather than indulging in furious battles with the sole purpose of shutting down the hydropower projects only.

In addition, the employment aspect in choosing the mode of construction should be looked into. Choice of masonry dams over concrete dams lead to larger employment at the construction site. Setting up Industrial Traning Institutes and commercial trade training centres; use of donkeys, mules, elephants and bullock carts in the handling of bulky construction materials; developing skilled technicians and project workers engaged in large water projects are some of the healthy practices often adopted in our country. The added advantage of adopting such practices is that it helps to bring the tribal and poor rural folks to the national mainstream.

3.5 Water Pollution

In many developed countries, storage projects are undertaken only for the purpose of diluting the discharged affluents. As per the studies conducted by the CWC to ensure the water quality, it has been ascertained that the quality of water in river basins, such as the Brahmaputra basin, where industrialisation has remained under developed yet, is able to meet the most stringent standards. Nevertheless, in some highly industrialised river basins, lacking the minimum desirable flow during the dry seasons to keep the affluents within permissible limits, building of upstream storage projects are essential to increase the lean-period flows and also to hold back water during the high-flow season for various uses which includes the flood-control.

3.6 Climatic Changes

The micro-climatic changes caused by the construction of large reservoirs are insignificant. However, moderating changes take place in temperature and humidity which are mostly beneficial. For instance, intense heat wave conditions that used to prevail in the Rajasthan deserts and Rihand dam areas have subsided signficantly following the construction of IGNP and Rihand dam. Anyone who has visited Ganganagar in the upper reaches of the IGNP will appreciate the greening of the desert that has taken place, In fact, this particular area has become "exporter" of foodgrains, which is an incredible achievement.

Such climatic changes occur in arid zones due to a high evaporation from reservoirs. This process drops the temperature during the night and cools the large evaporated water mass creating fog-like codition above the water surface. In the south of Brazil, frost has become less damaging to agriculture after the construction of a series of dams permitting growth of species sensitive to low temperature.

3.7 Health Impacts

WRD projects provide a dependable source for drinking water. People from irrigated areas in general enjoy better health and sanitation facilities, thus reducing the incidence of disease. A survey revealed that the Mula irrigation project has made substantial positive impacts on the incidence of illness, leading to a significant reduction in sick days. With IGNP, the percentage of reported relative incidences of water-borne diseases to all diseases dropped considerably over the last 12 years in the areas that received continuous irrigation facilities. As per the Institute of Vector-borne Diseases, there is no focii of Schistosomiasis snails in India. The increased incidences of malaria, reported in some irrigated areas, also occur in relatively naturally-wet areas. Vectoral risks can be reduced by removing sources of stagnant or slow-moving water and by ensuring continual maintenance of drains and canals. As per the UNICEF, Feb, '88 report, new water supply facilities created by large dams improved sanitary conditions which led to significant improvement in health conditions.

4. GUIDELINES FOR ENVIRONMENTAL IMPACTS ASSESSMENT

There was no proper mechanism for assessing the environmental impacts of the river valley projects (RVP) in India till 1978. The Central Water Commission and the Planning Commission had indicated the broad-lines for the preparation of the detailed project report and clearance of the Projects. The Department of Environment and Forests (DOEF), Government of India in 1978 prepared the guidelines for environmental impact assessment of river valley projects. It was aimed in these guidelines that each river valley projects should be subjected to rigorous assessment of their environmental impacts so that necessary mitigative measures could be incorporated in the project right at its inception stage. These guidelines for site selection as well as incorporation of mitigative measures were prepared and circulated among the project authorities through the Planning Commission. It was made imperative to analyse whether the adoption of environmental measures is going to result in any short or long term social and economic benefits or not. Special emphasis was laid on the health effects, plant genetic resources, aquatic life, water logging, salinity of irrigated soils, deforestation and soils conservation for considering the techno-economic viability of the project. It was made mandatory in the DOEF guidelines that the ecological considerations should be incorporated at every stage of planning and construction phases.

The Union Ministry of Irrigation, had set-up a working group in 1979 to prepare the comprehensive guidelines for the preparation of project reports for major RVPs. The working group considered the importance of soil conservation measures in the catchment of RVPs and was of the view that they should be implemented as a complement to river valley projects but should not form part of the project. Aspects recommended for consideration in site selection included immediate and long-term impacts on human settlements, flora and fauna in the vicinity, impacts on monuments and mineral resources, ground water levels, etc. The items mentioned in the Guidelines include most of the important effects on the environment such as (i) involuntary settlement of project affected persons, (ii) sedimentation, (iii) water-logging, (iv) salinity, (v) monuments/shrines, (vi) submergence of flora and fauna, (vii) aquatic life, (viii) health, (ix) climate, (x) water quality, etc.

Later, in 1980, the Forest Conservation Act was passed by Indian Parliament. Under this Act all the States concerned have to get the proposed river valley projects cleared before commencement of construction, in case any submergence or deforestation of land is involved.

The Ministry of Environment and Forests (MOEF) prepared a list of the type of data and basic information on the environmental impact studies in Jan, '85 which were sent to the project authorities along with the Guidelines for Environmental Impact Assessment of River Valley Projects.

Central Water Commission has published "The Guidelines for Sustainable Water Resources Development and Management (Sept. 1992)" to meet the Indian requirement for data collection and analysis, impacts assessment and management plans. Together with EIA for water resources projects, Environmental Management Plan (EMP) is also essential to ensure sustainable development and to limit the stress/load on the system within its carrying capacity. Special attention has to be given for the aspect of rehabilitation and resettlement, compensatory afforestation and catchment are a treatment (of directly drainage areas) in preparation of EMP.

The overall effort while preparing such gudelines and the study of the environmental impacts in various river valley projects is in the direction of rapid economic development on a sustained bassis with due care for safeguarding the environment.

5. ENVIRONMENTAL & FOREST CLEARANCE AND MONITORING

The National Committee on Environmental Planning and Co-ordination was constituted in 1972 by the Government of India for detailed appraisal of the various environmental aspects of development projects. On the recommendation of the Committee, a separate Department of Environment was created by the Central Government with the responsibility of providing costant vigil on the environmental requirements of development projects. It is serving as the nodal agency for environmental protection and eco-system development in a co-ordinating role. It has the responsibility to carry out environmental appraisal of development projects. It also has the direct administrative responsibility for pollution monitoring and regulation and conservation of critical eco-system.

At present, all new development projects and/or proposals for expansion of existing projects in the public sector are reffered to MOEF for environmental appraisal. The Environmental Impact Assessment (EIA) of the projects referred is carried out by the Impact Assessment Division (IAD) of the Department. The IAD, which has a multi-disciplinary base, carries out EIA of projects with the help of inter-ministerial appraisal committee, comprising of experts from related disciplines.

The Feasibility Reports along with Questionnaires and Environmental Management Plans (EMP) furnished by the project authorities are first scrutinised by the IAD and the project proponents are advised to plug the information gaps, if any. The project is then considered by the Expert Environmental Appraisal Committe (EAC). Site visits are undertaken by the sub-committees constituted as and when necessary to supplement the information submitted by the project authorities. The MOEF has developed guidelines, checklists and questionnaires to help the project authorities to collect the relevant information for EIA.

India's National Water Policy (1987) has also emphasised the need for an integrated and multidisciplinary approach to the planning and implementation of water development projects, catchment area treatment and environmental aspects of the catchment and the command areas. In the planning, implementation and operation of the projects, the preservation of the quality of the environment and the ecological balance are the primary considerations. Adverse impacts, if any, on the environment should be offset by adequate compensatory measures. An High Level Environmental Management Committee (EMC) headed by Member (WP & P), CWC with inter-ministerial/inter-agencies participation constantly monitors the environmental status reviews the mechanism established by project authorities to monitor the ecology of the project area and suggests additional compensatory measures. To date, 84 projects have been selected out of which 17 important projects are being closely monitored by the national level EMC. States and projects level EMCs have also been setup for review, co-ordination and ensuring the implementation of environmental safeguards in WRD projects.

6. COMMON ENVIRONMENTAL SAFEGUARDS

The agenda of EAC meeting generally brings out the nature and extent of scrutiny carried out by MOEF and the main observations, recommendations, etc. with regard to safeguards to be suggested to the Planning Commission when accepting project proposals. Some of the commonly suggested safeguards are summarised as follows:

(i) necessary arrangements for the supply of fuel wood by the project authorities to the labour force during the construction period.
(ii) restoration of land in the construction area and prevention of soil erosion;
(iii) compensatory afforestation;
(iv) drawing up of a master plan for the resettlement of displaced persons;
(v) identification of initially eroded areas in the catchment and undertaking of catchment treatment works at projects cost in at least 20% of the catchment;
(vi) a mechanism for the free movement of fish upstream and downstream of the structure across the river,
(vii) preparation of reports on command area development;
(vii) covering of river banks with a 10 m vide green belt;
(ix) provision of drainage-system in the irrigated areas;
(x) measures for the prevention of endemic health problems;
(xi) setting up of monitoring units for the implementation of the suggested safeguards;
(xii) alternatives in case of adverse effects on flora and fauna etc.

7. IMPORTANT SUGGESTIONS FOR ACCELERATED HYDRO DEVELOPMENT

The progress of most of the projects has been slow mainly due to low priority accorded to the project, by the States due to inadequacy of funds year after year, delays in land acquisition, getting forest and environment clearances, identification of non-forest land for compensatory afforestation and ressettlement & rehabilitation problems etc. Since many hydro power projects could not be commissioned within the projected time frame in the past, it sent erroneous signals to the affect that hydro power stations take very long time for completion. This perception further discouraged allocation of funds to hydro projects and most of the States preferred to take up the thermal or gas/liquid fuel based projects with comparatively lower gestation periods to secure benefits in the shorter time frame to meet the rapidly increasing demand for power. This is despite the fact that the development of thermal power plants, in turn, puts added pressure and burden with related additional costs on development of coal mines and rail transportation infrastructure not directly reflected in the project costs. If these costs on development of connected infrastructure are also taken into account, hydro power development would work out to be cheaper. It is therefore necessary that apprehensions that hydro projects involve higher investments need to be properly dispelled.

Although over 245 proposals aggregating to over 93,000 MW capacity are reported to have been received from private investors, very few serious proposals have been received for taking up hydel schemes, Perhaps investment and time involved in making a precise estimate of cost, based on not so thorough investigations made so far for these projects, prolonged time taken for acquisition of land, forest and environment clearance, resettlement problems, not so attractive financial incentives commensurate with the risk involved in taking up hydel projects, etc, could be some of the factors, discouraging the private entrepreneurs from ivesting in hydel projects in a big way. In a number of instances, inadequate investigations of the project formulation had resulted in Geological surprises leading to long delays in completion of the projects and the resultant cost over runs. Private investors cannot afford to take such risks. Some time ago it was felt that with the further liberalisation of government guidelines, more entrepreneurs would show interest, but this has not happened so far despite further liberalisation of incentives on hydro projects in January 1995. The lukewarm response from the private sector may be partly due to gaps in policy guidelines which still inhibit the private investors from investing in this sector.

In order to attract private investors for taking up hydro projects, the government must streamline the procedures for expeditious requisite clearance apart from the need to provide them with complete data and information on hydrological, geological and topographical aspects of the projects to enable the private promoters to formulate the project profiles and cost estimates in a fairly realistic manner. It is learnt that private investors also find it diffcult and furstrating to interact with numerous agencies and virtually impossible to bring even some of them together to consider common problems and obtain necessary sanctions/clearances within a reasonable time frame. Time invariably means money to them and private sector can hardly afford to lose time. Setting up of a High Level Empowered Committee for hydel projects at the cental level is a welcome step to expedite matters. The State Govts. should also help the project authorities by themselves getting all requisite clearances, acquiring the land and building infrastructure, as a part of their contribution to the development of hydel resources. This would enable the private investors to take up the construction of the project without much loss of time. The cost incurred could be debited to the cost of the project and revcovered from the private promoter. In order to tap the vast hydel resources in North Easterm Region, a special incentive by way of additional return on Equity could also be considered.

It is high time that all the concerned agencies responsible for hydro development work out a strategy for development of hydro resources as a National priority. Union Ministries of Power and Water Resources, CEA, CWC and Ministry of Forests & Environment should take a lead in the development of strategy for tackling the causes which are slowing down the pace of progress of hydro projects. availability of adequate financial resources is the foremost requirement for development of hydro resources. Way and means have to be found out to raise additional resources for hydro development. The States should take up construction of only such projects for which it would be able to provide full funds on a year to year basis till the commissioning of the projects. All other projects could be taken up with the assistance of private promoters. Adoption of new design and construction technologies and use of modern construction equipment available elsewhere in the world could help in reducing the construction period of hydro projects. If necessary, technical assistance of reputed and competent foreign companies could also be secured for this purpose. Leasing arrangements for high cost construction equipment could also be considered as it may not be feasible for project authorities to obtain such equipment individually.

For a capacity addition target of about 18,000-20,000 MW of capacity in the Ninth Plan, hydro capacity of hardly 3,000 MW-4,000 MW might come through the private sector route even by liberal estimates. A capacity addition of about 15,000-16,000 MW could still need to be planned in the public sector for which an investment of around Rs. 60,000-70,000 crores may be required. This kind of money would be difficult to comprehend from the Plan resources under the existing resources scenario. Some of the avenues that could be considered to raise the financial resources for hydro development are:

(a) Earmarking 50-60% Plan funds for hydro development.
(b) Joint ventures backed by financial institutions for construction of hydel projects.
(c) World Bank and other International agencies assistance for hydro power projects.

CONCLUSION

Consequently, dams must be built wherever it is technically feasible to store water for domestic, industrial, agricultural purposes and for the generation of power. It becomes evident that the benefits accrued from large river valley projects are so immense that they substantially outweigh the costs of immediate human and environmental disruptions. On the other hand, long-term adverse effects of not utilising the water resources would be catastrophic due to recurrence of floods, droughts and the resulting unemployment which further the backwardness within the people. The cntroversy of the large varsus small dams is irrelevant. The basic fact is that the medium and small water projects cannot substitute the large water storages but can at best complement the larger projects. This, too, depends upon the hydrological, geological, topographical, ecological and regional conditions.

Mankind is at the centre of environment which cannot be protected unless the basic needs of food, drinking water and energy for an all-round development are expeditiously met by accelerated vital improvement of environmental condition through irrigation, water supply, hydro and multi-purpose river valley projects. The pressure groups, lobbies and activists guided by vested interest have already created an environment of hatred opposing large hydropower and irrigation projects in particular. It is time once again to re-emphasise the requirements of these large river valley projects in order to ensure a better and stronger future. It is also important that the projects are evaluated simply not bacause of their size but what they would deliver to the people for years to come. It is essential that water resource development should not be blocked any longer through overstatements of adverse environmental impacts or irrational approach to such an important issue.

Hydro power plants, with their inherent capacity of quick start and stop operation most economically provide an ideal combination with thermal power plants to meet the peaking shortages. More hydro capacity would therefore need to be inducted into the power systems to meet the increasing peaking shortage economically. In view of the obvious advantages of developing hydro resources for power and the fact that a large number of thermal power projects are being set up by private promoters, atleast 50-60% of the total plan outlay for generation should be earmarked for hydro projects only, to be taken up by the States having large untapped hydro potential for benefits from Ninth Plan onwards. Unless the hydro development is pursued with persistent and dedicated mission, hydro development will not take off in a big way. The Union and State Governments, must act in unison to overcome the problems being encountered by the agencies keen to develop hydro projects.

REFERENCES

Beg M.I. and Srivastava R.N., "National Power Plan & Environmental Assessment of River Valley Projects with special reference to Hydro Power Development, "Published in Volume; `Environmental Impacts of Water Resources Development', Edited by Goel R.S., Tata McGraw Hill Publishing Co., Ltd., New Delhi, June '93.

Beg M.I., Srivastava R.N. and Gupta M.L., "Power Planning—An Overview of CEA's Experience" Third National Convention of Electrical Engineers on Planning for Power Development–Issues & Options, New Delhi, Oct. '87.

Central Electricity Authority, `Perspective of National Power Plan Development upto 2006-07', New Delhi, Aug '91.

Central Water Commission, "Guidelines for Sustainable Water Resources Development and Management", New Delhi, '92.

International Association on Electricity Generation, Transmission & Distribution (Afro-Asian Region), "Status Paper-Technical Committee No. 8- Bridging Demand & Supply Gap in Power Sector in India", New Delhi, 1994.

Goel R.S. (editor); `Environmental Impacts of Water Resources Development' Tata McGraw Hill Publishing Co. Ltd. (ISBN-0-07-462127-01), New Delhi, June '93.

Goel R.S.; "Large Dams-Needs and Controversy", Published in Volume; 'Environmental Impacts of Water Resources Development' Tata McGraw Hill Publishing Co. Ltd., New Delhi, June '93.

Goel R.S. and Kamta Prasad, `An overview of Environmental Impacts of Indian WRD Projects' Published in Volume; Environmental Impacts of Water Resources Development', Tata McGraw Hill Publishing Co. Ltd. New Delhi, June '93.

Goel R.S., Sinha K.N. and Chandpuri G.S. `Hydropower Development in India-Prospects and Problems' International Conference on Power Development in Afro Asian Countries, Volume I, New Delhi, 10-14 Dec '90.

Goel R.S. and Ghosh R., `Public Perceptions and Need of Large Dams for Power Generation', National Seminar on

Electrical Energy and Environment, INAE, New Delhi, April '90.

Goel R.S. and Maitra Ramtanu "Large Dams, Media and Ecology' National Seminar on Larger Dams–Environmental Loss or Gain? IWRS, Nagpur, 5-8 Feb ' 92.

Goel R.S. and Maitra Ramtanu, `An Overview of Environmental Impacts, Privatization and Public Perceptions of Hydro-Power Generation', Tenth Indian Engineering Congress, Jaipur, Dec. 20-23, '95.

Goel R.S. and Maitra Ram Tanu, "Large Hydro Power Projects and Environment", XI National Convention of Electrical Engineers, Institution of Engineers (India), Roorkee, Nov. 1995, Published in Volume, Environment Friendly Electric Power Generation", M/s Allied Publishers, New Delhi.

Goel R.S. and Maitra Ram Tanu, "Scientific Facts, Public Perceptions and Environmental Pseudoism Related to Water Resources Management", VI National Water Convention, Bhopal, January 1996.

Goel R.S. "A Critical Review of Environmental Impacts of River Valley Projects" International Workshop on Environmental Legislation & Management for SAARC Countries, New Delhi, October 23-25, 1996.

Institution of Engineers (India), "Proceeding of Third National Convention of Electrical Engineers on Planning for Power Development–Issues & Options", New Delhi, Oct. '87.

Karry W Canter, "Environmental Impacts of Water Resources Projects", Lewis Publishers Inc., Michigan, USA, 1985.

Ministry of Environment and Forests, "Guidelines for Environmental Impacts Assessment of River Valley Projects", New Delhi, '85.

Ministry of Power," India's Eletricity Sector - Widening Scope for Private Participation", New Delhi, Sept. 1996.

Tata Energy Research Institute & Canadian Energy Research Institute, "Planning for the Indian Power Sector - Environmental and Development Considerations", New Delhi, Study No. 62, June 1995.

World Bank, "Country Experiences with Water Resources Management", WBTP No. 175, Washington, '92.

WIND AND SMALL HYDRO ENERGY IN THE INDIAN CONTEXT

6

S.S. MURTHY and BHIM SINGH
Department of Electrical Engineering
Indian Institute of Technology
New Delhi - 110016

ABSTRACT

The paper discusses critical issues, technical and non-technical, relating to harnessing wind and small hydro energy available in different geographical locations in India. Both grid connected and grid independent systems are discussed as regards siting, mechanical and electrical aspects, control systems, indigenous capacities, R&D potential, technical manpower etc.

1. INTRODUCTION

Wind and small hydro energy are leading in the race in exploiting non-conventional energy sources in India, leading to sustainable energy technologies for power generation. Inspite of intense research on solar energy, major breakthrough is not achieved to effectively augment the energy requirement of the society and to reduce dependence of depletable sources. Gadgets using solar-thermal process have not been able to make a major dent in society. Solar PV panels are still prohibitively expensive and water pumping system still needs to be perfected for operational efficiency. Electricity generation using solar energy is still in demonstration stage nowhere near technological adoption. Geothermal and tidal energy are also in demonstration stage. Biomass and ocean (wave) energy need to be further researched for power generation.

While several wind and small hydro units are installed in India, it is necessary to identify various issues such as technology status and future forecast, indigenous potential, operational problems, obsolescence, maintenance, R&D, testing, HRD etc., so that effective actions can be pursued for utilization of these energy resources.

2. WIND ENERGY

Based on the wind mapping supported by MNES, several favourable locations are identified for wind electric farms with typical wind speeds of 5-10 m/s. While Tamilnadu and Gujarat are leading in installation of wind electric systems feeding to local grids, several other States such as Karnataka, Andhra, Maharashtra and Madhya Pradesh are also active.

2.1 Grid Connected System

Typical grid connected wind system is shown in Fig. 1. Wind turbine will generate power proportional to cube of wind speed and swept area, and transfer the same to the induction generator through a gear box. It is important to note that the power generated for each unit varies randomly as wind speed varies. There is a cut-in speed of about 4m/s below which the system has to stall, as also at very high gust speeds. Converted electrical energy is fed to the grid through a transformer and transmission network. System draws reactive power (VAR) from the grid and feeds active power to the grid. Terminal capacitors reduce the VAR drain.

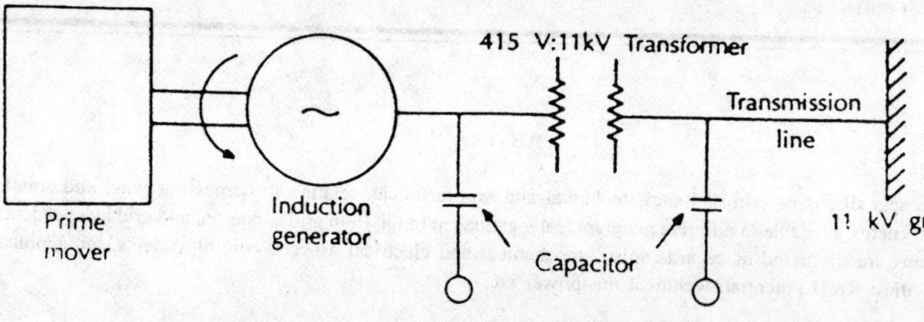

FIG. 1: *Grid Connected Wind Electric System*

All the systems installed in India employ horizontal axis wind turbines fixed to the necelle mounted on a tower. There is a motor driven yaw mechanism to orient the necelle to the wind direction to extract maximum power. Elaborate instrumentation system is incorporated to monitor wind speed, wind direction, temperature of critical points, rotational speed, generator voltage, frequency and current, active and reactive power etc. Appropriate data acquisition and computer control form indispensable subsystem for reliable operation, and to take care of continuous starting, running and stalling of the system.

Each wind system is connected to an electrical terminal box on the ground connected through a cable. The Electrical substation of the local utility located in the vicinity is connected to all these terminal boxes through transmission systems.

Initially 22 to 55 KW units were installed and 55 KW units stabilized and operated for a long time till recently. The Lamba wind farm in Gujarat (Pawan Shakti) then employed 110 KW units. Very recently higher capacity units of 110, 220, 300 KW are finding acceptability and getting installed in new wind farms. Factors deciding the unit sizes are—wind regime, tower height reliable equipments and operational issues. Bulk of the subsystems are imported although phased indigenisation is planned. Present developments in wind farms are subsidy driven needing efforts towards economic viability in the long run.

Based on the experience of operating the grid connected wind systems in India, the following issues can be identified.

2.1.1 Indigenous Manufacturing Capacity

A few companies are active in India in installing wind plants with international collaboration, where bulk of the subsystems are made by foreign partners. The first author was involved in the design of the first indigenous 55 KW Induction generator for wind plants. Based on this experience India is now in a position to produce Induction Generators of any rating for wind electric systems. India also has the capacity to make the hub, tower, gear box and electrical system. Local manufacture considerably eases after sales, service, maintenance and operational problems.

Major effort needs to be made to manufacture blades locally, as they are very critical components. Local expertise in aeronautical engineering in making aircrafts can be utilized to design and manufacture the blades. Instrumentation, electronics, transducers, computer control and data acquisition are other subsystems to be made locally. While some critical components may have to be imported, the subsystem can be built locally.

2.1.2 Grid Interphase

Grid interphase is a critical area on which depends the success of wind systems. The authors have studied this aspect intensively and reported in published literature[1,2,3,4,5]. Grid failure means no power from wind. Voltage/frequency variations, single phasing and supply unbalance considerably affects the power fed to the grid. Therefore a healthy grid is a prerequisite for efficient operation of wind systems. Unfortunately Indian grid condition is far from satisfactory. Drawing a dedicated grid is expensive. Close cooperation of the electricity boards and the wind energy producers (energy agencies) is very essential.

2.1.3 Energy Monitoring and Economics

It is essential to monitor the energy fed to the grid. This requires accurate and reliable instrumentation. Often there are disputes between energy agencies and electricity boards on the quantum of energy supply. To assess the economics of wind power generation, energy monitoring becomes important. Proper agreements need to be reached on sharing revenues and expenses.

2.1.4 Human Resource Development

It is found that proper knowledge of different aspects of wind systems is lacking among personnel dealing with them from policy makers to the operator. It is essential to conduct refresher training courses to personnel at different levels to impart necessary skills. Continuing education programmes for engineers and technicians for design, analysis, manufacture, operation and maintenance of wind systems are essential. Leading Institutions must play an active role.

2.1.5 R&D, Testing and Quality Control

Continuous R&D is required for improving the grid connected wind system to achieve improved efficiency and reduced cost. Analysis, design and simulation studies are required to understand the system. R&D encompasses different areas—aeronautical, mechanical, electrical, electronics, control, instrumentation, computers. Several research topics can be identified based on field experiences.

Testing of wind equipments is another critical area. Suitable test centres need to be established or identified to ensure quality control. These laboratories can also be used for prototype development. Special tests to simulate Indian condition (environmental, electrical) need to be conducted. RISO laboratory in Denmark is a role model. Some action is taken in India in this regard to establish wind energy centre by Government and other agencies.

2.1.6 Obsolescence and Retrofitting

Due to rapid improvements in technology some of the systems are getting obsolete fast. The rapid increase in unit sizes of wind systems make the old low capacity units obsolete and uneconomical to operate. The original installed 22 and 55 KW units may fall in this category and need to be replaced. Therefore action needs to be initiated for retrofitting higher capacity units in same locations. Further the low capacity units can be shifted to remote areas for operation independent of the grid in hybrid mode as described in the following section.

2.2. Grid Independent System

In view of the problem with the Indian electricity grid and the fact that large parts of the country still have unelectrified (or less electrified) habitats with favourable wind regime, grid independent wind energy systems to energize the areas appear attractive. For example, it is reported that the wind farm at Talkaveri in Karnataka is facing a lot of operational problem due to weak and unstable grid since the grid used is a normal 11 KV distribution feeder line. Due to the varying nature of wind energy it is impossible to meet the consumer needs by wind alone necessitating a back-up energy source in parallel. Wind battery system of a few hundred watts have been tried for defence and meteorological applications. In effect we are looking for hybrid energy systems with wind in combination with other energy source to feed a common load. Different possible options of grid independent wind electric systems are given below. Current technology status and actions needed to make them viable are also indicated.

2.2.1 Wind Diesel System

A schematic is shown in Fig. 2 where wind driven induction generator is connected in parallel with diesel engine driven conventional alternator, both feeding a common customer load. A dump load is introduced to absorb excess power above customer load under high wind speeds. Diesel unit will provide customer power above wind power and also compensate the reactive power needs of induction generator which is also partly met by the capacitor. A central control unit effects active and reactive power balancing.

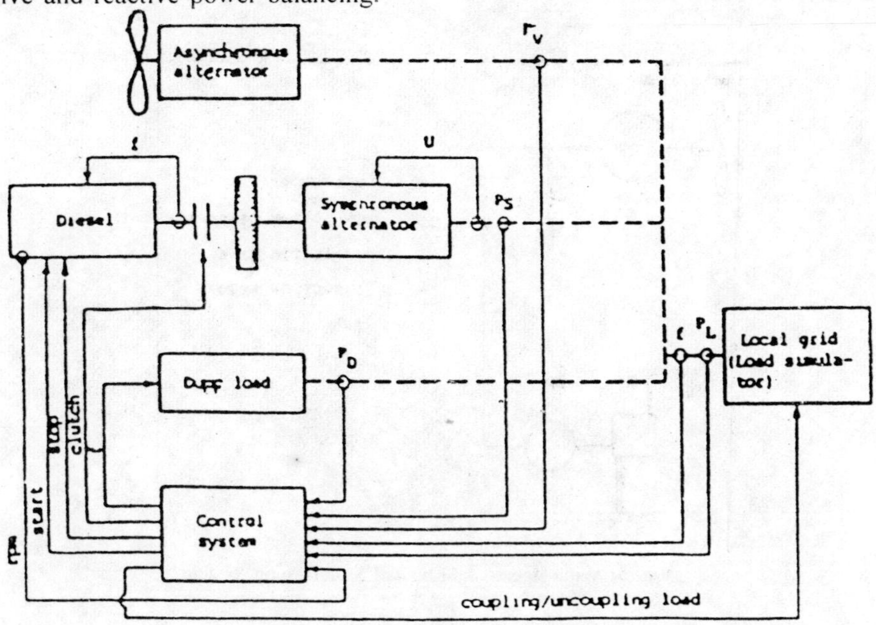

FIG. 2 : *Wind Diesel System*

This system is highly researched all over the world[6,7]. There is enough expertise in India to develop such a system. One may visualize a generalized scheme which can operate with the grid and switch over to wind-diesel under grid failure.

It is necessary to support a mission oriented consortium project involving Academic/R&D Institutions, Industries and User agencies to develop a system for installation in an identified location for field trial and demonstration. This is suitable for isolated locations such as Bet-Dwarka and Andaman-Nicobar, which are presently energized by diesel. Such hybrid systems may affect saving in diesel when wind energy is available. The technology need to be perfected before large scale installations. Perhaps one of the existing 55 KW unit can be used for operation in parallel with a diesel set to create a local grid. Needless to say that very effective control system is essential for the success of the unit.

2.2.2. Wind-Electric-Pump

It is common knowledge that large number of direct wind operated water pumps are not operated in India. Mechanical problems are attributed for large scale failure apart from lack of maintenance and owner accountability. Alternate system proposed is a wind operated induction generator energizing a standard motor pump system (Fig. 3). Depending on the wind speed generator produces variable voltage, variable frequency output which appears across the motor, which in turn runs at variable speed. Since all the components are "off the shelf", the system can be readily installed using a low power wind turbine. This idea needs to be pursued for technical perfection and field trials.

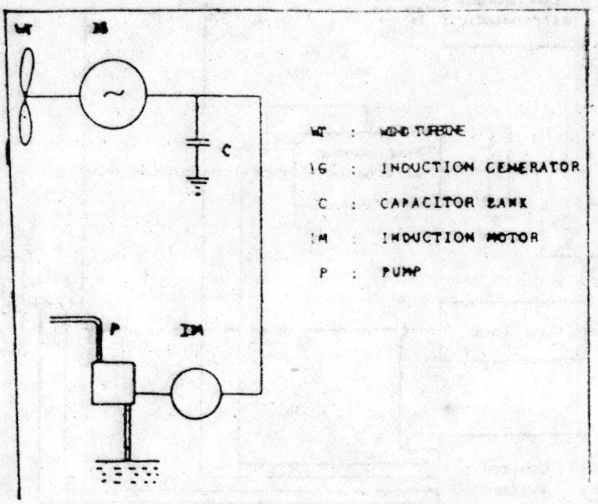

Fig. 3: Wind operated SEIG fed Motor-Pump System

3. SMALL HYDRO SYSTEMS

India has vast small hydro potential which avoids large dams and connected ecological hazards. Several favourable sites are identified in Western Ghats, Himalayas, Kashmir and North East region. The system to be chosen is site specific depending on the available hydro energy in a season. In small hydro system minimum civil works are envisaged. Considerable technological and manufacturing potential exists in India, specially on turbines and generators. Unlike wind systems, Hydro systems operate at fixed input power decided by the "Head" and "Discharge" of the turbine, unless there is a governor control of turbine blades or controlled water inlet.

3.1 Grid Connected Systems

Bulk of the small hydro systems installed in India are feeding to the local grids although a few autonomous systems are also in operation feeding isolated loads. Typical unit sizes range from 100 to 1000 KW. The hydro turbines can be locally manufactured as per the site requirement. Both synchronous and induction generators are employed. Synchronous generator can supply both active and reactive power (through field control), while induction generator has to draw VAR from the grid. Synchronous generator is a complicated machine requiring more maintenance and synchronizing system. Thus induction generator can be a strong competitor to the synchronous generator, and are locally manufactured. Several issues concerning wind system [section 2.1] are also relevant to hydro system. Ideally the hydro system delivers constant power to the grid, unless the turbine input is controlled, while wind system delivers randomly varying power dependent on wind speed. However, seasonal variation of power output is natural due to varying water levels. Normally unit sizes of grid connected hydro systems are larger compared to wind systems requiring appropriate design of the electrical system from generator terminal to the grid including protection. Electronic instrumentation and control are however simpler in hydro system.

3.2 Grid Independent System

In view of steady power from small hydro systems, they can also be operated independent of the grid to feed local loads. Typically 5 to 100 KW unit sizes are recommended. Since the system must be simple uncontrolled turbines (or reverse pumps) can be employed as prime movers. Presently, all the systems installed employ synchronous generators although self excited induction generators (SEIG) can be attractive alternatives [6,7,10,11,12]. Due to near constant power output of the generator, electronic load controllers have to be employed to absorb the power in excess of the customer load. Typical SEIG scheme is shown in Fig. 4. Considerable R&D work is being carried out at IIT Delhi supported by DST and some industries to develop a viable stand alone small hydro scheme using SEIG. It is hoped to achieve a breakthrough so that a practical scheme would be evolved for field demonstrations. Suitable systems to feed single-phase and three-phase loads need to be developed. Intensive experimental and theoretical investigations are carried out at IIT Delhi to develop suitable electrical systems. A single phase 2-winding SEIG is one option for 1-phase loads [13,14,15]. Alternatively standard 3-phase induction motor can be used [16,17], using a combination of capacitors. A demonstration unit of 1-phase and 3-phase SEIG is available in the laboratory at IIT Delhi. Due to the difficulty in supplying power to rural and remote areas through the utility grid, stand alone hydro system would be a boon to such backward locality, which can considerably improve the quality of life of the people through energisation, industrialization and improvement in irrigation for agriculture. Several rural based industries (e.g. food processing cold storage, timber) can be established.

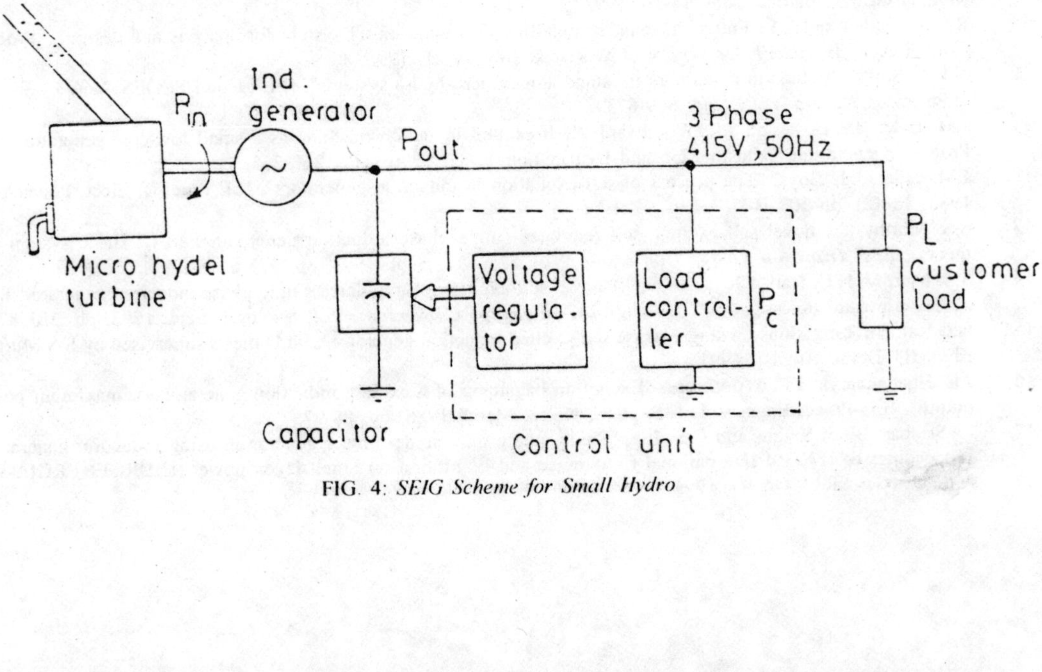

FIG. 4: *SEIG Scheme for Small Hydro*

4. CONCLUSION

Wind and small hydro energy systems are technically viable and can be exploited fully as sustainable energy technology. Grid connected wind systems have become workable in India although several technical operational, industrial and economical issues need to be looked into for better use of available potential. In view of poor grid conditions in several locations, grid independent wind systems need to be developed and installed. Considerable R&D efforts are needed in this direction. Hybrid system including wind and diesel appears to be feasible. Small hydro energy system is a mature technology to generate power and to feed to the grid. Induction generators are recommended for grid connected applications. For remote and hilly regions grid independent system are desirable. A major R&D project is in operation at IIT Delhi to evolve a scheme employing self excited induction generators.

REFERENCES

1. A.H. Ghorashi, "Investigations on grid connected induction generators relevant to wind/hydro energy systems" (Ph. D. thesis supervised by S.S. Murthy), Department of Electrical Engineering, Indian Institute of Technology Delhi, 1993.
2. S.S. Murthy, A.H.; Ghorashi; B.P.; Singh and Bhim Singh; "Analysis of wind driven grid connected induction generators under unbalanced grid conditions", *IEEE Transactions on Energy Conversion*, Vol. EC-9, no. 2, 1994. pp. 217-23.
3. S.S. Murthy, P.S.N. Rao and C.S. Jha, "Analysis of grid connected induction generator under realistic system constraints", *IEEE Transactions on Energy Conversion*, Vol. 5, no. 1, 1990, pp. 1-7.
4. S.S. Murthy, "On grid interface of wind energy conversion systems", TERI, *Energy Environment Monitor*, Vol. 11, No. 1, March 1995, pp. 21-39.
5. A.H. Ghorashi; S.S. Murthy, "Field studies and transient analysis related to wind electric systems", *IEEE Conf. PEDES*, New Delhi, Jan. 1996, Conf. Record Vol I, pp. 271-79.
6. Bhim Singh, "Induction Generators-A Prospective", *Journal of Electric Machines and Power Systems*, (USA). Vol. 23, no. 2, March/April 1995, pp. 163-77.
7. S.S. Murthy and Bhim Singh, "Capacitive VAR controllers for induction generators for autonomous power generation", *IEEE Conf. PEDES*, New Delhi, Jan. 1996, Conf. Record Vol II, pp. 679-86.
8. S.S. Murthy, "Development of wind-diesel electric energy system", Paper presented at Research Review Workshop of GEDA, Baroda, May 15-16, 1992.
9. R.C.A. Cullen and L.L. Freris, "Dynamic modelling of a wind-diesel system for analysis and design", Proc. of Conf. Record BWEA17, University of Warwick, July 19-21, 1995.
10. N.P.A. Smith, "Induction generators for stand alone micro-hydro systems", IEEE Conf. PEDES, New Delhi, Jan. 1996, *Conf. Record* Vol II, pp. 669-673.
11. J.M. Elder; J.T.Boys and J.L. Woodward, "Self excited induction machine as a small low-cost generator", IEE Proc. C, Generation, Transmission and Distribution, 1982, 129(6), pp. 260-265.
12. J.M. Elder; J.T. Boys, "The process of self-excitation in induction generators", IEE Proc. B, Elect. PowerAppl. 1983, 130(2), pp 103-108.
13. S.S. Murthy, "A novel self-excited, self regulated single phase induction generator, Part-I : Basic system and theory, *IEEE Trans. on Energy Conversion*, Vol. 8, No. 3, Sept. 1993, pp. 377-82.
14. S.S. Murthy; H.C. Rai; A.K. Tandon, "A novel self-excited self-regulated single phase induction generator, Part-II, Experimental investigation", *IEEE Trans. on Energy Conversion*, Vol. 8, No. 3, Sept. 1993, pp. 383-87.
15. H.C.Rai, "Investigations on single phase self-excited induction generator", Ph.D thesis supervised by S.S.Murthy, EED, IIT Delhi, August, 1991.
16. J.L. Bhattacharya; J.L. Woodward, "Excitation balancing of a excited induction generator for maximum power output", IEE Proceedings, Vol. 135, pt. C, No. 2, March 1988, pp. 88-97.
17. L. Sridhar; Bhim Singh; and C.S. Jha, "Single phase microhydel power generation using induction generator". Proceedings of IEE 3rd International Conference and Exhibition on Small Hydropower (HIDROENERGIA-93), Munich (Germany), Oct.4-6,1993.

POLICIES FOR COGENERATION DEVELOPMENT IN INDIA

7

PRADEEP DADHICH,
Fellow
TATA Energy Research Institute
INDIA

ABSTRACT

The advantages of cogeneration of power are mentioned. The factors likely to shape the future potential for cogeneration are dealt with. The constraints for promoting cogeneration such as legislative constraints, technological constraints and financial constraints have been touched upon. The incentives provided by the Government of India and the present policy for cogeneration have also been presented.

India's industrial sector is unique because of its vast size and variations in the geography of the country. The industrial sector in the country today consumes about 50 percent of the total commercial energy. The non-commercial energy consumption in the industrial sector cannot be accounted for because no definite market exists. However, in some sectors like sugar, rice mills and other agro-based industries the biomass generated in the manufacturing process is accounted for and consumed partly within the factory for energy purposes. Many of the high growth sectors like iron and steel, sugar, pulp and paper and others do account for over 70 percent of the energy consumption in the sector. Traditionally, petroleum refineries, fertilizer plants and sugar plants have cogenerated power and heat requirements for the process but without exporting to the grid. There are several Chemical and petrochemical processes which generate large quantums of waste heat which could be utilized for cogenerating power in the bottoming cycle mode in addition to topping cyle cogeneration. The estimated potential of cogeneration is between 17000 to 20000 MW on a conservative estimates. It will be very hard to give an accurate figure as many small scale industrial applications and industrial estates can also cogenerate and no estimates are available in these areas.

The power sector in the country is unable to meet the current demands and have a peak shortage of about 23 percent and an energy shortage of about 7 percent. The situation is likely to be aggravated if the IPPs do not come on in time. Some studies have indicated power. The total demand for power will be 120000 MW by the year 2000. The current installed capacity is around 82000 MW which means that an annual capacity addition that will be required is a little above 12000 MW for the 3-4 years. The industrial sector has projected an annual growth rate of around 12 percent. To sustain such a high growth rate high quality stable power will be required to fuel the industrial growth as well as the overall economic growth. So the industry has relied on captive power generation which is primarily diesel based generation and is estimated to be around 10500 MW. This is type of captive generation is not only expensive but also inefficient when compared to cogeneration. Several plants are now exploring the possibility of cogenerating and also exporting to the grid because of the changes in the current policies in the Electricity Supply Act for private power generation.

Research has indicated that the following factors are likely to shape the future potential for cogeneration, most of these currently exists in India:

— Future energy prices, both for cogeneration plants and for competing systems.
— The need for new electricity capacity, based on existing supplies, scheduled retirements and demand growth
— The potential transformation of electric industry structure will likely influence the development of cogeneration in ways which are presently difficult to predict.
— The presence and expansion of those industries which represent the best candidates for cogeneration, as well as the continued electrification of the associated processes for some of these industries.
— Technical advances in cogeneration, competing electric and thermal processes.
— State and central governmental consideration of environmental taxes.

Constraints for Promoting Cogeneration as applicable to India

While the extent of conservation potential through the adoption of cogeneration systems has been well appreciated, as also the numerous advantages one would derive from the concept, there are several barriers to cogeneration which restrict the adoption of such systems by the industries. The identified barriers were determined as a result of direct discussions with end users, equipment manufacturers, and government officials.

Legislative Constraints

Although most States have announced policies for bagasse based cogeneration presently there is no legal framework for the purchase of surplus cogenerated power is compulsory. Very few State Electricity Boards (SEBs) have encouraged cogeneration. In fact the self-generated sales tax imposed by some of the States have acted as a disincentive for the industries to opt for cogeneration. The policies on wheeling and banking announced by the SEBs are not very encouraging. The policies of the various SEBs on cogenerated power is given in the table which is shown as appendix at the end of the paper. It is important to have long term purchase contracts between the SEBs and cogenerators for purchase of surplus power to encourage industries to opt for cogeneration under power maximization mode. In fact cogeneration by industries other than sugar are encouraged to export power to the grid. Cases exists where surplus power generated through cogeneration by chemical and petrochemical industries where the power purchase price is around 80 piase.

Technological Constraints

Cogeneration under the power maximization mode requires high pressure and high temperature system parameters. High pressure boilers, turbines and accessories are not available in sufficient quantities indigenously. Turbines at high pressure are presently imported. There is also a considerable technological gap between the various advanced configurations and systems available in the developed countries. Major technological gaps exists in the area of steam and gas turbines. Mulitiple extraction steam turbines for cogeneration applications are now being manufactured by only two manufacturers in the country. Gas turbines in ranges 1 to 5 MW range capable of using multiple fuels are now being sold in the country.

Financial Constraints

One of the major problems for a firm interested in cogeneration is the relative capital intensity of a cogeneration system. The capital cost of a cogeneration system may range from Rs 1.8 to Rs 2.8 crores per megawatt of installed capacity for oil and gas systems to Rs 2.0 to Rs 3.5 crores per megawatt for coal, wood, or biomass systems. When in times of high interest rates and credit restrictions, such as those we are recently experiencing, this capital intensity makes the financing of such systems a major consideration in their successful development. Where the firm's overall economic picture limits its ability to use the tax benefits associated with such projects, third-party ownership is the answer to this problem in India.

Hence, the industries find it difficult to raise resources for the incremental cost of setting up a cogeneration system with the objective to sell surplus power to the utilities. The tax and duty structure on the capital equipment used for the cogeneration facility are not as attractive as in case of other non-conventional energy sources. Liberal credit facilities such as concessional interest rates, larger repayment period etc. should be provided to the industries to encourage cogeneration.

Incentives provided by the Government of India

The Government of India has targeted a capacity in addition to the extent of 2000 MW through renewable energy including 300 MW through bagasse based co-generation. In order to achieve this target of power generation, the Government of India has announced several measures to promote independent power production. These measures include:

- Upto 100 % foreign equity participation
- Flexible power sales arrangements in which private power companies can act as licensees, supplying and distributing power to a given region or as generating agencies supplying power to the grid at large.
- 4:1 Debt-Equity ratio on investments.
- Single window clearance through a new investment promotion cell.
- ROI of 16 % based at 68.5 % PLF.s
- 30 years initial investment period, with subsequent 20 years extensions.
- 5 years tax holiday, with partial holiday through tenth year of project.

Fiscal Incentives

The other fiscal incentives offered for this sector are mentioned in the table below:

COMPONENT	EXPLANATION
Income Tax	
01. Depreciation	100% depreciation in the first year on all the following co-generation equipments
02. Tax Holiday	First 5 years - Nil, Next five years - 70%.
Customs Duty	Duty leviable on renewable energy power projects of less than 50 MW capacity (under Project Import Category) is 20% ad valorem.
Central Excise Duty	Exempted for renewable energy devices, including raw materials, components and assemblies.
Central Sales Tax	-do-
General Sales Tax	Exemption is available in certain States.

Pricing Policy and Cost Considerations of the Cogeneration

One of the most crucial factor in promoting cogeneration relates to agreement on the tariff policy. Different States have adopted different pricing policy for purchase of power. Utilities while carrying out their long term planning exercise, must consider cogeneration option as a source of firm power supply in relation to conventional power generation options. The tariff for cogenerated power should therefore be set at the avoided cost for new capacity added for meeting base load demand.

Ideally, the utilities must pay the marginal cost or the avoided cost of their own power generation or purchase from any source other than the cogenerator. Utilities however, under the principles of grid discipline, consider the short run marginal costs (fuel and operating costs). This makes cogeneration option inviable and therefore deters investments by the industries. However, under the power maximization option, incremental capacity is added by the industry with the sole objective of supplying to the grid, the cogenerator must be paid both capacity as well as the variable costs. Also, utilities while carrying out their long term planning exercise, must consider cogeneration option as a source of firm power supply in relation to conventional power generation options.

MNES Task Force for Biomass Based Co-generation

The MNES had constituted a task force of experts in August, 1993 to formulate a policy for the Government of India for a National Programme on Biomass Based Co-Generation. The task force recommended the following steps to be taken by the Government to propagate co-generation in the Indian sugar industry.

- — Announce a policy on fiscal/financial incentives for cogeneration systems and provide subsidies/grants for few demonstration projects.
- — Formulate norms for co-generation schemes which would be eligible for incentives.
- — Formulate standard procedures for calculation of rate of return and selling price of power from co-generation plants, keeping in view opportunity cost of bagasse cost of supplementary fuels, if any, and minimum capacity utilization.
- — Formulate model agreement with State utilities to receive power from the sugar industry co-generation.
- — Evolve means of compensating co-generation schemes where the price of power offered by the State utilities does not give minimum 16% post tax return on equity.
- — Appoint suitable agencies to do detailed survey Statewise to assess co-generation potential and shortlist prospective mills for demonstration projects.

National Programme on Bagasse Co-generation

Based on the recommendations of the task force, the MNES had launched a National Programme on Bagasse Based Co-Generation in January, 1994.

The incentives offerred under the programme include:

I. Non-recurring capital subsidy (upto 30% of the equipment cost or upto Rs. 70 lakhs/MW of net surplus power, whichever is lower) for projects that envisage the generation of at least 5 MW of surplus power using boilers which generate steam at a pressure of 60 bar or higher. However, the maximum amount of subsidy to any cogeneration project will be limited to Rs. 60 million in case of existing sugar mills. In case of new sugar mills the Subsidy will be at half of the above mentioned levels. A maximum of 12 demonstration projects are proposed to be taken up in sugar mills in major sugar producing States in the country. Out of the 12 projects planned to be supported under the programme, 9 will be in the co-operative/public sector. The projects will be taken up in those States which have announced a general policy for purchase of surplus power generated.

II. Soft-semi commercial loans for projects which envisage production of at least 4 MW of surplus power. Under this activity MNES will provide grants (upto max. of 20 lakhs/MW of surplus power) to FIs like IDBI, IFCI, ICICI, NCDC, IREDA, etc to enable them to reduce interest rates by 2% on their lending (upto 60% of the project cost) to eligible sugar mills.

III. Assistance in the form of grant to the extent of upto 50 % of the cost of preparation of Detailed Project Reports, subject to a maximum of Rs. 4.0 lakhs.

IV. Support to Research & Development projects in the areas of bagasse drying, bagasse storage and baling, fluidized bed combustion and gasification of bagasse.

V. Support to information dissemination, training and assessment studies.

Multi-Lateral Investment Programmes

The United States Agency for International Development, (USAID) is presently operating a 19 million US $, five year Greenhouse Gas Pollution Prevention Project which is considered a part of U.S contribution to the Global Environment Facility. The project aims at reducing greenhouse gas emissions by catalyzing Indian development of efficient coal and alternative bagasse/biomass energy approaches for power sector expansion within India's economic development programme. The programme for cogeneration is known as Alternative Bagasse Co-Generation (ABC).The ABC programme focuses on options for off-season power generation in sugar mills, so that sugar mills can supply firm power year round.

The highlights of the ABC Investment Component (US $ 9 Million), meant to buy down the cost of the commercial line to be established by the IDBI are as under:

— Aims at demonstration and commercialization of bagasse based co-generation plants that rely on supplemental biomass fuels rather than fossil fuels. The biomass fuels recommended are cane field residues, stored bagasse which has been saved by the adoption of energy conservation measures, biomass such as rice husk, rice straw, groundnut shells etc. and wood from energy plantations.
— This component will work closely with the sugar industry to provide the industry with information, technical assistance and training on all feasible options.
— The component will also cover the incremental costs and risks associated with the initial adoption of technologies and development of skills and knowledge needed for the introduction and use of supplemental biomass fuels by bagasse co-generation systems.
— The ABC projects will have a debt-equity ratio of 3:2.

The objective of the ABC TA Component (US $ 3 Million), is to provide the pre-investment support needed to help stimulate private sector investment in biomass/bagasse based co-generation projects.

Conclusion

In past several studies have established that significant cogenration potential exists in India and the importance of cogeneration has also been recognised by the Government of India but the progress has been tardy. A study carried out by the IEA has expressed that the requirements for development of cogeneration systems in industry as follows:

— Clear and equitable interconnection specifications to the electric grid.
— Stable fuel supply.
— Unbiased standby rates and demand charges.
— Stable regulatory environment.
— Predictable and sufficient revenue stream.
— Power purchase agreements that are financeable (without open-ended liabilities).

Unless these issues are tackled and the vaiability of the SEBs established assuring long term payments to the cogenerators the various cogeneration programmes will not take off. However, some States have taken a substantial lead in the bagasse-based cogeneration, which shall pave the way for other cogenerators all over the country.

APPENDIX

State Government Policies and Incentives for Bagasse Based Co-generation

Description	MNES Guideline	Tamil Nadu	Karnataka	Maharashtra	Uttar-Pradesh	Madhya Pradesh	Punjab
Power Wheeling Charges	2% of energy fed of the grid, irrespective of the distance from the generating station	2% of energy generated for 3rd party 2% for sister concern within 25 km radius.	6% of the energy generated for 3rd party. (Likely 5%)	Nil	12.5% of the energy generated	2% of the energy generated	2% of the energy generated.
Power Banking Period/Charges	One Year/Nil	2% of energy generated	6% of the energy generated.	—	One year	Not specified	Not specified
Buyback rate by SEB	Rs. 2.25/kWh, with 5% escalation without any restriction on time or quantum of electricity supplied	Rs. 2.25 kWh, with 5% escalation every year for 5 years.	Rs. 2.25/kWh	Rs 2.00/kWh for export upto 4 MW. Rs 2.25/kWh for export above 4 MW. (Above rates for 6 years).	Rs. 2.25/kWh or at a rate which ensures a RoE of 16% and 68.5% PLF, whichever is lower.	Rs. 2.25/kWh	Rs 1.50/kWh
3rd Party Sale	Allow to HT consumers within the state at mutually agreed rate	Allowed to HT consumers at mutually agreed rate.	Allowed	Not Allowed	Allowed to HT consumers at mutually agreed rate within a radius of 5 km.	Allowed to HT consumers at mutually agreed rate.	Allowed at mutually agreed rates.
Capital Subsidy	Subsidy given to industrial units in backward areas.	Nil	Rs 25 lacs/MW (export)	For export > 4 MW, participation in cooperative mills by MSEB - upto 60% of equity.	Nil	Not specified	As extended to other industries.
Other Concessions	* Exemption from electricity duty.	1) Exemption from electricity	1) Exemption from electricity tax for 5	—	Nil	(1) Exemption from electricity duty for 5	(1) Exemption from electricity duty.

CLEAN COAL TECHNOLOGIES FOR POWER GENERATION

8

Dr. Y. P. Abbi & A.K. Ghosh,
Bharat Heavy Electricals Limited, New Delhi.

ABSTRACT

Power is a vital input for the development of the country. In this paper, the quality of Indian coals, and technological developments in coal fired power stations are presented, clean coal technologies, fluidised bed boilers, pulverised coal fired power plant and integrated gasification combined cycle are dealt with.

1. INTRODUCTION

Power is a vital input in the process of development of the country. While the gross domestic product of the country grew from 90.4 thousand crore rupees in 1970-71 to 269 thousand crore rupees (1980/81 prices) in 1995-96, the installed power generation capacity rose from 14709 MW to 81973 MW during roughly the same period. Out of this, the installed capacity of thermal generating stations, which largely means coal-fired power stations, even though gas-based power plants have made their appearance in the power scene of late (6268 MW) increased from 7900 MW in 1970-71 to 59160 MW in 1995-96. Thus, 72 percent of the installed generation capacity at present is based on thermal generation. These figures illustrate the dominant role played by coal fired power stations in the power as well as the economy of the country as a whole. This situation will continue to prevail in the foreseeable future given the relative magnitudes of the coal reserves of the country vis-a-vis oil and gas reserves, and the uncertainties associated with nuclear power, while large scale commercial exploitation of non-conventional energy sources is still some years away.

The above scenario calls for a critical look at the current technologies for coal utilisation in the power industry as well as the development trends which will have an impact in the future.

2. QUALITY OF INDIAN COALS

Some of the important physical and chemical characteristics of coal which are relevant to power station use are; the calorific value, ash content and ash characteristics, sulphur content, volatile matter, moisture content etc.

Indian coals generally have a high ash content. Coals meant for power station use have ash content of around 40 per cent or more. This percentage has been rising over the years due to the increasing adoption of open cast mining (while 20 per cent of total coal production in 1970-71 was through open cast mining, this percentage went up to about 75 per cent in 1995-96). The ash in Indian coals is highly abrasive due to the high silica content which exists mostly in the form of Alpha-quartz. The sulphur content is usually low (less than 0.5 per cent) except in the coals found in Assam. The calorific value is in the range of 3000 to 4000 kcal/kg. The washability characteristic of the Indian coals is poor due to the dispersion of the inert matter throughout the coal body.

These specific characteristics of Indian Coals call for tailor-made designs of coal handling and coal firing equipment, boiler furnaces and heat transfer surfaces, electrostatic precipitators, ID fans etc.

3. TECHNOLOGICAL DEVELOPMENTS IN COAL-FIRED POWER STATIONS IN INDIA

Since independence and till 1969, all the thermal sets were imported and were generally in the range of 30 to 60 MW. BHEL commissioned the first indigenous 30 MW and 60 MW sets in 1969-70. This was followed by the indigenous manufacture of 100/110/120 sets in the early seventies. Late seventies saw the introduction of 200/210 MW sets, to be followed by the 500 MW sets in the early eighties. As on date, 119 nos. of 200/210/250 MW BHEL sets and 14 nos. of 500 MW BHEL sets amongst a total of 219 BHEL thermal sets are operating in the country producing about seventy-five percent of the total thermal electricity produced in the country.

Both the 210 MW and 500 MW thermal sets employ sub-critical single reheat cycles with main and reheat steam temperatures of 540 degree C. BHEL have acquired substantial field experience on the operation of these units over the last one and a half decades. This experience has been fed back into the design of the units which is reflected in the significant improvement in the availability of these sets. This is indicated by the increase in the Plant Load Factor (PLF) of thermal power stations supplied by BHEL from 45 per cent in 1980-81 to 63 per cent in 1995-96 (78.7 per cent for 500 MW sets and 66.4 per cent for 200/210 MW sets).

In so far as the exploitation of Indian Coals is concerned BHEL,on the basis of its unique experience of handling a wide range of low grade coals, has developed the expertise to evolve the designs of different components of the boiler for its reliable operation for coals of calorific value as low as 3200-3500 kcal/kg.

4. CLEAN COAL TECHNOLOGIES

In the initial years of growth of the power sector in the country, the emphasis, naturally was on establishing the conventional coal-fired power plant technology so that the growing power demand could be met as quickly as possible. In course of time, as the consumption of coal grew rapidly to meet the rising power demand, its adverse effects on the environment began to be perceived. Advanced countries, where the per capita use of energy is ten to twenty times that of India, had, by seventies, already initiated measures to curb the emission of pollutants such as oxides of sulphur and nitrogen, and solid particulates. In India, the situation is somewhat different. Though SOx pollution is not an immediately significant issue due to the generally low sulphur content of coal, lower SOx content has the effect of reducing the conductance over the fly ash particles; thus leading to larger size of electrostatic precipitators. On the other hand, the high and abrasive ash content of the coals is a matter which needs careful consideration from the designer of coal combustion and ash collection/handling equipment. Regarding CO_2 emissions, it is a matter of global concern since it contributes to the "Greenhouse effect" which leads to global warming.

In the mid-seventies, BHEL had initiated a R&D programme with a view to develop technologies to utilise Indian Coals more efficiently and cleanly. Efficiency improvement also translates into less emission of pollutants per unit of electricity generation. Almost all approaches for efficiency improvement lead to thermodynamic cycles involving a gas turbine in topping cycle and a steam turbine in a bottoming cycle; the so-called combined cycle. While modern gas turbines are capable of operating in the temperature range of 1100/1200°C. to 550°C, the steam turbine takes over from the lower limit to the cooling water temperature. As a matter of interest, the latest heavy duty gas turbines are claimed to be capable of operating with inlet temperatures of almost 1300°C. The thermal efficiency of a combined cycle based on such a gas turbine is projected to reach almost 60% as against the level of 40% achieved by the conventional steam cycle. However, gas turbines need clean fuels such as natural gas or naphtha. Direct firing of high ash coal in a gas turbine combustion chamber is not technically feasible yet. Therefore, use of coal in gas turbines/combined cycle calls for conversion of coal to a clean fuel such as coal gas. Thus it is a fortunate coincidence that the high efficiency combined cycle for coals also leads to high efficiency clean coal technology.

4..1 Fluidised Bed Combustion Boilers

Before we consider combined cycle based clean coal technologies, an alternate technology to pulverised coal fired conventional power plant may be mentioned. This is fluidised bed combustion technology, which can be used for effective utilisation of low grade coals and washery rejects. Among the advantages of the fluidised bed combustion boilers are, compactness, ability to burn fuels with calorific value as low as 1800 kcal/kg while producing ash which is least erosive and combustion is independent of oil support. It is a clean technology since NOx emission is less due to low combustion temperature and sulphur can be retained in the bed itself by the introduction of suitable quantity of limestone into the bed alongwith the coal feed.

The fluidised beds are of two types – bubbling and circulating. While bubbling beds have low fluidisation velocities to prevent solids from being elutriated, circulating beds employ high velocities to actually promote elutriation. The solids are re-cycled while the hot gases pass through the convective sections.

BHEL has been carrying out R&D work on bubbling fluidised bed since 1975 and has developed boilers upto capacity rating of 30 MW for high ash coals and washery rejects. For units of capacity higher than 30 MW, it has been found that circulating fluidised bed combustion technology is more economical for high ash coals and/or high sulphur coals. BHEL has set up a one m^2 cross-section CFB test rig in association with Technical University of Nova Scotia, Canada and IDRC/Canada. Tests have been successfully carried out with representative high a.

Indian coals. In order to meet the immediate market requirements for higher capacity CFBC boilers, BHEL has entered into a tie-up with a leading international company in CFB technology. BHEL is currently executing an order for two units of lignite-fired CFBC boilers of 390 tph steam flow (125 MWe) each, and one unit of 175 t/h steam flow while firing high ash coal.

4.2 Pulverised Coal-fired power plant with Ultra Super Critical (PF-USC) Parameters

During the seventies and eighties, it appeared that the conventional pulverized coal-fired conventional power plant had reached a plateau in terms of thermal efficiency. The efficiency achieved was of the order of 40% in US and Europe. The corresponding figure for the Indian power stations would be in the range of 36-37.5%. The difference can be attributed to factors such as use of double reheat cycle as against single reheat cycle, low-ash coal as against high ash coal and lower cooling water temperature etc. The main steam temperature employed in these plant is limited to 540°C, which was dictated by the techno-economics of materials used for the high temperature piping and boiler pressure parts. Until a few years ago, use of steam temperatures higher than 540°C required the employment of expensive austenitic steels. With the development of high temperature ferritic steels, there was a revival of interest in ultra super critical parameters.

USC technology starts above a main steam pressure of about 250 bar and steam temperatures over 560°C. As an example, ELSAM of Denmark is setting up two coal/gas fired units of 385 MW each, having steam parameters at turbine inlet of 285 bar, 580°C with double reheat to 580°C. With cooling water temperature of 10°C, the thermal efficiency is projected to be 47 per cent (Net plant efficiency, LHV based). These plants are scheduled to be commissioned in 1997 and 1998.

The boiler is a once-thru sliding pressure tower boiler, with tangentially fired low-NOx burners. To control high temperature corrosion on the outer tube surfaces and steam oxidation on the inner tube surfaces, the last parts of the final super heater use austenitic tube material. Thick-walled components are made of the new 9% Chromium Steel P91. The spirally wound water walls are made of 13 Cr Mo 44 material.

In our country, the steam parameters remain sub-critical, though BHEL has introduced the once-thru boiler technology for sub-critical parameters in the Talcher 500 MW sets. Regarding the use of high temperature cycles, BHEL has developed and supplied an industrial steam turbine with live steam temperature of 565°C. Unit 6 of Trombay TPS has a 500 MW set with reheat temperature of 565°C.

4.3 Coal based Combined Cycles

As mentioned earlier, the Combined Cycles offer the highest thermal efficiency of the presently used technologies. However, the coal cannot be directly fired in combustion chamber of a GT; the reason being the high particulate loading in the combustion products. Two technologies have evolved to use coal in gas turbines/Combined Cycles, viz. (a) Integrated gasification Combined Cycle and (b) Pressurised fluidised Combustion boiler/combined cycle.

4.3.1 Integrated Gasification Combined Cycle (IGCC)

Gasification is a process to convert coal into a clean fuel for power generation. Gasification can be viewed as a partial combustion process in which air and steam or oxygen and steam react with coal to produce a fuel gas containing CO and H_2 as the main combustible constituents. Use of oxygen results in a medium calorific value product gas while the use of air produces low calorific value gas. Gasification can be carried out under atmospheric pressure or under higher pressures. For Combined Cycle applications, it is techno-economically advantageous to adopt the pressurised process. Three different gasification processes have been developed:

— Moving Bed,
— Fluidised Bed and
— Entrained Bed

In the moving bed gasifier, sized coal of 6 to 25 mm moves downwards in a counter-current flow to the gasifying media. The product gas leaves the gasifier at a temperature of around 540°C. The major limitations of the moving bed technology are that it has low throughputs and it generates tar during gasification. BHEL has designed and established a 150 tonnes per day (tpd) pressurised moving bed gasification system at its Tiruchi Complex. It has been operated since 1989. Three different types of Indian coals have been tested.

In the fluidised bed gasifier crushed coal of minus 6 mm size is gasified in a fluidised bed. The advantages of this process include the elimination of rejects during coal preparation, elimination of tars and oils in the product gas due to the operating temperature of around 900°C and a higher throughput than the moving bed gasifier. BHEL has designed and setup a 18 tpd pilot scale pressurised Process Equipment Development Unit (PEDU) at Hyderabad. This facility is being operated since 1990. The data generated through this facility has been used to design and set up a 150 tpd pressurised fluidised bed gasifier. This gasifier is under commissioning at Tiruchi Complex of BHEL alongwith its 6 MWe IGCC.

In the Entrained bed gasifier, coal is fed in the pulverised form. The process temperature is maintained at around 1500°C and ash, in the form of slag, is withdrawn from the bottom of the gasifier. Though its throughput is the highest, the entrained bed gasification process is not suitable for Indian coals with high ash fusion temperatures. Flux has to be added to generate free flowing molten ash. The overall efficiency is drastically reduced due to the loss of energy through the molten ash which is not recoverable.

Among the three processes for coal gasification, the Pressurised Fluidised Bed Gasification (PFBG) is considered the most suitable for the production of coal gas from Indian coals for power generation.

In an IGCC plant, the hot raw gas leaving the gasifier is passed through a gas cleaning system where the gas is cooled by generating steam in a heat recovery boiler. The steam thus generated is fed to the downstream steam turbine thereby constituting one element of integration. The cold gas is cleaned by scrubbing to remove particulates, halogens, ammonia etc. The clean coal gas is burnt in the gas turbine to generate power at the first stage. The hot exhaust gases from the gas turbine produce steam in a heat recovery steam generator and this steam is also fed to the steam turbine to produce power at the second stage. The air for gasification is bled from the gas turbine compressor, thus completing the integration of the gasifier system with combined cycle plant.

The efficiency of the IGCC is the product of the gasification system efficiency and the combined cycle efficiency. The combined cycle efficiency is directly related to the gas turbine inlet temperature. Thus the IGCC efficiency, being dependent of the GT inlet temperature, is expected to continue to rise significantly in future since there is a continuously rising trend in the GT inlet temperatures for the past few decades.

Recognising the potential of the IGCC concept, BHEL had conceived, designed and set up a 6.2 MW IGCC Demonstration Plant at its Tiruchi Complex, based on the 150 tpd moving bed gasifier, mentioned previously. Integrated operation of the plant on high ash coal (46%) has been achieved as envisaged in the design.

Currently, BHEL is engaged in commissioning the 150 tpd pressurised fluidised bed gasifier at the same site. This gasifier will be coupled to the combined cycle part of the plant, by substitution of the moving bed gasifier. This plant has been commissioned and run for two weeks. The plant is under repairs now because of fire on the coal conveyor belt. It is likely to be restarted by October/November 1996.

There is considerable world-wide activity in the IGCC arena, though a majority of the projects are based on the entrained bed gasification process due to the low-ash content of US and German Coals. One 250 MW plant is planned for operation in 1996 at Florida, USA. Another unit of similar capacity has been commissioned at Buggenum, Netherlands in 1995. A fluidised bed gasifier based IGCC of 100 MW rating is also planned for operation in 1996 at Pinon Pine, U.S.A.

4.3.2 Pressurized fludised bed combustion combined cycle (PFBC-CC)

PFBC is another combined cycle concept for coal based power generation in an efficient and clean manner. In this concept the conventional combustion chamber of the gas turbine is replaced by a pressurised fluidised bed combustor. The products of combustion pass through a hot gas clean-up system before entering the turbine. The heat of the exhaust gas from the gas turbine is utilised in the downstream steam turbine. The fluidised bed combustor for generation of steam could also have coils to extract heat from the combustor for generation of steam for use in the steam cycle.

The development of this technology is dependent upon the compatibility of the hot gas clean-up system with the gas turbine inlet requirements in terms of maximum particle concentration and size (typically 5 mg/cu.m. and 5 micron). The efficiency of the system is also limited by the process requirement that the fluid bed temperature be kept below the ash fusion temperature, thus placing an upper limit on the GT inlet temperature. Thus PFBC-CC operates with a GT inlet temperatures in the range of 850°C to 900°C. The development of this technology is crucially dependent upon the development of high efficiency hot gas clean-up system as well as the development of gas turbines which are more tolerant of the inlet particulate loading. So far four PFBC-CC demonstration plants in the capacity range of 70 to 80 MW are operating in the world. All these plants are from only one technology source in the World. The advantages of fluidised bed combustion, such as fuel flexibility and ability to burn a wide range of low grade coals make the PFBC-CC a potentially attractive option for India. Recognising this, BHEL has established a 6 tonnes per day PFBC test facility at Hyderabad and the suitability of the PFBC process has been proved for different Indian coals. Considerable test data has been generated on Combustion which will be useful for designing next higher size PFBC plants. This would, however, be possible only after suitable hot gas clean-up system is available commercially

5. CONCLUSION

Out of the clean coal technologies considered here, the atmospheric fluidised bed technology, both the bubbling and circulating type have already been commercialised. Of the technologies of the near future, the IGCC appears to be the most suitable in the Indian conditions in view of the availability of technology and equipment in the country/abroad, suitability of the process for Indian coals and experience of operation of pilot/demonstration plants. BHEL, in association with CSIR have already prepared a proposal for setting up a 45 MW IGCC retrofit. The potential site for this has also been located. The IGCC demonstration plants project has also been identified as one of the projects under the clean coal technologies mission of the Government of India. However, the funds required for the demonstration plant have not been sanctioned. It is felt that the time has come to set up such a technology demonstration plant with our coals as the final step towards commercialising a clean coal technology in the country.

NATURE OF LOW VOLATILE COALS OF JHARIA AND THEIR PREPARATION ASPECT

9

KALYAN SEN, S.G. CHAUDHURI and K.S. NARASIMHAN
Central Fuel Research Institute, Dhanbad (INDIA)

ABSTRACT

The coking coals of lower seams of Jharia coalfield are generally high ash, high inert and difficult-to-wash. The large reserves of such coals in India demand development of aggressive coal preparation technologies, specially when the reserves of good quality coking coals of upper seams are depleting. These high rank (mean Ro% - 1.25 to 1.35) low volatile coking coals have extremely poor liberation characteristics as indicated by petrographic studies. It has been concluded that these coals are not amenable to a simple, single-stage washing at reasonable recovery, so a multi-stage approach for washing these coals is necessary for the better utilization of reserves.

1. Introduction

With the increasing demand of coking coal for steel plants, the emphasis is to produce more coking coal with mechanisation, either in the underground or at the opencast which resulted in mixing of the coal so produced with extraneous materials. As good quality coking coals of upper seams (XII to XVIII) of Jharia is gradually depleting BCCL has been forced to go for low volatile coals of lower seams to produce more coking coal so that the demand of the steel plants could be met. This change in production technology desires change in technology of preparation of these coals as well. Hence a serious thought has to be focussed to modernise the existing washeries as well as the future washeries to take care of the feed which will come mostly from the lower seam coals whose physico-chemical vis-a-vis micropetrographic characteristics are completely different from the coals of upper seams of Jharia coalfield.

Before discussing the characteristics of lower seam coals, it will not be out of place to put forward comparison between Gondwana coals of Permian age and Carboniferous coals of Europe and North America. This will throw light why coking coals of India in general and lower seam coals of Jharia in particular are difficult-to-wash. India has large reserves (about 15.000 mt) of high ash, difficult-to-washing characteristics low volatile coking coals[1].

2. Geological Comparisons

The Carboniferous coals of Europe and North America are very closely related in their physical,

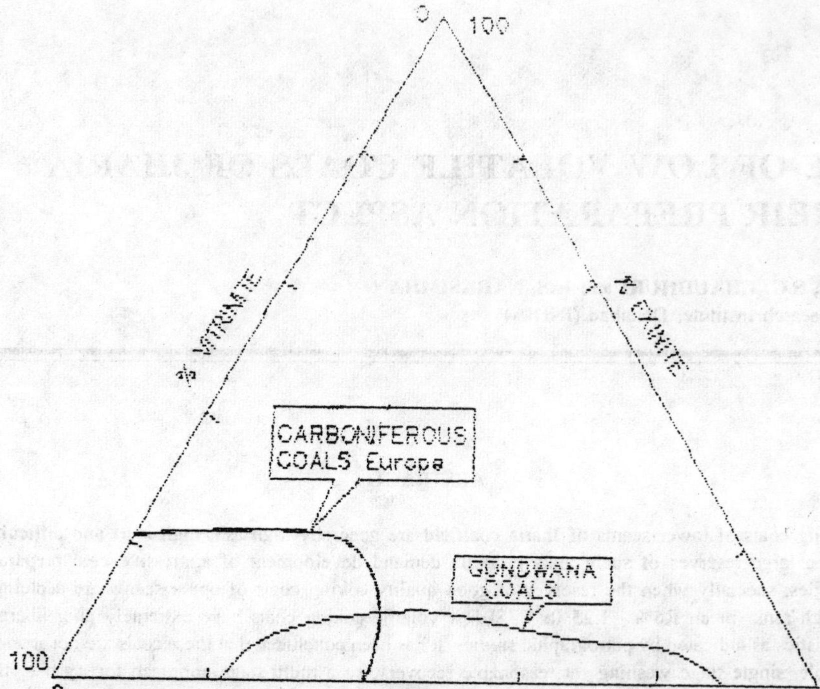

FIG. 1 : *Maceri Composition of Carboniferous and Gondwana Coals (after Sanders et. al.)*

chemical and petrographic properties. While the Gondwana coals are in many respects markedly different from the Carboniferous coals [2, 3]. Major differences existed in the geotectonic position, palaeotopography climate and flora and each of these had its effect upon the resulting petrography and mineral matter content. Fig. 1 depicts that the Carboniferous coals are predominant in vitrinite content while Gondwana coals are in general inertinite rich[4]. Petrographic composition explains significant differences in coal chemistry and has a direct effect on utilisation.

3. Petrographic and Other Characteristics of Low Volatile Coals

The lower seams coals of Jharia in general and V/VI/VII seam coals in particular have the following characteristics[5-7] which are due to their mode of origin and nature of deposition as already cited.

(a) The seams are very thick upto 25 and highly interbanded.

(b) The inertinite content may reach as high as 60 - 70 %, sometimes even more, mostly forming vitrinite type of microlithotypes.

(c) Mean reflectance of vitrinite varies between 1.25 and 1.35 percent (Fig. 2).

(d) One of the most significant characteristics of these coals is in their relatively high proportion of syngenetic mineral matter content (>10%) which is particularly important from the technological point of view when considering traditional methods of coal beneficiation. The most important syngenetic mineral is the various clays (70% or more) which are usually intergrown with the organic mass on a microscopic scale.

(e) The relatively low volatile matter of these coals is directly attributable to their low vitrinite content.

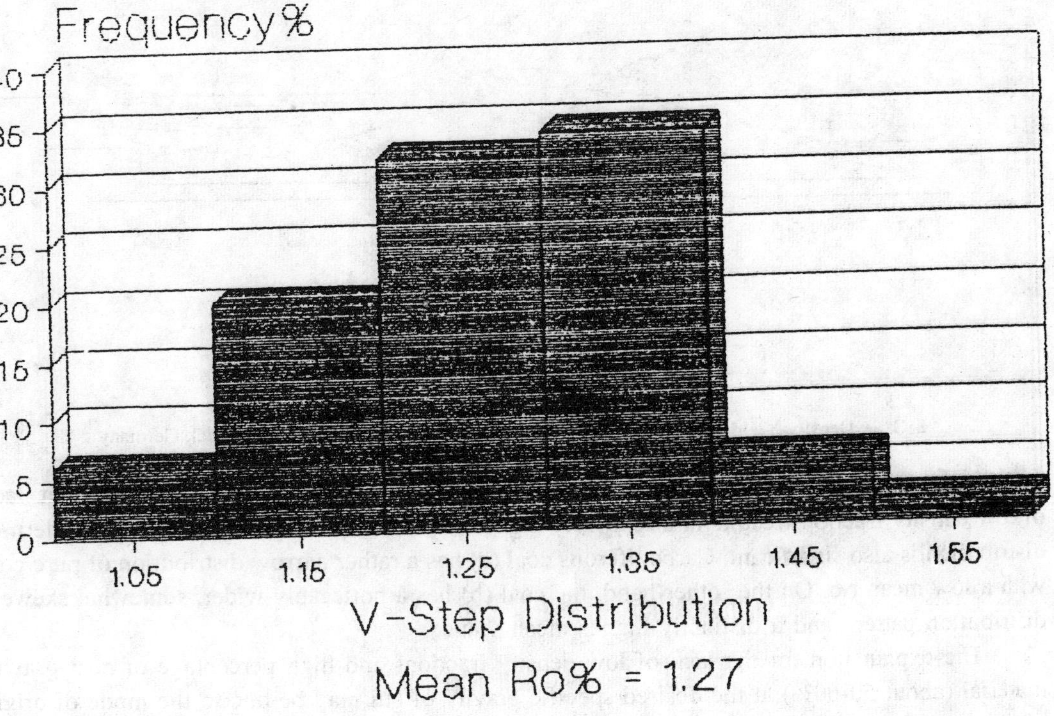

FIG. 2 : *Reflectogram of a Low Volatile Coal, Jharia Coalfield*

4. Particle Size and Density Distribution Characteristics

The work of Bennet[8] and Raab[9-10] had indicated that the particle size distribution patterns of Gondwana and Carboniferous coals are similar and they lie within the fairly narrow range of n=0.62 to 0.87 Rosin-Rammler distribution coefficient. From a wide variety of data, no detectable difference due to geological origin could be observed.

Density distribution of a typical Carboniferous coal and V/VI/VII seam coal of GOCP. Jharia coalfield are shown in Fig. 3. The main difference between Carboniferous and this coal is in the quantity of low relative density (r.d.) material. The coal (a) consists of overwhelmingly of material less than 1.4 r.d. with little between 1.4 - 1.6. Basically, this coal consists of pure and free coal particles and free dirt with little middlings, while the coal (b) shows a completely different density distribution pattern with almost complete lack of low relative density fractions.

The density distribution explains various significant properties that have been known to

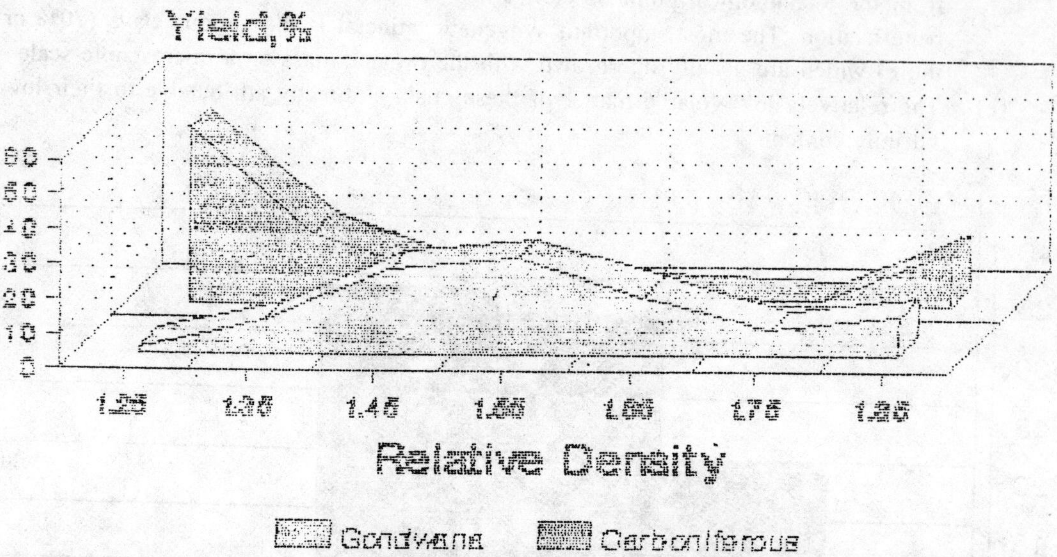

FIG. 3 : Density-Yield Distribution of a LVC Coal, Jharia and a Prime Coking Coal, Germany

characterise this lower seam coal of Jharia. Since r.d. is normally indicative of ash content, lack of low density fractions results in a relatively higher ash at comparable cut points. The pattern of distribution is also significant. Carboniferous coal (a) has a rather narrow distribution of pure coal with a low mean r.d. On the other hand, the coal (b) has a noticeably wider, somewhat skewed, distribution pattern and a distinctly higher mean value.

The explanation for the lack of low density fractions and high percentage of near gravity material (about 50-60%) at the desired specific gravity of cut may be due to the mode of origin of these coals which is focussed in the maceral/microlithotype composition and the higher amount of very finely disseminated ash-forming minerals. The light macerals, vitrinite and exinite is generally abundant in Carboniferous coal, while its amount in Gondwana coal is rather low. Durainous nature with the higher ash content and relative density, predominates in the lower seam coals of Jharia.

Density distribution determines the yield of clean coal for a given ash content. It therefore has a major influence on economics. Lack of low density material has a dramatic effect on product yield, as seen in Fig.4 where the cumulative clean coal yields are plotted against cumulative ash.

From the foregoing discussions it is clear that the lower seam coals are not amenable to a simple, single-stage washing at reasonable recovery rates. For the better utilisation of reserves, a multi-stage approach as depicted in Fig. 5, is extremely necessary.

5. CONCEPT OF OPERATING PARAMETERS

One of the inherent properties of coal from beneficiation point of view, is the relation between characteristic ash content and relative density of coal. In fact, these are the first set of data made

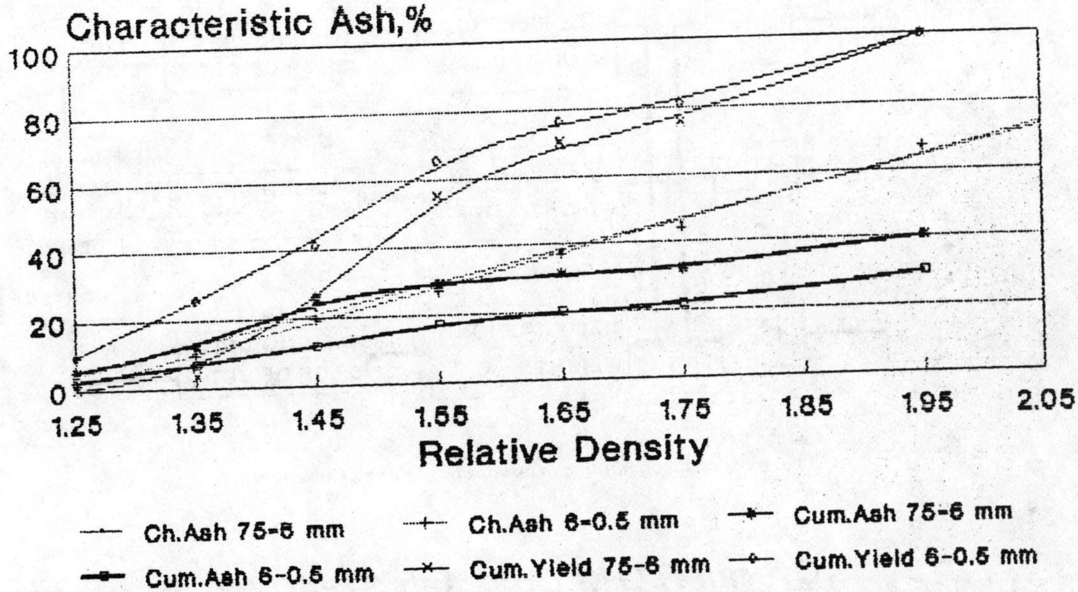

FIG. 4 : *Ash Content and Yield % vs Relative Density for Different Sizes in Composite Feed*

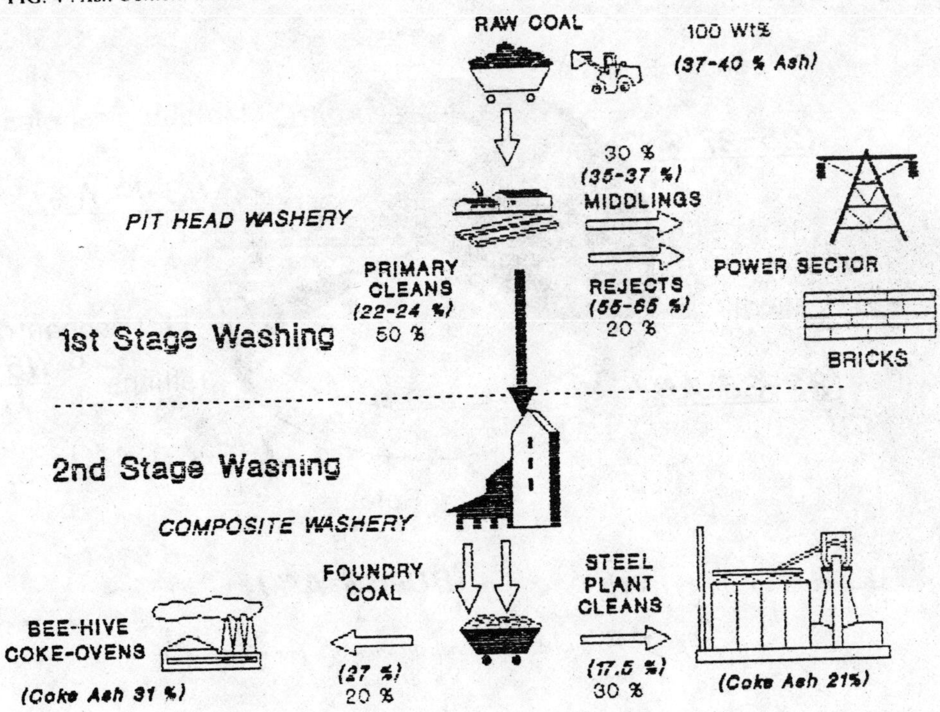

FIG. 5 : *CFRI Concept of Multi-stage coal washing*

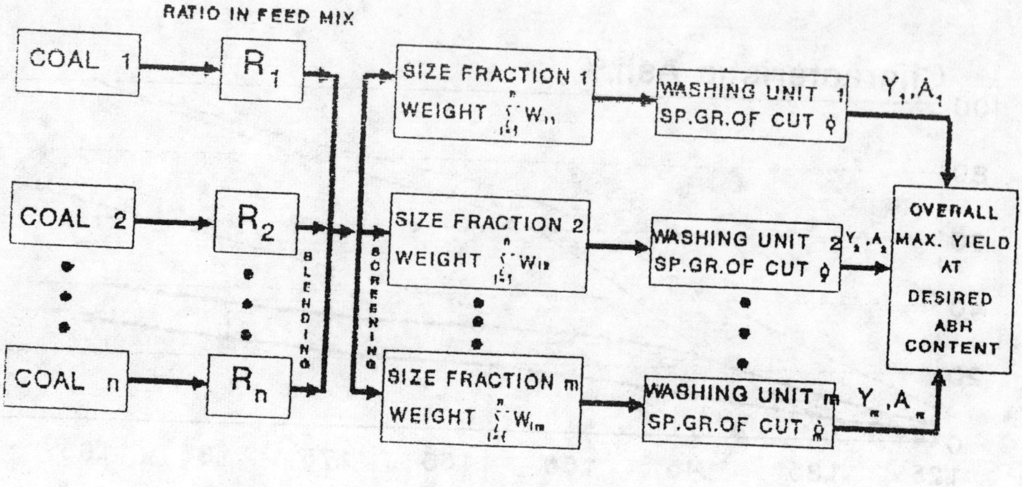

FIG. 6 : *Global optimization in composite washing of mixed coals*

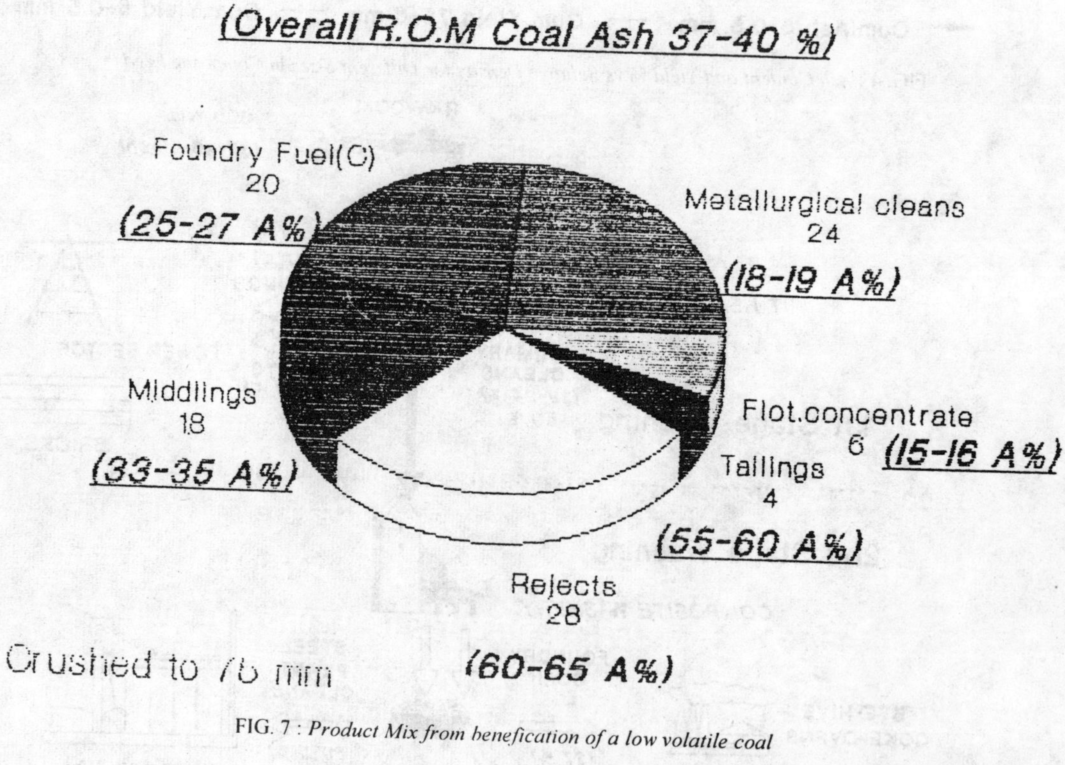

FIG. 7 : *Product Mix from benefication of a low volatile coal*

available after the basic float and sink tests. It is observed that this relation is (i) general or universal in nature and (ii) linear, particularly in the range of interest (1.45 to 1.65)[11].

Thus, it is stressed that this characteristic ash vis-a-vis characteristic reactive property may be used to maximize the recovery of cleansat the best possible carbonization properties, CFRI has developed software to optimize "composite washing of mixed coals" on this concept (Fig. 6)[12]. It is based on the fact that if the elementary coal fraction at the very point of cut has the limiting property regarding carbonization, combustion, etc. then other lighter particles will automatically be acceptable as such.

In fact, this explains why a carboniferous coal (having large amount of vitrinite) requires washing to 8-10 per cent (overall) ash content, but for Permian Gondwana coals it can be as high as 17 percent. In fines the limiting characteristic ash content in both the cases should be about 25±2 per cent. This may now be considered when coals from different origins may have to be blended for better economical benefits, under the new world order, which guaranteed carbonization properties. The product mix of such a composite washing of V/VI/VII seam coal of Jharia has been shown schematically in Fig. 7.

6. CONCLUSION

From the foregoing discussions the following significant conclusions may be drawn :

(1) Lower seam coals of Jharia coalfield are highly interbanded in nature, with high ash and overwhelmingly high amount of inertinite content.

(2) No detectable difference in particle size distribution patterns could be found between Gondwana and carboniferous coals.

(3) The density distribution in lower seam coals of Jharia is characterized by a low proportion or complete lack of low relative density fractions (i.e. < 1.3 or < 1.4).

(4) Lack of low density fractions accounts for lack of low ash fractions.

(5) Washing of these coals presents a serious challenge to processing of the large amount of near gravity material.

(6) Characteristic ash and reactive properties should be used to maximize the recovery of cleans at the best possible carbonization properties.

(7) The lower seam coals are not amenable to a simple, single-stage washing at reasonable recovery. For the better utilisation of reserves, a multi-stage approach for washing these coals is a must.

REFERENCES

1. Beneficiation of V/VI/VII seam coals of Goluckdih OCP. Jharia coalfield. - Rep. No 22.0104. Jan. 1993 (Restricted)
2. Chandra, D. & Taylor, G.H., 1982: 'Gondwana coals' Stach's Textbook of Coal Petrology, 3rd edition, Ch. 2.25. Gebruder. Borntraeger. Berlin-Stuttgart.
3. Mackowsky, M.Th., 1968: Coal & Coal Bearing Strata. (Edited by D.G. Murchison & T.S. Westoll), Ch. 14, Oliver & Boyd, Edinburgh.
4. Sanders G.J. & Brookes, G.F., 1986: *Int. Jour. of Coal Prepn.* 3.

5. Washability Investigation on Block II coals for Chasnala Washery. - Rep. No. TR/CFRI/1.04/93, July 1993 (restricted).
6. Ghose, S. & Chaudhuri, S.G., 1987: *Trans, MGMI,* 84, 1, 1987.
7. Studies on the suitability of LVMC coal available in Jharia coalfield for manufacture of coke of required quality for the pig iron industry through washing and carbonization. - Rep. No. TR/CFRI/1.07/93, 1993 (Restricted).
8. Bennet, J.G., 1936: *Jour. Inst. of Fuel.* 10.22.
9. Raab, G.A., 1971: Fuel Res. Inst. South Africa, Tech. Rep No. 11, Pretoria.
10. Raab, G.A., 1971: Fuel Res. Inst. South Africa, Tech, Memorandum No. 25, Pretoria.
11. Sen, K. & Chaudhuri, S.G., 1971: 12th Int. Coal Prepn. Cong.. May 23-27, Cracow, Poland.
12. Sen, K. & Choudhury, A. Dasgupta, R. Ghose. S. & Haque. R., 1994: *Int. Jour. Min. Processing*, 41, 1994.

10

TECHNOLOGIES FOR UTILISATION OF HIGH ASH COALS – INDIAN PERSPECTIVE

O.P Rao, Deputy Adviser
Council of Scientific & Industrial Research,
India.

ABSTRACT

The paper discusses various coal utilisation technologies in the power and steel sectors with reference to their status of implementation in India and global developments and their suitability to Indian coals. The technologies include clean coal tehnologies for power generation, direct ignition of pulverised coal, low NOX burners, baghouses, solvent refined coal, formed coke, coal dust injection into the blast furnace and COREX process for alternate iron making etc. Technologies suitable to Indian Coals have been indentified.

INTRODUCTION

The present assessment of coal resources in India according to the Geological Survey of India is around 200 billion tonnes, out of this 70 billion tonnes are proved reserves. Total estimated coal resources in India account for about 1% of the global resources and the proven reserves account for 6% of the global proven reserves. Coal desposits in India are mostly confined to Estern, southern and central parts of the country consisting of 27 major coal fields.[1] The Category wise resource position is as given below.

Coking Coal, billion tonnes	30 (15%)
Non Coking Coal, billion tonnes	170 (85%)

The consumption of coal in the country during 1995-96 (April-March) has been 273 million tonnes.[2] Power sector is the largest consumer of coal accounting for 67% followed by steel sector 13%, cement industry 4% and the balance 16% in other industries like textiles, refractories, brick kilns etc.[1] Indigenous Coal production by and large has been able to meet the coal requirement of various consuming sectors except the steel sector. About 9 million tonnes of low ash Prime Coking Coal is currently being imported annually for blending with indigenous coals for steel making.[1]

About 77% of the non coking coal reserves are of inferior grade characterised by high ash content.[1] The mineral matter is intrinsically mixed with the organic matter in coal. The ash is of inherent nature and has high presence of near gravity material. These characteristics make washing of Indian coal rather difficult. In addition to this grinding of coal to finer sizes is required to liberate mineral matter. This results in higher energy consumption in grinding and dewatering becomes difficult after washing. Excess amounts of fines are generated during mining, transportation and handling. Sulfur content is generally low. The ash fusion temperature is high >1400° C. The technologies for utilisation of coals must take care of the above properties for efficient conversion with economy and with least possible pollution.

In this paper, the Indian perspective on coal utilisation technologies relevant to power and steel sectors are discussed since they are major consumers of coal.

TECHNOLOGIES FOR POWER SECTOR

Coal plays a vital role in the power generation scenario of the country. About 64% of total coal produced in the country is presently consumed by the power sector alone and the same trend is likely to be maintained atleast for another 10 years. The coal based power generation dominates the power scenario in the country accounting for about 60% of power generation, and same trend in projected to continue for another 10 years.[1]

	1994-95	2001-02	2006-07
Coal Consumption, mt for power generation	167 (64%)	350 (70%)	500 (71.4%)
Installed power generation capacity (Coal Based), MW	58100 (71.5%)	81068 (55.5%)	111068 (53%)

Coal based power is generated through Pulverised Coal fired (PC) technology which is well established in the country. The PC technology has reached the maximum efficiency attainable with possibilities only for marginal improvements. Environmental regulations are becoming more stringent and the control measures specially the Flue Gas Desulfurisation (FGD) For sulfur dioxide removal is highly capital intensive. Indian coals in general have low sulfur content hence sulfur dioxide emission is not of immediate concern. When a cluster of thermal power plants are setup in a close vicinity specially in mega cities, control of sulfur dioxide emission could be necessary. FGD technology is commercially established in USA, Europe and Japan. At the end of 1993, more than 132 GW capacity had FGD units world wide. The efficiency of SO_2 removal is about 80-90% and the capital cost in US $ 150 – 270/Kw for retrofit and US $ 120 - 210/Kw for new installations.[3] Dry scrubbers are also used. The technologies based on catalytic and non catalytic reduction are under development abroad for control of emission of oxides of sulfur and nitrogen. These technologies may be available commercially after 10 years.

Particulate emission from thermal power plants is a serious concern for India. These emission are already high in many places. The situation worsens with deterioration in the quality of coals supplied to thermal power plants. The ash content in coals meant for power generation is likely to go upto 50%. The removal of ash is through electrostatic precipitator (ESP). Improvements have been made in the design and operation of ESPs resulting in higher efficiency of particulate removal upto 99.7%. The cost of new ESP varies from US $ 40 to 60/Kw. Bag filters or baghouses have been commercially used in USA, Japan and Europe in place of ESPs to the extent of 30 GW capacity. Baghouses are efficient in dust collection even for particle in the range of 0.05 to 1 micron and are cost effective - US $ 50 to 70/Kw.[3] The technologies of new ESP and baghouse are suitable to India. These may be implemented in the thermal Power Plants located in major cities or densely populated areas to begin with. Even 0.1% reduction in emission of particulate is quite susbstantial for 40% ash coal.

Pollution due to particulate matter can be reduced to a considerable extent by washing the coals. In addition to reduction in ash content the washing of coals for power generation have many other advantages such as: increase in thermal efficiency, plant availability and plant output and reduction in the operating and maintenance costs, load on transport system and oil support in the boiler. It has been estimated that if all power coals transported over 1000 Kms are washed to bring down the ash content from 40% to 34%, the likely bebefits in the year 2001-02 are: reduction of 11 MT in coal transport, reduction in flyash by 8 MT and reduction in CO_2 emission by 23 MT.[4] The washing of power coals has been taken up in India. The economics of use of washed coal and motivation for its use for power generation depend on technoeconomic attractiveness of washing and of course on the policy of the Government.

The PC technology is being improved. The recent developments include: thermodynamic cycle improvements by utilising super critical and advanced super critical steam conditions (steam pressure 25-35 MPa and steam temperatures of 560° - 650° C). The net cycle efficiencies can be in the range of 41.1 - 43.5% [5]. Significant improvements have also been made in power plant design, instrumentation and maintenance which can also result in higher efficiency and economy. Improvements in PC technology appear to be suitable to India. The coal based thermal power generation installed capacity is projected around 110,000 MW by the year 2006. Even 1% improvement in efficiency results in substantial increase in the number of units of power generation for such a high installed capacity.

The **Clean Coal Technologies** are alternate technologies for power generation which are more efficient and environmentally benign compared to PC technology. They are technologies of 21st cetrury to meet the environmental regulations and higher efficiencies. They can utilise unwashed coal, yet deliver higher specific power with very low pollution levels. These technologies are: Fluidised Bed Combustion (FBC), Integrated Gasification Combined Cycle (IGCC), and Coal Water Fuel (CWF).

Fluidised Bed Combustion

The atmospheric fluidised bed combustion (AFBC) boilers are commercially available world over for modules upto 200 MW and capacities betwen 250 to 350 MW are under planning stage in America. Europe and China. In AFBC boilers limestone can be injected to capture sulfur from coal and remove it as a dry byproduct. AFBC boilers are capable of burning inferior grade coals and also rejects from coal washeries. AFBC boilers can remove upto 90 to 95% SO_2 and solid wastes have been classified as non hazardous and therefore can be used for road construction, lightweight aggregate. NOX emissions are low in the range of 100 to 300 PPM. AFBC technology capital costs for a 150 to 200 MW are in the range of US $ 1300 to 1600/Kw.[3]. AFBC boilers are commercially available in India. They are suitable for burning rejects from coal washeries.

Pressurised Fluidised Bed Combustion (PFBC) Combined Cycle

The PFBC systems replace the conventional coal combustors and are capable of insitu sulfur removal which eliminates post-combustion FGD units. The lower combustor termperatures reduce NOX emission. A dry ash is produced which can be used safely. Hot flue gases after cleaning in a hot gas cleanup system drive a gas turbine. Steam is generated utilising the heat available in gas turbine exhaust to drive a steam turbine. Thus power is produced in a combined cycle mode. The efficiency achievable is in the range of 40-42%. More than 90% sulfur removal is possible and NOX emission is 100 to 200 PPM. PFBC is in the demonstration phase at a scale of about 70 to 80 MW in USA, Japan and Spain and 350 MW demonstration plant is under construction in Japan. Projections of capital costs for PFBC are in the range of US$ 1200 to 1550/KW[3] The isssues which need particular attention are:

- Coal and sorbent preparation and feed systems
- Effects of gas contaminates on gas turbine performance, efficiency and life,

Integrated Gasification Combined Cycle

Coal gasification converts solid coal into a synthetic fuel gas composed mainly of carbon monoxide and hydrogen. The fuel gas leaving the gasifier is cleaned to remove sulfur compounds and particulates are removed either in a hot gas cleanup or by scrubbing with water. The cleaning of gas in hot condition enables 2-3% higher efficiency compared to water scrubbing. IGCC can achieve upto 45% efficiency, greater than 99% sulfur removal and NOX emission below 50 ppm. NOX formation is suppressed in the reducing atmosphere of gasification. The IGCC costs are in the range of US$ 1500 to 1800/KW.[3] IGCC technology is adequately demonstrated in America, Europe and Japan in the range of 80-100 MW. The IGCC technology is moving from demonstration stage to commercialisation. IGCC plants worldwide which are under construction or planning stage are aggregating to 8000 MW and the plants under evaluation are 21,000 MW. For example, a 250 MW greenfield IGCC plant based on TEXACO gasifier belonging to Tampa Electric Company is under Construction at Florida, USA. The IGCC technologies demonstrated so far are for low ash coals. IGCC for high ash coals is yet to be demonstrated. The capital costs are high and they need to be brought down.

IGCC and PFBC are both relevant technologies to India to utilise large reserves of inferior coals for power generation efficiently and cleanly. IGCC is not handicapped due to non-availability of hot gas cleanup system since water scrubbing can be adopted for particulate removal. PFBC essentially needs a hot gas cleanup or a gas turbine to withstand dust loading in the gas. IGCC can be adopted on priority, but it needs demonstration with high ash coals before commercialisation.

Gas Turbines

The efficiency and economics of IGCC and PFBC technologies would improve with the advancements in gas turbine technology. The net heat rate (KCal/KwH) of gas turbine would decrease with increase in the firing temperature. Compared to a net heat rate of 3077 Kcal/KwH and a firing temperature of 1093° C in the late eighties, the gas turbines available now (GE9FA) have a net heat rate of 1560 KCal/KwH and a firing temperature of 1288° C. The GE's H technology gas turbine is projected to have an IGCC efficiency of 60% after the year 2000. Another development is the intercooled aeroderivative gas turbine which is projected to have efficiency in the range of 55-60% for combined cycle operation (based on natural gas). The application of Fog Evaporative Cooling system for gas turbine inlet air would increase the Power output considerably 10 to 15% for every 20° F drop in temperature.

Coal Water Fuel

It is a homogeneous mixture of 70% solids (coal) and 30% water. Additives are added to maintain homogeneity, control overall viscosity and for drag reduction. CWF can be stored, handled, transported and combusted in an environmentally benign way. The fuel can be used in a variety of applications in conventional coal fired or oil fired furnaces, advanced power generation systems of PFBC and IGCC and coal fired diesel engines. Coal needs to be cleaned for use in CWF. Dewatering which is a costly step in coal cleaning is not required when the fuel is used as CWF. CWF preparation processes are commercially mature with large scale production plants in Japan, Italy and other countries. The largest plants in service are 100,000 t/y in Italy, 70,000 t/y in Japan 3 million t/y in Russia. The world's first CWF project is established in Russia that integrates slurry production (3 million tonnes/year) long distance pipeline transport (262Km) and direct combustion in a power station of capacity 1320 MW. R&D work is in progress in USA, China, Italy and Japan[6].

Coal water fuel technology depends to a large extent on the economics of washing. Development of technoeconomically attractive cleaning methods for non-coking coals are very crucial for CWF. Pipeline transport of fuel may pose political and social problems such as transport from water deficient to water surplus locations and laying of pipelines and prevention of pilferage. CWF technology could be applied to low ash coals from north eastern sector, but they contain high sulfur.

Direct Ignition of Pulverised Coal

Thermal Power sector is one of the major oil consumer, although pulverised coal is the main fuel. Oil support is required in the PC fired boilder for startup, ignition of coal and low load stabilisation. It is estimated that average consumption of fuel oil per million units of power generation is 9.23 kiloliters[7]. A new technology known as "Direct Ignition of Pulverised coal (DIPC)" for reduction of oil consumption in boilers has been developed by BHEL, India through use of a DIPC burner. The DIPC system has been succcessfully retrofitted on commercial scale in 210 MW boilers in India. This technology has a scope for wider application in the country.

Low NOX Burners

They are designed for the combustion to take place in stages in the boiler/furnace which results in the reduction of NOX formation. Reduction in NOX emission to the extent of 30 to 55% can be achieved with low NOX burner compared to the conventional burners without any appreciable adverse side effects on corrosion, burnout and heat transfer. Further reduction in NOX emission is achieved through combination of low NOX burner, overfire air and gas or coal reburning. The overfire air system introduces 10 to 25% of the combustion air above the main combustion zone. The combination of low NOX burner and overfire air can reduce NOX by 40 to 60% and gas or coal reburning can reduce NOX by upto 60 to 70%. In reburning the rate of flow of coal into the main combustion zone is reduced by upto 20% and replaced by equal input of natural gas or coal introduced above the main combustion zone[3]. At present NOX is not controlled, but will be necessary when the regulations become stringent. This technology should be applied in India on wider scale since coal based power capacity is likely to cross 100,000 MW in the next decade.

TECHNOLOGIES FOR STEEL SECTOR

Steel sector is the second largest after the power sector in consumption of coal, only 15% of the Indian coal reserves are coking variety suitable for use in steel making, the consumption and demand of coking coal for steel sector is as follows :

	1994-95	2001-02	2006-07
Coking coal, MT	33	50	60
% of coal production	(13%)	(12%)	(11%)

Even the meagre deposits of coking coal predominantly contain high ash and are difficult to clean. There is wide fluctuation of ash content in the coals supplied to steel plants. The high ash and variation affects the performance and smooth operation of the blast furnace. The ash content in blend for feeding to blast furnace varies form 20 to 22% whereas the steel plants are designed for the consistent quality of 17%. As the limited reserves of good coking coals are getting depleted fast, the quality of mined coal is getting deteriorated. Every 1% increase in the ash content result in the increase of coke rate in blast furnace by 23-25 Kg/thm[8]. Thus ash content in coal is very critical for steel sector.

The imports of prime coking coals are steadily increasing on the ground that the domestic coking coals are deficient in quality and quantity. The imports are affecting the foreign exchange reserves. Therefore technologies are required to utilise indigenously available coking coals and also non coking coals for steel making to reduce the dependence on imports. The washeries commissioned during sixties are designed to use raw coal of 22-25% ash content. Over time the ash content in raw coal has gone up to 30-35%. The existing washeries obviously need modernisation to wash the available high ash coals. The full capacity of washeries need to be utilised[1].

It is a standard practice to combine dense media separation for washing coarser size of coal (+25/13mm) with improved jigs for washing stack coal (–25/13mm) and treatment of fines (0.5

mm) by froth flotation technique. Many Indian washeries which have adopted froth flotation are reported to have problems with regard to filtration and drying of cleans. CFRI has developed a technology by combining heavy medium cyclone for coarser fraction and oil agglomeration for finer fraction. A combined dewatering and upgradation of fines has been carried out by CFRI through oleoflotation technique. The moisture content in final product is about 8% compared to froth flotation which is around 10-13% which needs drying before feeding to coke ovens. Another technology successfully demonstrated for upgrading the coal fines is column flotation technique. The amount of fines generated in washeries is around 4.75 MT per year. The fines have to be utilised to improve the economics of washing and also steel making. Therefore the fine coal beneficiation technologies are needed in the country and oil agglomeration may be preferred choice. However the oil consumption of 8-10% needs to be brought down to improve the economics of the process[9].

Partial Briquetting of coal charge (PBCC) enables to use higher proportion of semi coking coal in the coal blend without affecting the coke strength. SAIL is adopting PBCC technology. Stamp charging helps in obtaining better coke strength using inferior coal charge. TISCO has adopted this technology utilising 60-70% of medium coking coal. Selective crushing is another technology in which coal is homogeneously crushed. If pneumatic classification is applied ash bearing particles being denser get separated. The coal makes a good blend which helps in improving the coke strength The technologies of PBCC, stamp charging selective crushing may be beneficial to India.

Solvent Refined Coal

This is a coking agent prepared from non-coking coals/wahery middlings as feedstock. Addition of about 8-10% solvent refined coal in the coking blend has shown significant improvements in coke strength by way of decrease in M10 value apart form reduction in ash content of the coke by 1-2%. This technology is developed upto pilot scale of 0.5 tpd by CFRI, India[9]. This coal can replace imported coal but it is yet to be demonstrated with Indian coals.

Formed Coke

A number of processes have been developed abroad for the production of formed coke on commercial scale in Germany, Poland, Japan and France etc. based on non-coking coals and lignites. In India CFRI has developed a process based on char and tar produced form non-coking coal and briquetting the char with binder prepared by processing the tar and thereafter curing it. The formed coke produced has been successfully tested in low shaft furnace. The adoption of formed coke in metallurgical application can lead to conservation of coking coals[9].

Coke Dry Cooling and Non-Recovery Coke Oven

In coke making wet quenching of hot coke causes atmospheric pollution by emitting about 500 grams of dust (40% of total dust emission in coke ovens) and 156 grams of chemical pollutants per tonne of coke. Coke dry cooling eliminates this pollution, recovers energy form hot coke and improves coke quality. This results in the reduction of coke rate in the blast furnace upto 3%[10].

In comparison to by-product recovery ovens, the modern non-recovery type have several advantages such as pollution free, flexible, profitable and no handling of byproducts. Two large scale non-recovery coke oven plants of capacity 0.7 mtpa and 0.24 mtpa are operating in USA and Australia. In India a oven called Kumbhraj is developed by CFRI[10].

Coke Rate Reduction in Blast Furnace

Metallurgical coal is scarce in India and consumption of coke made form it constitutes the most significant cost factor in the production of hot metal. Energy accounts for nearly 35% of total cost of steel production and coking coal provides around 80% of energy input. Therefore efforts are continuously being made to reduce the coke consumption in blast furnace. The coke rate in the blast furnaces in U.K., Germany and Japan range from 450-525 Kg/thm as against 675-750 Kg./thm in India. The advanced technologies, relevant to coke/coal in the reduction of coke rate are,[11]

* Upgradation of coke quality.
* Injection of coal dust or coal gas production from non coking coal into the blast furnace, resulting in the reduction of coke requirement upto 40%. The rate of coal dust injection practised in the advanced countries is in the range of 80-90 Kg/thm as against 50-70 practised in India.

This is a suitable technology for India. The rate of coal dust injection can be increased. The ash content in coal need to be below 15% which is a constraint for India in popularising this technology.

Alternative Iron Making Processes

Blast furnace iron making, the most widespread iron making technology is capital intensive and dependent on metallurgical coal. These limitations have led to renewed interests in the following alternative iron making processes which are based on non-coking coal.

— Direct reduction.
— Smelting reduction.

Direct reduction includes a large number of processes in which iron ore or pellets are reduced in the solid state by either solid or gaseous reducing agents. In these processes reformed natural gas or non coking coal is mainly used as the reductant and primarily source of energy. The final product after reduction is solid which must be melted during steel making like scrap, mostly in electric arc furnace. The processes used for gas based reduction are retort, shaft furnaces and fluid beds and for coal based are rotary kilns. Gas based processes account for 90% of Words's installed capacity because of certain advantages of gas based over coal based.

Smelting reduction as the name suggests involves both reduction and smelting like in a blast furnace where both steps occur in a single reactor. A number of selting reduction processes have been developed which can be classified into 3 categories: shaft furnace processes, electrical processes and converter processes. Out of all the smelting reduction processes, COREX is the only one which has reached the commercial stage. A commercial plant of 300,000 tpa was tarted in 1989 in South Africa. A second commercial plant of 600,000 tpa is scheduled to start soon in South Korea. The capital cost of a typical 300,000 tpa COREX plant is estimated as US $ 250 million. A plant of 1.25 million tpa is under construction in Inda. The main advantages of COREX process are;

— Lower emissions than blast furnace route
— Wide range of non coking coals can be used
— Lower investments and production costs.

The smelting reduction technology has a lot of potential for the production of ferroalloys. Japanese are working on a process similar to COREX for the manufacture of ferrochrome. This process is less energy intensive compared to the submerged arc process normally employed for making ferrochrome and other ferroalloys[12].

ROMELT or Ferrous Liquid Phase Recovery Process is another smelting reduction technology developed using non coking coal. This is developed by Moscow Institute of Steel & Alloys, CIS. A semi commercial plant of about 300,000-350,000 tonnes of liquid iron per annum has been set up there. This process is considered unique for the following reasons[12].

— The process can use iron ore in the size range upto 20 mm including fines without any pretreatment of beneficiation and agglomeration.
— The process can use run of mine non-coking coal in the size range upto 20 mm.
— Design and operation of the process are simple
— Cost of production of hot metal is projected to be 10 to 30% lower compared to blast furnace.

The drawbacks and technical issues yet to be addressed are,

— The Russian plant used non coking coals of low ash and low volatile matter. Performance with high ash and high VM coals such as those available in India have not been demonstrated.
— The process is oxygen intensive around 700 NM^3/Thm.
— The waste gases from the furnace carry away 25-30% of the energy input upsetting the economics of the process.

Mini Blast Furnace

The large blast furnaces are associated with high investment, large capacity for economy of production, stringent quality of raw materials etc. While the mini blast furnace is characterized with lower investments, smaller capacities, lower gestation periods and market oriented locations.

A large number of mini blast furnaces are operating in Brazil, China and India. Operation of the mini blast furnaces with charcoal and coke has been extensively tested. China uses metallurgical coke and the coke rate varies between 550 to 630 Kg/thm. Injection of pulverized anthracite coal to the extent of 60 Kg/thm has been tried which has brought down the coke rate by about 40-50 Kg/thm. A mini blast furnace of 175 cubic meter capacity at a capital cost of US $ 10.72 million was commissioned in Goa (India) in 1992. The coke rate of the plant is 600 Kg/thm.

Direct Steel Making Process

Conventionally steel making from iron ore is a two stage process i.e. iron making and then steel making. Attempts are being made to develop a process wherein steel could be produced out of iron ore/coal without converting them to Pig iron as in the case of blast furnace. Such a process has been classified as direct steel making. RESINESS process is one such process. The name RESINESS combines three metallurgical operations: RE—signifies reduction, SIN—signifies sintering and ESS—signifies electroslag smelting. This process uses iron ore fines and non coking coal which are preheated to 300 to 350° C. The preheated mixture is charged continuously in an externally heated retort maintained at 1000 to 1020° C to simultaneously reduce the iron ore particles and sinter the reduced iron which is then electroslag smelted to directly produce steel. This technology is at laboratory stage of development.

CONCLUSIONS

Though India has abundant reserves of coal compared to oil and natural gas, majority of them have high ash content. Fortunately sulfur content is low. Power and steel are the core infrastructure sectors for economic growth. They are also major consumers of coal. The traditional technologies of coal utilisation in power and steel production are still being practised in India in a big way for example pulverised coal fired technology for power generation and blast furnace route for steel making. Depletion of good quality coals, environmental considerations, and economics of operation are forcing the country to deviate from traditional technologies. Beneficiation of coals for power generation is becoming necessary and the country is taking right steps in this direction. The clean coal technologies specially IGCC is required in the country and its introduction on commercial scale needs to be expedited. In the steel sector, the technologies for substitution of coking coals with non coking coals, specially coal dust injection in blast furnace needs to be given priority. Technologies for non-blast furnace route of steel making needs careful examination for adoption to Indian coals.

REFERENCES

1. Report on Integrated coal Policy 1996, Chari Committee.
2. Economic Intelligence Sevice, Monthly Review of Indian Economy April 1996, CMIE
3. E. Stratos Tavoulares and Jean—Pierre Charpentier Clean Coal Technologies for Developing countries, The World Bank, Washington, 1995, Chapters 1 to 7.
4. Sachdev R.K. Indian Coal Industry: Growth Perspective to 2010, Workshop on the Indian coal Perspective to 2010, June 1996, New Delhi.
5. John Sligar, New Coal Utilisation Technologies, UNDP/PACEE Workshop on Coal Technology & Environment, New Delhi, 1995.
6. Kelly Thambimuthu, Developments in coal Liquid Mixtures, IEA -CLM Workshop, Japan 1994.
7. BHEL's Brochure on DIPC
8. P.N. Roy and A Chatterjee, Journal of the Instituion of Engineers, 71 (1190), p 50.
9. TIFAC studies, Economic Reduction of Ash in coal while conserving Scarce Coking Coal Resources, Sept. 1991.
10. Ajay Kumar, Dipak Chaudhari and Ashok Kumar, MECON Journal, 14, (1995), pp. 8-18.
11. TIFAC studies on steel, summary & Recommendations, New Delhi-1991.

ENERGY TECHNOLOGY & SOCIO-ECONOMIC DEVELOPMENT: DST PROGRAMMES AND ACTIVITIES

11

MALTI GOEL
Department of Science & Technology
Government India,
India

ABSTRACT

The paper briefly discusses the importance of Energy Technology and its role in socio-economic development. The potential for longer term Energy R&D in different sectors is described. The purpose is to highlight the coordinating role of the Department of Science & Technology in promoting Energy R&D, which can lead to develop a policy framework for an industrial ecology. The background of programmes of Department of Science & Technology in promoting research and development in energy, in cooperation with socio-economic ministries, industry and academic institutions is discussed.

The approach in the field of energy technology, major activities, various studies commissioned as well as publications are described.

INTRODUCTION

Energy consumption is a vital parameter of economic growth. The intensity of a nation's energy use and pattern of energy mix depend on many factors, viz. climate, geography, population and energy resource endowment. Rapid industrialisation, rising living standards and global competitions demand higher energy intensity in all sectors of economy. So far, the most analysis for energy development has relied on availability of resources, both material and financial, as was in the case of petroleum shortage in 1970s led to global energy crisis. This scenario is, however, changing. Energy-Environment nexus and increasing concern for greenhouse gas emissions from energy related activities and consequent climate change have led to a second global debate on Energy in the 1990s.

GREENHOUSE GAS EMISSIONS AND CONTROL

The greeenhouse gases, viz. carbon dioxide, methane, nitrous oxide, ozone and CFCs, are accumulating in the atmosphere as a result of industrial/economic activity. Increasing greenhouse gas emissions in the atmosphere are giving rise to threat of global warming. Their dispersion in the atmosphere, their impact on earth's environment are subject of intense scientific investigations. A large component of greenhouse gas emissions is linked to use of energy by industries. The largest single contributor is carbon dioxide and the large source of those emissions is burning of fossil fuels (coal, oil and gas). Countries, who use a lot of energy, top the list of carbon dioxide polluters. Whereas, carbon dioxide emission per capita in India is about 4% of a US citizen, and the energy consumption per capita is 1/6 of the world average. Both China and India produce small amount of carbon dioxide pollution per person, but, because their population is large, when they reach an avarage energy consumption comparable to the world average, their emissions are likely to increase. This has given rise to the need to develop energy system models and applying greenhouse gas emission control technologies in various industrial processes[1].

WHAT IS ENERGY TECHNOLOGY?

Control of greenhouse emissions demands application of specific technology measures in various sectors of economic activity. It is being realised that a more quantitative approach to identify significant energy related activities where technology development could substantially help to reduce these emissions trends in a longer term is possible, whereas in non-energy sectors options are more qualitative in nature. The energy wealth of India in comparison to different countries is shown in Table 1. Process conversion of it in the usable form of energy can be termed as energy technology. The energy technologies are those aiming at pollution control and greenhouse gas emission control, i.e., processed concerning energy availability, energy conservation and efficiency.

TABLE 1 : How the world's energy wealth is distributed ? (estimate)

Country	Coals (BT)	Oil (MT)	N. Gas (Baft)	Uranium or Thorium Oxide (K Tons)	Hydro power (BKW/Year)
Australia	107.8	31	0.011	23.5	23
Brazil	1.6	47	0.38	196	17
Canada	82.6	54018	300050	479	105
France	4.4	29	7060	11.7	67
India	196.8	765	707	571	7.7
Japan	20.6	7	20668	1.6	130
UK	166.3	-	-	-	8.2
USA	1479	161140	1249620	304	643

Source : Energy by Mitchell Wilson, time life International (Netherland) N.V., 1970 except for India.

Inter-Governmental Panel on improvements in Climate Change has drawn an exhaustive list of technologies which relate to industrial activity in socio-economic development. Those concerning energy are grouped into energy production, energy intensive manufacturing, and energy end use.

SOCIO-ECONOMIC DEVELOPMENT AND ENERGY R&D

Realisation of energy service at a place depends upon the energy resource, efficiency of energy conversion process and distribution as well as end use efficiency. Pattern of energy consumption

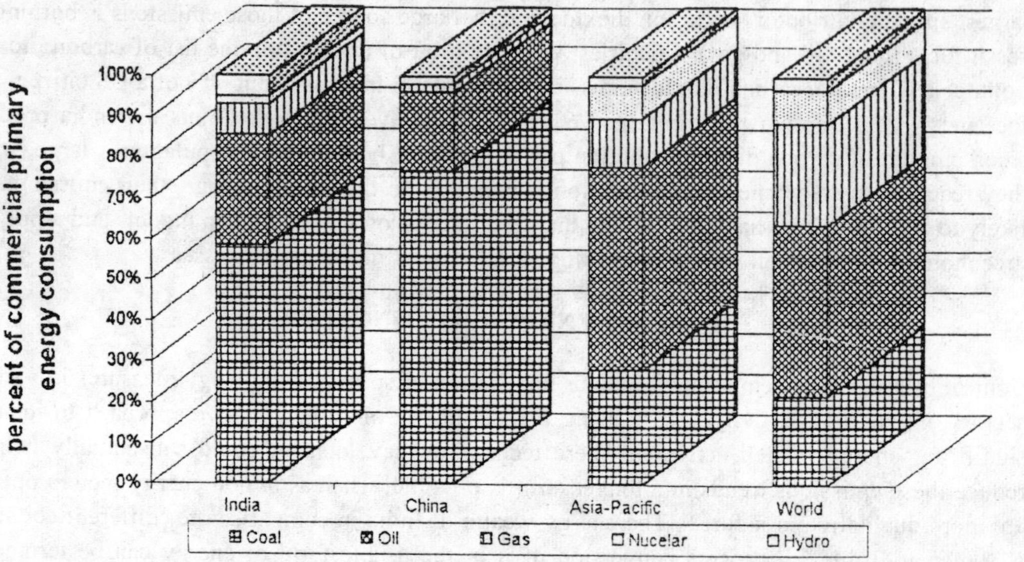

FIG. 1 : *Pattern of Fuel consumption in commercial energy for India, China, Asia-Pacific and World. (The values for Asia-Pacific and World do not include China, Ref. 2)*

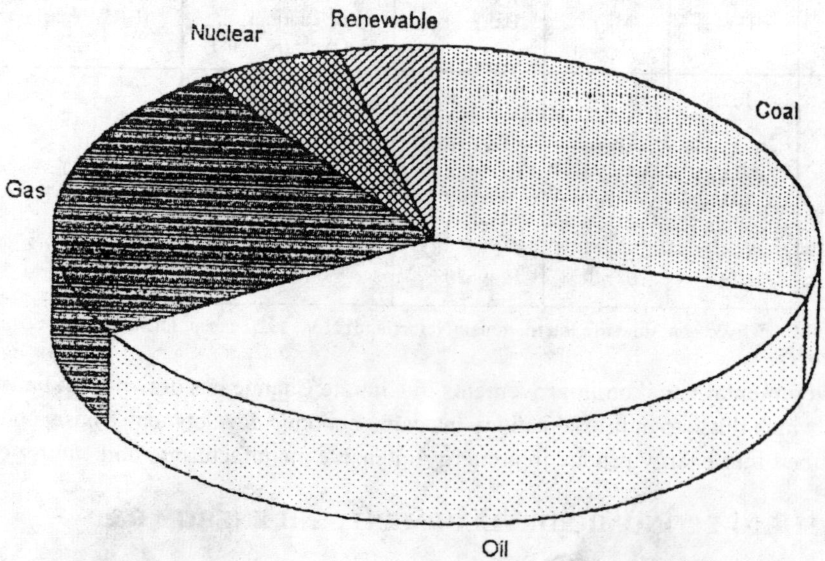

FIG. 2 : *IEA Outlook on Fuel Mix for the World in 2010*

is an indicator of priorities in economic development. In India, during the last four decades the share of commercial energy in the total energy consumption has increased steadily from 28.4% in 1952-53 to 63.2% in 1992-93. The total energy consumption is dominated by four sectors: Industry (50.4%), Transport (24.5%), Household (13.8%) and Agriculture (9.0%). The largest increase has taken place in industry, consuming almost half of the commercial energy. Several energy conservation measures have also been initiated by the industry.

In commercial energy, coal has contributed largest share of 58.3%, followed by oil (27.1%), natural gas (7.53%) and hydro electric power (4.76%), in 1993-94. The percentage share of different fuels in India in comparison to China,[2] Asia Pacific and world is shown in Fig.1. The world projections for 2010 in pattern of fuel consumption is shown Fig.2. According to IEA outlook[3], the global share of energy from renewable sources would be about 4%, in the year 2010.

The linking of climate change problems with challenges of economic growth has become an integral part of energy planning process. The future scene for energy demand and supply will thus be governed by not only the availability of resources, the fuel mix, the consumption patterns and life styles but also the availability of new technologies in the concerned sectors. It has led to the recognition of the importance of new and improved energy technology development in socio-economic sectors[4]. On the basis of R&D and evolution of individual technologies, further improvements in energy efficiency would be possible in different sectors of society. All these will have a profound impact of future energy development strategies and policies.

The Energy R&D technologies are grouped into following categories.

(i) Energy availability-Conventional source base technologies

The technologies concerning commercial energy using fossil fuels viz. coal, oil, natural gas, fall into this category. The approach to development include reduction in emissions through efficient utilisation of the fuel and use of alternate fuel. Development and adoption of technologies in the field of coal based thermal power plants, such as pre-combustion clean coal technologies and introduction of emerging coal conversion – coal conversion technologies like IGCC, PFBC and fuel cells as well as improved design of the plant equipment such as boilers and compressors with a view to increase energy efficiency, are some of the options. The use of alternate fuel such as oil and natural gas in place of coal in electricity generation also contributes to reduction of emissions. The technologies concerning efficient transmission, improvement in quality of the power and management of distribution of electricity, are also very important.

(ii) Energy availability – Non-conventional resource based technologies

Use of renewable energy sources to produce pollution free electricity has been a scientific goal for past several decades, in many countries. Technologies for solar thermal and solar chemical power generation, wind energy, micro-hydel, tide and ocean energy, biomass and geothermal energy conversion are some of the targets of this group. Efficiency improvements of large size solar cells panels, solar collectors, wind energy and field evaluation and market potential of renewable energy devices are major developments taking place for increasing share of renewable energy.

(iii) Energy in Manufacturing and Production

Industrial processes leading to manufacturing of steel, other metals and non metals have been more or less unchanged since the beginning of industrial revolution. In today's crisis the new demands on manufacturing sectors are being put to achieve higher efficiency in production. This can be attained through application of process control technologies and introduction of efficient electric motors, drives, pumps in the existing systems. For steel making newer technologies in production such as continuous steel casting, COREX process are emerging. New techniques of metal forming, increasing use of aluminium scrap as an energy conservation measure, are some examples being adopted worldwide. Non metal energy intensive industry such as cement and pulp and paper are also transforming with the introduction of new technologies, namely improved crusher technology and continuous pulp digestor, respectively.

(iv) Energy End Use

After the industrial sector, transport sector is the next major energy consuming sector and also major environmental polluter. Reduction in emissions through use of alternate fuel is governed by the fact that carbon content of the oil and natural gas is much less than that of oil and coal. Technologies are emerging for compressed natural gas operated vehicles, battery operated vehicles and for more efficient two stroke engine designs as well as through use of lighter material in vehicle designs. Mass rapid transit system and system control of vehicular traffic are other means of saving energy.

The other major end user is household sector, already having about 1/7th share of the commercial energy. This would increase rapidly as the share of non-commercial energy decreases in the coming years. The technologies for this sector include not only efficient household appliances, efficient lighting systems, efficient heating and cooling systems, but also the intelligent/smart buildings, energy conserving building designs and use of energy saving efficient building materials in construction, etc. The municipal waste utilisation technologies form part of this group.

DST's PROGRAMMES AND ACTIVITIES

The Department of Science & Technology (DST) as part of Ministry of Science and Technology, is unique department in networking available scientific and technological strengths and promoting development of new areas of science and technology in the country. Concerns of renewable energy development led to the establishment of a high powered Commission for Additional Sources of Energy (CASE) in the Departmernt of Science & Technology in 1981. This has now become a separate Ministry of Non Conventional Energy Sources.

DST fulfills a major task of interacting with socio-economic ministries on matters of science and technology. The Department assists in the formulation of policy and guidelines for S&T

inputs to programmes of various ministries. Different economic ministries have set up Science and Technology Advisory Committees (STACs) with thė objective of formulating long term S&T plans and programmes relevant to concerned sectors. Under its Inter-Sectoral S&T Advisory Committee (IS-STAC) providing a framework for catalyzing and coordinating technology development projects in academic institutions, government and industry establishments in collaboration with STACs.

The IS-STAC is highlighting energy conservation measures adopted by economic sectors[4]. The multi-partnership projects with socio-economic ministries in emerging areas of energy technology include development of column flotation at Hindustan Zinc Limited as more compact and efficient process, it is expected to reduce power consumption as compared to conventional circuits of ore beneficiation. Under another joint project in the coal sector, Regional Research Laboratory at Bhopal is developing a vorsyl separator for LVMC coal for its induction in cleaning circuits near coal washeries. Microprocessor controlled STATCON development at IIT Delhi at electricity sub-station, in cooperation with Energy Management Centre, is expected to improve the quality of power.

Engineering Science Programme in DST covers emerging concepts applicable to applied research in dfifferent disciplines of engineering. Some of the energy oriented schemes relate to development of Cross Flow turbine for micro-hydel power generation, conversion kits for oil and natural gas in automobiles and inter-conversion computers in load despatch centres. Through R&D efforts at Indian Institute of Science, Bangalore, a few demonstration power micro hydel plants have been set up with capacities ranging from 5 KW to 5000 KW in the states of UP, Kerala and Mizoram. Not exclusive to energy management, these have contributed to improvement of supply of energy or energy conservation through considerable savings compared to conventional methods or research on the fuel for energy for making it more efficient and less polluting. Under Science & Society programme of the DST, several energy related technology projects have been supported for the benefit of weaker sections of the society.

Recognising the problems in Energy R&D cannot be looked at in isolation, an integrated approach to promote energy R&D is being developed, in which selected projects as Research, Development for increasing utilization of indigenous coal resources and minimising pollution are identified, with concerned sectors on Coal, Steel and Power taking the lead. DST conducted a workshop on Greenhouse Gas Control Technologies jointly with Confederation of Indian Industry in June, 1995. The important recommendations of the workshop concerning energy included:

— a coal demonstration plant based on PFBC technology
— promotion and deployment of renewable energy technology
— formation of separate energy management cells in industry for encouraging R&D for energy efficiency and energy conservation
— introducing project selection criteria in R&D funding for promotion of energy efficiency
— subsidising investment in energy efficient equipment
— incentive to industry for meeting the norms, which they can trade off for loan benefits
— increased funding to renewable energy R&D and support to development of solar-hydrogen technology

ROLE OF TECHNOLOGY INFORMATION AND ASSESSMENT

Technology information is going to be a major promoter in determining the technological progress at a place. Unlike the scientific information which is considered an international property and is universal, the information on technologies has been kept closely guarded because of economic reasons. The sources of information are merely through company advertisements and patent informations. Implementation of new technologies is not only hampered by the lack of resources, infrastructure and access to technology know how but also by serious lack of informations on new developments by the management. The key issue in innovation and technology change is also the assessment capabilities of the management.

Technology Information, Forecasting and Assessment council Technology Information Forecasting and Assessment Council under the aegis of Department of Science & Technology has carried out studies as Techno-Market-Surveys, Technology Assessment and Technology Forecasting in the emerging areas of technologies. Among others, improving energy efficiency are discussed for various applications. A Technology Menu Project started in 1993, is intended to aid the designing and implementation of Industrial Energy Efficient Improvement Programmes and Projects. It provides a concise, user friendly, reliable data and analysis tools relating to the performance & costs of energy, use of technology and options for improving the efficiency of components such as Electric Motors, Pumps, Fans, Compressors and Variable Speed drives etc. The major stake-holders are professionals in Planning and Implementation of Energy Efficient Programmes, especially industry engineers/managers and utility engineers working in conjunction with their industrial customers.

In the following section, some publications of TIFAC, DST on energy are summerised.

— *Energy Efficient Technologies for oil extraction* - The study highlights the technology options and process steps for yielding better efficiency in oil extraction namely high yielding modern expellers, better desolventizer and micell distillation systems. The vacuum systems ejector steam is utilised for desolventizer storage steam leading to lower steam consumption (TMS:072, 1993)

— *Rice Husk as energy source in rural areas* - The report deals with Energy from rice husk, systems of combustion, boilers for energy extraction and describes technical aspects of pyrolysis and also provides an analysis of the cost and revenue between pyrolysis and gasification (TMS:065, 1993)

— *Energy Conservation in Cement Industry* - This techno-market survey presents the energy consumption scene in cement industry and discusses major conservation measures. Energy technology trends for various operations and processes are described (TMS:049, 1992)

— *Energy Scenario and Emerging Technological options for Effective Utilisation of Coal* - The study recommends adoption of emerging technologies, with particular applications to power sector. The coal beneficiation, fluidized bed combustion IGGC, coal water slurry combustion are some of the suitable options for different scales of power generation (P:004, 1990)

— *Economic reduction of ash in coal while conserving scarce coking coal resources* - The important problem in this study include the intrinsic quality constraints of our coking coals and iron ore, and technological options including solvent refined coal and carbonisation with a focus on steel industry (S:037, 1991)

— *TIFAC-STAC Study on Precombustion Clean Coal Technology* -This technology assessment report covers profiles of clean coal technologies and global trends. The problems and key issues in the optimal utilisatioin of Indian coals for application in steel and power sectors and criteria for choice of projects as well as profiles of selected projects based on detailed analysis are described (EIS:002, 1995)

— *Solar photovoltaic Systems for rural applications* - This report describes technology trends in solar cells as well as the status and viability of this technology in Indian and global markets (TMS:066, 1993)

— *Mini/micro hydro power stations* - The study gives the definitions and classification of mini/micro hydel power stations, technology status, and techno-economic appraisal, specific recommendations for mini/micro hydel stations, suggests an integrated approach in agro processing industries and communication facilities (TMS:070, 1993)

— *Techniques to improve operational efficiency of thermal power stations* - A study of coal based thermal power stations has been conducted. There is scope for improving availability of power and large scale energy conservation through increasing plant load factor. The need for monitoring performance, secondary oil consumption and maintenance planning has been highlighted (TMS:099, 1995)

— *Energy Recovery Systems including hot gas clean up* - This study discusses various possible energy recovery systems in an integrated steel plant. Use of combined cycle power generation for efficient use of by-product fuels, utilisation of low temperature waste heat for cooling are suggested. The projects have been identified (S:030)

— *Indian Energy Scenaro : Technology Status* — This report brings out briefly some significant areas in power generation technologies on which the country has to concentrate in the coming years. The future technology requirements for transmission, distribution and instrumentation for improving higher efficiency in generation are discussed. (TSR : 001, 1992)

— *TMS in Technologies and Systems to make photovoltaics economical* - Improvement in silicon ingot cutting technology have led to further advancements in PV systems. This study aims to identify the most cost effective ways and means of utilising PV technologies and systems for power generation (TMS : 092, 1995)

— *Energy from bio-mass production utilisation* - The future requirements as well as biomass resources are identified. Present utilization of biomass resources for the purpose of energy generation are stated. Major facilities for the manufacture of equipment for gasifiers, biomass engines and plants have been detailed. The impact of various technology options are highlighted. (TMS : 012, 1990)

CONCLUSIONS

A unified national energy development strategy is required to be developed. Policy thrust has varied during past several decades. The gap between demand and supply is increasing. In order to achieve a balance between energy security, environmental protection and economic growth, an increasing support to Energy R&D need to he given. This is more so, as the traditional energy path, the path followed so far for economic growth in the industrialised countries are being questioned and alternative paths are to be identified and implemented that are better for the environment, provide more and more energy and are also suited to resource realities.

Finally, there is need to introduce energy technologies in our socio-economic development priorities. The examples of energy technology for concerned sectors have been cited. Programmes of DST for promoting R&D in energy and some publications of TIFAC on energy have been described. An Industrial ecology framework need to be developed with simultaneous description of both economic and environment aspects.

REFERENCES

1. Malti Goel, Greenhouse Gas Emissions and Technology Development, in Proceedings of the Workshop on Control Technologies for Greenhouse Gas Emissions, CII, New Delhi, 1995.

2. Z.X. Zhang, Energy Conservation in China, An International Perspective Energy Policy 29, 159-66, 1995.
3. Energy and Environment technologies to respond to global climate change concerns, IEA Publication, 1994. STAC-Scan Newsletter, DST Publication, Vol 3. No 3 and Vol 4, No 1, 1995-96.

R & D NEEDS IN ENERGY SECTOR FOR THE NEXT 20 YEARS

12

C. S. MOORTHY

Formerly, Professor, IIT, Kanpur (India)

ABSTRACT

The possible power mix and the investment required during the next 20 years have been projected. Based on these estimates, the nature of short and medium term, and long term R & D problems are described. The role of manufacturing industries, and the plan of action for energy and material management are presented. Finally, the outlay on R & D and manpower requirements are projected. As the power projects get completed, and the corresponding invesments for industrial growth and the energy conversion are put in place, the total manpower demand will be few millions per year.

Introduction

The estimates of power demand in the country very. However, it would appear that, by the year 2020, the additional installed capacity is likely to be 250,000 MW to satisfy the several national needs. This is a gigantic task. There is an urgent requirement to develop hydropower, including mini hydel potential. The non-conventional energy potential is estimated to be high, and at least 30,000 MW have to be developed during the next 15 years.

The total investments on various sectors will be more than Rs. one million crore. These investments call for ambitious and indigenous R & D efforts. The outlays on R & D will be high, as also the required manpower. These problems are discussed in the paper.

2. Conservation of Energy vs Energy Efficiency

The concept of conservation of energy is based on First Law of Thermodynamics. This also includes wasted energy and energy used in poor designs. instedad, we should talk about energy efficiency in various operations based on Second Law of Thermodynamics. This would create a more positive approach and be able to compare with a model for positive energy audit. Efficiency implies that we should use less energy without compromising on quality and quantity of output.

3. Estimated Power Demand

The Planning Commission document[1] envisages an increase of nearly 300,000 MW over the current installed capacity of about 80,000 MW in the period up to the year 2020. This estimate is perhaps based on the current load factors and the efficiency increase in production and consumption during the same period. If the load factors improve by 10% then, the additional capacity required would be around 250,000 MW.

On the other hand, it is estimated by some that the additional capacity to be installed by the year 2010 is about 60,000 MW. If this is the correct position, then, it is difficult to increase the capacity by nearly 200,000 MW in a period of 10 years after 2010. The addition can only be perhaps another 40,000 MW by 2020. This will be grossly inadequate, considering the fact that by that time many power stations will have become old. Therefore, the increase in capacity will have to be nearer to 150,000 MW from all sources by the year 2010, if this is achievable.

4. Possible Power Mix

The estimated hydel power potential in the country is nearly 84,000 MW at 60% load factor. Only, about 22% of this has been developed or at various stages of development. At least, another 50% of the potential has to be developed. This gives clean and Pollution free power. The mini hydel potential is estimated to be 10,000 MW, out of which only 250 MW has been developed. The full potential of this source has to be exploited during the next 20 years, and another 25,000 MW of medium and large hydel sources has to be developed and installed. The hydel power production efficiency can be at least 80% and this can be increased to more than 85%.

Since coal is a major source of thermal power in the country, the efficiencies in the use of coal and thermal efficiency of production has to be consideraby improved. The boiler efficiency can be improved to about 95%. At present, the thermal efficiency of conversion in the turbines is not more than 40%, resulting in an overall efficiency of only 32%. This can be improved to at least 40 to 45%, which would result in a considerable amount of saving in coal use. The coal based thermal power generation has to be increased by 90,000 MW.

The gas based power generation has to be increased to at least 45,000 MW with an efficiency of 45% using modern technologies. The oil based power generation has to be decreased as a percentage with improved efficiency.

The installed nuclear power is not as much as was expected and an additional capacity of 5000 MW has to be developed.

The non-conventional energy potential is estimated to be more than 140,000 MW. The immediately promising sources are wind energy, biogas and biomass based plants, solar energy and ocean thermal energy. Using these various sources, at least, 30,000 MW of power had to be installed and generated during the next 20 years.

5. Investments during the Next 20 Years

Based on the above proposed distribution of power mix for development, the estimated investment and efficiencies are given in Table, 1, taking into account the varying investment rates for different sources of generation.

TABLE 1 : **Additional Power Mix and Investments.**

Sources	Power (MW)	Current overall efficiency %	Possible efficiency %	Investments Rs. in crores (Thousands)
Coal based Thermal power	90,000	32	40 to 45	360
Oil Based	45,000	40	45	135
Gas Based	45,000	40	45	180
Nuclear	5,000	35	40	25
Hydro	35,000	80	85	175
Nonconventional	30,000	-	60	150
Others	10,000	20	30	30
Total	260,000			1,055
Transmission and Distribution cost @15%				160
				1,215

The investment on Power Sector comes to over Rs. 1.0 million crores and in addition the investment on the transmission and distribution equipment will be at least 15% of the basic investment. Thus, the total investment will be over Rs. 1.2 million crores for only power sector.

There will be an additional Rs. 4-5 million crores investments for efficient energy conversion for industrial growth.

6. Estimated Energy Conversion or Consumption

In addition to improvement in energy production there is a greater need to improve the energy consumption efficiency. On the basis of possible increase in the efficiency of energy use, the estimated energy consumption in various sectors is shown below.

	Percentage
Industry	36
Transport	16
Agriculture	16
Domestic	20
Feed stock and others	12
	100

7. Nature of R & D Problems

(i) Clarity About R & D

We must be clear as to what we mean by research and what is meant by development. In olden days, R & D meant generating basic concepts and proving principles on a small scale. Basic and applied research, and limited development work can be done in educational institutions, unless large and sophisticated production facilities are created for large developmental studies. Even then, continuous development work cannot continue after the students leave the institution. Further, large and interdisciplinary manpower will be required.

But, development has to be made on hardware, This work involves development of components, simulation, development of heat treatment, castings and forgings, surface coatings, system design, integration and testing. Such activities cannot be undertaken by educational institutions. in modern times, a considerable amount of applied research and development are being undertaken in National Laboratories and industries themselves.

The criteria for development are knowledge, technology, information flow, coordination, monitoring, manpower training, decision-making, commitment, finances and implementation. The efficiency of the whole system is less than the weakest link in the chain.

(ii) Short and Medium Term R & D

Research can be software development, development of sensors and their response under various conditions, change or replacement of materials or processes by new ones which require less energy, development of small components, reduction of thermal losses, failure and reliability analysis and simulation. Some such projects can be given to institutions and laboratories. Special emphasis may be given to non-conventional resource utilization. In all cases, there must be an institutional commitment to complete the projects in time. Even if some members leave the organization, the concerned authorities must ensure the completion of projects. Penalties may be imposed for non-fulfilment of commitments. These projects required only moderate capital inputs.

(iii) Long-tern R & D

Some large developmental projects may involve building suitable size prototypes for process and manufacture, recycling, redesign of inefficient equipment such as fans, compressors, pumps, motors, air conditioning equipment improvement of thermal efficiency, of transport equipment and vehicles, life extension studies, safety measures, realiability, improvements in boiler efficiency, steam usage, furnaces, adoption of cleaner technologies to improve the efficiency of existing plants through use of washed coal and introduction of technologies like PFBC and IGCC new blends, retrofit improvements in quality of power to meet the requirements of industries like data management and information technologies and prototypes for renewable energy development.

The R & D projects in this category are much more demanding in terms of technolgoy, development costs, space, prototypes building and testing facilities without dismantling the existing plants, large and interdisciplinary manpower. These require large capital inputs. These new plants may ultimately help revamp the existing ones on economic basis in which energy and material efficiency will be a major criterion.

It appears that some modern technologies are to be imported in these capital intensive industries and it is also possible that these advanced technologies can be obtained from some countries.

8. Government Sensitivity

(i) Concern for sensitivity to imported technologies is a matter for national policy. Government and policy makers are to be concerned with implications of resources utilization. They have to consider also whether a museum of technologies from several countries is to be permitted. Their standards differ and even the spares are to be imported and guarantee clauses are not adequate in actual practice.

(ii) If technologies are to be imported in some areas, then we must be clear as to the role of indigenous R & D in those areas. These must be debated and decisions made.

(iii) For hydroelectric power, there must be an understanding on water sharing, height of dams, environment and rehabilitation. This is also the concern of the government.

9. Manufacturing Industries

(i) Energy Efficiency

These fall in almost all sectors such as coal, petroleum products, recovery of various grades of oil/petroleum, paper and pulp industries, steel making, and other matals production, cement, chemical industries, textiles, fly-ash utilization, waste disposal, transport equipment and vehicles and so on.

The industries must aim at profitability by minimizing the use of energy and materials. These depend on better product quality, better production performance, recycling and elimination of rejected parts. This brings us to efficient design of processes, better plants and better labour after retraining. This in turn leads to careful recording of information in each production centre, cost centre and energy centre.

Those who can invest in energy efficiency measures will benefit and make more profit, and those who cannot invest ultimately will have to close down by sheer competition.

(ii) Transport Sector

This sector is much less efficient than the thermal power generation sector. The thermal efficiency is 22% to 25% at best. The vehicles are capable of at least 2% to 5% improvement in thermal efficiency by redesign of thermal conversion equipment. After all, transport vehicles need power to overcome air drag and road friction. The air drag is not important at speeds less than 60 km/h. But, road friction coefficient can be reduced by more smooth roads consistent with road grip. The condition of the roads is to be improved.

10. Plan of Action for Energy Management

The following steps/strategies are to be initiated for quick returns in the energy sector.

(i) Energy and material audit must be conducted on a national level in each major industry in each sector in each energy and material centre. These reports must be published.

(ii) From these audits, some R & D projects may emerge to improve the energy efficiency.

(iii) Preferably, these audits must be conducted by groups of persons from industries, academics, professional bodies, R & D laboratories and government agencies. Then, these audits can be compared with possible and achievable standards.

(iv) A SWOT analysis is to be made so that we are clear about our strengths and weaknesses.

(v) Some short and medium term R & D problems may also be identified from the long term projects cited in 7 (iii).

(vi) After identifying the projects, and also the centres to undertake R & D, the short and medium term projects may be given to two or three institutions/laboratories for comparing the results.

(vii) Urgent decisions are to be made on long-term projects after identification of the same. The same are to be given to suitable industries for development within a stipulated time. Standing committee may be constituted for the purpose.

(viii) Some incentives must be given to industry for improving on energy and materials efficiency against a standard.

(ix) At least, few maintenance Engineering Institutions must be opened.

11. Outlay on R & D and Manpower

(i) Gigantic Task

The total R & D problems described above constitute a gigantic task in which the whole country is to be involved. Especially the renewable and non-conventional energy develoment requires considerable investments and manpower. The R & D inputs in these sectors should be of the order of at least 2% of the total investment. The investment on the short and medium term projects is nearly Rs. 400,000 crores as indicated in paras (4), (5) and (7). Then, the R & D inputs comes to about Rs. 8,000 crores.

However, the bulk of the investments would be on the longterm projects mentioned in 7 (iii) and the R & D inputs in this category will be at least 5% of the investments which are about Rs. 0.8 million crores. Then, the R & D expenditure comes to Rs. 40,000 crores. Thus, the total R & D funds required will be

	Rs. crores (× 1000)
Short and medium term projects	8
Long term projects	40

For the results to be usefu!, the R & D projects must be completed in a period of 10 to 12 years. Then the R & D expenditure on short and meduium projects would come to Rs. 700 crores per year, and that on long term projects comes to about Rs. 4000 crores per year.

Conditions must be created to absorb this much yearly outlay. The educational institutions must be requested to re-orient their programmes to suit the energy sector.

(ii) Manpower

Assuming that 20 persons can be employed per Rs. one crore, the manpower for short and medium term projects will come to 14,000 persons. Perhaps, only about 1/2 of this number is available currently in the country. Assuming that for quick R & D results the additional number will have to be recruited and trained over the next 10 years or less, the manpower comes to 700 per year. Ways and means have to be found to generate the numbers quickly.

Similarly, for long-term projects, about 80,000 persons will be required and in this case also only about 1/2 of this number is perhaps available currently. The balance requirement comes to about 4000 persons per year. These are the requirements for R & D. The total for R&D comes to 4,700 per year.

Additionally, in order to absorb the capital investments in power sector and the corresponding industrial growth amounting to about Rs. 6 million crores, a manpower of over 100 million will have to be positioned over the next 20 years. This would mean 5 million persons per year.

12. Recommendations/Conclusions

(i) The power requirements and investments during the next 20 years are described in paras (4) and (5).

(ii) The R & D needs are briefly described in (7)

(iii) The steps for plan of action are given in (10). These are recommended to be implemented.

(iv) Various committtees are to be appointed to implement the recommendations in (10).

(v) The R & D investments envisaged are given in (11). The recommended R & D outlays must be released immediately.

(vi) The additional manpower requirements are high. The technical and educational institutions are to be requested to augment the personnel in the energy sector.

(vii) For proper maintenance training, some maintenance engineering institutions are to be opened.

REFERENCE

1. J. N. Maggo, "R&D Needs of Energy Sector" paper presented at the workshop on R&D Needs for Energy Sector on 23rd Sep 1996, at IIT/Delhi.

RECENT TRENDS IN RENEWABLE ENERGY EDUCATION AND TRAINING

13

H.P. GARG and **T.C. KANDPAL**
Centre for Energy Studies
Indian Institute of Technology
INDIA

ABSTRACT

A brief review on the recent trends in renewable energy education and training is presented alongwith some important inferences and recommendations for future work.

Introduction

Formulation and implementation of suitable strategies for balanced and accelerated adoption of renewable energy technologies, is a direct consequence of first oil crisis of early seventies. Efforts were made, on a global scale, towards the development and dissemination of a variety of renewable energy technologies to reduce the consumption of conventional fuels. However only a few technologies successfully penetrated into the consumer market and in view of the internationally falling oil prices development of appropriate renewable energy technologies became a low priority activity. In recent years, once again, the interest in the development of renewable energy technologies has renewed primarily due to their environmental friendliness and also due to the realization of the fact that it is not at all possible to meet the global energy needs on a sustained basis with conventional fossil fuels alone.

Design, development and dissemination of sustainable renewable energy technologies to meet the growing energy requirement for improvement in the quality of life of their large population is a major challenge for most of the developing countries. One of the serious handicaps in the implementation of any strategy in this direction is the unavailability of trained manpower at all levels[1]. Moreover, the attitudes and preferences of the common public as well as of the decision makers have to be changed to effect wider acceptance of renewable energy technologies. This necessitates that sincere efforts be made in the area of renewable energy education and training not only to provide the required technical manpower at all levels to design, develop, fabricate, install, operate and maintain the various renewable energy systems but also to motivate the common public towards a large scale utilization of these systems.

Global efforts have been made in the past towards the establishment of education and training programmes in the field of renewable energy and other relevant disciplines[2-6]. In this paper an attempt has been made to briefly review the recent trends in renewable energy education and training. Attempt has also been made to summarize the emerging recommendations for strengthening renewable energy education efforts.

Levels of Renewable Energy Education

It is now widely accepted that for successful and sustained implementation of renewable energy technology development and dissemination programmes, education and/or training at the following levels (Table 1) may be provided[7].

(i) Primary and secondary school level, (ii) University level (both undergraduate and post-graduate) teaching/training programmes, (iii)Training of technicians and mechanics, and (vi) Educating policy makers and common public.

A brief review of efforts undertaken, so far, in each of these categories is presented in the following sections.

Primary and Secondary School Education

The future well-being of our younger generation primarily depends upon our individual and collective abilities to make the best possible use of presently existing nonrenewable resources of energy and to harness new and renewable sources of energy. As the problems involved are both immediate and long range, many of the viable solutions inevitably will have to come from the scientists and engineers of tomorrow – the youth of today. Therefore, one of the most important responsibilities of the present generation of energy educators is to impart renewable energy education/training to our younger generation so that they appreciate all the energy related complex issues and are equipped to seek proper solutions. While the basic principles of energy conversion and utilization may be introduced in the primary school curriculum, it may be possible to explain the principles of operation of various energy devices at the secondary school level. Availability of appropriate teaching learning resource materials is an essential prerequisite for successfully introducing energy related new concepts in primary and secondary school curriculum.

Several groups around the globe have developed curricula and packages for teaching renewable energy in secondary schools. Activities undertaken in this direction include:

(i) development of lecture-laboratory curriculum base for the teaching of alternate and renewable sources of energy within the secondary school industrial arts laboratory at the Montclair State College in New Jersey (USA).

(ii) solar energy education activities undertaken by Florida Solar Energy Centre, at school level in the State of Florida (USA).

(iii) development of solar energy experiments for school level students at the Centre for Energy Studies of the Indian Institute of Technology Delhi (A list of some of these experiments is given in Appendix 1)[8].

(iv) primary and secondary school teaching packages developed for Victorian Renewable Energy Education Programme by Victorian Solar Energy Council (Australia)[9].

(v) Educational solar laboratories and courses for secondary school students at Moscow Institute of New Technologies.

TABLE 1. Different Levels of Renewable Energy Education.

Age Group	Institution	Type of Course(s)	Remarks
5-10 years	Primary School	Introducing simple concepts in the course on environmental studies	
10-13 years	Junior High School	Introduction of relevant concepts and demonstration experiments in the science curriculum	
13-16 years	High School/Secondary School	(a) Introduction of relevant concepts, demonstration experiments in the science and biology curricula.	Separate chapters can also be introduced at appropriate places
		(b) Introduction of pre-vocational course(s) in the area of renewable energy technologies	
15-18 years	Senior Secondary School	(a) Introduction of relevant concepts, technologies, demonstration and laboratory experiments in the Physics, Chemistry and Biology curricula.	
		(b) Introduction of vocational course(s) in the area of renewable energy technologies	A properly planned full fledged vocational course may be offered in this area.
Above 17 years	Technical Colleges Polytechnic Engineering Colleges Institutes of Technology	(a) Certificate and diploma courses for technicians and mechanics	
		(b) Undergraduate and postgraduate degree courses	
		(c) Short term/on the job training courses for updating knowledge and skills	Useful for policy makers and administrators as well besides teachers
Any age	Non-governmental and Voluntary organisations Any other appropriate organisation	Awareness programmes for common public	

Renewable Energy Education at University Level

Efforts to offer university level courses in the area of new and renewable energy sources have already been initiated in several countries. Postgraduate level teaching/training programmes have, so far, been given priority as compared to the full fledged undergraduate level programmes in this area[10-12]. The international postgraduate level teaching/training programmes being offered presently include those at AIT Bangkok, University of Reading, University of Oldenburg, IIT Delhi and Anna University Madras. In India many universities and engineering institutions have recently started degree/ P.G. diploma level courses in the discipline of energy/renewable energy (Table 2). It is widely appreciated that development of structured curricula for university level courses in the area of new and renewable sources of energy is crucial for their successful and effective implementation.

Preliminary efforts towards the development of a postgraduate level teaching programme in the area of energy engineering have been made at the Centre for Energy Studies of the Indian Institute of Technology, Delhi[6, 13-15]. Details of the developed postgraduate level course and its proposed adaptation for a 3-semester M.Tech. programme are given in Appendix II and III respectively. For efficient and effective communication between the teacher and the learner development of appropriate teaching learning aids, laboratory experiments etc. is quite important.

Training of Technicians and Mechanics

Training of technicians to supervise, fabrication, manufacture, installation etc of technologies and of mechanics to actually work on the shop floor and to provide repair and maintenance back-up is very necessary for sustained development and dissemination of new energy technologies. By forecasting the future manpower requirements for the solar industry and identifying the tasks that should be performed by technical and skilled workers, suitable curricula to train these workers may be developed.

Educating Policy Makers and Common Public

It is necessary to educate and convince policy makers about the merits of the new innovations in the area of energy. Similarly educating the common public about the energy related issues and exposing them to the state of the art energy technologies is also of immense importance as the utilization and maintenance of the energy technologies is the responsibility of the end user. Women must be provided training in the use and management of renewable technologies.

Challenges and Problems in Developing Countries

In the last two decades some efforts have been made to provide renewable energy education and training in the developing countries. However, most of these efforts have been made in the area of postgraduate level courses in universities, engineering colleges and institutes of technologies[10-12]. Introducing relevant inputs into the school curricula has not yet been seriously considered. Similarly the programmes for technicians and mechanics are also not available. Sometimes refresher courses are organised. Professionals from majority of the developing countries usually go abroad for education and training in this area. Moreover, a variety of problems are yet to be resolved to provide renewable energy education in an efficient and effective manner in developing countries. Some of the important issues are briefly described in the following paragraphs[16].

(a) Employment potential and curriculum development

Large scale unemployment and underemployment is a rather common characteristic of many developing countries. In a country like India it is highly probable that, unless carefully planned, the manpower with certificates/diplomas/degrees awarded in the field of renewable energy also do not get suitable employment. Therefore, prior to the development of any course-curriculum it is necessary to identify and analyse the potential job opportunities in the field of renewable energy. Depending on the job requirements of each type of skilled manpower the necessary inputs to satisfy the cognitive, psychomotor and affective domains of education must be provided. This necessitates an indepth analysis of the job requirements for each job opportunity (self or wage employment) envisaged with a specific teaching/training programme in the field of renewable energy. Only then the syllabus for the respective courses should be formulated.

(b) Courses of different levels

Realizing the importance of imparting renewable energy education at all levels it is desirable that such an attempt is made simultaneously at several levels. It is therefore necessary to ensure consistency and continuity in the inputs provided at different levels to maximize the overall efficiency and effectiveness of the renewable energy education strategy. While the inputs given at a certain level must match with the expected job responsibilities it is also necessary to maintain a clear distinction between the learning objectives of two different level courses.

(c) Selection of inputs

The course-structures of the renewable energy education programmes often have the subjective bias of the educators/professionals developing and implementing the curriculum. Rather than the inputs ought to be given at a particular level the inputs which can be given with the available expertise of educators often dominate the course curricula. Such a strategy, though unavoidable at times, may lead to a mismatch between the knowledge and skills required for the jobs to be undertaken by the student (when employed) and those acquired by him in the course.

(d) Teaching-learning materials

Being a relatively new interdisciplinary area, renewable energy education largely suffers from the unavailability of appropriate resource materials for teaching-learning (text books are not available even for the postgraduate level teaching/training which has received maximum attention in developing countries). However, in view of the latest developments in the fields of computers and electronics effective audio-visual aids and appropriate materials for computer aided instruction can also be developed to impart renewable energy education besides a variety of other teaching learning aids.

(e) *Financial constraints*

One of the major constraints in promoting renewable energy education activities on a large scale is the unavailability of funds with most of the schools, colleges and universities in the developing countries. Being interdisciplinary and predominantly hardware intensive, renewable energy education and training efforts require a considerable amount of funds to provide the bare minimum inputs. In fact, to a large extent, the unavailability of appropriate teaching learning materials can also be attributed to the lack of adequate funds.

Recommendations

The following recommendations may be made for strengthening energy education programmes in general and renewable energy education in particular at all levels :

(a) Concepts and courses dealing with renewable energy sources and technologies must be introduced from primary classes through all formal and informal stages of education. All sections of the society should be made aware of the existing energy related problems and should be motivated to seek appropriate solutions.

(b) There is an urgent need of developing and implementing structured course-curricula for all types of teaching/training programmes on renewable energy. The curricula may be standardised at the national, regional or even global level to facilitate technology transfer and exchange of know-how.

(c) Renewable energy education programmes must offer a mix of academic as well as hands-on skills training to the student. The latter can be accomplished by conducting laboratory experiments, practical demonstration of operational systems, field visits and field installation of an actual working system. Wherever applicable, the students may be encouraged to undertake hardware oriented projects during the course of the teaching/training programmes.

(d) Preparation of text books, laboratory manuals, teacher's guides and other teaching-learning aids for all levels and modes of education in the field of energy is very important for successful implementation of the teaching/training programmes. If appropriate, modern techniques of communication and information processing as well as computer aided instruction techniques may also be used for this purpose. The books on new and renewable sources of energy being published by John Willey & Sons under UNESCO Series of Learning Materials in Engineering Sciences may be quite useful if their cheaper international editions are made available.

(e) Sincere efforts must be made to educate all the members of the society at large about the basic concepts of efficient energy utilization and also about the various renewable energy options available to them. Use of mass communication media, organisation of exhibitions and trade fairs etc. and offering community college courses and adult education programmes are some suggested mechanisms for this purpose.

(f) It is necessary to organize short term courses/workshops etc. for providing basic input about energy and relevant topics to the policy makers and other associated administrative personnel.

(g) For providing an efficient and effective renewable energy education at the school level it is necessary to establish facilities and device mechanisms for in-service training of existing teachers.

(h) The objectives of the teaching/training programmes on renewable energy should be consistent with the development strategy followed by the country/region.

(i) The university level teaching/training programmes should be designed to provide ample job opportunities and/or should be capable of providing self employment.

(j) A close interaction among the organisations imparting energy education at national, regional as well as international level is quite necessary for increased efficiency, economy and effectiveness of the teaching/training programmes[17].

(k) The employment aspects of renewable energy education programmes which are offered beyond the secondary/high school level must be realistically analysed before starting the programme.

REFERENCES

1. Boris Berkovski and Charles M. Gottschalk Strengthening Human Resources for Energy Engineers of the 21st Century. In Renewable Energy Engineering Education (editors H.P. Garg and T.C. Kandpal), Omega Scientific Publishers, New Delhi (1996).
2. Lars Broman. On a Possible Worldwide Programme for Solar Energy Education Progress in Solar Energy Education 3, 11-14 (1994).
3. Garg H.P. and T.C. Kandpal. Renewable Energy Education: The Global Experience. Proc. of the Workshop on Technology Transfer (ed GD Sootha, SK Gupta, Jugal Kishore) pp. 196-214. Solar Energy Centre, Ministry of Non-conventional Energy Sources, New Delhi (1991).
4. Garg H.P. and T.C. Kandpal. Renewable Energy Education in Asia : Survey of Existing Graduate Teaching/ Training Programmes. Progress in Solar Energy Education 1, 26-28 (1992).
5. Konrad Blum. Courses and Packages for the Renewable : The State of IASEE's Survey. Progress in Solar Energy Education 1, 7-8 (1992).
6. H.P. Garg and T.C. Kandpal. Renewable Energy Engineering Education. Omega Scientific Publishers, New Delhi (1996).
7. J.C. McVeigh. Training in Solar Energy: Curriculum Development Report submitted to UNESCO, Paris, (1982)
8. S.S. Mathur and T.C. Kandpal. Development of Laboratory Experiments of Solar Energy Final Project Report submitted to Tata Energy Research Institute, New Delhi, India (1988).
9. W.W.S. Charters. Development of Primary and Secondary School Teaching Packages for Renewable Energy Education. Proc. 1st World Renewable Energy Congress, Reading, U.K., 23-29 Sept. 1990, pp. 2845-2849, Pergamon Press (1990).

10. S.C. Bhattacharya. The Status of International Postgraduate Courses in Renewable Energy. Progress in Solar Energy Education 2, 3-5, (1993).

11. Garg H.P. and T.C. Kandpal. Global Survey of Resource Material for Energy Education at University Level Report submitted to UNESCO/ROSTSCA, New Delhi (1995).

12. H.P. Garg and T.C. Kandpal. Energy Education at University Level: Results of a Questionnaire Based Survey Applied Energy. 52, 649-654 (1995).

13. H.P. Garg and T.C. Kandpal. A Proposed Postgraduate Teaching Programme in Energy Engineering. Progress in Solar Energy Education, 2, 15-19 (1993).

14. H.P. Garg and T.C. Kandpal. Energy Education at Postgraduate Level: Issues Involved, Course Structure and its Proposed Adaptation. Renewable Energy 5, 1406-1412 (1994).

15. H.P. Garg and T.C. Kandpal. Resource Materials for Renewable Energy Education. Renewable Energy 5, 2362-2366 (1994).

16. H.P. Garg and T.C. Kandpal. Renewable Energy Education: Challenges and Problems in Developing Countries Renewable Energy 9 (1-4), 1188-1193 (1996).

17. Konrad Blum, Lars Broman and S. Niwong. International Association for Solar Energy Education IASEE-Five Years of Experience Renewable Energy 5, 2377-2379 (1994).

APPENDIX - I

The following school level experiments on Solar Energy have been developed at the Centre for Energy Studies of the Indian Institute of Technology, Delhi[8].

1. To measure the altitude of the sun.
2. To measure the azimuth angle of the sun and to study its variation during the day.
3. To study the absorption of solar radiation by different colours (I).
4. To study the absorption of solar radiation by different colours (II).
5. To demonstrate the greenhouse effect.
6. To compare the insulating properties of insulating materials used in solar energy appliances and systems.
7. To study the effect of the thickness of a glazing material (glass) on its performance.
8. To measure the reflectance of commonly used reflecting materials.
9. To study the movement of the focal line of a linear Fresnel lens solar concentrator with the apparent movement of the sun.
10. To study the rise in temperature of water filled in a rectangular channel absorber placed at the focus of a laboratory model composite parabolic trough.
11. To determine the melting point of an organic compound (paraffin wax) using a composite parabolic trough solar concentrator.
12. To study the rise in the temperature of water filled in the absorber placed at the focus of a conical solar concentrator.
13. To study the effect of a roof overhang on the air temperature in an enclosed air space.
14. To study the effect of a movable insulation on the performance of a laboratory model passive solar house.
15. To study the effect of the orientation of the transparent wall of an enclosed air space on the enclosure air temperature.
16. To study the characteristics of a silicon solar cell.
17. To study the effect of the series and/or parallel combinations of solar cells.
18. To study the performance of a laboratory model of a simple flat plate solar water heater.
19. To study the performance of a laboratory model of a box type solar cooker.
20. To study the performance of a laboratory model of a solar still.

APPENDIX-II

PROPOSED POSTGRADUATE TEACHING PROGRAMME IN ENERGY ENGINEERING

(A) BRIDGE COURSES

S.No.	Course	Contact Hours	S.No.	Course	Contact Hours	S.No.	Course	Contact Hours
1.	Thermodynamics	20	4.	Applied Physics	20	7.	Principles of Electrical Engineering	20
2.	Heat and Mass Transfer	20	5.	Engineering Mathematics	20	8.	Principles of Chemical Engineering	20
3.	Fluid Mechanics	20	6.	Natural Resources Economics	20	9.	Engineering Laboratory	20

(B) CORE COURSES

S.No.	Course	Contact Hours	S.No.	Course	Contact Hours	S.No.	Course	Contact Hours
1.	Energy Resources	20	6.	Energy Systems	20	11.	Energy Storage	20
2.	Advanced Heat and Mass Transfer	20	7.	Economics of Energy Technology	20	12.	Advanced Energy Systems	20
3.	Advanced Thermodynamics	20	8.	Basic of Energy Conservation	20	13.	Advanced Engineering Laboratory	20
4.	Advanced Fluid Mechanics	20	9.	Energy, Ecology and Environment	20			
5.	Combustion Engineering	20	10.	Instrumentation and Control in Energy Systems	20			

(C) SPECIALIZATION COURSES

Conventional			Energy Management and Planning			Renewable Engineering		
S.No.	Course	Contact Hours	S. No.	Course	Contact Hours	S. No.	Course	Contact Hours
1.	Energy Engineering Equipment-I	30	1.	Energy Auditing	30	1.	Solar Energy Thermal Utilization	30
2.	Energy Engineering Equipment-II	30	2.	Energy Analysis	30	2.	Biomass Energy	30
3.	Coal Technology	30	3.	Industrial Energy Conservation	30	3.	Wind and Microhydel	30
4.	Petroleum Technology	30	4.	Rural Energy Conservation	30	4.	Passive Solar Building	30
5.	Gas Technology	30	5.	Energy Management in Building	30	5.	Direct Energy Conversion	30
6.	Nuclear Technology	30	6.	Energy from Wastes	30	6.	PV Systems Engineering	30
7.	Hydroelectric Power	30	7.	Energy Planning-I	30	7.	Commercial and Industrial Utilization of Renewable Energy	30
8.	Energy Aspects of Environmental Control	30	8.	Energy Planning-II	30	8.	OTEC, Wave, Tidal and Geothermal energy	30
9.	Power Systems Engineering	30	9.	New and Renewable Source of Energy	30	9.	Integrated Energy Systems	30
10.	Energy Laboratory-I (Compulsory)	60	10.	Energy Laboratory-II (Compulsory)	60	10.	Energy Laboratory-III (Compulsory)	30

APPENDIX-III

ADAPTATION OF THE PROPOSED POSTAGRADUATE TEACHING PROGRAMME IN ENERGY ENGINEERING FOR A THREE SEMESTER M. TECH. PROGRAMME.

(A) BRIDGE COURSE

S. No.	Course	Contact Hours	S. No.	Course	Contact Hours
1.	Heat and Mass Transfer	40	5.	Principles of Electrical Engineering	40
2.	Fluid Mechanics and Thermodynamics	40	6.	Principles of Chemical Engineering	40
3.	Advanced Engineering Mathematics	40	7.	Engineering Laboratory	80
4.	Natural Resource Economics				

(B) CORE COURSES

S.No.	Course	Contact Hours	S. No.	Course	Contact Hours
1.	Energy Ecology Environment	40	4.	Basis of Energy Conservation	40
2.	Economics of Energy Technologies and Systems	40	5.	Instrumentation and Control in Energy Systems	40
3.	Energy Systems Technology	40	6.	Advanced Heat and Mass Transfer	40

(C) SPECIALIZATION COURSES

Conventional Energy Engineering			Energy Management and Planning			Renewable Engineering		
S.No.	Course	Contact Hours	S.	Course	Contact Hours	S.	Course	Contact Hours
1.	Energy Engineering Equipment	40	1.	Energy Auditing and Analysis	40	1.	Solar Energy Thermal Utilization	40
2.	Energy Aspects of Enviornment Control	40	2.	Industrial Energy Conservation	40	2.	Biomass Energy	40
3.	Power Systems Engineering	40	3.	Rural Energy Conservation	40	3.	Wind, Microhydel and Tidal Energy	40
4.	Coal Technology	40	4.	Energy Management in Buildings	40	4.	Passive Solar Buildings	40
5.	Petroleum and Gas Technology	40	5.	Energy Planning	40	5.	Photovoltaics:Principles and Applications	40
6.	Hydroelectric Technology	40	6.	Energy and Waste Management	40	6.	Energy Storage	40
7.	Principle of Energy Conversion	40	7.	New and Renewable Sources of Energy	40	7.	Integrated Energy Systems	40
8.	Energy Engineering Laboratory-I	80	8.	Energy Enginering Laboratory-II	80	8.	Energy Engineering	80

WORLD WIDE REGULATIONS FOR SUSTAINABLE DEVELOPMENT IN PETROLEUM SECTOR

14

Dr. A. Bhattacharyya
Economic Adviser, Contract Cell,
MOP&NG, New Delhi (India)

ABSTRACT

Sustainable development offers countries around the world a different future, one in which improved energy efficiency and environmental performance leads to improved economic performance. It seeks to balance finite resources with efforts to grow and expand, that is, to harmonize competing economic and ecological aspirations. The present paper describes the following issues: (i) existing world-wide regulations/legislations required for sustainable development in petroleum sector; (ii) experiences of Asia-Pacific countries in implementing environmental regulations/legislations; (iii) energy efficiency measures pursued in South Asian countries including India; (iv) environmental management and sustainable development in oil and gas sector in India—a case study of national oil company i.e., Oil & Natural Gas Corporation Limited (ONGC).

Indian petroleum industry is to deal with environmental problems like many other energy-intensive industries. In this regard, our Ministry of Petroleum & Natural Gas (MOPNG) has set up Petroleum Conservation Research Association (PCRA) in 1978 and Oil Industry Safety Directorate (OISD) in 1986 and Directorate General Hydrocarbons (DGH) in 1993. As far back as in 1983, ONGC decided on a statement of corporate environment policy. A lot of importance is attached to environmental and social obligation. The organisation has set up a full-fledged department of Safety and Environment Management and introduced a well conceived system of environmental information and awareness as well as monitoring implementation of standard industrial practices for maintaining a clean and safe environment. ONGC has also established an Institute of Petroleum Safety and Environment Management (IPSEM) to provide R&D support to environmental aspects. By using various energy efficient measures like waste heat recovery, use of soft-starters, variable frequency controllers, solar energy, efficient lighting fixtures etc., savings to the tune of Rs. 107.67 crores has been achieved during 1995-96.

India, as a signatory to the Rio Conventions, is deeply concerned about degradation of human living environment. Our legislation requires that rigorous environmental impact assessment be prepared for all projects including petroleum, during the planning and approval process. More recent guidelines of Ministry of Environment of Forests cover the conservation of biological natural resources and their eco-systems and water pollution control. One of the key areas of focus in the formulation of Indian Energy Sector's long range development plans is to ensure that environmental considerations are fully integrated into sustainable development.

1. Introduction

Sustainable development offers countries around the world a different future, one in which improved energy efficiency and environmental performance leads to improved economic performance. It seeks to balance finite resources with efforts to grow and expand, that is, to harmonize competing economic and ecological aspirations. In the past, one of the ways in which industrialised countries realized economic growth and well-being by consuming more and more energy with little concern for energy efficiency. The process of significantly improving the effectiveness of energy use in advanced countries, initiated in the early 1970s and now taking place in all OECD counties, reduced energy demand in relative, but not absolute terms. Still, higher energy consumption largely meant burning more fossil fuels, which can have a negative effect on the global environment unless appropriate safeguards are employed. Society tolerated this development because it translated into economic prosperity and are improved standard of living. Today, people around the world still demand development and economic prosperity, but in a manner consistent with the preservation of the environment. In short, society demands sustainable development in both developed and developing countries.

2. Approach to Sustainable Development

The present paper describes the following issues; (i) existing world-wide regulations/legislations required for sustainable development in petroleum sector; (ii) experiences of Asia-Pacific countries in implementing environmental regulations/ legislations; (iii) energy efficiency measures pursued in South Asian countries including India; (iv) environmental management and sustainable development in oil and gas sector in India—a case study of national oil company i.e., Oil & Natural Gas Corporation Limited (ONGC).

3. Energy and Economy

Energy is a crucial input in the productive process. Energy consumption and the level of economic activity are closely linked. Any expansion of the economy requires a proportional increase in energy supply. While the industrialised countries have managed to decouple economic growth from proportional input of energy through improved energy technology, the traditional view of linking growth and energy input largely remains valid when applied to most developing countries like India. The factors which have the greatest influence on energy consumption in developing countries are : (i) population and urbanisation growth rates, (ii) economic and industrial growth (iii) efficiency of energy use and advent of improved technologies.

4. World Oil and Gas Scenario—Role of Asia-Pacific Region

At present 75 per cent of world oil and 40 per cent of gas reserves (138 billion tonnes of oil and 148 TCM of gas) is with OPEC countries, whereas they produce about 40 per cent of total oil/gas production (3.2 billion tonnes of oil and 2.1 TCM of gas). In the early 1970s oil accounted for about 50 per cent of world total energy demand which has fallen to about 40 per cent in 1994. The global primary energy consumption is expected to reach a level of 11,560 MTOE by 2010 as against 7850 MTOE in 1994 whereas the share of oil in global energy mix is expected to remain relatively stable at 40 per cent level during the same period. The share of natural gas in the world's total primary energy mix is expected to increase from 22 per cent today to just about 24 per cent in 2010. The growth rate in the Asia-Pacific region is expected to be around 4 to 5 per cent as against on average rate of around 2 per cent for USA as also the world as a whole. India and China together account for 47.5 per cent of the total Asia-Pacific demand. Thus, growth in demand has shifted the centre of world's oil demand from USA to Asia-Pacific region.

5. Energy Intensity and Efficiency in South Asia

Given the environmental impact of energy use, there is a wide range of potential and currently available technologies aimed at decreasing the energy intensity of end use. The energy intensity measures how much energy is consumed per unit of real GDP created. Energy efficiency is only one aspect of energy intensity. In order to analyse trends in energy intensity, it is necessary to separate energy intensity changes into structural and behavioural changes on the one hand and technical energy efficiency changes on the other. From 1971 to 1993 world energy intensity has improved by 0.8 per cent per annum, in OECD countries this has improved at an average rate of 1.3 per cent annually, whereas South Asia including India has showed a deterioration in energy intensity during the same period. For example, in the United States, there is a substantial increase in the proportion of steel produced in electric arc furnaces and this has reduced energy intensity from 0.48 TOE in the manufacture of each tonne of crude steel in 1981 to 0.34 TOE in 1993 whereas India still relies on the open hearth furnace which is much less efficient compared with other furnaces for about one quarter of its steel produced and energy intensity is around 1.0 TOE per tonne. For modern steam crackers, the specific Energy Consumption (SEC) per tonne of ethylene (excluding feedstock) varies between 13 GJ (Giga Joules) for ethane and 25 GJ/tonne for gas oil (hydrocarbon processing) in U.S.A., whereas average plant (steam cracking of naphtha) consumes 73 to 90 GJ/tonne in China and India. For ammonia, modern steam reforming plants consume 30 to 40 GJ/ tonne in Europe whereas in China and India ammonia production is still dominated by small and medium size plants, unit energy consumption run 20 to 25 per cent higher than plants of modern design. The USA pulp and paper industry's SEC decreased from 26 GJ/ tonne in 1972 to 22 GJ/tonne in 1993, whereas overall SEC in Indian pulp and industry range from 40 to 55 GJ/tonne. The high SEC is mainly due to small mill size, use of old technology, lack of energy efficient devices and lack of cogeneration. Primary SEC in building material industry (particularly Cement industry) in South Asian countries including India is in the range of 5 to 6 GJ/tonne, compared to 2 to 3 GJ/tonne in European countries, mainly due to use of less efficient wet process instead of dry process technology. Similarly, fuel efficiency of automobile vehicle (passenger cars on the road between 1985 and 1993) differs across different countries. Passenger cars and engine sizes are smaller and of old vintage and less efficient in India, China and Korea compared with Venezuela and fuel consumption per kilometre is on the higher side. The lower energy efficiency in India as compared to other countries indicate two things. One we can achieve higher output for same level energy of input and two we may achieve this with much lesser impact on environment.

6. Experience of World's Leading Energy Companies

As the world's leading suppliers of coal, oil, gas, electricity and energy services, companies (e.g., Electricite de France, Tokyo Electric Power Co., Enel, Spa, Italy Hydro Quebec, Canada, BP, U.K, RWE Energic AG, Germany, EXXON, Chevron, Southern California , Edison Co., USA, recognize that they have a significant role to play in society achieving this goal. They are pursuing policies that contribute to this goal for many years. They have developed expertise on both the supply and demand side and are involved in the development of a number of new infrastructure concepts. They have developed expertise in assessment design implementation, operation of dispersed power systems, integrated pollution prevention plant using end-of-pipe control technology, photovoltaic generation, fuel cells, small hydro, solar, wind farms in isolated areas that match country requirements while fostering optimal primary energy use, energy efficiency, cost effectiveness, environmental compatibility. Actions on the demand side include (i) design of tariff structures, load management techniques, use of electro technologies; (ii) clean air coating, thermal energy storage energy efficient lighting, infra-red heating in refineries, petrochemical plants, infra-red dying etc. which not only reduces emissions, but offers other non-environmental benefits like better workplace, icreased flexibility, improved product quality.

7. World Wide Regulations

The Paris Commission (for Prevention of Pollution from Land Based Sources, 1975) after reviewing its permitted levels of oil content for water discharged by North Sea offshore installations and pollution caused by the discharge of oil-contaminated cuttings from wells, decided that the level of oil on cuttings from exploration and appraisal wells, and new development wells drilled with oil-based mud, must be less than 1 per cent by 1 January 1994 (other wells must achieve this target by 1 January 1997). Tankers are also subject to the MARPOL Convention (International Convention for the Prevention of Pollution from Ships 1973). Platforms have to be equipped with an oil discharge monitoring system which measures and records the oil content of effluent and automatically stops discharge into the sea where the oil content exceeds permitted levels.The Offshore Pollution Liability Agreement (OPOL) covers direct loss or damage arising out of the escape or discharge of crude oil or natural gas liquids from offshore facilities located in the jurisdiction of the United Kingdom, Denmark, Germany, France, Ireland, Netherlands, Norway, Belgium, Greece, Italy, Portugal and Spain. In fact in general OPOL is considering to provide more prompt and effective environmental compensation than any other system of law.

7.1 UK Experience

In the United Kingdom the Petroleum Act of 1987 provides for safety zone around an offshore installation, infringement of which is an offence where if a vessel enters a protected area without consent, its owner or master can be criminally liable as it happened in June 1988, when a vessel struck an oil rig in the North Sea causing extensive damage. Oil and gas pipelines are among the projects covered by the Environment Assessment (EA) Directive. Danger to navigation and fishing clearly influence pipeline routes.

7.2 Thailand Experience

One of PTT's (Petroleum Authority of Thailand) major policies has been the development of quality oil product for environmental protection despite increasing investment cost. The first such attempt started in 1984, when PTT tried to reduce the lead content by mixing pure alcohol in premium gasoline. The mixture, called "gasohol", was put on trial with PTT's fleet of cars for two years. But the pure alcohol, as weighed against gasoline, was very expensive, compared with a then downtrend in world oil prices. The new fuel, may not be developed further, but information from the experiment in terms of usage, mixture and storage is still maintained to prepare for any future oil crisis. In 1987, PTT introduced methyl tertiary butyl ether (MTBE), another octane booster in its premium gasoline. The new products, "PTT HI-Octane", was marketed in early 1988 as the gasoline with the lowest lead content in the market, i.e. 0.40 grams per litre, while the products of other traders have 0.45 grams per litre, the highest content allowed according to the law at that time. This new product has contributed to the reduction of pollutants from fuel combustion, as it decreases not only the lead content in the atmosphere but also 40% of carbon monoxide (CO). PTT continues to maintain its leadership role in product development by being the first to introduce unleaded gasoline for distribution on May 1, 1991. This supported the government policy which requires the nationwide use of unleaded gasoline in 1996. In addition, PTT is also the first to import diesel oil with low sulphur and low distillation point, which can decrease the black smoke and sulphur dioxide (resulting from combustion of fuel with sulphur). Another project recently launched in cooperation with the Bangkok Mass Transit Authority(BMTA) is the natural gas for vehicles (NGV) project, using natural gas for buses instead of diesel in order to ease pollution problems in Bangkok.. The laying of natural gas distribution lines around Bangkok, called the Bangkok Gas Ring, is another feasible project to encourage industrial plants to use natural gas in place of fuel oil.

7.3 Indonesian Experience

Indonesian Legislations passed in 1992 and 1986 require that rigorous Environmental Impact Assessment be prepared for all projects, including petroleum, during the planning and approval process. More recent legislations, covering the conservation of biological natural resources and their ecosystems and water pollution control, are included in the requirements for the environmental impact assessment. Agencies have been established to review the Environmental Impact Assessment and to continuously monitor its performance.

7.4 Indian Experience

In India, amongst all forms of energy, hydrocarbon continues to be the major source of energy today. Consumption of petroleum products is expected to grow from the present level of 74 million tonnes in 1995-96 to over 100 million tonnes by the turn of the century and to 150 million tonnes by 2010. India being a net importer of hydrocarbon, the management of oil resources is of vital importance to our economic development. ONGC produce over 32 million tonnes of crude and 20 billion cubic metres of gas. Considering that more than 70 per cent of our oil production is coming from offshore areas, environmental protection is the key to successfully manage our business as environmental regulations are more stringent for offhsore oil operations.

8. Energy and Environment Regulations in India

Indian petroleum industry is to deal with environmental problems like many other energy-intensive industries. In this regard, our Ministry of Petroleum & Natural Gas (MOPNG) has set up Petroleum Conservation Research Association (PCRA) in 1978 and Oil Industry Safety Directorate (OISD) in 1986 and Directorate General Hydrocarbons (DGH) in 1993. Major activities of PCRA include field surveys, energy audits, diagnostic studies, R&D work, training and education of the energy consumers in various sectors of the economy. OISD lays down the procedures and guidelines in design, operation, maintenance fire, safety aspects of all installation of PSU oil companies. Similarly, DGH has been promoting sound management of hydrcarbon resources having a balanced regard for the environment safety, technology and economic aspect of the petroleum activities. In recent years, environmental awareness has increased in India and we are reacting strongly to public concern about the environment. The construction and installation of drilling rigs and production platforms are affected by many provisions of the environmental law of our Ministry of Environment and Forest (MOEF). The environmental law requires prior assessment of proposed installations before consent is given for various projects taking in view likely significant environmental effects.

9. Environment Management : A Case Study of ONGC

As far back as in 1983, ONGC decided on a statement of corporate environment policy. A lot of importance is attached to environmental and social obligation. ONGC published the first ever five year (1985-90) "Environment Management Plan" for environmental safeguards, enabling advance planning in this regard. Now, one of the corporate objectives of ONGC is environment protection. The organisation has set up a full-fledged department of Safety and Environment Management and introduced a well conceived system of environmental information and awareness as well as monitoring implementation of standard industrial practices for maintaining a clean and safe environment. ONGC has also established an Institute of Petroleum Safety and Environment Management (IPSEM) to provide R&D support to environmental aspects.

10. Energy Conservation Measures

Energy consumption for various activities of like drilling, production, cargo operations, LPG production, etc. are monitored, operation-wise and region-wise. The consumption patterns of operations are analyzed and the strategies to cut down the consumption rate are evolved from time to time. The fuel-wise consumption pattern, for 1995-96, is as follows natural gas 87.0 per cent (1227 MMSCM), HSD/Petrol 11.0 per cent (201.2 Mltr) and electricity 2.0 per cent (337.5 MU). Emphasis is being laid on maximizing the usage of natural gas thereby reducing the consumption of liquid petroleum products. The liquid petroleum products consumption in ONGC was 18.0 per cent less in 1995-96 (201.2 Million Lts) when compared to base year 1989-90 (245.28 Million Lts). Energy audits are being conducted for identifying and locating energy intensive units and plants. Specific improvement programmes are being designed suiting the needs of particular site/equipment. Various conservation measures have been adopted like use of lube oil analysis kits, conversion of diesel engines into dual fuel (gas + HSD) operation, by using gas engines in place of diesel engines, CNG for automobiles and gas turbine generator sets. Pilot project using CNG in place of diesel/petrol in various types of vehicles and generators have been conducted successfully in Bombay, Tripura, Gujarat and Southern Region. Various automobiles converted are Ambassador car, jeeps, trucks and buses, which have been successfully trial tested. natural gas based, 90 ton Absorption chilling A/C systems have been installed for centralized air conditioning at Nazira, ERBC. They are substituting the consumption of electrical energy by natural gas. ONGC is also using waste heat recovery systems at its major installation like offshore platforms for well fluid heating, Glycol heating, PPD storage tanks heating and other process heat requirements, offshore drilling rigs for generation potable water through waste heat recovery water makers and cogeneration is used at onshore processing Complexes of Hazira and Uran, through waste heat recovery boilers installed in conjunction with two 20 MW gas turbines at each complex. By using various energy efficient measures like waste heat recovery, use of soft-starters, variable frequency controllers, solar energy, efficient lighting fixtures etc., savings to the tune of Rs. 107.67 crores has been achieved during 1995-96.

11. Environment Protection Measures

Various steps has been taken by ONGC to protect the environment and different pollution control measures have been implemented to maintain ecological balance during land acquisition. Drilling locations sometimes fall on forest land necessitating felling of trees and bushes, and sometimes displacement of inhabitants. In such cases ONGC acquires equivalent land area somewhere (preferably nearby areas) and hands it over to the forest department for development purpose. During the development stage of a drilling location, attempts are made to plant trees, bushes and grass of compatible and fast growing nature so as to minimise the effect on flora and fauna. Upto the year 1995-96, around 22 lakh trees have been planted. Bunds are made wherever necessary to avoid soil erosion. Steps have been taken to minimise use of land by drilling a number of wells (usually 3 or 5 wells) as far as possible from a common platform i.e. same location. Along with the development of drilling sites ONGC also takes steps and implements social programmes like construction of roads, drains, bridges and socio-economic development of the area. Waste pits are constructed to receive various streams of effluent, Care is always taken to minimise use of water for operational purpose. Effluent water collected in pits is treated with lime and alum in a specified pit to make it fit for operational use through recycling. Modified filtering arrangement in pits is also practiced in some drill sites. Here, alum and lime treatment is given to a treatment tank placed in between waste pit and other pits having arrangement of filter compartments. The treated and filtered water which is very clear is used for operational purpose. To control the pollution by the associated gas of crude oil, the flaring of the same at Group Gathering Station (GGS) is accomplished in a flaring pit near GGS. During winter, oil floating on water in oil pit is converted to semi solid state and collected through skimming manually and transported to GGS for disposal. All the pits of drill site are initially filled with rock cuttings followed by earthen materials such that level of earth work becomes upto normal ground level. Leaving the area earmarked for X-mass tree and oil pipe line connected to GGS, all other area of left over drill site is cleaned and levelled properly, and made easily approachable.

12. Concluding Remarks

India, as a signatory to the Rio Conventions, is deeply concerned about degradation of human living environment. Our legislation requires that rigorous environmental impact assessment be prepared for all projects including petroleum, during the planning and approval process. More recent guidelines of our Ministry of Environment of Forests cover the conservation of biological natural resources and their ecosystems and water pollution control. During the last decade, public concern on the environment has grown tremendously towards energy efficiency, conservation and potential negative environmental impact associated with the development of petroleum resources. Natural gas is recognised as having less pollutance and considered more environmental friendly. One of the key areas of focus in the formulation of Indian Energy Sector's long range development plans is to ensure that environmental considerations are fully integrated into sustainable development.

RENEWABLE ENERGY SOURCES WITH EMPHASIS ON SMALL HYDRO

15

S C BHATIA, FIE
Dy. Chief Engr. (Design) El
SAIL, Bokaro Steel Plant
India

ABSTRACT

Electrical energy is one of the most important Index of economic development and growth of a nation in modern times. Its generation from various resources, in an optimum manner causing least impact to the environment is a matter of concern for all of us. Any discussion on our electrical energy options have to be based on the projected population growth and developmental strategy that will be adopted during the decades to come. Indications are that India's population will corss 1000 million mark by the turn of the century and will continue to grow further before stabilizing at a higher level. It is this large population base that will govern our developmental strategy and determine our electrical energy options.

Introduction

Any discussion on our electrical energy options have to be based on the projected population growth and the developmetal strategy that will be adpted during the decades to come. The large population base of 1000 million people by the turn of the century is most likely to govern our developmental strategy and determine our electrical energy options, out of Thermal (including gas), Nuclear, Hydro (Including Small Hydro) and Non-conventional sources for adequate Power management to ensure sustainable growth in the 21st century. The SHPs offer a promising source of energy which could be exploited beneficially with short gestation period and with least capital investment for deriving quick results in order to sustain the growth of Industrial development. Any delay in tapping this source leads to wastage of 'zero cost' electrical energy in the form of free running water.

Economics of Hydro Power

Hydro electricity comes from non-consumptive use of a natural and renewable source of energy without polluting the environment. Water presenting the 'zero cost' input, hydro plants produce, perhaps the cheapest energy, once installed. While the capital cost/MW of Hydro project may appear equal to that of thermal plants, they are far cheaper in absolute sense. Usually hydro projects are multi-purpose projects with associated benefits like flood-control, irrigation, water supply, navigation, fisheries development, recreation etc. The cost/benefit ratio, therefore, works out to be highly favourable in case of hydro projects. They use high efficiency (90 to 95%) Prime movers of longer life capable of providing maximum flexibility in operation, thereby providing excellent peaking partners in an integrated power system.

Environmental Concern

Its associated environmental impacts being the least, hydropower has been ranked as an attractive energy source by the World Commission on Environment and Development in their Report "Our Common Future" which recommends its development to the extent possible to help create sustainable society. The International Conference on Global Warming and Climate Change: Perspectives from Developing Countires' held at New Delhi during 21-23 February, 89 has favoured higher priority for non-fossil renewable energy sources and nuclear power. At the World Energy Conference held at Montreal in September 89, it has clearly emerged that the renewable energy sources, including hydro power, have an unquestionable edge over nuclear power in their quality and safety value to a sustainable energy development.

Global Perspectives

The gross theoretical world hydro power potential amounts to an annual electric power generation of about 40,000 billion units. Asia is endowed with 28% of the total potential. India has a potential worth 600 billion units. USSR has highest potential in the world. Developed countries like USA, Canada, Japan, Soviet Union etc have significantly harnessed their hydro power potential. Enrope has exploited moe than 60% of its hydro power potential. Many countries like Norway, Sweden, Austria, Brazil, Canada, Swtzerland etc meet their major power requirement from hydel sources. Norway has achieved 100% hydro power. During the last 30 years, nearly 150 underground power houses involving more than 3000 Kms of hydro power tunnels have been built in Norway. Great potential remains, particularly in the third world countires. Development seems to be hindered by the financial problems of the South countires and the scarcity of the sites in the North countries.

Indian Scenario

Nature has endowed India with high mountains and gigantic rivers. Himalayas, spreading across a length of about 3200 Kms and a witdh of about 640 Kms, with falls ranging from a couple of 100 mtrs to a couple of 1000 mtrs, abound in hydro energy potential, the snow clad giving rise to perennial flows. The Great Indus, Ganga and Brahmputra system with their tributaries emanating from Himalayas constitute about 70% the country's major assessed hydro power potential which according to the recent studies carried out by CEA, stand at 84,000 MW at 60% load factor. The hydro energy potential has been assessed as 600 billion units including secondary energy.

Electrical Energy Demand

As per the 14th Power survey conducted by the Central Electricity Authority, the demand of electrical energy is projected as follows :

Assessment year	Energy Requirement MW	MKwH
1994-1995	72,711	359,612
1999-2000	112,319	517,005
2004-2005	174,190	726,096
2009-2010	210,000	986,777

Electrical Energy Forecast

The above demand projection shows that in 15 years, the power demand would multiply itself by four times, both in respect of peaking demand as well as energy requirement. Studeis by Planning Commission and other groups also indicate that the power demand may increase by twice in each cycle of 10 years, though the absolute figures may bc slightly less than the above figures projected by CEA. The national economy which is maintaining a growth rate of 4 to 5% is poised to grow at a much faster rate in future Plans. Keeping in view all these factors and extrapolating the findings of the 14th Power Survey, the requirement may be of the order of 626,000 MW. Even considering on a conservative scale the demand could be expected to be in the region of 500,000 MW.

Small Hydro Power

Power generation through Small Hydro Power (SHP) stations is not only cheap but also environment friendly. Despite its dual advantage, the SHP has however, remained untapped all over the world. Things are now changing slowly with more and more focus on SHPs. China has set an example in this field, worth emulating. At present the total power generation in China through SHPs is to the tune of 15,000 MW. The advantages of SHPs are as follows :

1. Reliable and simple to operate.
2. Can be exploited in water flows of small streams, rivers etc.
3. Does not involve setting up of large dams.
4. There is no problem of deforestation or submergence.
5. There is no problem of rehabilitation of people.
6. Much less initial investment is needed.
7. Power availability with short gestation period.
8. Ideally suitable for hilly regions.,

India has a total installed capacity of SHPs at around 295 MW only — 110 from 48 schemes of 3 MW each and 185 MW from 27 plants of 3 to 15 MW Capcity each. Considering the large potential of a Small Hydro Power Plant and its overriding advantages, the Govt. of India is putting more thrust on SHPs. A target has been set to generate 600 MW of electricity through SHPs by the ned of Eighth Plan. The Govt. of India has also announced various incentives to atract private sector participation in the development of SHPs. Hydro power being the most clean source of electrical energy from environmental considerations also, needs to be tapped by the end of the century. Any delay in tapping this source leads to wastage of century. Any delay in tapping this source leads to wastage of this 'Zero cost" electrical energy in the form of free running water. The SHPs offer one of the most suitable alternative in order to tap the Hydel source of electrical energy.

Non-conventional Energy

Though there has been much discussion about adequate development of Non-conventional energy sources like Solar, Wind, Biomass etc, any real breakthrough is yet to be made in harnessing these sources as major alternatives of electrical power. These sources of energy are free gift of nature to mankind. During the past one decade the use of non-conventional sources of energy for power generation has come in a big way to stay as commercial viable source. It has great relevance for India where a large population lives in remote villages, where the reach of electrical power through the long distance conventional transmission system has not been realised to the required level. Non-conventioal sources of energy could provide the answer, at least to supplement the system, if not a total solution to the problem.

Wind Energy

The existing capacity to generate about 40 MW of electrical power has been commissioned under the demonstration programme of government and a further 29 MW of electrical power generation is encouraged for commissioning in the private sector. In addition to this a total of 35 MW of wind farms are under installation at present. The future planning envisages a power generation of approx 500 MW by 2000 AD against the total Wind energy power generation potential in India estimated at 20,000 MW.

Solar Energy

Solar energy is one of the naturally available energy source, available in abundance in India and absolutely free of cost as a gift of nature to mankind. The scenario on Solar power generation is as follows :

1. A 30 MW Solar thermal power plant established in Rajasthan.
2. Proposals invited for 10MW solar-thermal power projects.
3. Proposals invited for large photovoltaic generation plants.

4. Special scheme of capital subsidy for grid interactive plants.
5. Estimated potential of solar-thermal generation from each 350 hectares of arid land in the sunny states is around 100 MW.

Biomass Energy

An overall potential of 17,000 MW of power generation through the use of various types of Biomass materials has been estimated. The government has planned to emphasize co-generation based on bagasse from sugar mills for which a potential of 3000 to 4000 MW has been established, out of which 300 MW capacity addition has been envisaged in the Eighth Plan through these systems.

Incentives for power from non-conventional sources

1. 100% accelerated depreciation on equipment.
2. World bank soft loan through IREDA.
3. Five year Tax holiday.
4. Concessions/exemptions from Excise duty.
5. Concession/exmptions from customs duty.
6. Exmption from Excise duty/Sales Tax.
7. Facilities for Wheeling, banking, buyback and Sales by SEBs.
8. Capital subsidy by certain State Governments.
9. Abolition of licencing.

Incentives for SHPs

1. Availability of World Bank soft loan credit of $ 70 million for disbursement to private sector entrepreneurs through Indian Renewable Energy Development Agency (IREDA) for the SHPs upto 15 MW capacity.
2. Extending capital subsidy upto 50% of the capital for implementation of Projects for hilly regions and non-grid projects and 25% of capital for others.
3. Extending support upto 50% for DPR preparation and upto 100% for survey and investigation for all catagories of projects.
4. Power generation schemes involving capital outlay upto 25 crores need no prior clearance from Government.
5. Five Year Tax holiday for new power campanies.
6. Concessional Customs duty @20% for import of equipment.
7. 100% depreciation under Income Tax Rules—Under Consideration.

Conclusion

For a developing country like India the demand of electrical energy is fast growing to meet the day-to-day domestic, Institutional, agricultural and industrial requirements. To cope up with the growing demand, the Government has chalked out capacity addition programme of about 38,000 MW of power generation during the Eighth Plan period alongwith associated transmission and distribution system.

The SHPs offer a promising source of electrical energy which is perennially renewable with 'zero cost' input, could be beneficially exploited with short gestation period and with least capital investment for deriving quick results in order to sustain the growth of Industrial development.

REFERENCES

1. Electrical enrgy and Environment, INAE Journal March 90.
2. Energy Options, MA Hai, INAE Journal Apr 90.
3. Energy Perspectives, SP Prothia, IEI Journal Dec 89.
4. Energy Options into 21st Century, SC Bhatia, IEI Journal Dec 94.
5. Elect Energy and Power, Mecon Journal March 95.

DELPHI ANALYSIS FOR THE UTILISATION OF RENEWABLE ENERGY SOURCES IN INDIA

16

S. Iniyan, L. Suganthi and **T.R. Jagadeesan**
Department of Mechanical Engineering
College of Engineering, Anna University
Madras - 600 025, INDIA.

ABSTRACT

The use of renewable energy technologies in developing countries has steadily increased over the past few decades. The widespread use of renewable sources require a greater understanding of the available options. In order to ascertain the quantum of acceptance of renewable energy sources in the context of possible deterioration of the environment, on account of the increased use of fossil fuels, Delphi study has been conducted. The feedback of the study was collected and analysed, so as to arrive at a general consensus. By the year 2020, the renewable energy contribution is effected to be 20% of the total energy use in India. At that time, the main resources utilised would be biomass, wind and solar in the order of their quantum of use. This study will help in the formation of strategies which will ensure the development of the optimal integrated energy systems for continuous power supply.

1. Introduction

Energy is an important input for the process of development of any country. As population expands exponentially, demand for energy increases. To meet this growing demand renewable energy sources have to be utilised to a large extent. The advantages of renewable energy sources are (i) they are locally available, (ii) they help in generating non-polluting energy and (iii) they help in sustaining a clean environment. Also their development and utilisation will improve the quality of life and provide dependable power supply to the rural and urban areas. In the energy front it is clear that renewable sources will play a significant role in the long-run.

2. Delphi Characteristics and Application

The characteristics of Delphi are Anonymity, Iteration and Feedback. Several research works have been initiated in this area. Delphi is not only an instrument for conducting technological forecasts, but it is also used in technology assessment and in research on development strategy and policy making. Chen Yuxiang *et al.*, (1990) state that Delphi is chiefly used in the following fields in China : Forecasting of technological developments, research on development strategy, technical evaluation, analysis of impact factors and estimation of model parameters. Jan (1990) makes a comparison case between individual and group interviews. A methodology for Science and Technology Planning Mario *et al.*, 1990 was developed using Delphi techniques. Jan (1990), reports on the potential transformation of the Delphi method into an instrument to control technological change by employees. Fred (1991) has reviewed the literature concerning the quantitative applications of the Delphi method.

3. The Process of Delphi

A Delphi study in the present research is conducted to determine the social acceptance and the percentage use of renewable energy sources. The economical, technological social and environmental factors were taken into consideration while framing the questionnaire. The Delphi questionnaire was given to ten experts as a preliminary test to check its validity. The first round questionnaire was then prepared and sent to 700 respondents. The participants were Secretaries for Ministry of Non Conventional Energy Sources (MNES), Scientists, Researchers, University professors, Agriculturists, Manufacturers and users of renewable energy systems in industrial, transportation, agricultural and domestic sectors. The mean value of the responses was determined and the second round questionnaire was prepared and sent to the first round respondents. Around 50% responded in both the rounds.

4. Analysis of Delphi Study

The Delphi questionnaire was analysed to determine whether there is any change in the responses from the first round to the second round, whether uncertainty has decreased and whether a consensus has been reached. Mean, Standard Deviation, Confidence and Stability of the participants in answering the questions were determined.

5. Mean

The mean response is the average response of the participants. The responses are given a scaling and then the value is determined. The mean response nearer to a certain option is found and that option is taken to be the mean value. The score is given for the relative importance of the different items.

$$\text{Mean score} = \Sigma \frac{(\text{number of responses for option i} \times \text{score})}{(\text{Total number of responses}) - (\text{questionnaires not answered})}$$

where i indicates different options.

6. Renewable Energy Distribution

The predicted percentage distribution of renewables and commercial energy sources for the year 2020 is given in Fig. 1. The experts' opinion indicate that solar, wind, biomass and commercial energy sources will be utilised to an extent of 7.12, 7.9 10.49 and 74.49 per cent of the total respectively.

7. Solar Energy Application

The extent of Solar energy utilisation for various end-uses was analysed. The Delphi study indicates that, as in Figure 2, the percentage utilisation of solar photovoltaic and solar thermal electric conversion for lighting end-use for the year 2020 are 9 per cent and 6 per cent of total lighting consumption. Solar cooker, solar thermal electric conversion and solar photovoltaic electrical conversion system are preferred for cooking and the expected percentage utilisation are 9, 4 and 4 per cent. Solar photovoltaic, solar thermal and solar thermal electric conversion have been suggested for pumping with the following percentage utilisation of 8, 6 and 5 per cent of total demand of energy for pumping. In the case of heating end-use the solar thermal, solar thermal electric conversion and solar photovoltaic electric conversion are preferred and the expected percentage of utilisation are 10, 6 and 4 per cent respectively, Solar thermal (absorption systems), solar photovoltaic and solar thermal electric conversion can be used to the extent of 7 per cent, 6 and 5 per cent respectively for cooling purposes. An amount of 6 per cent and 5 per cent can be used for solar photovoltaic and solar thermal electric conversion respectively for transportation. The policy should be framed to substitute the suggested quantum of renewable energy sources to meet the future energy demand. As the percentage of substitution expected is high, efforts should be taken to strengthen R & D in classified areas of importance. To be successful, incentive programmes to stimulate the increased use of solar energy must include a well-planned integration of technology-push and market-pull strategies.

8. Wind Energy Application

The wind energy distribution for different end-uses have been analysed and it is plotted in Figure 2. Wind energy electric conversion distribution for lighting , cooking, pumping. heating, cooling and transportation end-uses are 21, 12, 27, 13, 15 and 12 per cent respectively of the total energy demand respectively. More reliable and cheaper wind energy systems are the need of the hour. An integrated approach should be followed for optional use of wind, considering the location, nature of output and operational characteristics.

9. Biomass Energy Application

The extent to which the biomass resources should be used in various end-uses are given in Figure 2. The utilisation pattern are : for lighting biogas 6 per cent and biomass gasifier 6 per cent; for cooking biomass direct combustion 9 per cent, biogas 9 per cent and biomass gasifier 6 per cent; for pumping end-use biomass gasifier 6 per cent biogas 6 per cent and ethanol 6 per cent ; for heating purposes biomass direct combustion 8 per cent, biomass gasifier 7 per cent and biogas 6 per cent ; for cooling end-use biomass gasifier 4 per cent and biogas 3 per cent; for Transportation ethanol 8 per cent, biomass gasifier 5 per cent and biogas 5 per cent. In the traditional biomass sector, research and development should be encouraged in Energy Farming for sustainable use of bioenergy. Also Increased co-operation is needed to bring together bioenergy research, development

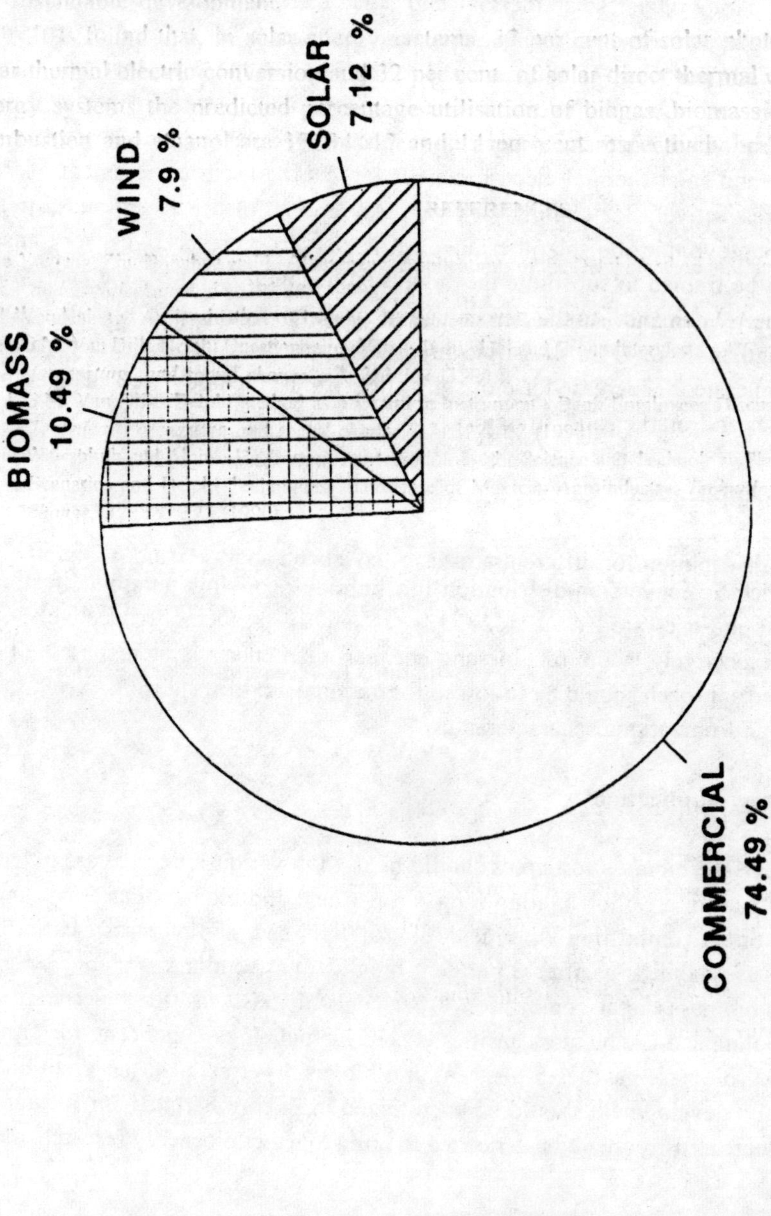

FIG.1 : *Predicted Energy Distribution Pattern for India in the Year 2020-21*

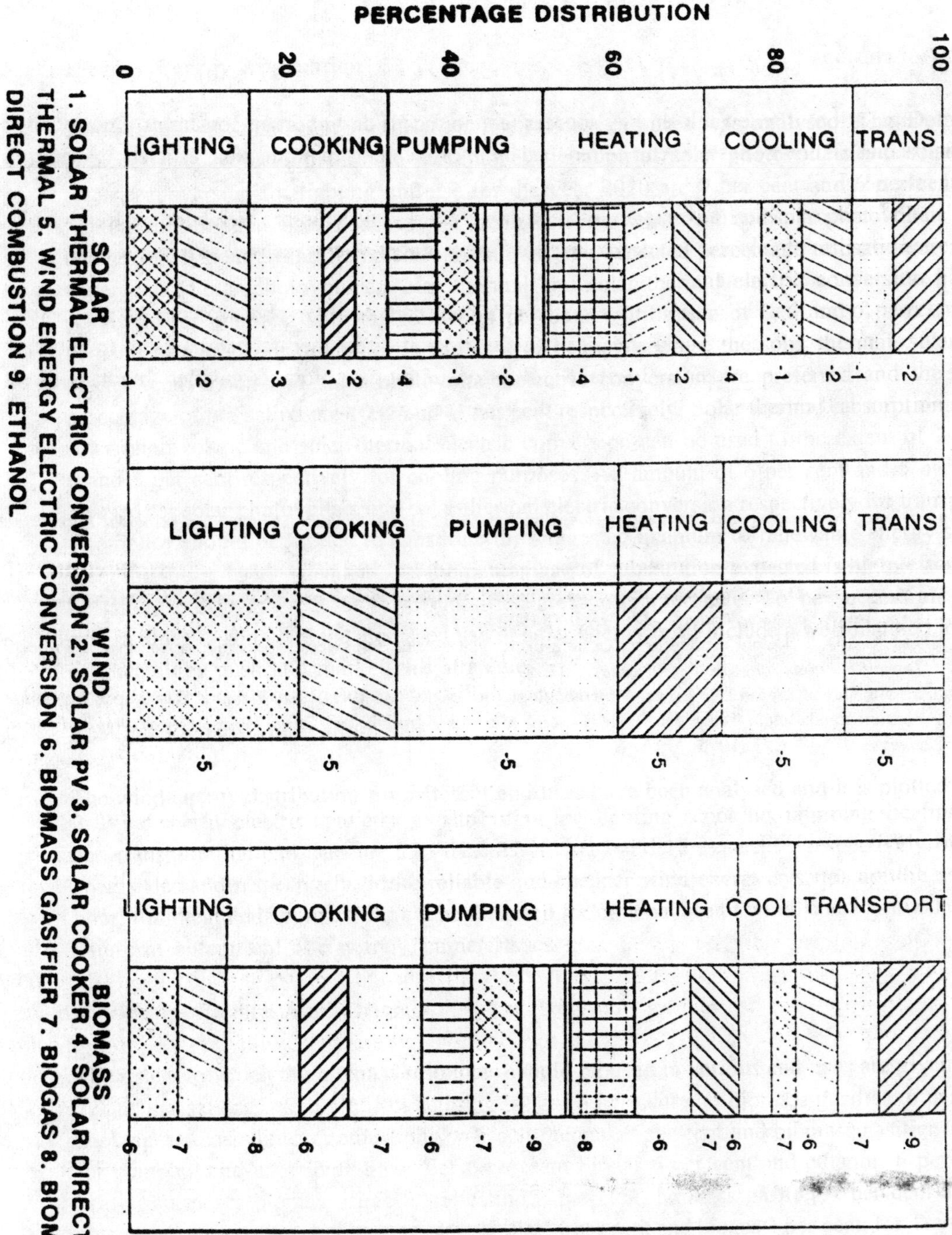

FIG. 2 : ***Predicted Renewable Energy Distribution for Different End-Uses***

and demonstration work already carried out in different regions and countries. Efforts should be initiated to implement afforestation programmes in the country.

10. Conclusions

In the utilisation of renewable energy sources, economics is an important constraint. Energy costs from renewable energy systems can be reduced by improvements in production, efficiency, materials and reliability.

The environmental advantages of renewable energy systems are tremendous. Renewables are seen as a part of the total solution to our environmental problems, as well as a major component for sustainable development.

It is found that, in solar energy systems, 37 per cent of solar photovoltaic, 31 per cent of solar thermal electric conversion and 32 per cent of solar direct thermal can be used. In Biomass energy systems the predicted percentage utilisation of biogas, biomass gasifier, biomass direct combustion and ethanol are 35, 34, 17 and 14 per cent respectively.

REFERENCES

Chen Yuxiang, Zhu Donghua and Li Changgeng, Applications of the Delphi method in China, *Technological Forecasting and social change, 38, 293-305*(1990)

Fred Woudenberg, An Evaluation of Delphi, *Technological Forecasting and social change, 40, 131-150* (1991).

Jan A.G.M. Van Dijk, Delphi Questionnaires Versus Individual and Group Interviews : A Comparison Case, *Technological forecosting and social change. 37, 293-304* (1990.).

Jan A.G.M. Van Dijk, 'Delphi method as a Learning Instrument : Bank Employees Discussing an Automation Project', *Technical Forecasting and social change, 37, 399-407* (1990).

Mario Waissbluth and Alonso De Gortari, A Methodology for Science and Technology Planning Based upon Economic Scenarios and Delphi Techniques : The Case of Mexican Agroindustry, *Technological Forecasting and social change, 37, 383-397* (1990).

ECONOMIC RAMIFICATIONS OF POWER AND DEVELOPMENT STRATEGIES

17

S.R. Mohnot,
Executive Chairman, Centre for Industrial & Economic Research,
New Delhi

ABSTRACT

The losses suffered by the economic system are narrated the critical issues confronting power development are discussed, and the policy frame work is presented.

1. Introduction

To comprehend fully the present and future economic implications of energy scenario in India, one must go back. The last four-and-a-half decades have witnessed, paradoxically though, both a phenomenal growth in power generation (installed capacity of 1,360 MW in 1947 rising to over 83,000 MW at the end of 1995-96 besides 11,000 MW installed under non-utilities) and equally staggering avoidable economic cost with a high cascading multiplier.

2. Economic Cost

The multiplier is generated by the cost of alternative sources and a still higher cost in terms of loss of effective productive capacity in all primary, secondary and tertiary sectors. The low growth rate of GDP (3.5% over a period of three decades until the onset of 1980s) and highly adverse trend in the incremental capital output ratios (ICORS) have been due substantially to short and erratic power supplies. The power famine continues to be a perceptible bane in mid-1990s and is likely to continue beyond the turn of the century, thanks to weird policy development processes and their applications. The shortages persist despite the fact that the Indian per capita consumption is far below comparable international standards (China 550 KwH to India's 360 KwH). "Power in plenty is the requirement of the day to prevent India from becoming (in V.S. Naipaul's words) an area of darkness (literally) and to place the Indian economy right at the top of the resurgent economies". (Mark the words at the end of the five years of reform from one who is intimately engaged in a large power organisation).*

* Rajendra Singh, Chairman and Managing Director, National Thermal Power Corporation in "Shifting Indian Paradigm in Survey of Indian Industry 1996. *The Hindu,* Madras:.

As the GDP growth is affected by the power supplies in quantum and quality, it undermines the real power elasticity of GDP. The divergence is due to underutilisation of power capacity (both generation and transmission) itself, underutilisation of capacity in the other sectors and low productivity resulting from deficient and erratic power supply. Power is not merely an essential component of GDP, it is an engine of growth. Larger power supplies stimulate growth beyond itself.

It is indicated that as of March 1996, the energy shortfall in India is of the order of 9.0 per cent (peaking shortage even higher at 18.0%). Ignoring the energy elasticity for a while, even at 9 per cent – and delibitating peaking shortages – the loss of GDP is of the order of Rs 900 billion annually (at the 1995-96 rate), more than what the country proposes to invest in the coming years – by all, the Central and State Governments and the private sector, Indian or foreign.

India's energy load factor in utilities has been claimed to be around 60 per cent subject to positive and negative errors of estimation. Should this be raised to 75 per cent which is an attainable goal, this would tantamount to 25 per cent increase (15 percentage points) in utilisation. The shortage is estimated at between 12 to 15 per cent after providing for peak shortages. By implication, only by dint of achieving the realisable utilisation level, the country can wipe out the shortages without adding an additional MW. In other words, what is required a saner management, to use a provocative phrase, which on analysis might be found to be a realistic assessment. The losses suffered by the economic system and its dynamics have taken the following forms:

1. Direct loss, suffered by the under-utilization of capacities created with massive investments for generation, transmission and distribution of electricity.
2. Loss of GDP due to non-availability and erratic power supplies.
3. Loss of productivity due to deficient and erratic power supplies.
4. Incremental investment in captive power supplies most of which is grossly unutilized (since most of this is standby).
5. Incremental operating cost of alternative or captive power generation.
6. Cost of equipment to rectify erratic power supplies.
7. Cost of increased maintenance due to grossly low standard of maintenance of power equipment - for generation, transmission and distribution.
8. Incremental maintenance cost or loss of power-driven equipment at the consumer-end (industrial and bulk consumers and domestic consumers).
9. Multiplier effects of the foregoing and in particular (1) and (2). Three major areas which ought to invite urgent attention and bold moves which have been relatively neglected are:
   (a) better maintenance of plants and transmission and distribution systems leading to higher capacity utilization and higher consumption coefficient.
   (b) Improvement in the quality of power.
   (c) Conservation of energy.

Seemingly insignificant area of energy conservation has the potential of wiping out the present shortage. The poor quality of energy supply is a dimension which calls for as much concern as power generation itself. The cost of work in these three areas have an incredible potential to meet the yawning gap in energy supply quantitatively, qualitatively and in terms of total cost.

3. The Critical Issues

The multivariate economic issues confronting power development are:

1. Opportunity cost of supply-demand imbalances
2. The location size and environmental aspects (costs) of power projects
3. The dimension of growth and consequent capital resource requirements
4. Sourcing of financial resources
5. Relative economics of fuels for power generation – solid, liquid, gaseous, nuclear, hydel and wind power (large and small)
6. Cost overruns
   (a) due to time overruns
   (b) changes in design, size, fuel resources
7. Cost of dithering
   (a) in policy formulation
   (b) in going through multi-point clearances.
   (c) in implementing projects
   (d) in operating the systems.

8. Viability of the development process from the three standpoints, IPPs, transmission/distributions and the consumer.

4. The Policy Framework

Even after the iteratative realisation of the critical role of power supplies in the development processes, the performance of the Eighth Five Year Plan, soon to terminate in March 1997, is just dismal. The basic issues which were known continue to plague the scene and are being debated, thanks to the contrived processes of federalism and democratic systems. Some of the critical issues confounding the position are:

1. Restructuring of SEBs (State Electricity Boards, the main source of utility supplies throughout the country). Delhi Electricity Supply Undertaking, operating so close to the seat of Central political power, only highlights non-action in an activity where the need for immediate action is universally admitted.
2. The entry of the domestic and international private sector in the area of power generation, transmission and distribution.
3. The identification of the areas/cases for (but the denial/procrastination of) sovereign guarantees.

4. Identification of the fuel for energy production and the conditionalities governing their usage.
5. Finding structure, mechanisms and modalities of power financing.
6. Fixation of the financing norms such as debt-equity and cost of capital.
7. Determination of the cost price parameters including the price of fuel, rates under PPAs.
8. Company structures (with or without alliances).
9. The structure and modalities of foreign participation.
10. The fuel policy and its nuances – supplies, sources, prices.
11. Policies regarding ownership and control – outright ownership, BOT (Build, Own, and Transfer) BOOT (Build, Own, Operate and Transfer), BOLT (Build, Own, Lease and Transfer).

What is now happening is that each State Government and 'the lowly credit - rated' SEBs are entering into long term, often repetitive negotiations analysing, assessing, raising, moderating and modifying PPAs and project related agreements with or without counterguarantee or sovereign guarantees and with or without clarity about environmental issues and policies. The result is that the progress has been slow and halting. It is not the price the economic development process is paying for the dilatory and iterative negotiations which is a matter of concern, the greater casualty is the lost time. It will have dilabitating impacts on future growth of the entire economic system.

It is enough to quote two top stories published by the *Times of India,* New Delhi, on the same day as late as October 2, 1996:

> "A power crisis looms large over Delhi as the Union Cabinet's decision that the railways will not carry coal to any power station unless freight charges are paid in advance, came into force on Tuesday.
>
> "Coupled with this came the threat of National Thermal Power Corporation (NTPC) Chairman that the Corporation would cut down supplies to the Delhi Electric Supply Undertaking (DESU) from the Northern Grid if immediate steps were not taken towards payment of Rs 744 crore owed by DESU to the corporation as well as for regular payment of the current monthly bills.
>
> "The powerhouse, managed by NTPC, is required to pay the railways Rs 22 crore per month. However, it pays only Rs 11 crore per month as it receives only Rs 31 crore of its monthly bill of Rs 45 crore from DESU.
>
> "Of Rs 1,000 crore owed by various powerhouses and state electricity boards to the railways, the Badarpur plant accounts for about Rs 750 crore.
>
> "The World Bank has given a notice to the National Thermal Power Corporation (NTPC) threatening to cancel loans worth about $1.2 billion if the Corporation fails to bring down its dues on State electricity boards to the level of two months bills. The issue is slated to be taken up during the forthcoming visit of the World Bank President.
>
> "NTPC's dues on SEBs have gone up to Rs 4,955 crore, the major defaulters being the boards in the eastern and northern regions.
>
> "He said the corporation would be forced to cut down supplies to the defaulting SEBs if

they did not take steps to pay back the arrears.

"The UPSEB was the largest defaulter with a total outstanding of Rs 1,006.58 crore followed by the Delhi Electric Supply Undertaking (DESU) with arrears of Rs 743.89 crore".

It is stated that the government has designed a package of incentives to promote participation of the private sector in the energy sector. It includes:

— Freedom to set up projects of any size.
— Freedom to set up thermal, hydel, wind and solar energy based projects for generation or for distribution as licencees; 11 per cent of outlay to come from promoters.
— Generating companies to sell power on a suitably structured two-part tariff mechanism.
— 60 per cent of outlay to come from sources other than Indian public financial institutions.
— Foreign capital participation upto 100 per cent.
— A 16 per cent return on Equity.
— 5-year tax holiday for both power generation and distribution.
— Customs Duty on import of equipment to be charged at 5 per cent.
— Rate of return to be escalated upto 0.7 per cent on Equity for every percentage point increase in PLF beyond 68.5 per cent.
— Hydro-tariff improved providing cover for hydrological/geological risks and long gestation periods.
— Competitive bidding made compulsory.

Now examine all the incentives and concessions. Except that a 5 year tax holiday is permitted – which existed – what is precisely the incentive? Incidentally, given the long gestation and lactation periods and capital intensity, the 5-year tax holiday is a highly eroded premise. Is a 16 per cent return on a massive investment with the risk factors an incentive? Even with 16 per cent return, what will be the IRR (internal rate of return), the real measure of profitabiiity . And why is it necessary at all – only because the dispensation permits the companies to access the consumer through the SEBs who for no fault of theirs have lost their own credibility. The customs duty of 20 per cent is not a concession only because there is a reduction; it is one of the highest levels. What does it imply: if you bring equipment worth Rs 1000 crore for your project, hand over a hefty Rs 200 crore to the government even before a KwH is produced!

There is no integrated, focussed, goal-oriented approach to achieve the goal. So long as focussed attention is not paid to the infrastructural development as a whole, in particular to the energy sector, Indian growth goals to match that of far eastern tigers would turn into empty dreams.

The 1996 Budget, for example, while reducing the customs duty on non-coking coal from 35 to 25 per cent, counterbalances it by disallowing carry forward of depreciation beyond eight years. Even the reduction in customs duty is a measely measure. The duty of 25 per cent is much too high by itself on any product, much more on the use of coal for power production. What is worse, disallowing full depreciation is wrong as a canon of fiscal finance. Depreciation is a legitimate cost. The rates at which it is provided could be a matter of debate, not the total amount. What it really means is that even if you have continued to make losses, you pay taxes. This is worse than MAT (Minimum Alternate Tax), newly introduced, which is payable only if you show profits. The imposition of 2 per cent import levy will also cost a lot to those power companies who will have to import fuels.

Now that the issue of sovereign guarantees is now on hold, there could be a real setback. In fact, the policy was never clear ever. It was encumbered by a number of road-blocks of 'clearances' and 'approvals'. No masive investment of the order desired for the power sector can be forthcoming unless a crystal clear, consistent policy is not only adumberated but also applied.

In a notable study (Michael Chia, Rob Mallick, 1996), has found that all the 200 new would-be power producers in China, India and Indonesia need to change the way they conceive and deliver projects*. It points out that forecasters had predicted that worldwide demand for new capacity would increase during the 1990s, an increase that exceeds the entire installed capacity of Western Europe. Of this, developing countries were to account for 70 per cent of the incremental capacity.

On the other hand, the projects get installed and do not take off. Of the 258 projects in developing countries that had received letters of intent or won competitive bids, about half had not concluded financing bids, while a further quarter either had failed or were experiencing problems.

In being able to attract the best international power producers and financing, one must recognise that the profitability of power utilities in India had been the poorest in Asia: Singapore and Malaysia (2.5 cents/KwH), Thailand and Korea (0.8 cent/KwH), Pakistan, China and Indonesia (0.1 cent/KwH) and India a negative (–) 0.7 cent/KwH. India's poor performance is not due to low rates. Industrial customers pay higher rates in Bombay than do in Germany, UK, US, Seoul or Shanghai, Kuala Lumpur or Colombo.

What is needed is the change in the mindset. Like the policy on DFI (direct foreign investments) for four decades, the policy-frame should recognize that the freedom to operate and to follow the normal rules of the game is not a 'concession' which will tempt private capital to rush and set up projects involving very large investments without being assured that the there is protection against exogenous factors. To provide that protection is not a concession, it is an essential conditionality.

The past has been a phase of missed opportunities. With the almost unlimited incremental demand for energy in the emerging countries worldwide, especially in Asia, there are competing demands on the resource supply. India will have to create hospitable environment for the flow of private capital to the energy sector – domestic or foreign. The demand for commercial energy in the Ninth Plan is likely to grow at a rate faster than the Eighth Plan. It is estimated that the capacity requirement in 2011-12 works out to 256,600 MW, thereby necessitating an addition of

---

The McKinsey Quarterly, 1996, No. 2.

175,400 MW of capacity in the next 17 years at an average annual rate of 10,300 MW. That costs a hefty Rs 400 billion annually to instal the desired capacity.

The vacillating policies and complex operating systems for evaluation and implementation of projects punctuated by recurrent emergence of diverse decision-centres, will not lead to the desired promotion of the power programme.

There has been in the recent period a shift from the central to decentralized decision-making processes. This is a welcome sign. But there is so much of interlocutory mechanisms which make the decision making apparently decentralized, it is more interlocutory, complex and dilatory. The discretionary authority has received premium over transparent and objective criteria. This is not obviously conducive to the demands of the critically urgent energy programme. A total transformation of the contents and processes of policy is the prime need if the acceleration of development is not to be stalled and sacrificed.

SUSTAINABLE ENERGY SYSTEMS FOR RURAL AREAS

18

Er. P. P. DUA
Principal
Dayalbagh Educational Institute
and
Dr. S. B. RAO
Reader,
Mech. Engg. Deptt.
Faculty of Engg.
Dayalbagh Educational Institute
(Deemed University)
Dayalbagh, Agra
(INDIA)

ABSTRACT

The Paper deals with the potentral of biomass energy for development of rural areas. Role of biomass production and waste have been separately dealt.

Introduction

Energy is a primary input in the production of goods and services. The role of Energy in contributing towards human welfare has increased in stature alongwith industrialisation and the organised development of the service sector of modern society. The far reaching changes in the form of energy and their respective shares in supporting human activities. Various attempts have been made since early 1980s to incorporate biomass in energy planning activities. There is a remarkable shift in the perception of the present and potential role of biomass in the national and rural economy.

It has become increasingly important to recognise the problems and potentials of biomass energy and the production and the rural incomes generated in the process of production and the environmental effects need to be considered as well.

Why Biomass Energy Development ?

In the 'Energy Crisis' perception of the 1970s, the negative effects on the rural subsistence economy, deforestation effects and impressive 'National Fuel Gaps' between supply and demand predominated. Remedies for the rural fuel wood problems were thought to be relatively simple: biogas production from different kinds of wastes, more trees in the form of every plantation, better stoves for rural families, tapping of energy stored in wind, water and solar energy conversion for heating are widely being studied.

As alternative to fossil fuels the growth and economic utilisation of biomass has to be seriously considered. Although traditionally all rural energy sources have been basically derived from biomass in the form of wood and waste (65% fire wood, 20% agricultural waste and 15% dung cake), it is only after the primary energy crisis that attention has been drawn to the wastefulness of direct burning of wood and the ecological and environmental problems created through massive uncontrollable deforestation. In fact it is said that India may produce sufficient grains for everyone, but unless measures are taken, it cannot provide the necessary fuel to cook the same.

TABLE 1 : Energy-Development Nexus

	Per Capita GNP- $ 1992	Per Capita Energy Consump. Kg oil eq. 1992	Average Growth Rate-Energy Consumption 1980-92 (%)	Energy Imports as % of Merchandise Exports-1992
Part A				
World Average	4280	1447	3.9 (2.5)	10
High Income Countries	22,160	5101	1.5	10
Low Income Countries	390	338 (376)	5.4	9
Sub Saharan Africa	530	258 (135)	2.9 (3.5)	7 (12)
South Asia	310	209 (289)	5.6 (7.2)	23
East Asia and Pacific	760	593	6.8	10
Middle East and North Africa	1950	1109	5.7 (7.3)	5
Latin America and Caribbeans	2690	923 (1051)	2.7	10 (13)
Part B				
Bangladesh	220	59	8.5	21 (26)
India	310	235 (337)	6.8 (7.2)	26
China	470	600	5.1	4
Philppines	770	302 (218)	3.1 (1.9)	22
Singapore	15,730	4399 (6178)	6.7 (5.6)	15
Egypt (UAR)	640	586	6.1 (4.6)	4 (5.6)
Saudi Arabia	7510	4463 (4866)	5.5 (9.3)	Nil
Costa Rica	1960	566 (507)	3.7	22
Brazil	2770	681 (908)	3.9 (4.7)	14 (19)
Part C				
Japan	28,190	3586	2.6 (2.2)	16
USA	23,240	7662	1.7 (1.4)	15
Canada	20,710	7912 (9390)	1.6 (2.0)	4
Highest Energy Consumption— United Arab Republic 14,631 kgoe.				

Source : World Bank: World Development Report, 1994.
N.B. Figures in parenthesis pertain to the year 1991.

Since there exists a positive nexus between energy consumption and food production, this low energy consumption impinges upon their long drawn battle to achieve self sufficiency in food production. Table 2 shows the consumption levels of the top 22 most populous countries – having poulation of more than 50 million and their level of self-suffiency in food in 1979-81 and 1988-90. The situation was almost self-sufficient in food in India, the situation continues to be most grim in Bangladesh.

Until and unless some reliable measures are adopted and a dynamic new approach in energy planning and management is vigorously promoted, majority of developing countries will continue to languish in slow growth, poverty, ill health and growing unemployment.

TABLE 2 : Energy Consumption and food self-sufficiency of the most populous countries (Population more than 50 million)

Country	Per Capita Energy Consumpton-1992 (1)	Food Self-sufficiency (%) 1979-81 (2)	1988-90 (3)
USA	7662	168	131
Russian Federation**	5665	–	–
Germany	4358	82	96
France	4034	172	215
Ukraine**	3885	79	85
U.K.	3743	94	110
Japan	3586	28	26
Italy	2755	75	81
Mexico	1525	83	74
Iran*	1256	76	68
Turkey	948	102	97
Brazil	681	83	89
Thailand	614	153	140
China	600	95	98
Egypt*	586	55	56
Indonesia	303	90	95
Philippines*	302	91	83
India	235	96	106
Pakistan	223	112	97
Nigeria	128	79	97
Bangladesh*	59	88	87
Ethopia*	21	99	87

Sources : No. 1 from World Development Report, 1994 and
Nos. 2 & 3 from FAO's Agriculture: Towards 2010.

* The most vulnerable countries, with sizeable food imports.

** USSR combined.

The most preferred fuel electricity is plagued with heavy slippage in India, the targetted production always falling short, the Table 3 shows the planwise slippage in power production in addtion, heavy transmission and distribution losses averaging 22%, thus adding to the growing energy starvation. Although, there has been an increasing allocation for the power sector–now

accounting about 30% of the total plan funds— the rural areas gain only marginally. So far, only 27% of the households in the villages have been electrified.

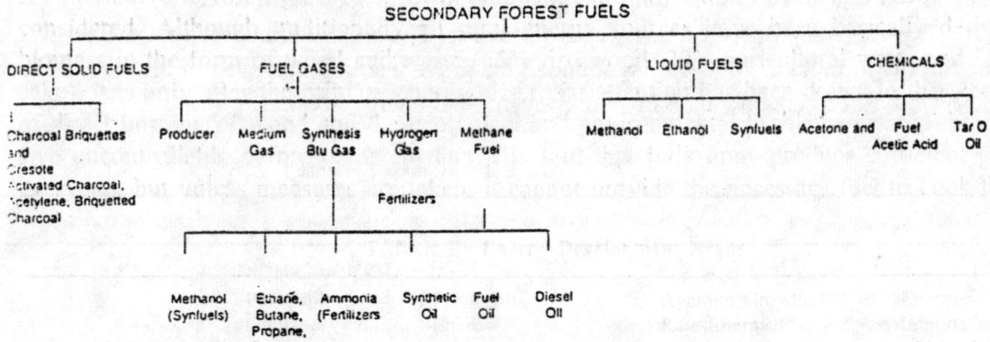

Some Yardisticks ofr Bio-fuels:

1 ton Wood	–	0.55	Ton Methanol
5 tons Cellulosic Wastes	–	1	Ton Ethanol
107.5 tons Sugarcane	–	1	Ton Ethanol
1 ton Ethanol	–	400	kg Butadiene or 600 kg Ethylen or Single Cell Proteins

TABLE 3 : Plan-wise slippages in the power sector in India

		Target (MW)	Achievement (MW)	Slippages (%)
First Plan	**1951-56**	**1300**	**1100**	**15.4**
Second Plan	**1956-61**	**3500**	**2250**	**35.7**
Third Plan	**1961-66**	**7040**	**4520**	**35.8**
Annual Plan	**1966-69**	**5430**	**4120**	**24.1**
Fourth Plan	**1969-74**	**9264**	**4579**	**50.5**
Fifth Plan	**1974-79**	**12,499**	**10,202**	**18.4**
Annual Plan	**1979-80**	**2945**	**1799**	**38.9**
Sixth Plan	**1980-85**	**19,666**	**14,226**	**27.7**
Seventh Plan	**1985-90**	**22,245**	**21,402**	**3.8**
AnnualPlan	**1991-92**	**4,212**	**2,776**	**34.1**

CMIE : Current Energy Scene in India. October, 1991.

Oil Production in India is constantly falling short of demand. From a peak of 34.09 million tonnes in 1989-90,.the production has slipped down to 26.95 million tonnes. The condition is targetted to group to 44.45 million tonnes in 1996-97. By then the total demand for oil and petroleum product is expected to touch 80 million tonnes.

Preponderance of non-conventional energy sources

The 'World Energy conference' held at Delhi in 1994 in its report had categorically highlighted the excessive dependence on the non-conventional fuels in Asia and Africa. It had further pointed out that there is not going to be any letup for a long time to come.

The Advisory Board on energy, in its report 'Towards A Prerspective On Energy Demand and Supply in India in 2004-05 has sounded the grave warning that even in the early 21st century there would be preponderance of non conventional energy sources.

TABLE 4 : Biomass Systems

	Advantages		Problems
1.	Stores energy for use at will	1.	Land use competiton
2.	Renewable	2.	Land areas required
3.	Dependent on technology already available with minimum capital input	3.	Supply uncertainty in initial phase.
4.	Can be developed with present manpower and material source	4.	Fertiliser, soil and water requirement
5.	Reasonably priced		
6.	Ecologically inoffensive and safe		
7.	Does not increase atmospheric CO_2		

Adapted from D.O. Hall 1979.

Biomass from Plants — The Photosynthetic Pathway

Most of the developing countries, being located in/around the tropics are well endowed with sunshine. This bounty of nature makes the photosynthetic pathway as the most desirable and efficient energy option for the Third World Countries. Amongst all the new and additional sources of energy, it is the biomass and bioenergy derived there from, when promoted and produced in a decentralised manner, could eventually turn the tide in favour and lead to speedy growth, with the added benefit of year-round employment for the needy millions.

Green plants is a naturally available machine to fix atmospheric carbon dioxide into carbohydrate from which ultimately we get energy as shown in the equation

$$6\ CO_2 + 6\ H_2O \xrightarrow{\text{Light}} C_6H_{12}O_6 + 684\ \text{Kcal.}$$

Generally in common fuels such as wood, the carbon (C) and hydrogen (H) are available in the form of cellulose, hemicellulose and lignin and the former is the material which burns easily. When wood is directly burnt, at the most 13% of total energy is utilised and with improved chulhas the efficiency is raised. However the volatile matters going out as smoke and soot are wasted and become pollutents. In the 1970s, Gunnerman from America developed small wood pellets called "woodex" from wood waste. Woodex is nothing but a concentrated form of carbon and can be burnt throughly. It is environmentally clean particularly with reference to potent pollutent sulphur.

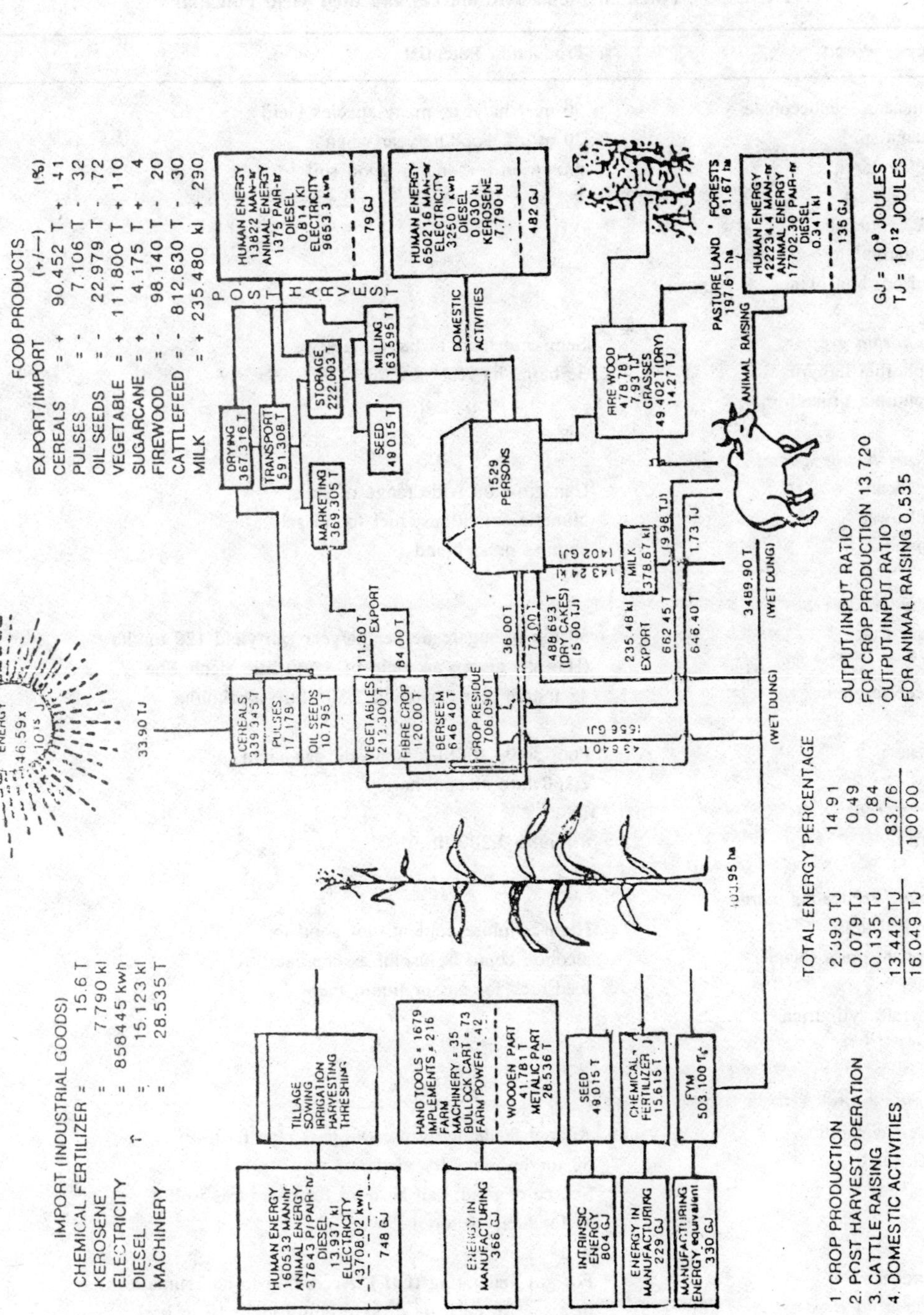

Annual Energy Flow in Village Ecosystem

TABLE 5 : Potential Biomass Resources and their yield Potential

Trees / Wood	Production Potential
Leucaena leucocephale Acacia sp. Colliandra sp. Sesbania sp. Eucalyptus sp. Casuarina sp. Hydrocarbon plants	50 m^3 / ha/year; many species yield 20 m^3 of wood/ha/year when grown on reasonably good soil
Euphorbia group Euphorbia lathyrus Euphorbia tirucali	Semi-arid land 17/barrel/ha/year, 36/barrel/ha/year
Oil producing species Soyabean Sunflower Groundnut	Can grow on wide range of soils, climatic conditions; fuel for diesel engines or as blend
Sugar/starch crops	
Sugar cane	50 ton of sugarcane per ha/year can yield 120 ton/ha (Hawali) giving an yield of 3,600 litre alcohol/ha.
Sweet sorghum	45 ton/ha/year can yield 3,500 litre alcohol/ha.
Cassava	Poor soils, 10-20 ton/ha/year can yield 2,160 litre alcohol/ha
Other crops	
Corn	6 ton/ha–2,200 litre
Green herbaceous plants	
Tropical grass Elephant grass (Napier grass) Imperate cylindrica (Asia)	High cellulose content, not good for alcohol, could be useful as charcoal, feedstock for gas or liquid fuel.
Aquatic crops	
Water hyacinth	Animal feed/energy use (biogas) plus fertilizer.
Duckweed	50 ton/ha/year (dry matter)
Seaweeds	Source of food; can be used for energy as well as for fuel; 90 ton/ha.
Seakelps	For gas generating (CH_4). According to an estimate, in a sea for each ha of kelp could yield 10 m kcal of food.

The global biomass production of 200 billion tonnes can yield 3×10^{21} joules of energy per annum, which is about ten times the total world energy consumption. Thus, this energy source NULLI SECUNDUS, biomass can become a resource that is far cheaper, easily renewable, pollution free and something that easily conforms to their very ethos.

Given meticulous planning, biomass can definite help micro regions become virtually self sustaining 'Energy Republics'.

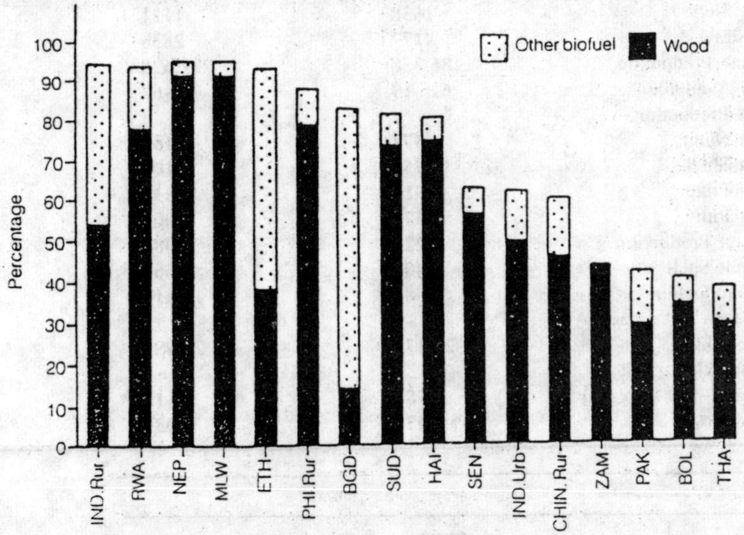

FIG. 1 : *Bio and wood-fuel dependency in selected developing countries.*

The fuel ladder shown in Fig. 2, illustrates the general point of "upward shifting" consumer preferences for more convenient source/devices of energy. Most energy policies seem to focus at possibilities to influence the transition at, or towards, the top of the ladder.

Making an inventory of biomass resources, one is amazed at the rich diversity that is to be found in any and every situation, ranging from agro-horticultural wastes, animal wastes to forestry and fishery waste, besides the agro industrial wastes. The diverse bioresources and their potential for energy is shown in Table-5.

Since it is the crop residues and the agrowastes which form the most easily available and economically renewable resource. Table 6 lists the availability of major crop residues in India during last decade.

TABLE 6 : Potential availability of agro-wastes in India"
('000 tonnes)

	Crop	1981-82	1989-90
1.	Rice	53,284	74,053
	Rice Straw	26,624	37,027
2.	Wheat	37,452	49,652
	Wheat Straw	49,811	66,037
3.	Maize	6897	9409
	Maize Cobs	2069	2823
4.	Bajra	5537	6620
	Bajra Straw	[illegible]	[illegible]

TABLE 6 : Potential availability of agro-wastes in India" ('000 tonnes)

Contd.

	Crop	1981-82	1989-90
5.	Ragi	2961	2781
	Ragi Straw	14,213	13,349
6.	Small Millets	1638	1121
	Millet Straw	4177	2836
7.	Sugarcane Production	186,358	222,628
	Bagasse Available	62,119	73,467
8.	Coconut Production		
	(Million Nuts)	5573	8163
	Coconut Shell	752	1102
	Coconut Fibre	914	1338
	Coconut Pith	1371	2008
9.	Groundnut Production	7223	8088
	Groundnut Shell	2407	2669
10.	Cotton Production	7884	11,414
	('000 Bales–70 kg each)		
	Cotton Stalks	24,171	21,993
11.	Jute Production		
	('000 Bales–180 kg each)	6788	7112
	Jute Sticks	2478	2037

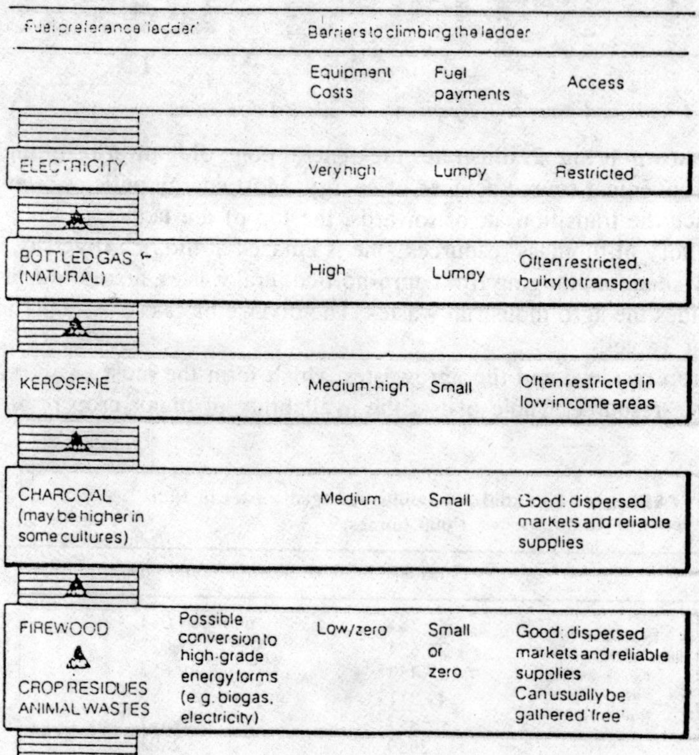

FIG . 2 : ***Urban fuel preferences and constraints***

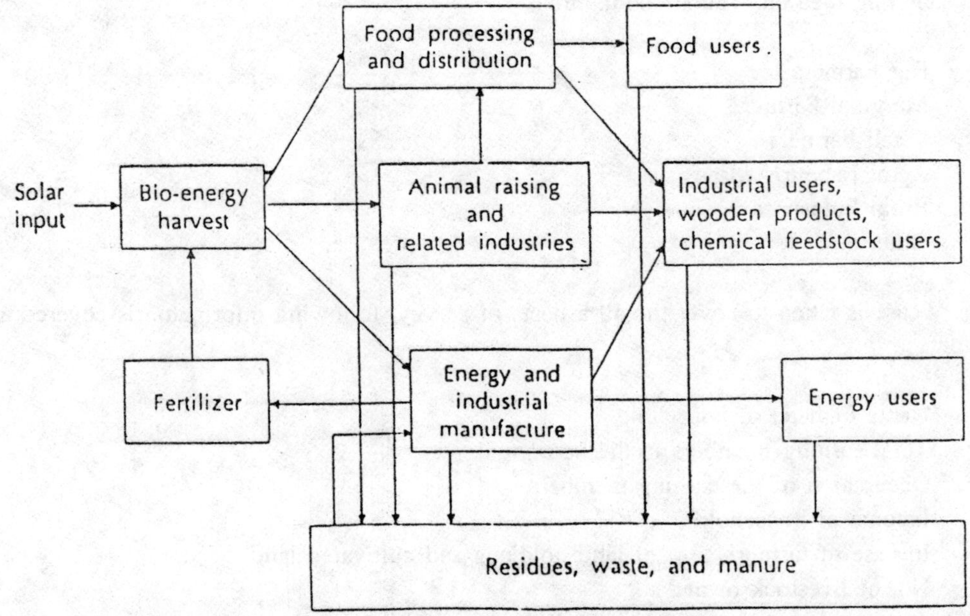

FIG. 3 : *The biomass conversion system*

Assessment of Existing Pattern of Energy Consumption

For evolving an appropriate strategy for energy in the rural areas it is essential to have an assessment of existing availability of energy resources and every consumption so as to make projections of energy demand and supply. Energy consumption in rural areas are in the main four sectors.

1. Domestic Sector — Cooking, Heating, Lighting
2. Agricultural Sector — Irrigation, Operation of Implements
3. Service and Industry Sector — Cottage Industries and Commercial Activities
4. Transport Sector — Transportation of produce and progress.

Data regarding consumption were obtained under the classification as -

1. Resourcewise consumption of energy
2. Sectorwise consumption of energy
3. Income groupwise consumption of energy

The level of consumption of energy and type of source is more or less proportional to the economic level of household.

A sample size of 10% of the total villages in the block is ideal for case study. On random basis, in each village only one-fourth of the household were surveyed that too on the basis of stratified sampling method. The six strata are

1. Big Farmers
2. Marginal Farmers
3. Small Farmers
4. Rural Labour/Artisans
5. Rural Industries
6. Others

Every case is taken to cover the all aspects of survey, following informaton is covered as primary data.

1. Name of head of household
2. Total earning members in the household
3. Occupation of the earning members
4. Income of household
5. In case of farmers, size of land holdings and cultivated land
6. No. of livestock owned
7. Types and members of domestic, agricultural, Industrial, Transport implements owned by household.
8. Basic energy consumption data

   Purpose *Type of fuel used*

(i) *Domestic*
    Cooking
    Heating
    Lighting
(ii) *Agricultural*
    Irrigation
    Cultivation
    Operation of Implements
(iii) *Industrial Purpose*
(iv) *Transport Purpsoe*

Information regarding the awareness of villagers towards the importance of energy, its efficient utilisation through improved and appropriate means of technologies are also collected.

The behaviour and approach of the villages in general towards adaptability of new technologies is assessed, which is also a requisite for case study and thus a part of survey.

The survey data, collected from village/block Barauli Ahir of Agra District in U.P., reveals that mostly non-commercial fuel is used by the households in the domestic sector. About 80% fuel is consumed for cooking purpose alone out of the total household energy requirement of which 90% comes from non-commercial energy sources. Among non-commercial fuels used for cooking, dung cakes contribute about 55% requirement rest is met by firewood. Sesbania Bispinosa

(Dhancha plantation) is used as firewood in the village the present survey was undertaken to bring out a composite energy supply-demand pattern, keeping in view the following variables.

Fuelwood Consumption pattern
Amount of fuelwood collected in each season
Sources of fuelwood in different seasons
Other energy sources

In summer the weather in Agra is hot and sunny with strong winds blowing throughout the day and calm periods for about 42% of the time. The winter is cold, with sunny days and stable meteorological conditions prevails.

Nearly 48-51% of total fuelwood is collected in winter
17-21% in rainy season and 29-42% in summers

Fuelwood and kerosene are normally used for cooking, coal is sparingly used and LPG is rarely used because its prohibitively high expense.

On an average 220 Kg. fuelwood is consumed annually per capita (5.5 persons per household). Share of kerosene consumption is low, about 33% and is used mainly for lighting, the total projected rural population of the block is estimated to be 1,58,436. Therefore the fuelwood consumption is 35,000 tonnes. The socio-economic survey indicators of the study area shows that it is one of the backward blocks of the district, literacy is also the minimum i.e. 26% as against 36% for the district.

REFERENCES

1. Bent Sorensen, 1995, Renewable Energy and Environmental Policy : Pacific & Asian Journal of Energy 5(1) : 79-95.
2. Bungay H. 1981, Energy, The Biomass Option, New York : Whiley.
3. EC 1994, Biofuels. DG XII report EUR 15647 EN Brussels : European Commission.
4. Hall D. O. 1979, Plants as an Energy source, Nature (London) 278, pp. 114-17.
5. Hall D.O., Rosillo-Calle F. 1991, Why biomass matters energy and environment. Biomass Users Network, Network News, 5(4): pp 4-15.
6. Jenson J., Sorensen Bent 1984, Fundamentals of Energy Storage, New York : Wiley.
7. Kalyana Sundaram S. 1984, Biomass Energy Technology in Bioenergy convention India, New Delhi.
8. Rajvanshi A. K. 1994, Biomass Energy - A panacea for Developing Countries in Bio-Resources 1994 Conference India, Bangalore.

A CRITICAL EVALUATION OF CLEAN COAL INDUSTRY AS A SOURCE OF ENERGY IN INDIA.

19

R.K. SACHDEV,
Ex-Adviser, Minister of Coal,
Government of India
MALTI GOEL,
Director, Department of Science & Technology,
Government of India, (INDIA)

ABSTRACT

Coal will continue to be a sustainable source of energy until 2020 AD. This paper describes the importance of coal beneficiation vis-a-vis coal production in India. A study on pre-combustion clean coal technology was commissioned by Technology Information Forecasting and Assessment Council under the aegis of Department of Science & Technology. The report traces the global trends and highlights the developments in coal industry, making an assessment of various projects for their techno-economic viability. Here we examine the emerging trends and recent developments in Indian coal industry for applications in the power, steel and future metallurgical industry.

Thrust areas and the initiatives to take up clean coal projects in the Research, Development and Demonstration mode as inter-sectoral programmes are highlighted.

INTRODUCTION

The coal will continue to be an attractive preposition as a source of global energy supply, not only in near future but in the decades to come. The projections for the year 2020, indicate a slight decreasing share of oil and nuclear fuels in the global energy mix, while the share of Lignite and Coal will remain stagnant at 29%. Consequently, world coal consumption is expected to rise at 2.1% per annum and electricity generation is expected to be the most rapidly growing form of energy. SWOT analysis carried out by Osborne *et al.*[1] also supports sustainability of coal as cost effective option in comparison to other fuels. The major threat in coal use, as energy resource however, is environmental pollution, which can be effectively handled through value addition from cleaning of coal as well as from use of advanced coal conversion technologies.

India is third highest producer of coal after China and the USA. Endowed with coal deposits of approximately 202 billion tonnes (of which 68 billion tonnes are proven), the projected Indian scenario is that by the year 2020, the demand for coal consumption would be around 1300 MT. The power and steel sectors at present are the major consumers taking 80% of share of the total coal produced. In the electricity production, installed capacity of coal based generation has increased from 8,000 MW in 1973 to 50,000 MW in 1994-95 and is expected to go up by another 50,000 MW in the next 15 years and further 50,000 MW is to be added in the next 10 years. To cater to this growing —energy demand, it is necessary to develop low cost techniques to reduce extraneous inert matter in coal, which is due to contamination during large open cast mining, where separation of coal from overburden material at the seam junctions is not feasible.

With over 85% of country's coal resources having ash content varying for 24 to 45%, the problem of combustion of high ash coal in an environmentally friendly manner will continue to be the focus of discussions in years to come. In this paper we examine the historical and recent developments in cleaning of high ash coals for applications to steel and power industry. Section 2 deals with the importance of beneficiation for non-coking coals. Suggested areas of development from the TIFAC-STAC study on Precombustion Clean Coal Technology are enumerated in section 3. Section 4 and 5 describe the same recent developments in the industry and the current programme of Technology development as the Coal Research, Development and Demonstration projects.

IMPORTANCE OF COAL BENEFICIATION

Indian Coal Industry has undergone considerable technical changes in the last two decades. However, the conditions after the new Industrial Policy of 1991 are of a different nature compared with earlier decades in post nationalisation. The Industry faces a much stronger competition from other fuels which are more environment friendly as well as from imported coals. With more than 60% of the coal reserves in India within 300 m depth, contribution of opencast mining has increased its share in coal production by about 8 folds in the last two decades. The underground coal production on the other hand has been more or less same i.e. 58.22 MT in 1974-75 and 56.85 MT in 1992-93. As a result share of inferior grade coal has increased from 9% in 1972-73 to about 75% in total coal production (Fig.1).

The non-coking coal has been graded into 7 categories (Table I) on the basis of their Useful Heat Value (UHV) which is computed as

$$UHV = 8900 - 138\,(A+M)$$

Here A and M represent percentage ash and moisture contents respectively (Table 1). A major share of total production comprises of E, F and G grades. Beneficiation of inferior coal is important from several considerations. For the thermal power plants, removal of non-coal materials leads to several advantages, viz.,

(i) Reduction in transportation costs—In India, a large part of coal is transported to regions located at more than 1000 Kms. Estimated savings in transport by washing of coal to different ash levels has been calculated and is shown in Fig. 2.

(ii) Reduction in carbon dioxide emissions and fly ash emissions

(iii) Improvements in efficiency of coal utilisation

(iv) Improvements in plant efficiency and overall plant utilisation factor.

(v) Elimination of support fuel oil and power savings in operation

TABLE 1 : Grading of non-coking coal & useful heat value

Grade	Useful Heat Value K. Call/Kg.	(A+M) % age	Calculated Gross Calorific Value Cal./Kg.
A	>6200	Upto 19.6	>6454
B	5600-6200	19.6 to 23.9	6049 to 6454
C	4940-5600	23.9 to 28.7	5597 to 6049
D	4200-4940	28.7 to 34.1	5089 to 5597
E	3360-4200	34.1 to 40.1	4524 to 5089
F	2400-3360	40.1 to 47.1	3869 to 4524
G	1300-2400	47.1 to 55.1	3113 to 3869

The demand for coal in power sector will increase at a rate of 7 to 8 % per annum, the total demand in 2010 is projected to be 550 MT against present demand of just over 200 MT. Steel plants deploying new technologies, such as COREX process which uses non-coking coals, thereby reducing dependence on import of prime coking coal, are to be benefited from the value addition in the inferior coals.

TABLE 2 : Requirement of Value addition in inferior grade Coal for major applications

Application	% Ash	% Moisture	Volatile Matter	Fixed Carbon	UHV
Power	34-35	6-7%	22-29	30	3000-3500
Cement	26-58	6%	30	30-34	4000
Sponge Iron	25	5%-7%	25-35	42-50	5200
Corex	20	5%	20-30	60-70	5500-6000

The requirements of value addition of different grades of coal in major application sectors are summarised in Table-2. Meting the growing requirements for future applications will need careful control of various activities in the coal preparation/beneficiation chain. The selection of an appropriate process for an application would depend upon the technology feasibility and its flexibility for use of coal from different mines, as well as its environmentally acceptability.

THRUST AREAS IN COAL BENEFICIATION

Tests on washability of coals in India date back to around 1920 when laboratory investigations were started. In 1951-52 the first two coking coal washeries were set up. However, the efforts till date have been on washing of coking coal for the steel industry alone. It is only recently that washing of coal for power sector has assumed critical significance in the light of benefits that arise from use of coal cleaning. The need for cleaning power grade coals has been emphasized by the fact that more than 70% of thermal capacity in power stations operates at efficiencies below 30%. Estimates so far show[2] that by enhancing the operating efficiency to the design gross efficiency level of 32%, would increase aggregate power consumption by nearly 15% and decrease CO_2 emissions per kilowatt hour (KwHr.) by 12%. A study of 210 MW Satpura Thermal Power Station, carried out in November 1988 revealed that beneficiated coal marked a reduction

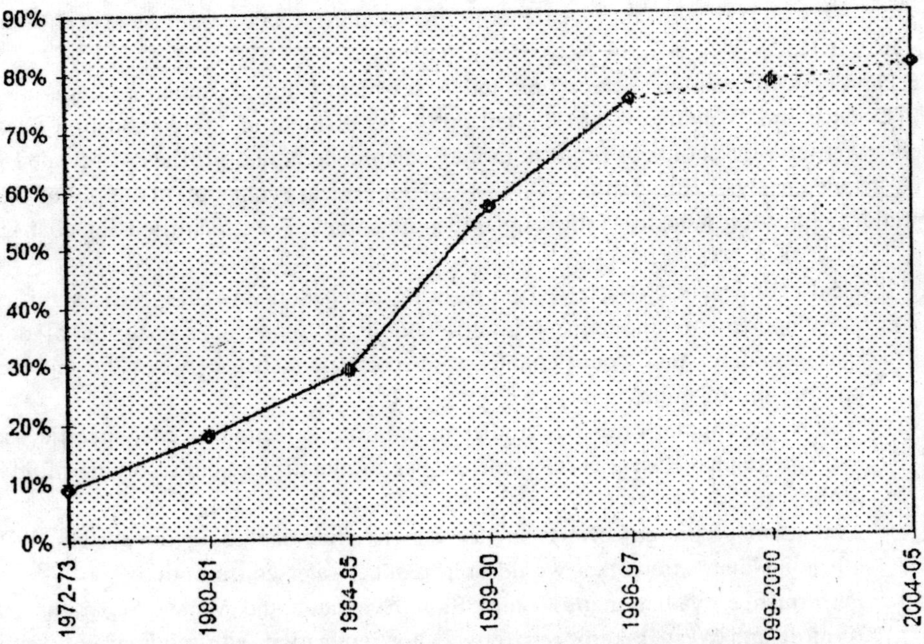

FIG. 1 : *Percentage Share of Inferior grade coal in total productions*

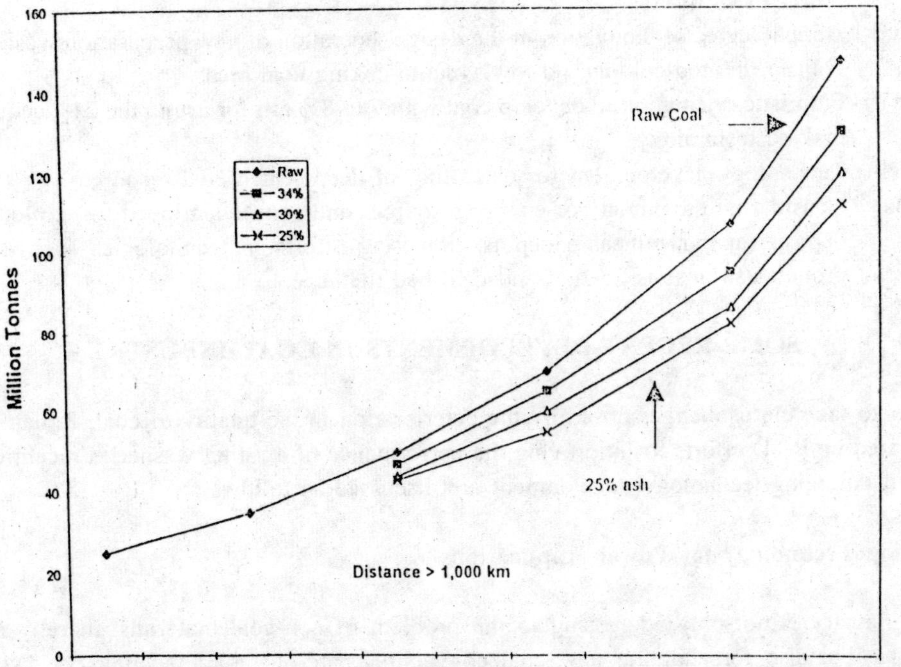

FIG. 2 : *Savings in Transport by Washing of Coal to Different Ash Levels*

in ash from 44% to 31% with improvements in gross calorific value from 3948 Kcal per Kg. to 5064 Kcal per Kg.[3]

Technology Information, Forecasting and Assessment Council, under the aegis of Department of Science and Technology, has carried out an assessment of Pre-combustion Clean Coal Technology. criteria for choice of projects in the Indian context and profiles of selected projects based on detailed analysis is presented[4]. On the basis of critical evaluation of supply of clean coal to the major application industry, following thrust areas have been identified for coal industry.

(i) Coal endowment characterization to bring out parameters of specific significance.

(ii) Development of innovative mining technologies as well as ensuring of descipline during mining that would enable closer correspondence between the *in-situ* quality and the run-of-mine quality.

(iii) Intensive computer oriented studies to optimize selection of coal specific washeries and Operational Research (OR) techniques in identifying and resolving multivariable problems.

(iv) Evaluation of present efforts in dry beneficiation including ore-sorters, ROM and other jigs-innovations that would help reduce water consumption.

(v) Performance evaluation of Contra Flow Separator and Vorsyl Separator set-up at Kymore and West Bokaro, respectively for replication with modular concept as well as capacity upgradation. The target market for these are, small and medium size non-power users of coal.

(vi) A high degree of innovation in the design, operation of new preparation/washing units both in the non-coking and low volatile coking coal areas.

(vii) Futuristic orientation to develop coal with 7 to 8% ash for use in the 21st century steel making technology.

(viii) Technology development for extracting of deep seated coal resources.

(ix) Considering the paucity of energy resources and consequent need for exploiting coal deposits at non-mineable depths, mounting efforts in technologies such as underground *in-situ* gasification and coal bed methane.

SOME RECENT DEVELOPMENTS IN COAL INDUSTRY

In order to face the challenges posed by the deterioration in the quality of coal, Indian industry has greared up R&D efforts for improving the performance of existing washeries recent developments in emerging technology development are discussed as follows :

Emerging Technology for Power Applications

Power industry is not only addressing to the problem of non-coal materials, thereby reducing emissions, but also for adopting new coal conversion state-of-the-art technology. A Gasifier driven combined cycle is an emerging coal utilisation technology for power generation. IGCC and PFBC are evolving coal based power generation technologies already commercialised abroad or at the verge of commercialisation. An IGCC demonstration plant of 250 MW has come up in short time at Buggenum, Netherlands[1] and has started producing electricity since 1994, to be upscaled to 600MW commercial plant by 1997. The technique essentially requires improved quality non-coking coals. The initial testing began for coal with 14% ash content, but it has a comprehensive testing programme for 24 different coal types. A 135 MW PFBC plant operating in Stockholm, Sweden not only has demonstrated greater thermal efficiency leading to cogeneration of 220MW distinct heating, but also produces lower levels of pollution. A lignite based power plant is another option to utilise the huge lignite resource in our country. The 300 MW demonstration plant at Goldenberg power plant in Germany would be worth examining for possibility of implementing this gasifier technology. Among the environmentally benign future energy options from coal, worth pursuing is the molten carbonate fuel cell, yet in the demonstration stage.

In India, the lead role in PFBC development was taken up by Bharat Heavy Electricals Ltd. at Trichy eight years ago. BHEL had set up a 6.2 MW Coal Based Combined Cycle Demonstration Plant during 1988 and conducted tests with different grades of coal. The plant has a 150 TPD coal pressurised moving bed gasifier. Based on these developments an Expert Group of the Ministry of Energy (1987) studied the global developments in gasifier technology and proposed a design stage plant at North Karanpura Coal Fields using unwashed non-coking coal. Indian Institute of Chemical Technology (IICT), Hyderabad and Central Fuel Research Institute (CFRI), Dhanbad have conducted preliminary tests with low grade, high ash coals available in India on one TPH moving bed pressure gasification plant.

Major breakthrough in improving the quality of non-coking coal through cleaning and deshaling of coal is ROM Jig Technology. This technology is proposed to be adopted for beneficiation of grade 'G' coal produced from Bina Opencast Mines. To meet the requirements of thermal power plants, ROM Jig has been specially designed for Indian conditions giving a yield of 67% with ash content of 31.6% from the input of coal with 45% ash and being recommended as potential technology after some more testing with different coals. The recent advancements in coal washing are summarised in Table 3. The input to all these processes is 'G' grade coal , except for Kymore unit, where the preferred input is 'F' grade coal. Detailed economic analysis of coal preparation was recently conducted by scientists at Pittsburgh Energy Technology centre USA, from the perspectives of a preparation plant investor, and a power plant operator. The analysis for Talcher coal showed that cleaning is profitable for a preparation plant investor for achieving clean coal with ash content of 32-34% using the conventional processes. By inclusion of advanced fine coal cleaning circuits, the coal preparation would become much more attractive. In power plant operation, four plants are evaluated to see whether it would be gainful to use beneficiated coal. Again high gross profits have been suggested depending on the plant capacity and transportation distance from the mine. Under the Programme for Acceleration of Commercial Energy Research (PACER), a collaboration between USAID/ India and ICICI, a 500 ton/hr washery project is launched to be located at Bilaspur for supply of washed and deep cleaned coal to the Dahanu Power Plant.

Table III : Recent Developments - Some notable examples in non coking coal washing

Unit	Process	Capacity/Year	Yield	Output		
				a*	M**	UHV***
Kallinga Coal Preparation plan	Improved Jig	8 MT	74.00%	34.00	7.00	3242.00
Bharatpur OCP & Ananta OCP	-do-	2.6 MT	76-77%	34.60	6.50	3300.00
Piparwar Project	-do-	6.5 MT	80.00%	34.00	8.00	–
Bina Project	ROM Jig	4.5 MT	67.00%	31.60	–	3435.00
Kymore ACC Beneficiation Plant	Contra flow separator	8 MT	70.00%	23.20	–	–

* Ash %, ** Moisture %*** Kcal/Kg %

Developments in Coal Preparation for Steel Industry

For the steel industry studies have been conducted to examine the potential of low volatile medium coking coal (LVMC) available in substantial quality. Recommendations of Sachdeva Committee (Ministry of Coal) stresses on High Utilisation Factor of Clean Coking Coal increasing from 67.6% in 1992-93 to 92.3% in 2001-02, through fructification of LVMC based capacity. The programme relies on modernisation of existing plants, qualitative improvements in clean coal and implementation of new technology for LVMC fields. The work of LVMC has been carried out at CFRI, TISCO and Central Mine Planning and Designing Institute, (CMPDI), Ranchi.

The recent developments include design of Kumbhraj Oven for production of blast furnace coke from LVMC at CFRI. In a two stage process, HM Cyclones is expected to produce Clean Coal with 26% ash, from the input as 'E' grade coal. In this direction, a promising development is Vorsyl Separator at West Bokaro Plant for Coal Cleaning. Under a Government sponsored project, Testing of Vorsyl Separator Laboratory Unit developed at RRL, Bhopal is being carried out at two coal washeries of Coal India Ltd. Techno-economics of this process indicates considerable benefits in coal output and less energy consumption compared to HM Cyclones.

The global trends in advanced technologies for steel sector indicate larger share of non-coking coal in the next 25 years. Indian industry has responded to those developments. The production of steel by Sponge Iron route by smelting in electrical furnace requires non-coking coal. This process is an alternative to Blast Furnace (BF) route, which is entirely dependent on prime coking coal blended with imported coal. Laboratory scale and pilot plant trials have been carried out for improving the quality of coal to the desired level. Six samples of coal drawn from 3 different mines of M/s. Singareni Collieries Company Ltd. have been tested to bring down the ash to 25% with corresponding percentage of recovery of useful fuel to 70%. Further field trials are needed for utilisation of indigenous resource.

COREX is another process of Steel making which does not require scarce coking coal and emerging as an eventual replacement of a significant portion of BF technology. It consists of a melter gasifier and a reduction furnace containing lumped iron ore. COREX plants are coming in operation at South Africa and Korea. In India also there are plans to have units based on utilisation of non-coking coal, which can meet stringent requirements of COREX Technology. To establish this process cost effectively it is essential that indigenous coal resources are used and the off-gas generated is simultanously utilised in power generation.

Among the successful demonstrations made on quality improvement of non-coking coal, worth mentioning is development at ACC Kymore. A Contra Flow Separator Pilot Plant of 100-150 TPH capacity functioned extremely well for cleaning of coarse coal and deshaling from the tests conducted on seven samples having varying ash content of 31.3 — 47.7%, it turned out that 77% output as ash content of 23.2% can be obtained. The Techno-economics of this process has been worked out in the TIFAC-STAC Study[3]. The process is graded as lower power consuming and water consumption is also low, of the order of 3 to 5 cum/hr. This plant is highly recommended for steel and cement industry.

Coal preparation for future applications in Metallurgical Industy

It is well known that coal can be beneficiated by physical methods as well as chemical methods[5]. The physical methods have limitations in ash reduction and yield coals with upto 20-30 % ash content. Chemical leaching methods such as Solvent Refining process (SRC) or Organo solvo-refining have been tested for developing coals with much lower ash, for applications in metallurgical industry. While a feasibility study of SRC carried out for low grade coals was evaluated it as highly costly and requiring high pressure reactions and temperature conditions, was not found suitable, IIT Delhi has developed organo-refining for production of zero ash coal. The product from this process, not requiring extreme temperature and pressure conditions is said to have very high calorific value of about 8000 Kcal/Kg. Technology upgradability of this laboratory technique is being tested under a sponsored project at IIT Delhi, for applications to graphite and aluminium industries.

COAL R & D PROJECTS-SELECTED PROJECTS IN TECHNOLOGY DEVELOPMENT

Recognizing the need for a strong science and technology (S & T) base for sustenance of the Indian Industry in a globally competitive environment, Department of Science & Technology (DST) has been emphasizing on formulation of an Inter-sectoral Science & Technology Programme with the participation of concerned stake-holders. The Science & Technology Advisory Committees (STACs) constituted in individual ministries have addressed to the S & T related issues for the concerned sector. The areas of joint interest were discussed in one of the inter-ministerial STAC Group with Coal Ministry having a lead role and steel and Power actively participating. Several areas of joint interest related to coal applications have been spelt out. Clean Coal Technology Development for inferior grade coal is identified as the priority area. The selected Research, Development and Demonstration type projects, with emphasis on increased utilization of low grade coals available in India will be taken up. A programme on Advanced Coal Utilisation Technologies (ACUTE) is being conceptualised integrating with concerned sectors. Under the programme intense research and development efforts would be directed towards developing a clear understanding of which coal will be needed for a particular process, utilising the existing knowledge base in the country. The programme will aim to develop most effective advanced coal cleaning techniques for power and steel industries. It will also deal with the advancement in coal technology for application in cement and metallurgical industry.

CONCLUSIONS

The developments in new technologies for coal as sustainable energy source have been discussed. In the coal industry, which is experiencing unprecedented technical changes, new environmental consideration have led to more severe threats and opportunities. India having large low grade coal resources, problem further pronounced as a result of open cast mining, is gearing up to face the challenge for their increased and effective utilisation in the Power and Steel industry. Thrust areas have been identified. The extensive knowledge base created over the years, can be exploited to achieve rapid developments in selected areas. Recent efforts in the industry vis-a-vis some examples of rapid developments abroad have been cited. Finally, in order for coal to be fully utilised for meeting future energy demands, the efforts need to be mounted through further research and developments and linking them with the stake holders. The utilisation of low grade coal in an efficient and environmentally friendly manner would form the basis of our future Research Development and Demonstration programmes.

REFERENCES

1. D.G. Osborne, J.M. Graham and L.K. Elliott, New Coal, p. 215-233, 1996.
2. R.K. Sachdev, Coal as Sustainable Source of Energy in India concerns and issues, Int. Conference of Energy, Nov. 17-18, 1994, New Delhi.
3. D. Biswas, Technologies for Control of Greenhouse Gas Emissions, Proceedings of the Workshop on Control Technologies for GHG Emissions, CII, N. Delhi, p.1, 1995.
4. Precombustion Clean Coal Technology Assessment, A TIFAC-STAC study, TIFAC Publications: EIS:002, 1995.
5. Malti Goel, Perspectives in Coal Utilisation with suitable options for reducing emissions, in the Seminar on Clean Coal Technology, 15-16 April, 1994, IICT, Hyderabad.

EFFICIENT METHOD OF SIZING SOLAR ARRAY FOR A BATTERY BANK

20

KAMALA DEVI, V (Scientist)
Agency for Non-conventional Energy and Rural Technology
India.

ABSTRACT

The size of the SPV module for a battery operated SPV device is normally chosen by the peak current capacity of the battery rather than taking into account the efficient charging capacity of the battery. If the design of an SPV device is done based on certain useful parameters of the load related with the SPV module, it is possible to reduce the array or module size to minimum . The paper discusses about the design concept for efficiently sizing an SPV module for a battery operated SPV device. The methodology used is an iterative procedure.

Introduction

Solar cells or modules when exposed to sunlight generate electric power which acts as input source for the SPV devices. Response of the solar modules towards sunlight and ambient temperature is significant in deciding the power and voltage rating of SPV modules to suit a particular device.

The most common load for a PV module or array is battery. Normally a solar module is sized for a battery depending upon the peak current required by the battery. This method of sizing is not a perfect matching solution. A criteria is to adopted by which the module load performance characteristics and certain load parameters has to be considered before deciding the size of an array for battery bank. The design of the power system should be based on the electrical characteristics i.e., current-voltage relationship of the array under various levels of radiation and at various temperatures.

Methodology

A battery has a voltage, depending upon its state of charge or discharge and other factors. This voltage determines the voltage of everything connected to the battery, including the module or an array. The battery voltage will set the operating point on the array characteristic curve. For a 12V battery bank, the battery voltage varies within a fairly narrow range usually between 11.5V with average full charge operation around 13-14 Volts. To size an array for a battery bank, the current sourcing capacity of the array is to be found out and matched with certain criteria of the battery. The array for a 12V battery bank be so sized that its operating point at maximum condition of irradiance and ambient temperature lie at a point where worst conditions are best satisfied. Since the battery voltage determines the operating point, the array size should be such that at 11.5 Volts, the array should be able to source current only in the range of C/10 and not a larger excess of it. The charging current is purposefully limited to this value in order to maintain the charging efficiency of the battery which may deteriorate if the charging current is kept above C/10 as it is clear from the experimental results shown in Fig. 1. and Fig. 2. These charge-discharge curves shows that the battery capacity is deteriorating with higher charging rates.

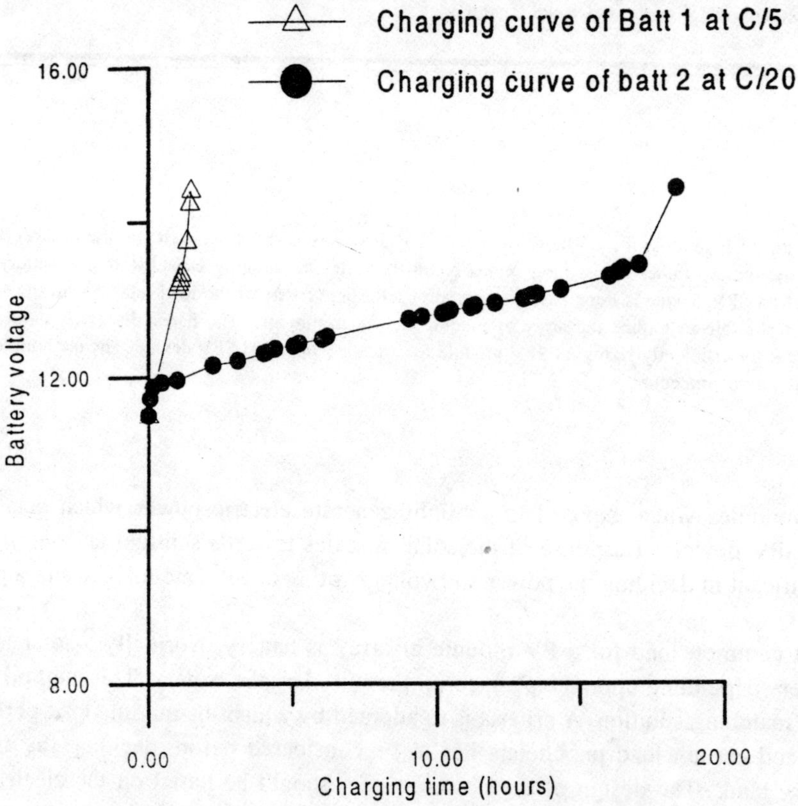

FIG. 1.

The classical solar cell equation is

$$I \times \left(1+\frac{Rs}{Rsh}\right) = Il - \left(\exp\left(Ko(V+I\times Rs)\right)-1\right) - \frac{V}{Rsh} \quad ....(1)$$

$$Il = I\left(1+\frac{Rs}{Rsh}\right) + Io\left(\exp\left(Ko(V+I\times Rs)\right)-1\right) + \frac{V}{Rsh} \quad ....(2)$$

where power is $P = V \times I$ ....(3)

The sizing is done at maximum power point to satisfy the worst conditions.

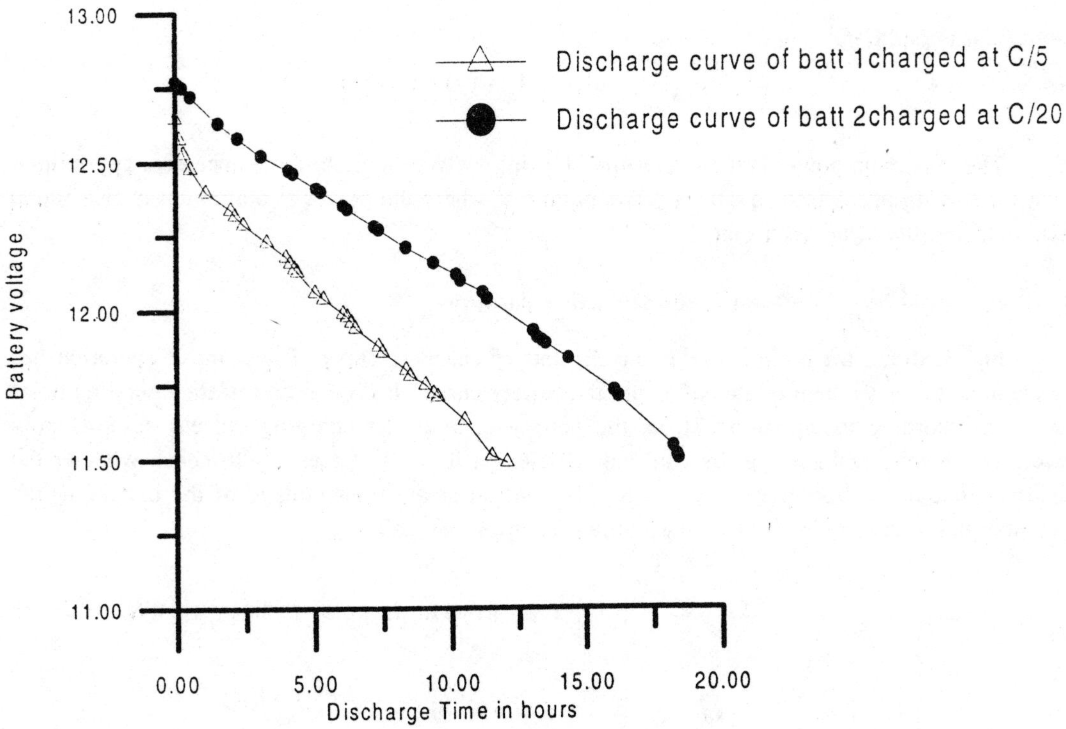

FIG. 2.

From equation (1) which is the solar cell equation, the five parameters (I*l*, Io, Ko, Rs, Rsh) which describe the equation can be determined by using numerical analysis technique for solving the equation and an iterative trial and error least squares approach for finding the equation parameters. The I-V curve data are first obtained using measurements of current and voltage of the module at any condition of irradiance (Hm) and temperature (T). The parameters are determined and I*l*, I0 and K0 are then corrected for irradiance = 1000 watts /m^2 and temperature = 298°K.

$$\text{where corrected } Il = = \frac{Il \times 1000}{Hm} \mu_{isc}\,(T - 298) \quad \text{....(4)}$$

$$\text{corrected } I0 = I0 \times \left(\frac{298}{T}\right)^3 \Big/ \exp\left[\frac{\varepsilon N_s K0}{1}\left(1 - \frac{298}{T}\right)\right] \quad \text{....(5)}$$

$$\text{corrected Ko} = \frac{Ko \times 298}{T} \quad \text{....(6)}$$

where ε is the band gap energy, N_s is the number of cells in series in a module times the number of modules in series and μ_{isc} is the temperature coefficient of short circuit current and is measured at standard condition.

$$\mu_{isc} = \frac{dI_{sc}}{dT} = \frac{I_{sc}(T_2) - I_{sc}(T_1)}{T_2 - T_1}$$

The maximum power is then determined using Newton's method of successive approximation for solving non-linear equations. The point I-V where the power is maximum is also found out and this should be such that

$$P_{max} = 11.5 \times \frac{C}{10} + 20\% \text{ for well matched solar array} \quad ....(7)$$

Fig. 3. shows the module curve and the battery charging curve. The point of operation lies at the intersection of the module curve and the battery curve. It is clear that if the operating point is fixed according to equation (7), as the battery charges the charging current automatically decreases thereby maintaining the charging efficiency. It is also necessary to check whether the selected module will be able to source trickle current at the float voltage of the battery which sets the minimum requirement of the module open circuit voltage.

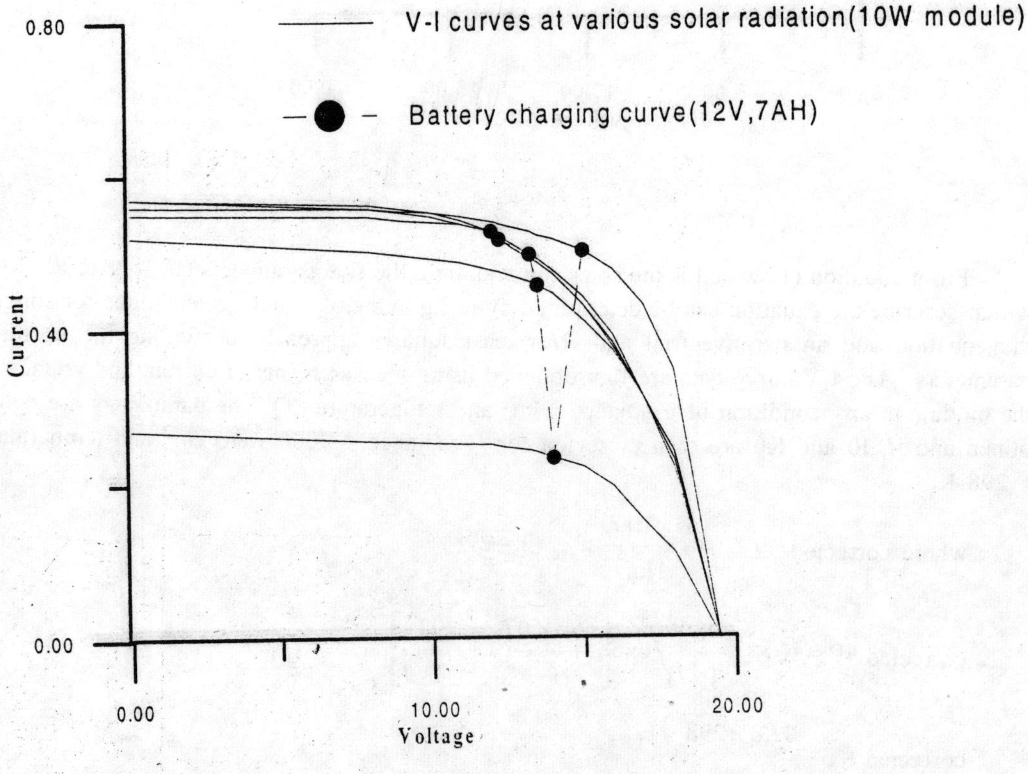

Fig. 3 : *A typical battery charging curve for a day. The operating point is shifting on fluctuation in solar insolation.*

A Typical Example

A typical example for determination of parameters and maximum power point of a 10W module for charging a 12 V, 7 AH lead acid battery is given below. I0 and Il are assumed constant to variation in temperature in this example. Fig. 4. gives the measured and calculated I-V curves of the SPV module at 700 and 1000 W/m^2 respectively. The data chart is also given below which shows that this module has a maximum power of 10 W with open circuit voltage not less than 15 V having trickle current capacity of not less than 10% of the rated charging current at this voltage. Hence this module is suitable for a 12 V, 7 A H battery.

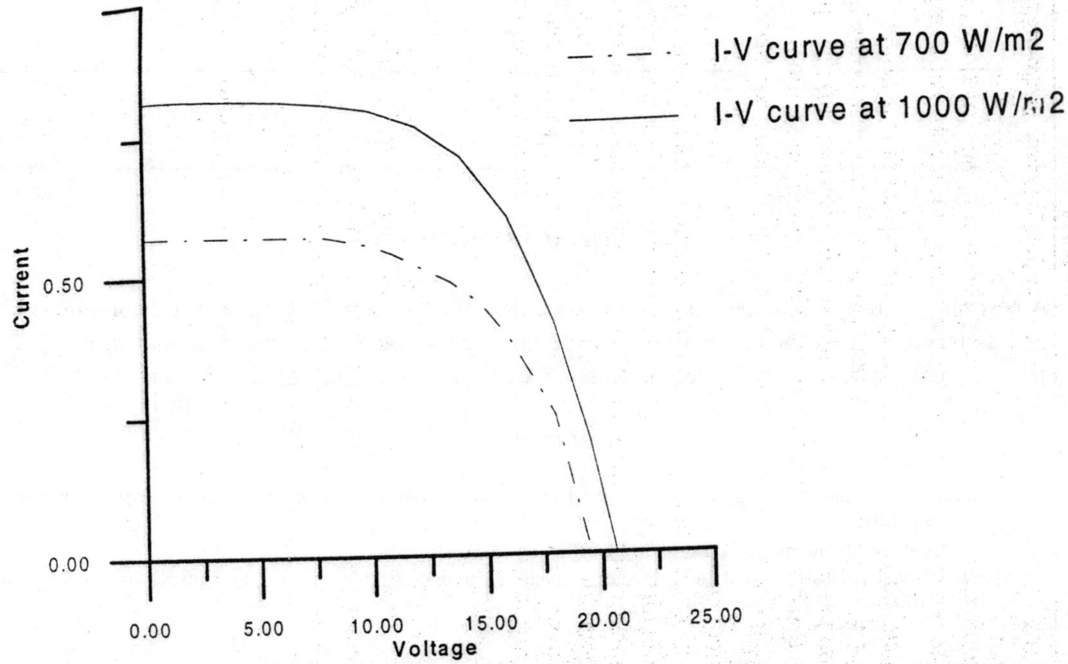

FIG. 4.

Measured V_R, I_R points at H=700 W/m^2 and T=30°C	V_R	0	8	10	11	11.5	12.5	13	13.5	14	14.5	15	16	18	19.5
	I_R	.57	.569	.555	.54	.53	.51	.5	.49	.475	.46	.44	.39	.25	0
Calculated values of parameters	Il	I0	A	K0	Rs	Rsh									
	.57023	1.1527E-04	87.75	.4361	2.4	10,000									
Corrected values of the parameters at H=1000W/m^2 and T=25°C	Il	I0	A	K0	Rs	Rsh									
	.8146	1.1527E-04	87.75	.4289	2.4	10,000									
Calculated I_C values at H=700W/m$_2$ and T=30°C		.57	.563	.553	.544	.539	.523	.512	.498	.483	.464	.442	.384	.203	8.48E-08

Calculated Ic, V_c values at H=1000W/m^2 and T=25°C	V_C 0 2 4 6 8 10 12 14 16 18 19.5 20.62 I_C .814 .8137 .8127 .81 805 .795 .769 .766 .607 .0415 .204 .776E
Calculated maximum power	Pm=10 W, Vm = 14.62 V, Im = 0.689 A
Required maximum power	(11.5 × C/10) + 20 % (depends upon the battery) = 9.66 W (approx 10 W) for a 12 V 7 AH battery.
Calculated current at V=11.5 volt	.69926 A , power = 8 W
Calculated current at V=14.4 volt	.69926 A, power = 10 W

CONCLUSION

A best match always operates the module such that at maximum power the worst conditions are best satisfied. The charging current and voltage are to be best adjusted at this point. This selection criteria may be extended to other SPV devices for sizing of solar arrays

REFERENCES

1. "Solar Photovoltaic lighting systems–Design Trends", by V.Kamala Devi published in National Solar Energy Convention-94.
2. Solar Energy of Thermal Processes – by Duffic and Beckman.
3. A numerical analysis method for computer modelling photovoltaic I-V curves - by Don Kitfoyle–ISES Solar World Congress, USA 1991.

DEVELOPMENT OF ALTERNATIVE ENERGY SOURCES

21

ASHISH K. JAIN
AEE (Electrical) MSG TO GGM (EM)
Department of Equipment Management,
Oil & Natural Gas Corporation Ltd.
Dehradun, (India)

ABSTRACT

Energy is an essential component for economic development and social progress of any nation. The economic and political security of developing countries greatly depends on their access to a secure supply of fossil fuel energy at reasonable costs. Energy is at present derived largely from fossil fuels. The major sources of conventional energy at present are crude oil, natural gas and coal, which are constantly depleting.

The present available resources are capable of meeting a small part of this demand. Under the circumstances to achieve part of the goal we must adopt energy conservation measures. For the future power generation the energy shortage will pose a recurring problem due to fast depletion of fossil fuels, and the increase in demand of energy due to increase in population and the growth of industry.

Developing countries are going through a challenging task of addition of generating capacity to bridge the ever-increasing gap between supply and demand of electric power, with the constraint of resources and the conflicting demands of protection of the environment.

India is a developing economy with a large agricultural base and a growing industrial infrastructure. Energy sector had more than 30 % of the plan allocation in recent years, of which oil had the second largest allocation. With the rapidly growing demands for liquid petroleum products, India is faced with the prospects of substantial increase in imports, despite an increase in domestic production of crude oil and gas. The export earnings are not sufficient to cover the increasing oil imports. We have to take some urgent steps to reduce our dependence on oil through energy conservation, development of alternative/ renewable energy sources and energy research and planning.

The objectives of the energy planners should be to assure the development and application of these low cost non-conventional/renewable energy technologies which offer the potential of making a significant contribution to our energy needs, and which can help in reducing long term dependence on imported oil. To face this situation, a need to search for renewable sources of energy has emerged as an option to the conventional energy sources. Development of renewable energy technologies, therefore should be at an economically and environmentally acceptable cost.

The paper thus aims to cover a few such conservation measures along with the commercially viable alternate sources of energy in the country.

Energy Crunch

Energy crunch is the talk of the day and poses a recurring problem due to fast depletion of fossil fuels, and the increase in demand of energy due to increase in population and the growth of industry. In fact, energy has assumed such an important role that it is one of the prime requirements in whatever development strategies we plan and employ. It is one of the most important factor to be considered for the economic development of the country and to improve the standard of living.

The requirement of energy for our nation is considerable and is always on the increase. Worldwide the present mode of generation of electricity is largely dependent on the fossil fuels. The consumption of fossil fuels for power production is expected to continue steadily in our country. Experts have estimated that the availability of coal will be for another 100 to 200 years. In the case of oil it is estimated that the sources will last for another three decades or so. Petroleum products like diesel and oil are to be imported from other countries.

With the fast developments in this industrial world, the energy demands are also increasing at a fast pace, while the conventional energy sources are diminishing at a much faster rate thereby creating energy crisis in the entire globe. Energy shortage is expected to double to 14.7 % in the next decade. A shortfall of nearly 50,000 MW by the year 2004 has been forecast.Our installed power capacity is around 72,000 MW against projected demand of around 1,72,000 MW by 2010 i.e. addition of 6500 MW per annum for next 15 years. This in turn means requirement of around Rs. 26,000 crores per annum. The present available resources are capable of meeting only a small part of this demand.

There is sufficient evidence to show that there is a direct correlation between economic growth and availability of energy sources. It is universally accepted that the continued and acute shortage of power prevailing in almost all the corners of the country will have far reaching consequence in crippling the future economic development of the nation.

The present energy scenario has become so fragile that even an external event in far away areas like the recent Gulf War have adversely affected the economy and development of our country. The developed countries have recognized these threats due to oil crisis and energy crunch and have taken immediate steps not only to reduce their energy consumption but also develop alternate energy sources.

The long gestation periods for establishment of conventional energy projects, dwindling natural resources and rising import bill for oil will severely affect energy supply, unless sustained efforts are made to supplement energy supply through non-conventional energy sources. The inflation and continuous rise in cost of fossil fuel will make the conventional power much more costly. To face this situation, a need to search for an alternative source of energy arose and the renewable sources of energy have emerged as an option to the conventional energy sources.

Need for Alternate Energy Sources

During 1960-61, commercial energy contributed 40 % towards the national energy requirements and balance 60 % coming from non-commercial energy, whereas in 1994-95 the share of commercial energy is estimated to have reached a level of 67% and is further projected to reach to a level of 79.6% by (2004-05).

This growing need of commercial energy will have in turn the pressure on individual sources and hydrocarbons will continue to contribute as a major portion of commercial energy. They are estimated to provide 47 % of commercial energy by the turn of century.

Although the Indian energy situation has been subjected to severe stress in recent years, there are several areas of energy sector in which significant progress has been made. There are several simultaneous challenges we have to overcome if we have to progress, particularly when market economy is becoming the fashion world-wide. We have to be globally competitive in cost, quality and delivery schedules.

Industry is fast realizing the pressing need to be energy efficient. Energy generation and conservation are buzzwords in industry today and can rightly promise a win-win situation for all concerned. There is a need to promote greater understanding and potential of alternate energy sources and their overall impact on the economic development.

Eventhough considerable importance is given for the power development by the Government, there is a shortage of power. In India traditional fuels characterized by low levels of efficiency still account for as much as 41 % of country's energy supplies. India is relatively poorly endowed in terms of commercial energy resources. Proven reserves of hydrocarbons are small accounting for less than 0.6 % of global reserves.

Our country is not blessed either with high quality fuels or with a wide range of energy choices; these adversely affect our energy sufficiency and energy security. The conventional sources of energy that is oil and coal are exhaustible. Coal continues to be predominant source of commercial energy in India accounting for about 60 % of its energy needs. In the power sector, at present, around 65 % of capacity is coal based.

With the increasing industrial and agricultural activities in India the demand for energy is also rising. It has now been realized that the exploitation of alternate energy sources could not only contribute, but also supplement meaningfully alongwith the further development of conventional energy sources.

Sharply rising energy costs have given impacts to novel and more efficient utilization of valuable energy sources. Many methods of conventional and non-conventional energy production are under focus as alternatives to future capacity additions. With the price shocks of 1970's, efficiency of fuel use took a renewed importance and development of alternative fuels has been given a special attention worldwide to use fuel efficiently.

The gap between demand and supply of energy can be bridged to a certain extent by supplementing conventional sources of energy by alternate sources of energy and above all by energy conservation.

Potential of Energy Conservation in the Country

Energy is vital to all of us as, no modern economy can function successfully without energy being available. Availability of adequate energy is a pre-requisite for economic development. To ensure a transition from developing economic state to developed economic state energy supply has to be more than the demand. Efficiency of use of energy must be there so as to keep the demand under control and ensure that energy goes where it can most profitably be used. By adopting various energy conservation measures such as waste heat recovery system, energy efficient devices considerable amount of savings have been achieved.

The growth in the demand of primary energy has been accompanied by a shift towards the increasing share of commercial energy in total energy. The share of commercial energy in total energy demand has grown up from 40 percent in 1960-61 to 67 percent in 1989-90 and is expected to grow further 80 % by turn of the century (year 2001-05).

India was consuming less than 3 MT of petroleum products per annum at the time of independence. By 1994-95 it touched to about 65 MT. It is expected to grow to about 103 MT by 2001. While the demand has been increasing rapidly, indigenous production of crude oil has not kept pace with it. As per the plan projection the production of crude oil would be about 57 MT in 2001. This calls for a judicious use of this precious resource.

Our country presently is undergoing a phase where the industrialization is going on at a rapid pace. With the ongoing liberalization of economy, the demand for energy is further going to increase. It is anticipated that the per capita consumption of energy would increase considerably. Pressure on fossil fuels is mounting due to new power plants coming up. Due to scarce availability of these, it is essential that most efficient use is made of our natural resources.

The present day awareness for the future assurance of adequate fossil fuels in view of their limited stocks, open up the campaign for energy conservation. With the growing demand and dependence on commercial energy and specially on oil and gas, it is imperative that these resources are used efficiently. It professes that wasteful energy expenditure must be stopped.

In view of the resource crunch hovering on our heads, the need of energy conservation has acquired now a much deeper and urgent emphasis than ever before. However, despite of this energy intensity continues to be very high and with the increased development in the country this is further widening gap between energy demand and supply. Energy Conservation is thus an obvious solution to the problem of energy crisis. It has been indicated that about 8 to 10% of energy savings can be obtained in all the sectors with little or no investment and simply by awareness good house keeping, energy auditing and energy budgeting. Additional 5% savings can be obtained by small investments and with a pay back period of around 2 years.

In our organization the conservation of energy has resulted in savings of about Rs. 107.69 crores during 1995-96, yet there is ample scope for improvement.

Development of Alternative Energy Sources

The overall consumption of commercial energy has increased tremendously in the last decade. The per capita consumption is around 337 Kgoe. This is quite low when compared to per capita consumption of some of the developed and developing countries. The share of various sectors in commercial energy consumption is indicated in Table 1 :

TABLE 1

SECTOR	% SHARE
Agriculture	4.3
Industry	51
Transport	22.4
Residential	10
Others	12.3

Industrial sector is the largest consumer of commercial energy followed by the transport sector. Since the dependence on fossil fuels is fairly large for meeting the requirements of these two sectors, there are several constraints we have to tackle in the course of time, such as depletion of deposits, quality of fossil fuels, heavy price to be paid for basic minerals and above all the environmental degradation.

In this context development of alternate energy sources have emerged as an alternative to supplement the conventional energy sources. Thus, the challenges to be met are :

(i) Judicious utilization of available fossil fuels
(ii) Implementation of Conservation Measures, and
(iii) Exploration and exploitation of Alternate Energy Sources.

We need to concentrate our resources for full development of alternative sources of energies, while we continue using fossil fuels. India is endowed with potential for large scale use of solar, tidal, wind and hydel energies, hence availability of energy is not a constraint but only its harnessing and storage is important.

The overall potential of renewable energy has been estimated to be :

TABLE 2

Sources/Technology	Potential/Availability
Solar Energy	5 × 10 15 WHr/Year
Small Hydro	10,000 MW
Wind Energy	20,000 MW
Tidal Power	9,000 MW
Urban, Municipal & Industrial Waste	1,000 MW

Alternate energy sources cannot completely replace the primary energy sources but can supplement meaningfully alongwith the further development of conventional energy sources. The demand for energy must be met by alternative source, increased efficiency and conservation of available sources of energy.

Alternative Fuel Substitution to Petroleum Fuels

Road transport in India mainly uses diesel and petrol as a fuel. HSD and petrol constitutes about 40% of the total demand of petroleum products. Further nearly 62% of the total 23 million ton of HSD and 50% of lubricants used is accounted for transport sector. Even a marginal efficiency by 10% in oil consumption in transport sector can provide 1000 crores in foreign exchange.

The number of automobile vehicles has increased from 3 lakhs in 1951 to 203 lakh's in 1995 and is projected to increase to a level of 424 lakhs by the year 2001. Accordingly the demand projections of HSD have been estimated to be about 35.7 million tonnes and that of petrol to be about 11.6 million tonnes by the Planning Commission, by the year 2000.

These high requirements of HSD and petrol demands for substantial efforts in checking its consumption by adopting greater efficient technologies along with use of alternative fuels in the automobiles such as Compressed Natural Gas, Methyl alcohol, Electricity etc.

Using Waste Heat and Cogeneration to Improve Efficiency

It is estimated that 30 to 50 % of fuel energy is lost to atmosphere as thermal energy. If this rejected thermal energy is used in a process, then the system efficiency becomes very high as this can become an additional source of energy in the plant. Major sources of waste heat could be furnace, incinerator, regenerator in refineries; turbines and generator exhaust gases etc. The waste heat available can be used by many schemes such as waste heat boiler, process steam requirement, Air preheater, Heat pumps, absorption chilling and space heating etc.

Many methods of conventional and non-conventional energy production are under focus and cogeneration is espoused as one of the many alternatives to future capacity additions. Cogeneration and combined cycle power plants have improved the efficiency to a level of around 70 to 90% compared to 25% to 33% in open cycle system in case of gas turbine power generation, using natural gas as a source of energy. It is estimated that the cogeneration potential is over 5000 MW in sugar plants itself. Cogeneration power plants has drawn attention of todays energy planners, because of following inherent characteristics.

— Maximum fuel efficiency.
— Increased reliability.
— Use of various types of fuels.
— Low Nox emission.
— Low incremental capital cost.
— Low gestation period.
— Reduced transmission losses.
— Low maintenance cost.

Cogeneration and Waste Heat Recovery in ONGC

In ONGC, waste heat recovery systems have been installed at various places wherever, in the operations, the utilization of waste heat is feasible. Some of the major installations in ONGC are described below :

(i) Waste Heat Use on Offshore Process Platforms

In the offshore process platforms, waste heat is available as the exhaust of Gas Turbines, being used for power generation and pump and compressor drives.

The heat recovered is being used for :

— Well Fluid heating
— Hot water trough heating
— Glycol heating etc.

On an average 347 million of electrical equivalent units of heat is being recovered on process platforms which in turn provides ONGC saving of Rs.52 crores per annum. The most of the heating requirement are being met through waste heat recovery on these platforms.

(ii) Waste Heat Co-generation Plants Gas Processing Complexes at Uran Bombay and Hazira Surat.

In the gas processing complexes at Uran (Mumbai) and Hazira (Gujarat) waste heat recovery boilers are installed in conjugation with each 20 MW, capacity gas turbines. The steam so produced is used for processing of Crude oil and production of LPG in the process plant.

Due to recovery of heat from exhaust gases of gas turbines, energy savings of 36 × 10 KCAL/Hour approximately can be achieved in each waste heat recovery boilers. Cost of these HRSG units has been recovered over a period of two years of operation. Use of waste heat recovery has provided a savings of Rs.18 crores on account of waste heat recovery.

(iii) Waste Heat use on Offshore Drilling Ships

Generation of potable water from sea water for operational and other requirements, is an essential requirement on any offshore installations. Many of the offshore drilling rigs are having waste heat recovery water makers. In the process about 27 million units worth Rs.4 crores is being recovered annually from these water makers.

(iv) Waste Heat for Vapour Absorption Refrigeration System

Waste heat is directly being used in vapour absorption refrigeration system as a source of heat energy to produce chilled water. The chilled water can be used for meeting the air-conditioning requirements. Natural Gas based vapor absorption technique has been adopted at onshore installation in our Eastern region.

Due to the distinct advantages of using cogeneration power plants such projects should be preferred. The potential of cogeneration in India which is more than 10,000 MW mainly in sugar, paper, fertilizers and petrochemical industries needs to be further exploited. Generally lack of information on economic benefits dithers industries venturing into the area of cogeneration. If Government enacts a law making mandatory to the local electricity boards to purchase the surplus power generated from these plants with priority, cogeneration plants need not be designed exclusively for internal consumption of a particular industry and they can be designed for optimum generation.

Cogeneration, as a source of energy production is highly efficient, environmentally sound, low cost method of producing electrical and thermal energy. It is going to play a significant role in reducing our dependence on imported fuels and reducing cost of energy to a step towards better environment.

Conclusions

India is heavily dependent on the fossil fuels for meeting its power requirements. In view of the limited availability of fossil fuels it is essential to open up the campaign for energy conservation and development of alternate energy sources. In order to cope up with the ever increasing demand of energy pressure on these vital resources is going to be there. Hence, these precious resources need to be conserved as they are not going to last forever. The wasteful use of energy needs to be checked.

The future energy requirements are forecasted to be of very high order. Achieving a 5% annual compound growth rate in GDP would depend mainly on our ability to fulfil the energy needs. The situation calls for concerted and innovative efforts collectively in all sectors and by all concerned to evolve suitable strategies, technological break through, measures for energy efficiency, inter fuel substitution, commercialization of renewable sources of energy, cogeneration and use of waste heat etc.

Energy efficiency has to grow as people's movement and every section of society will have to contribute towards its achievement.

SUSTAINABLE DEVELOPMENT OF ELECTRICITY INDUSTRY IN INDIA

22

Dr. UDESH KOHLI
Chairman Director
Power Finance Corporation, New Delhi
Dr. K.K. GOVIL
Director (Projects), Power Finance Corporation, New Delhi.
P.K.SINGH
Dy. Manager (Environment), PFC.
P. MALIK
Asst. Manager (Environment), PFC.

ABSTRACT

Sustainable development is assuming greater importance in view of the limited resources on the planet and increasing demand on these resources due to growth in population. Growth of Science and Technology has opened new horizons in converting the resources into products and environment, so as to change the quality of life towards higher satisfaction levels.

Energy is one of the primary needs for human existence. The energy is needed in various forms depending on the human needs. These needs could be seen to belong to some primary categories to work out alternative models of sustainable development. The energy needs of human race could thus be divided into:

(a) Cooking
(b) Transport
(c) Motive Power for work and Industry
(d) Lighting and Environmental control
(e) Communication and Entertainment

Convertibility of electricity into many forms for satisfying various forms of human needs, make it one of the most desirable forms of energy. Electricity industry is one of the largest single industry in terms of capital employed and resources used. International systems of electricity generation, transmission and distribution have been developed with technology growth and have reached maturity in terms of large capacity of generating units and high transmission voltages for increasing efficiencies and to desire benefits of economies of scale.

Environmental concerns, and high level of fossil fuel consumption already reached are cited to convince the developing economies to move towards smaller projects and decentralised power generation. Rising costs of energy may also lead them towards life-styles needling lower energy consumption. In a way this would help the developing nations to adopt an optimal path towards limiting their energy consumption to sustainable levels.

The paper examines the human energy needs futuristic scenario and the options available to meet energy demand. Emphasis would be laid on analysing the environmental impacts due to electricity

Note: Views expressed in this paper are those of the authors and do not necessarily match with those of the Corporation.

industry and what kind of policies and strategies should be adopted to move towards the sustainable level of development. Comprehensive action plan to move towards a sustainable level of development cannot be built without linking it to the current stage of socio-economic development and the financial and technological possibilities. Such a model of development would have to link all the resource inputs needed to change the life styles and methods of production. The resources and needs which would require balancing for such a growth are capital and investments, science and technology, population and labour, minerals, fuel resources, life styles and consumption patterns.

The paper attempts to identify possible scenarios in growth of human energy needs. It suggests possible options to meet them in a sustainable scenario of energy demands and environmental impacts with the current level of scientific knowledge. The subject is vast and impinges upon a vast strata of scientific, industrial, socio-economic and political alternatives which are controlled by conflicting class interests. The policies and strategies emerging out of the paper shall have to be converted by the policy makers and the decision makers in National policies and priorities.

1. Introduction

Sustainable development is assuming greater importance in view of the limited resources on the planet and increasing demand on these resources due to growth in population. Growth of science and technology has had twofold effect on human societies. On the one hand it has opened new horizons in order to convert resources into newer products and services that improve the quality of life and on the other hand by-products of modern societies are increasingly threatening survival by having a negative impact on the environment.

Sustainable development ensures meeting present needs without compromising the ability of future generations to meet their own. It requires that societies meet human needs both by increasing productive potential and by ensuring equitable opportunity for all. There are many critical development issues that are related to uneven growth between urbanised and rural regions of the country, poverty and population growth; as all these factors place unprecedented pressures on the land, water, forests and other natural resources particularly in a heterogeneous and diverse country like ours.

Energy being one of the primary needs for human existence, future development crucially depends on its long-term availability in increasing quantities from sources that are dependable, safe and environmentally sound. Concern about dependable future for energy is only natural since energy provides essential services for human life – heating and cooling for comfort, cooking and lighting in homes and workplaces, power for industries, transport and mechanical work etc.

Data over the past decade and future projections of growth in annual GDP, primary energy supply and electricity generation in India are shown in Table-1.

Since, industrial and service sectors are major contributors to overall GDP, this has resulted in tremendous rise in commercial energy consumption. Indian per capita commercial energy consumption was 9.5% of World's average in 1970 which has moved up to 25 per cent in 1991. However, it is worth mentioning that these figures do not include the traditional, non-commercial source of energy (viz. fuel wood, dung cakes etc.) which form a large part of India's total energy consumption.

2. Energy and Electricity Consumption Patterns

The industrial sector continues to be the single largest commercial energy consuming sector, although its share is declining gradually. Agriculture sector has seen a gradual rise

TABLE -1 : Estimates of Selected Economic Parameters from 1970-2010

Year	1970	1980	1990	1993	2000	2010
Population (Million)	541	679	837	890	1000	1140
GDP (US$ Billion	110	149	256	296	420	700
Rate of growth %	3.5	3.1	5.6	5.0	5.1	5.2
GDP per capita (US $)	203	219	306	333	420	631
Rate of growth %	1.0	1.0	3.4	3.2	3.2	4.2
Primary Commercial						
Energy supply (Mtoe)	59	96	175	210	310	510
Rate of Growth%	5.0	5.0	6.2	6.3	5.7	5.1
kgoe/Capita	109	141	209	210	310	447
kgoe/$ GDP	0.54	0.64	0.68	0.71	0.74	0.73
Electricity Gen. (Twh)	61	119	289	355	540	986
Rate of growth %	11.6	6.9	9.3	7.1	6.2	6.2
Per Capita kwh	113	175	346	400	540	865

Note : oe=Oil Equivalent (1kg oil=10200 Kcal) GDP=Gross Demestic Products
1 US $=Rs. 35.0
(*Source* : Energy for a billion people – H.N Sethna: Technorama; June 1996)

in the intensity of commercial energy consumption. This reflects the trend towards mechanisation of Indian agriculture, both for land preparation and for lift irrigation. Although agriculture is not a major commercial energy consumer at the national level yet, it may account for a significant share of commercial energy consumption in certain States and in certain seasons. Commercial energy intensity of the transport sector has also increased regularly. A fall in share of traffic handled by the railways and consequent increase in the road traffic is the major reason. Goods have been carried over increasingly longer distances by trucks. Growth of public modes of transport has been lagging behind population growth resulting in increased use of private modes of transport. The Changing Pattern of Utilisation of Electricity in various Sectors is presented below in Table-2.

TABLE-2 : Changing Pattern of Utilisation of Electricity in the End Use Sector (%)

Year	*Industry*	*Agriculture*	*Household*	*Others*	*Total*
1960-61	74.3	5.0	9.0	11.7	100.0
1970-71	70.7	9.2	7.9	12.2	100.0
1980-81	61.7	16.2	10.3	11.8	100.0
1990-91	49.8	23.8	15.1	12.3	100.0
2000-01(Anticipated)	44.8	18.8	25.5	11.2	100.0

Source: (Indian Member Committee of the World Energy Council).

It is clear from the Table-2 that bulk of the electricity consumption has also been in the production sectors of the economy viz. industry and agriculture although it is showing somewhat declining trend due to fast growth in the household sector and railway electric traction. Non-commercial energy consumption has also increased from 143 mtoe (million tonnes of oil equivalent) in 1972-73 to 161 mtoe in 1982-83. However, the percentage share of non-commercial energy consumption of the total energy consumption has declined from 74 per cent to 66 per cent. In 1994-95 electricity generation through all the power sectors viz. SEBs, EDs, Central Utilities and other sources grossed 351 BUs. The 15th Electric Power Survey of India (CEA) has computed pattern of utilisation of electrical energy in various categories namely domestic, commercial, irrigation, industry and others. According to this survey, industry continues to be the major energy consuming sector in India, although the share of percentage domestic electric energy consumption is increasing gradually. This is further illustrated in Table-3.

TABLE-3: Pattern of Utilisation of Electrical Energy (Public Utilities)

	ACTUAL					ESTIMATED	
Categories	*1980-89*	*1989-90*	*1990-91*	*1991-92*	*1992-93*	*1993-94*	*1994-95*
Domestic	15.54	16.44	16.71	17.23	18.00	18.09	18.76
Commercial & Misc.	7.44	6.97	7.31	7.28	7.16	7.05	7.04
Irrigation	24.21	25.23	26.35	28.12	28.69	28.12	27.54
Industry	45.87	44.60	43.19	41.05	39.81	40.49	40.37
Others	6.94	6.76	6.44	.32	6.34	6.25	6.29
Total	100.00	100.00	100.00	100.00	100.00	100.00	100.00

	ESTIMATED						
Categories	*1995-96*	*1996-97*	*1997-98*	*1998-99*	*1999-00*	*2000-01*	*2000-02*
Domestic	19.37	19.90	20.61	21.91	22.91	22.61	23.52
Commercial & Misc.	7.01	6.98	6.96	6.94	6.95	6.90	7.03
Irrigation	27.04	26.58	26.13	25.61	25.14	24.20	24.10
Industry	40.31	40.18	40.01	39.93	39.64	39.30	39.00
Others	6.27	6.20	6.29	6.30	6.36	6.35	6.35
Total	100.00	100.00	100.00	100.00	100.00	100.00	100.00

Source: 15th Electric Power Survey of CEA

3. Basic Household Needs and Sustainability

Household sector is the largest consumer of energy in India, accounting for 40-50% of the total energy consumption when both commercial and non-commercial energy sources are considered. Bulk of the energy consumed by the household sector consists of traditional fuels such as fuelwood, animal dung and agricultural residues. However, the supply of non-commercial fuels has not kept pace with the increasing population in the last two decades.

The rapid pace of urbanisation and diverse urban growth patterns has resulted in a shift in energy consumption mix in the past few decades. The decline in the use of firewood is clearly evident. The use of softcoke has also declined but this may be due to its decreasing supplies. The use of kerosene and LPG has increased substantially. The daily consumption of the poor households is between 1256 and 2093 KJ of useful energy per capita. This is much lower than the minimum requirements of 2847 KJ worked out by the Advisory Board on Energy. Even in Urban areas, poor households use biofuels which are burnt in low efficiency devices and cost per unit of useful energy is much higher than that obtained from modern fuels.

Energy requirements for meeting the needs of household sector by the years 2004-05 has been depicted in Table-4.

TABLE-4: Energy Requirements for Meeting Household Sector Need by 2004/2005

Fuel	*Energy Requirement*
Fuelwood (10^6 Tonnes)	300
Softcoke (10^6 Tonnes)	20
Kerosene (10^6 Tonnes)	16
Agriculture Residue (10^6 Tonnes)	90
Animal Dung (10^6 Tonnes)	199
Gobar Gas (10^6 Cu. m)	169
LPG (10^6 Tonnes)	2
Charcoal (10^6 Tonnes)	3
Electricity (BU)	96

Source: TERI Energy Data Directory and Yearbook, 1994-95.

The assumption for the extrapolation is no change in fuel mix and no change in the efficiency of utilisation. The population of India by 2004/2005 has been assumed at 1003 million (Scenario 1 of Table-5).

The demand of fuelwood at the level of 300 million tonnes cannot be sustained even with large scale afforestation programme, realising the low productivity of Indian forest lands and increasing requirement of wood by other sectors like paper and pulp industry etc.

3.1 Classification of Human Needs

The paper attempts to look at the futuristic demands of electricity and energy in consideration of the effect of improved standard of domestic living and increased population by the year 2004 A.D. The human needs have been classified into five main categories to estimate energy/ electricity requirements with a view to reach sustainable levels of consumption and to identify possible policy initiatives required for this purpose.

Cooking
2. Transport
3. Motive power for work and Industry
4. Lighting and Environmental Control
5. Communication and Entertainment

Table-5 gives an overview of the population trends with low, medium and high growth rate scenario (ref. Advisory Committee for Energy, 1984). These three possible population scenarios have been used in this paper to evaluate future energy needs for Indian households in a simplistic manner to evolve the system of planning and policy making.

TABLE -5 : Forecasted Population Values used for forecasting Energy/Electricity Requirements (millions)

Scenario	I	II	III
1999	939	969	1022
2004	1003	1046`	1115

Source: Towards a perspective on Energy Demand and Supply in India in 2004/05, Advisory Board on Energy.

In these assessments, 1994 is taken as the base year and population is assumed to be 900 million in the base year. Table-6 gives the assumed population distribution for each of the three different classes which have been used for arriving at total energy/electricity consumption values for the nation.

TABLE - 6: Population Distribution assumed in different income classes

Income Class	Population
High	50 Million
Middle	200 Million
Low	650 Million

For the sake of calculating total energy consumption for the classified human needs Indian population is visualised in three distinct classes i.e. 'high', 'medium' and 'low' income classes. Household size is kept uniformly at 6 persons for the purpose of estimation in each of the categories. High income class household is assumed to have all modern amenities such as a car, air-conditioner, food processor, microwave oven, dish washer, fully automatic washing machine, a T.V set, video, high-fi stereo music system, refrigerator etc. The medium class is assumed to have a 2-wheeler automobile, cooler, semi automatic washing machine, mixer-grinder, two in one music system, one T.V, Basic lighting, Refrigerator etc. The low income class household although struggling for survival is also assumed to have possessions like a bicycle, a heater, some basic devices for illumination, a radio, B&W T.V set and predominantly relying on public transport system.

3.2 Cooking Activities

Cooking is one of the most important energy consuming activity in any house. Table-7 presents a rough estimate of cooking energy needs for various income classes of society. An attempt has been made to tabulate the major energy consuming cooking equipments, their equivalent power load, utilisation time and thereby overall energy consumption for cooking has been estimated.

TABLE -7 : Estimated Cooking Energy Needs for 1994 (Base Year)

Class	*Cooking Mode*	*Energy Requirement (W, hr)*	*Total Annual Electric Reqt. (KwH/Household)*
High Income	ElectricityLPG	Refrigerator(150W,10 hrs.) Cooking Range/Microwave (1000 W, 1/2 hrs.) Mixer/Grinder (500 W, 1/4 hrs.)	776
Medium Income	Electricity LPG Kerosene	Refrigerator (150 W, 10 hrs.)Toaster/ Mixer (500 W, 1/4 hrs.)	593
Low Income	Kerosene Fuelwood, Charcoal, Coal/Soft coke, Dung Cake	—	—

Based on the assumed population distribution in Table-6, the present cooking energy needs per household, and total energy consumption for the country works out to:

$$776 \times 50 \times 10^6 / 6 + 593 \times 200 \times 10^6/6 = 26.2 \text{ BUs}$$

This energy consumption is extrapolated for the year 2004 for the three different population estimates given in Table-5 based on the average consumption of 593 KwH per household for the total population as given below

$$593 \times 900 \times 10^6/6 = 89 \text{ BU}$$

TABLE-8: Projected Cooking Energy Needs in 2004

Population Scenario	*Cooking Energy Needs (BUs)*
I	$\frac{1003}{900} \times 89 = 99.2$
II	$\frac{1046}{900} \times 89 = 103.2$
III	$\frac{1115}{900} \times 89 = 110.3$

With the increase in standard of living of the low income class population, their energy requirements for cooking are presumed to go up also. Even though desired level of energy consumption may be that of high income class, it would be unsustainable. So, suitable incentives and taxes may have to be incorporated in the energy pricing and fuel/energy availability systems to limit the national average energy requirement for cooking to the present middle level energy consumption figure of 593 KwH.

This energy requirement could additionally be supported by alternative energy modes. Continuance of subsidies for high efficiency stoves, solar cookers, biogas and implementation of these devices/techniques for meeting household cooking needs could help in easing conventional electricity energy demands for cooking purposes.

3.3 Transportation Needs

Large amount of energy is consumed in meeting the transportation needs. Energy consumption in the transport sector increased @ 4.8% a year between 1980-81 and 1991-92. This has been on account of greater use of private modes of transport especially cars and two wheelers. According to the report of the working group on energy policy, the share of passenger traffic through private modes increased gradually to 23% by 1979. The consumption of petroleum products in the year 1992-93 was 3615 and 2058 thousand tonnes of petrol and diesel respectively.

It has been assumed here for computation that the high income class household uses one car, middle income class household has one scooter and lower income class uses either bicycle or public transport (bus). It has been also assumed that each household commutes 4,000 kms per year (national average). Table-9 presents the estimated energy needs for transport.

TABLE-9: Estimated Energy Needs for Transport (1994)

Class	*Transportation Mode*	*Fuel Requirement per year*	*Total Reqt.(Million ltrs. of oil)*
High	Private Car	400 Ltrs.	3332
Medium	Private Scooter, Public Transport (Bus & Rail)	100 Ltrs.	3332
Low	Public Transport (Bus & Rail)	-	-
			Total=6.0 mtoe

This energy consumption is extrapolated for the year 2004 for the three different population estimates given in Table-4 and is presented below in Table-10 based on the average consumption of 100 lts. per household for the total population for which the total transportation energy needs for the present population works out to be 13.5 mtoe.

TABLE-10: Projected Energy Needs for Private Transport in 2004

Population Scenario	Energy Requirement (In million tonnes oil equivalent)
I	15.0
II	15.7
III	16.7

The calculations shown above take into account only passenger transport into consideration in a very simplistic manner assuming that every household needs to travel 4,000 kms. yearly.

With increase in the standard of living of population, the energy requirements for transport are also likely to go up for low income group people. At the same time, the energy requirement by higher income group may not be sustainable in future and would have to brought down either by policy intervention or technological innovations.

Excessive consumption by personalised use of automobile may be taxed in suitable manner to reduce congestion in urban cities centres and fuel consumption and by giving incentive for using efficient public transport system. Provision of dedicated cycle paths that encourage safe-cycling can also help in a manner to meet transportation needs of a portion of urban population in a sustainable way.

3.4 Motive Power and Work related Miscellaneous Activities

Whereas all the manual work, food processing, washing, laundry, cleaning, scrubbing, washing dishes etc. was done manually by the women in the household and/or servants earlier, now due to consumer boom, working women, modernisation and improvement in standard of living increasingly there has been a shift in life style towards adoption of gadgets and commercial energy for operating them.

For the purpose of calculating electricity required by all the sections of society (viz. high, middle and low) for meeting motive power and work needs in a household, we assumed that a modern high income group category household possesses a fully automatic working machine, a vacuum cleaner and a dish washer. Middle income household owns a semi-automatic washing machine. For the lowest income household it is assumed that all the household work is done manually.

TABLE-11: Estimated electricity needs for performing miscellaneous activities (1994)

Class	Energy Requirement (W, hr)	Total Electricity Reqd. (KwH/yr)
High Income	Laundry (1000 W, 1/2hr) Vacuum cleaner (200 W, 1/2hr) Dish Washing	219
Medium Income	Laundry (500 W, 1/2hr)	91
Low Income	–	–

Total energy requirement for performing miscellaneous activities in all households in nation (at 1994 level) based on population distribution in Table-6 is estimated at:

$$219\times50\times10^6/6 + 91\times200\times10^6/6 = 4.9 \text{ BU}$$

However, based on the average consumption of 91 KwH per household for the total population, the total electrical energy requirement for miscellaneous activities works out to be 13.7 BU.

TABLE-12 : Estimated energy need for Motive power and work (Miscellaneous Activities) in 2004

Population Scenario	*Energy Requirement (in BU)*
I	15.3
II	15.9
III	17.0

Alternative/Appropriate technologies for meeting human needs would have to be encouraged. In rural households where members do not have to spend a lot of time commuting to and fro offices and factories it may be more suitable to go in for more labour intensive processes and energy efficient technologies rather than use of electricity based gadgets for performing various work related activities. Air conditioned, wall to wall carpeted homes using electrical cooking range, vacuum cleaners, electric water heater and washing machines for 900 million people of India is an unsustainable situation. As the same time, expecting the whole country not to use any energy for miscellaneous household activity is not a desired situation. The sustainable consumption level may be at that of the present middle income class.

3.5 Lighting and Environmental Control

There has been sharp rise in requirement of energy for lighting, cooking and heating in households in the last few decades. An attempt has been made to assess the present energy needs in meeting this requirement and estimate future needs in terms of electrical energy. A high income household is assumed to be making use of energy efficient tubes/bulbs, Airconditioner, Geyser, Heater etc. to meet their lighting and comfort needs. Middle income household may afford cooler, basic lights, fans and heater for the same need. A low income household may have access to only basic lighting or kerosene for lighting. Table 13 depicts the estimated lighting, cooling and heating needs of household at 1994 level whereas Table 14 presents the future projection for year 2004 with different growth in population rate.

TABLE 13 : Estimated Energy Need for Lighting, Cooling and Heating (1994)

Class	*Energy Requirement*	*Total Electricity Requirement*
High Income	Energy efficient lighting (100 W,10 hr)Air-conditioner/Heater (2000 W, 1/2 hr) Geyser (1000 W, 1/4 hr)Inverter/ DG Set	821 KwH
Medium Income	Cooler/Heater (500 W, 1/4 hr) Basic lighting/fans (200 W, 8 hrs) Emergency lamps/Candles	630 KwH
Low Income	Basic lighting/fan (80 W, 1 hr)Kerosene lamps/Hurricane Lanterns Natural Lighting/Ventilation	29 KwH

The total energy requirement for lighting, heating and cooking needs of household. India (at 1994 level) based on population distribution at Table-6 would be:

$$821\times50\times10^6/6 + 630\times200\times10^6/6 + 29\times650\times10^6 = 30.9 \text{ BU}$$

However, based on the average consumption of 630 KwH per household for the total population, the total electrical energy requirement for miscellaneous activities works out to be 94.5 BU.

TABLE 14: Projected Energy needs for Lighting, Cooling and Heating in 2004

Population Scenario	*Electricity Requirement (BU)*
I	105.3
II	109.8
III	117.1

As a matter of policy, we in India ought to exploit natural light particularly for meeting the lighting needs in houses and offices in day time. Solar passive houses for the cold climatic regions, low energy evaporative cooling system and natural draught based ventilation system for all regions also need to be encouraged for sustainable consumption of electricity in the household and offices.

3.6 Entertainment and Recreation

In pre-industrial Indian society means of healthy recreation were quite straight forward and simple. Physical activity like folk music and dance provided the recreation. These activities have been partially replaced by satellite television, video, computer games and radio in this information age.

For the sake of our calculations we have assumed that per household these are 2 colour televisions, 1 video, 1 hifi stereo and one computer in high income group. In the middle income group category these are 1 colour television and 1 music system and for the low income group 1 Black and White television and radio has been assumed.

TABLE-15: Estimated electricity need for entertainment (1994)

Class	*Energy Requirement (W, hr)*	*Total yearly Electricity Required (KwH)*
High Income	1 Colour TV (100 W, 4 hr) 1 Video (50 W, 2 hr) 1 Hi-fi Stereo (25 W, 4 hr)	219
Medium Income	1 Colour TV (100 W, 3 hr) 1 Two-in-one (20 W, 5 hr)	146
Low Income	1 Radio (10 W, 5 hr)	18

Total daily electricity requirement for entertainment and recreation in India (at 1994 level) based on population distribution of Table-6 would be:

$$219 \times 50 \times 10^6/6 + 146 \times 200 \times 10^6/6 + 18 \times 650 \times 10^6/6 = 8.6 \text{ BU}$$

However, based on the average consumption of 146 KwH per household for the total population, the total electrical energy requirement for entertainment activities works out to be 21.9 BU.

TABLE 16: Projected electricity needs for entertainment and recreation (2004)

Population Scenario	*Electricity Requirement (BU)*
I	24.4
II	25.5
III	27.1

Keeping energy consumption level low to meet human needs in this sector is not very challenging. Number of TV and radio sets would not lead to excessive demand on energy supply situation and number of channel beamed would reach equilibrium level simply due to the overall economic and sustainability of advertising revenue. However, as a national policy we may aid inculcation of healthy values through these media and motivate the billion odd population in near future to excel in the sports and cultural arena that would help produce better, responsible and healthy citizen who can be justly proud of their country.

3.7 Impact of Population on Energy Needs

Present rate of population growth has put excessive pressure on available resources and cannot be sustained. Poverty and illiteracy are two major reasons of unsustainable population growth in India. Hence, there is a need for policy makers and implementing authorities to work on several fronts to limit population growth, to enlarge the range of resources and their productivity, to convert population into productive resource and to provide people with forms of social security other than large numbers of children.

The role of women is very crucial in affecting fertility. Fertility rate falls as women's opportunity for employment, their access to education and their age at marriage all rise. Improvement in health and education can have profound effect in bringing down the existing population growth rate in India. A population policy incorporating lessons learnt from previous population control efforts and adopting pro-active approach aiming at positive tangible results is needed.

3.8 Effect of Energy Conservation on Human Needs

While the use of energy per US $ of GDP has declined in the US and elsewhere in developed countries since the seventies, the same has increased in India from 0.53 kg oe in 1980 to about 0.7 kgoe in 1990. Energy requirement in typical energy intensive industry of India has been compared with that of some developed nations below in Table-16. Based on the comparison given in Table-17 it can be clearly seen that tremendous potential for energy conservation exists in industrial sector. Energy conservation has already started showing results. While the intensity of electricity consumption in Indian industry has increased the rate of growth of intensity has declined.

TABLE 17: Energy requirement in energy intensive industries (kgoe/tonnes)

Country	Cement	Steel	Fertiliser
India	950	200	1113
UK	607	130	762
USA	606	95	970
Japan	418	120	–

Source : *The Economic Times* 13.3.94.

The energy and environment audits could result in more energy efficient processes that reduces emissions, effluent and wastes. There is a need for enabling atmosphere that makes energy efficiency and resource conservation the top priority for the industry.

Energy conservation also has enormous scope in the rural household energy usage. Efficiency of cooking especially in the rural areas is less than 10%. Women who cook in an earthen pot over an open fire use almost eight times more energy than their affluent counterpart with a gas stove and aluminium pans. The poor who light their homes with a wick dipped in a jar of kerosene get one-fiftieth of the illumination of a 100 watt electric bulb. These examples illustrate the tragic paradox of poverty. For the poor, the shortage of money is a greater limitation than the shortage of energy. This issue is vital for Indian economy and needs to be addressed carefully by policy making and enabling/implementing authorities.

Buildings offer enormous scope for energy savings and perhaps the most widely understood ways of increasing energy efficiency are in the home and work place.

4. Policy Interventions for the Sustainable Energy needs of India

1. *Ensure a Sustainable Level of Population*

   A population policy needs to be formulated integrating various economic and social development programmes such as education, health care and expansion of livelihood base for poor. Increased access to family planning services and women education will be helpful in controlling the population growth rate.

2. *Conserve and Enhance the Resource Base*

   Conservation of natural resources such as clean air, water, forest and soils and using energy, water and raw materials efficiently is required to ensure sustainability. Improvement in efficiency of production will ensure reduction in per capita consumption of natural sources. The long term industrial policy should encourage shift to non-polluting products and technologies particularly in densely populated areas.

3. *Re-orient Technology Development*
   Greater emphasis is required on policy level regarding orientation of technological development which is environmentally benign. Non-conventional renewable energy resources need to be exploited to the maximum extent. Wind energy, small hydro and solar power have already started becoming popular. The energy policy should aim at mounting further stress on sustainable forms of renewable energy.

4. *Change in Quality of Life*
   Policy intervention is required to evolve price mechanism in order to ensure a life style which uses less resources in a more efficient manner. Improvement in life style need not necessarily be through consumption of more energy. Many communities in USA, Japan and other industrialised countries have already adopted environment friendly life style that are based around re-cycling, depend on renewable energy like solar for heating and cooking needs besides meeting some of their other energy needs by storing renewable energy in efficient batteries.

5. *Urban Strategy*
   In order to reduce the exodus from villages to cities and metropolis, policy intervention are required to accelerate development of secondary urban centres. India needs to avoid policy intervention that increase the attractiveness of major cities beyond their carrying capacity particularly subsidies on food and energy, overly generous provision of urban infrastructure and excessive concentration of administrative power in cities. The need of the day is to encourage growth of secondary centres by building on the natural economic advantages of their regions, specially by creating infrastructure in resource processing and marketing areas that would create job opportunities. Rural and urban development strategy and approaches should be complementary rather than contradictory.

6. Public transportation in India needs to be improved and strengthened in order to reduce usage of cars and other private means of transport. This will not only help in saving precious foreign exchange but also save energy to a large magnitude, besides lowering the level of air pollution by reducing automobile exhaust.

7. Policy and legislative measures need to address the issue of thermal characteristics of buildings. Encouragement needs to be given to natural ventilation and lighting, increased usage of solar energy for heating and cooling by popularising solar power storage for miscellaneous uses, photovoltaic lighting and solar cooking.

8. A multipronged strategy to make effective use of power capacity should include rehabilitation of existing power plants, integration of regional grids for effective hydro-thermal optimisation, reduction in T&D losses, setting up of pump storage plants at existing hydro sites and introduction of "time of day" tariff.

5. Conclusion

Energy efficiency should be the cutting edge of national energy policies for sustainable development. The cost effectiveness of "efficiency" as the most environmentally benign `source' of energy is well established. The energy consumption per unit of output from most efficient processes and technologies is one-third to less than one-half of typically available equipment. The older and the less efficient equipment has sunk energy costs and also the sunk capital which can have severe negative impact on a national economy if a such equipment has to be discarded. Thus energy efficiency can be maintained and extended, through judicious and explicit policy which minimises combined cost of capital and energy in a long term perspective weighing the costs and benefits in the means of production with gradual shift towards energy efficient technologies.

Holistic understanding suggests that a low energy intensive path is the best way towards a sustainable future. Within the next 50 years, nations have the opportunity to produce the same levels of energy services with as little as half the primary energy supply currently consumed. This requires transition to a safer, more sustainable energy era. Both the routine practice of efficient energy use and development of renewables will help take pressure off traditional fuels, which are most needed to enable developing countries to realise their growth potential world-wide. This should not mean that dependence on electricity and energy and life styles do not need any change. Developing countries will have to combine their technological growth, manpower skills, all forms of available energy sources so as to minimise their own resource inputs for a stronger and sustainable economy.

REFERENCES

Annual Report on the Working of SEBs and EDs by the Planning Commission, 1995.
"Energy for a billion" people - H.N. Sethna; Technorama, June 1996 issue.
Fifteenth Electric Power Survey Report of Central Electricity Authority.
"Our Common Future": World Commission on Environment and Development.
"TERI Energy Data Directory and Yearbook (TEDDY) 1994-95" - Tata Energy Research Institute, New Delhi.
"Towards a perspective on Energy Demand and Supply in India in 2004/2005" - Advisory Board on Energy.

MEETING ENERGY NEEDS THROUGH RENEWABLES:

23

Can it meet the Socio-economic needs of 930 million Indians

M. D. JOSHI
Director,
Gujarat Energy Development Agency (GEDA),
Vadodara - 390 005

DAKSHA VAJA
Documentaion Officer,
Gujarat Energy Development Agency (GEDA),
Vadodara-390005.

ABSTRACT

Nowhere is the role that Renewables can play more evident than in Third World countries like India, whee a chronic power deficit means there is a continuing need (which is also impossible to meet) to install generating capacities. Renewable Energy Technologies have a set of operating and financial attributes — modularity and flexibility, very low operating costs and the ability to create strategic options for the future — that is considerably different from those of traditional, fossil-based technologies whose attributes include investment shortfalls, long lead times and operating cost uncertainties especially regarding future fuel outlays.

Energy needs of our rural population (75% of the country's 930 million) could be met using modern bioenergy options. Additionally they make a significant impact/contribution in alleviating the stress on forests and the national electricity grid. They also offer an opportunity for sustainable, self-reliant and equitable development and provide tangible benefits of land reclamation, local employment and reduction of greenhouse gas emissions. The paper details the opportunities of modern bioenergy options that can provide loal and global benefits.

An attempt has been made to document the relevant experiences of Renewables in India with a special reference to Gujarat in terms of decentralised generation, its impact on growth and efficiency improvements. The basic premise of the analysis is to provide quality fuels to meet energy services required to improve the quality of socio-economic lives of India's teeming millions. The study concludes that wide use of solar and wind energy and bio-fuels, and other renewable energy sources can at least meet 20% of our basic energy needs, and reduce greenhouse gases. This coupied with a movement towards a long term objective of a unifed global energy consumption system would create a substantial reduction of both environmental damage and also the problems of security and sovereignty.

1. INTRODUCTION

There is little doubt that we live in an age of rapid and accelerating change. Indeed in today's world every thing at times seem to be in flux. The global environment is also rapidly—and unpredictably-spurred by unprecedented growth in the number of human beings and even more rapid expansion in the number of factories, buildings and automobiles.

Yet, thinking about the world's nergy future is surprisingly bereft of this kind of vision found in so many other areas. Outside a few pockets of innovation, planners and analysts at the worlds' leading energy institutions seem to be trapped in stagnation and confusion that began with repeated failures in the 1970s and 1980s to develop a viable energy strategy.

An energy economy marked by continuing heavy dependence on fossil fuels, made possible by a gradual shift away from conventional oil towards lower-grade and (more abundant) fossil resources viz. lignite and oil share in order to avoid the need to make significant technological or policy changes is anticipated. One area in which the forces of change have become visible in the past few years is in the host or more energy-efficient technologies that have emerged. Such technological advantages suggest that government and international agencies are still over stating future levels of energy use. The energy landscape has also been altered by the recent emergence of natural gas as the most rapidly growing major energy source. Recent developments are turning wind power, solar energy and a host of other renewable resourcees into economically viable energy options. The energy sector is no exception.

How quickly the world's energy system will change is one of the biggest uncertainities. History suggest that major transition -from wood to coal and coal to oil — take time togather momentum. But once economic and political resistance is overcome and the new technologies prove themselves, things can unfold rapidly. This is how today's energy systems emerged at the end of the last century, and it may be the way a sustainable energy economy begins to emerge.

2.0 The Energy Dilemma in India

The power sector is caught between the pressure to add new generating capacities to match the rapidly growing demand for power to fuel its growing economic and social development and environmental challenges arising from large-scale power generation plants. Estimates show that new generating capacity of 10,000 to 150,000 MW will be required during the next 15 years. Coal is and will continue to be, the mainstay in India. This brings into focus issues relating to the quality and composition of atmospheric emissions and issues relating to fly ash disposal from thermal power stations and the need to consider appropriate technologies to reduce such emissions.

According to estimates of the 14th Electric Power Survey (EPS), electicity demand will grow at a compounded annual rate of 8.3% during 1993 to 2012. Among the various consumer categories, demand is estimated to increase at a rate of 6.7% during this period while growth in electricity demand in the agriculture sector is forecast to decline significantly after 2002.

High T & D losses is the major issue before electric utilities in India. Losses have increased from about 15% (1966-67) to 21.74% (1986-87) and since then to about 23%. The impact is two fold: (1) they result in under utilization of the total energy being generated. This is unpardonable considering the energy shortages and staggered supplies that the country faces, and (2) energy wasted due to T & D losses must be compensated for new generating stations in all probability (fossiled fuel-based thermal) with environmental degradation problems created due emissions of greenhouse gases and other associated problems of exhausting the feedstocks – that have serious environmental repercussions. However, the actual rate of consumption is expected to multiply with exponential growth of population and their aspiration. Long-term energy security cannot therefore be ensured by conventional sources only.

But despite the grim power scenario manifested by shortfall in generation, increase in demand and huge losses by the State Electricity Boards (estimated at about Rs. 27,000 Cr. plus) the dependence on conventional sources continues, leaving the vast potential of renewables untapped.

3.0 Development of Renewable Energy Resources

The setting up of the Commission for Additional Sources of Energy (CASE) and the creation of the Department of Non-Conventional Energy Sources (DNES) which was later elevated to a full-fledged Ministry (MNES) reveals the fact that the Government of India has attached significant importance to the development and promotion of renewable energy. India is perhaps one of the few countries in the World to have drawn up a comprehensive Strategy and Action Plan and set specific goals for renewable power generation-based capacity additions. The goals have been redefined and increased from 600 MW to 2000 MW. While the extent to which renewable is used in this country is still below the economic potential even at the current levels of technology in several use, renewable energy development has fared quiet well. In addition, the establishment of the Indian Renewable Energy Development Agency (IREDA) and the announcement of fiscal incentives and financial packages have placed renewables on to the commercialisation roadway.

We have over two decades of renewable energy development and promotion experience that is not economically viable in most cases but provide options that are sustainable. Thus far, their insignificant contributions (not accounting for direct biofuel use in rural area) in the economy's over all energy mix is mainly because or various technological constraints. Other factors that has been detrimental to its growth : the low budgetary allocations — between 1980-92 the cumulative government expenditure is over Rs. 11 billion compared to over Rs. 1300 billion provided to the power, petroleum and coal sectors.

Even in the 8th Plan allocations to renewables represent only 0.8% of the total Energy Plan outlay..... The existing inactive linkages with Industry and R & D or its applications which otherwise could have mooted private sector participation into renewables are other factors that are responsible for its slow development.

The Biogas Programme has traditionally been the single largest programme accounting for over half of the allocated funds. Other individual programmes received less than 10% of the total funding provided to the renewable sector. In number terms, however, Improved Cookstoves have enjoyed a fair amount of success and acceptability among rural households. Scope for considerable improvement in implementing this programmes exists.

Among other technologies Windfarms have emerged as a viable power option. Windfarms of over 800 MW capacity are operational. The performance of these Windfarms have shown an average reliability of 98% with a capacity utilization factor as high as 30%. India is likely to be second largest wind power producing country with the growing interest of the private sector.

Small hydro has evolved as another promising option for power generation especially in remote hilly areas where the cost of grid electricity is prohibitive. Almost 110 MW have been harnessed from 148 small hydro stations/projects. However, this capital intensive programmes with adoption of canal or cluster appraisal to achieve short gestation periods will lower the cost.

The Gasifier Programme in India was launched in 1986 and so far 1000 wood-based systems have generated 3.7 MW, Blomass Gasifiers offers significant potential to meet rural energy needs, particularly in rural and remote areas. Future promotion of biomass gasification depends on technology development, improved operation reliability, O&M cost and biomass prices.

Among Solar Thermal Technologies, Water Heating Systems for industrial applications and Solar Cookers are gaining popularity. However, attention needs to be paid to product development in both the technologies. The technologies also must be made more compatible with users requirements.

In general, not all Renewable Energy Technologies can be defended purely on economic grounds. But environmental and social concerns, which are equally important in view of the need for sustainable development make a strong case for their widespread dissemination and penetration.

4.0 Modern Biomass Options

Biomass is a renewable fuel used in nearly every corner of the developing world as a source of heat, particularly in the domestic sector. Biomass energy includes energy from all plant matter (tree, shrub, crop) and animal dung. Currently biomass energy is characterized by a low efficiency use and a low quality of life due to drudgery associated with its gathering and use. However, biomass, unlike other renewables use versatile source of energy which can be converted to modern forms viz. liquid and gaseous fuels, electricity and process heat. Bioenergy also permits operation at varying scales. For example, small-scale (5-100 kW), medium-scale (1-10 MW) and large-scale (about 50 MW) electricity generation systems or biogas plants of few cubic metres (Indian and Chinese family plants for cooking) to several thousand cubic metres (Danish systems for heat and electricity generation). This variety of scales are useful for power generation for decentralised applications at the village level as well as for supply to the national grids. Unlike solar, wind or micro hydro electric systems, modern biomass energy systems could be set up in virtually in any location where plants can grown or domestic animals are reared. Renewables like solar, wind and micro hydro require spare or additional capacity to produce adequate energy when to conditions are right viz. sunny days, water flow or wind speed. The Intermittent features of such renewables energy systems necessitate electricity storage facilities, especially with small and local systems sustained demand requirements are to be met. Bioenergy sources viz. producer gas systems do not require electricity storage. In short, biomass energy systems offer an opportunity for sustaiable, self-reliant and equitable development.

According to WEC (1993) under the minimum case scenario biomass is the most important renewable resource and is projected to account for 45% of the new contribution by renewables to World energy by 2020.

The US-DOE suggest that biomass power will be the most important renewable energy option for a next quarter of a century and has projected installations cumulating about 25 GW capacities in the US alone by 2010. This clearly establishes the dominant role biomass energy can play.

A study of potential role of biomass gasifiers in India has shown that electricity needs for lighting and stationary shaft power of half million villages could be met sustainably through decentralised biomass electricity while requiring only part of the existing degraded land for biomass production. Biomass-based village energy systems where all the energy services of a village ecosystem are met by bioenergy technologies have shown to be technically feasible. In addition, to meet heating and electricity requirements at micro level an enorous potential to supply electricity to the national grid like the existing 9000 MW in USA, and liquid fuel transportation of 12 billion litres of ethanol used annually in Brazil are successful programmes.

Biomass Energy Technologies are currently available off the shelf (biogas, producer gas, etc.) and those *in-situ* that will become available soon are (advanced gasifier/gas turbine) technologies. Additionally the large-scale utilization of biomass can provide a basis for rural development and employment in both developed and developing world. A further significant benefit is that, if used in a sustainable way, bioenergy could lead to no net emission of CO_2, infact, could lead to a reduction of greenhouse gas emissions when used as a substitute for fossil fuels. We readily recognize net ultimate choice of fuels or devices which must be left to the individual farmers, housewife or entrepreneur in rural areas, but it is the responsibility of the development planner to create affordable and accessible energy alternatives to improve the quality of life.

Thus modern bioenergy systems offer potential to meet the heating, cooking, shaft power and electricity requrements of villages through a combination of biogas and producer gas systems. These technologies are available indigenously in India and their feasibility in villages have been demonstrated. Bionergy options that use locally produced feedstocks lead to self reliane, local employment, reclamation of degraded and waste lands, enhancing biodiversity substitution of fossil fuels with concomitant reduction of carbon emissions.

4.1 Land and Sustainable Biomass Supply

The need for dedicated Energy Plantations for generating about 122 MT of woody biomass annually for decentralised electricity generation is stressed. It is not desirable to use the existing natural forests (even for supposedly sustainable modes of removal) given the low area under forest (64 Mha or < 0. ha of forests/capita) and the potential threat of biodiversity in the procss of extraction, Currently 181 MT of woody biomass is used nationwide annually in the rural domestic sector for cooking. With the provision of biogas and liquid fuels, the fuelwood requirement for cooking would become redundant.

A land-use strategy to produce food, fodder, timber and fuel (for bioenergy) shows that land is not a constraint in rural India to meet all these biomass needs of the growing population. A part of farmers crop land and much of village commons (pasture land) and forest lands (< 10% tree crown) are degraded and not suitable for profitable cropping and this non-crop area has remained nearly constant over the past 20 years. Such land have not been converted to crop land as one would expect with rising populations. Even at a low energy conversion efficiency of 18% only 15-30 Mha productivity (8-4 t/ha/year) are adequate for meeting energy needs of 75% of rural settlements. Energy Plantations with appropriate species mixes and density could also provide other products, viz. grass for livestock, leaves for biogas, and other valuable tree products (truits, seed etc). The land required would decrease with increase in energy conversion efficiency as well as plantation productivity, the potential exists for both. The leafy litter by-product from village Energy Plantation, is suggested as feedstock for biogas generation along with cattle dung for cooking. The main premise is that bioenergy programmes could provide the necessary incentives to reclaim currently degraded land, which have little incentives to otherwise reclaim and rehabilitate through conventional afforestation or fodder development programmes. Thus bioenergy programmes could become part of rural development packege involving reclamation of degraded and wastelands, generate employment at local level (production and processing of wood feedstock operation of biogas and producer systems) promotion or self-reliance, and improving the quality of life specially for women and rural poor.

4.2 Large Bioenergy Systems

The potential for megawatt-scale biomass electricity generation and liquid fuel systems cannot be ignored since there could be significant economic benefits to farmers who own degraded lands that could lead to reduce oil imports and replacement of coal in power plants. We calculate that not all the existing degraded lands are required for village-level decertralized producer-gas electricity systems or timber production. Similar to sugar co-operatives, where growers in a given area set up a sugar mill to process their cane, farmers owing degraded lands could set-up megawatt-scale electricity generation units and supply feedstock. Potential of 16000-32000 MW of installed capacity is shown to be feasible by dedicating only 5% of India's geographical area, depending on the efficiency of conversion. For a one mega watt system, biomass feedstock needs to come from a radius of only 8 km, at a productivity of 8t/ha/year assuming a conversion efficiency of 18%. The motivations of farmers would be the same as in the case of a large mill i.e. profits and employment from largely unused lands. The large potential for co-generation in sugar mills has been recognised by the goverment to be at least 3500 MW. This option needs to be pursued on a prority basis since bagasse in already available in centralized location and is a by-product that is presently used inefficiently.

For convenience a tonne of wood/capita/year currently used it is assumed that 0.5t/capita/year will be used for electricity generation and the same amount for liquid fuel generation. Assuming conversion at 18% and 36% efficiencies for 0.5t of wood, the potential of electricity generation using producer gas route is 384 kWh and 768 kWh/capacita/yr respectively. The methanol fuel production potential using 0.5t/capita/yr of lignocellulose feedstock via thermal-chemical route is 217 kg/capita/yr.

Producer-gas technology is available at various scales, but the technology for liquid fuel (methanol and ethanol) manufacture using lignocelluiose at small and moderate scale is not commercially available. However, it could bcome available in near future. Liquid fuel production from woody biomass is assumed as it is not considered feasible to dedicate good land for maize, sugar cane or other crops for energy generation.

4.3 Biogas : An Environment - friendly Alternative

By far the most promising alternative is biogas from dung, which is no longer given much impetus as more sophisticated technologies ar floating around the market place. The technology is cost-effective and the know-how is relatively simple allowing local variations and what is most important can be repaired *in-situ*, without many spare parts and by local villagers who are trained. Experience has proved that the family unit is more socially viable than the community set, which is more efficient. The problem with the community sets is that they require dung from several heads of cattle, and those who possess such animals (the ownership of cattle is even more skewed than the ownership of land) are reluctant to part with waste for those who do not own any. This of course leaves families altogether but, in as much as it relieves the pressure on wood for fuel, it is environmentally benign. There is the bonus of the slurry left over proving to be an excellent sourse of fertiliser.

Thus Biogas very well qualifies as an energy source for meeting cooking and lighting energy needs of rural India. Estimates indicate that biogas generated by cattle dung available in the country would be equivalent to nearly 195 billion kWh annually of equivalent to 24 billion litres kerosene. At present biogas, available from more than 2 million family size plants (1-4 m^3 capacities) is generally used for cooking and lighting. Cooking energy needs even today, are mostly met by burning agriculture residues in the form of wastes, twig, cow dung and fuelwood, all these fuels are available at zero cost since they are collected tree by women and children.

The opportunity cost for collection of biomass for domestic energy requirements is also negligible as vast numbers are unemployed/underemployed who have surplus time for this activity. Economics of biogas plants is not straight forward since the pricing of the inputs and outputs of uncertainties are involved in their quantification. Biogas used for production activities in rural industries can replace a considerable amount of commercial energy directly which is otherwise not available from the utilities or is inadequate and irregular whenever available. The problems faced in managing Community Biogas Plants are reviewed on account of difficulties of dung collection and their managment could be integrated to dairies which is part of the rural Industrial complex; which can use the energy obtained from the biogas complex. Various backward-forward linkages and combination between the two systems can emerge as a very effective and productive venture. Biogas produced by dairy cattle dung can be supplied to common rural industries viz. milk processing, food processing, nursery raising etc. These activities are closely linked with each other and can be carried out-in one complex to make them economical. There will be no dependence on any external sources of energy supplies. Biogas would be able to meet all the energy demand and if needed additional energy requirement can be met by supplementary biomass gasification or animal power. The rural industrial complex could be made to function for a cluster of 20-25 villages that uses local resources for decentralised energy applications. These efforts could diversity rural activities and generate employment. Sustained rural enterprises with decentralised systems will develop the rural areas and make them competitive in all respects that can contribute significantly to alleviating poverty levels.

4.3.1 Biogas : Application in Rural Industries

4.4 Biomass Gasification

The basic principles of gasifier technology have been studies and developed since the early 19th century. A technology to use solid fuels for Internal combustion engines was developed and extensively used in European countries during World War II. Tractors, railway engines, fishing boats, saw mills, pumping stations and other industrial machinery were run on gas produced from partial combustion of charcoal and wood. For instance, Sweden had 70,000 cars, tractors and buses using producer gas during the Second World War. Japan had 100,000, France 110,000 and Germany 350,000.

Under the combined pressures of rising oil prices, balance of payments difficulties and unreliable fuel supplies, the incentive to develop alternative fuel sources in developing countries is strong and gasification of biomass is amongst a range of possible choices. Producer gas, the product of gasified biomass, can replace fossil fuels in a number of applications : direct heat application, where the gas is burnt directly in boiler furnace or kiln to provide heat; shaft power applications the gas is used to run engines; and large-scale electricity generating applications using gas turbines.

4.4.1 Direct Heat Applications

The producer gas utilised as engine fuel in stationary applications is a technology which has termendous potential. The technology is used in a few direct heating plants and timber-kilns for pulp mills in Sweden and Finland capacities range from about 5 MW to 30 MW thermal.

Direct-heat systems can be used in industries where the gas is burnt directly in furnace or oven, or used to fire boilers for steam generation. Oil and gas equipment already in use can often be converted by replacing the existing burner with produces gas burners generation.

4.4.2 Shaft Power Application

Most of the work on gasifiers over the past 100 years has been direct towards the production of engine fuel. One of the great advantages of producer gas is that it can be used virtually in any type of internal combustion engine.

The much lower sulphur content of producer gas reduces the SO_2 emissions, lowers the risk of soil quand water acidification. And with all the fears of increasing the pressure on existing wood and charcoal supplies and the heavy dependence of biomass supplies in fuel scarcity areas, properly designes systems with adequate safety procedures, the impact of producer gas would be minimised.

An appropriate use of shaft power gasifiers in industry can be found in rural areas where grid electricity is either expensive or unavailable and diesel or gasoline power are the conventional alternatives. For Instance, a village of 400 people with a fairly low per capita demand for electricity say 200 kWh/year, would require 140 tonnes of wood per year to run a gasifier powered generator. The system can operate on wood harvested from 14 ha. wastelands at a rate of 10 tonnes per ha/year.

In developing countries irrigation pumps can be operated on shaft power. A reasonably efficient gasifier system with duel-fuel diesel engine uses approximately 1.4 kg of wood per kWh of shaft power generated. For irrigation, an area between 3% - 11% of the irrigated land would have to be planted with trees to supply enough wood to run pumps. Eventually gasifiers may be perfected to use crop residues. A study by Foley and Barnard shows that high yielding maize crop can produce as much as 6-8 tonnes of residues per ha. Waste generated from 1 ha of an irrigated area can systems generated from could be effectively fuel an irrigation system 10-15 hp capacity.

4.4.3 Biomass Integrated Gasification Systems

Biomass integarated engine gas turbine technology for co-generation on stand-alone mode power for generation in many instances looks promising to produce electricity at a lower cost than most alternatives. The sugarcane industry is a favourable target for near term applications. Depending on the choice of gas turbine technology and extent to which cane tarsh can be usd, the amount of electricy that can be produced could be up to 44 times of site power needs of a sugar factory or an alcohol distillery. Globally a conservative potential of 50,000 MW has been estimated from cane sugar residues.

5.0 Bioenergy Options : Socio-economic Feasibility

Technical feasibility is only one of the several factors required to promote rational energy strategies. Technology or appropriate designs are already available or will soon become available in many developing as well as industrialized countries. Bioenergy devices promote many loal, social, economical and environmental benefts apart from the global benefits. At national level it will reduce fossil fuel consumption and alleviate some of the foreign exchange needs. The social benefits of using biomass energy are : creating market and local employment opportunities for enterpreneurs and development of skills.

6.0 Successful Case Studies

6.1 Surendranagar

When Lalbuben of village Bamanbore, near Surendranagar, Gujarat, stated, "The value of biogas cannot be calculated. It has added 20 years of my life" she was not making an over-statement. Wherever rinstalled and working well, biogas plants have economically reduced the drugery of rural women and children from the daily rigours of searching for fuelwood and head loading numerous baskets of dung, cooking in smoky kitchens cleaning blackened vessels and the frequent recurreness of eye and lung diseases due to smoky chulhas. As a technology, the biogas plant makes optional use of locally available resources like animal dung and human excreta and other organic matter, which is digested in the biogs plant to produce an easy-to-use and convenient fuel gas (biogas) for cooking, lighting, or motive power purpose. The process also produces an enriched manure from the digested solid material as organic fertilizer which can use in agricuture, silviculture mushrooom cultivation etc. As against dung cakes which can only be used as fuel in the *chulha*, the dung in a biogas plant produces cooking fuel and can still be used as excellent fertilizer. Biogas, therefore, makes sound economic impact and cost studies indicate that the levelized annual cost comes to Rs. 0.53 kWh comparing very well with that of a traditional *chulha*–at Rs. 0.54 kWh. As a versatile fuel. Biogs is cheeper than fuelwood and dung except when they are collected free and compares closely with kerosene.

5.2 Methan

In this village in Sidhpur taluka of Mehsana district, Gujarat, a Community Biogas plant (630 m^3) operational since 1987 provides cooking fuel to 54% of 600 odd households. Households pay for the biogas supply. The plant is managed by a registered co-operative society. Its efficient working has led to regular supply of gas to the users daily for two hours, besides making available rich manure which is returned to the beneficiaries in the proportion of their contribution of dung. Besides the benefits of cooking and nutrient rich manure the society in that initial years of operation, generate a net profit of Rs 3000 p.a. The society out of this profit purchased a tractor and trailer to transport dung and manure. And during 1988 declared dividends from its profits. This Community project — the largst in Asia — has notched the highest success ratio in managing energy co-operatives.

5.3 Kothara

A village about 80 km from Bhuj (Abdasa taluka, Kachchh district) of Gujarat state recently performed the *Bhumi Pujan* ceremony. The first large-scale grid connected wood-based gasification system will annualy provide 26 lakh units of electric power to the State Electricy Board and generate a revenue of over Rs. 65 lakh annually @ Rs. 2.25 per lakh. Funded by MNES and the state government for Rs. 187 lakh when operational can meet the electricity needs of about 10-12 villages. The systems will operate on dual-fuel mode and require about 10 tonnes of wood and 1000 litres of diesel. These activities will change the landscape, green the land, invite rainfall, ease fuelwood crisis of this dry desolate region and uplift the economic life by providing 20,000 mandays employment. The project is expected to improve the socio-economic status of the region and encourage wild life to return.

6.4 Kachchh

Energy Plantation Programmes have covered over 2200 ha wastelands spread over 17 locations in different agro-climatic zones of the state. The Energy Plantation activities in Kachchh have offer local villages with employment of 80000 mandays besides providing 40 tonnes of free fodder for their cattle. The expected yield of about 10-12 tonnes/ha/year can easily sustain a power plant of 4

MW capacity at 60% load factor. The plantation has thrived against all odds-adverse environment of saline soil and water scarcity with a survival rate of 70%.

8.5 Vadodara

Direct heat applications experiments in a Bakery in Vadodara (Gujarat) daily producing 500 kg of bread have improved the end-product's quality, hygiene and the efficiency of the wood-fired oven. Extensive trials carried out at this bakery with retrofitted wood gas system have proved an improved quality of bread (without smell), uniformed backing throughout the charge and browning all round. The rejection rate ranging from 3-5% in the wood-fired over was brought down to 1% in the gas-fired oven. The simplicity of obtaining gas from the gasifier, lighting up and controlling the burner was found to be attractive and above all increased the output on a continous basis. A country-wide application to the bakery sector in rural, semi/para urban and urban areas can offer advantage of hygiene quality and raise efficiency. And site specific locations where a sustained biomass supply is available and electric supplies are diffcult to reach new simple designed wood-based oven can eliminate dependence on electricity, PLG or diesel.

8.6 Ungra

A community irrigation system powered by a 20 kW produces gas using mulbery crop residues and has been commercially managed since July 1993 in Karnataka state. About 30% of the annually pruned stalks provide the required feedstock. Farmers pay full commercial charges for its operation.

6.7 Hosahalli

Hosahalli a non-electrified village, about 8 km from Ungra, where a supported mix tree plantation on 2 ha of common land provides fuelwood for a 5 kW wood-based genset servicing the village's energy needs, supply domestic water and illuminate the village. The system has been run by two youth of the village and managed by a village committee since 1988. The people in the village in turn pay Rs. 5 for a 40 W fluroscent lamp per installation which meets the running costs for diesel input at 70% replacement.

Ungra and Hosahalli the two villages in South India — demonstrate the socio-economic viability of the whole community that have uplifted rural life.

Nowhere is the role that Renewables can play more evident that in Third World countries like India, where a chronic power deficit means there is a continuing need (which is also impossible to meet) to install generating capacities. Renewable Energy Technologies have a set of operating and financial attributes — modualarity and flexibility, very low operating costs and the ability to create strategic options for the future - that is considerably different from those of traditional, fossil based technologies whose attributes include investment shortfalls, long lead times and operating cost uncertainties especially regarding future fuel outlays.

Energy needs of our rural population (75% of the country's 930 million) could be met using modern bioenergy options. Additionally they make a significant impact/contribution in alleviating the stress on forest and the national electricity grid. They also offer an opportunity for sustainable, self-reliant and equitable development and provide tangible benefits of land reclamation, local employment and reduction of greenhouse gas emissions.

Wide use of solar and wind energy and bio-fuels, and other renewable energy sources can at least meet 20% of our basic energy needs and reduce greenhouse gases. This cupled with a movement towards a long term objective of a united global energy consumption system would create a substantial reduction of both environmental damage and also the problems of security and ....

REFERENCES

El-Ashru, M.T., Meeting Energy Needs in Developing world: Growth, Efficiency, Environmental Sustainability. PAGE, 1994 (1): 7-14-pp.

Flaim, S.J. and Elkis, J.R.A. Environmental consequences of Renewable Energy Development in India. *PAGE,* 1995, 5(1), 43-52 pp.

Hulscher, W., *Socio-economc consideration in the Power Guide: An International Power Guide*, in Intermediate Technology Publications, 1994.

Kishore, V.V.N. ed., *Renewable Energy Utilization: Scope, Economics and Perspective*, New Delhi, TERI, 1993.

Mahajan, V.S. ed; *Studies in Energy and Economic Development,* New Delhi Deep and Deep Publications, 1991.

Mortimer, N.D., Kellett; J. E. and Graat. J.F., *Rural Renewable Energy use for Urban Sustainbility. PAGE,* 1995 (1), 55-66 pp.

Naidu, B.S.K., Indian Scenaro of Renewable Energy for Sustainable Development. *Energy Policy*, 1996, 24 (6).

Rajagopal, R. and Chitra, N. Environmental impacts of Bioenergy. *ENVIS,* 1992 (1) , 1-5 pp.

Sheer, H., The Socio-economic Importance of Solar Energy. Speech delivered at World Solar Summit, Unesco, 1993.

TERI (New Delhi) and CERI (Calgary), Planning for the Indian Power Sector: Environmental and development considerations. *Executive Summary Series*, 2, 1996.

Tinker, *et al;* ed, by Desai, *A Human Energy* New Delhi, Wiley Eastern, 1990.

Who, our plant, our Health: Report the WHO Commission on Health and Environment. Geneva, WHO, 1992.

DIVISION - II

ENERGY SECTOR DEVELOPMENT IN SOUTH ASIA

SUSTAINABLE ENERGY SUPPLY OPTIONS

24

Dr. W.S. HULSCHER
FAO Regional Wood Energy Development Programme
Bangkok, Thailand

ABSTRACT

The most economical option for supplying renewable energy is still to grow a tree and burn it. Wood and other biomass supply about one-third of the national energy consumption in India. Some common misconceptions about wood and biomass fuels are cited and examined. Challenges for wood energy planning, policies, and institutions have been covered.

Role of Wood and Biomass as a Renewable Energy Source

The cheapest and most reliable option for supplying renewable energy is still to grow a tree and burn it. This is actually what is commonly done in Asia, even though tree growing is seldom meant for the sole purpose of supplying energy. Unfortunately, the important roles and associated problems of wood and other biomass as a sustainable energy supply option are largely undervalued by energy planners and policy makers. A few basic data will suffice to substantiate the point[1]. Table 1 overviews the share of wood and biomass energy consumption in 10 countries in South and South-East Asia. Table 2 shows the trend in biomass energy consumption (including wood) in 14 countries. In Table 3 the total value of woodfuel consumption per annum is estimated in terms of dollar equivalent. Table 4 summarises some figures on estimated employment per fuel type. These data lead to the following important observations:

— Wood and other biomass supply about one-third of the national energy consumption in India, and some 37% of the energy consumption in the region. Further comparisons show that amongst the various renewable energies wood/biomass provides by far the largest share in the region.
— In all the named countries the consumption of wood and other biomass fuels is still increasing in absolute terms, even when their share in national energy consumption is decreasing.
— The total value of wood and biomass fuels amounts to more than 9 billion US$ per annum in India (which was 30-40% of export earnings in 1991), and as much as some 30 billion US$ per annum for the named countries in the region together.

TABLE 1: Share of Biomass and Wood Energy Consumption

Country	Year	Share of Biomass (%)	Share of Wood (%)	Share of Biomass in Domestic Sector (%)
Bangladesh[2]	1992	73	13	89
Bhutan[3]	1991	82	-	-
China[4]	1992	10	-	25
India[4]	1992	33	-	78
Indonesia[5]	1992	39	31	73[1]
Laos[3]	1991	88	-	-
Malaysia[5]	1992	7	2	15[1]
Myanmar[3]	1991	74	-	-
Nepal[2]	1992-3	92	68	97
Pakistan[2]	1993-4	47	27	83
Philippines[5]	1992	44	26	66[1]
Sri Lanka[2]	1990	77	-	93
Thailand[2]	1994	26	9	65[1]
Vietnam[3]	1991	50	-	-

— No data available, 1. Domestic and commercial sector, 2. National energy balance

3. World Resource Institute, 4. United Nations, 5. AEEMTRC

TABLE 2 : Biomass Energy Consumption (PJ) in Member Countries

Country	1981	1986	1991
Bangladesh	243	262	277
Bhutan	7	9	12
China	1,541	1,820	2,018
India	2,165	2,441	2,824
Indonesia	1,181	1,320	1,465
Lao PDR	29	33	39
Malaysia	69	78	90
Myanmar	156	175	193
Nepal	113	197	206
Pakistan	192	233	296
Philippines	308	327	382
Sri Lanka	70	78	89
Thailand	484	546	526
Vietnam	197	222	251
Total:	6,755	7,742	8,666

— Woodfuel use generates at least 20 times more local employment than energy from oil products (per unit of energy). Hence, national interests like balanced growth, employment and equity are linked to energy supply policies.

Unfortunately, common misconceptions about wood and biomass fuels can still be observed. They are of the type: woodfuel is a traditional commodity which will phase out in the near future as society is modernising; the consumption of wood and biomass fuels is confined to poor and rural households; fuelwood originates from forest lands and their consumption is responsible for destroying natural forests. The last assumption conflicts with many survey results revealing that some 70% of all woodfuels originate from non-forest land (e.g. agricultural land, waste land, etc). Reference is also made to the discussion summarised by Ravindranath and Hall on the presumed relationship between woodfuel use and deforestation[2]. Few people look at fuelwood as a commodity which can be, and indeed partly is, produced and harvested like rice or wheat, though with a gestation period much longer than the named grains. Furthermore, looking at wood energy as only a traditional commodity contrasts with today's policies of many industrialised countries to deliberately increase wood energy use, mainly for environmental reasons. For instance in the Nordic countries in Europe, modern bio-energy has already been increased to 10% of total energy consumption, and it will be further increased considerably in the years to come. The most significant contribution to bio-energy comes from wood and the increase is due to a combination of: better use of wood residues, wood energy plantations, improved wood energy conversion efficiencies, higher forest productivity and lower production costs[3]. The policies include combined heat and power generation and co-combustion of wood and coal with modern technologies, whether stand alone or grid-connected. It is interesting to note that in Europe income and job generation in rural areas is being recognised as an additional justification for promoting bio-energy. Unfortunately, planning institutions of governments at centralised and decentralised levels in Asia often display a lack of interest in what is considered 'traditional' fuels, including wood and biomass energy.

Obviously there is an imbalance between, on the one hand, the development and planning of wood energy and, on the other hand, the major efforts which are directed, for instance, to planning of electricity and fossil fuel supplies as well as incentives for their use. Furthermore, trying to overview the numerous efforts worldwide to further develop renewable energies like solar, wind, hydro power and other options, it occurs to me that wood and other biomass energy is also undervalued amongst renewable options in terms of R&D, development, policy making and planning. Finally, even within the range of efforts directed to wood energy itself, also imbalances are noted. My estimate is that R&D on e.g. gasification received, may be, ten times more funding than R&D on traditional combustion, whereas in fact combustion is at least a thousand times more common than gasification. Probably, the rationale of this imbalance is the assumption that at present gasification would pose more serious challenges to R&D than combustion does. The assumption is false, unless challenges to R&D are defined in a narrow conceptual manner.

Challenges to Wood Energy Planning

Wood energy qualifies as a modern option when it is efficient, clean, convenient, reliable, and economically and environmentally sound. We know that numerous wide-spread practices associated with wood energy are still far away from these qualifications. This poses many challenges to the ingenuity of not only scientists and engineers, but also specialists in natural resource management, forestry, agriculture, economy, planning, environmental management, social sciences, extension, etc., as well as the integration of elements of the named disciplines. For instance, the use of wood and biomass in rural small and medium-scale industries (SMI) is wide-spread in India and other countries. The SMI are vital for the economy in terms of product output, employment, rural sustenance and entrepreneurship. However, fuel use in SMI suffers from many problems like inefficient

TABLE 3: Woodfuel Values in Million US$ using Average Woodfuel Prices

Country	Fuelwood	Charcoal	Dom. FW	Dom. Char	Ind. FW	Ind. Char	Total
Bangladesh	306	-	255	-	50	-	306
Bhutan	37	3	33	-	3	3	40
China	9,320	-	9,320	-	-	-	9,320
India	9,080	-	8,440	-	640	-	9,080
Indonesia	2,317	-	2,317	-	-	-	2,317
Laos	88	-	88	-	-	-	88
Malaysia	31	49	31	49	-	-	80
Maldives	3	-	3	-	-	-	3
Myanmar	914	213	914	213	-	-	1,127
Nepal	469	-	451	-	17		469
Pakistan	1,318	-	1,317	-	-	-	1,318
Philippines	618	491	618	491	-	-	1,109
Sri Lanka	436	5	363	-	-	-	440
Thailand	432	1,595	432	1,595	-	-	2,027
Vietnam	1,139	139	1,055	133	-	1	1,278
RWEDP member countries	26,506	2,494	25,637	2,481	710	4	29,000

Fuel prices in	US$/Ton	US$/GJ	End-use efficiencies of stoves		Fuel heating value GJ/Ton	
Fuelwood	40	2.67	Fuelwood	20%	Fuelwood	15 GJ/Ton
Charcoal	250	8.62	Charcoal	30%	Charcoal	29 GJ/Ton

TABLE 4: Estimated employment per fuel type

Fuel type	Tons of Fuel per Tera Joule (TJ)	Estimated Employment per TJ Energy Consumed in Person Days[1]
Kerosene[2]	29 Kilo Litre	10
LPG[2]	22 Tons	10-20
Coal[3]	43 Tons	20-40
Electricity[4]	228 MWh	80-110
Fuelwood[5]	62 Tons	110-170
Charcoal[5]	33 Tons	200-350

Source: Estimates by the World Bank mission members for the Philippines Householes Energy Study.

1. Where applicable employment covers growing, extraction, production, transmission, maintenance, distribution and sales, including reading meters. It excludes employment generated outside the country for fuels that are imported in semi-finished or finished state
2. This assumes that crude oil (for refining), kerosene and LPG are imported.
3. Varying according to capital intensity of the mine, seam thickness, energy value of the coal as well as the distance from demand centres.
4. Varies according to production method ranging from hydro to traditional oil/coal fired units and the efficiency of electricity generation, transmission and distribution.
5. Depending on productivity of the site, efficiency of producers and distance from the market.

combustion, unhealthy working conditions, unreliable supplies, poor financial viability, lack of technical and managerial skills, etc. The response to these challenges should not be to ignore these energy-related problems, or bet on alternative forms of energy, or establish a few subsidised SMI show cases, or blame the SMI themselves in the context of today's market oriented economy. The response should rather be to engage in the cumbersome work of inventorising the many problems, analysing them, and developing feasible solutions. This type of work requires integrated efforts and skills of many of the above named specialists.

This is not the place to specify the many and varied problems associated with wood energy at large. Rather, as an illustration the challenges to data availability for planning and policy making will be elaborated a bit. First, to be feasible at all, wood energy planning should be based on decentralised structures, e.g. linked to area-based planning and/or rural development planning. Decentralisation allows close interaction between planning and implementation, whether in projects, programmes or policies at large. Experience has shown that such interaction is vital for successful interventions which address local and site-specific issues like wood energy. Wood energy planning has both long-term and short-term components, i.e. from 1 year to 10 years or more.

For wood energy planning, systematic data are required as a time series, which implies more than some casual survey. Data availability is a problem in most countries. The relevant data are scattered over various institutions and they are not as accurate as the data readily at hand for fossil fuels*, but they need not be. Usually, data are sufficiently accurate if at least they allow a correct understanding of problems at stake and provide a grasp of trends over time. These two requirements can be met by a combination of macro-studies at certain intervals (e.g. base-line surveys linked to nationwide censuses) and well-selected micro-studies. Relevant micro-studies would be on wood energy consumption patterns and trends (in households, industries, institutions, commerce, etc), and on wood energy supply patterns and trends, including studies of woodfuel sources, flows and markets.

As usual in the energy sector, planning is taken as matching supply and demand. On the supply side wood energy planning calls for quantitative data on supply patterns and trends of wood and other biomass fuels. Such data include the sources of supply, their productivity and sustainability, and the patterns of woodfuel flows from sources to markets and finally to consumers. Related socio-economic data are on employment (or income generation) and gender aspects. Linkages with geo-information on land use from forestry, agriculture and environment sectors backs the interpretation of data obtained from supply surveys. Part of the data can be obtained from aerial surveys, but many aspects also need ground surveys. The data are to be compiled in data bases providing overviews for planning and policy making.

On the demand side wood energy planning requires quantitative data on consumption patterns and trends of wood and other biomass fuels in a systematic way. Such data, again, can only be obtained by surveys. The data need to be linked to information on the local rural and national economy,

***The accuracy and reliability of national fossil fuel data can be questioned. Reference [1] presents country cases where ambiguities of 30% and more have been detected. Furthermore, it is quite revealing to study linkages (or rather the lack of them) between development planning and actual decision making in various sectors of the economy, including the electricity sub-sector. Such analyses undermine the argument that wood energy could not be incorporated into energy planning and policy making.**

and key information from relevant sectors like energy, forestry, agriculture, and environment. Relevant social conditions, gender related information, and distributional aspects must be studied to help interpret consumption data. The surveys should lead to quantitative overviews to be installed in data bases which are accessible for purposes of planning and policy making.

Policies and Interventions

The most direct adverse effects of neglecting wood energy planning and policy making can be severe scarcities and drudgery for weaker groups, over-exploitation of local resources, lost opportunities and, as we often observe, ineffective government interventions. In many countries, typical institutional barriers to the development of policies for wood and other biomass as a renewable energy supply are of the following kind.

— The energy sector lacks appreciation and adequate data, particularly with regard to the supply side, for incorporating woodfuels into energy planning. The interests of energy institutions are often focussed on new technologies or small pilot projects rather than on understanding and upgrading main stream use of woodfuels.

— The forestry sector largely considers woodfuel a low value by-product of trees, and foresters are not used to thinking in terms of Joules or Calories.

— The agriculture sector is not sufficiently aware of the importance of agricultural land for woodfuel supply. Agro-forestry options meet scepticism amongst agriculturalists, and experts still know very little about allopathy of trees and agricultural crops[4].

Hence, wood energy seems to be nobody's land, whereas it should actually be centrally located in the triangle energy-forestry-agriculture, with linkages to other policy areas. National policies for wood and other biomass as a renewable supply of energy can be based on a combination of several important policy considerations, e.g.:

— The factual importance of wood and other biomass in terms of energy quantity, money equivalence, saving foreign exchange, employment, balanced regional development and equity, as indicated above;

— For many decades to come woodfuels will remain a basic commodity for more than half of the population, particularly for the rural poor, as well as numerous small-scale industries and enterprises;

— The supply of woodfuels is diverse and locally becoming increasingly scarce;

— Sustainable wood energy use poses no threat to the global environment in terms of carbon emissions;

— The proper use of wood energy links to additional public interests like health and social security, acknowledging that the woodfuel business provides a 'safety net' for impoverised people;

— Energy demand in transitional economies can be quite dynamic, whereas developing energy supply options requires long lead times, e.g. up to 20 years. This calls for long-term scenarios.

Effective energy policies require adequate wood energy data and planning, as stated above. Wood energy planning 'works' if it directs resource allocation and leads to further sound government interventions amongst other energy options. Therefore, wood energy planning should be part of energy planning.

Interventions and Institutions

Interventions in the wood energy sub-sector largely depend on government actions like public resource management, pricing, taxes, subsidies, economic incentives e.g. fuel substitution, and licensing. Such interventions can only be prepared and implemented when the country avails of adequate institutions and a sufficient number of qualified experts. Furthermore, interventions with regard to wood energy are to be supported by the establishment of effective public-private partnerships for implementing new supplies and managing existing ones, as well as private sector organisations for production, conversion and utilisation technologies.

Adequate institutions include inter-ministerial coordination between policy areas as referred to above. At present this may be the largest single constraint to providing wood energy its proper place amongst the renewable energy options. In virtually all other forms of energy, whether renewable or not, a ministry or department in charge of energy is at least in principle well equipped to deal with the supply side. However, this is much more complicated in the case of wood and other biomass energy sources. The situation obviously calls for proper coordination mechanisms witch match with established government procedures. The development or selection of methodologies which are most suitable for wood energy planning in a particular country, is subject to evaluation by government experts. I have observed that countries in Asia can benefit a lot from each other's experiences.

Next comes the development of skills for units dealing with developing, planning and management of wood and biomass energy. In order to gain momentum, a critical mass of educated and well-trained people will be necessary. The present situation does not seem satisfactorily to me. Surprisingly, forest training institutes still spend very little attention to woodfuel as a tree product, and agricultural training institutes probably not at all. Furthermore, most institutes which educate energy engineers, focus on fossil fuels or even nuclear energy. As far as education in renewable energy is concerned, I guess that in a country like India more technologists have been trained, and are being trained at this very moment, e.g. solar photo-voltaics than in wood energy. Generally, it cannot be taken for granted that a forester, thermal engineer or physicist who is educated in present mainstream education, will naturally develop an interest or skill in wood energy issues. The situation may be slightly better in disciplines of energy economy and natural resource management, because a properly trained development economist should be able to deal with wood energy problems when required. However, many economists could still benefit from basic technical familiarization courses. In the present educational situation, it is hard to imagine how the above named imbalances in the energy scene will be corrected, and how a critical mass of trained bio-energy experts in various institutions will develop.

Further important aspects of institutional development relate to the roles of state and local government authorities, NGO's, community participation and private sector organisations, as referred to above. Joint efforts will be required to develop prognostic tools and indicators for wood energy. This will not be elaborated here.

CONCLUSION

It is concluded that wood and other biomass constitute the main source of renewable energy in India as well as in many other countries in Asia, and this situation will not and cannot change in the near future. Woodfuel can and should develop as a modern and sustainable energy supply option. Notwithstanding increasing appreciation of wood energy development, the importance of supporting planning activities still remains generally not recognized by national policy and decision makers from the energy, forestry and agriculture sectors. Problems and opportunities associated with this form of renewable energy are not yet fully taken up. It is recommended that wood energy planning and policy making will be fully incorporated into energy planning.

REFERENCES

1. Wood Energy News, Vol.11, No. 2, FAO-RWEDP, June 1996
2. N.H. Ravindranath and D.O. Hall, Biomass, Energy and Environment - A Developing Country Perspective from India, Oxford University Press, 1995, pp 61-62.
3. 9th European Bioenergy Conference and 1st European Energy From Biomass Technology Exhibition, 24-27 June 1996, Copenhagen, Denmark (as reported by M. Trossero, FAO)
4. Wood Energy News, Vol. 11, No.1, FAO-RWEDP, March 1996

SOUTH ASIAN COOPERATION FOR ENHANCING ENERGY SUPPLY IN THE 21st CENTURY

25

TOUFIQ A. SIDDIQI
Regional Adviser on Energy
Environment and Natural Resources Management Division
United Nations ESCAP
Bangkok 10200

ABSTRACT

The energy sources in South Asian countries is discussed and their environmental impact. Regional cooperation in the use of natural gas and hydro power has been emphasised. The need for clean and pollution free energy is emphasised for South Asian countries.

1. INTRODUCTION

The use of primary commercial energy in South Asia increased at an annual rate of 6.8% between 1971 and 1993. Global studies that have been undertaken in recent years, for example those by the International Energy Agency (IEA, 1996) project continued growth during the next several decades. As a convenient reference point for the magnitude of energy requirements up to 2010, we shall use the results of the IEA study, which projected the energy demand for all the regions of the world using a number of assumptions. The assumptions underlying key variables such as economic growth, energy prices, technological progress, consumer behaviour, and government policies on energy, were treated independently of each other, with full realization that links between them do exist.

One set of assumptions is designated the "Capacity Constraints" (CC) case. It combines baseline GDP and population assumptions with rising energy prices and historical trends in energy efficiency. The second set of assumptions represents the "Energy Savings" (ES) case. In this case, baseline GDP and population assumptions are combined with higher rates of improvement in energy efficiency and energy prices that remain flat. The energy demand projections for the two cases are given in Tables 1 and 2. The Tables also provide projections of the contributions of different energy sources to the total energy supply, as well as to electricity generation, which is increasing at a faster rate than overall energy growth.

The "Solids" category in Tabes 1 and 2 includes coal and lignite only, and biomass contributions are not included in the demand projections. Although the demand for biomass is expected

TABLE 1: **Energy Demand - South Asia - Capacity Constraints Case**

	Levels				Fuel Shares (%)			
	1971	1993	2000	2010	1971	1993	2000	2010
Primary Energy (MTOE)	**72**	**251**	**355**	**658**	**100**	**100**	**100**	**100**
Solids	39	129	153	270	53.5	51.2	43.2	41.1
Oil	27	82	128	233	37.6	32.7	36.0	35.4
Gas	3	31	57	125	4.5	12.2	16.2	19.1
Nuclear	0	2	4	5	0.5	0.6	1.0	0.8
Hydro	3	8	12	23	3.9	3.3	3.4	3.5
Geo/Others	0	0	1	1	0.0	0.0	0.2	0.2
Final Energy	**55**	**162**	**245**	**461**	**100**	**100**	**100**	**100**
Solids	25	44	58	93	45.5	27.4	23.8	20.1
Oil	23	72	112	207	41.5	44.6	45.8	44.9
Gas	2	18	31	74	3.6	10.8	12.5	16.0
Electricity	5	28	44	88	9.4	17.2	17.9	19.0
Electricity Output (TWh)	**76**	**419**	**650**	**1241**	**100**	**100**	**100**	**100**
Solids	32	250	363	710	42.5	59.8	55.9	57.2
Oil	6	25	43	65	7.3	6.2	6.6	5.2
Gas	4	40	90	178	5.4	9.5	13.9	14.3
Nuclear	1	6	14	19	1.7	1.4	2.1	1.5
Hydro	33	97	139	268	43.2	23.1	21.3	21.6
Geo/Others	0	0	1	1	0.0	0.0	0.1	0.1

Source: International Energy Agency (1996).

TABLE 2 : **Energy demand — South Asia — Energy Savings Case**

	Levels				**Fuel Shares (%)**			
	1971	**1993**	**2000**	**2010**	**1971**	**1993**	**2000**	**2010**
Primary Energy (MTOE)	**72**	**251**	**344**	**625**	**100**	**100**	**100**	**100**
Solids	39	129	147	248	53.5	51.2	42.8	39.7
Oil	27	82	125	227	37.6	32.7	36.3	36.3
Gas	3	31	56	120	4.5	12.2	16.2	19.2
Nuclear	0	2	4	5	0.5	0.6	1.0	0.8
Hydro	3	8	12	23	3.9	3.3	3.5	3.7
Geo/Others	0	0	1	3	0.0	0.0	0.2	0.4
Final Energy	**55**	**162**	**238**	**440**	**100**	**100**	**100**	**100**
Solids	25	44	56	83	45.5	27.4	23.3	18.9
Oil	23	72	110	202	41.5	44.6	46.2	46.1
Gas	2	18	30	71	3.6	10.8	12.5	16.1
Electricity	5	28	43	83	9.4	17.2	18.0	19.0
Electricity Output (TWh)	**76**	**419**	**634**	**1180**	**100**	**100**	**100**	**100**
Solids	32	250	350	659	42.5	59.8	55.3	55.9
Oil	6	26	41	61	7.3	6.2	6.5	5.2
Gas	4	40	89	170	5.4	9.5	14.0	14.4
Nuclear	1	6	14	19	1.7	1.4	2.2	1.6
Hydro	33	97	139	268	43.2	23.1	21.9	22.7
Geo/Others	0	0	1	3	0.0	0.0	0.2	0.3

Source: International Energy Agency (1996).

to grow as population and per capita energy consumption increase, the availability of biomass is not expected to increase much, if at all. Thus, in the time frame up to 2010, we shall assume that the amount of energy supplied by biomass in South Asia remains constant, and increases in total energy supply will come from commercial energy sources, including a relatively small amount supplied by new and renewable energy sources such as wind power and solar energy.

A comparison of Tables 1 & 2 shows that, by 2010, the total energy demand for the Energy Savings case is 33 million tons oil equivalent (MTOE) less than for the Capacity Constraints case. There is, however, very little change in the share of energy supplied by the different energy sources. Solids (coal and lignite) continue to supply the largest share of energy, closely followed by oil. Natural gas supplies about 19% of the total energy, and hydro power only about 3.5%. nuclear and renewable energy sources contribute only modest amounts to the total commercial energy supply in 2010.

From an environmental perspective, the heavy dependence in South Asia on coal and oil for energy is unfortunate, since these are generally considered to be the most polluting of the enrgy sources. We shall briefly discuss in the next section the environmental impacts associated with different sources of energy.

2. Environmental Impacts Associated with Energy Sources

There are environmental impacts associated with all energy sources, even the renewable ones. The magnitude and nature of the impact, however, varies considerably from one sorce to the other. these impact can be short-term, or extend over a longer time horizon. Further, the impact can be at local, regional, or global levels, and affect the terrestrial, aquatic, and atmospheric ecosystems. These impacts are summarized in Figure 1. It should be pointed out that the impacts shown in the figure are from routine operations only. In the case of accidents, for example the collapse of a hydropower dam or a nuclear accident as at Chernobyl, the impacts can be regional or global.

In the limited time available here, it will not be possible to discuss the environmental impacts in any detail. Those interested in more comprehensive discussions may wish to consider a number of excellent surveys, e.g. those by ADB (1991), IEA (1989), and UNEP (1985). We shall only summarize the impacts associated with the coal and oil systems. These are presented in Figure 2 and Table 3.

In *routine* operation, the environmental impacts associated with natural gas, hydro power, and nuclear power are smaller than those associated with coal or oil. We underline the word "routine" here, since there can always be a major catastrophic accident associated with nuclear power, natural gas, or hydro power. As far as the other renewable sources are concerned, there are substantial emissions of local air pollutants as well as carbon dioxide (CO_2) from biomass combustion. The production of photovoltaic cells and panels, as well as of the materials for wind turbines, also has environmental effluents associated with it. The aggregate impacts, however, are much smaller than those from the use of coal or oil.

From the environmental perspective, it is therefore desirable to increase the shares of natural gas, hydro power, and renewable energy. In the next section, we can discuss ways in which this could be done.

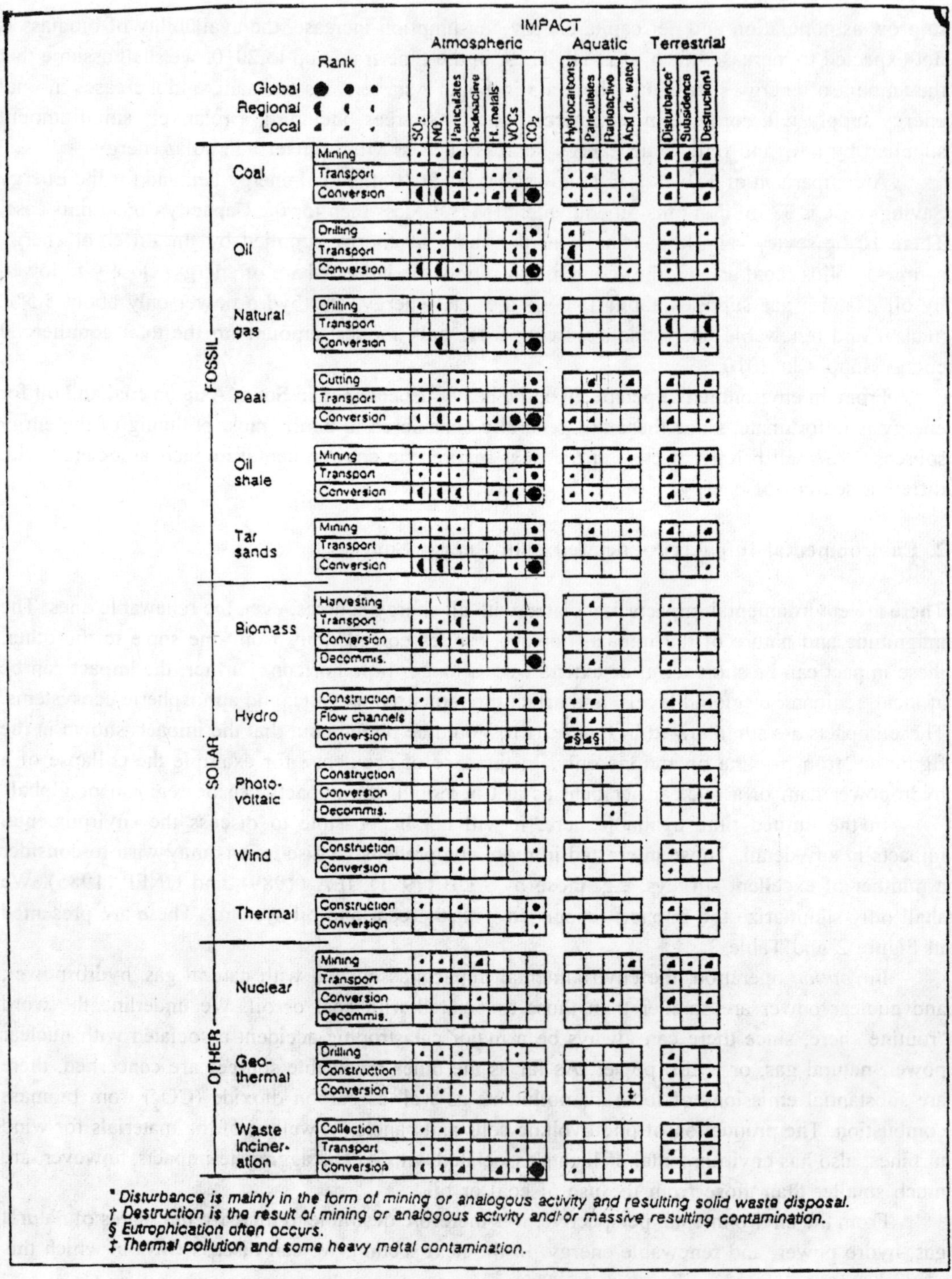

** Disturbance is mainly in the form of mining or analogous activity and resulting solid waste disposal.*
† Destruction is the result of mining or analogous activity and/or massive resulting contamination.
§ Eutrophication often occurs.
‡ Thermal pollution and some heavy metal contamination.

FIG. 1 : *Ranking of Environmental Impacts*

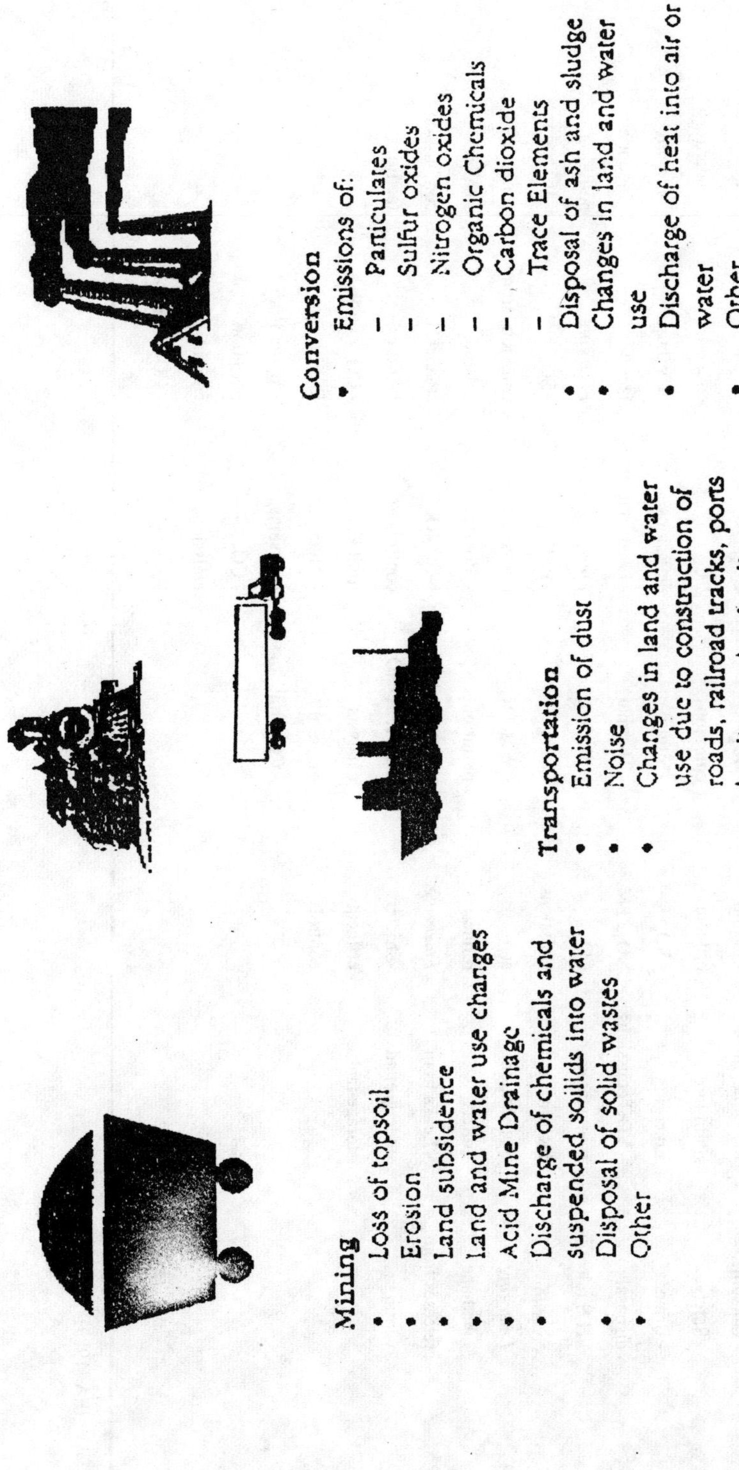

Source: Toufiq Siddiqi

FIG. 2 : *Environmental Impacts Associated with the Coal Systems*

Table 3

Fuel Cycle	**Exploration Production**	**Transport Storage and distribution**	**Treatment**	**Electricity Generation**	**Transport end-uses**	**Industrial end-uses**	**Resiaential and commercial end-uses**
Oil	Ongoing: - Oil spills. - Safety. - Brine disposal, - H_2S releases, - Siting, *Emerging:* - methane releases from venting	Ongoing: Maritime oil pollution, Emerging. - VOC emissions from storage facilities and distribution outlets - Leaks from underground storage tanks,	Ongoing: - Liquid effluents. - SO_x NO_x,. PM. CO emissions at refineries, *Emerging:* -Stricter effluent controls.	Ongoing: - SO_x, NO_x emissions. - Siting. - Thermal pollution, *Emerging:* - Waste from pollution control, FGD sludge, de-NO_x catalysts disposal - CO_2 emissions	Ongoing: CO_x, NO_x IIC emissions, - Lead content, *Emerging:* - PM emissions - Stricter NO_x/VOC controls lo limit O_x - Aldehyde emissions, - CO_2 emissions	Ongoing: SO_x. NO_x PM cmissions for new, large facilities. *Emerging:* - SO_x. NO_x. emissions from existing, smaller facilities. - CO controls facilities. - Waste from pollution control - CO_2 emissions	Ongoing: - Fuel quality standards. *Emerging :* - Indoor air pollution. - NO_x. SO_x emissions from smaller combustion facilities

Source: IEA (1989)

3. Regional Cooperation in the Use of Natural Gas and Hydropower

In this section, we examine ways in which the share of energy supplied by natural gas, hydro powre, and renewable energy can be increased in countries of South Asia. It is worth pointing out here that there are large differences between the countries of the region in terms of the share of energy supplied by each source. Tables 1 & 2 had provided aggregate shars for the whole region, but the large energy use in India may give a misleading picture of the relative contributions of the energy sources in the other countries. In Figure 3, we provide an overview of the percentage of energy supplied by each of the commercial sources in each of the member countries of the South Asian Association for Regional Cooperation.

In Bangladesh, natural gas is the largest source of energy. In Pakistan, oil and natural gas contribute roughly equal amounts of energy. The supply of commercial energy in India is still dominated by coal, although the energy supplied by petroleum products (oil) and natural gas has increased substantially in absolute amount. Nepal still derives more than 90% of its energy from traditioonal fuels, but its use of commercial energy is based mainly on oil and hydropower.

It may be useful to clarify that for the next decade, at least, theuse of all commercial sources of energy – coal, oil, natural gas, hydro power, and nuclear energy – is likely to increase in South Asia. What we shall be discussing here are ways in which the use of natural gas and hydro power can be increased.

3.1 Natural Gas

Starting first with natural gas, its use in the region has increased a great deal during the past two decades. As a result, the Reserves/Production (R/P) ratios in Bangladesh, India, and Pakistan have declined to below 40 years. Thus large increases in domestic production cannot be anticipated, unless major new discoveries are made. The countries have not yet been adequately explored for natural gas, and a high priority could be to encourage large increases in exploration, both onshore and offshore.

A parallel approach that could be implemented is for india and Pakistan to import natural gas from the Middle East of Central Asia, where several of the countries have very large reserves and R/P ratios. The proven reserves and R/P ratios of Bangladesh, India, and Pakistan are compared with those of several potential suppliers of natural gas to the region in Figure 4.

Several options for importing natural gas have already been examined by India and Pakistan. Pakistan has signed Memoranda of understanding with the Islamic Republic of Iran, Qatar, and Turkmenistan, for the possible supply of natural gas. India has been exploring the possibility of importing natural gas via pipeline or in liquefied form (LNG) from Oman, or via pipeline from the islamic Republic of Iran (Hamso *et al.,* 1994).

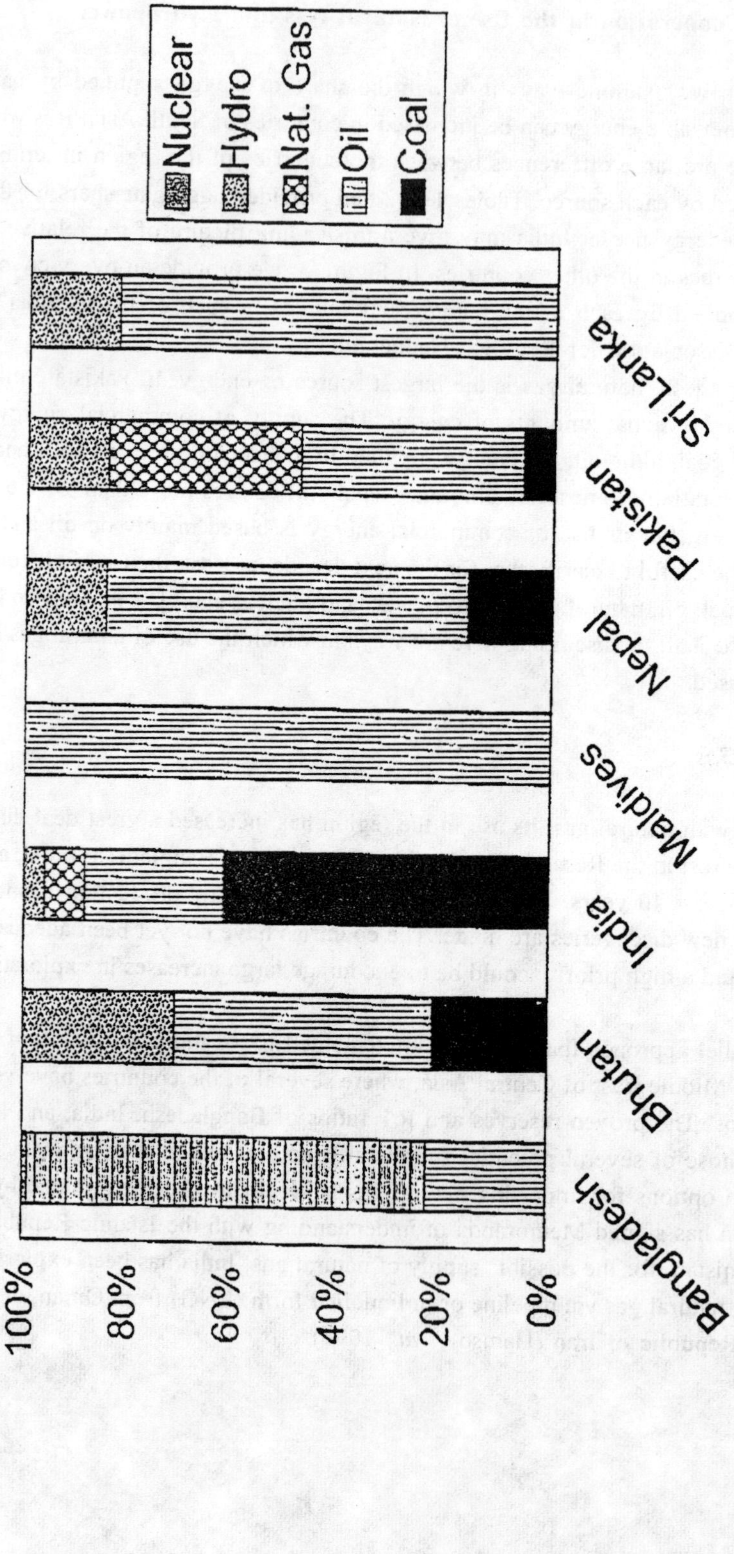

Sources: United Nations, except for India (TERI) and Pakistan (HDIP)

Toufiq Siddiqi (1996)

FIG. 3 : *Share of Primary Commercial Energy Supplied by Different Energy Sources in South Asia (1994)*

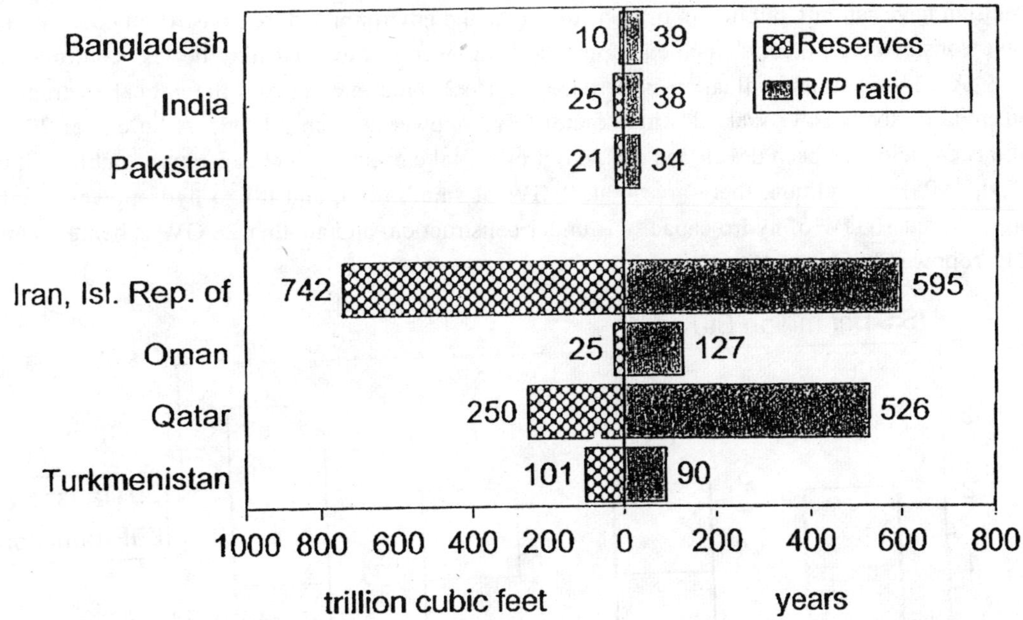

FIG. 4 : *Proven Reserves of Natural Gas in Selected Countries of South and West Asia, 1995*

Although India and Pakistan could make independent arrangements for importing natural gas, economies of scale suggest that each of the countries could benefit a great deal by cooperating in the construction of one or more pipelines that could supply natural gas to both countries (Siddiqi, 1995). The anticipated demand for natural gas in India and Pakistan could exceed 8 billion cubic feet per day in each country by 2010 (Raza and Mehta, 1995), of which only about 25% could be supplied from domestically proven reserves.

Natural gas supplied via pipeline is expected to cost about 35% less than LNG in Pakistan and Western or Northern India. Cost comparisons for Pakistan are given in Figure 5. Given the large amounts of natural gas the might be imported, and the trade balance of each of the two countries, a saving of 35% would represent a substantial amount. In addition, india may still require to import LNG for the Southern part of the country, where the length of the required pipeline to bring natural gas imports may be too long to make it economically attractive.

3.2 Hydro Power

Only a few decades ago, Hydropower was the largest source of electricity in most of the South Asian countries, including India. The rapid growth in electricity demand, combined with the long construction times for major hydro projects and availability of domestic coal (India) and natural gas (Pakistan, India, and Bangladesh) have led to the construction of a great deal of fossil-fuel based electrical capacity. Further, some of the planned large hydro power sites in India and

Pakistan have run into public opposition for social and environmental reasons. Both countries still have considerable potential for developing hydro power at sites that may be less controversial.

According to an evaluation undertaken in 1992, india has a gross theoretical hydropower potential of about 84 GW at 60% load factor (Hydropower & Dams, 1996). A little over 20 GW of this capacity has been developed so far, out of a total current 77 GW capacity of public utilities (Teri, 1995). in addition, there are about 2 GW of small, mini, and micro hydropower installations. About 10 GW of hydro capacity is under construction, and another 28 GW is being planned (Hydropower & Dams, 1996).

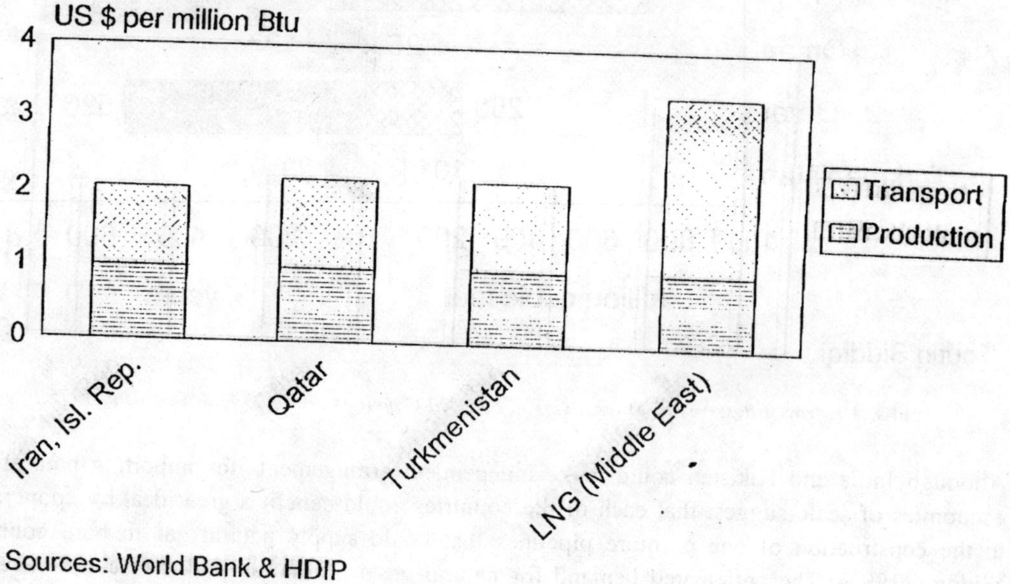

FIG. 5 : *Estimated Cost of Natural Gas Delivered to Pakistan by Pipeline (2.0 bcf/d) or as LNG (0.6 bcf/d)*

Pakistan has an installed hydro capacity of about 5 GW out of a total installed capacity of 12 GW (HDIP, 1996). The potential that could be utilized amounts to about 27 GW (EUAD/IUCN).

Quite early during the next century, India and Pakistan are likely to have fully utilized their hydro power potential. Before that happens, it would be to the mutual benefit of Nepal and India to cooperate in the development of the vast hydro power resources of the former. The estimated theoretical potential for hydropower in Nepal is about 83 GW, of which 25 GW could be economically developed at present (Shresta and Bhattarai, 1993). Nepal is at present generating only about 250 Megawatt (MW) of electricity from water resources, and its own demand in the intermediate term could be met by a capacity of just a few GW. A great deal more of additional capacity could be developed for export, initially to India with possible extension to Bangladesh at a later time. There have recently been discussions between India and Nepal which enhance the likelihood of cooperation for the development of hydro power.

Bangladesh, Bhutan, and Sri Lanka also have hydropower reserves, but only Bhutan is in a position to have capacity that is surplus to its own needs. Bangladesh, for example, generates only about 10% of its electricity needs from hydro power at present, and this share is expected to decrease to about 7% by 2020 (GOB, 1996). If satisfactory financial arrangements can be worked out, electricity from Bhutan can also be exported to India and Bangladesh.

4. Conclusion

Energy policies in South Asia have traditionally been based on making use of indigenous energy resources to the maximum extent possible. The repid growth in the transportation sector necessitated imports of oil and petroleum products. The rising demand for electricity, and increasing concerns about the serious deterioration in air quality in the countries of the region necessitate the imports of "clean" energy by many of the countries in thr region.

Air pollution in many of the largest cities in South Asia greatly exceeds (Brandon and Ramankutty, 1993) the levels considered acceptable by the World health Organization. Calcutta, Delhi, Karachi, Kathmandu, Lahore, and Mumbai are amongst the cities with the highest levels of particulates or other air pollutants. In addition to these local pollutants, the contribution of the region to emissions of greenhouse gases (GHGs), particularly carbon dioxide (CO_2), is also rising. Asian emissions of CO_2 now exceed those of North America, Western Europe, or Eastern Europe including the Russian Federation, as shown in Figure 6. (Siddiqi, 1996).

On a per capita basis, of course, CO_2 emissions from South Asia are only a small fraction of those from the industrialized countries, and are likely to keep increasing well into the 21st century. However, all the South Asian countries have signed the Framework Convention on Climate Change (FCCC). The Convention does not require countries to set limits on emissions of GHGs, but signatories from the developing are expected to identify policy options that could reduce future emissions of greenhouse gases, while meeting their development objectives.

The options discussed in this paper, such as the import of natural gas by Pakistan and India from the Middle East and Central Asia, and the import of elecricity by India and Bangladesh from Nepal and Bhutan, could make a substantial contribution to lowering the future levels of air pollution in the region. They could also be instrumental in reducing the future rate of growth of greenhouse gas emissions from the region.

Due to time constraints, we have not discussed the role of new and renewable energy sources in the region. India is already well along in the use of wind power, particularly in the Southern part of the country. This experience can be utilized elsewhere in the region. Solar photovoltaic (PV) systems are also being used in some locations that do not have access to the electrical grid.

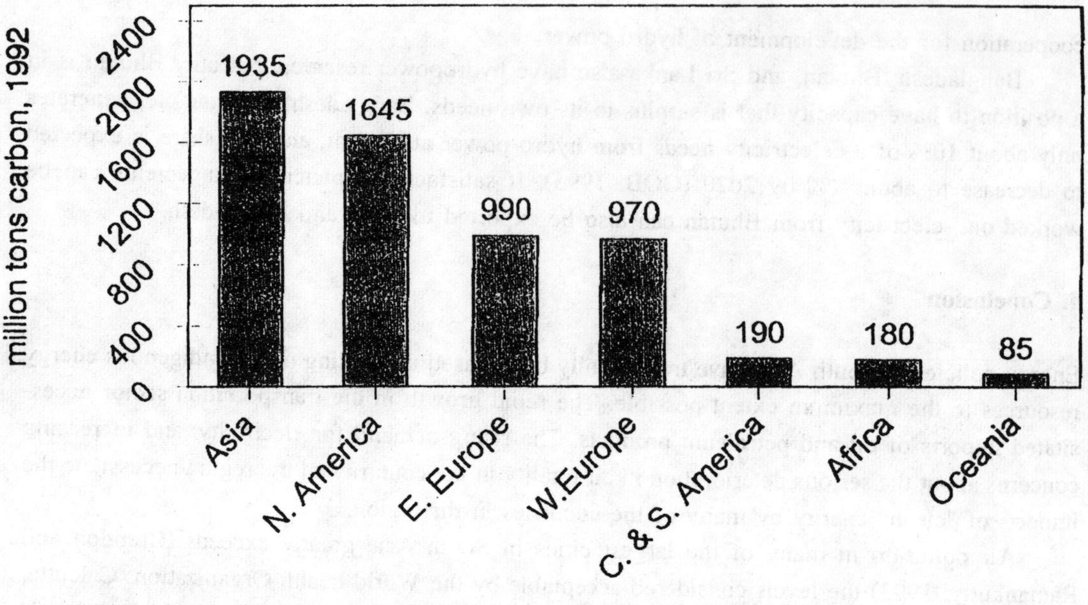

FIG. 6 : *Estimated Emissions of Carbon Dioxide from Fossil Fuel Use in Different Regions*

As the costs of PV systems come down further, we can anticipate their use in many locations throughout South Asia.

The countries of South Asia have the natural and human resources, including well trained manpower, necessary to meet the twin goals of energy supply and environmental quality. Although the countries could undertake most of the efforts on their own, the pace of development could be accelerated, and the costs lowered, by increased cooperation between them. A modest start has been made through the initiation of a project on "Enhancing Regional Cooperation in South Asia through Collaboration in Energy and Environment". The project is being supported by the United Nations Development Programme, and involves participants from several South Asian countries (UNDP, 1996). Potential projects dealing with cooperation in natural gas imports, development of water resources for electricity, new and renewable energy technologies, energy efficiency improvements, and air quality managment, are being considered.

REFERENCES

Asian Development Bank (1991). *Environmental Considerations in Energy Development*. Manila : Energy and Industry Department, ADB.

Brandon, Carter, and Ramesh Ramankutty (1993). *Toward an Environmental Strategy for Asia.* World Bank Discussion Paper No. 224. Washington, D.C.: World Bank.

Environment & Urban affairs Division, Government of Pakistan, and the World Conservation union (1991). *The Pakistan National Conservation Strategy* Karachi: IUCN.

Government of Bangladesh (1996). *National Energy Policy.* Gazette Notification Number DA-1. Dhaka: GOB.

Hamso, Bjorn, Afsaneh Mashayekhi, and Hossein Razavi (1994). "International Gas Trade: Potential Major Projects". *Annual Review of Energy and Environment,* Vol. 19, pp. 37-73.

Hydrocarbon Development institute of Pakistan (1996). *Pakistan Energy Yearbook 1995.* Islamabad: HDIP.

Hydropower & Dams (1996), Issue One, pp. 129-131.

International Atomic Energy Agency (1991). *Senior Expert Symposium on Electricity and the Environment: Key issues Papers.* Vienna: IAEA.

International Energy Agency (1989). *Energy and Environment: Policy Overview.* Paris: IEA, Organization for Economic Cooperation and Development.

International energy Agency (1996). *World Energy Outlook: 1996 Edition.* Paris: IEA, Organization for Economic Cooperation and Development.

Raza, Hilal, and Y.R. Mehta (1995). Papers presented at the Workshop on *Enhancing Regional Cooperation in South Asia through Collaboration in Energy and Environment,* held at Singapore from 2-4 December, 1995. Sponsored by the United Nations Development Programme.

Shresta, Ram M., and Gopal B. Bhattarai (1993). "Electricity Planning with Demand-Side Management in Nepal". *Energy Policy,* Vol, 21, pp. 757-767.

Siddiqi, Toufiq A. (1995). "India - Pakistan Cooperation on Energy and Environment" To Enhance Security", *Asian Survey,* Vol. 35, pp. 280-290.

Siddiqi. Toufiq A. (1996). "Carbon Dioxide Emissions from the Use of Fossil Fuels in Asia: An Overview", *Ambio,* vio. 25, pp. 229 - 232.

Tata Energy Research institute (1995). *Teri Energy Data Directory & Yearbook 1995/96.* New Delhi: Teri.

United Nations (1996). *1994 Energy Statistics.* New York : United Nations, Statistics Division.

United Nations Development Programme (1996). *Enhancing Regional Cooperation in Sough Asia through Collaboration in Energy and Environment* New York: Regional Bureau for Asia and the Pacific, UNDP.

United Nations Environment Programme (1985). *The Environmental impacts of Production and Use of Energy.* Nairobi: UNEP.

CONVENTIONAL AND RENEWABLE ENERGY PROSPECTS IN IRAN

26

AMIR HOSSIEN GHORASHI
Center for Renewable Energies
Research and Applications (CRERA)
Atomic Energy Organization, Tehran, (IRAN)

ABSTRACT

This paper presents the recent energy scenario in the Islamic Republic of Iran, with an emphasis on utilization environmentally friendly energy resources in spite of very rich conventional sources of energy in Iran.

The paper includes future scenario in the Iranian context, technology gaps and some critical issues involved in some renewable energy conversion systems like wind, hydro feeding power to the local grid. This is supported by relevant data based on the extensive field experience in some wind farms in India.

The paper, besides being informative about energy prospects in Iran, it may also provide useful guidelines to energy planners in other Asian countries.

Introduction

In all the bulk of life, including energy processes, it is vital to work with less irreversibility. The more irreversibility or entropy generation, the more possibility of hostile environmental reaction, fossil fuel combustion was one of man's historical mistakes. During combustion process, high quality chemical energy, which all can be transferred to electric energy through ideal "cold combustion" is changed to low quality thermal energy. While, as the second law of thermodynamics states, it is impossible to transfer all these thermal energy to mechanical work or electricity. The difference as we know is high irreversibility, high rate of entropy generation, and consequently will result in waste of human lives and wealth, environmental disorder and disasters. Therefore it is unwise to continue in this way and certainly "Hot-combustion" must be replaced by "Cold-combustion" and meanwhile, conventional fuel substitution is unavoidable.

Today, the sustainable realistic development and growth is one which, does not have diverse effects on the environment and it should not damage the natural resources. The recent unrealistic developments by industrial countries has resulted many changes in the choice of mankind. Hence it pollutes the earth surface by injecting yearly more than 6 billion tonnes carbon dioxide into the atmosphere. With this, the world faces a very real and irreversible climate changes, whose consequences are very well-known and need no further comments.

On the other hand, the real value of fossil fuels is much more precious than simply burning them for their heat. Moreover, the life of oil reserves are limited and the share of the next generations is to be considered so that they may have better option to utilize these badly treated treasures. The present best option of utilizing fossil fuels is their Bio-products which can pay many times with compared to their hot combustion.

Iran is the second middle east country with respect to having various energy resources. Besides abundant conventional sources of energy like oil, natural gas and coal the country is blessed with large renewable energy resources like solar, wind, geothermal, hydro etc. Based on the government policies after the Islamic revolution in Iran, i.e. minimum extraction of fossil fuels and also cope with the technological progress taking into account the environmental considerations. That is why, the government in its Second Five Year Plan largely emphasizes on the developments of renewable energies in Iran.

The presented information in the paper regarding recent energy situation in Iran indicate reasonable energy balance in the country. However, in spite of such achievements in power generations based on mainly conventional energy resources, efforts are extensively carried out towards new achievements in utilization of renewable energies in Iran.

In this paper recent scope, status as well as future scenario of both conventional and nonconventional energies are presented referring to the elements of national programmes on the Iranian context with the emphasis on renewable energy programmes.

Conventional Energy Resources

There is no coordination between discovery of oil reserves and the consumption factor. Population growth parallel with economical developments both add to the energy requirements, hence increasing the Energy Intensity. For instance the Energy Intensity in USA, Japan and Iran is 0.8, 0.5 and 1.5 respectively. The main conventional energy resources in Iran are; Crude Oil, Natural Gas, Coal and some non-commercial energies.

Oil

The present known oil reserves in Iran are estimated to be around 39 billion barrels[1]. Iran's economy is to some extent dependent on oil export and the local consumption of oil is around 700,000 barrels per day. Therefore, the extraction from oil reserves is mainly related to the quantity of oil exported. Before the Islamic revolution the oil export was to extent of 6 million barrels per day. The present policy is entirely based on conservation of oil reserves for better utilization against non-calculative use of such great national economical and political backing. However, considering the increasing rate of the local consumptions the present oil reserves will last not more than 30 years from now.

Natural Gas

Iran is one of the richest countries in natural gas resources. The present known gas reserves are estimated to be around 17.7E12 m³ which is around 12.5% of the total world's reserves[1]. The gas utilization including export, mainly covers the local domestic, industries and power plants consumptions through widely spread pipe lines throughout the country. Further future scenario, besides boosting the export programmes, the extension of pipelines are to be extended even to the most inferior parts of the country. Considering only the local consumptions, the life span of the natural gas reserves are estimated to be around 200 years.

Coal

The capacity of the coal mines in Iran is estimated as about 6 billion tonnes. These mines are located in much deep layers of the ground and the extraction process becomes quite expensive. Hence, at present only 10% of the coal mines is extracted which is mainly consumed by the industries[1].

Non-Commercial Energies

These include the energies obtained from animal wastes, forest wood, bushes and other motes and stalks. Such energies cover more than 50% of domestic heating/cooking requirements of the rural population in Iran. Iran is an agricultural country and more than 50% of the population live in villages. In order to boost the agricultural products, improve the life standard in villages and encourage local handicrafts, the supply of electricity to the rural areas is an essential task. Therefore implementation of renewable energy programmes in rural areas can substantially provide their energy requirements.

Electric Power Generation

Iran with a population of 61.2 millions and an area of 16,48,000 Km^2 is capable of annually generating 1388 KwH electricity per head with an average annual growth of around 9%[1]. Excluding some rural areas with no approach to the national grid, more than 80% of Iranian families are provided electric connections. The numbers of Thermal, Gas and Hydro power generating stations in Iran are 15, 35 and 15 respectively. The latest data shows that the electric supply and demand in Iran is to some extent balanced. Table 1 indicates some details of power generations in Iran during year 1995-96 which is self explainer.

Nonconventional Energy Resources

Although, fusion energy is certainly an important and fatal option, of course with great technological challenges facing mankind, but development of solar energy systems and technologies are much easier to achieve. Solar energy including hydro, wind, biomass, geothermal and ocean energy are environmentally clean and non-problematic sources of energy and due to their technical simplicity are certainly one of the interesting options left for men.

TABLE 1 : Annual Power Generation in Iran (1995-96)

Type of Power plant	Output MW	Units 1000 MWh	Efficiency Max/Min	Ratio %
Thermal	11557	55901	98% / 95%	43.5%
Gas/Combined cycle	7746	16145	89% / 61%	29.5%
Hydro	1953	7275	100% / 100%	7.5%
Diesel	658	723	77% / 67%	2%
Industries	831	2034	-	3.5%
Others	3826	2891	-	14%
Total	26571	84969	-	100%

Results of a project concerning the "possible renewable energy consumption in Iran" shows that target of 20,000 MW is feasible in year 2020, of course, 10 percent renewable energy contribution in 25 years needs determination, data acquisition, technology achievement and planning. For this purpose crucial elements of a national programme were defined. These elements are in the form of three different types of projects:

— Fundamental projects leading to detailed analyses of renewable energy data, techno-economical analyses and feasibility studies concerning the availability and difficulties of renewable energy conversion technologies.
— R&D programmes and prototype projects for better understanding technical achievements and efficiency improvement these projects were defined regarding the state of art technology, national and international bodies recommendations.
— Joint venture cooperation and research programmes at regional and international levels intended to reach further technical improvements and technology transfer.

As it can be seen these projects are inter-connected and therefore the precise definition and sequences are very important, the results of these projects in less than 4 years, are the essential elements for policy making and a national development programme, obviously the results will determine priority, economic viability, future investments and the share and contribution of each source.

Solar Energy

All the available energies on earth both conventional and nonconventional are originated from solar energy and the Asian and African countries are abundantly blessed with these energy resources. Iran in spite of being very rich in conventional energies, has also great potential for large-scale application of solar energy systems. The country is covered 60% barren land with maximum solar radiations. The yearly average sun radiations is about 20-30 Mega-joules per m in a day. The sunny hours during four seasons are: 700 hours during spring, 1050 hours during summer, 830 hours during autumn and 500 hours during winter season, in short, through the utilization of solar energy received by 15510 Km^2 or less than 1% of its land. Iran can produce enough hydrogen to meet its present annual energy demand of 3.6 EJ, and export 6.77 EJ of clean energy, at the same level of 1992[2].

However, the introduction of solar energy into our energy system will take place over decades, that is, extensive penetration of solar energy and solar hydrogen into energy markets will not happen quickly for several reasons, such as the time lag for technology development and the development of large manufacturing industry, and a limitation on total investment capital available. Even an ecology driven scenario proposed by world energy council set renewable energy contribution to the world energy demand at 12% in year 2020[2]. The programme to install several solar energy systems in Iran will enlighten the deprived people in rural areas spread over 60,000 villages throughout the country. There are already several installed photovoltaic plans which are providing electricity to number of villages and telecommunication posts. Moreover, Iran has got its own production of photovoltaic panels, manufactured by the fiber optic industry in Tehran.

Wind Energy

Wind Energy as an exploitable resource is now a reality and not a dream as related by hundreds of rotating wind turbines installed in the length and breadth of many countries producing electricity and feeding to the local/national grids.

Persians were the first people to install the first wind mill some 200 years BC. Still, some of the very old wind mills are operating in rural areas of Khorasan province. This indicates availability of locations with high wind energy potentials in Iran. According to the recent wind energy survey in 15 different provinces in the country, the wind energy potential of about 50,000 MW is estimated[3]. The availabilities of such vast wind energy potential in Iran insures the possibilities of establishing both large wind farms, feeding bulk amount of power to the national grid as well as autonomous units for isolated loads. Presently two Wind Energy Conversion Systems of 500 KW each are installed at Manjeel and Roodbar in northern Iran, where the converted energy is fed to the local grid. The capacity factor obtained by these wind turbines is more than 60% and each turbine generates more than 3500 MWh electricity in a year. The CRERA is in the process of augmenting the present 1 MW power production to further 100 MW of which, the work of 10.1 MW wind farm has already been started. This project is planned under the assistance of Danish company (Nordtank) granting transfer of technology to Iran.

Besides the wind electric systems, there are many multi-blades turbines, pumping water mechanically for irrigation purposes throughout the country. However, it is a far journey to go in order to fill up the already created gap and to utilize the large wind energy potentials. It is essential that Iran should not fall back in this new field of technology. Therefore, all efforts have to be made to bridge the technological gap and achieve self reliance in wind energy exploitation to augment power generation in the country. The question today is not "why" to install a wind electric system but how to make it successful. While it is one thing to import a few systems and install in selected places but to make it operate successfully against all odds require tremendous technological and organizational skills. If the weather-dependent supply of wind power were to be integrated successfully into already existing supply structures and those which are to be developed. Besides the energy supply, the influence on the grids as well as on the consumers of electricity are becoming more and more important. This is mainly due to the reactive power consumptions by the wind electric machines. On the contrary the adverse effects of a poor grid on the wind farms are of great concern. Presenting the analytical cum-practical studies regarding such problems is out of scope of this paper and are already published[4-6]. However, Table 2 as a sample, indicates the effects of grid abnormalities on the performance of Lamba wind farm in India. These data are recorded by the author during November 1993.

TABLE 2 : Abnormalities of the Lamba Wind Farm

Causes of Wind Turbine Failures	Occurrence No.	Duration (Hr.)	Energy (MWh)
Low Voltage	100	12.5	0.315
High Voltage	88	10.3	0.26
Current Unbalance	22	10	0.252
Power Down	336	35.8	0.902
Remote Stop	100	19.5	0.492
Manual Stop	30	44.3	1.12
Feeders Failure :			
Feeder 1	34	2.13	0.966
Feeder 2	46	2.83	1.141
Feeder 3	50	3.67	1.48

Hydro Power

Primary estimations on hydro energy potentials in Iran is about 16,000 MW. There are 15 hydro power generating stations of total capacity of about 2000 MW, annually feeding around 7300 million KwH electricity to the national grids. It is aimed at increasing this capacity to further 5000 MW[1]. Besides, power generation the water storage behind huge dams solves the irrigation problems in many parts of the country. There are also many widespread mini-hydro units installed for local loads at rural areas.

Geothermal Energy

Iran is located on the global geothermal belt. This is evident by many warm water streams, volcanoes and regular ground movements. There are four main locations with the total area of more than 31,000 Km² which contain great amount of geothermal potentials. The government have already approved the installation of two geothermal power plants with equal capacity of 20 MW at Sabalan and Khoy province in Iran. At each locations, at least 5 to 10 geothermal wells are dug with 1400-2000 m depths[7]. These schemes are to be furnished by assistance of Philippian exports.

Biogas Plans

The importance of the biogas is its site independence against all other renewable energies which are very much site specific. Domestic and industrial sewage/waste, animal dung/waste, 80% (valid

% in Iran) of garbages and left over of agricultural products are the main sources of biogas energy. All these matters are also great sources of pollutión and carriers of many infectious diseases in and around every city and village which, may become epidemic mainly during hot seasons. These are the serious problems mostly faced by the third world countries. If other renewable energy resources are converted by the non-pollution-making systems, the biogas plan will convert the pollution source to energy and non polluted useful matters like, fresh fertilizer and water for irrigation. Hence, the focus is not to be on the production of energy or its quantity and in order to have a healthy leaving environment, the biogas systems are the best options. The CRERA is investigating the planning of a biogas systems at Seveh town 150 Km south of Tehran. This will be followed by another biogas reactor to be constructed in Isfahan with a purposed capacity of 10 MW[7]

Conclusions and Suggestions

The conventional energy systems are ecologically problematic and must be replaced by a clean renewable energy system. Meanwhile it is gradual and hard task due to the lack of recognition of the external cost of conventional energy use. This fact must be considered in microeconomic decision making. However it is noticed that ecônomic availability and capital constraints are the most important factors in any energy substitution programme, therefore, at first stage every renewable energy resources must be carefully examined and their potential and energy intensity for all locations should be determined.

Further, it is vital to work hard for technial achievements. In other words, for implementation of an effective programme and widespread application of renewable energy systems, essential elements of the required technologies must be made inside the country (Iran/third world countries). For this reason R&D programmes as well as joint venture projects leading to technology transfer are fundamental elements of a national renewable programme.

In short, energy substitutions need determination, thinking, planning and last but not least capital investments. It is a duty of the intrnational bodies to give proper supports to the implementation of renewable energy conversion programmes. Besides the development need more technical break through by increasing worldwide technical cooperation and fair joint venture research programmes.

Finally, fossil fuel subsidies is a universal barrier to the development of renewable energy systems, and its gradual elimination is accepted in Iran.

REFERENCES

1. Annual reports on "Electric Power Generation and Electrical Industries in Iran, 1995-1996 (in Persian).
2. Fundamental Elements, A report on "renewable Energy Programmes in Iran", Office of deputy minister, Energy Affairs, Ministry of Energy, Tehran Iran, March 1996.
3. Report on "Estimations of Wind Energy Potentials in Iran" Atomic Energy Organization Tehran Iran, 1994.
4. A. H. Ghorashi, "Investigations on grid connected induction generators relevant to wind/hydro energy systems", PhD thesis, IIT Delhi 1993.
5. A.H. Ghorashi, S.S. Murthy, B.P. Singh, Bhim Singh, "Analysis of Wind Driven Grid Connected Induction Generators Under Unbalanced Grid Conditions", IEEE/PES Summer meeting 1993, Vancouver, Paper No. 545-4 EC.
6. A.H. Ghorashi, B.P. Singh and S. S. Murthy, "Analysis of Wind Driven Grid Connected Induction Generators in Single Phasing Condition", Proc. ELROMA-1992 (International conference on electrical rotating machines). II, Session IV-A, Bombay, India.
7. Annual Report by CRERA, Atomic Energy Organization, Tehran, Iran, 1995.

DEMAND FORECAST FOR BIO-FUELS IN RURAL HOUSEHOLDS

27

I. NATARAJAN, Chief Economist
National Council of Applied Economic Research, India

ABSTRACT

Bio-fuels are the most dominant sources of energy in rural India. The energy surveys conducted are presented in the paper. Demand forecast, population projection and pattern of fuel consumption are dealt with.

Introduction

Bio-fuels are the dominant source of energy in rural India. In a survey done in India by the National Council of Applied Economic Research, NCAER[1] in 1978-79, it was estimated that bio-fuels accounted for over 90% of the heating energy used by the rural households. The major bio-fuels used are firewood, crop wastes and dung cakes. Firewood alone accounted for more than 55% of the total bio-fuel consumption.

The survey also showed that the rural households consumed over 125 million tonnes of bio-fuels in 1978-79. Of these 67 million tonnes were dung cakes and 30 million tonnes were crop residues. The balance of 79 million tonnes was firewood made up of 20 million tonnes of logs and 59 million tonnes of twigs.

Since 1978-79 the energy demand has grown due to both income and population increase and all the available indications show that the rural households continue to depend heavily on bio-fuels to meet their energy requirements.

During these years there had been an impressive increase in the availability of LPG and kerosene, the two petroleum fuels essentially used in the household sector. These apparently had not made any dent in the pattern of fuel consumption in the rural areas. The Council had conducted two urban surveys[2, 3] in the last 10 years to estimate the potential of LPG in urban India. The results of these studies show that the increased availability of kerosene and LPG was mostly absorbed in the urban areas.

Unfortunately, no comprehensive rural energy surveys at the all-India level were done since the 1978-79 survey of the NCAER. There were, of course, a number of regional or small area surveys done by different institutions to understand the pattern of energy consumption in different areas. Most of them are in the nature of case studies and therefore their results do not provide the basis for generalisation.

Nevertheless, various researchers and government bodies had in the past tried to forecast the demand for bio-fuels in general and firewood in particular making use of all the available material. The range of estimates provided by them was wide. For example, the estimates of firewood demand by the turn of the century varied between 160 million tonnes to over 350 million tonnes. Curiously, a large number of them used the results of the NCAER survey of 1978-79 as the base. Obviously, the methodology and the underlying assumptions were different.

In 1993, the NCAER completed a study[4] to evaluate the "National Programme of Improved Chulha". The findings of this study were based on data collected from a sample of 16,000 households spread over 1600 villages selected from all the districts in the country. As a part of this study data was collected from the sample households on the energy use and the pattern of fuel consumption. Based on this data an estimate of rural energy consumption and the pattern of fuel use was made and these were presented in a paper[5], to be published shortly. An attempt is made in this paper to use the results of these two NCAER studies to provide yet another estimate of demand for bio-fuels in rural India by the turn of the century.

According to the Improved Chulha study, the estimated consumption of energy for heating by the rural households in 1992-93 was 153.8 million tonnes of coal replacement (mtcr).

This is about 72% more than what was used in 1978-79. The pattern of fuel consumption for these 2 years is as follows.

Energy Consumption for Heating in Rural Households

Fuel	Unit	1978-79			1992-93		
		Qty	CR	Percent	Qty	CR	Percent
Coal/Soft coke	000't	1143	1,713	1.92	429	584	0.38
Kerosene	000'kl	414	2,279	2.55	1103	6,831	4.44
Dung cake	000't	66,755	20,087	22.51	86,732	26,151	17.00
Firewood, logs	000't	20,109	16,905	18.95	57.956	49.989	32.49
Firewood, twigs	000't	58,742	31,774	35.62	73,418	44,785	29.11
Crop wastes	000't	29,529	15,531	17.41	24.955	20,530	13.35
Others			913	1.03		4.971	3.23
Total			89,202	100.00		153,841	100.00

It is seen that the share of bio-fuels in the 2 years has remained unchanged at about 95%. Firewood consumption has gone up both in quantity and share. An interesting feature is the increase in the use of logs in 1992-93 compared to the earlier year. The survey data also showed that more people collected logs in 1992-93 compared to 1978-79.

Forecast for 1999-2000

Both the total heating energy demand and the likely pattern of fuel consumption need to be assessed for the year 1999-2000. It is relatively simple to forecast the total energy demand as its growth is more uniform. The approach adopted in this paper is, therefore, to forecast the total heating energy demand and then apportion this between different fuels based on available trends with a few assumptions

A number of factors affect the level of energy consumption. Two of the most important among these are income and population. To take care of the effect of population increase the following procedure is adopted. The likely average household energy demand is projected for the year 1999-2000. This figure is multiplied with the projected number of households in 1999-2000 to arrive at the demand for total energy in that year.

To estimate the average household energy demand for any future year, the income effect is to be taken into consideration. In order to gauge the effect of a rise in real (household) income on average energy consumption, it is necessary to know the income elasticity of demand for energy.

In the chulha study, data on household income was not collected. The Central Statistical Organisation (CSO) has estimates of personal disposable income at constant prices on a time series. However, these data are not suitable for the present purpose as the income data presented by the CSO is for both urban and rural areas put together whereas for the present exercise data on growth in rural household incomes is required. Moreover, a single elasticity of demand encompassing all levels of income for the entire population neglects the fact that income elasticity of demand for any product varies at different levels of income and also between different segments of the population.

In the absence of data on energy consumption at different levels of income, the following procedure is adopted to forecast the average household heating energy demand for 1999-2000.

Between 1978-79 and 1992-1993, the average household heating energy demand grew at 1.9% per annum. This growth may be attributed mostly to the increase in the annual average household income between 1978-79 and 1991-92. It is assumed that the period between 1992-93 and 1999-2000 also, the average household energy demand would grow at 1.9% per year. This essentially means that the income effect is taken to be the same in the two periods under consideration.

It is likely and there is also evidence for it that the income would grow faster in the coming years compared to the past. But what is relevant is how the income distribution would get affected through this growth. If bulk of this increase accrues to the affluent where the income elasticity for heating energy is likely to be low, the average household energy consumption may not grow faster than what was observed in the past. The results of some of the consumption surveys[6] done by NCAER in the recent past indicate that income in equality has increased in the rural areas. This suggests that the income growth is likely to be relatively more for the affluent.

The demand for average household energy for heating 1999-2000 estimated as above comes to 1550 kgcr.

Population Projections

The rural population of India was estimated at 628.7 million in the 1991 Census. The growth rate between 1981 and 1991 was estimated at 1.8% per year. The census report also reckoned the growth rate for the rural and urban areas put together at 2%. The report of the Expert

Committee on Population[7], estimated the growth rate of the Indian population between 1991 and 2000 under different sets of assumptions giving rise to 3 levels of projections namely low, medium and high. These projections were made prior to the 1991 Census. After the 1991 Census results were released, it was found that the committee's "High" projections were closer to the observed census growth rate. In view of this, it is proposed to use the growth rate under this alternative to estimate the population for 1999-2000. The report adopted an overall growth rate of 1.9% per year between 1991-2000. Rural and urban growth rates were not estimated separately for this alternative. It was therefore decided to adopt the same differential between urban and rural as evidenced between 1981 and 1991 for the years beyond 1991 also. On this basis the average annual growth rate for rural population between 1991-2000 come to 1.7%. The projected rural population in 2000 was estimated at 744 million.

Some of the recent surveys done by the NCAER, indicated the average household size in the rural areas to be 5.7. Using this average size, the estimated number of rural households in 2000 comes to 130.5 million.

With an average household heating energy demand of 1550 kgcr, the demand for heating energy in rural households in 1999-2000 is estimated at 202.4 mtcr. Pattern of fuel consumption

The next step in the exercise is to apportion this total energy demand between different fuels. It is assumed that the fuels used would be soft coke, kerosene, dung cake, firewood, crop wastes and bio-gas. Commercial fuels like LPG and electricity and other non-conventional energy sources are not considered as their share in the total energy demand is likely to be insignificant.

To assess the likely pattern of fuel use the following procedure is adopted. The demand for all the fuels other than firewood is estimated based on the likely availability and past trend. The residual is attributed to firewood. The method of estimation and underlying assumption are spelt out below.

Coal/Soft Coke

Soft coke is not a preferred fuel in rural India because of its high cost. Between 1978-79 and 1991-92 its consumption had dropped to a third. Its use is mostly in the eastern region near the coal mines where it is collected. This fuel therefore is not likely to have a future as a rural energy source. In view of this it is assumed that demand for this fuel in 1999-2000 would be at the same level as in 1992-93 at around 400,000 tonnes.

Kerosene

The use of kerosene for heating in rural areas is limited. Past studies have shown that it is not a preferred fuel for cooking. Its use is mostly confined to water heating for making tea, coffee etc. and bathing. Being a commercial fuel it has not made inroads into rural areas where households invariably obtain their fuel free of cost. Yet, the consumption of kerosene increased from 404,000 kilo litres in 1978-79 to 1103,000 kilo litres in 1991-92, a growth of around 7.5% per annum. Its share in total energy is still small at 4%.

The question is whether this growth can be sustained beyond 1992-93. The answer probably is "No". The use of kerosene as a heating fuel is mostly among the affluent. With the expansion of bio-gas programme, many are likely to switch over to the cleaner bio-gas. Further, with the introduction of parallel marketing of petroleum fuels, the price of kerosene is also likely to go up making this fuel unattractive to the consumer.

In view of this, it is assumed that the demand for kerosene for heating in rural areas would grow at 5% per annum. It may be noted here, that any change in this assumption is not likely to affect the estimated pattern of demand for different fuels as kerosene's share in the energy basket is very small. The estimated demand for kerosene under this assumption comes to 1700 thousand kilo litres.

Bio-gas

There is a growing demand for bio-gas in rural India. From a modest beginning in the early eighties, the bio-gas programme has made rapid strides and it is estimated that upto the end of 1995 around 2.2 million plants were installed in rural India. It is envisaged that this would reach a figure of 3 million by the turn of the century.

In an evaluation study[8] done by NCAER, it was found that around 85% of the plants installed are functional and each of them produce around 400 cu.m. of gas in a year. On the basis of these estimates, the likely availability of bio-gas by 2000 would be 1020 million cu.m.

Dung Cake

Dung cakes are an important source of energy in the rural households. Animal dung is also used as manure. The availability of dung is directly related to the cattle population in the country. The rate of growth of animals has been uniform around 1% per annum during the last three decades. It is assumed that the number of animals would continue to grow at the same rate in future also. Further, it is assumed that the proportion of dung used for making dung cakes would not change significantly during the next few years. With these assumptions, the likely use of dung cake in 1999-2000 was estimated using the same average annual growth rate of the animal population. No allowance was made for the possible diversion of dung to bio-gas plants. For the dung requirements to feed the additional plants (about 1 million) is very small (above 10 million tonnes of wet dung, equivalent of 2 million tonnes of dung cakes). Moreover, it is more likely that the dung used as manure is diverted to bio-gas plants, as these plants in addition to providing a cleaner fuel, also supply enriched manure. The demand for dung cakes estimated as above comes to around 93 million tonnes.

Crop Wastes

Burnable agricultural residues are the second major source of energy widely used by cultivators and farm labourers. The agricultural production witnessed impressive growth during the last

2 decades thanks to the green revolution. But the growth in the yield rates was essentially in foodgrain crops. The by-producers of these crops are mostly used for feeding the farm animals. The burnable wastes normally come from crops like cotton, pulses, mesa, oil seeds etc. There has not been any major breakthrough in the yield levels of these crops. The outputs of these crops continue to grow at the historical growth rate. It is assumed that this rate would continue till the turn of the century. The likely availability of crop wastes is therefore projected on the basis of growth rate observed in the past. The estimated demand for crop wastes in 1999-2000 is 38,000 tonnes.

Firewood

The likely contributions of all the other fuels estimated as above is 65.8 mtcr. From the total energy demand of 202.4 mtcr, this quantity is deducted to arrive at a residual of 136.6 mtcr which should be met through firewood.

To convert this energy into physical units of firewood, it is necessary to know the composition of firewood demand in terms of logs and twigs. The twigs burn faster and therefore are less efficient than logs. In other words one kg of log would give more useful energy than one kg of twigs.

The NCAER studies observed that the share of logs had increased in the total firewood from 25.6% in 1978-79 to 44.3% in 1992-93. This trend may continue further and it is assumed in 1999-2000 that logs and twigs will have equal share in the total firewood demand. Under this assumption the demand for fire wood in 1999-2000 would be around 182 million tonnes.

In the following table, the likely pattern of demand for different fuels estimated as outlined earlier is presented.

Demand for heating energy and likely pattern of fuel use in rural households 1999-2000

Fuel	Unit	Qty	CR '000't	Percent share
Coal/soft coke	000't	400	600	0.30
Kerosene	Th.k.l.	1700	10,540	5.21
Bio-gas	Mill.cn.m.	1020	3,570	1.76
Dung cake	000't	93,000	27,900	13.79
Crop wastes	000't	38,000	23,180	11.46
Firewood	000't	182133	136485	67.47
Total		202,275		100.00

Adjustment for Improved Efficiency

The estimates provides so far, have not made any allowance for any likely increase in the efficiency of fuel use in the future. Under the National programme of Improved chulhas fuel efficient wood stoves are propagated in the rural areas. Till the end of 1994, around 15 million chulhas were constructed. The last few years saw additions of around 2 million chulhas a year. If this trend continues, by 1999-2000, around 30 million chulhas would have been installed in rural homes.

These chulhas are smokeless and fuel efficient. It is claimed that an improved chulha is three times as efficient as a traditional chulha. If this is true and when all the 30 million chulhas are used in 1999-2000 then a dent would certainly he made in the demand for bio-fuels.

However, recent studies by NCAER show that the improved chulhas installed by the rural households are only 40% more efficient than the traditional chulha. Further hardly 50% of the chulhas are operational. Faulty construction and poor maintenance were cited as the major reasons for the poor performance. The study also noted that around 20% of the existing chulhas become non-functional each year. As a result the new constructions mainly replace the ones that go out of order each year leaving very little for net addition to stock. It is unlikely that this system would improve in the near future. Based on these it is estimated that out of the 30 million chulhas only 9 million would be in use in 1999-2000.

On the basis of a 40% higher efficiency, it is estimated that an improved chulha would save fuels capable of yielding useful energy of around 620 kgcr. For the 9 million chulhas in operation, the total saving would be to the tune of 5.5 mtcr. In terms of firewood it is around 7 million tonnes. Thus the total demand for firewood after allowing for likely improvement in the efficiency of use of fuels would be around 175 million tonnes in 1999-2000.

REFERENCES

1. Natarajan, I. - Domestic Fuel-Survey with Special Reference to Kerosene, NCAER, Feb. 1985.
2. NCAER - Demand for LPG in Urban Areâs, 1985 (unpublished).
3. NCAER - Demand for LPG in the Northern Region 1995 (unpublished).
4. NCAER - Evaluation Survey of National Programme of Improved Chulha, 1994.
5. Natarajan, I - Trends in Firewood Consumption in Rural India, 1995 (To be Published).
6. NCAER - Market Information Survey of Households (1986, 1987, 1988, 1990, 1992, 1993).
7. Planning Commission, Government of India - Report of Expert Committee on Population Projections, 1987.
8. NCAER - Evaluation Survey of Household Bio-gas Plants set up during Seventh Five Year Plan, 1992 (unpublished).

DECENTRALISED RURAL ELECTRIFICATION FOR SUSTAINABLE LONG TERM POLICY OPTION

28

I.S. ANAND AND S. N KAMRA
General Manager (PI), Dy Chief Engineer (PI)
Rural Electrification Corporation,
New Delhi, INDIA

ABSTRACT

The rural electrification policy is closely linked to rural development policy. The government has stressed social rather than financial criteria in the feasibility of rural electrification. The strategy to meet the lighting loads calls for midcourse corrections in the policy of rural electrification. Financial and organisational models are presented.

INTRODUCTION

Electricity forms on an average only about 10-15% of the total energy supply in the rural areas in most of the countries, yet rural electrification is virtually synonymous with modernisation of the rural energy sector. This is due to the most convenient form and versatility of electricity in meeting the varied demand of rural population when compared to other forms of energy. Most of the countries have given national mandate to power utilities to extend electricity to the rural areas as a part of their overall socio-economic development strategy. Thus rural electrification policy is closely linked to rural development policy and is strongly influenced by welfare objective of the planners.

In keeping with the political aims to ensure the basic energy needs of the rural population, who happen to predominate the urban population, the government has stressed social rather than financial or economic criteria in assessing the feasibility of rural electrification projects under the programme. Centralised grid development, which preceded the social objective of rural electrification, got prominence as it helped the country to develop its individual potential and urban electrification. The rural electrification, mostly was done through extension of grid network, even though in many cases, in remote areas, the decentralised power generation and distribution took prominence in initial stages, in absence of availability of grid nearby and prohibitiveness of

The views expressed are of the authors, not necessarily of the organisation they represent.

its extension for small load potential. Nevertheless, with coming up of big industries and other development programmes, the grid extension was justified and decentralised generation and distribution pushed over to oblivion.

In India, out of about 5,79,132 inhabited villages as per 1981 census, about 87% have access to electricity, mostly grid based with a few exceptions in North-East and other difficult and remote areas. The balance over 80,000 villages are posing challenges to the utilities forcing them to have a second look on the strategy of rural electrification. Majority of these villages, would still hope to get grid extension in next few years but as per rough assessment made recently, about 18,000 inhabited villages will offer themselves as candidates for decentralised generation and distribution system through various alternatives available.

ELECTRICITY FOR ALL

With the electrification of 5,01,784 inhabited villages out of 5,79,132 as per 1981 census as on 31.3.1996, (As per 1991 Census, there may be about 90,000 villages left for electrification), country has made electricity available at the doorstep of the rural areas,. Yet, only about 45 million out of 112 million households in rural areas are estimated to have the facility as on 31.3.1996. The strategy to ensure that all the households in the country are provided with grid quality electricity initially to meet their lighting loads in a definite time frame, calls for mid-course corrections in the policy of rural electrification. Based on the resources availability, Statewise strategies are required to strengthen the existing network, extend network where possible, employ decentralised power generation system based on locally available sources, ensuring that each household is provided the benefit of electricity at an affordable cost; comparable to the existing tariff or cost of saving effected in place of kerosene, candle etc. being used normally – but it should not exceed the tariff for the urban areas.

Incidentally, the balance villages remaining unelectrified are predominantly in lower population groups of upto 500 i.e. households around upto 100. The Statewise details are in Annexure-I. While rural electification will continue its course during the next 14 years with 2010 AD as the terminal year for cent per cent electrification, (as recommended by NDC Committee on Power) the funds required for electrification of 30,000 villages per plan, on an average would be of the order of Rs. 1,200 crores during 1997-2002 for distribution system in 24,000 normal villages and Rs. 900 crores in 6,000 remote villages with matching funds for Generation and Transmission. Similar amounts would be required for Tenth and Eleventh Plan at the present day cost data. A beginning could be made in the Ninth Plan with a target of atleast 3000 remote villages, 15 million household electrification which would require an allocation of Rs. 450 crores for villages and over Rs. 3000 crores for household electrification, commercial connections, LT agro based industries, street lights etc. through conventional and alternative sources. The major role would be played by the decentralised generation and distribution system in remote villages for bringing grid quality electricity expeditiously. The targets are achievable, provided funds are forthcoming with matching funds for sub-transmission. The generation and transmission would be separately provided. It can be expected that after 2010 AD– load intensification would be the thrust areas in all the villages, with experience gained in the already electrified villages in various States over the years. New and innovative usages, energy conservation methods, efficient and economic material and equipment etc. would lead to acceleration of the process; per capita consumption increasing with availability of additional generation to meet the demand.

EXPERIENCE GAINED

In the initial demonstration phase, in the 1980s, the programme focused on field deployment of some rural based applications of PV such as street lights, community lights and TV, water pumps and small capacity (2-5 KW) village power plants. About 40,000 systems were deployed in various parts of the country, implementation being done by State level nodal energy development agencies for renewable energy supported by State level utilities (State Electricity Boards) for street lights projects with loans from Rural Electrification corporation for items other than modules which were fully financed by MNES and for other systems with the support of manufacturers and users . However, as reported , regular tariff of utility is not being charged and ad-hoc charges are levied to sustain the maintenance and operation cost.

In a few States, the decentralised generation system have come up based primarily on diesel and small hydel. In these cases, utilities provide the operation and maintenance support and collect revenue as part of State Govt. programme.

The planning commission had identified about 10,000 villages to be provided with non conventional system for electrification during the Eighth Five Year Plan 1992-97. The State level energy development agencies who have so far been promoting the decentralised generation system have so far not generated enough confidence to approach State utilities and REC for taking up the programme in an organised manner on commercial basis.

INSTITUTIONAL FRAMEWORK FOR DECENTRALISED RURAL ELECTRIFICATION

The experience gained in the field of SPV system over the years led to a policy decision in 1993, to give new orientation to the programme. it was decided:

(i) To throw open the SPV industry to the private sector.
(ii) To remove barriers to foreign investment and foreign technology ;
(iii) To make budgetary support available to all suppliers of SPV device both public and private sector ;
(iv) To insist that State Government. agencies procure "complete system" from manufacturers who would give warranties and offer after sales service;
(v) To try to promote a non-subsidised market for SPV products while continuing to give financial support to certain applications in specific geographic areas.

For remote villages which are yet to be electrified there is thus a need to develop " Decentralised Rural Electrification (DRE) policy for each State. It should consider developing the financial and management models to suit the requirement of each group of remote villages within the overall Rural Electrification Policy for 10-15 years.

Financing Models

In the context of the emerging scenarios, two programmes need to be developed for next 10-15 years period :

(i) Socially oriented
(ii) Market oriented

Under the socially oriented programme, PV systems are promoted through subsidies and other measures, especially in rural areas to enable household electrification, street lighting and pumping systems helping the people to improve quality of life. MNES provides 50% subsidy for the Solar Photovoltaic (SPV) system for village electrification. Grid interactive programme with SPV systems, provides two-third of the total cost, subject to a maximum of Rs. 2.0 crores/100 KW. Balance amount have to be provided by SEB. The balance amount could also be provided by REC.

MNES provides interest subsidy for small Hydro Projects upto 3 MW station capacity in the hills, North Eastern region and andaman and Nicobar Islands. The maximum capitalised interest subsidy amount upto Rs. 38.3 lakhs/MW or the actual capitalised interest subsidy amount whichever is lower to bring down the interest rate by the financial institutions to 10% in hilly areas and 12.5% in other areas.

It is desirable to have special promotion scheme to be continued over 10-15 years period to give impetus to the development of small-Hydro schemes in the country with a policy decision on line with PURPA Act of USA under which all the generation of such projects would be mandatorily accepted by the utilities and paid for at agreed rates. This will help India to exploit about 10,000 MW of small hydro potential by the end of Tenth Five Year Plan.

While MNES is providing capital or interest subsidies and other incentives for project formulation, IREDA and REC provide loans for various decentralised system. Banks and non-banking financial institutions should also come forward to finance such projects. The financial assistance for decentralised power systems for rural electrification by various Government institutions are given in Annexure-II.

In all cases, wherever required REC can consider providing loan for the uncovered portion as also for evacuation of power and distribution lines.

Organisational Models

Ministry of Non-Conventional Energy Sources (MNES) is responsible for overall planning, formulation, implementation and evaluation of the programme of non-conventional sources of energy in India. The Ministry of Power (MOP) which is the core Ministry, is responsible for power generation, transmission and distribution including Rural Electrification. The responsibility of implementation of various programmes in case of non-conventional energy systems is primarily that of State level nodal agencies promoted by MNES.

In case of Rural Electrification, the implementation is with State Electricity Boards (SEBs)/

Power Departments, with finances made available primarily by Rural Electrification Corporation. The electrification of remote villages, with the availability of commercially viable based on decentralised non-conventional power systems, could be operated and maintained by the local people/organisations. There is an institutional understanding between, MNES, REC and Indian Renewable Energy Development Agency (IREDA), and the utilities. Joint Sector companies, Panchayat Raj institutions, Energy cooperatives, NGOs etc. could be considered at the grassroot level with adequate support from the Nodal agencies and financing institutions.

In order to develop the strategies leading to commercial operation of the decentralised power systems, the viable organisational models need to be identified and encouraged. The leasing of equipment, down payment of partial cost and servicing fees per month model could be considered.

ROLE OF UTILITIES

The Electric power Utilities are tuned to the concept of project approach promoted by Rural Electification Corporation over the last 27 years. The programme that would be acceptable immediately, would be where the villages have low tension network compatible with existing standards governed by Indian Electricity Act. As the ultimate goal is to provide grid quality electricity to each household may be by 2010, the Electric Power Utilities find that there are number of localities which defy grid-extension due to various problems and constraints and, therefore, would welcome decentralised energy systems which can be introduced as part of integrated RE in the area.

The decentralised generation source may be mini/micro hydel or village level solar packs or biomass gasifier systems. These should be managed by a dedicated team/cell and look after the problems of such generation-cum-distribution systems. At par with grid system, they could release connections to domestic and commercial categories or others if solar photovoltaic panels are provided with sufficient capacity and recover commercial tariff prevalent in the State, providing services as per O&M plan. Such villages, would be several Kms. Away from the grid lines, with low load potential, very few houses and identified to be served through photovoltaic rural electrification atleast for 5-7 years period. Studies have shown least cost compared with Grid Electricity when usages are little and distances are manifold.

It is visualised that apart from LT network fed through solar power pack, the introduction of solar home packs and solar lanterns in the households could be entrusted to the utility on commercial basis. The consumers located away from the feeder lines both in the villages being electrified and those already electrified in remote areas, could avail the services. The battery charging stations at a central place, leasing of the solar lanterns on monthly rental basis and other innovative financing could be thought off.

The State utilities could encourage small local cooperatives or societies formed to provide service to the consumers in the project area. These service societies could be developed into an "economic model" where other activities leading to the income generation for the community and individuals would be encouraged so that societies are self-independent and could slowly take over the extension of systems on a technoeconomic viability basis. The main strategy would, therefore, would be to undertake the following:

(i) "Carry out field survey and collect data for each such village as under:

Sl. No.	Name of village/ locality	Census Code No.	Population at present	No. of household at present	Distance from electrical grid in Km.		Tentative requirement		
					HV	LV	No. of H/H	Other loads	Cumulative System loads
1	2	3	4	5	6		7	8	9

No. of hours of operation required	Optimum Array required	No. of LT feeders required	Total Dist. network in Km.	Approx. cost (Rs. in lakhs)
10	11	12	13	14

(ii) Indentify and select the villages which are not covered under regular RE programme; which are far away for 11 KV and LT lines; and where utility does not plan extending network in 5-7 years horizon.

(iii) Obtain the detailed 11 KV and 440 volt line network for the area and the plans of utility for short term and long term.

(iv) Obtain data for feasible non-conventional power systems including solar photovoltaic system for the village to meet the identified loads.

(v) Develop low tension 220 volt - AC network for the village keeping the configuration of loads into view and providing isolated houses, stand alone systems.

(vi) Develop management model for the village to operate and maintain the systems on sustainable basis.

(vii) The State Electricity Board's services be utilised for formulating a viable project for a group of villages, as part of PV RE in the State.

(viii) The consumers necessarily will have to have grid compatible internal wiring and the network laid in the village also as per conventional system so that the village is ready to receive grid extension, as and when incomes and aspiration of people rise, creating a demand for higher level of energy services and the utility is in a position to serve the same. In such a case, the existing low tension network would be available and no expenditure need to be incurred.

(ix) The project be posed for being financed by REC and MNES.

The utilities would undertake the programme as part of decentralised rural electrification policy to supplement Rural Electrification. It would be appropriate to lay the tariff policy under the PV and other decentralised power systems as distinct from the grid system to ensure that people in remote villages also avail reasonable services by paying marginally more but at the same time having the satisfaction of meeting at least minimum basic electricity needs in the house.

The project approach, envisaged by REC would ensure back-up for the services through the turn-key contractor who would implement the project and maintain it for 5 years initially to train local agencies to take over. Initially the utilities would be slow in taking up the programme, as such, the innovative methods of financing under a package of subsidies, grants, low cost funds, pilot projects would be introduced and utilities encouraged to develop their own models for sustainable implementation, operation and maintenance, coordinating with the efforts of other developmental agencies to provide health sevices, drinking water, school class room lighting etc. In fact, while formulating the project, the financial inputs of various agencies would be identified to meet the load requirement and to develop tariff structure.

PILOT PROJECTS

In view of the electrification of about 18,000 villages through alternative generation and distribution systems being contemplated, it would be appropriate to launch pilot projects with active involvement of REC, SEBs and bilateral and other agencies so that the models developed are replicated on large scale after gaining confidence.

The pilot projects will have to be preceded by sample energy surveys in typical areas such as hilly, deep forests, desert, riverine and such other difficult and remote areas,. The detailed project reports, formulated thereafter for financing would identify the organisational and maintenance models backed up by financial arrangements so that the projects basically implemented by SEBs or other such agencies could cater to the consumers at or similar pattern as of the existing consumers of the SEB with innovative and better service model because of remoteness.

Rural Electrification Corporation is in touch with agencies to bring in low cost funding for the pilot project initially and to develop long range policies based on the same.

CONCLUSIONS

Decentralised rural electrification of selected villages in remote and difficult areas in India, is being embarked, as part of the overall Rural Electrification policy, under which all the households are proposed to be provided basic household lighting facility by 2010 A.D. The efforts made by State level energy development agencies are required to be upgraded by synergetic merger with the programmes of State level utilities, who would have to adopt project area approach as distinct from dispersed provision of PV and other systems, to ensure that villages have a regular operation and maintenance programme, justifying the techno economic selection of such villages for sustained overall development in the years to come. The low tension lines for A.C. Operation would be welcomed by utilities, who can bend a little to accommodate decentralised power systems, entrusting the operation of dispersed systems to State level energy development agencies for the time being, who could involve NGOs and other agencies for operation and maintenance. The systems could be upgraded, formed as a mini-grid, integrated as a hybrid system or integrated with local electricity grid, depending upon the aspirations of the people by virtue of their economic upliftment, load growth and expectations of higher quantity and quality of electricity supply leading to better standards of living in the rural areas, matching with urban standards.

ANNEXURE-I

VILLAGE ELECTRIFICATION IN INDIA

(Population Groupwise)

As on 31.3.1996

Village classification by population group	INHABITED VILLAGES AS 1981 CENSUS						
	Total No. of villages	Elecfd. as on 31.3.94 as per 1981 census	%age	Elecfd. as on 31.3.96 (projections)*	%age	Remaining villages as per 1981 census	%age of remaining vill. to total villages
1	2	3	4	5	6	7	8
Below 500	283067	233476 + 1792@	83.1	236106 + 1792	84.0	45169	58.40
500 - 999	141878	124659	87.9	126557	87.9	15321	19.81
1000 - 1999	97498	84166	86.3	85684	88.2	11814	15.27
2000 - 4999	47626	41521	86.8	42660	88.6	4966	6.42
5000 - 9999	7228	6770	93.7	7150	98.9	78	0.10
1000 and above	1835@	1807	98.5	1835	100.0		
	579132 (583003)+	494191	85.3	501784	86.6	77348 (80074)**	100.0

Note: As per "General Review" of Central Electricity Authority (CEA) (1993-94) based on 1981 Census. Data as per 1991 Census is under compilation.

* Based on projection made by the author. (CEA has reported electrification of 5,01,784 villages as on 31.3.1996.

\+ Figures of totals in col. 2 in brackets include Tripura State's 1971 Census figures of 4,727 villages. even though there are only 856 villages as per 1981 Census.

@ Breakup not included by CEA for 1792 non-census villages reported electrified in Assam. Included in below 500 category by author.

** Taking into account 1151 non-feasible villages left under 100% States/UTs and 6 villages excess reported by Mizoram net villages to be electrified comes to 80,074 as per 1981 Census.

HIGH SULPHUR COALS OF INDIA

29

S.K. Srivastava & S.K. Verma
Central Fuel Research Institute,
India

ABSTRACT

Presently, high sulphur coals are being used in industries and in power generation. The paper deals with the recovery of sulphur, desulphurization, and R & D processes.

Introduction

In the Indian context, coal forms an important source of fuel and it will continue to dominate the fuel scene for several decades. The studies on the reserves[1] and availability of Indian coals reveal that broadly (i) gondwana coals have 5300.00, 22600.53, 1799.71 and 165460.67 million tonnes of reserves of prime coking, medium coking, semi/weakly coking and non-coking coals respectively, (ii) tertiary coalfields of North-Eastern (NE) region have 325.89 and 538.89 million tonnes of reserves of coking and non-coking coals respectively, and (iii) tertiary lignite fields of Neyveli, Rajasthan, Gujarat, Kashmir and Kerala have 4900.00, 564.00, 324.00, 90.00 and 100.00 million tonnes of reserves respectively. The proportion of sulphur in coal varies widely from as low as 0.2% to as high as 14% but Indian coals are usually low in sulphur content, generally well below 1%. Amongst the above regions, coals from NE region coalfields, Wardha valley Coalfield, part of Godavari Valley Coalfield and sizeable quantity of lignite from Gujarat, Neyveli, Rajasthan and Kashmir have high sulphur, mostly in inorganic forms, except in the case of NE region coals of India where the organic sulphur content is almost 75-90% of the total sulphur, which is unique[2].

High Sulphur Coals

Presently, the high sulphur coals are being utilised in thermal power stations, brick making, tea drying, making iron and steel, cement, paper and textile industries, and for industrial/domestic heating. Although NE region coals are of good quality, but they are not being utilised efficiently because of the detrimental effects due to the presence of high sulphur. Since the rate of coal utilisation for heat and power generation besides coal conversion processes, is increasing enormously, the problems caused by the sulphur in coal are also expected to be immense.

Recovery of Sulphur

Although high sulphur coals are less in quantities in comparison to the total reserves of coal in India, its utilisation and recovery of sulphur in useful form are of utmost importance in view of the following[3]:

(i) Conservation of prime coking coal by blending low ash high sulphur coking coals for iron and steel industries.
(ii) Reduction in the import of sulphur in India by recovering elemental sulphur.
(iii) Pyrites recovered from physical cleaning of high sulphur coals for manufacturing sulphuric acid.
(iv) Recovery of sulphur (solid/gaseous) in the form of chemicals (e.g. gypsum, ferric sulphate, etc.) for their utilisation in different industries.

Although a large number of methods have been tried to remove organic sulphur in coal, so far the maximum sulphur removal achieved was only about 25%, that too under stringent conditions of temperature and pressure, primarily because the sulphur functionalities present in the coal structure are not clearly known[4].

Sulphur content in coals and lignites is variable and is not rank dependent like other elements such as carbon, hydrogen and oxygen. Depending upon their utility, sulphur content in the high sulphur coals/lignites have to be defined. It is known that all over the world the coals/lignites mined are mostly being used for heat and power generation. In India to, except for the high sulphur coking variety coals of NE region, all other high sulphur coals are being used for heat and power generation. The coking variety NE region coals may be desulphurized and blended with other coking coals for coke making in iron and steel industry. Except the high organic sulphur coals of NE region, all other coals have high inorganic sulphur which can be removed by beneficiation.

Thus, a knowledge of organic sulphur distribution in coal is a must for prediction of extent of desulphurization of coal. Since no method was available for organic sulphur distribution with quantitative sulphur material balance, a method for estimating all the organic sulphur functional groups present in coal using a combination of temperature programmed reduction and controlled nitric acid oxidation was successfully developed by CFRI[5,6]. Table-1 summarises sulphur evolution temperatures, amount of organic sulphur functional groups present and sulphur material balance for Tipong coal. Based on above studies on Assam Coals, it was concluded that (i) Sulphur is present in Assam Coals in the forms of disulphide, thiol, thioether, thiophene, pyrite, sulphate and thioketone, (ii) as the total sulphur content enhanced, percent thiophenic sulphur increased, (iii) thioether content in various coals were almost similar, (iv) thioketonic sulphur was found to be more in coking variety coals in comparison non-coking coals, and (v) disulphide was present in less amount in coking variety coals than in non-coking coals.

TABLE 1 : Organic Sulphur Distribution of Tipong Coal

Sl. No.	Sulphur Evolution Temperature Range (°C)	Sulphur Evolved (%)	Sulphur Functionality Assigned	Material Balance for Sulphur
1	190-220	0.36	Disulphide	Total of Sl. Nos 1 to 4 & Sl. Nos. 7 to 9 = 5.8
2.	260-290	1.70	Thiol	Total Sulphur by IS Method =5.7
3.	360-390	0.51	Thiother	
4.	460-490	0.88	Thiophene	
5.	510-540	0.35	Part of Pyrite + Part of Complex Thiophene	
6.	590-620	0.24	Part of Complex Thiophene	
7.	-	0.46	Pyrite	
8.	-	0.11	Sulphate	
9.	-	1.80	Thioketone	

Sl. Nos. 1 to 6 from Temperature Programmed Reduction (TPR), Sl. Nos. 7 & 8 from IS Method, Sl. 9 from Controlled HNO_3 Oxidation Method.

The understanding of redistribution of sulphur among the gaseous, liquid and solid products of the reactions is very important as far as the sulphur removal/fixation is concerned[3]. The variables that affect the redistribution of sulphur during the reactions among coal and gases may be groupd into two categories: (1) variables that are a function of the inherent properties of the coal such as volatile matter, ash and its composition, sulphur content and its forms, and (ii) variables that depend on the processing conditions such as time and temperature programme through which the coal passes, type of gas in the reactor and its pressure, coal particle size, reactor type and the superficial gas velocity.

Desulphurization

Desulphurization of high sulphur non-coking coals can be done before, during and after combustion. There are various methods available viz. Physical, Microbial and Chemical including pyrolysis, to desulphurize coal prior to use. While physical and microbial cleaning processes remove pyritic sulphur, chemical cleaning processes remove part of organic sulphur also.However, from economic point of view dry/wet beneficiation of coal/lignite would be more useful. Separation is done on the basis of specific gravity difference between coal/lignite and mineral matter as well as on difference in hydrophobic/hydrophilic character of the coal/lignite and mineral matter[3]. The other physical cleaning processes commonly used are Froth floatation, Oil agglomeration, Magnetic separation, Electrostatic separation etc. Microbial cleaning process is a slow one and requires high selectivity i.e.one kind of bacteria can remove only one type of sulphur functional group. Although upto 19% of organic sulphur has been claimed to be removed by microbial means[7] but because of its very slow rate of desulphurization and high selectivity this process has not reached to a commercial scale. Chemical cleaning processes are divided into 6 categories viz. Oxidation,

Reduction, Pyrolytic, Solied-gas reactions, Microwave and Radiolytic. Desulphurization of coal prior to use offers the following potential advantages: (i) Wear and tear as well as corrosion of equipments could be minimum, (ii) Additional investment in the form of flue gas desulphurization unit in power plants can be avoided, and (iii) Chances of spontaneous combustion could be minimised in storage.

High sulphur coking coals of NE region have to be desulphurized in such a way that the coking property is not destroyed and the desulphurized coal can be used as blend constituent for coke making in iron and steel industries. All oxidative processes are known to destroy coking property, hence hydrodesulphurization process may be suitable. In case of NE region coals, pyrites cannot be separated by an wet/dry beneficiation processes as it is highly disseminated in the coal maceral. A method for coal desulphurization using a combination of selective crushing and screening and wet beneficiation has been developed by CFRI[8]. This worked very well with effective desulphurization of high sulphur Central India coals.

The sulphur content of coal/lignite may not come down to below 1% even after wet/dry beneficiation of high sulphur coal with very good yield. Therefore it is necessary to have desulphurization during combustion. With the addition of lime, while carrying out combustion in Atmospheric fluidised bed/Pressurized fluidised bed/Circulating fluidised bed about 80-90% sulphur is removed and is fixed in ash. The flue gas contains only small amounts of sulphur dioxide and nitrogen oxides due to lower temperatures being maintained in above processes. Various flue gas desulphurization processes viz. wet scrubbing, dry scrubbing, catalytic reduction etc. are post combustion methods for sulphur/nitrogen removal. Any process my be economically viable by recovering and selling sulphur in useful forms and other by-products. During the process of wet/dry beneficiation, pyrites are recovered which is a source of Sulphuric Acid/Ferric sulphate. During combustion, desulphurization generates "Suphur fixed" ash which has potentiality to be used as fertilizer for sulphur deficient soils, citrus plants/fruits and also in cement industry because of formation of gypsum. The oxides of sulphur and nitrogen can also be converted to fertilizers by lime-sulphur reactions. The hydrogen sulphide and sulphur dioxide evoloved during hydrodesulphurization and oxidation processes can be converted into elemental sulphur through known processes. Threatment of high sulphur coals/lignites with air and steam would produces elemental sulphur in addition to desulphurized products[9].

Use of Coals in Industry

High sulphur coals are being used in thermal power stations, cement, paper, textile and iron and steel industries, brick making, tea drying, reilways and for industrial/domestic heat generation. The higher the sulphur content in a coal, the greater are the problems of its utilization and more are its adverse effects on atmospheric pollution. Thus, a prior evaluation of the nature, total sulphur content and distribution of sulphur in coal as well as the available practical techniques for desulphurizations are most essential for knowing the magnitude of problems in utilising high sulphur coals of India. For example, thermal power stations may tolerate sulphur content upto 0.7% but in cement industry high sulphur coals upto about 2% can be beneficially used. Sulphur compounds in industrial gases cause corrosion in industrial equipments. Sulphur also poisons the catalysts wherever used in the processes. Similarly in coal carbonisation industries, 30-40% of the total sulphur in coal is generally found to be converted to hydrogen sulphide and 1-2% to organic sulphur compounds. So the sulphur content in coal and of its coke, to be used in iron and steel industry, are nearly of the same order. Similarly to be within the limits of Environmental Pollution Control Act, the sulphur content in coal to be used in industries for heat and power generation should be of the order of 0.4–0.7%.

The magnitude of the problems in utilising high sulphur coals/lignites in different industries can be reduced if the sulphur content of the feed coal is reduced to the minimum. Central India coals are high in inorganic sulphur which can be removed to a great extent through beneficiation. In cases of Eastern and Western parts of Wardha Valley Coals, blending is necessary[8] for rational utilization of coals because western part of Wardha valley coals are of high sulphur content (1.5–4%) while Eastern part coals are having upto 1% sulphur content. NE region coals contain very high amounts of organic sulphur which cannot be removed by beneficiation. Blending is the only option left for coking variety coals of NE region. Non-coking coals of NE region may be utilised for heat and power generation using Fluidised Bed Combustion technologies in presence of lime. High sulphur lignites of India can be desulphurized using combination of dry beneficiation and treatment of lignite with air and steam[9]. Problems, such as dust pollution, spontaneous combustion during storage of high sulphur coals/lignites, water pollution due to wet beneficiation of high sulphur coals cannot be avoided because these are linked with utilisation of these coals.

R & D Processes

The various R & D processes on coal/lignite desulphurization[3, 10] are LOL Process, AMES Process, PETC Oxydesulphurization process, ARCO Promoted Oxydesulphurization Process, Meyers Process, JPL Chlorinolysis Process, KVB Process, BHC Process, TRW Gravimelt Process, Magnex Process, IGT hydrodesulphurization process, Chemical Comminution, Microwave treatment process, Radiolytic Process and Combined Gamma Ray-Microwave treatment process. In the TRW Meyers Process, only pyritic sulphur is removed, but other oxidation processes convert various organic sulphur functional groups into sulphones and sulphoxides depending upon the degree of oxidation. Radiolytic process is basically an oxidation process while microwave treatment[10] is a heat treatment of coal but if it is done in presence of sodium hydroxide, it again becomes an oxidation process. The processes using alkali in different concentrations lead to oxidation of sulphur groups. IGT hydrodesulphurization process is basically heating of high sulphur coals in presence of hydrogen to eliminate sulphur as hydrogen sulphide. Air and steam treatment of high sulphur coals/lignites at higher temperatures would be suitable for reducing pyritic sulphur as well as part of the orgainc sulphur content of coal. Most of the processes mentioned above are as yet limited to R &D stage of development. Therefore, lots of efforts are needed for making some of them cost effective.

Studies on chemical desulphurization of some high sulphur coals were carried at CFRI. Results on treatment of coal with distilled water led to the conclusion that about 90% of the sulphate sulphur present in Assam coals are water soluble and by such treatment the moisture content (air dries/60% RH, 40°C) of the residual coal can be brought down[11]. When such a work was extended to lignites/coal fine slurries, the results were encouraging. Treatments of high sulphur coals/rejects with ferric sulphate solution removed about 90% pyritic sulphur from original coal and ferric sulphate was recovered[12]. Under optimum conditions at 900°C and 30 min. time coals from Ledo, Baragolai, Tipong, Dilli and Jeypore Collieries of upper Asaam were pyrolysed. The coke yields and the total sulphur reductions obtained were 63.7%, 56.4%, 60.5% & 60.5%, 58.0% & 58.4%, 56.5% & 59.5% and 55.5% & 70.9% respectively. Distribution-wise about 80% sulphate sulphur, 85% pyritic sulphur and 50% organic sulphur were removed[13]. Ten coal samples viz. Jeypore, Ledio, Tipong, Bellora, Bapung, Majri, Tikak, Bapung (middle), Baragolai and Dilli, and one reject (Sp.gr.>1.8) from Wardha Valley viz. Rajpur were treated with chlorine in presence of water resulting in reduction of about 65%, 36%, 42%, 50%, 40%, 76%, 22%, 27%, 23%, 31% and 78% sulphur respectively. Simultaneous reductions in ash and moisture contents were also observed[15].

When high sulphur coals were subjected to Hydrodesulphurization (HDS) in presence of (i) tertralin at 200°C, and (ii) anthrancene oil at 280°C, under normal pressure, about 20-25% sulphur could be removed from coal. Similar treatment (in tertralin) under pressure (and in presence of Co-Mo Catalyst) from 40 to 310 atm and at temperatures of 300 and 350°C led to the conclusion that the temperature of the reaction should be below 300°C and the hot pressure should be in the range of 40-60 atm for removing sulphur from these coals in such a way that the destruction of the coal structure is minimum and the coking properties are not destroyed. Further experiments

TABLE 2: Coal/Lignite Desulphurization using various methods

Reagent	Reaction conditions	Sulphur Removed (%)	Ash Removed (%)
Cupric chloride	10% Concn, 200°C, 1.2 atm, 1 hr.	33-72	37-86
H_2O_2	Room Temp. (RT), 30 min.		60-65
	3% Concn.	17-41	
	6% "	19-50	
	15% "	28-55	
	30% "	36-95	
NH_4OH	RT, 30 min.	22-24	60-61
$K_2Cr_2O_7$	Reflux, 30 min.	24	61
NaOH	Reflux, 30 min.	23	63
CH_3COOH	RT, 30 min.	23	14
NO_2	RT, 5 min.	29	65
Pyrolysis	900°C 3 mm size, 30 min.	57-71	-
Chlorinolysis	100°C, 3 hrs.	23-78	-
Ferric Sulphate	Reflux, 1 hr	more than or equal to 90 (Pyrite only)	-
Water	Reflux, 1 hr.	82-93 (Sulphate only)	-
Hydrodesulphurization,			
Tetralin	200°C, 1 atm. 30 min.	21-48	-
Anthracene Oil	270°C, 1 atm, 30 min.	25-33	-
Tetralin	300-350°C, 1 hr, Co = Mo Catalyst, 41–310 atm.	27-100	-
Anthracene Oil	270°C, atm, 1 hr.	25-40	

on HDS of 10 high sulphur Assam coals in anthracene oil at 280°C and 40 atm hot pressure revealed that except for Ledo coal, in all other cases the total sulphur content of the desulphurised coals are upto 3.3% with coking properties[2]. Thus, the desulphurized coals can be used as a blend constituent for coke making in iron and steel industry.

Studies on simultaneous desulphurization and demineralisation of high sulphur Indian coals/ lignite were carried out using various inorganic reagents[16] viz. Cupric Chloride, various concentrations of H_2O_2, NH_4OH, CH_3COOH, $K_2Cr_2O_7$, NaOH and NO_2. Amongst them H_2O_2 was found to be most suitable in removing 50-90% of sulphur from coal at room temperature. The organic sulphur was found to be difficult to remove. About 50-55% of mineral matter originally present in coal can also be removed simultaneously. Coking properties were destroyed while using all the reagents except cupric chloride.Table-2 presents the data on extent of desulphurization and demineralisation on treatment of high sulphur coals with various reagents under different raction conditions.

A radiolytic method, involving irradiation of acidic aqueous coals slurry by gamma-rays for simultaneous desulphurization and demineralisation of 4 high sulphur coals, was tried at CFRI. It was observed that the removal of inorganic sulphur, organic sulphur and mineral matter were in the range 40-75%, 35-75% and 20-35% respectively[17].

REFERENCES

1. Annual Report on Reserves of Indian Coals, Geological Survey of India, Calcutta, 1995.
2. S.K. Srivastava, Technical Report No. TR/CFRI/3.01/93, "Chemical Desulphurization of high sulphur Assam Coals"
3. S.K. Srivastava and S.K. Verma, "Monograph on High Sulphur Coal of India". Ed., P.C. Kumar, Allied Publishers, New Delhi, 1996 (to be published).
4. S.K Srivastva, A. Kumar and R. Haque, Fuel Sc. Tech., 11, 111, 1992.
5. S.K. Srivastava and A. Kumar, Fuel Sc. Tech., 13, 71, 1994.
6. A. Kumar and S.K. Srivastava, Fuel, 71, 718, 1992.
7. D. Chandra and A.K. Mishra, "Removal of Sulphur from Assam Coals by bacterial means" in Bioprocessing and Biotreatment of Coal, Ed. D.L. Wise, Marcel Dekker Inc., New York and Basel, 631-652, 1990.
8. "Studies on the possibilities of Removal of Sulphur in Coals of Western Parts of Wardha Valley Coalfield", Report submitted to WCL, Nagpur, 1980.
9. S.K. Srivastava, "State of Art on Desulphurization Technologies", Report submitted to NCB, Ballabhgrah, 1996.
10. S.C. Tasai, Ed., "Fundamentals of coal Beneficiation and Utilization", Elsevier Scientific Publ. Co., Amsterdam 1982.
11. S.K. Srivastava, A. Kumar and A. Ali, Fuel Sc. Tech. 12, 35, 1993.
12. S.K. Srivastava and J. Sinha, Fuel Sc. Tech. Intl., In Press.
13. S.K. Srivastava, J. Sinha and A. Ali, "Pyrolytic Desulphurization of High Sulphur Assam Coals of India". Proc. of 6th Australian Coal Science Conf., New Castle, Australia, 1994.
14. S.K. Srivastava and J. Sinha, "Hydrodesulphurization of High Sulphur Assam Coals of India", *ibid.*
15. A. Ali, S.K. Srivastava and R. Haque, Fuel, 71, 835, 1992.
16. S.K. Srivastava and J. Sinha, "Aqueous Chlorinolysis of High Sulphur Coals of India", Fuel Sc. Tech. Intl., In Press.
17. K.K. Mishra, P.S.M. Tripathi, L.C. Ram, S.K. Jha, R.R. P. Roy and G. Singh, "Adhigandhakiya Koyle Ka Vigandhakan: EK Audyogik Awashyakata Rashtriya Vikas, CFRI, Dhanbad, 18 Nov., 1995.

ACADEMIC SPECIALISATION IN INDIA IN THE FIELDS OF POWER AND ENERGY

30

BALAKRISHNA K. NAYAR
Vice President
Indian Association for the Advancement of Science
New Delhi (INDIA)

ABSTRACT

With its wide network of educational institutions, India has at present an intake of 2,000 persons at the post graduate level in electricial engineering and connected fields, 15,000 at the graduate degree level and 22,000 at the diploma level. Specialisation at the Master's Degree includes 17 subjects in the field of power engineering and 13 subjects in energy engineering. A number of multi-disciplinary schools and departments have been set up in a dozen or so of the leading technological institutions in the country which produce highly qualified personnel in a wide range of specialities. India is today in a position to provide academic specialisation of a very high order to talented students and key personnel from foreign countries.

Introduction

The planning, production, distribution and effective utilisation of energy involves the participation of wide range of expertise. Engineers, particularly electrical engineers, have a pre-eminent role. The present paper is concerned with the status of academic training of elctrical and allied engineers at various levels in India.

Degree and Diploma Levels

Electrical engineering continues to be a major branch of study in technical education in spite of its branching into communication and electronics engineering, and blossoming more recently into computer science and engineering. The sanctioned intakes into Electrical engineering at the diploma and degree levels in India in 1996-97 are given in Table 1.

The sanctioned intake into electrical and connected courses at the diploma level was slightly over 22,000. Nearly three-fourth of the intake was in electrical engineering proper, spread over 370 institutions. Electrical and Electronic engineering with a sanctioned intake of about 6,000 in 114 institutions had a share of 27%. There were no other branches of electrical engineering at the diploma level.

TABLE 1: **Intake at the Diploma and Degree Levels into Electrical Engineering and Connected Courses 1996-97**

Subject	Level	
	Diploma	Degree
Electrical Engg.	16,055	6,153
Electrical & Electronic Engg	6,007	8,064
Electrical & Communication Engg.		300
Electrical Engg. (Power)		60
Electrical Power System Engg.		90
Electrical Engg. (Power)		10
Power Plant Engg.		
Power Electronics		90

At the graduate degree level, the sanctioned intake into electrical and connected courses was over 14,500. Electrical engineering proper, taught in about 125 institutions with a sanctioned intake of over 6000, constituted only about two fifths of the total. Electrical and Electronic engineering with an intake of over 8000 in about 140 institutions had the major share. Electrical and communication engineering was offered in half a dozen institutions and had an intake of about 300. It is worthnoting that there are six institutions offering courses connected with power and energy at the graduate level through the individual intake was less than 100 each.

The ratio of the sanctioned intake at the degree level to that at the diploma level in electrical and connected courses in India in 1996-97 was about 1:1.5. The sanctioned intake into electrical and electronic engineering at the diploma level was only about 75% of the intake at the degree level. The ratio of degree to diploms level intake in electrical engineering proper was about 1:2.6.

Post-graduate Level

The sanctioned intakes at the post-graduate degree level in electrical and connected fields are given in Table 2.

There are 17 specialities related to Power and 14 related to Energy in which post-graduate degree programmes are being conducted in India today. The sanctioned intake in about 1700 for PG degrees related to Power Engineering and about 300 for those related to Energy, making a total of 2000.

The intake at the post-graduate level for courses related to Power and Energy is about one seventh of the degree level intakes into courses related to electrical engineering and connected fields.

Power Systems Engineering with an intake of 581 in 34 institutions, Power Electronics with an intake of 267 in 11 institutions, Heat Power Engineering with 236 in 15 institutions and Thermal Engineering with 231 in 13 institutions constituted three-fourths of the intakes in Power-related fields. High volage Engineering with an intake of 67 in three institutions, Electrical Power Systems with an intake of 61 in three institutions, Control Systems with an intake of 37 in three institutions and Design & Production of Heavy Electrical Equipment with an intake of 37 in two institutions were the other major areas of post-graduate specialisation.

TABLE 2 : **Intake at the Post-Graduate Level into Electrical and Connected Courses**

Subjects	Number	Institutions
A. Power related		
Power System Engineering	581	34
Electrical Power systems	61	3
Integrated Power systems	18	1
Industrial Electrical systems	19	1
Power and Energy systems	17	1
Electrical Machines & Power systems	26	2
Heat Power Engineering	236	15
Thermal Engineering	231	13
Thermal Science & Engineering	18	1
Hydro Electricity	19	1
High Voltage Engineering	67	3
Measurement & Control	12	1
Control Systems	37	3
Power Electronics	267	11
Power Apparatus & Electrical Drives	18	1
Industrial Power Control Drives	26	1
Design & Production of Heavy Electrical Equipment	37	2
B. Energy Related		
Energy Systems	64	4
Energy Engineering	46	3
Energy Conservation & Management	18	1
Energy Management	18	1
Energy Technology	17	1
Energy Science & Technology	10	1
Energetics	18	1
Electrical Energy Systems	18	1
Thermal Science & Energy Systems	18	1
Alternate Hydro Energy systems	18	1
Ocean Engineering	18	1
Nuclear Engineering	18	2
Renewable Energy Systems	45	2
Energy Systems and Pollution	13	1

Energy systems with an intake of 64 in 4 institutions, Energy Engineering with an intake of 46 in three institutions and Renewable Energy Sources with an intake of 45 in two institutions were the three major subjects of specialisation in the field of Energy. There were 11 other specialisations in which the intakes varied from 10 to 18.

Intake at the Post-graduate Diploma level in electrical and connected fields was restricted to two specialities. There was an intake of 50 each in four institutions for operation and Maintenance of Thermal Power stations. There was also in intake of 18 P G. Diploms course in Integrated Rural Energy Technology.

Outturns in the Late Eighties

Academic specialisation in the fields of electrical and connected areas has been going on at a substantial pace in the past also in India.[3]

Two hundred and forty-eight and qualified for the doctorate degree in electrical engineering in India during the period 1984-89. The major contributions were made by the Indian Institute of Technology (IIT) Delhi with 61, IIT Madras (42), IIT Bombay (39) Indian Institute of Science, Bangalore (31), IIT Kanpur (23) University of Roorkee (18), IIT Kharagpur (17), and VJP Institute.

At the post-graduate degree level, over 2500 persons had taken their Master's Degree in electrical engineering and connected fields during 1984 to 1989 India. Another 330 had obtained their Master's Degrees in the field of thermal power form departments of Mechanical engineering.

Energy Engineering

One of the most significant features in the development of high level personnel through academic specialisation in the field of power and energy is the growing emphasis on the study of energy as an interdisciplinary subject. Specialisation in the field of energy engineering has been gathering momentum from the middle of the eighties. IIT Delhi led the field with the award of 44 PhD's, 63 M Techs and 2 post-graduate diploma in energy studies during 1984 - 89. IIT Bombay had an out turn of 35 post-graduates in energy systems and engineering. Regional Engineering College, Tiruchirapalli, produced 11 M Techs in energy engineering. Three persons had taken the MTech in Renewable Energy Sources, one from the Division of Agricultural Engineering, IARI in 1984, and two from the college of Technology and Agricultural Engineering, Udaipur in 1989. Post-graduate Diploms in Solar Energy has been awarded to 5 persons form the Birla Viswakarma Mahavidyalaya, Vallabh Vidyanagar.

The formation of multidisciplinary groups or schools for the pursuit of excellence in the field of energy engineering has been making substantial progress. Some of the groups, and the specialisations offered, are given in Table 3.

Of these active groups, two deal with Power and six deal with Energy in all its aspects, including renewable sources. Among those which deal with limited areas are two concerned with Hydro-energy, two with ocean engineering, one with Nuclear engineering and technology, and one with Rocket Proplusion. They thus over a very wide range of fields in the area of Power and Energy.

It may be mentioned that most of these are among the top twenty engineering institutions in the country. With increasing interaction with industry, these institutions are today in a position to provide academic specialisation in a wide range of areas in power and energy.

Discussion

India has at present an intake of 2,000 persons at the post-graduate level in Electrical engineering and connected fields. Academic specialisations at the Master's degree level include 17 subjects in

TABLE 3. : **Multidisciplinary Departments in Indian Universities dealing with Energy Education and Research**

	Department/Schools	Subjects of Specialisation
1	Centre for Engineering Studies, IIT Delhi	Power Science & Technology.
2.	School of Electrical and Electronics Engineering Anna University, Chennai.	Power Electronics and Drives.
3.	Multidisciplinary Group IIT Mumbai	i. Energy Systems Engg. ii. Systems & Control Engg.
4.	School of Energy Studies Jadavpur University, Calcutta	Energy Science & Technology.
5.	School of Energy, JNTU College of Engg. Hyderabad	Energy Systems
6.	Schools of Energy, Bharatidasan University, Tiruchirapalli	Energy Conservation and Management
7.	Engineering Science, Devi Ahilya University, Indore	Energy Management
8.	Alternate Hydro-energy Centre Roorkee University, Roorkee	Alternate Hydro-energy systems
9.	Water Resources Engg. and Management, M.S. University, Vadodra	Water resources Engg.
10.	Ocean Engineering, Indian Institute of Technology Chennai	Ocean Engineering
11.	Ocean Engineering and Naval Architecture, IIT, Kharagpur	Marine Technology
12.	Neclear Engineering and Technology, Indian Institute of Technology, Kanpur	Nuclear Engineering and Technology
13.	Space Engineering Birla Institute of Technology, Mesra	Rocket Propulsion

the field of Power engineering and 13 subjects in Energy engineering. Over a dozen multi-disciplinary schools and departments have been set up in some of the leading technological institutions.

Of these active groups, two deal with Power, six deal with Energy, two with Hydro Energy, two with Ocean Engineering, one with Nuclear Engineering & Technology and one with Rocket Propulsion. They cover a wide range of fields in the area of Power and Energy. Greater emphasis on certain areas of special importance to India appears necessary in the promotion of academic specialisation.

The balance of recoverable reserves of natural gas are presently estimated at over 700 billion cubic metres. Fresh reserves are being discovered form time to time. The production of natural gas has increased from 1.5 billion cubic metres in 1970 to 19.4 b.cu.m, an increase of 250%. In spite of an installed generating capacity of about 5000 MW, commercial energy production form natural gas in 1993 was 16.86 mtoe, and had recorded an increase of only 2% during 1990-93. Nearly 4 decades ago, the National Council of Applied Economic Research (NCAER) had considered the feasibility of setting up a national gas grid based on gasification of low grade coal. Toady, the promotion of a natural gas grid based on available supply in India may appear chimaeric. Development of indigenous technologies for the generation of electricity using natural gas is of paramount importance. Organisation of appropriate courses and R& D in the area have become imperative.

India has a coastline of about 6000 KM. It is estimated that the approximate potential for ocean Thermal Energy is 50,000 MW,[5] of sea Wave Power is about 20,000 MW, and Tidal Power about 10,000 MW, making a total of about 80,000 MW. Very little spectacular success has been achieved in exploiting this enormous resource. Only two institutions have specialised courses in Ocean Engineering, though a few others have courses dealing with marine structures. Concerted efforts appear to be necessary for gaining mastery in this field.

Exploitation of Nuclear Energy[6] has been making scant progress. Through growth in electrical generation form 2.42 billion KwH in 1970 to 5.40 bkwH in 1993 shows a 2.25 fold increase, the share of Nuclear energy in the Indian Commericial Energy Mix has come down from 0.3% in 1970 to 0.2% in 1993. Only one institution outside the DAE has a PG course in Nuclear Engineering.

The enormous deposits of Thorium have to become the subject of much greater study, research and development.

While there is at least one PG Course specifically concerned with the problems of Energy Systems and Pollution, the exploitation of the mounting pollutant-Urban Municipal and Industrial wastes with a potential of 1,000 MW is not yet the subject of academic specialisation in any PG level course.

Commendable work is being done in a number of institutions on the exploitation of Renewable sources of Energy. Research and Development have also been making good head way. Two centres, one at IIT, Delhi and the other at Anna University, Chennai have been recognized as UN University Centres. Ten to fifteen students form other countries have been attending courses of about 8 months duration each.

India is today in a position to provide academic specialisation of a very high order at the post-graduate and doctoral level students to talented students and key personnel form foreign countries in a wide range of areas of Power and Energy.

REFERENCES

1. All India Directory of Approved Institutions (Degree and Diploma Programmes) 1996-97; AICTE, New Delhi.
2. Directory of AICTE approved Institutions–Postgraduate Programmes as on October 31, 1995; AICTE, New Delhi (Mimeographed).
3. Sushil Kumar *et al*, Outturn of Scientific and Technical Manpower in India 1984-89, Vol 2, Engineering and Technology, CSIR, New Delhi 1993.
4. Maggo J N, R & D Needs for Energy Sector- A paper presented at a Brainstorming Workshop on Thrust Areas in R&D for Energy availability, Energy Conservation and Energy Efficient Technology, 23 September 1996; Indian Member Committee of the World Energy Council, Indian Member Committee of the World Energy Council, Institution of Engineers (India), New Delhi, 1996.
5. National Paper on Energy scenario in India 1995- Prepared by Indian Member Committee of the World Energy Council, Institution of Engineers (India), New Delhi, 1995.
6. Chaturvedi, Pradeep and Narayan, M P—Energy Conservation Perspectives for India, Paper presented at a National workshop on Energy Conservation held under the auspices of Rajiv Gandhi Foundation and Institution of Engineers (India), New Delhi, 1995.

A MATHEMATICAL MODEL PREDICTING ENVIRONMENTAL DEGRADATION FROM ENERGY UTILIZATION IN INDIA IN 2020-21

31

L. SUGANTHI, S. INIYAN and T.R. JAGADEESAN
Faculty of Mechanical Engineering
Anna University, India.

ABSTRACT

Importance of efficient energy utilisation has been recognised the world over. During the early fifties the major contributor to the energy requirements of India had been the non commercial sources of energy. In the past decade the trend has changed with heavy reliance on commercial sources. A glimpse of future commercial energy is vital for planning a proper energy strategy. Time series and econometric models have so far been used to predict future commercial energy requirement. In this paper a modified model has been developed taking into consideration factors such as previous year's energy consumption energy price, national income, population and environmental quality. The model links environmental quality and technological efficiency with energy economic factors. The Model is tested for its validity as against time series and econometric models using Squared error, Regression Coefficient and Durbin Watson statistic. The coal, oil and electricity requirement in 2020-21 is 298 Million Tonnes of Coal Replacement (MTCR), 2177 MTCR and 2497 MTCR. If this quantum of energy is to be utilised the amount of emission that would be added to the atmosphere is then determined. The amount of CO_2 as compared to the rest of the emissions is very large in magnitude and is expected to be around 25 BT in 2020-21. This is because of the emission from coal based thermal power stations. Hence coal utilisation need to be reduced and to that extent renewable energy have to be substituted.

Introduction

The distribution of energy resources over the earth is lopsided. Most energy resources are concentrated in certain areas, for example oil in Middle East Countries. Such concentrated reserves of energy have resulted in a situation where 20% of the population uses 80% of the energy. The energy problem is further complicated by the general non-substitutability of many energy sources. Exponential increase in the population growth rate has resulted in energy being consumed at an alarming rate. It has been established that technological and social progress necessitates an increased energy usage and development of new and more efficient energy sources. But simple exploitation of the available resources will lead to environmental degradation. Hence, advancement of any country will depend on efficient energy utilisation with minimum pollution.

Energy demand is linked and interlinked with several factors which need to be considered while developing an energy model. Several parameters like price, availability and efficiency affect its utilisation and in turn causes change in the economy and environment. All these aspects need to be considered to arrive at an overall integrated approach. Also the production and utilisation of energy, strongly and most often adversely affect the environment. Tremendous atmospheric pollution is caused by fossil fuel combustion. In future, energy utilization should be planned, designed, developed and operated keeping in view, the impact on economy and environment.

Environmental Quality

Environmental analysis and least cost energy models are applied to United States Energy structure (OECD 1983). Regional/national energy mix and environmental impact scenario evaluations are determined for energy systems. Torren (1983) has addressed the energy environmental problem from an end-user perspective. Direct and indirect emission, air pollution, impact of conservation are studied. The case studies include a local energy environment model, a regional environment energy analysis using a computer optimization model and a national environment energy analysis using an economic energy model.

The importance of emission abatement was realised and was analysed in depth. The emission from fossil fuel utilisation was found to be Carbon dioxide (CO_2). Sulphur dioxide (SO_2), Nitrogen dioxide (NO_2), Total Suspended Particulates (TSP), Carbon monoxide (CO) and Volatile organic compounds (VOC). Accumulation of CO_2 in the atmosphere increases the atmospheric temperature resulting in global warming. The effects of global warming are already being felt-increased occurrence of flooding in coastal areas, general temperature increases, northerly shifts in climatic patterns, hot dry regions becoming hotter an drier, warm wet regions becoming warmer and wetter, increased severity and violence of weather patterns and greater fluctuations of temperature, wind and precipitation.

Energy Environment Nexus

There is considerable debate over the issue that the presence of CO_2 in the atmosphere causes the temperature to increase. Keeling and Bacastow (1978) has carried out research and found the concentration of CO_2 to be steadily increasing and state that this in turn will increase the average atmospheric temperature. Brooks (1989) states that the temperature and CO_2 concentration have remained almost stable over the past hundred years and unnecessary caution is imposed on energy usage. But he agrees that there may be a small percentage of rise in temperature.

There are several econometric and optimization models to study the effect of changes in energy prices, changes caused in the economy, improvement in efficiency of the appliances. In recent years emissions and its effect has caused concern and hence it was felt an environmental quality factors has to be introduced.

Leinweber *et al.* (1980) developed a model to assess the environmental effects of different energy policies and gauge the validity of such assessments. The model calculates the amount of air pollutants and nuclear wastes. Nordhaus and Yohe (1983) determine global GNP as a function of labour productivity, population and energy consumption. Substitution of fossil fuel by non-fossil fuel is considered. Probabilities scenario analysis is carried out. Ten variables whose value are uncertain is given a high medium and low value. The model is then used to analyse CO_2 emissions for these scenarios., In addition to sensitivity analysis, Nordhaus and Yohe attempted model validation. Though model validation gave positive results, Nordhaus and Yohe model did not consider income and end use efficiency which are found to be sensitive parameters in other models. Their omission in Nordhaus and Yohe model is significant.

Darmstadter and Edmonds (1989) have studied the impact of technological innovation with reference to energy models. In this global model the world is divided into nine regions and studied. The CO_2 emission is then determined after balancing the energy demand with energy supply.

Diepstraten *et al.* (1989) state that the only short term option available to reduce CO_2 emission is to reduce fossil fuel utilization. They are of the opinion that the problem cannot be solved either by energy conservation, carbon tax or fuel tax, but energy efficiency should be improved and henceforth energy analysis concept has to be used in policy making.

Harvey (1989) has found out how the atmospheric temperature would vary for different scenarios. His results indicate that the air borne fraction of CO_2 falls rapidly once emission begins to decrease, so that the total emission can be reduced to half the present value to stabilize atmospheric CO_2. Hence to stabilize atmospheric CO_2 it is sufficient to reduce 50 % of the existing value. Also deforestation has a disproportionately large extent on net emission. The changes in temperature, CO_2 concentration in the terrestrial biosphere are analysed. He agrees that CO_2 emission can be reduced by improving efficiency, by replacing fossil by non fossil fuels.

Boyle (1990) feels 60 % reduction in CO_2 emission and significant reduction in other greenhouse gases may be required to stabilize the climate.

It is found by vector optimization that the application of energy saving technologies together with CO_2 removal techniques may reduce the overall emission of CO_2 by more that 70 %. This will increase the costs of energy system by 100 % and reduce energy consumption by 25 % (Groscurth *et al.* 1990). Alfsen *et al.* (1990) have calculated the CO_2 emission for power plants, industry and transport as a function of consumption, fuel efficiency and emission coefficient.

The large scale utilization of fossil fuels to generate heat and electricity for a growing population has increased the level of carbon dioxide in the atmosphere. This has resulted in global warming. It was hence realised that energy models have to be built taking environmental quality into consideration. From the literature survey it was found that very few have attempted to develop an econometric model with environmental factor. Hence to constrain the rise of CO_2 it has been included in the energy consumption model. The CO_2 emitted per tonne of coal and oil is multiplied by the energy consumption figures to obtain CO_2 emission. About 65 % of the total electricity generation is at the thermal power station and this results in tremendous quantity of CO_2 emission. Hence CO_2 emitted at the thermal power plants is taken as the CO_2 emission from electricity. If CO_2 emitted from energy sources is curtailed, the amount of energy that needs to be utilised sourcewise can be determined. The development of such an energy environment model is likely to be useful for planners, for evolving suitable energy resource planning to meet future energy demand.

Forecasting Model

Energy planning deals with energy economy interactions. The energy prices on one hand and rise in national income due to inflation on the other have made the energy problem opaque. Inclusion of energy prices, national income, population, technological efficiency indicates its effect on consumption. It involves two stage least square technique.

Seven time series equations, linear, power, exponential, logarithmic, quadratic, double exponential smoothing and triple exponential smoothing have been used. Past consumption data with time are considered. The best fit using squared error is determined. As time series models are used as first approximations, more sophisticated econometric models with previous year's consumption, energy price and national income have determined.

In the Modified model, refined energy consumption (Y) at time 't' is a function of energy consumption (X) at time 't' obtained using regression analysis, price (P) at time `t' corresponding to a base year, GNP (G) at time 't' corresponding to a base year, technological efficiency (TE) at time 't' and CO_2 emission (EQ) at time 't'. K is a constant, θ is the time response parameter, α, β, τ and δ are elasticities of price, income, technological factor and environmental quality. Modified model (Suganthi and Jagadeesan, 1992) can be represented as

$$Y_t = K\,(X_t)^{\theta}\,(P_t/P_o)^{\propto}\,(G_t/G_o)\delta\,(TE_t)^{\tau}\,(EO_t)^{\delta}$$

The results obtained are given in Table 1. In the econometric model, the price elasticity of coal, oil and electricity are 0.03718, 0.00582 and –0.10227. The elasticity coefficient is negative only in the case of electricity. For the other two, coal and oil it is positive. This indicates that econometric models do not seem to fit Indian conditions.

In the Modified model, the price elasticities of coal, oil and electricity are –.001, –0.01 and –0.4486. It can be seen that in the case of coal, the effect of coal price on coal consumption is negligible. The income elasticities of coal, oil and electricity are 0.00108, 0.03002 and 0.64544 repectively. This indicates that any minor change in consumption will immediately cause changes in the national income. The elasticities of technological efficiency are –0.00465, –.0022 and –0.01 for coal, oil and electricity. Of the three, the elasticity for electricity is high and hence changes in the efficiency in the case of electricity will have a greater bearing on its consumption than coal or oil. The elasticity for environmental quality is very high compared to the other elasticities which indicates the vital impact of consumption on emission. The elasticities are 1.0032, 0.9744 and 0.7703 for coal, oil and electricity respectively.

The model is validated with other model, the squared error (SE), square of correlation coefficient (R^2) and Durbin Watson (DW) statistic. The square of the difference between the actual and the predicted is called the squared error. The model which has the least squared error will hence mean that the deviation of the predicted from the actual in that model will be the least. Square of the correlation coefficient R^2 explains the strength of the relation of the independent variable with the dependent variable. If the R^2 is near the value of 1, the independent variable will explain a large part of the variation in the dependent variable. When there is a natural sequential order in the error term the correlation is referred to as autocorrelation. Autocorrelation occurs when a variable is omitted in the model on the right hand side. In such cases, a cluster of errors in noticed. Presence of autocorrelation is always checked by the Durbin Watson statistic.

TABLE 1 : Comparison of the Three Models

COAL

Time Series

$$X_t = 29.50189 [e^{0.03401432\,t}]$$

$SE = 555.8188 \quad R^2 = 0.963375 \quad DW = 0.544809$

Econometric Model

$$Y_t = 1.38429 \left[Y_{t-1} \right]^{0.8729} \left[\frac{P_t}{P_o} \right]^{0.03718} \left[\frac{G_t}{G_o} \right]^{0.01781}$$

$SE = 279.7885 \quad R^2 = 0.981564 \quad DW = 1.6027$

Modified Model

$$Y_t = 181.8 \left[X_t \right]^{-0.00266} \left[\frac{P_t}{P_o} \right]^{-0.0001} \left[\frac{G_t}{G_o} \right]^{0.00108} \left[TE_t \right]^{-0.00465} \left[EQ_t \right]^{1.0032}$$

$SE = 0.0868 \quad R^2 = 0.999994 \quad DW = 1.7856$

OIL

Time Series

$$X_t = 25.55914 [e^{0.06527355\,t}]$$

$SE = 1928.383 \quad R^2 = 0.988846 \quad DW = 0.3218$

Econometric Model

$$Y_t = 0.6022 \left[Y_{t-1} \right]^{0.60065} \left[\frac{P_t}{P_o} \right]^{0.00582} \left[\frac{G_t}{G_o} \right]^{0.33761}$$

$SE = 577.126 \quad R^2 = 0.996661 \quad DW = 1.50212$

Modified Model

$$Y_t = 180.5 \left[X_t \right]^{0.02156} \left[\frac{P_t}{P_o} \right]^{-0.01} \left[\frac{G_t}{G_o} \right]^{0.03002} \left[TE_t \right]^{-0.0022} \left[EQ_t \right]^{0.9744}$$

$SE = 2.8533 \quad R^2 = 0.999983 \quad DW = 0.72517$

ELECTRICITY

Time Series

$$X_t = 12.50911 - 0.45366 [t] + 0.120885 [t^2]$$

$SE = 2256.781 \quad R^2 - 0.967714 \quad DW = 0.320241$

Econometric Model

$$Y_t = 0.35597 \left[Y_{t-1} \right]^{0.8621} \left[\frac{P_t}{P_o} \right]^{-0.10227} \left[\frac{G_t}{G_o} \right]^{0.47393}$$

$SE = 453.258 \quad R^2 = 0.993515 \quad DW = 1:979349$

Modified Model

$$Y_t = 39.86 \left[X_t \right]^{0.1481} \left[\frac{P_t}{P_o} \right]^{-0.4486} \left[\frac{G_t}{G_o} \right]^{0.64514} \left[TE_t \right]^{-0.01} \left[EQ_t \right]^{0.7703}$$

$SE = 432.0231 \quad R^2 \; 0.993819 \quad DW = 0.9682$

The squared error is lesser in the case of Modified model for coal, oil and electricity than in that of the econometric or time series model. The R^2 value is greater in the Modified model than that obtained in the econometric and time series model for coal and oil. These validations indicate that the Modified model gives results with a higher degree of accuracy.

The best fit from among the seven regression equations is obtained for GNP index and price index for coal, oil and electricity and is given in Table 2. The CO_2 emission from coal and oil is determined based on the predicted percentage consumption value. Coal to be used at the thermal power station is predicted to determine the CO_2 emission from electricity. The best fit equation is given in the same table. The table indicates that in most of the cases, quadratic fit results as the best fit.

TABLE 2

EQUATIONS FOR PREDICTING THE PRICE INDEX, GNP INDEX AND COAL REQUIRED AT THERMAL POWER STATIONS

Price Index

Coal	PI_t =	$43.3852 - 7.469512\,[t] + 0.36137709\,[t^2]$
Oil	PI_t =	$23.6328 - 3.224693\,[t] + 0.2036408\,[t^2]$
Electricity	PI_t =	$38.5103 - 0.279637\,[t] + 0.2033762\,[t^2]$

GNP Index

$$GI_t = 42.06208 + 0.2796369\,[t] + 0.06940838\,[t^2]$$

Coal Required at Thermal Power Stations

$$EC_t = 4.040288\,[\,e^{0.1017569\,t}]$$

The coal, oil and electricity requirement is predicted using Modified model to the year 2020-21 in Fig. 1. The amount of coal, oil and electricity requirements in 2020-21 are 298 MTCR, 217 MTCR and 249 MTCR respectively. It is found that the present oil consumption which is almost double that of electricity is slowly changing and in 2020-21 it is found to exceed the oil consumption rate. If this trend continues then the emission that would result from coal based thermal power station would be tremendous.

Environemntal Concern

About 1550 million tonnes of CO_2 was added to the atmosphere during the year 1988-89 in India. To reduce the addition, curtailment of fossil fuel usage and afforestation measures are required. Growing of trees not only helps reduce CO_2 emission but also helps generate bioenergy. Using the predicted consumption values from the long term Modified model the amount of emission that would be added to the atmosphere to the year 2020-21 is determined. The amount of CO_2 that would be emitted from different sources to the year 2020-21 is plotted in Fig.2. The amount of CO_2 emission from coal and oil is less compared to the emission from electricity. This is due to the exponential growth expected from electricity. The amount of CO_2 emission from electricity is 13.6 times that of direct use of coal and 22.7 times that of oil.

About 90 % of the SO_2 emitted into the atmosphere is because of fossil fuel combustion. Usage of low sulphur containing fuels would result in lesser sulphur emission. Sulphur content in the atmosphere can be controlled by going in for coal washing, coal gasification and oil desulphurization. These techniques have to be made cost effective if they are to be utilized. Policy decisions are mostly based on cost. The emission of SO_2 is given in Fig.3. The emission from oil likely to be around twice than that emitted from coal in 2020-21 while the emission due to electricity gêneration will be around fourteen times that emitted from coal.

Of the total Nitrogen Oxides (NO_x) present in the atmosphere, only 50 % is because of natural consequences. Of the remaining, 75% is due to fossil fuel combustion. NO_x combines with CO, resulting in ozone formation which causes photochemical smog. NO_x reacts with the water vapour forming nitric acid which leads to acid rain. Nox is formed because of inadequate supply of air during the usage of oil stoves, gas appliances and other combustion process. The usage of catalytic converters will help reduce NO_2 emission from vehicles. Emission from other sources can be reduced by going in for flue gas treatments or by using low NO_x burners. The emission of NO_2 from coal, oil and electricity is given in Fig. 4. The emission from oil is around thrice than that emitted from coal and emission due to electricity generation is around five times than that emitted from oil in 2020-21. The tremendous rise in NO_2 is because of steep rise in the electricity growth rate while for oil, the emission is because of the steep rise in oil requirement in transportation sector.

Particulate matter arises from different sources. The major contributor are from the industrial and agricultural sectors,. These minute particles are blown by the wind and remain suspended in the atmosphere. Sometimes sand particles that are blown into the air get accumulated. Around 5 to 50 % of TSP in the atmosphere is caused by human beings. Coal washing, usage of natural gas where possible will help reduce this emission. Installation of particle removal equipment in power stations along with efficient combustion techniques will bring down the TSP emission to a large extent. TSP is found to cause several hazards. Its presence in the atmosphere needs removal. The TSP emission from different sources are given in Fig. 5. The emission from coal and electricity are very high compared to oil. This is because of the high ash content prevalent in the coal used in India. The emission from coal is around 2.5 times than that emitted from oil and the emission from electricity is almost five times the emission from coal in 2020-21.

Incomplete combustion of carbonaceous fuels in industries and vehicle result in CO formation. CO is emitted from the stoves because of the absence of sufficient quantity of air. Also burning of fuelwood and hydrocarbons oxidation result in CO emission of the total prevalent CO in the atmosphere 60 % is because of these above factors. Reduction in CO emission is carried out by reducing the vehicular CO emission, though improvements in engine design and exhaust emission controls. CO is found to cause several harms. The CO emission from coal and electricity is negligible compared to oil as seen in Fig. 6.

VOC results from inefficient combustion of hydrocarbons (HC). HC are the essentiats ingredients of photochemical smog,. Most of the HC emission are from vehicles and hence reduction is carried out by controlling vehicular emission. The emission of VOC is also maximum in the case of oil with negligible emission from coal and electricity as shown in Fig.7. This is also

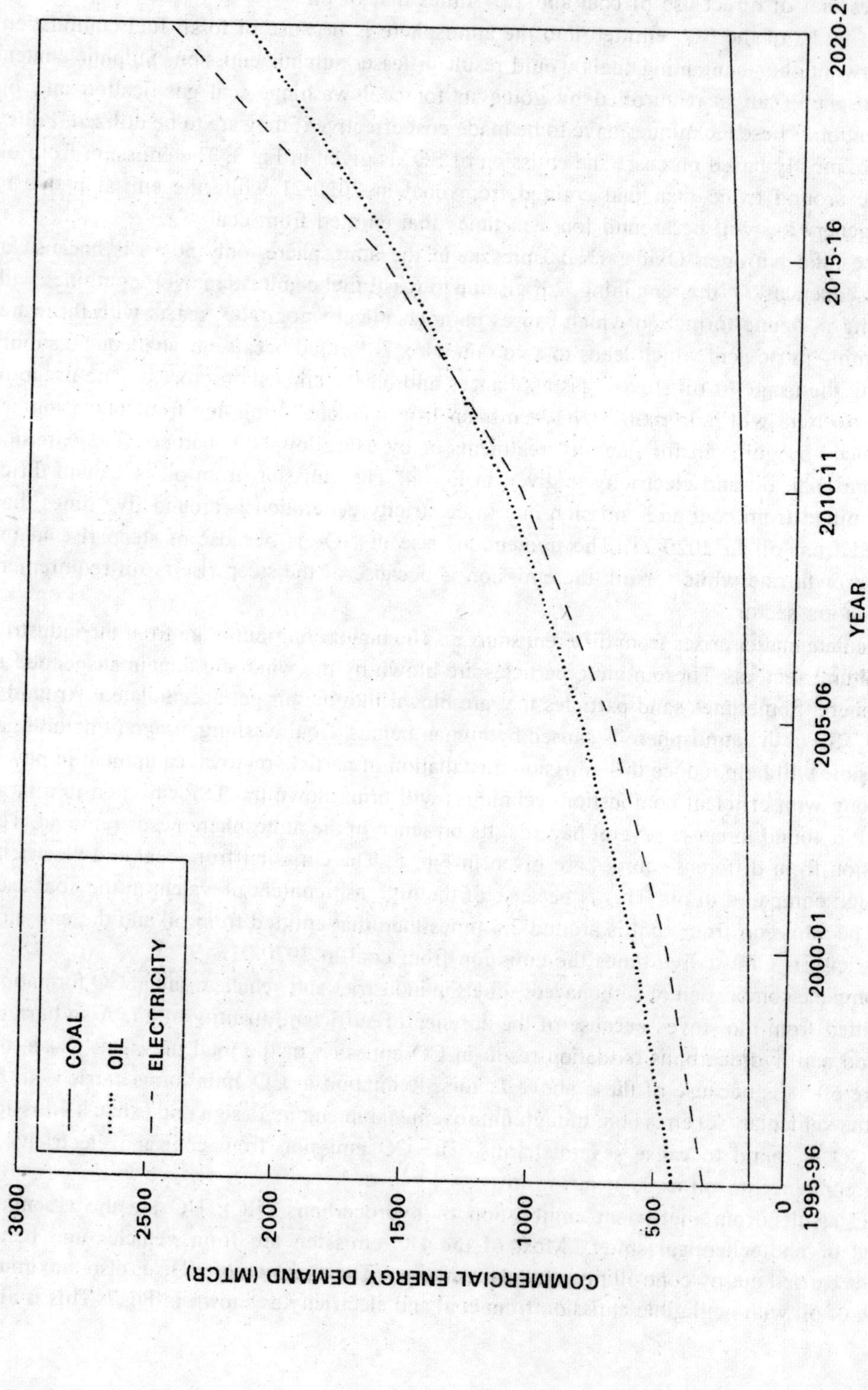

FIG. 1 : *Predicted Commercial Energy Demand using Modified Model*

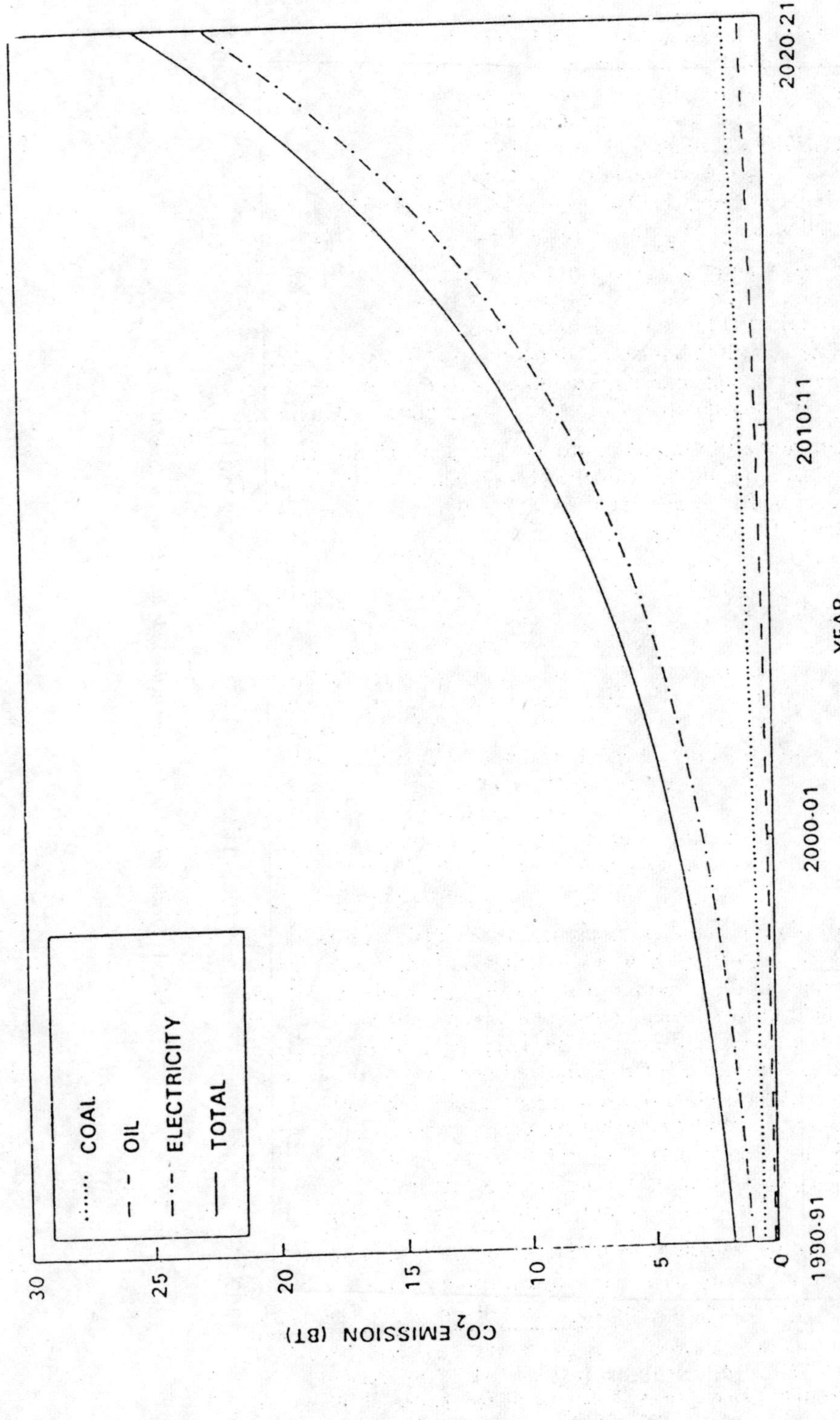

FIG. 2 : Prediction of CO_2 Emission to The Year 2020 Using Modified Model

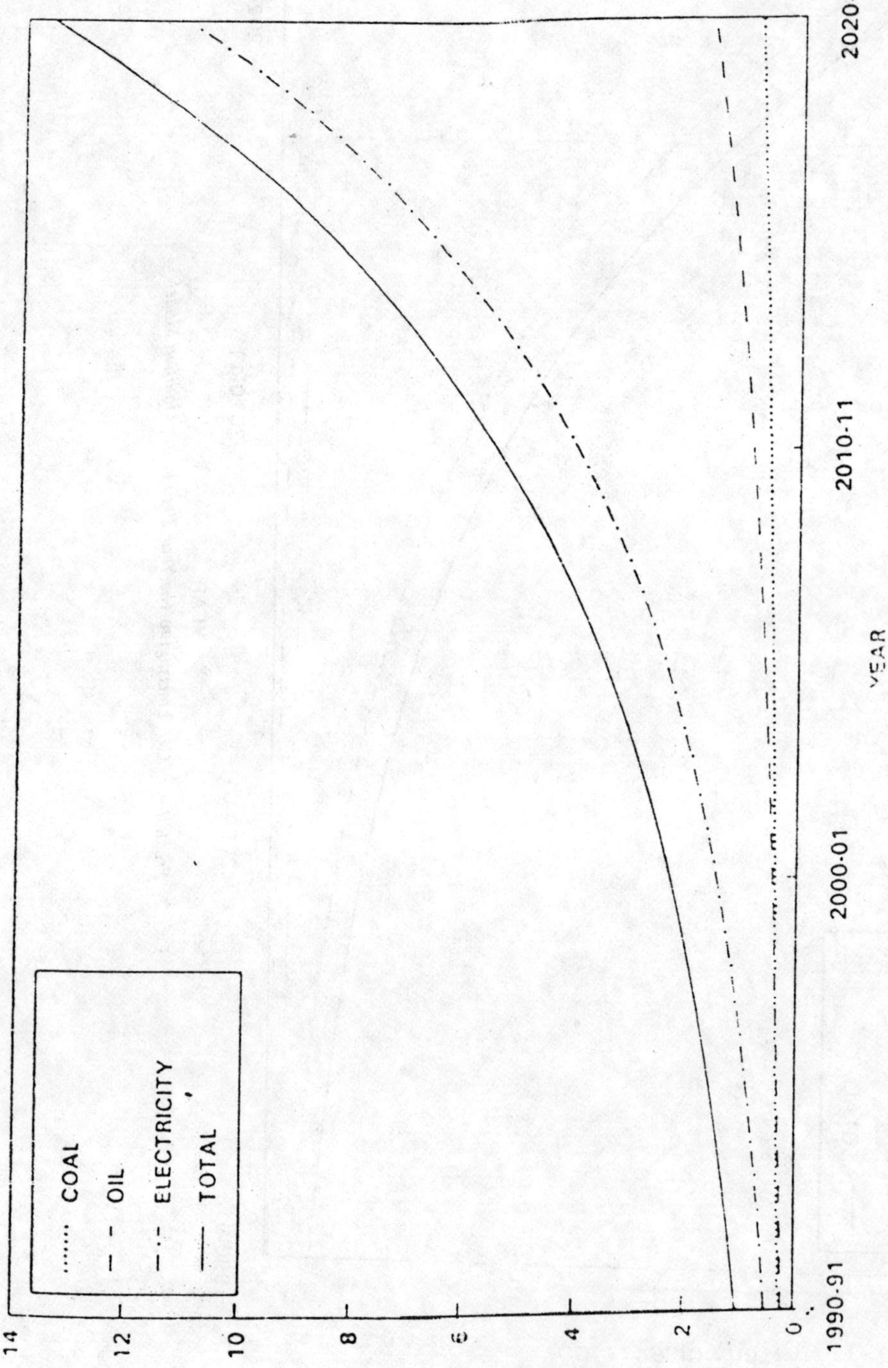

FIG. 3 : *Prediction of SO_2 Emission to the Year 2020 Using Modified Mode.*

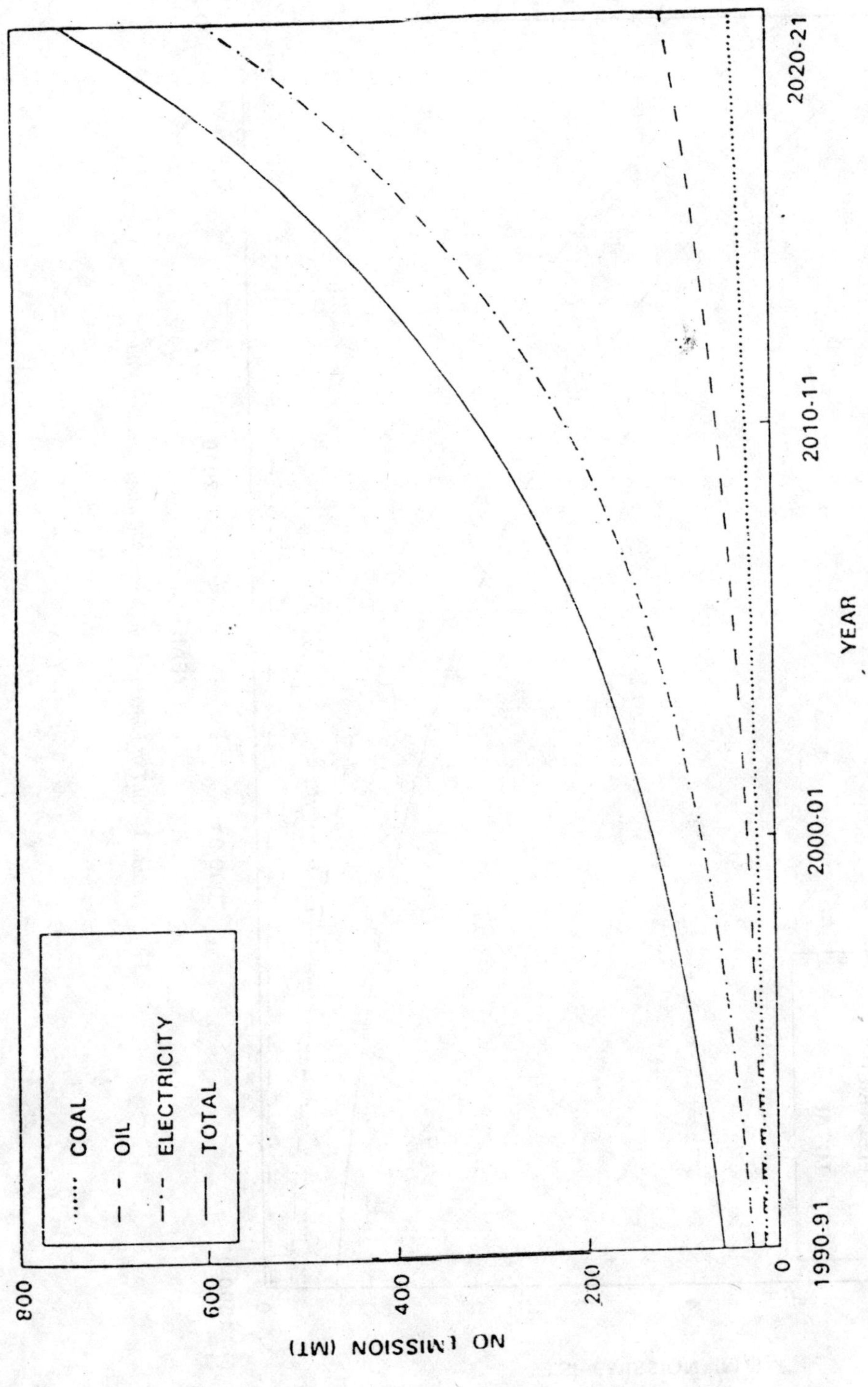

FIG. 4 : *Prediction of NO_x Emission to the Year 2020 Using Modified Model*

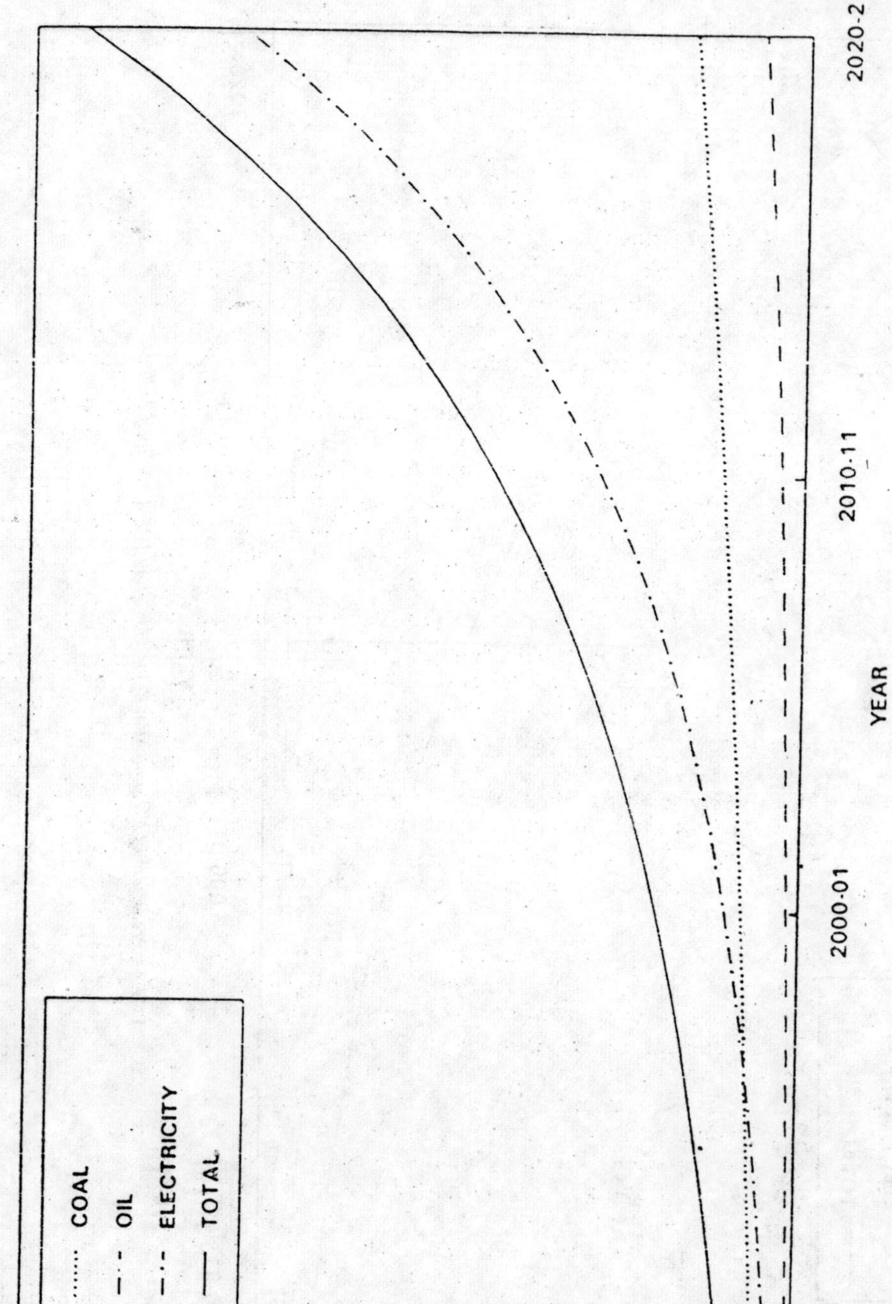

FIG. 5 : *Prediction of TSP Emission to the Year 2020 using Modified Model*

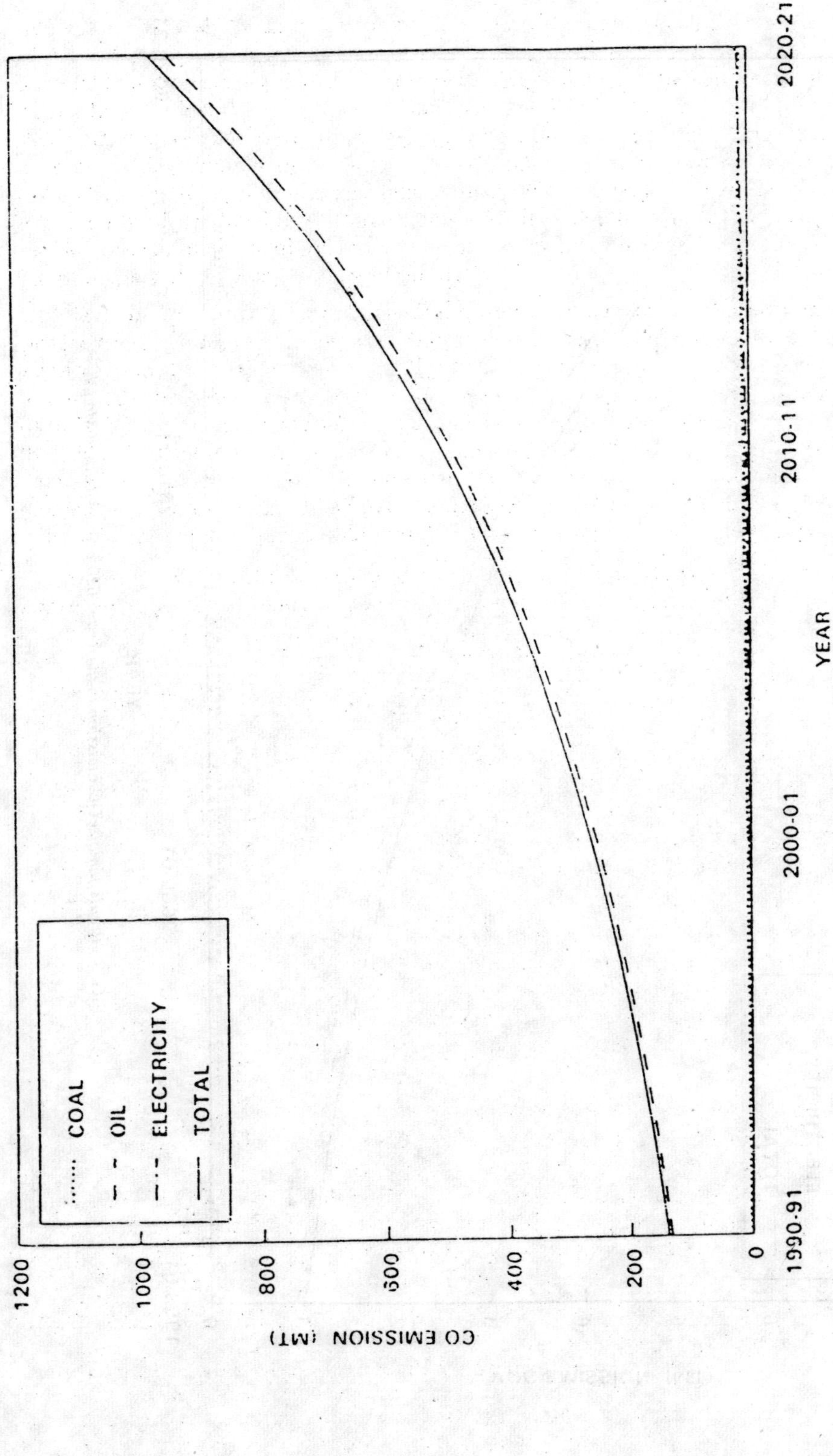

FIG. 6 : *Prediction of CO Emission to the Year 2020 using Modified Model*

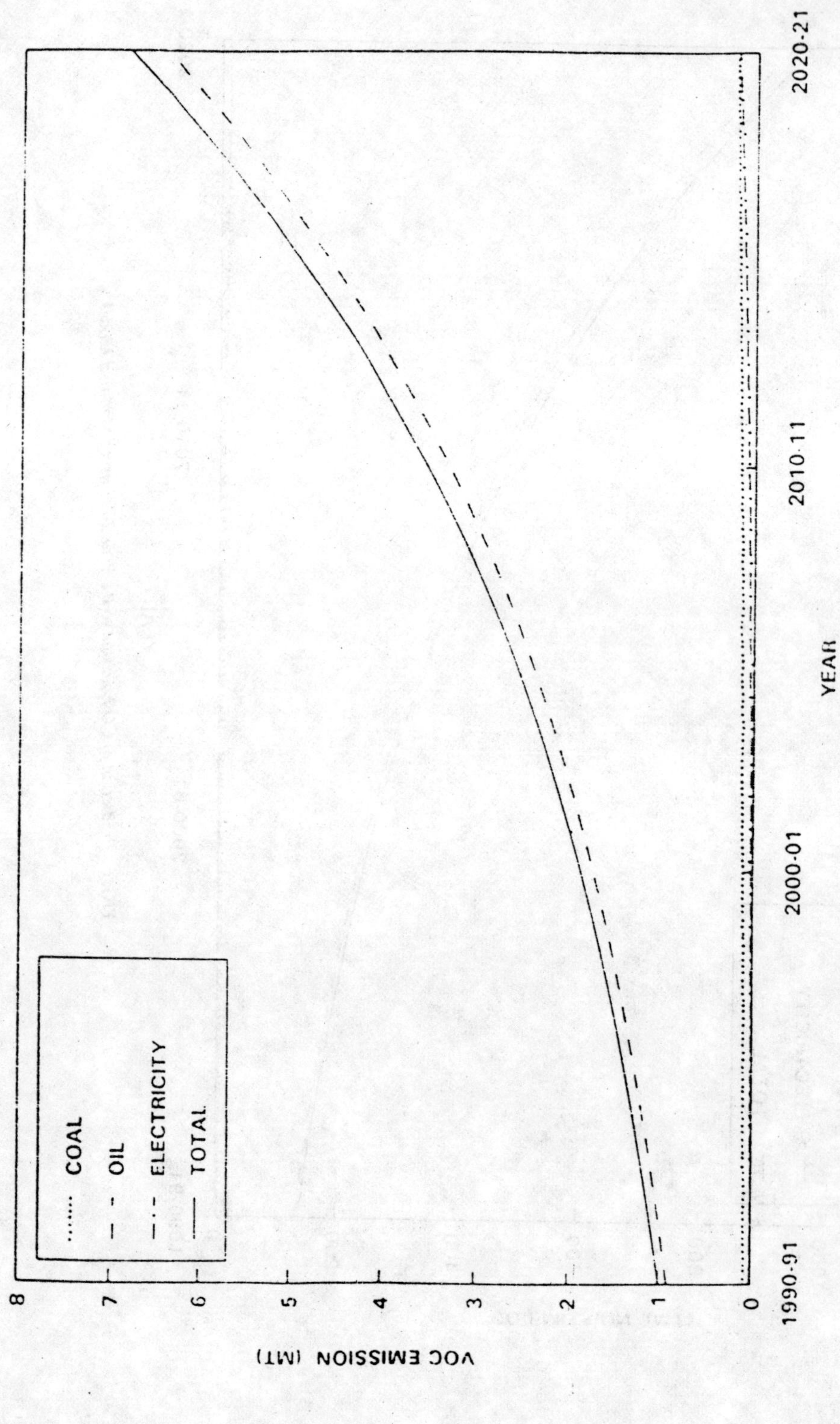

FIG. 7 : *Prediction of VOC Emission to the Year 2020 using Modified Model*

because of the increase in vehicular transport expected by 2020-21. It has been found that NO_2 and VOC present in the lower atmosphere results in ozone formation especially during the mornings in places where there is a heavy traffic. This ozone is detrimental to health.

Conclusion

A comprehensive energy model to predict the future energy requirement of India incorporating all essential and desirable parameters have been developed and the results have been analysed. The model developed has been adequately tested for its validity and can therefore be used effectively for assessing our future energy requirements keeping in view technologial and environmental factors.

Environmental factor has an influencing role to play. The elasticity of environmental factor is quite high when compared to the elasticities of all other parameters. This can be noticed by the elasticity term for environment factor for oil demand which is 0.9744 while all other elasticities lie below 0.03.

The emission arising out of electricity generation constitutes around 52.2 % of the total emission as of 1988-89. To bring down the emission level, electricity generation should be reduced which is possible by increasing the efficiency or by utilizing alternate enrgy system. As the CO_2 emission from coal based thermal power plants is several times more than that from coal, energy utilisation should be encouraged for direct utilisation of coal.

REFERENCES

Alfsen K.H, Lorentsen L. and Nyborg K. (1990), `Environmental effects of a transition from oil and coal to natural gas in Europe', Recent Modelling Approaches in Applied Energy Economics, Ed Bjerkholt O., Olsen O. and Vislie J., Chapman and Hall, pp. 87-100.

Boyle S.T.(1990), `Power Generation and Global Worming: The role of advanced combustion systems, energy efficiency and fuel switching in UK', International Conference on Environmental Impact of Electrical Power Generation, United Kingdom.

Brooks W.T. (1989), `The Global Warming Panic', Forbes, Dec., pp. 96-101.

Darmstadter and Edmonds (1989), `Human Development and Carbon dioxide emissions: The current picture and the long term prospects', Greenhouse warming : Abatement and Adaptation, Proceedings of a workshop held in Washington in 1988, Resources of the Future, pp. 35-51.

Diepstraten F.M.J.A and Van Gool W. (1989), `Prospects for carbon dioxide emission reduction', Climate and Energy, Ed. Okken P.A., Stwart R.J. and Zwerner S., pp. 60-77.

Groscurth H.M. and Kummel R. (1990), `Thermoeconomics and CO_2 emissions', Energy, Vol. 15, No. 2, pp.73-80.

Harvey (1989), `Managing atmospheric CO_2', Climatic change, vol. 15, pp. 343-381.

Keeling and Bacastow (1978), `Impact of Industrial gases on Climate', energy and Climate, National Research Council, Geophysics Study Committee, '77 national Academy of Sciences, pp. 72-95.

Leinweber and Squtieri (1980), 'Linking uncertainty and environmental models, 'Rand Corporation, Report No. R-2609/ 1-2-DOE/RC.

Nordhaus and Yohe (1983), 'Future paths of energy and carbon dioxide emissions', Changing Climate, Washington D.C., National Academy of Sciences.

OECD Report 138 (1983), 'Environmental effects of enrgy systems: The OECD Compass project'.

Suganthi L. and Jagadeesan T.R. (1992), 'A Modified Model for prediction of India's future energy requirement - International Journal of Energy and Environment, Vol. 13, no. 4, pp. 371-386.

Torren (1983), 'Environmental effects of enrgy systems : The OECD Compass projects', OECD Observer, No. 122, pp. 25-28.

PETROLEUM EXPLORATION POLICY FOR THE ASIA-PACIFIC REGION

32

Kuldeep Chandra
Regional Director
Keshav Dev Malaviya
Institute of Petroleum Exploration
Oil & Natured Gas Corporation

ABSTRACT

The paper deals with the main sedimentary basins and its hydrocarbon endowment, current load, self sufficiency of total energy requirement, future envisaged gap between demand and supply and strategic exploration policy.

1. Introduction

It is a well known fact that the economic growth of any country is primarily governed by the availability of energy and its pattern of consumption. Presently hydrocarbons contribute over 60% towards the commercial energy requirement of the world and this scenario is likely to continue well into early part of 21st century. The hydrocarbons are formed in the natural habitats "Sedimentary Basins". The evolution history of such basins dictate the richness of this non-renewable source of energy. The impact of liberalised economic era is mainly responsible for rapid economic growth globally which is manifested to a greater extent in the Asia-Pacific region more because of the fast economic growth taking place in many of the developing countries in the region like China, India, Malaysia, Indonesia, Thailand etc. The region as a whole is not self sufficient in oil and gas production. Currently about 21 per cent of crude oil is imported, which is likely to go up to 30% in the year 2010 (Mackenzie, 1995).

2. Sedimentary Basins of the Region

World over there are over 600 sedimentary basins out of which 400 are of different stages of exploration and only about 160 basins are under active exploration. only 25 basins contribute more than 80 per cent of the hydrocarbon demand. Although, the Asia- Pacific region has about

Note : The views expressed are those of the author and not necessarily of the organisation to which he belongs.

100 sedimentary basins (Fig. 1) and consumes about 25 per cent of the world primary energy requirement, it is not self sufficient in hydrocarbon requirements. Among major players in the region like China, Indonesia and Malaysia, which were oil surplus countries till recently, China has now become net importer and Indonesia and Malaysia will also follow the trend by early part of 21st century.

It is evident that bulk of commercial energy requirement is likely to be met by hydrocarbon alone All the countries of the region and their sedimentary basins do not hold similar hydrocarbon richness, primarily because of the evolution of the basins under different tectonic settings. Although the prolific Tethyan and South Gondwana realms predominate in the region, basins of this region are not as prolific as they are in the Middle East and Gulf Coast of America.

There are four independent facets of prime interest in the evolution of sedimentary basins. They are their geometric shape, size and basin configuration; sedimentary fill; tectonic setting and lastly the amount, nature and location of hydrocarbons within the basins. These facets, how they occur in basin, govern its richness. The incidence of hydrocarbon occurrence in a basin depends upon the richness and its type, thermal maturation, suitable reservoirs and entrapment mechanism.

Broadly the sedimentary basins are classified as rifted or orogenic basins. In case of rifted basins early subsidence decreases with time and the basins are characterised by prominent continental rise, gentle continental slope, broader shelves and broad coastal plains. These basins are characterised by high geothermal gradient which favours better source maturation and hydrocarbon generation. The black euxinic mud provides excellent source, shoreline sands, platform carbonates and reefs provide favourable reservoir and shales and evaporites provide excellent seals in the rifted setting. Whereas, in orogenic basins the subsidence rate tends to build up towards final climax. Such basins are characterised by deep oceanic trenches, steep continental slope, narrow continental shelves and coastal plain. In the Asia-Pacific region these orogenic basins are convergent margin which include normal faulting and formation of small oceanic basins. The convergence between oceanic and continental crust results in high geothermal gradient and generation of oil in sediments as young as Pleistocene due to rapid burial with subsidence. However, many of the basins are polycyclic and have undergone changes in their tectonic framework and setting during their evolution and are called transgressional basins. Rifted and orogenic basins are found in Indian subcontinent, China and South East Asia.

3. Exploration Maturity of Proven Basins

The recorded history of exploration dates back to late 1860 immediately after Col. Drakes discovery in USA, although, crude oil has been used for medicinal purpose and lighting since time immemorial in China and Burma. It is a misnomer to conceive a basin as mature. The exploration maturity may be achieved for certain plays, however, still some untested plays remain in a basin for exploration i.e. the Michigan basin of USA was considered matured for Devonian plays , but still, the basin is being explored for Silurian plays. The main producing basin of Asia-Pacific region have good amount of seismic coverage and exploratory drilling resulting in discovery of giant and large fields, but still quite good amount of reserves are being discovered annually. However, the discovery size in these basins has considerably reduced.

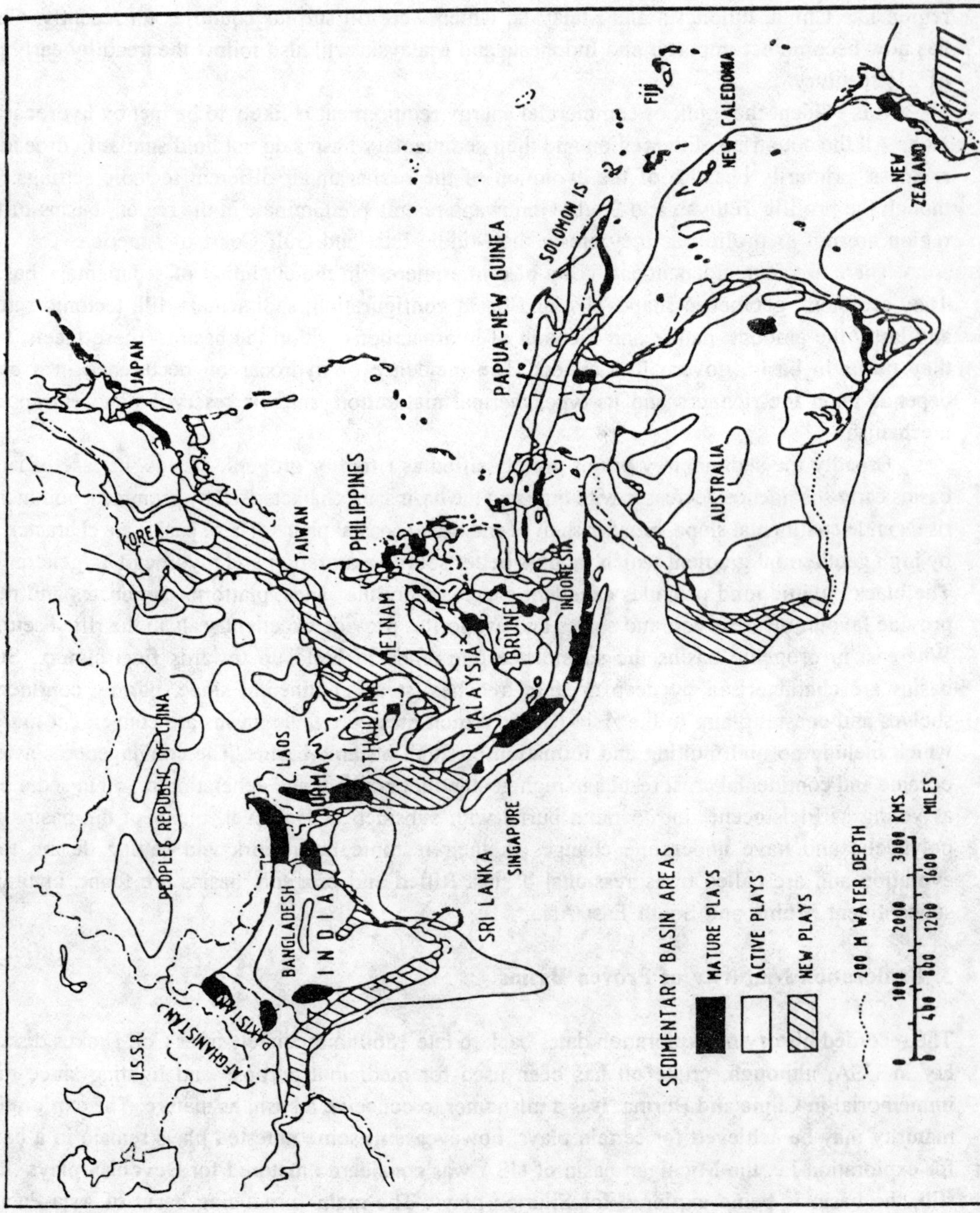

FIG.1: Prospective Sedimentary Basins of Asia-Pacific

4. Exploration Potential of Virgin and Frontier Basins

The discovery trend in a basin follow a lognormal trend resulting in discovery of large fields during initial phase of exploration followed by smaller and smaller discoveries. This trend is manifested in the main producing basins of the region like Indus, Bombay Offshore, Cambay, Upper Assam, Burma, Mahakam delta, Malaysian (Sarawak and Sabah), Indonesia etc. However, still there is a large tract consisting of deepwater and frontier basins having high risk and high to medium reward potential awaiting intensive exploration. The pertoleum legislations in Asia-Pacific have been evolved so as to attract large number of companies to work in this region. The acreages hitherto, reserved for national oil companies in countries like China, India, Burma etc. have been thrown open for private participation. The production sharing contract initially developed by Indonesia is being practised with certain modifications to attract more private investments in exploration from multinationals. In this context the countries like Vietnam, China and to some extent India have attracted private participation by multinational companies in their exploration ventures in addition to prime areas of Indonesia, Malaysia, Papua New Guinea and Australia.

5. Status of Exploration

The ultimate potential of different countries of the Asia-Pacific region is well documented by Masters *et al.*, 14th WPC, 1994 (Table-1 & 2) although, the figures indicated especially for India are not correct. However, for a comparison the figures quoted in the World Petroleum Congress are retained. The total ultimate potential estimated by Masters *et al.* is 23.37 billion tonnes oil and 936.1 TCF of gas, which clearly indicates that still there is significant amount of undiscovered potential requiring upgradation. The exploration in deepwater and frontier areas is gaining momentum in this region.

6. Current Demand and Future Projections

Asia-Pacific region is likely to exhibit an average energy demand growth rate of about 3.7% per annum to 2010. This is mere reflection of the energy supply that is likely to be available rather than the demand, which could be much higher (Fig. 2)

The current production from the region is estimated to be about 350 MMt of crude oil per annum with R/P ratio of 22 (Table-1). The oil production is likely to grow slowly reaching a plateau of 360 MMt O+OEG. The estimated proved reserves for oil and gas in the Asia-Pacific region in 1995 is 6.09 billion tonnes of oil and 350.65 TCF. China will continue to contribute about 40 per cent of the region's production of 150 MMt O+OEG per annum. This production in China comes mainly from onland Daqing, Shengli and Liaohoe basins located in the northeast China and account for 70 per cent of the production. Production is expected to stabilise in China around 2000 at 170 MMt O+OEG against demand of 300 MMt of O+OEG around 2010.

Indonesia, which is the second largest producer of the region will see a steady fall of about 3.5 per cent per annum in production in the next 15 years to reach about 41.49 MMt per annum by 2010.

TABLE 1

ESTIMATES OF RESERVES, UNDISCOVERED RESOURCES, PERCENTAGE UPGRADED AND R/P RATIO FOR OIL: ASIA-PACIFIC REGION (IN BILLION BBI)

Country	Cumul. Prod	Reserve	Undisc. Oil	Ultimat. Resour.	% upgra-ded	R/P Ratio
Afganistan	0.0	0.1	0.4	0.5	20	–
Australia	3.7	3.1	2.1	8.9	76	19
Bangladesh	0.0	0.0	0.2	0.2	–	–
Brueni	2.4	1.8	0.5	4.7	89	30
China	16.9	36.6	30.5	83.9	64	25
India	3.5	6.5	1.5	11.5	88	25
Indonesia	15.5	13.1	6.1	34.8	82	19
Malaysia	2.7	7.3	3.7	13.7	73	16
Myanmar	0.5	0.2	0.9	1.6	44	54
New Zealand	0.1	0.1	0.1	0.4	75	29
Pakistan	0.3	0.4	0.2	0.9	78	12
Papua	0.0	0.3	0.0	0.3	100	26
Thailand	0.1	0.2	1.0	1.3	29	15
Vietnam	0.1	0.9	3.7	4.7	21	14
Asia-Pacific	46.3	71.2	53.2	170.6	69	22

TABLE-2

ESTIMATES OF RESERVES, UNDISCOVERED RESOURCES AND PERCENTAGE UPGRADED FOR GAS : ASIA-PACIFIC (IN TRILLIONS OF CUBIC FEET)

Country	Cumul. Prod.	Reserves	Undisc. Gas	Ultimate Resource	% Upgraded
Afganistan	2.0	3.1	14.7	19.8	26
Australia	8.5	77.1	19.0	104.8	82
Bangladesh	1.7	10.8	13.7	26.2	48
Brunei	5.9	12.2	4.4	22.5	80
China	15.2	38.9	172.9	227.0	24
India	3.5	23.1	14.7	42.1	65
Indonesia	19.7	105.9	63.3	188.9	67
Malaysia	5.5	64.4	40.6	110.5	63
Myanmar	0.4	9.2	5.8	15.5	62
New Zealand	1.7	5.7	1.0	8.4	88
Pakistan	7.8	23.9	28.2	59.8	53
Papua	0.0	12.2	17.3	29.5	41
Thailand	1.7	10.8	21.2	33.7	37
Vietnam	0.0	3.0	8.6	11.6	26
Asia-Pacific	77.2	411.2	447.8	936.1	52

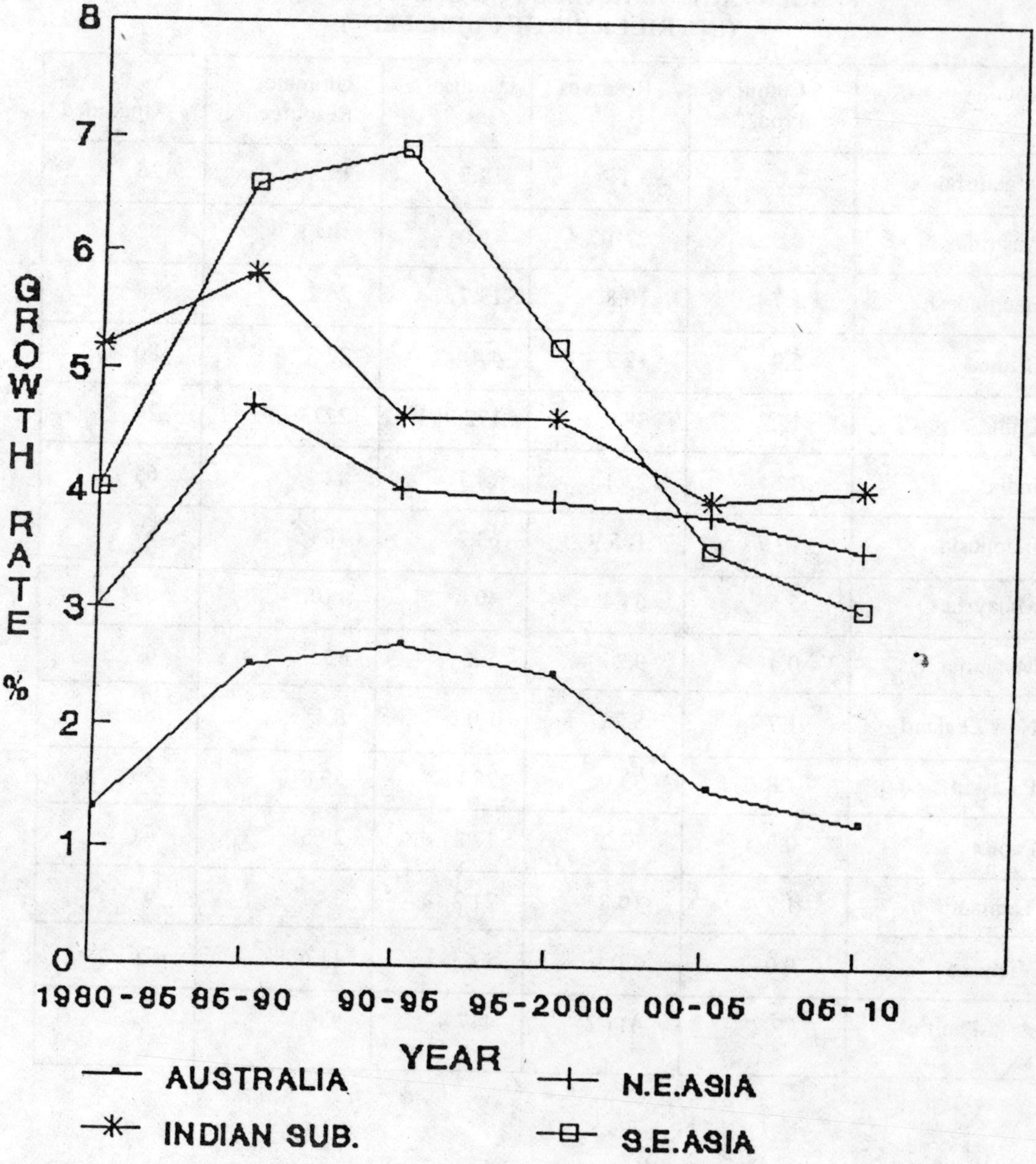

FIG. 2 : *Annual Growth of Energy Demand*

It is expected that Indonesia may be the first OPEC country to become a net importer by the year 2003.

India produced about 32.5 MMt of oil during 1995 which is about 9.3 per cent of the region's production. Wood Mackenzie (1995) have forecasted a drop in oil production by 2000 and there after an increase of about 4 to 5 per cent per annum and will reach a production of 45 MMt 2010.

Total demand for the region is set to rise from 800 MMt in 1995 to 1200 MMt in 2010 (Fig-3) averaging a growth rate of 2.9 per cent per annum for the region as a whole. Although most of the governments of the region formulated policies to reduce their dependence on this source of energy. However, in many cases despite the reduction of oil consumption for power generation, industrial and residential settings, demand for petroleum products in transport sector has led to a continuing increase in overall demand. Oil demand in India is predicted to be more than double from an estimated 78 MMt in 1995 to 160 MMt by 2010 with growth rates varying between 5 to 6 per cent per annum. it is expected that Japan will see a very low growth of 1 per cent China will require less than 20 per cent of the total energy demand for crude oil by 2010. However, it is expected to require about 300 MMt of oil by 2010 (85 per cent growth over 15 years period)

Oil Net Import : With oil production plateauing and demand increasing dramatically the regions oil import are forecasted to rise by over 90 per cent by 2010. The Asia-Pacific region is already importing over 450 MMt of oil per annum and it will rise to 870 MMt by 2010 (Fig-4). Thus the region is expected to be about 30 per cent self sufficient in oil by 2010 (Figure-5).

Gas production is increasing dramatically in the region and is likely to replace to some extent oil because it is eco-friendly. Increased exploration activity has provided many countries with newly discovered gas reserves which is used in the domestic market increasingly replacing imported fuels. Gas production is forecasted to rise by an average of 6 per cent per annum to 290 BCM by 2000 and further increasing to 360 BCM by 2010 (Fig. - 6A). Demand for gas is forecast to rise by about 7.3 per cent in the Asia-Pacific region over the next five years to reach 300BCM by 2000 and further rise to 464 BCM by 2010 (Fig.-6B). In terms of share of gas in primary energy demand, it is likely to increase by 3 per cent from 1995 through 2010 while the demand for oil is likely to drop by 9 per cent to 3 per cent (Fig.-7)

Indonesia will remain largest producer of gas followed by Malaysia. The countries of the Indian sub-continent will experience large growth of gas production over the next 15 years. India is already planning to import gas from Oman and Iran through trans-regional pipelines.

7. Exploration Scenario

The Asia-Pacific region is considered to have a very large 'Yet to Find (YTF) hydrocarbon potential'. Most of the major oil companies consider Asia-Pacific as the most prospective region for exploration. The region has an abundance of pertoliferous sedimentary basins and significant rate of economic growth to boost demand for oil and emerging market for natural gas making the region most attractive for exploration ventures. However, there are challenges for exploration companies since the acreages available for exploration in the region are mostly in logistically

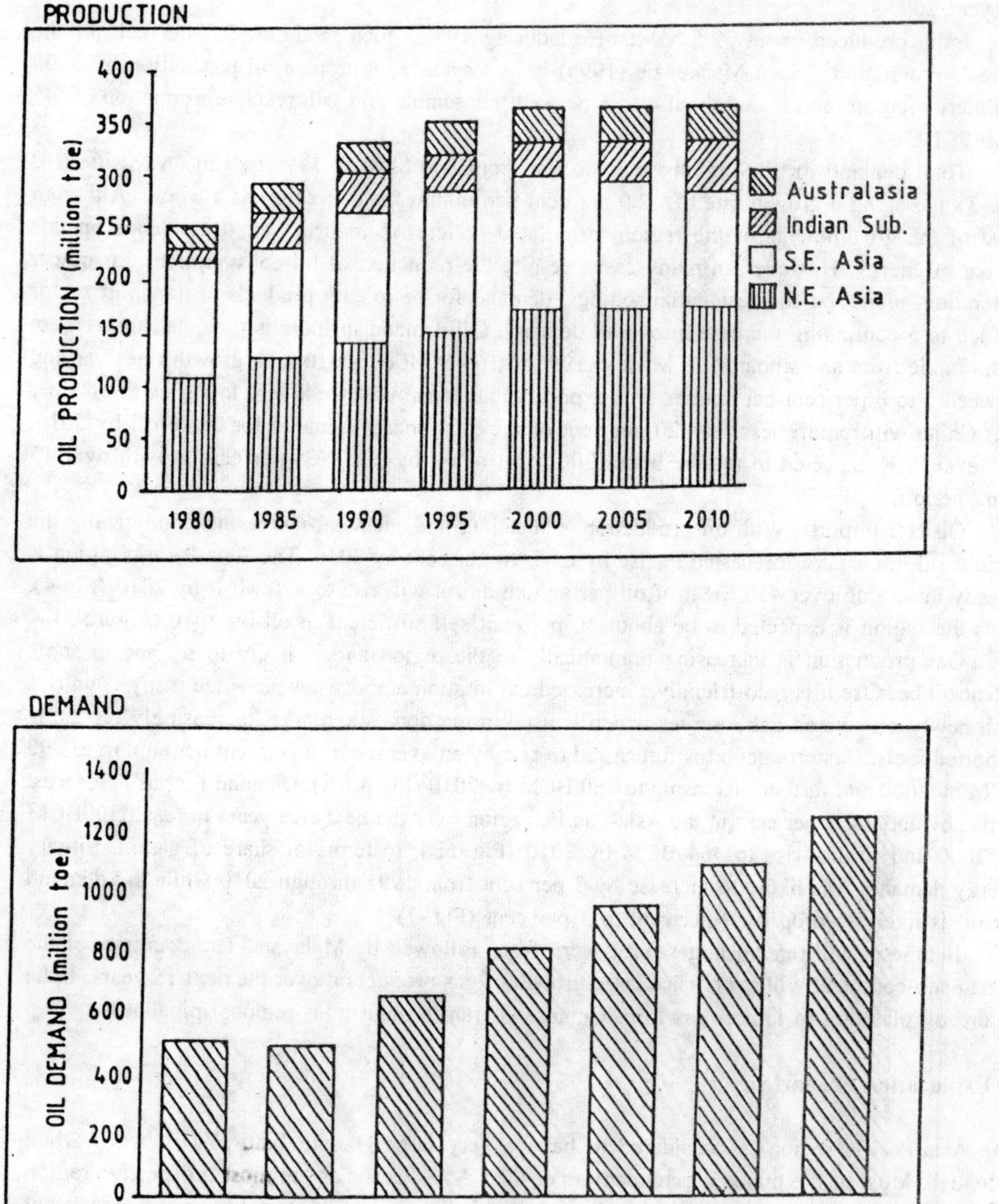

FIG. 3 : *Oil Production and Demand, Asia-Pacific Region*

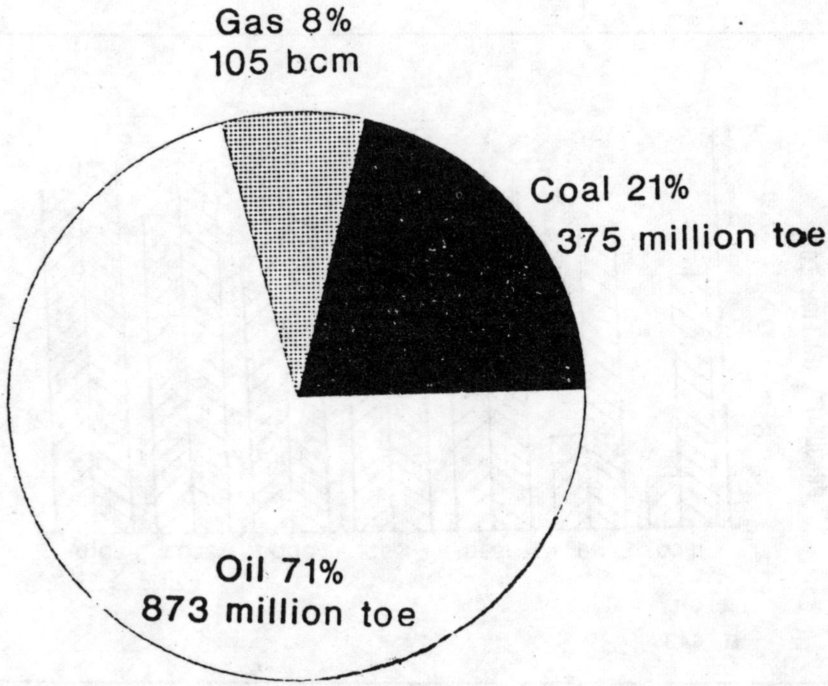

FIG. 4 : *Net Imports of Energy - 2010*

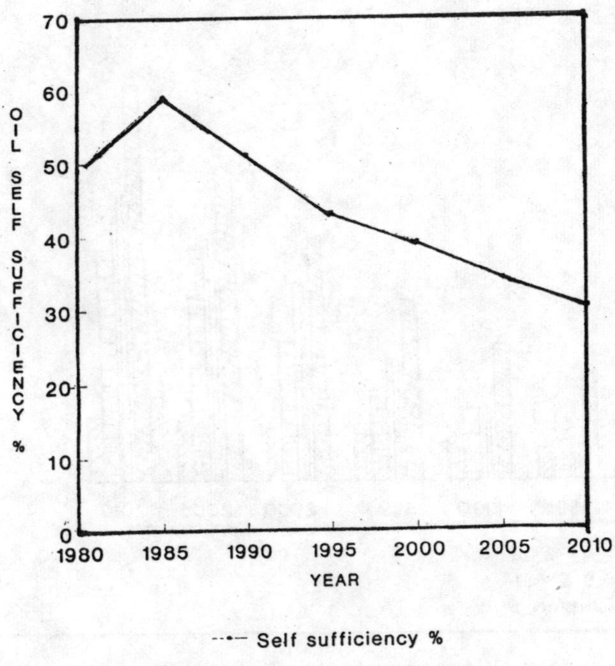

FIG. 5: *Oil Self Sufficiency, Asia-Pacific Region*

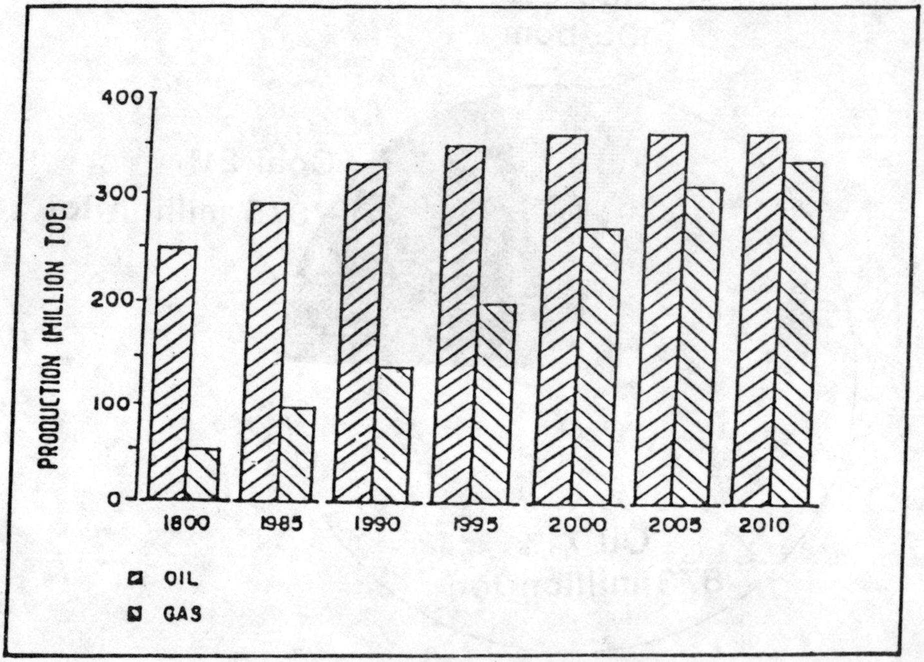

FIG. 6 A : *Gas vis-a-vis Oil Production, Asia-Pacific Region*

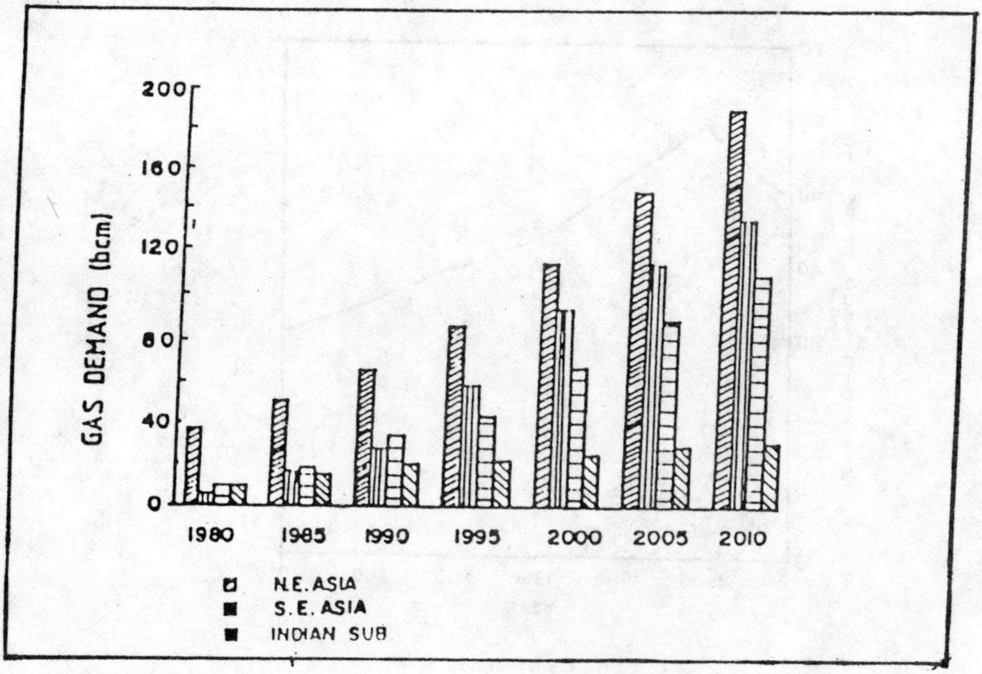

FIG. 6B: *Gas Demand in Asia-Pacific Region*

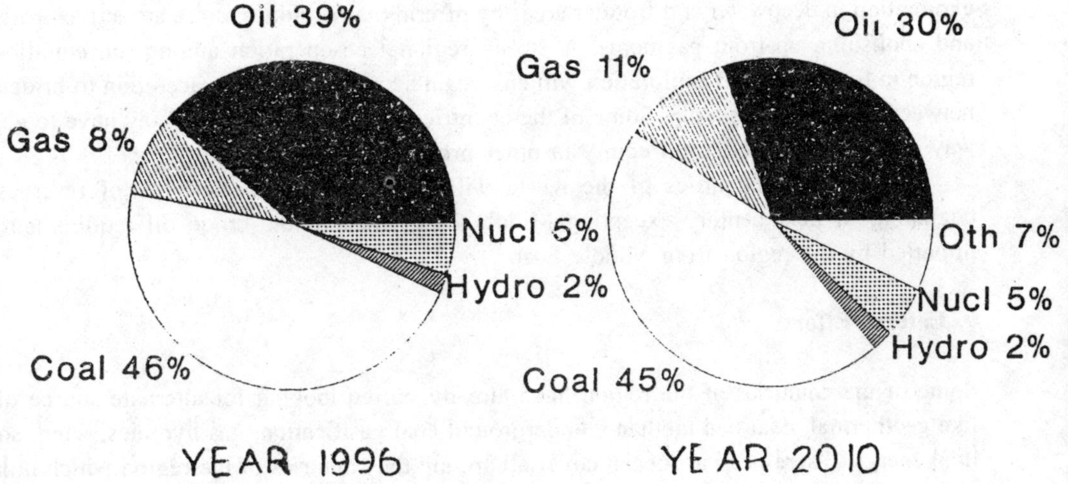

FIG. 7 : *Primary Energy Demand, Asia-Pacific Region*

difficult, highly tectonically disturbed (Fold belts), deserts (Tarim basin of China) and in deepwaters. Easily accessible and shallow pools have already been discovered. The new targets are likely to be at deeper depths, less obvious subtle traps and frontier areas (deepwaters and fold belts) Seismic imaging of the sub-surface features especially in the tectonically disturbed areas in the thrust fold belt and deeper targets will provide a major challenge to explorationists. In the well explored area the average field size has considerably declined and will warrant innovative technology to make them economically viable. Enhanced oil recovery will play a very major role in boosting the production form the existing fields, Innovative drilling technology involving drain hole, horizontal and cluster drilling concepts will reduce the cost and increase productivity of such smaller discoveries. 3D seismics may also be vital tool for future exploration to reduce the cost and increase success rate.

As already mentioned the self sufficiency of crude oil production will go down in the region to 30 per cent by the 2010 and to bridge the gap between demand and indigenous supply it would require a very large investment on exploration. Even India alone will require an investment of about US $50 billion to US $ 95 billion on exploration by 2010 to meet its crude requirement.

8. Policy issues

The future areas available for exploration lie in high risk-high reward category where success rate are to be quite low, more attractive fiscal regime will be required to promote exploration by the host governments. In some of the countries, still large investments are being made by many oil companies in the form of signature bonuses, work commitments etc. Already a very large progressive fiscal regime has been developed by Government of India wherein there is no clause for signature bonus, 100 per cent cost recovery and profit shares are linked with post tax rate of return and/

or on investment multiples. In my opnion, still the governments of this region can encourage exploration in deepwater and frontier areas by offering tax holidays, more attractive profit-sharing and abolishing up-front payments. A strong regional co-operation among the countires of the region in the field of oil exploration will ensure an accelerated reserve accretion to bridge the gap between demand and supply. Some of the countries like India and others may have to go in a big way overseas to generate oil equity in other prospective acreages.

Most of the countries of the world will be reaching their half life of reserves by the beginning of next century excepting Middle East. Most of the crude oil requirement will be imported by the region from Middle East.

9. Lateral Efforts

Some of the countries of the region have already started looking for alternate source of energy like geothermal, coal-bed methane, underground coal gasification, gas hydrates, wind, solar and tidal energy. There is a vast ocean cover all around the countries of the region which hold a very large gas hydrate prospect. It is estimated that in India alone there are likely to be about 142 billion tonnes of OEG of gas hydrate deposits around the coastal tract. It is expected such gas hydrate deposits may be located around shelfal regions of Indian and Pacific ocean and may reduce considerably the dependence on conventional crude oil and gas.

10. Conclusions

— Asia-Pacific region consumes about 25 per cent of the world's total energy requirement of which hydrocarbon contribute about 60 per cent.

— The region as whole is self sufficient in its hydrocarbon requirements. The region's major player like China has become a net importer and Indonesia and Malaysia will follow the trend by beginning of twenty-first century.

— The region is expected to be about 30% self sufficient in oil by 2010.

— As the existing onland and shelfal discoveries are getting smaller and smaller, the frontier and deepwater basins are likely future targets for exploration.

— To promote speedy exploration the government of this region can encourage exploration in difficult areas by offering tax holidays, more attractive profit sharing and abolishing up-front payments.

— The venture for alternate source of energy like geothermal, coal-bed methane, underground coal gasification, gas hydrates, wind, solar and tidal energy may considerably reduce the dependence on crude oil.

Acknowledgements

The author is highly thankful to CMD, ONGC for according kind permission to present this paper. The Asia-Pacific Energy Report by Wood Mackenzie (1995) and Petroconsultants publications have been used in preparation of this paper and the same are acknowledged.

FIREWOOD : THE NEED FOR AN ACTIVE POLICY

33

J.B. LAL
Former Director,
Indian Institute of Forest Management,
Bhopal (India)

ABSTRACT

The detailed survey made by Forest Survey of India in the late eighties revealed that India consumed as much as 157 million tonnes of firewood annually against the sustainable level of production of 58 million tonnes per annum. The enormous gap between demand and sustainable level of production of firewood is one of the major causes of depletion of forests in the country. In the rural India over 60 percent of energy needs in the domestic sector are met from firewood. Even in the urban India the share of firewood in energy consumption in the domestic sector is well over 25 percent. The attempts made to bridge the gap between demand and supply of the firewood in the last few decades have been based on the philosophy of increasing the production. The solution however lies in managing the demand. The way we use firewood currently, we obtain only 5 to 8 percent thermal efficiency. We can reduce the annual demand of firewood to less than 50 million cubic meter, if we increase the thermal efficiency to 40 per cent. Unfortunately, we did not have an active forest policy in regard to firewood. The paper advocates an urgent adoption of an active policy in regard to firewood.

INTRODUCTION

Though firewood accounts for nearly 60 per cent of the energy consumption in the rural domestic sector in India, the institutional aspects in regard to firewood have drawn little or no attention of the decision-makers. The discussions on the subject of firewood have been mostly confined to silvicultural and socio-economic aspects. Even the technical aspects have been ignored.

FIREWOOD CONSUMPTION IN INDIA

The share of firewood in energy consumption in rural household in India has not decreased over the years. On the contrary, as Table 1 indicates there has been a marginal increase.

It should be admitted, however, that as most of transactions in relation to firewood, either in cash or in kind, take place in informal sector, not much authentic data is available about the demand and supply of the commodity. Nor is available any convincing study about the consumption of firewood at the national level. A few studies that have been made cannot stand a thorough statistical scrutiny. The one study which is statistically reliable, not totally but substantially, is the one made by Forest Survey of India (FSI) in 1987.

TABLE 1: **Energy Consumption in Rural Household in India**

Fuel Type	% share 1978-79	% share 1992-93
Coal	1.92	0.38
Kerosene	2.55	4.44
Dung Cake	22.51	17.00
Firewood	54.57	61.60
Crop Waste	17.41	13.35
Others	1.03	3.23
Total	100.00	100.00

Source : Natarajan (1995)

The FSI study revealed that in middle eighties, India, on average, consumed 157 million tonnes annually against the sustainable level of production of only 58 million tonnes, 28 million tonnes from forest lands, and 30 million tonnes from other lands. The gap of nearly 100 million tonnes in legitimate production and consumption was obviously met by illegitimate cuttings in government forests. Indeed, the big gap in consumption and sustainable level of production of firewood, has been one of the major causes of depletion of forests, and consequent environmental degradation.

In his comprehensive study, Lal (1989) came out with the stunning fact that during the period 1953 to 1987 the country consumed 3244 million cubic meters of firewood in excess of sustainable level of production, which amounted to total destruction of 3 million hectares of forests and depletion of 27 million hectares to various degrees.

INEFFICIENT USE OF FIREWOOD

The sad part of the situation is that the gap between consumption and sustainable level of production would not be as big as it is, if firewood were to be used rationally and efficiently. The State of Forest Report, 1987 (FSI 1988) reveals that per capita annual consumption of firewood varies in various States and Union Territories of the country from 0.2 cubic meter to over 0.5 cubic meter. There is sufficient evidence to indicate that consumption was more related to availability of firewood than to the actual need. In absence of an active firewood policy, a kind of Parkinsonian law seems to have been operating in the country, that is, demand rising to equal supplies, legitimate, or illegitimate.

The inefficiency in the use of firewood is proved by the figures in Table 2. The way firewood is till used by most of the people, that is by using it without drying, and burning it in open hearth, it yields only 5 to 10 per cent thermal efficiency. If firewood were to be dried before use, and some kind of efficient stoves were to be used it is estimated that consumption would go down by about 30 per cent. One of the reasons for irrational and inefficient use of firewood is lack of a policy.

TABLE 2 : **Thermal Efficiency of Various Fuels**

Fuel	Approximate thermal efficiency
	Per cent
Firewood	5-10
Charcoal	20-35
Kerosene	35-50
Propane Gas	50-60

Source : Planning Commission, 1982.

THE ECONOMICS OF FIREWOOD

This should be remembered, however, that even if rationality and efficiency is introduced in its use, firewood, for the reasons of its basic economics, is not a commodity which can be treated at the national, or even regional level for planning supplies. Presently, the domestic user gets firewood either free, or at a nominal price in terms of money, though he pays a price in terms of labour. If firewood supplies involve long distance transport, the price can only be in terms of money, and this may cause suffering to the poorer section of the rural Indian society. The economics of firewood, makes it mandatory that regular studies are made at local level, which be integrated to derive a national scenario.

Valued in terms of money, the woodfuel sector in India amounts to some 9080 million US dollars per annum (Wood Energy News, June, 96). Although most of firewood supplies is gathered by the users themselves, the woodfuel is also important as a source of main cash income to a percentage of Indian rural population. The exact figures are not known, but it is estimated that local employment generated by wood energy sector is much higher than in the oil sector, if we consider the situation per unit of energy (wood energy News, June 96.)

WOOD ENERGY IN INDUSTRY, TRANSPORT AND POWER SECTOR

The current trend that use of firewood reduces with the increase in per capita GNP will reverse sooner or later, Fossil fuel supplies are limited, and are bound to exhaust at some point of time. The man will have to resort to renewable forms of energy. The choice is between nuclear and solar energy. In solar energy, the wood is most convenient form. The day is not far off when wood energy would be used in industry, transport and power, if wood energy is available in adequate quantities that day onwards. The supplies of wood energy cannot be increased in weeks, months or even years. It takes decades. To increase supplies of wood, we have to plan now, and to plan we have to have a policy.

In Sweden the wood energy has already been used in transport. Lorries, tractor, cars and buses have all been used on producer gas obtained from wood (FRI, 1943). There are examples of wood energy being economically used for production of power. It was worked out by Earl (Earl, 1975) that an input of 5,60,000 tonnes of wood can run a power station of size of 150 megawatt. The researches at Indian Institute of Chemical Technology have proved that oil extract from widely available Jatropha plant could be cracked to generate a wide range of hydrocarbons as an effective substitute of fossil fuels.

Even if we ignore the likely future demand of wood energy in industry, power and transport sector, we still have to plan, and have a policy for adequate supplies of firewood to domestic sector. If a simplistic assumption is made, and a linear relationship between population and firewood consumption is accepted, the firewood requirement at the turn of century is likely to be 210 million tonnes, equivalent to 300 million cubic meters (Lal, 1989). If efficient use reduces the requirement by 30 per cent, and use of biogas and solar cookers by another 20 percent, we still need to produce 150 million cubic meters of wood annually. To produce this quantity we have to bring nearly 20 million hectares of land under energy plantations, as in the given environmental conditions, even with improved technology we cannot expect to produce more than 8 cubic meters of wood per hectare per annum on average.

CONCLUSION

The discussions made in the earlier paragraphs underscore the need for an active policy on firewood. A major handicap, however, in integrating wood into national energy planning is non-availability of reliable information, primarily because firewood related activities take place in the informal sector. The data presently available is derived from secondary sources and could be misleading. To be reliable, it should be collected by surveys and case studies. Presented with adequate and convincing data, the decision makers would easily recognise wood energy as major policy and programme concern.

NOMENCLATURE

The term firewood describes all fuels containing wood fibres that are used for direct combustion, such as branches, twigs, stems and roots.

Wood fuels, in addition to firewood, also include all wood-derived fuels such as charcoal and producer gas.

REFERENCES

Earl, D.E. (1975). Forest Energy and Economic Development. Clarendon Press, Oxford

Forest Survey of India, FSI (1988). State of Forest Report, 1987

Forest Research Institute, FRI (1943). Indian Forest Leaflets, No. 35

Lal, J.B. (1989). India's Forests : Myth and Reality. Natraj Publishers, Dhera Dun

Natarajan, I. (1995). Wood Energy News Vol 10, No. 4

Planning Commission, Govt. of India (1982) : Report of the Fuel Wood Study Committee.

POWER THROUGH ENERGY PLANTATION—A RENEWABLE ENERGY SOURCE

34

M. SRINIVAS, U.C. TRIVEDI, AND M.RAMAMOORTHY*
Electrical Research & Development Association,
Baroda (India)

ABSTRACT

Since 1950, the energy consumption in the world has been rapidly growing. The wrong perceptions in the use of renewable energy technologies are no longer valid especially in the developing countries. Since the fossil fuels are always limited, energy plantation may be the only massive alternative to relieve the pressure on the consumption of fossil fuels. The contribution of waste lands in power generation is because of the natural growth of species like *Prosopis Juliflora, Eucalyptus, Acacia nilotica, Alnus Rubra, Leucaena Leucocephala*, etc. Since Asia is having more desert areas and unused waste lands, adoption of this kind of power generation may reduce the difficulties to be encountered to meet future Asia energy demands. The generation cost and hence, the power cost per unit may ultimately cheaper than the conventional power. Due to the power shortages and increasing environmental concerns in Asia, the new technologies related to efficient utilization of non-conventional energy resources, are going to play very important role in the future energy scenario.

Introduction

Besides land, capital and labour, energy is the key input to the economic development of any nation. Since the year 1950, the energy consumption in the world has been growing. The fossil fuels are always limited. The growing environmental concerns will limit its use in the country. So we have to ultimately depend upon the renewable energy sources.

Domestic economies in many parts of the developing world rely upon renewable sources of energy to meet their needs. Developed countries have come to rely upon increasing consumption of finite fuel resources to maintain and increase their relative position in the world economy. The rationale for growth based upon the utilization of under-priced, non-renewable resources is now being challenged because it inevitably must result in irreversible depletion of the world's stock of finite energy in accordance with the dictates of the Law of Entropy.

There has been a perception that renewable energy technologies can only give small amounts of energy in certain locations and therefore their contribution to the total energy requirements would be marginal. The second bottleneck to the growth of renewable energy has been the perception that they are rather expensive as compared to the conventional alternatives. Both of these perceptions are no longer valid especially in the developing countries.

Waste lands are considered as an indirect renewable energy resources as it can be put to use for production of energy plantation. Of course, the same land could be put to other alternative uses, like bio-resources, if it cannot be assumed to be totally available for energy purposes.[1]

The contribution of waste lands in power generation is because of the natural growth of species like *Prosopis Juliflora, Eucalyptus, Acacia*, etc. The present paper discusses the economics involved with a 100 MW wood fired power plant based on *Prosopis Juliflora* plantation and the indirect social and environmental benefits that accrue by the proposed power generating plant.

The Plantation

In Asia, most of the waste lands are either saline, alkaline or desert sands. The major part of Western Asia is highly covered with desert sands and middle and North-East Asia is rich in unused islands. All this land is eroded by either wind or water. These lands could be made use for energy generation by growing plantation. *Prosopis Juliflora* is one of the best species suited for any of the soils like saline and sodic soils, degraded soils and desert sands.[4]

The following are some of the features of *Prosopis Juliflora*.

1. The survival rate is 100 % for any soil and climatic conditions.
2. Due to fast growth, hardiness and salinity tolerance it would be an ideal plant to harness waste lands.
3. A good understorey crop for optimal site utilization with species like *Teak, Eucalyptus* and *Subabul (Leucaena Leucocephla)*.
4. It can tolerate fairly heavy amount of shade. The yield rate is as much as 25-30 tonnes per hectare per annum.
5. The recycling period is 4 years.

A power station of reasonable size i.e. 100 MW requires one lac hectares of land for growing plantation. The plantation can be brought under aforestation of waste lands. This will provide employment for a large number of people round the year. Since continuous fuel supply is essential, the plantation is playing an important role in power generation. The forest area needed to sustain production would depend upon the amount of wood residues available and on the increment from natural forest or energy plantations.

The quantity of fire wood can be increased by providing improved varieties of species like *Alnus rubra, Eucalyptus Saligna, Leucaena Leucocephala*, systematic planting and developing agricultural practice. But these species cannot fully survive in desert or saline lands. The above species can be used in areas where the average rainfall is 200-400 mm per year.[4]

The Power Plant

Electrical energy is very expensive to produce compared with energy supplied directly from fuels but it has many advantages which could make it desirable to convert wood energy to electricity in areas where the supply of wood is not limiting. The low specific gravity of wood does not affect the calorific value that is based on unit weight. The ash content of wood is usually low (vary from 0.5-5 %) and is of little consequence.

The most important controllable factor influencing the efficiency of wood as a fuel is the moisture content. Moisture in very fresh wood may amount to more than 100 % of the dry weight substance that reduces its value as fuel because of the absorption of heat required in the evaporation of water. It is customary in the timber trade to estimate moisture as a percentage of the dry material.[5]

The working of wood based power station is almost similar to a thermal power plant. The only differences are coal handling unit and boiler. In this system, the coal handling unit is replaced with a series of millers for cutting the wooden logs and the boiler is wood fired boiler. The following assumptions are made for estimating the land requirement.

1. The air-dried wood yields 4000 kcal/kg
2. The thermal electric conversion is 35 %.
3. The plantation yields on an average of 25 tonnes per hectare per annum.

At 35 % conversion, one ton of wood gives 1625 KwHr.
1 MW-year = 8.76×10^6 KwHr.
1 MW requires $8.76 \times 10^6 / 1625$ = 5390 tonnes of wood
Hence, total wood required per MW per year ≈ 5,400 tonnes
Land required for plantation for 1 MW power generation = 216 ha.
Therefore, land required by 100 MW power plant to sustain wood supply is 21,600 ha.

Apart from the factors that influence the plant site selection, the fuel transportation distance should be as minimum as possible. Because continuous lifting of fuel wood from the areas where it grows is required, the plant should be located not more than 30-40 Km from any place of plantation growing areas.

With the total utilization concept embodied in a project of this type, it should be comparatively easy to arrange for the ash residues in the furnace to be returned to the waste land, thus reducing the need for input of Calcium, Potash and trace elements.

Cost Benefits

It is reported that, for setting up a thermal power station of 100 MW size, nearly Rs. 400-500 crores of money is required. Since the millers for cutting logs and wood fired boilers are relatively cheaper than coal handling unit and coal fired boiler, the initial establishment cost will come down to a reasonable extent. Hence, the unit power cost may ultimately cheaper than the conventional unit power cost.

The following are other benefits from the above system:

1. It provides employment to a large number of people round the year which leads to better economic status of local people.
2. Since the fuel is available at nearby places, major transportation is not required.
3. Fuel is available continuously round the year and no fuel shortage problems.
4. The CO and CO_2 in the exhaust gases will be absorbed by the plantation and wood is free from sulfur content. SO, SO_2 content is nil in the exhaust gases. Therefore the system is pollution free and eco-friendly.
5. The power generation cost is relatively cheaper than other conventional methods.
6. Brings better economical status to the local people. *Prosopis Juliflora* acts as a good fodder for animals that will be helpful to the local people.

Conclusions

Due to the shortfall in energy supply and increasing environmental concerns in Asia, the new technologies related to efficient utilization of available fuel resources and increased utilization of non-conventional energy resources, are going to play more and more important role in the energy scenario of the future.

In this respect, it should be noted that some renewable sources of energy such as fuel wood, agricultural residues, biomass etc., have been used, especially in rural areas of the developing countries.

Future production is influenced by new discoveries, improved production techniques, development of new energy resources, economic and environmental constraints and other factors. In view of the difficulties to be encountered to meet future world energy demand, technological developments will be increasingly necessary to maintain sufficient energy supplies.

REFERENCES

1. Proceedings of Bio-Energy Society First Convention and Symposium, 14-16 October, 1984, New Delhi.
2. Dennis Anderson, The economics of afforestation - A case study in Africa, World Bank occasional papers, Number one, New Series, 1987.
3. Proceedings of First National Workshop on Role of Prosopis in Wasteland Development, 18-19 February, 1986, edited by Patel, V.J.
4. Energy Plantation in Wastelands - The case of Kutch, An ORG report, 1988.
5. Mathur, A.N., Iqbal Hussain, Rathore, N.S., Economics of Energy Plantation, Himanshu Publications, 1992.
6. Earl, D.E., Forest Energy and Economic Development, Clarendon Press, Oxford, 1975.
7. Dunn, P.D., Renewable Energies: Sources, Conversion and Application,
8. Zalenskii, O.V., Photosynthesis and Solar Energy Utilization, Amerind Publishing Co. Pvt. Ltd., 1977.

RENEWABLE NON CONVENTIONAL ENERGY — BIOGAS FROM AGRO BASED INDUSTRIES

35

S. Vaidyanathan
Additional Chief Engineer Navy
INDIA

ABSTRACT

The alternative sources for augmenting energy production have been briefly covered. Waste treatment for power generation in process industries offers on alternative option. The cost economics of a biogas generation systems for a distillery waste has been presented.

Introduction

The availability of a reliable and economic form of energy is a pre-requisite for economic and social development. It is clearly demonstrated by the close relationship between energy consumption and economic growth in industrial countries and between persistent energy deficiency and the low standard of living in many developing countries.

In spite of annual growth rate of 10% in the capacity in the last seven Five Year Plans, the country experiences shortage of power. India currently faces peak electric generating capacity shortage of 25% and approximately 12% demand is left unserved. This power supply shortfall of 15 thousand mega watts may become worse as demand growth exceeds the abilities of utilities to finance the new capacity.

With this background, it is necessary to explore alternatives for augmenting our energy production. These alternatives sources for energy generation should be environmentally friendly, technically viable and based on proven technology, waste treatment for power generation in process industries offer an alternative option. This has led to identification of bio-degradable resources which satisfy the above conditions and can also be used commercially for generation of energy.

Technology

Industrial bio-degradable wastes for power generation comprise the following :

(a) Municipal solid wastage (MSW)
(b) Effluents of agro based industries.
(c) Distillery spent wash.
(d) Agro-residues.
(e) Sewage, cattle and poultry waste.
(f) Black liquor from paper mill.

Though there is the problem of safe disposal of these wastes, today, technologies are available for waste treatment and simultaneous energy generation. Hence the disposal of waste is no more a problem but an opportunity. Technology is fully established and available commercially in the market. In India, Western Bio-Systems Ltd., Pune and western Paques India Ltd., Pune are offering expertise to harness liquid effluent using continuous feed Anaerobic sludge digestion called bio-methanisation to generate energy. The basic steps for the generation of power from the liquid effluent are as under.

The liquid effluent is collected in a tank or pit, suspended solids are allowed to settle down and the effluent is then pumped to the digester through a Heat exchanger. The effluent is cooled so that the reactor temperature is maintained at 36-38 degree C. The effluent is allowed to be digested anaerobically in the digester for a period of 12-45 days during which biogas is generated. The digestor is provided with a central agitator and three lateral agitators for proper mixing of feed with the fluid, inside the digestor and to provide homogeneous distribution of fluid inside the digestor. The bio-gas produced in the digestor accumulates holder acts as an intermediate gas storage as well as pressure control device. The treated effluent overflows into the degasing pond where traces of biogas are released. The effluent then goes to Lamella clarifier where active biomass is separated and recycled to digestor with the help of recycling pump. The excess sludge collected at the bottom of the Lamella clarifier is removed and sent to sludge Drying Beds for its further use as form manure. The treated effluent from the lamella clarifier can be used as irrigation water after proper secondary treatment. The biogas from gas holder is blown out with the help of blower system to the boiler house or to the Generating sets. The flow diagram indicating typical process is given in Figure 1.

Cost Economics of Bio Gas Generation System for a Distillery Waste

A. For Generation of Steam:

I Technical Data :

(i)	Distillery capacity	:	30,000 Litres / day
(ii)	Amount of spent - wash	:	450 cum/day
(iii)	COD of spent wash	:	90,000 mg/l
(iv)	BOD	:	50,000 mg/l
(v)	PH	:	4.0-4.5

Assuming COD reduction as 70%

Total organic load : 28350 kg/day
Amount of bio-gas generated : 0.5 cum per kg of COD destroyed
= 05. × 28350
= 14175 cum/day

(b) Average methane content of biogas would be 60% by volume and calorific value as 5400 cal/m³

Calorific value of Bio gas = 5400 × 14175
= 76.55×10^6 K cal/day

II. Price Gain Per Year

(i) Calorific value of coal = 400 Kcal/kg

(2) Coal equivalenrt of this = $\dfrac{76.55 \times 10^6}{4000 \times 10^3}$

= 19.14 T/day
(say, 20 Tons/ day)

(3) Price of coal considered = Rs. 1600 per ton
Savings per anuum = Rs. 1600 × 20
= Rs. 32000 per day

Taking average working period as 300 days in a year,
Price gain per year = 32,000 × 300
= Rs. 94 lakhs.

III. Investments

The total investment of the project is divided into two categories viz. capital investments and Operational expenditure.

Keeping detention time as 15 days, volume of digestor needed = 15 × 450 = 6750 cum

Provide a steel digestor of 7000 cum.

$$\text{Organic loading} = \dfrac{\dfrac{450 \times 90{,}000}{10^6}}{7000}$$

= 5.76 kg/cum/day

(i) Capital Cost

(1) Plant, Machinery including Royalty, Technology fee, erection and commissioning

= Rs. 200 lakhs

(2) Civil work, subject to soil condition	15
(3) Freight, Insurane, Octroi etc	20 lakhs
Total	235 lakhs

(ii) Operational Expenditure

(1) Interest at 15.5 on 75% of the cost of Biogas plant : Rs. 27.32 lakhs

(2) Maintenance at 1% on plant and maintenance : 2.00 lakhs

(3) Salaries of 1 chemist in-charge 2 Lab Chemists, 3 operator, 6 persons at Rs. 24,000/- per annum : 1.44 lakhs

(4) Chemicals : 1.00 lakhs

(5) Power required 50 HP i.e. 38 KWH.
Daily power required = 912 KWH : 5.472 lakhs
Daily electrical charges at Rs. 2
per KWH = 912 × 2 Yearly expense = 1824/-
1824 × 300

Total 37.232 lakhs

IV. **Savings**

1. Rs. 96.00 lakhs 37.24 lakhs = 58.76 lakhs
2. Pay Back Period = 4 to 5 years

Similarly the economics and the pay back period in respect of other agro based industries have been worked out and is given in Table 1.

TABLE 1 : Pay-back Periods of Various Agro Based Industrial Effluents for Biogas Recovery

Name of Inudstry	Capacity (Litres)	Operating days/yr	Operating Digestor volume cum	Digestor Detention Time days	Capital cost Lakhs	Power required HP	O&M Cost	Revenue gain per year	Pay back perior (yrs)
Distillery	30000	200	7000	15	200	50	15	54	11
		250	7000	15	200	50	17	67.5	7
		300	7000	15	200	50	19	81	5
Sugar	2500mt	200	1500	3	40	30	5	2	14
Pulp and paper	20mt	250	1500	5	80	40	8	13.5	6
		300	1500	5	80	40	9	16.5	4
Dairy	50,000	3	450	3	20	20	4	1.5	13

(B) For Power Generation

I. Technical Feasibility

1. Distillery Capacity : 30 KLPD
2. Quantity of Effluent : 450 m^3/hr
3. COD (mg/1tr) : 90,000
4. BOD (mg/1tr) : 50,000
5. PH : 4.0-4.5

II.	Outflow Characteristics		
	1. Effluent Quantity	:	450 m^3/day
	2. COD Reduction	:	70% ± 5%
	3. BOD Reduction	:	85% ± 5%
	4. pH	:	7.0-7.5
III.	Biogas Production		
	1. Biogas Generated @ 0.5 Nm^3 /kg of COD destroyed	:	14, 175 Nm^3/ day i.e. 590 Nm^3/d (approx)
2.	Methane content in biogas	:	60% approximate
3.	Net calorific value of Biogas	:	5400 kcal / Nm^3
IV.	Savings		
	1. Power generated @ 2 KWH per cu. mtr of biogas	:	1181.25 KWH (1476.56 KVA)
	2. Cost of Electricity	:	Rs. 2 per unit (likely to be revised to Rs. 3)
	3. Savings in terms of power (For 300 working days)	:	Rs. 170.10 Lacs/yr
V.	Manpower Required		
	1. Chemist Incharge	:	1
	2. Operators	:	3
	3. Lab Chemists	:	2
VI.	Economics		
	1. Plant & machinery including Royalty, Technology fee, Erection & Commissioning	:	Rs. 350.00 Lacs
	2. Civil work & oil storage & handling facilites	:	Rs. 25.00 Lacs
	3. Freight, Insurance, Octroi Duties, Taxes, etc.	:	Rs. 30.00 Lacs
	Total	:	405.00 Lacs
VII.	Operational Expenditure (For 300 Working Days)		
	(A) Fixed Cost...		
	1. Interest @ 15.5% on 75% of the cost of Biogas Plant	:	Rs. 47.08 Lacs/yr
	2. Maintenance @ 1% on Plant and machinery	:	Rs. 3.50 lacs/yr
	3. Salaries & wages of chemist Incharge & Operators 6 persons @ Rs. 24,000/- per annum	:	Rs 1.44 lacs/yr
	Total		Rs. 52.02 Lacs/yr

(B) Variable Cost :
1. Chemicals : Rs. 1.00 lacs/yr
Grand Total (A+B) Rs. 53.02 lacs/yr

After stabilisation of the process normally no chemicals are required to be added regularly.

VIII. Saving
Rs. 170.10 lacs (—) Rs 53.02 lacs : Rs 118.08 lacs/yr
Pay- back period : 3–4 years

IX. Means of Finance
1. Own Contribution : 25%
2. Loan from F.I.s : 75%
100%

Potential

A survey conducted of various industries in the country has established that about 1500 MW of Power can be generated from liquid effluent. The main agro based industries covered are sugar, distillery, pulp and paper, food processing industries, dairy and poultry. The agro based industrial effluents have high organic content and solids and are completely bio-degradable and can be treated biologically and bio-gas recovered as a byproduct.

Extensive research and studies have been carried out in the existing Molasses based Distilleries, Black Liquor paper mills, effluents from sugar and Dairy industries for establishing the design and performance parameters of Anaerobic digestors.

Extensive trails and work have been carried out in India to involve a appropriate technology to utilise the biogas as a source to generate electric power. 2 KW can be generated per cubic meter of biogas. Thus the biogas produced through biogas methanisation of spent wash from a 30 KLPS distillery can generate about 1 MW of power for its captive use. The biogas can be fed to the Generating set to obtain best efficiency on dual fuel Engines, using biogas and diesel economically. Bio gas will be used as primary fuel and HSD as pilot fuei. The thermal efficiency vary from 35 to 40%

TABLE 2 : Biogas Production From Various Agro Based Industrial Wastes.

S.No	Parameter	Distillery	Sugar	Paper & Pulp	Dairy
1.	Production capacity	30,000 ltrs	2500MT	20MT	50,000 ltrs
2.	Waste generated cum./day	450	500	300	150
3.	COD mg/l	1,00,000	3000	30,000	4000
4.	Organic load kg/d	45,000	1500	9000	600
5.	Bio gas cum/day	14625	485	2925	195

PRESS MUD FROM SUGAR INDUSTRIES

Among the various organic substrates, sugarcane filter cake or press mud (obtained from clarification of mixtures of sugar cane juice, lime, sulphur dioxide and phosphoric acid) is one of the material which is abundantly available in our country.

Two demonstration bio-gas plants were set up in one existing sugar factory in Maharashtra for Bio-gas generation from sugarcane filter cake. The plants were receiving a daily charge of 1.4 tonnes of filter cake along 3.5 cum of water in order to have 35 days Hydraulic Detention time. The results indicated the biogas production of 80 litres per kg of fresh mud. The characteristics of influent and effluent slurry of press mud is given in Table 3.

TABLE 3. Characteristics of influent and effluent slurry of press mud and Daily Bio gas production

Parameter	Influent	Effluent	Biogas
pH	5.91	7.3	120 cum/day
TS	98500	158000	
TVS	95850	50150	
COD	1,44,000	71,000	
BOD	21,957	7383	

It has been estimated that 99 Mt/day of filter cake is produced in a sugar Mill. Taking average size of an Indian Sugar mill plant as 2500 TCD (Tonnes of cane per day) and 421 such units, the potential of power generation in this biogas generation is considerable if fully harnessed. This is in addition to power generation from liquid effluent from sugar factories.

STRATEGIC PLAN

Since solid/liquid effluent is available in every region of India both in town / cities, such power plants can be installed in respective places. Of course sizing of the plant will depend upon the potential availability of the waste. Installed capacity of such plants can vary from 2.5 MW to 40 MW. Since these plants work on resources available on 24 hour basis and are close to the load centre, they can offer high PLF. In spite of these developments the participation of the private sector in this industry is essential because waste treatment for power generation has not been given the recognition it deserves. Moreover there is lack of communication between the various concerned departments of the government, State Electricity Boards, municipal corporations, Agro based industrial sectors and financial institutions. Fiscal and financial support is available to this activity in the form of debt, sales tax and excise exemptions and the policies formulated by the Government further more, need to be unambiguous.

CONCLUSION

Bio degradable solid and liquid wastage is the cheap source for the sustainable generation of power and bio-fertiliser. In addition to this, it also helps in abatement of pollution and saves depleting natural resources such as fuel and land. In future this offers great opportunity of sustainable energy generation.

FLY ASH AS BUILDING MATERAL IN INDIA

36

S.N. Mukherjee
Gentral Fuel Research Institute, India

ABSTRACT

Fossil fuel, being abundantly available, has always been the single largest source of commercial energy. By 2010 AD in India too, 30 per cent of the total energy will be produced by thermal power sector (coal and gas) where coal and lignite are to play the major role when total installed capacity in thermal power generation will be almost twice the present capacity. Fly ash generation will naturally increase to about 100 million tonne or more per annum which is to add to the pollution of environment as well as the acute disposal problem. On the other hand, to execute the programme of the National Housing Policy there will be huge shortage of conventional building materials. In view of the above, the tremendous potentialities of fly ash to produce different types of building materials and their present status in India have been discussed.

Introduction

It has been forecasted by the Chief Executive, World Coal Institute, U.K. that by 2010 AD coal will again be the world's single largest primary source of energy with the invention of new eco-friendly conbustion technologies[1]. In Indian energy scenario, coal and lignite are already playing a major role in spite of quest for cleaner energy source driven by the concern for environment. At present, near about 70 percent of the total energy in being produced by the solid fossil fuel which is expected to go up by another 8 per cent or so by 2010 AD when installed capacity of thermal power generation in the country will be around 1.1 lakh MW assuming 15 per cent annual addition as per the programme of the Planning Commission[2]. General run of Indian power coal being of high ash content, the fly ash generation will naturally go up and be somewhere between 100 and 110 million tonne per annum[3] during the period as shown in Figure 1. This huge quantity of ash is to add to the pollution of environment causing serious health hazards to human beings and vegetation[4] as well as acute disposal problem. Annual land requirement for disposal of fly ash alone is minimum 2 acre per MW power generated. In ideal condition i.e., with 1.0 PLF for dumping of fly ash, a minimum of 2.2 lakh acre of land has to be arranged for 1.1 lakh MW installed capacity for a life span of a thermal power plant, of 30 years only[5]. This will create a very hazardous situation not only in respect of eco-system but also wastage of land for the country. But power has to be generated as because power is the prime pre-requisite for advancement of any country.

On the other hand, human settlement in rural and urban areas, not only in India but also in most of the under developed countries consume earthy resources for construction of their dwelling units. The indiscriminate use of earthy material i.e. mainly top soil for shelter as a whole becomes critical in any effort to arrest and reverse the current alarming trends in reserve-depletion and environmental degradation[6]. Moreover, the National Housing Policy adopted by Govt. of India clearly defines its objective that "to help every family to own an affordable shelter by the year 2001 AD". Considering the rate of growth of population, the National Building Organisation (NBO) has estimated that the requirement of dwelling units during the current plan period alone is 21.8 million. This is to result in huge shortage of conventional building materials for which the annual cost of building materials has been increasing by 13-14 per cent even when the annual inflation is less than double digits. Here again, if the clay brick, as per present rate @ 55 billion/ year), is produced, it will consume 1 foot deep soil of about 1 million acre of land[7].

Therefore, a search for alternative raw materials for new building components like bricks, cement, etc. is an immense necessity. The results of R & D carried out in India and abroad on utilisation of fly ash for the production of different types of building materials indicate its tremendous potentialities as a raw material for the purpose. Use of fly ash will not only save the land which would have been otherwise used for its storage, but will also save the eco-system from potential damage arising out of its dumping. Further, production of building materials, especially the bricks/blocks using fly ash would save the corresponding amount of precious top soil, which would have been used for clay brick making. In spite of the above, utilisation of fly ash in India is not at all encouraging which is only around 3 per cent of the ash generated per annum.

Fly Ash as Raw Material

The R & D carried out of different types of building and constructional materials using fly ash over the last 3-4 decades is quite voluminous considering the available literatures from National and International Insitutions/Organisations. The scheme of total utilisation of fly ash as enunciated in the Keynote address[8] in a Seminar in 1967 at Pittsburgh (USA) has been depicted in Figure-2, although its techno-economy has been found to be doubtful in some cases considering locational advantages/disadvantages and/or market potentiality. Based on such R&D activities, quality of some of the building components has been drawn up and presented in Table-I. All these have judiciously been modified with incorporation of fly ash by different countries. In Indian context also, considering the shortage of different building materials (Table-2) during the current plan period, utilisation prospects of fly ash by the turn of the century has been drawn up[3, 9] and presented in Figure-3 which indicate about 50 percent of ash generated could be utilised gainfully if proper steps are initiated. Out of this 50 per cent, building materials/components as well as other construction activities may consume fly ash around 70 per cent.

Fly Ash Based Technology

Number of process technologies for the utilisation of fly ash for the manufacture of building materials/components[3, 9, 10] have been developed in the country by various Research Institutions, Public and Private Sector Agencies and Thermal Power Stations. The present status of some of these technologies[11] has been set out in Table-3. The potential and prospective areas for use of fly ash as raw material for making building components are:

(i) Portland cement clinker
(ii) Portland pozzolana cement
(iii) Ready mix fly ash concrete
(iv) Sintered fly ash light weight aggregates
(v) Aerated light weight cellular blocks/slabs
(vi) Clay fly ash bricks
(vii) Fly ash sand lime bricks/blocks
(viii) Unsintered fly ash sand lime aggregates
(ix) Lime pozzolana mixture
(x) Masonry cement
(xi) Fly ash lime gypsum bricks/blocks (FAL-G)
(xii) Pre-cast fly ash concrete building units
(xiii) Tiles for floor and wall panels
(xiv) Sodium silicate bonded bricks.

Apart from the above, fly ash could also be utilised for other constructional activities, i.e.,

(i) Structural fills
(ii) Water dam concrete
(iii) Stabilisation of road base and sub-base

Salient features of some of the various above technologies developed so far were examined and evaluated by Building Material Technology Promotion Council, Ministry of Urban Development, Govt. of India[9] and the results of which have been set out in Table-4.

TABLE-1 : Normal Range of Properties of Building Bricks and Blocks

Property	Clay bricks	Concrete blocks (dense)	Light weight aggregate concrete block	Calcium Silicate bricks	Aerated concrete block (autoclaved)
Strength, compressive strength tested in a wet condition (kg/cm^2)	105-840	70-350	20-175	105-560	20-55
Drying shrinkage (percent of length)	0.0-0.015	0.02-0.05	0.04-0.08	0.01-0.035	0.05-0.1
Thermal conductivity (K) gram-cal/cm^2/hr/°C/cm	6-2-11-1	8-7-14.9	1-2-6.2	10.0-13.6	0.9-1.9
Durability under severe natural exposure	Excellent to very poor	Good to poor	Good to poor	Good if strength above 150kg/cm^2	Good

Fire resistance	Very good ro moderate	Good to poor	Very good to poor	Good	Very good
Efflorescence	Fairly common none too severe	None or slight	Generally none	None	None or slight
Density (g/cc)	1.44-2.4	1.76-2.24	0.64-1.6	1.68-2.08	0.40-0.83

Source : Taylor H.F.W. The Chemistry of Cement, Vol. 2, 1964

TABLE 2 : Requirement of Various Building Materials in the Eighth Plan Period of India

Sl. No.	Material	Unit	Requirement	Quantity available	Shortage	Remarks
1.	Cement	Million tonne	131.98	114.93	17.05	6% growth rate assured, 40% total available for housing sector
2.	Steel	Million tonne	31.94	11.24	20.70	5% growth rate assured, 20% available for housing sector
3.	Bricks	Millions	286945	231598	54898	3% growth rate assured, 80% for housing sector.
4.	Aggregated	Million tonne	555.94	270.35	285.59	6% growth rate assured 40% for housing sector.
5.	Timber	Million cu. m.	29.09	15.75	13.34 assured,	3% growth rate 40% for housing sector.

TABLE 3 : Status of flyash based building material units in India

S. No.	Product	Location	Source of Technology	Rated output	Utilisation of fly ash	Status
1. 2-	Fly ash sand—lime brick	M/s Jagadhatrl Brick Works Pvt Ltd. Barrackpore, West Bangal	CFRI/NRDO	9000 bricks per day	88%	In operation
2.	—do—	M/s Bright Bricks Pvt. Ltd., Golthan, Raichun Karnataka	—do—	1000 bricks per hour	80-82%	In operation
3.	—do—	M/s Sand Plast (India) Pvt. Ltd. Berar, Alwar Rajasthan	Dorstner, Germany	4000 brick per hour	30-35%	-do-
4.	Flyash sand-lime blocks/ bricks	M/s AEC Cements & Construction Ltd. Ahmedabad	Own source	10000 bricks per day	80-82%	-do-
5.	Fal-G Blocks	M/s Bhanu International Visakhapatnam,	M/s Bhanu Int. natl. AP	400 bricks per hour	60-70%	-do-

TABLE 3 Contd...

S. No.	Product	Location	Source of Technology	Rated output	Utilisation of fly ash	Status
6.	Clay flyash brick	Kolaghat TPS, Midnapur, WB	CBRI, Roorkee	30 lacs bricks/yr	40-50%	-do-
7.	Clay-flyash brick	CPWD, Delhi	-do-	20 lakhs brick/yr	25%	-do-
8.	Flyash-sand lime brick	Neyveli Lignite Corpn Neyvell, T. Nadu	CFRI/NRDC	1000 bricks per hour	82%	-do-
9.	-do-	S.G. Brick Field Pvt. Ltd. Bandel, West Bengal	-do-	30000 bricks per day	80-85%	-do-
10.	-do-	M/s Pulver Ash Projects Ltd., Bandel, W.B.	-do-	80000 bricks per day	80-82%	-do-
11.	-do-	M/s Thirumala Flash Tech (P) Ltd. Vijayawada, II AP	-do-	30000 bricks per day	80%	To be commissioned shorty.
12.	Fly ash-sand lime brick	M/s Sterling Kak Sand bricks Ltd Indore, MP	ATLAG BREMEN Germany	4000-5000 bricks/hr.	30%	To be commissioned shortly.
13.	Flyash-lime gypsum bricks	Ahmedabad Electric Co Near Ahmedabad, Gujarat	AEC, Ahmedabad, Gujarat	10000 bricks/day	80%	-do-
14.	Clay-flyash bricks	M/s Pushpa Builders, Gaziabad UP	CBRI. Roorkee	15 million bricks/yr	40-50%	Proposed
15.	-do-	M/s Tatanagar Bricks Ltd., Jamshedpur, Bihar	China	54 million per year	80%	-do-
16.	Flyash brick	M/s NUTECH Flyash Product Ltd., Calcutta	Czechoslovakia	50000 bricks/day	60%	-do-
17.	Flyash-sand lime bricks	M/s Gujarat Narmada Flyash Co. Gujarat	CFRI/NRDC	—	80%	-do-
18.	2 Nos -do-	Pvt. Sector, Rajghat Thermal Power Station, Delhi	Yet to be decided	3000-4000 bricks/hr each	70-80%	-do-
19.	-do- -do-	Indraprastha Power Stn.	-do- Pvt. Sector, DESU, Delhi	-do-	70-80%	-do-
20.	Flyash-sand lime bricks 4 Nos	Pvt Sector, Badarpur Thermal Power Stn. Delhi	Yet to be decided	3000-4000 bricks/hr.	70-80%	Proposed
21.	Clay flyash brick	DVC, Durgapur, WB	CBRI, Roorkee	2000 bricks per hour	70%	Under erection.
22.	Flyash based building centre	Chandrapur, Maharashtra	CBRI, Roorkee	30000 bricks/day	25-30%	-do-
23.	-do-	HUDCO, NTPC, Dadri, UP	CFRI/NRDC	—	70-80%	Proposed
		HUDCO, NTPC, Dadri, UP	Dutch Tech	—	—	—
24.	Flyash-sand-lime brick	GAIL, Durgapur	CFRI/NRDC	5000-10000	80-83%	Under process
Cellular Concrete (AAC)						
1.	Aerated light Weight concrete (AAC) Block/Slab	M/s Ballarpur Industries Ltd, (BILT), Palwal, Haryana	Y- Tong, Germany	560 m^3day	60% Operation	In

TABLE 3 Contd...

S. No.	Product	Location	Source of Technology	Rated output	Utilisation of fly ash	Status
2-	-do-	Tamil Nadu Housing Board, Ennore. Madras	Polymex-Cakov Poland	500m ^{3}/day	30%	-do-
3.	Callular Concrete Block	HPL, New Delhi	Indigenous	—	60-70%	-do-
	Cellular concrete blocks & slab	Punjab State Indl. Dev. Corpn (PSIDC, Bhatinda, Punjab)	Y-Tong Germany,	500m^3/day	70-60%	Proposed
	-do-	West Bengal Power Dev. Corpn Ltd. (WBPDCL), Kolaghat, WB	EPDC, Japan	5000 m^3/day	70/80%	-do-
Synthetic Aggregate						
	Synthetic flyash light weight aggregate (sintered process)	West Bengal Power Dev. Corpn., Kolaghat, WB	EPDC Japan	—	80-85%	Proposed
	Synthetic light weight flyash aggregate plant (cold process)	M/s TATA Elec. Co., Bombay	Aardelite Netherlands	Pilot Plant	70-80%	To be commissioned
	Synthetic flyash light weight aggregate (sintered process)	M/s Guardian Plasticode Ltd., Calcutta (WB)	Own source	50 ton per day	70-80%	Proposed
Portland Pozzolana Cement (PPC)						
	PPC	Associated Cement Co Thane, Bombay	Own source	—	20-25%	In operation
	PPC	Cement Corpn of India Haryana	Own source	—	20-25%	-do-
	PPC	M/s Madras Cement Ltd Hyderabad	Own source	—	1000-2000 MT/month	Reported
	V. Concrete Flyash concrete (FAl-G)	Institute of Solid Waste Research & Ecological Balance, Visakhapatnam	Salzjitter, Germany	—	20-25%	Reported
	Flay ash concrete	Neyveli Lignite Corpn Neyveli, T. Nadu	NLC, Tamil Nadu	—	10-30%	Reported
	-do-	Rihand Dam, Orissa Tenughat, Kakki	Basin Bridge Madras	Highways Research Station, Madras & SERC	10-30%	-do-
	Precast Component Lintels, Sunshades, Jallies, doors, windows frame, roof slabs, prulin, etc.	NLC, Tamil Nadu	NLC, TN	—	—	Reported

Considering the present position, some problems and prospects[12] related to production and marketing of fly ash based building materials/components are mentioned below:

— Very cautious and slow approach by the users towards the use of new building materials without valid scientific reasons.
— Construction agencies are disinterested in the production of building materials. As a result, the new technologies have been the victim of as to who takes the lead.
— The potential market of fly ash based building materials is generally far away from the fly ash generating plants.
— Total lack in R&D inputs for the development of cost effective plant and machineries for the new and emerging technologies.

To overcome the above as well as to augment ash utilisation for the production of building materials, the following steps may be considered:

— Positive interest is to be taken for promoting the new and emerging technologies for fly ash utilisation through suitable financial intruments like buy back guarantee, capital, land, power, water, etc.
— Action may be initiated for equity participation and buy back arrangements by the construction agencies.
— Excise duty on indigenous plant and machineries as well as sales-tax on product are to be exempted, at least for some years during intitial period.
— Use of precious top soil should be restricted for brick making where fly ash is available through suitable legislation with ultimate aim to ban completely.
— R&D has to be taken up for indigenous plant and machineries.
— Policy guideline on fly ash utilisation should not necessarily abide by economic consideration alone, but should also strategically consider the far reaching effects and impacts on eco-system and society.
— Adequate publicity for popularising alternative building materials and for creating interest among common people and users is the prime necessity.

If the actions are earnestly intiated as above right now, a substantial amount of fly ash, may be 50 per cent or so of what is generated per year, could be utilised for the production of different types of building materials/components alone before much damage is done. This may save proportionately the agricultural land otherwise wasted for dumping the ash and used for making building materials. Further, another point needs to be mentioned which is to take steps to minimise generation of fly ash at thermal power station (TPS) itself. The studies carried out and reported in 1981 on the techno-economics of ROM and beneficiated coal for power generation by National Productivity Council in collaboration with National Thermal Power Corporation reveal that even a reduction of 15 percent ash of feed coal used in TPS will tell upon the generation cost insignificantly whereas the Nation will produce 15 percent less ash which is not at all insignificant.

TABLE 4 : Evaluation of Technology for Conversion of Flyash into Building Materials and Components

S. No.	Building Materials & Components	Process Technology Investigators	Use of dry flyash	Use of wet flyash	Extent of utilisation (%basis on end product	Remarks
1.	Portland Pozzo lana cement	Central manufacturing unit, CBRI, NCB, BICP, CEA, APSEB, WVSEB, ACC	Essential	—	Max 30%	Commercially exploited, Quantum of utilisation does not match the cernent production due to various reasons.
2.	Portland cement clinker	-do-	Essential	—	Max. 30%	-do-
3.	Flyash concrete	CSMRS, CBRI, Cement manufacturing units HPL, NCB, AEC	Essential	—	Max. 25%	-do-
4.	Synthetic light weight flyash aggregate	CBRI, ACC, TATA Elec. Co. WBPDCL	Desirable	Possible	Max.80%	Pilot plant studies conducted. No commercial production so far.
5.	Light weight auto-claved Aerated Concrete (cellular concrete.)	TNHB, CBRI, BRL, HPL PSIDC, BILT	Essential	—	Max. 70%	Commercially exploited with foreign technology. Few machines not available indi-genously, require import.
6.	Lime Pozzolana Mixture	CBRI, NLC, CRRI, ACC pvt. Entrepreneurs, AEC, NBO	Desirable	Possible	Max. 75%	Has not become popular due to non-uniform quality and increased availability of cement.
7.	Clay-flyash brick	CBRI, CPWD, WBPDCL, Pvt. Entrepreneurs, AIBTMF,	Desirable	Possible	Max. 35%	Commercially exploited. Gradually gaining

		NTPC, CPRI				popularity.
8.	Flyash-sand-lime bricks	CFRI, NRDC, NCB, NLC, NTPC, TISCO, AEC, HPL, CBRI, ABC, several	Desirable entrepreneurs	Possible	Max. popularity.	80% Commercially exploited Gradually gaining
9.	Flyash-lime-gypsum block	M/s Bhanu International Pvt entrepreneurs	Desirable	Possible	Max. 70%	Has b n utilised in small project, Techno-economic feasibility is yet to be established for large scale application.
10.	Soil stabilisation	CSMRS, CBRI, CRRI	Desirable	Possible	Max. 10%	Has been used in few small scale projects and for pilot plant investigation. Not value added.
11.	Road construction	CRRI, University of Roorkee, IRC	Desirable	Possible	Max. 75%	Commercially exploited Quantum of utilisation not significant. Technically feasible, but not value added
12.	Land filling/ reclamation	CSMRS, CRRI, NTPC	Desirable	Possible	Max. 95%	Investigation pilot plant scale. Commercial use negligible. Not value added.

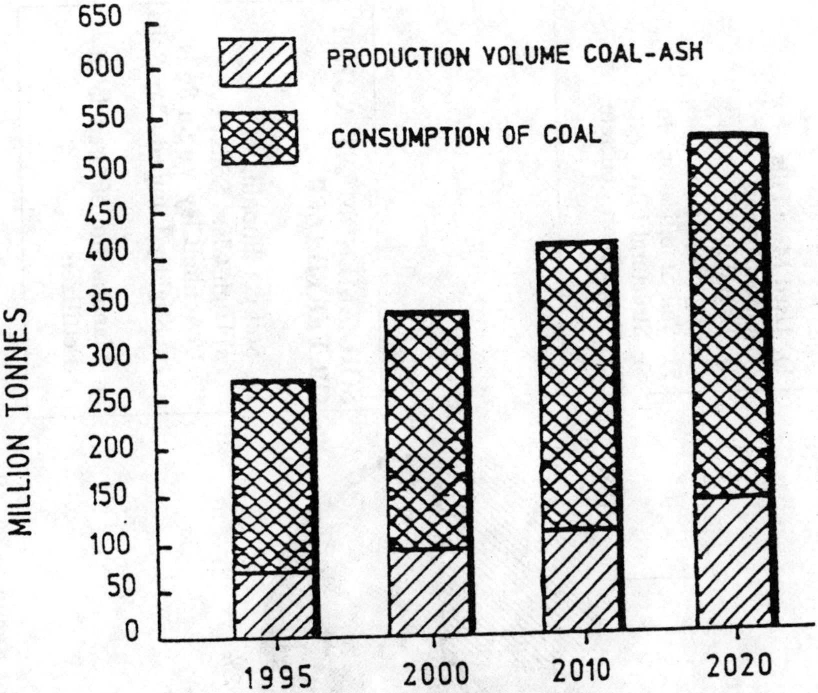

FIG. 1 : *Prospect of Production Volume of Coal Ash*

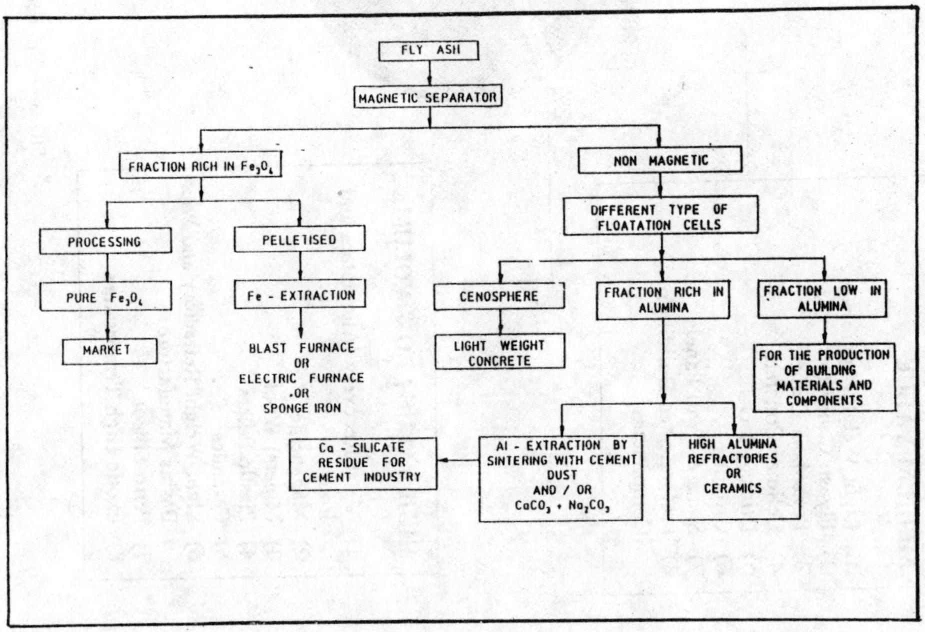

FIG. 2 : *Scheme for total utilisation of Fly Ash*

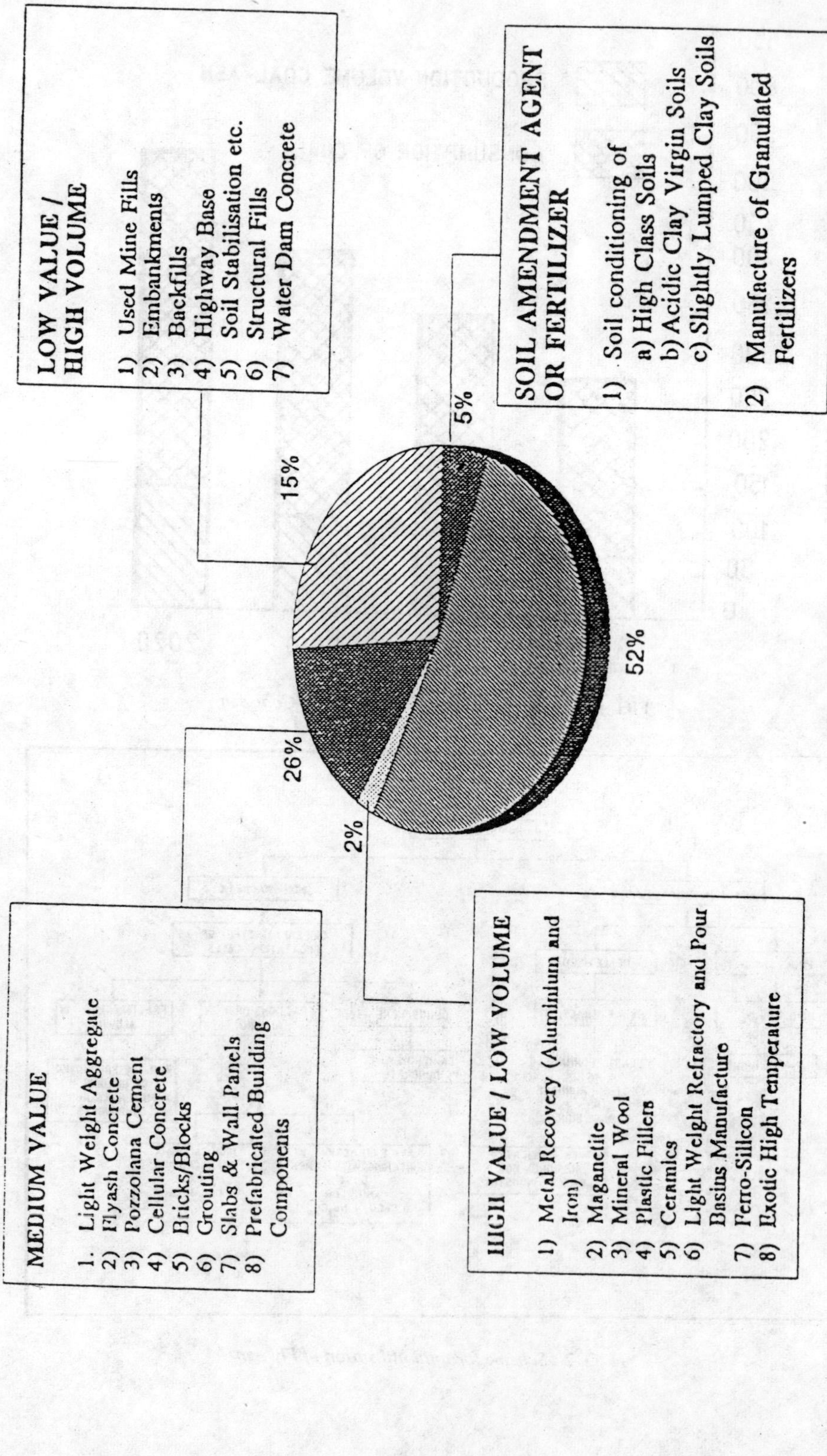

FIG. 3 : *Utilization Prospects of Flyash by 2000 AD*

Conclusion

Before drawing up the conclusion, it would perhaps be better to visualise what actually may happen in 2020 AD in the background of what has been stated herein above. Ash generation per year will be about 135 million tonne (Figure-1). Very optimistically, it could be foreseen that following the current trend of ash utilisation, maximum 10 percent of ash could be utilised which means about 120 million tonnes of ash per year has to be dumped. This will require about 8000 acres of land per year for storage and at the same time, top soil for 10 feet deep of around 1.5 lakh acres of land will be utilised for production of conventional clay brick per year. Further, the ash dump will contaminate environment in various ways and affect human beings, flora and fauna as well as acquatic life because of the toxic trace metals present in ash which will also pollute directly or by leachates of land, surface and underground water due to gradual accumulation over the years. To-day, for disposal of fly ash, the code as per RCR Act of USA for protection of environmentally sensitive areas is "Fly ash disposal should not be encouraged in (1) wet land areas, (2) 100 years flood plains, (3) critical habitats for endangered species, and (4) recharge zone for acquifers". To-day, it is an Act in USA, but tomorrow, it may be a case for India too. Therefore, the only option to minimise potential damage of eco-system is to convert fly ash into a product of utility. Fly ash has to be utilised.

In view of the above and considering the issue in the total perspective, it may be concluded that the problem will be disastrous by 2020 AD. Perceiving this, the concerned authorities are to sit and adopt a National pledge, i.e., Ash has to be utilised for healthy growth of thermal power sector as well as the country and initiate action without much delay.

Acknowledgement

The atuthor is grateful to the Director, Central Fuel Research Institute, Dhanbad for his kind permission to present the papaer in International Conference on Energy "Asia Energy Vision 2020" to be held on Nov. 15-17, 1996 at New Delhi organised by World Energy Council.

REFERENCES

1. Tallboys, R.G. - Proce. Fifth New Zealand Coal Conference, Vol. I, October, 1993, 2.
2. Chari, K.S.R.- Address on Golden Jubilee Celebration of Central Fuel Research Institue, Dhanbad (India), Nov. 17th, 1995.
3. Gupta, T.N., Mukherjee, S.N., Majumdar, S.K.,. Sundarajan, V. and Sen Gupta, J. - Proce. International Conference on Environmentally Sound Coal Technologies" held at IIT, Madras, 15-18th Jan., 1992.
4. Studies on the investigation of Trace Metal content in Neyveli Lignite, CFRI Technical Report No. TR/CFRI/ 1.09/90, Central Fuel Research Insititute, Dhanbad (India), September, 1990, 41-49.
5. Techno-Market survey on "Technologies for disposal of Thermal Power Station Fly Ash" - Technology Information, Forecasting-Assessment Council (TIFAC), Dept. of Science & Technology, New Delhi, Nov., 1990, 5.
6. Ramachandran, A. -Building Materials News, BMTPC, Oct., 1992, 2.
7. Mukherjee S.N., Biswas, D. and Nag, A.K. - Proce. on Seminar on :Fly Ash Brick", WBSEB, Calcutta, 23rd August, 1990, 1.
8. Dedman, W.G. - Proce. on "Fly Ash Symposium", Pittsburg. 1967, 1, Bur. of Mines, I.C. No. 8348.

9. Report of Building Material and Technology Promotion Council, Ministry of Urban Development, New Delhi, 1993.
10. Specification for "Guidelines for Utilisation and Disposal of Fly Ash" -No. IS: 10153-1982, Bureau of Indian Standards, New Delhi.
11. Dr. (Mrs.) Saroj, Ministry of Environment and Forest, Govt. of India-Workshop on "Ash Utilisation - Present Trends", NTPC, Noida, July 25-27, 1995.
12. Mukherjee, S.N. - Proce. of Workshop on "The Problems and Prospects of Coal Ash Bricks"- Coal Ash Institute of India & The Institution of Engineers, Calcutta, September 27th, 1995, 1.

BENEFICIATION OF POWER STATION COAL

37

Dipl. Eng. Jürgen Winckler
General Manager Sales–Coal Division
KHD Humboldt Wedag AG
Cologne, Germany
and
Gurudas Mustafi, B.Sc., Eng. IRIMME
Chief of Marketing & Sales
Humboldt Wedag India Ltd.
Calcutta, India

ABSTRACT

Growing mechanisation of coal mining drastically raises the amount of refuse as well as the amount of fine material in the R.O.M. coal. This leads to an increase and a wide fluctuation of the ash content and to a higher moisture content of the coal. Due to this the amount of raw coal, which is beneficiated, is permanently increasing worldwise. Process steps for the various size fractions of the raw coal have proven their feasibility for the beneficiation of powerstation coal and are presented herein.

Aside the beneficiation of the raw coal the improved technologies allows for a recycling of the dumped unwashed coal, which has been wasted in former times due to selective mining methods for example. Even slurries out of ponds, which very often contain material with an ash content of less than 40%, can be beneficiated to make optimum use of the energy, which is available.

1. Beneficiation of Lump Sized Coal

1.1 General

A great percentage of the power station coal mined throughout the world has been burned so far without preparation or refuse separation, because the preparation costs are too high or combustible substance is lost in the preparation process. Considering, however the lower wear in the boiler houses and the increased volume of transportation in the case of prior separation of refuse from the power station coal, then a preparation, at least of the lump sized portion should be profitable.

Test regarding the abrasive properties of different types of coal have shown that the abrasion and thus the wear are considerably limited due to a separation of refuse from lumpy run-of-mine coal.

The preparation of coarse coal is a process step frequently found in Europe. This has to be attributed to the fact that the proportion of refuse in the raw coal varies with respective grain size. Normally the percentage of refuse in coarse raw coal is higher than in fine granular raw coal. In West Germany about 25 per cent of the raw coal is in the grain size fraction plus 40 mm, and

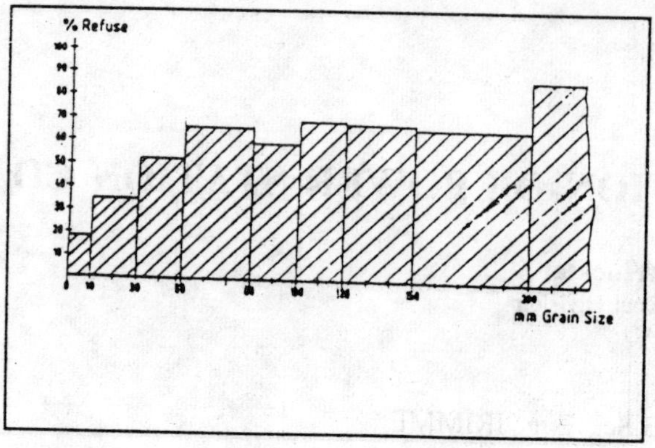

TABLE 1 : *Refuse content of a typical German ROM-Coal*

the refuse content of this fraction average 90 per cent, which is nearly one-half of all refuse contained in the raw coal (Table 1).

The removal of a certain protion of refuse form the uncrushed raw coal offers great advantages for the preparation process, such as for example.

- Decrease of power reqirements for size reduction.
- Increased yield of graded coal in the case of domestic coal production,
- decreased yield of slurry,
- decrese of ash content in the raw slurry,
- more favourable homogenizing properties of the raw coal,
- decreased wear in the concentration machines,
- more favourable utilization of the preparation capacity, i.e. increased efficiency of the preparation plant.

The development of the ROMJIG opened the possibility, to use a very simple process for the beneficiation of the lump size coal—the top size is 400 mm and the bottom feed size is 30 mm–by using only one technologically improved machine and some standard auxiliary equipment.

In fact the suggestion to reconsider an economic process for the beneficiation of ROM coal actually came from India. At the first joint meeting of the Indo-German Coal group, the German Beneficiation Industry was asked to propose a process, as far as possible dry, to solve the problem of preparing non-coking coal.

After many years of development such a plant has been installed in the Bina coal handling plant and is ready for operation. In this plant the raw coal is screened at 30 mm. The coarse coal 250-30 mm is fed into the ROMJIG and separated in refuse and clean coal.

1.2 Description of ROMJIG

First of all it has to be ascertained whether lumpy material could also be stratified according to density in water bath like the size fraction of minus 120 mm mentioned before. For this purpose sink tests with a size fraction of 400-40 mm were conducted in water bath which showed that after

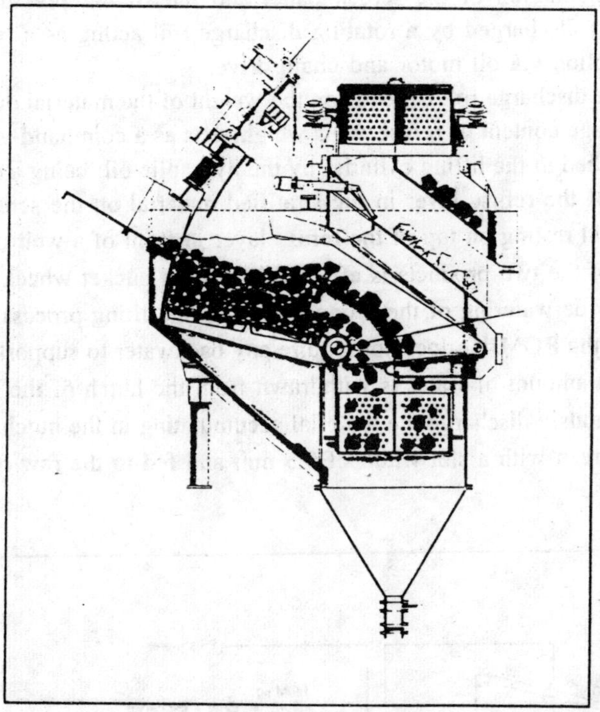

FIG. 1 : *ROM Jig*

repeated mechanical lifiting and free settlement of the material a stratification according to density is effected, with the refuse settling in the lower area while the lighter coal lumps are overlying.

This knowledge led to the construction of the Movable Screen Jig, designed as single cut jigging machine for the separation of stones from run-of-mine coal (Fig.1)

To find a proper name for this kind of jigging machine, we called in "ROMJIG' which is self-explanatory in defining the function of the machine. In Germany we call it also the "Indian Jig" because the initiation to develop it came from there. In the meantime the word "ROMJIG" has been registered to stand for this type of movable screen jig.

The ROMJIG consists of an unilaterally supported screen panel with a slot width of 15 mm which is hydraulically moved up and down in a water vesssel. The movement of the screen panel is effected via a hydraulic cylinder suspended on gimbals, by which the screen panel is lifted. Upon reaching its highest point the oil pressure in the hydraulic cylinder is released and the screen panel freely falls through the water to its lowest dead position.

The kinetic energy is absorbed by a hydraulic shock absorber connected with a central hydraulic plant preventing the building from receiving noticeable shocks.

The temperature of the circulating hydraulic oil is kept within permissible limits by a water-operated heat exchanger.

The material fed along the side of the maximum stroke length of the screen panel is stratified according to its density by the lifting movement of the screen panel. Transport is effected due to the inclination of the screen panel and the sliding pressure of the following material. The refuse is discharged by a rotating discharge roll acting as a retaining edge which is set into circular motion via oil motor and chain drive.

The speed of the discharge roll depends on the weight of the material dumped on the screen panel, i.e. on the average content of refuse. This weight acts as a command variable to determine the pressure to be applied to the lifting cylinder by the hydraulic oil; being a sufficiently accurate index for the height of the refuse layer in the stratified material on the screen panel.

The light material resting on top of the refuse layer in front of a weir moves over the weir edge. The discharge of the two products is effected by a twin bucket wheel, sufficiently dimensioned to allow proper de-watering of the products during the lifting process.Unlike the conventional coarse coal jigs the ROMJIG does not require any backwater to support the jigging process (Fig.2). Only a certain amount of water is withdrawn from the hutch of the ROMJIG through a gate in order to contnuously discharge the material accumulating in the hutch This hutch product is de-watered on a screen with a slot width of 0.5 mm and fed to the raw feed coal for further preparation.

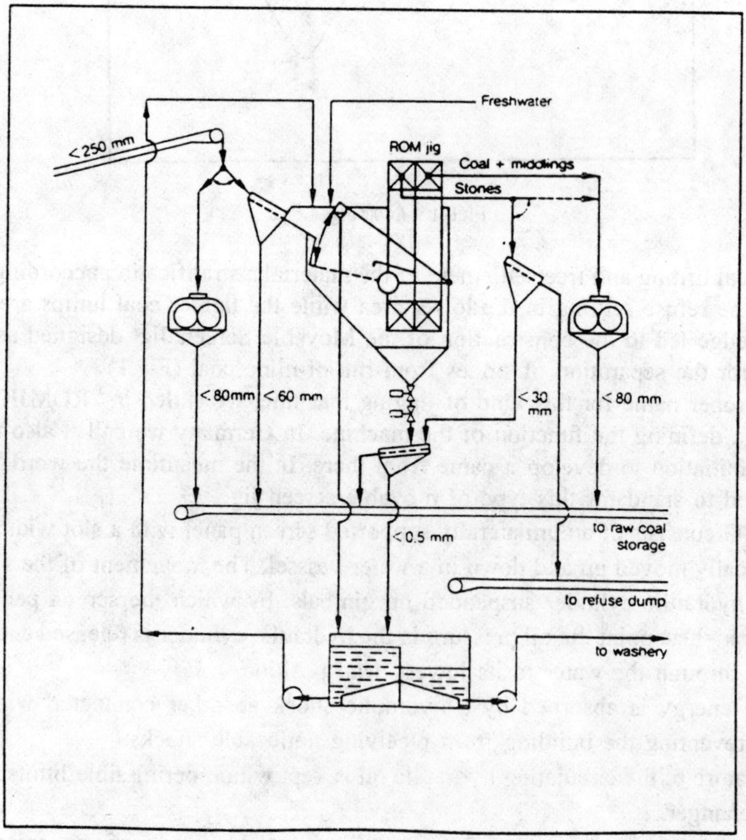

FIG 2 : *First Application of the ROM JIG*

The screen undersize does not need any additional clarification. It runs to a pump sump and is returned directly to the ROMJIG.

Quite generally one can say that the ROMJIG requires so much additional water as it is taken up and carried away by the particle surfaces of the product. As known from experience this comes up to 1-2 per cent or 3 to 6.5 m^3/h for the size fraction plus 30 mm.

1.3 Application of ROMJIG

Such a plant has been installed in the Bina coal handling plant and is ready for operation. In this plant (Fig. 3) the raw coal is screened at 30 mm. The coarse coal 250-30 mm is fed into the ROMJIG and separated in refuse and clean coal.

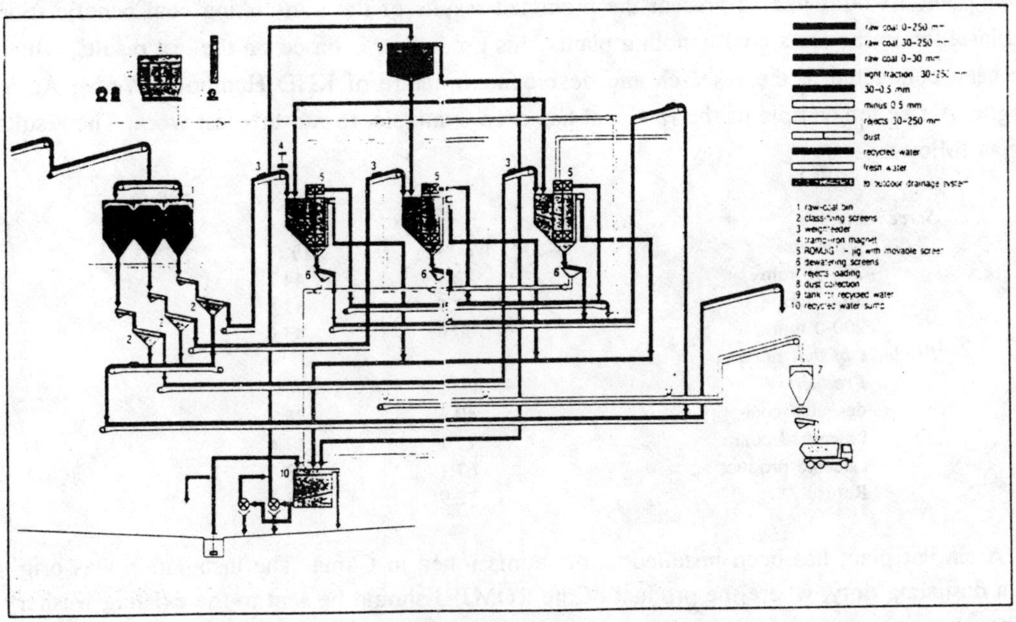

FIG. 3 : *Flowsheet of the Bina Deshaling Plant*

The refuse discharged from the ROMJIG is separated form the undersize on a stationary bar screen with a slot width of 30 mm and the oversize finally conveyed to the refuse tip.

The light product of the ROMJIG, consisting of coal and middlings, is crushed down to a washable particle size by a double-roll crusher and then conveyed together with the screened out undersize material to the two raw coal storages.

Fine material may accumulate in the water bath having passed through the opening of the jigging surface and will have to be extracted by an occasional introduction of water in order to avoid such accumulations.

This hutch product is de-watered on a screen and added to the light material.

The underflow of the de-watering screen is returned to the ROMJIG without any intermediate clarification. From operational data obtained it has been found that, depending on the feed rate, with its low proportion of fines, a balance of solids is formed which in no case exceeds 50 g. The ash content of the circulating solids came up to approximately 60 per cent.

This knowledge led to a considerable reduction of the freshwater requirement to 3.5 m^3/h. just sufficient for covering the losses of water absorbed by the products.

1.4 Results

At this point I would like to present the predicted results of the non-coking coal beneficiation plant installed in the Bina coal handling plant. This prediction is based on the test results, which have been conducted in the research and development centre of KHD Humboldt Wedag Ag in Cologne. A five ton sample of the raw coal has been available, to run this test work. The results were as follows:

Screen analysis

grain size	*Wt. %*	*Ash %*
200-30 mm	82.3	44.2
30-0 mm	17.7	37.6
200-0 mm	100.0	43.0

Products of that plant:

Product	*Wt.%*	*Ash%*
deshaled coal	49.4	28.3
unwashed coal	17.7	37.6
saleable product	67.1	30.8
Refuse	32.9	68.1

A similar plant has been installed and commissioned in China. The installation was originally a deshaling duty, where the product of the ROMJIG should be sent to the existing washery after having been crushed down to the required size. While commissioning this system the results were so excellent that the plant is selling the product of the ROMJIG as a final product. In fact the imperfection value of the plant has been determined to be 0.09.

1.5. Prospects

The successful first employment of a ROMJIG for large refuse separation of run-of-mine coal opens a wide field of future applications. In view of the favourable secondary conditions, such a low water and power consumption, low expenditure for necessary auxiliary equipment this process stands any economical comparison with other possibilities of large refuse separation and is the ideal process, in case people think of a pit head instalment.

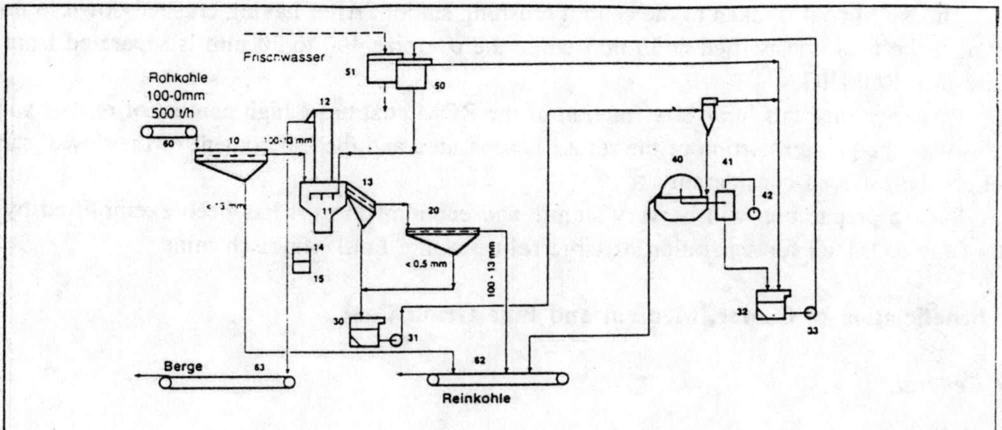

FIG. 4 : *Flowsheet of a Coarse Coal Beneficiation Plant*

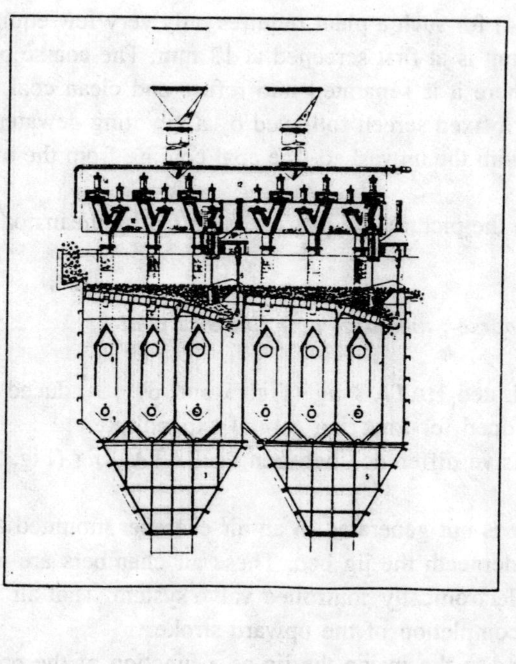

FIG. 5 : *Batac Coarse Coal Jig*

Such a preparation plant for power station coal be exemplified as follows:

The ROM coal is taken to the central crushing station. After having crushed down to minus 400 mm, the coal is classified at 30 mm while the oversize 400 to 30 mm is separated from the refuse in a ROMJIG.

By experience this lump size fraction of the ROM coal has a high content of refuse, so that in this way the greater portion of the refuse is separated and the ash content of the power station coal can be lowered considerably.

Such a preparation will be very simple and economical as it has been exemplified by the use of the ROMJIG for separation of large refuse at the Emil Mayrisch mine

2. Beneficiation of Coarse, Medium and Fine Grain Coal

2.1 General

Whenever the characteristics of the raw coal does not allow, to achieve the required product quality with the ROMJIG by only washing the lump size coal, the grain size of beneficiation has to be reduced. Reasons can be the grain size distribution. when the amount of material <30 mm is too small and the ash content of the fine material is too high. This leads immediately to the beneficiation of coarse coal by means of Batc® jig.

2.2 Description of the plant

A typical process (Fig. 4) for such a plant requires only very few equipments. The raw coal with a top size of 75 (120) mm is at first screened at 13 mm. The coarse product 75 (120)–13 mm is fed to the Batac jig, where it is separated into refuse and clean coal. The clean coal product is dewatered by means of a fixed screen followed by a vibrating dewatering screen. The dewatered product will be mixed with the unwashed fine coal coming from the underflow of the classifying screen.

As it is seen form the picture, the process is closed by means of a combination of cyclones and filters.

2.3 BATAC® jig for coarse-, medium- and fine-size coal

The under bed air, pulsated BATAC jig (Fig. 5 and 6) of reduced weight and constructional volume has been developed for lowering capital expenditure.

There is one decisive difference between an BATAC jig (Fig. 5 and 6) and a Baum type jig.

The water current is not generated in an air chamber mounted adjacent to the screen plate but in air chambers underneath the jig bed. These air chambers are intermittently supplied with compressed air by an electronically controlled valve system. That air is intermittently discharged from the system after completion of the upward stroke.

Motion is imparted to the inside the jig as a function of the pressure generated inside the air chambers.

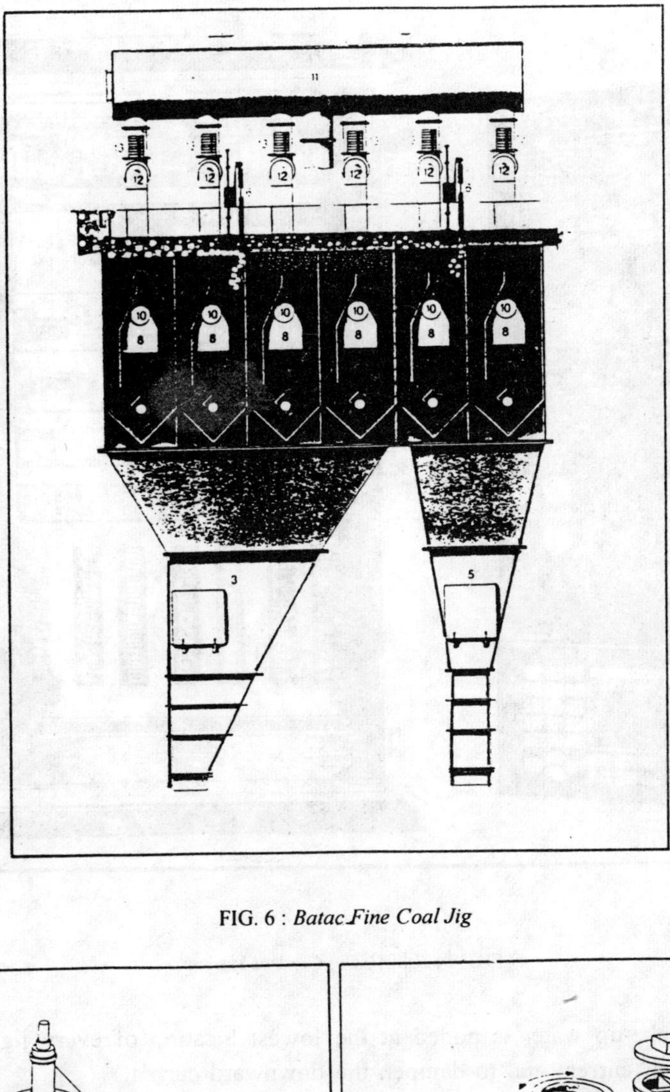

FIG. 6 : *Batac Fine Coal Jig*

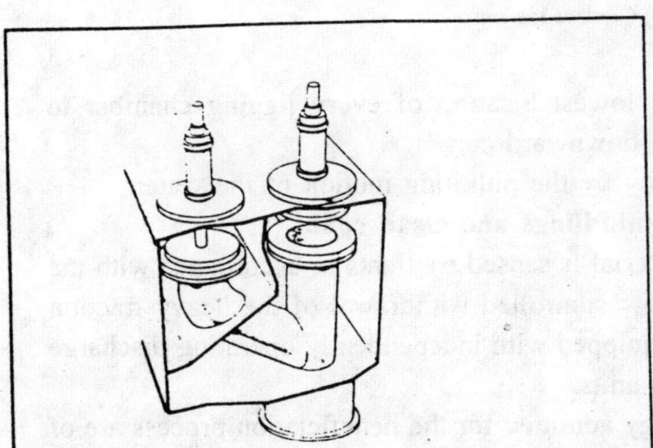

FIG. 7 : *Disc Valve*

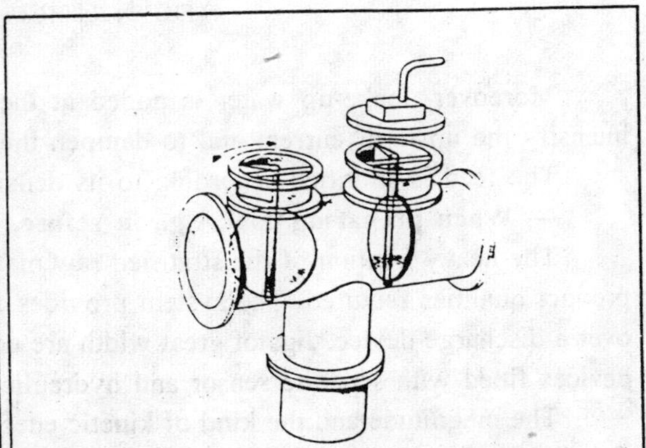

FIG. 8 : *Rotary flap valve*

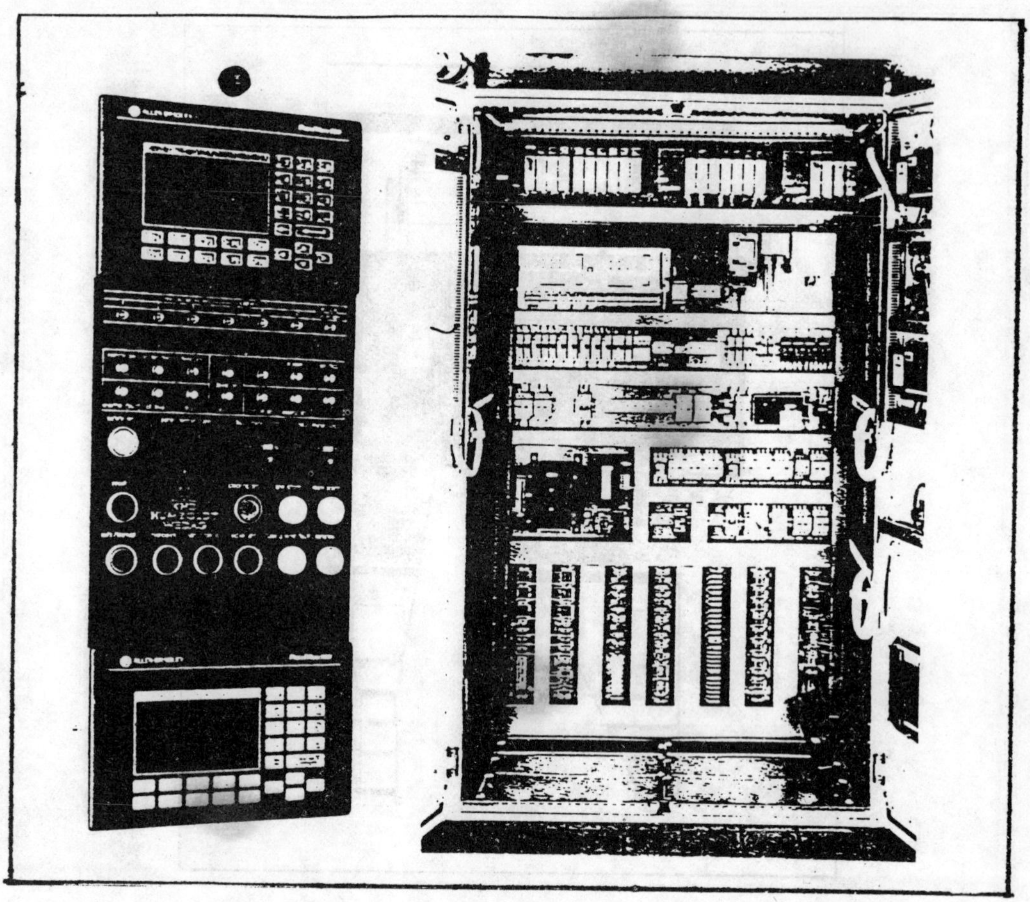

FIG. 9 : *BATAC Jig Control System*

Moreover, make-up water is added at the lowest location of every jigging chamber to intensify the upward current and to dampen the downward current.

The feed is stratified according to its density by the pulsating motion of the water:

— **When preparing coal, e.g., in refuse, middlings and clean coal.**

The heavy fraction of the stratified raw material is sensed by floats in accordance with the product qualities required. The system provides for controlled withdrawal of the heavy fraction over a discharge device. Jigs of great width are equipped with independently operating discharge devices fitted with separate sensor and hydraulic units.

The magnitude and the kind of kinetic energy admitted for the beneficiation process are of decisive importance for a successful separation .

Therefore, special attention has been given to the development of the pulse generators for the controlled admission of the compressed air (Fig. 7, 8).

The quantities of jigging air are controlled electronically. The amount of air can be set

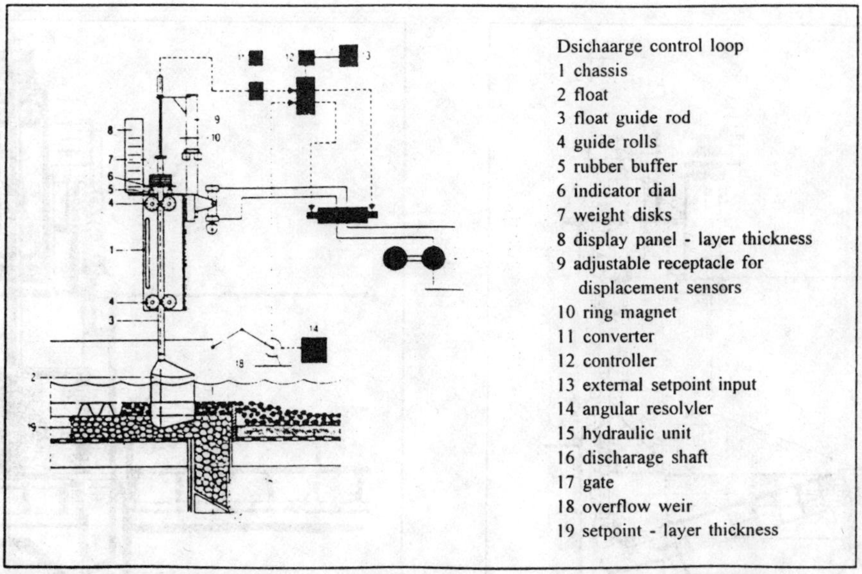

FIG. 10 : *Discharge control system*

individually for each jigging chamber and an identical pulsation frequency (40-120 strokes/min) is set for all jigging chambers in the jig PLC (Fig. 09).

The thickness of the coal layers stratified by specific gravity is sensed with the aid of a float equipped with a suitable linkage (Fig 10).

Formerly employed induced current metering units are today replaced by analog displacement measuring systems operating at ultrasonic velocity. The basic float setting is done by additional weights.

The hydraulic systems of the discharge gates (Fig. 11 and 12) are actuated from the. PID controllers which are integrated in the jig PLC (Fig. 09).

2.4 Prospects

The Batac jig is so flexible that the bottom washing size can be reduced from 13 mm to 6 mm or even to 3 mm, without a modification of the flowsheet. The only problem. is the dry screening at the lower cut points. This can be solved by using the right screen system under these circumstances.

Additionally the depth of beneficiation can be increased by adopting further processes, like the combination of coarse and fine coal jigging and at the end the addition of the flotation system, which is relatively seldom in the beneficiation of a power station coal.

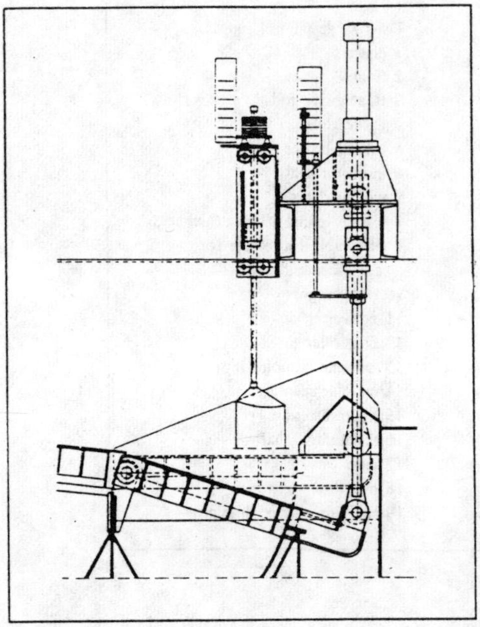

FIG. 11 : *Discharge Gate Coarse Size Jig*

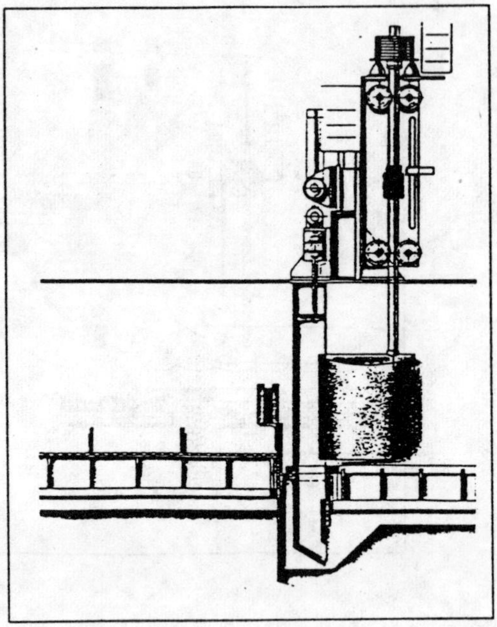

FIG. 12 : *Discharge Gate, fine size jig.*

3. Beneficiation of Dumped Coal

3.1 General

In good coal reserves – preferably in open cast mining – a relatively big amount of good quality raw coal is dumped, i.e. wasted, due to the selective mining method. This is the material close to the border between coal seam and shale band, which contains upto 50% of saleable coal.

Worldwide plants are being installed to process especially this so called interface material. The process flowsheet of such a plant appears very simple. The heart of this plant is the Batac jig and the depth of beneficiation is again defined by the raw coal characteristics and the product requirements. Due to this the flowsheets of such plants vary widely and I would like to present one system, which can be considered as a standard for this duty (Fig. 13).

3.2 Description of the Plant

The raw coal is fed in the grain size 50.0 mm to the beneficiation plant. Here it is screened at 0.5 mm first, especially to achieve a good distribution over the width of the following machine

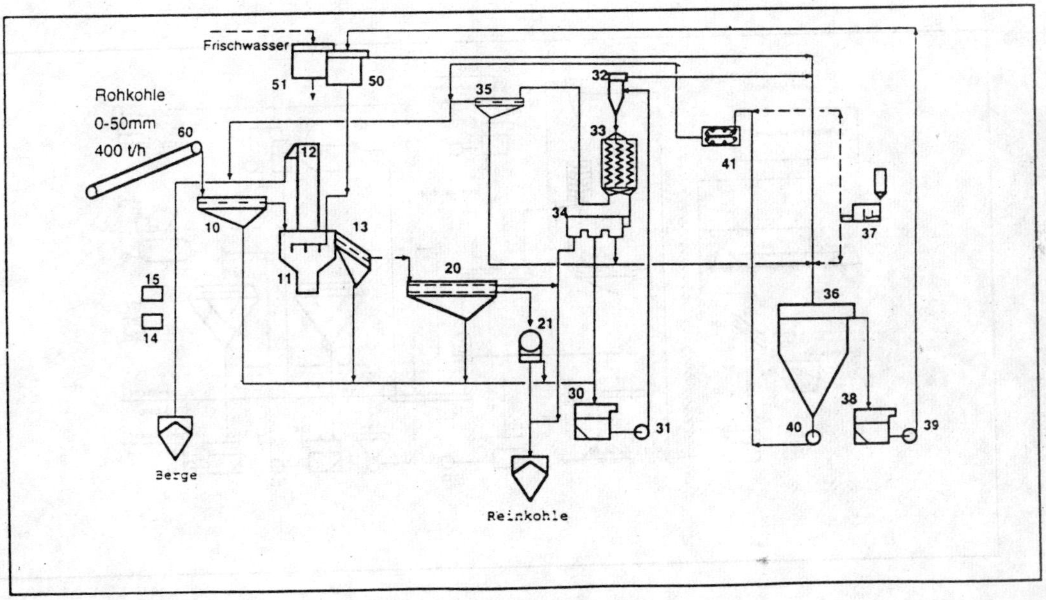

FIG. 13 : *Flowsheet of a Dumped Coal Beneficiation Plant*

and to wetten the material prior to the jig. The overflow of the screen is directly going into the Batac jig, where it is separated in refuse and clean coal.

The refuse are taken out of the jig by a dewatering bucket elevator and falls down onto a belt conveyor, which brings the refuse out of the washery.

The clean coal is overflowing at the end of the Batac jig and is dewatered by means of a fixed screen followed by a double deck vibrating screen. The overflow of the upper deck is dewatered enough and can be sent directly to the clean coal belt conveyor. The overflow of the lower deck is further dewatered in a basket centrifuge, whereas the underflow of the screen is classified together with the underflow of the desliming screen in cyclones. The underflow of the cyclones can be beneficiated in spirals. The spiral concentrate is dewatered together with the jig concentrate in the centrifuge, whereas the refuse are dewatered by small dewatering screen and then sent out of the plant together with the jig's refuse.

The overflow of the cyclone is fed to a thickener, from where the clarified water is recirculated back to the plant. The underflow of the thickener can be either dewatered on filtler presses or on centripress centrifuges.

4. Beneficiation of Pond Material

4.1 General

In many places in the world slurries are pumped into big slurry ponds. Aside the fact that this is

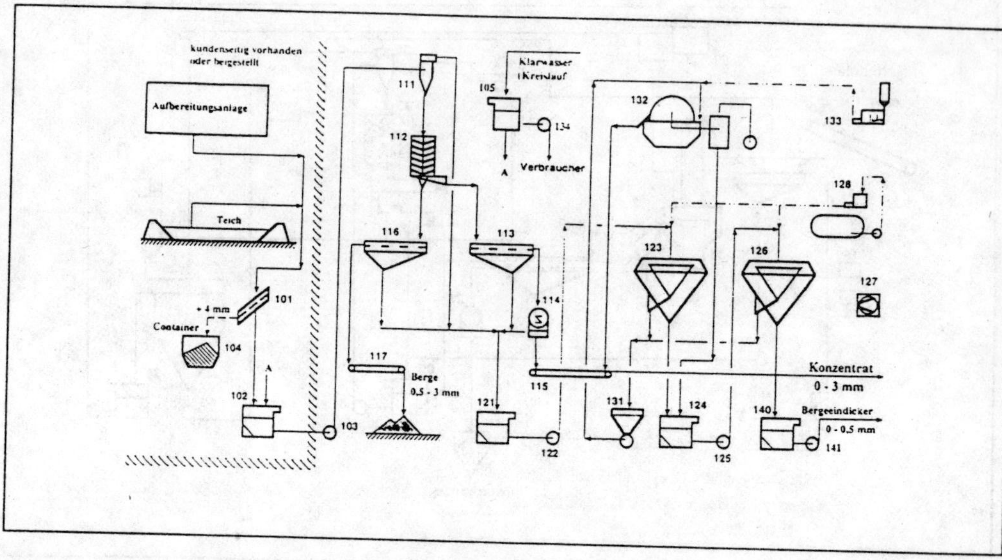

FIG. 14 : *Flowsheet of a beneficiation plant for slurry pond material*

wasting of energy, it also destroys the environment. In future the use of such ponds will no longer be allowed. Technically it is no problem, to close the water circuits of the mines or the preparation plants, which are the main reason, why the ponds exist at all.

This section of the beneficiation plant is unfortunately not only the most expensive — as well in investment as in operation — but at the same time the less valuable one due to the higher content of moisture and ash. In some countries even the thermal drying is required.

4.2 Description of the Plant

A typical flowsheet of such a plant is shown in Fig. 14 and consists of the following beneficiation steps.

The slurry is pumped into this beneficiation plant and is at first screened at 3 and at 0.5 mm. The overflow is a check screening to avoid oversize material to enter into this system. The overflow 3.0 - 0.5 mm is separated in spirals into rejects and flotation concentrate. The spiral concentrate is dewatered in screen bowl centrifuges, whereas the rejects are screened and sent out together with the coarse rejects.

The underflow of the screen is collected in a sump and is being fed to the flottion system. The flotation is the Pneumatic flotation (Fig. 15), a further development of the conventional cell flotation. As the cross-section shows, one single central aerator is located on top of the separating vessel. The aerated pulp flows through a pipe into a distributor box near the bottom of the vessel. Orifices in the distributor discharge the pulp vertically, with controlled speed, into the vessel.

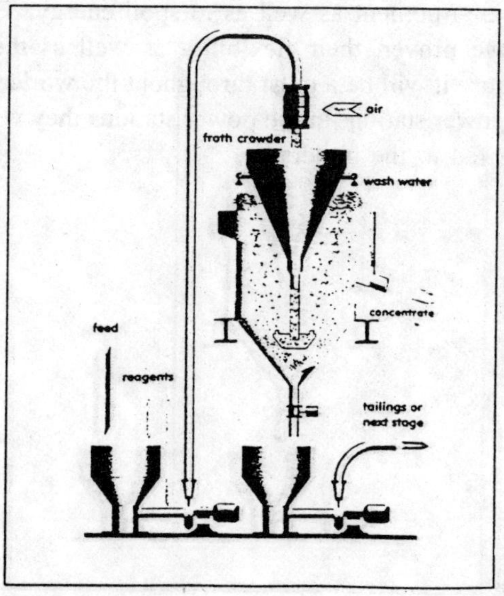

FIG. 15 : *Cross section of a Pneumatic Flotation*

A conical froth crowder can be moved up and down on the feed pipe via a hydraulic drive to adjust the size of the froth area to optimise it for various applications. Further control of the concentrate quality is obtained via a wash water system on top of the froth.. The economics of the system are further improved by a new type of aerator that does not need compressed air.

The proven design of the Pneumatic flotation has been further enhanced: The operating features are:

— high recovery, especially for course particles
— optimum selectivity due to washing device
— optimum flexibility due to froth crowder
— unbeatable economic efficiency because of
  — low investment costs
  — economical use of spares
  — excellent product quality

In the pneumatic flotation the raw slurry is separated into tailings and flotation concentrate. The flotation concentrate can be dewatered either together with the spiral concentrate in the screen bowl centrifuge or separately on vacuum filters.

The flotation tailings are sent to a thickener from where the overflow, i.e. clarified water. is pumped back into the circuit. The underflow of the thickener can be dewatered on centripress or filter presses.

5. Conclusions

According to today's standard it is no longer required or neccessary to work with open washery circuits and to damage the environment as well as to spoil energy. The described processes for the various applications have proven their flexibility as well as their acceptance in the coal beneficiation industry. In future it will be a must throughout the world, to beneficiate the raw coal before it will be sent to the power station. In the power stations they will appreciate the better and constant coal quality, being fed to the boilers.

ESTIMATION OF ELECTRICITY, ENERGY AND TRANSPORT DEMAND : FUNCTIONS OF MUMBAI AND ENVIRONMENTAL IMPACTS

38

BABU RAM
Energy Policy Division
Planning Commission, India

ABSTRACT

This paper reports a study on estimation of electricity, energy and transport demand functions of Mumbai.

Since planning by an electric utility depends on forecasting of peak electricity demand, the peak-time electricity demand functions of Bombay Electricity Supply and Transport undertaking (BEST) and Bombay suburban Electric Supply Ltd. (BSES) have been estimated taking price as a key variable.

In the case of energy, the demand functions of key petroleum products such as Liquefied Petroleum Gas (LPG) and Superior Kerosene Oil (SKO) have been estimated. The own-price and corss-price elasticities estimated from demand functions indicate potential for energy conservation. In order to assess demand of High Speed Diesel (HSD) and Motor spirit for road transport, the passenger transport demand, mix of vehicles and model split in the year 2001-02 have been estimated. It has been found that bus transport systems is likely to be a dominant mode in the year 2001-02.

Further, transport related air pollution loading has been estimated using the normative emission coefficients, utilization level of different types of vehicles on a daily basis during 2001-02.

1. Introduction

Mumbai is financial and commercial capital of India. Mumbai's contribution to Maharashtra economy is as high as 35-40% of the State Domestic Product. On the population front, the population of Mumbai has been growing at a rate of 33.90% per decade, employing annual increase of 2,80,160 persons. Mumbai with a population of 12.6 million (1991) is ranked as the Sixth largest city in the World. further, Mumbai's population has been projected to increase to 24.3 million by 2010 AD according to a UNDP study. Due to industrial expansion, Mumbai offers vast employment opportunities which correspond to more than 10% of employment in manufacturing, in addition to what it offers in small scale and unorganized sectors. Further, the compound annual growth rate of Net Domestic Product (NDP) of Mumbai at constant price is 5.5% per annum. The per capita income of Mumbai has been rising at a rate of 1.05% per annum. With the growing population, expanding economy and rising per capita inçome, demand of basic infrastructural services and municipal services have been rising rapidly at enhanced quality and reliability. On the other hand, development of infrastructure to meet growing demand of urban areas requires a major part of State outlays which leaves out inadequate resources for development of 'backward' areas of the State.

The 16th World Energy Congress (WEC, 1995) has expressed concern about problems degrading urban societies all over the world. uncontrolled growth of towns and cities generate production and consumption patterns, that are unsustainable.

In the long term, without a urban development strategy, a city cannot remain sustainable. Impacts of unsustainable pattern of production and consumption on environment are severe. The strategies have been thought to prevent such adverse impacts. The first and foremost is to minimize negative impacts of city based production and consumption pattern. Second is to implement urban management strategies based on the understanding that non-renewable resources and carrying capacity of eco-system to absorb emissions from use of non renewable resources are finite in a wider regional, national and international context. Consequently, the author of this paper had earlier suggested a programmed approach to reduce energy intensity of city based production and consumption patterns [Babu Ram, 1995] with particular analysis of Delhi and Calcutta. An extension of above theme has led to estimate demand functions of electricity, energy and transport services of Mumbai with price as a key variable. For managing demand, prices can be manipulated to effect inter-period and inter-fuel shifts in demand of electricity and transport, and energy respectively.

From the point of view of demand management, peak-time demand functions of electricity have been estimated as a function price and load connections.

So far energy demand is concerned, energy demand functions for domestic cooking gas, i.e. LPG, and Kerosene (SKO) have been estimated as a function of price of LPG, price of a close substitute, viz. kerosene, and per capita net domestic product of Mumbai.

The estimation of HSD demand function is linked to transport demand. In order to estimate HSD and petrol demands, passenger transport demand has been estimated. Impacts of transport on environment have also been studied by estimating the emissions of carbon monoxide, unburnt hydro-carbons, nitrogen oxides, etc. on a normative basis.

2. Electricity

The electricity is distributed by the Bombay Electric Supply and Transport undertaking (BEST) in Greater Bombay areas and by the Bombay Suburban Electric Supply Ltd. (BSES) in suburban areas. The Tata Electric Company is a major supplier to BEST and BSES. The BSES had commissioned a project of 500 MW to augment its capacity to meet load demand in its franchised area. The Maharashtra State Electricity Board (MSEB), further, supplies electricity to New Bombay areas.

2.1 Characteristics of Demand

The electricity demand characteristics of BEST, BSES and MSEB are shown in Figure - 1. It may be noted from Figure-1 that domestic and commercial consumers are dominate group of consumers receiving electricity supply from the BEST system. However, as opposed to BEST system, domestic and industrial consumers predominantly constitute the BSES consumer grid. On the other hand, in the New Bombay areas, industrial loads dominate the supply system of MSEB.

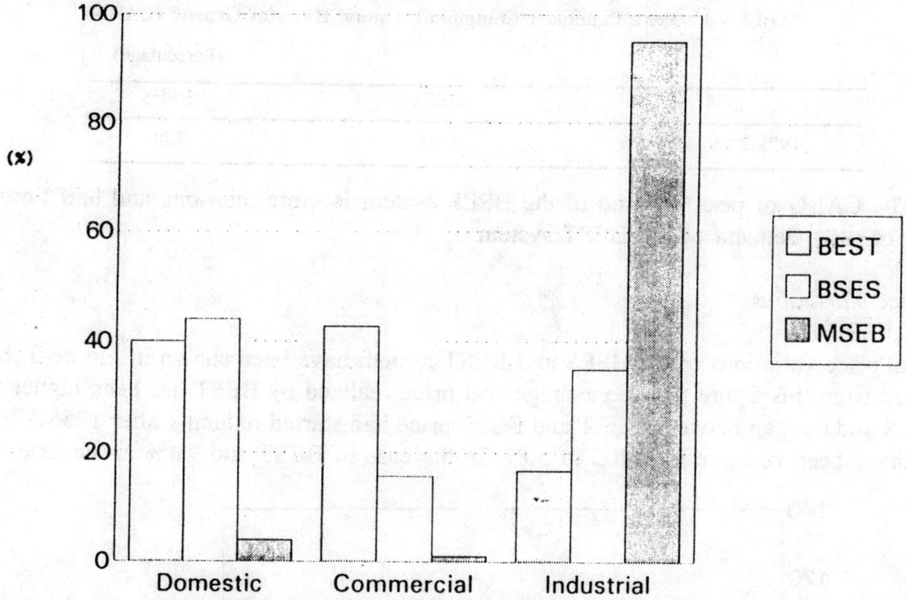

FIG. 1: *Consumer Grid of Utilities in Mumbai*

Further, the growth rate of connected load and consumption have been presented in Figure-2 for different categories of consumers of BEST and BSES systems. The rate of connection of average load on the BSES system is higher than the average rate of connection of load on the BEST system. Regarding consumption, consumption by commercial consumers of BEST is increasing at a faster rate than that of commercial consumers of BSES system. However, the consumption by domestic and industrial consumers of BSES has been rising at a rate faster than that of consumption of domestic and industrial consumers of BEST.

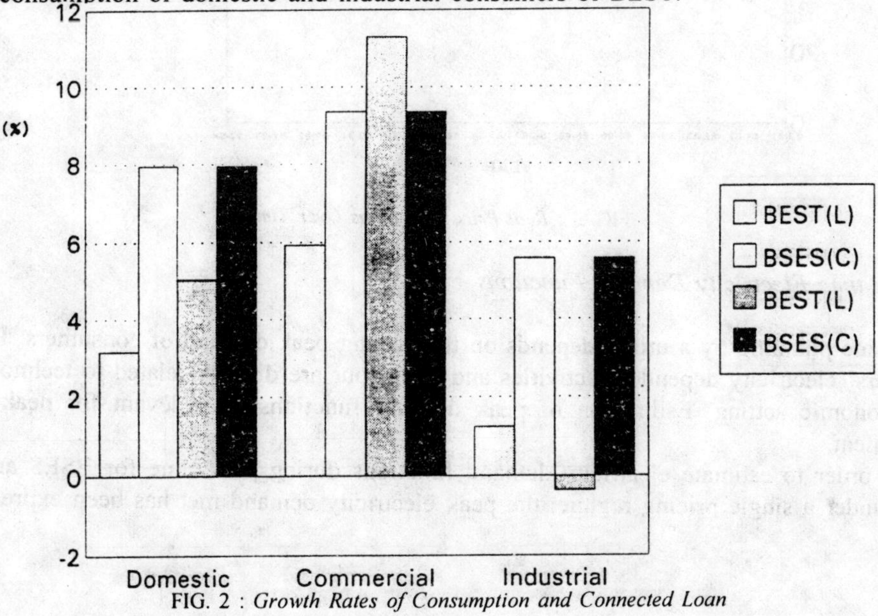

FIG. 2 : *Growth Rates of Consumption and Connected Loan*

Further, the compound annual growth rates of peak-demand for BEST and BSES are given below:

TABLE - 1 : **Peak Demand : Compound Annual Rates of Growth (CARG)**

(Percentage)

	BEST	BSES
1978-79 to 1995-96	4.44	7.06

The CARG of peak demand of the BSES system is more than one and half times that of CARG of peak demand of the BEST system.

2.2 Price Variations

The real price variations of the BSES and BEST system have been shown in Figure-3. It may be observed from this figure that the average real price realized by BEST has been higher than that of BSES and the gap between BEST and BSES price had started reducing after 1986-87. The real prices have been rising at a CARG of 5.8% in the case of BEST and 9.4% in the case of BSES.

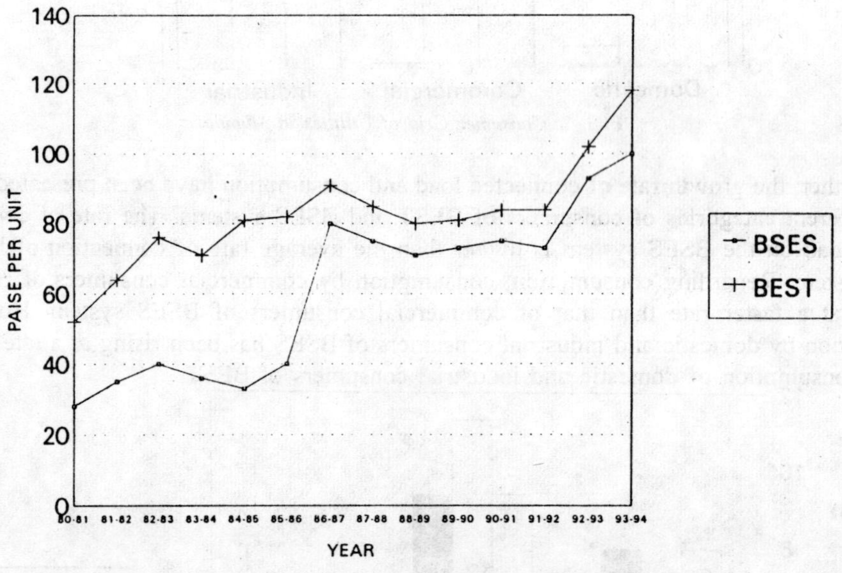

FIG. 3 : *Real Price Variations Over Time*

2.3 Peak-time Electricity Demand Functions

Supply side planning by a utility depends on forecasting peak demand of consumers. However, consumers' electricity dependent activities and behaviour are directly related to technology and socio-economic setting. Estimation of peak demand functions are relevant for peak demand management.

In order to estimate electricity demand functions during peak time for BSES and BEST system under a single pricing regime, the peak electricity demand met has been expressed as a function of the average price deflated by wholesale price index at 1981-82 base and connected load. The peak demand met in the previous year, in order to account for impact of change of technology of end-uses, has also been included as a explanatory variable. The time series data from 1980-81 to 1994-95, have been obtained from TEC, BSES and BEST. The Maximum Likelihood Estimator [Babu Ram, 1994] has been deployed. The estimated demand functions of BSES and BEST are given as under :

TABLE 2 : **Peak - Time Electricity Demand Functions**

Dependent Variables	Constant	Explanatory Variables Ln (Price)	Ln (load)	Ln (KVA (-1))
Ln [KVA (BSES)]	-1.71 (2.5)	-0.056 (0.34)*	0.97 (3.09)	0.12 (0.46)
	R - Squared = 0.98, DW = 1.69			
Ln [KVA (BEST)]	-0.20 (0.33)	-0.064 (1.66)	0.14 (0.84)	0.92 (7.24)
	R - Squared = 0.99, DW = 1.99	* T Ratios		

The own- price elasticities of peak demand met are (–) 0.056 and (–) 0.064 respectively for BSES and BEST system. Further, the connected laod elasticities are 0.97 for BSES and 0.14 for BEST.

In the case of BEST, low elasticity of peak demand with respect to connected load indicates the load that is being connected on the system does not contribute to peak demand as much as the load connected on the BSES system contributes to peak demand.

3. Energy

The percentage shares and CARG of petroleum products consumed in Mumbai are given below:

TABLE - 3 : **Consumption Pattern and Growth Rates**

	Percentage Share of 1993-94 Consumption	CARG (1980-94) (%)
LPG	6.71	7.13
SKO	11.58	1.25
Motor Spirit	7.98	3.35
HSD	15.50	2.47
ATF	17.15	1.73
LSHS	23.00	2.07
Others	18.08	

The CARG of LPG use is 7.13% per annum during 1980-81 through 1994-95 for Greater Bombay. The percentage share of LPG consumption of Greater Bombay was 35.92% of the LPG consumed during 1994-95 in Maharashtra. Along with LPG, SKO is used for cooking and other purposes in Greater Bombay. The CARG of SKO is 1.25% during 1980-81 to 1994-95.

3.1 Energy Demand Functions

The estimated demand functions of LPG and SKO of Greater Bombay are given in Table - 4.

TABLE - 4 : **Energy Demand Functions**

Dependent Variable	Constant	Explanatory Variables Ln (LPGP)	Ln (SKOP)	Ln (PNDP)
Ln (LPT)	13.52	–0.12	2.08	0.16
	(4.15)	(0.43)	(4.96)	(0.41)*
	R - Squared = 0.95; DW = 1.80; * T-Ratio			
Ln (SKO)	11.39	-0.1441	0.5894	0.2478
	(8.70)	(1.42)	(3.86)*	(1.58)
	R - Squared = 0.85; DW = 1.59; * T-Ratio			

The above demand functions for LPG and SKO have been estimated using Least Square Estimation technique. The data sources are Oil Coordination Committee for consumption of LPG and SKO, Petroleum and Natural Gas Statistics for prices of LPG and SKO in Mumbai and Directorate of Economics and Statistics, Government of Maharashtra for Net Domestic Product of Mumbai. The LPG stands for actual LPG consumed in the Greater Bombay area, LPGP indicates price of LPG in Rs/KCAL and SKOP is the price of kerosene in Rs/KCAL. The PNDP refers to per capta NDP of Mumbai. The demand function of LPG reveals that own price elasticity of LPG is — 0.12 and cross price elasticity of LPG with respect to kerosene is 2.06 showing substitution of kerosene by LPG.

It may be observed from the estimated demand function of SKO that demand of SKO is price inelastic with respect to its own price. Secondly, the cross-price elasticity of SKO with respect to LPG is — 0.14 which shows that SKO and LPG complement each other in Mumbai.

The demands of High Speed Diesel (HSD) and Petrol are dependent on various kinds of transport vehicles and their utilization rate. Before estimating demands of HSD and Petrol, it is prerequisite to estimate transport demand in the medium term.

4. Transport

The Sub-urban Rail and Bus transport system operated by BEST constitute Mass Transit System that handles about 88% of all passenger travel demand of Mumbai. During peak periods, 4000 passengers per train against normal train capacity of about 1700 passenger per train travel. The public transport system acts as a feeder service to the sbu-urban rail transport system. On an average 46 lakhs passengers travel daily by BEST bus and 50 lakhs travel by sub-urban railway system.

To estimate energy demand of transport and impact of transport demand on environment, only surface transport has been considered in this paper.

4.1 Forecasting of Transport Demand

The growth and composition of vehicles are shown in Figure-4. It appears from Figure-4 that, among others, over stretched public transport system has given rise to the growth of personalized mode of transport viz. two wheelers, etc. Of the total number of registered vehicles, 60% of vehicles are located in Central Business District.

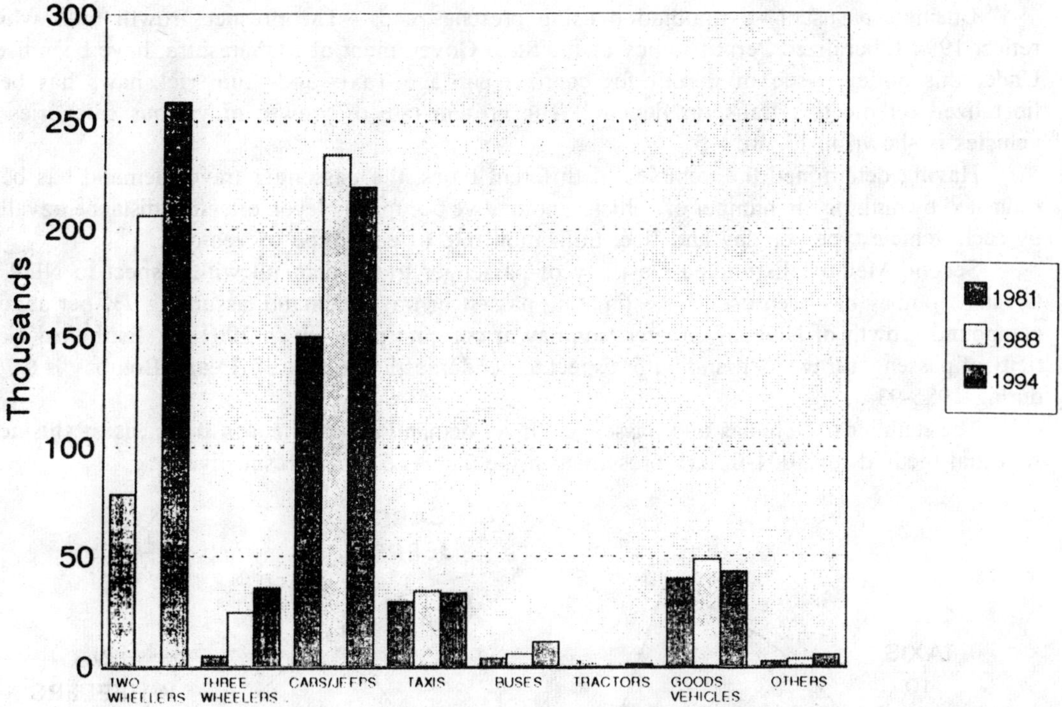

FIG. 4 : *Growth and Composition of Vehicles in Mumbai*

A number of Metro transport studies have been caried out viz. (i) Central Rail Research Institute (CRRI, 1983, 1986), New Delhi, (ii) Advisory Board of Energy Sponsored a study on urban Transport to Central Institute of Road Transport (CIRT, 1988), Pune; and (iii) Tata Energy Research Institute (TERI, 1993). Multi-modal Urban Transport Project-II of Bombay Metropolitan Region Development Authority is being implemented.

The demand for transport is a derived demand. The distribution of demand in space and time is dependent on location of economic and governmental activity centres and on location of residences. Consequently, land use becomes a key veriable in a demand forecasting model.

In a multi-stage transport demand forecasting model, (Gwilliam, 1975) Zones with different land use characteristics are considered. Subsequent stages are (a) a trip generation stage determines total number of trips to be made, (b) trips distribution stage connects origins to destination, (c) a modal stage generates choice of mode of transport and (d) assignment stage models the choice of route for each mode. The model finally assigns trips on the network.

Transport Engineers and Planners have traditionally assumed that cost of travel (Price) by transport remains constant and concentrated on exogeneous variables, viz. income, household structure, etc.

However, in the face of price-inelastic nature of passenger travel demand of public transport system, the methods use in this paper are described below :

4.2 Forecasting Methods

For forecasting passenger transport demand the following methods are used :

Business as usual –extrapolation using present trends—The presnet growth rates which reflect 1994 Liberalized Permit Policy of the State Government of Maharashtra, have been used. Under this policy, issue of permit for contract carriage Taxis and Auto rickshaws has been liberalized for meeting transport demand. The present rate of growth of various categories of vehicles is shown in Figure - 5.

Having determined the vehicles of different types, the passenger travel demand has been estimated by multiplying number of vehicles, normative occupancy level, effective distance travelled by each vehicle type per day and fleet utilization rate of each type of vehicle.

Second Method: Estimated elasticity of passenger travel demand with respect to NDP of Greater Bombay has been used for estimating passenger travel demand, assuming 7% per annum compound growth of NDP of Greater Bombay at constant price up to 2001-02. Estimated elasticity of passenger travel demand with respect to net domestic product of Greater Bombay is 0.797 during 1985-93.

The number of vehicles and passenger travel demand determined on the basis of Business As usual method for 2001-02 are presented in the Tables 5 and 6 respectively :

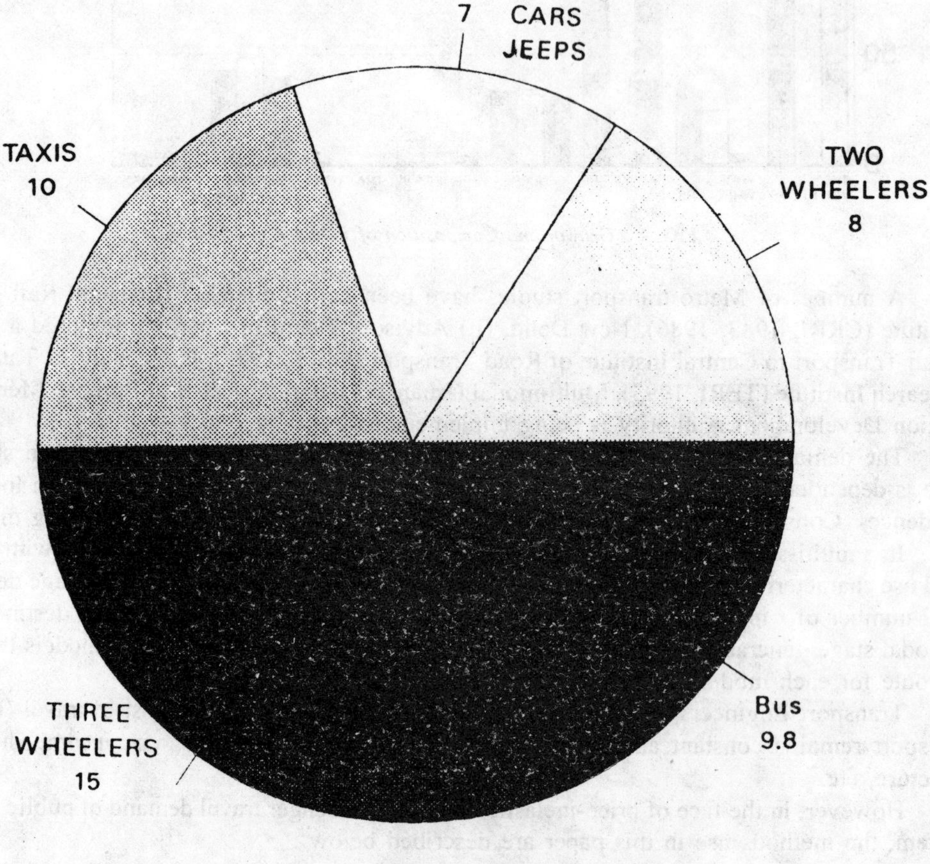

FIG. 5 : *Growth Rates of Vehicles Percent Per Annum*

TABLE - 5 : Population of Vehicles in 2001-02

	2001-02 (000) Business As usual
Two Wheeler	479
Car & Jeep	377
Taxi	72
Three Wheeler	109
Bus*	24*
Total	**1061**

* Indicates public & private buses.

TABLE - 6 : **Passenger Travel Demand**

	Business As usual	Based on Passenger Km. Elasticity w.r.t. NDP
Passenger Travel Demand (Billion passenger Km)	89	68

TERI (1993) had estimated Passenger Travel Demand of Mumbai in 2000-01. According to TERI (1993) which was conducted before Liberalized Permit Policy regime of Government of Maharashtra, projected passenger travel demand as 64 billion passenger km in 2000-01.

Model Split is given below :

TABLE - 7 : **Model Split**

	Two Wheeler	Car & Jeep	Taxi	Three Wheeler	Bus
Model Split (%)	6.03	8.33	2.58	4.32	78.74

Estimation of Energy Consumption

The fuel demand of a given type of vehicle is a product of travel demand catered by that vehicle and corresponding operating energy intensity. By summing up energy demand of different vehicles, energy demand of passenger transport sector is obtained and presented in the Table - 8 given below :

TABLE - 8 : **Estimated Energy Demand in the Passenger Transport Sector (2001-02)**

	(000 TOE)	
	Petrol	Diesel
Mumbai	429	58

5. Environmental Impacts

Mumbai's Road network per 1000 vehicles has been declining. Due to congestion, transport productivity declines. Consequently, energy consumption and transport related air pollution are on the rise.

Increase in personalized mode of transport is the main cause of growing transport related air pollution. The motorcycle, scooters, three wheelers, and four wheelers emit carbon-monoxide, oxides of nitrogen, unburnt hydro carbon, oxides of sulphur, etc.

The emissions from diesel powered vehicles are carbon monoxide,unburnt hydrocarbons and nitrogen oxides. The concentration of nitrogen oxides is very high in the exhaust of diesel vehicles compared to carbon monoxide and unburnt hydrocarbons.

The normative mass emission factors used in the study are taken from Central Pollution Control Board's (CPCB's) study on "Assessment of Vehicle Pollution in Metropolitan Cities, 1988".

The magnitude of vehicular emissions has been estimated by multiplying normative mass emission factors and normative vehicle Km travelled per day by number of vehicles of different type in the targeted year. Based on above, the pollution loadings of different vehicles are given below in case of Mumbai for Business as usual scenario on a dialy basis for 2001-02.

TABEL 9 : **Pollution Loadings of Vehicles in 2001-02**

(Tonnes)

	Vehicle	2001-02					
		CO	HC	NOX	SO_2	TSP	Total
1.	Two Wheeler	126.42	75.59	0.0	0.16	0.0	204.17
2	Three Wheeler	99.91	38.60	0.0	0.28	0.0	138.79
3.	Car & Jeep	398.05	58.82	31.79	8.00	0.0	496.65
4.	Taxi	119.61	17.67	9.55	2.40	0.0	149.23
5.	Bus	26.50	4.42	44.17	3.18	1.59	79.85
	Total	778.48	195.10	85.51	14.02	1.59	1068.71

Central Pollution Control Board (1988) estimated pollution loading for Passenger Transport as 510 tonnes per day for 1986-87. Using Business as usual scenario, pollution loading per day has been estimated as 1068.71 tonnes/day during 2001-02. The vehicles estimated for 2001-02 are 3.44 times of the registered vehicles of 1986-87 whereas the estimated pollution loading in 2001-02 is 2.1 times of the pollution loading estimated by CPCB (1988). Further, TERI (1993), on the basis of trends which prevailed before 1994 liberalized permit policy regime, estimated pollution loading as 876 tonnes per day in 2000-01.

6. Conclusions

1. This study has demolished the myth that peak electricity and energy demands are price-inelastic in the Indian context.
2. The study has established the relationship among demand of electricity during peak time, price and connected load. The price elasticity of peak demand indicates potential for conservation of electricity during peak period. This can be utilized for designing tariff structures for Demand Side Management.
3. The energy demand functions estimated for LPG and SKO indicate potential for coservation of LPG. The positive cross-price elasticity between LPG and SKO reveals that SKO is being substituted by LPG. It has been observed that the kerosene is price inelastic due to Administered Price Mechanism. The negative cross-price elasticity of SKO with respect to LPG indicates that SKO and LPG complement each other.
4. The demand of passenger transport in Mumbai has been estimated using Business as usual approach which reflect liberalized permit policy of the Government of Maharashtra. Along with estimates of the passenger travel demand, mix of population of vehicles and model split have been estimated. The bus transport system turns out to be a dominant mode of travel at the dawn of next century.
5. The pollution loadings of the emergent transport scenario have been estimated. The estimated pollution loading for 2001-02 is 2.1 times of the pollution loading estimated by CPCB (1988). On the other hand, the number of vehicles estimated for 2001-02 are 3.44 times of registered vehicles in 1986-87.
6. The findings of this study can be used to manage demand of infrastructural services of Mumbai. The demand management is an integrated element of strategy to alter unsustainable pattern of production and consumption in urban areas in a wider regional, national and international context.

7. Acknowledgements

The author expresses his gratitude to the following for providing valuable information and data:

1. Bombay Electricity Supply and Transport Undertaking.
2. Bombay Suburban Electric Supply Ltd.
3. Tata Electric Compnay.
4. Oil Coordination Committee, Government of India.
5. Directorate of Economics & Statistics, Government of Maharashtra.

Views expressed in the Paper are those of the author.

6. Transport Commissioner, Government of Maharashtra.
7. Association of State Trarnsport Undertakings.

REFERENCES

World Energy Council's 16th Congress, (1995), Energy for our Common World - What will the future ask of us? (Tokyo), October 8-13.

Babu Ram (1995), 'Investigating Energy Transition and Strategy for Sustainable Energy Supply and use in India with a focus on Delhi and Calcutta', Paper no. 3.2.04, World Energy Council's 16th Congresss, Tokyo; October 8-13.

Babu Ram (1994), 'Tariffs and Load Management : A Post Privatization Study of the U.K. Electricity Supply Industry', IEEE Transactions on Power System, Vol. 10, No. 2, pp. 1111-1117, May, 1995.

CRRI (1983), 'Planning of Road System for Bombay Metropolitan Region. New Delhi' : Central Road Research Institute. Volume I, traffic and transportation studies.

CRRI (1986), 'Traffic and transport flows for selected cities in India. New Delhi'.

CIRT (1988), Report on energy implications in urban transport. Report submitted to the Advisory Board on Energy, New Delhi, Pune: Central Institute of Road Transport.

Gwilliam, K.M. (1975), 'Economics and Transport Policy' Book, George Allen Unwin Ltd. (London).

Transport Commissioner's office, Government of Maharashtra.,

TERI (1993) 'Impact of Road Transportation Systems on Energy and Environment - An Analysis of Metropolitan Cities of India', Submitted to Ministry of Urban Development.

CPCB (1988), 'Assessment of Vehicles Pollution in Metropolitan Cities'.

LOGISTICS OF LOCATION OF PIT-HEAD COAL BASED POWER STATIONS

39

L. K. BOSE
Executive Director (Corporate Planning)
Coal India Ltd. Calcutta (INDIA)

ABSTRACT

There is need for integrated energy planning covering exploitation and utilisation of resources to its ultimate use. Coal continues to be predominant source of commercial energy in Inida accounting for about 60 % of its energy needs. Coal consumption by power sector has increased eightfold from a level of 20.3 MT in 1974-75 to 185.00 MT in 1995-96 registering an average annual growth rate of about 10%. This is likely to increase to 291 MT in 2001-02, 351 MT in 2006 -07 and 435 MT in 2011-12.

While coal reserves are concentrated in few places, the setting up of coal based stations in widely dispersed coal centres, at distances from coal sources, has led to irrationality in coal linkages and also constraints in movement of coal. As the fuel cost is an important component of cost of generation, the generation cost is significantly high in power stations located away from coal sources due to high incidence of freight. Since 1981, while annualised growth rate of pit-head coal price was around 9%, those for railway freight and Government duty were 14.8% and 21.85% respectively. As a consequence, in the landed price of coal for distant power stations, share of pit-head coal price has decreased from 65.5% to 45.6% over this period. The distant power stations also suffer from various operational problems such as coal quality, pilferage, disruption in rail movement etc. Performance of pit-head power stations have been found to be much better than distant power stations from the point of view of PLF, lower cost of generation, consistency in performance etc.

Pit-head power stations have also been observed to be more cost effective on the basis of various techno-economic studies. Issues related to long distance transmission of power to multiple States need to be sorted out so that more pit-head power stations could come up in future near the potential coalfields.

Introduction

Integrated energy planning is recognised as the essential element right from exploitation and utlisation of resources to its ultimate use. For future energy development of the country it is essential to ensure efficient use of fuel and ensure sustainability of long-term supply.

Coal remains the predominant source of commercial energy in India, acounting for about 60% of its energy needs. In the Power Sector at present around 65% of capacity is coal based. Coal based power generation (utility) accounts for nearly 2/3 of overall generation. This has been possible due to substantial improvement in performance of coal based stations which have proved to be more reliable and efficient compared to other power stations. The present indications reveal that share of coal based capacity and generation will continue and/or further improve.

Transmission of Power vis-a-vis Transportation of Coal

The relative merits of Pit Head Generation vis-a-vis generation at Load Centres have been studied by various Site Selection Committees for large thermal power projects. As per the report of the Site Selection Committee appointed by the Government in 1973, Pit Head Generation have been given preference over Load Centres based on Techno-Economic Studies. Based on such studies, a large number of Pit Head Power Stations have been constructed in various new coalfields. Many of these sites have already been utilised.

For major load centre power stations sizeable amount of power is absorbed there itself requiring very short transmission lines, thus ensuring reliability of power supply to the Load Centres enhanced stability and reduced transmission losses. Moreover, this avoids construction of long transmission line which involves high capital cost and long gestation due to problems of various clearances and right of way issues. However, various studies have revealed that the alternative of the Pit Head Generation using raw coal as well as beneficiated coal with transmission of power is cheaper than the Load Centre Stations.

The Pit Head Power Stations taken up so far were meant for catering to the need of the same region. The new Pit Head Sites or extension projects at the existing Pit Head Stations would be needed to meet the need of distant regions as well as additional need of the region.

In the present paper various issues related to coal availability, movement of coal, fuel cost etc. hae been elaborated to examine the merits of alternative location of power stations.

Power Scenario

Out of total installed capacity of 81,164 MW (as on 31.3.95) 65% is owned by the State, 31% by the Centre and 4% by the Private Sector. Steam based power stations account for 64.2% of total installed capacity. The total addition in installed power generating capacity during the first three years of VIII Plan (1992-95) was 12,674 MW, out of which addition in installed capacity of thermal based power station (coal and lignite) was 10,614 MW (i.e. 83.75%).

The compound annual rate of growth for installed power generating capacity has been around 7.5% during the last two decades, whereas the annual growth rate of thermal generating capacity has been about 9.4% out of which Coal-based generation was predominant.

Coal Scenario

The compound annual growth rate of coal production has been around 6.1% during the last two decades since nationalisation of coal industries as against about 1% in the IV Plan before the nationalisation. The overall country's coal production has increased from 72.94 MT in 1973-74 to 253.7 MT in 1994-95. The coal consumption by power sector has increased eightfold during the last two decades from a level of 20.3MT in 1974-75 to 185.00 MT during 1995--96 registering an average annual growth rate of about 10%. Thus the share of Power Sector (Utility) in coal consumption has increased from 23.41% in 1974-75 to 68.07% in 1995-96.

Power sector development is mostly dependent on coal as the country is well endowed in coal, while endowment of other resources like crude oil and natural gas are much less assured. As per the level of consumption the reserves of crude oil and natural gas will last for 21 years and 38 years respectively, whereas the reserves of coal is expected to continue over 245 years. *(Source - BP Statistical Review of World Energy* 1996).

As on 01.01.95, total reserves (proved and indicated excluding inferred) of power grade coal (Grade D, E, F & G) are estimated as 105,033 MT out of total reserves (proved and indicated excluding inferred) of all grades of non coking coal amounting to 129,480 MT. Thus in line with our future requirement of coal based power gestation we are fortunate to have major share (81%) of non-coking coal reserve in power grade. Most of these power grade coal deposits are in the form of thick seams which are amenable to Open Cast mining. However, the deposits of power grade coal are mostly concentrated in the eight major coalfields as given in Table below:

TABLE : Coal Reserves in Major Coalfields

(Billion Te)

State	Name of Coalfield	Coal Company	Power Grade Coal Reserve (Proved & indicated)
Bihar	Rajmahal	ECL	9.43
	North Karanpura	CCL	7.02
	South Karanpura	CCL	2.95
Orissa	Ib Rver	MCL	11.39
	Talcher	MCL	15.46
Madhya Pradesh	Mand Raigarh	SECL	7.45
	Korba	SECL	7.56
	Hasdeo-Arand	SECL	2.67
	Total		**63.93**

It can be noted from the above that 63.5% of power grade coal reserve (out of 105 Billion tonnes) occur in these eight major coal fields. The future Power houses will have to be fed by coal mostly from these coal fields.

Fuel Cost in Power Generation

The trend in the major components of cost in the unit cost of power supply for different SEBs is given in Table below :

TABLE: **Structure of Unit Cost of Supply (%)**

	1992-93	1993-94	1994-95	1995-96
Fuel	25.9	23.1	23.0	23.5
Power Purchase	27.9	31.7	31.1	31.8
O&M	4.7	4.5	4.4	4.2
Estt./Admn.	14.9	15.6	15.3	14.4
Miscellaneous	1.6	2.0	1.8	1.4
Depreciation	7.5	7.0	7.5	7.5
Interest	17.5	16.2	16.9	17.2
Total	**100.0**	**100.0**	**100.0**	**100.0**

The proportion of expenditure on fuel cost in the total cost of supply of electricity has varied in the range 23-26% in the above noted period. The trend in the fuel consumption and fuel cost ar given in Table below:

TABLE : **Consumption and Cost of Fuel in Coal Based Thermal Generation**

	Fuel Consumption per KwH		Fuel Cost per KwH	
	(Coal (Kg.)	Oil (ml)	Coal (Paise)	Oil (Paise)
1992-93	0.76	6.95	56.16	3.57
1993-94	0.77	5.73	64.67	3.48
1994-95	0.76	5.85	70.21	3.42
1995-96	0.76	5.76	71.94	3.5

It can be noted from Table above that while the average specific coal consumption in thermal plants has more or less remained same, there is gradual decline in the specific oil consumption in the coal based thermal units from a level of 6.95 MI/KwH in 1992-93 to 5.85 MI/KwH in 1994-95. However, the cost of coal per unit supply of power has increased from 5.16 paise per KwH in 1992-93 to 70.71 Paise per KwH in 1994-95. As per the report of the Planning Commission, the main reason for increase in coal cost is due to higher transport cost as most of the States where there is increase in coal cost are located away from coalfields.

While coal reserves are geographically confined to a few places, the setting up of coal based stations in widely dispersed Load Centres, at distances from coal source have led to rapid spread of coal production Centres in outlying areas. This has led to irrationality in coal linkages and also constraints in movement of coal. As the fuel cost is an important component of cost of generation, the generation cost is significantly high in power stations located away from coal sources due to high incidence of freight apart from other operational problems.

The coal cost is observed too much higher in the States where coal has to be transported to long distances such as in Delhi, Gujarat, Haryana Kerala (KPS), Punjab, Rajasthan and Tamil Nadu. However, cost of coel per unit of generation is much lower in the States like Bihar, Madhya Pradesh, Maharashtra, Orissa, Uttar Pradesh and West Bengal as coal is produced in these States thereby having lower cost of transportation. In coal producing State of Andhra Pradesh the cost is higher because of higher administered price of Singareni Coal.

Cost of Transport in Fuel Cost

It has been observed that over the years, there has been gradual decline in the share of FOB coal cost in the total cost at the destination end as compared to railway freight and Government duties. It has been noted that during the last 15 years since 1981 while the annualised growth rate of Pit Head Coal price was around 9%, the annualised growth rate for railway freight and Government duty were 14.8% and 21.85% respectively. As a consequence share of Pit Head Coal price has decreased from 65.5% to 45.6% over this period.

Coal Movement

With development of pit head power stations in different parts of the country at a fast rate, there has been steady rise in the movement of coal by captive mode viz. MGR, Belt, Ropeways etc. The trend in the mode-wise movement of coal in India is given in Table below:

Table: Modewise Trend in Coal Despatch

	1987-88	1995-96	*Annualised growth* Rate(%)
Rail	114.89	156.62	3.95
Road	42.17	54.46	3.25
MGR/Own Wagon	17.10	57.13	16.27
Belt	6.20	6.37	0.34
Rope	4.30	5.92	4.08

The growth in the rail despatch to power stations has shown a declining trend during the recent years. As against a growth rate of 21.9% in 1990-91 and 12.99% in 1991-92, the actual growth in rail despatch by wagon to power stations during 1994-95 was only 2.96%.

Analysis of Coal Cost in Power Generation

Analyses of these tables reveal the following:

(i) Coal cost is lowest at pithead stations and Coal cost goes up progressively with increase in distance of transport.

(ii) Distant power stations pay much higher for transportation which is increasing at a higher rate than that of coal price. Thus power stations located away from coal resources has a definite economic disadvantage due to higher cost of generation.

Pithead generation had been given some preference as per recommendation of Site Selection Committee appointed by the Government in 1973 based on techno-economic studies. A few sites were selected and the stations which have come up are giving high level of performance in terms of capacity utilisation and PLF. In view of the magnanimity of power supply situation in the economy considering constraints and costs of coal transport, it is desirable to have regional and inter-regional power grid long transmission lines, both from operational and cost of generation angles. From the data of transportation cost, it is very clear that while cost of fuel is a major component of generation cost, cost of transport in the fuel cost is high, and significantly high where long distance coal haulage is involved.

Operational Issues

Apart from price disadvantage, there are operational problems faced by long distance power stations as given below.

(i) *Coal Quality Problems:* There are frequent disputes between coal companies and SEBs about quality and tonnage of coal supply to mostly distant power stations resulting in withholding of payments of coal sale dues which severely affect liquidity of coal suppliers. Whatever may be the reason or fact, quality control could be adhered to for most of the pithead stations.

(ii) The distant power stations also suffer due to disruption in rail movement from time to time and loss in transit. In recent years, there has been gradual decline in growth of rail movement while the demand of power stations has increased steadily over the years.

Conclusion

Considering projected industrial growth of 12 to 14%, there is likelihood of substantial growth in commercial energy in future years. The electricity demand is likely to increase from a projected demand of 336 TWH to 725 TWH in 2009-10.

As per future energy policy, coal will continue to predominate in commercial energy scenario in future years. Demand of coal as worked out by Working Group on Coal and Lignite for power utilities is likely to increase from a level of around 210 MT for 1996-97 to 291 MT in 2001-02 and 435 MT in 2011-12.

The power grade coal reserves are concentrated in major eight coalfields in the country. Thus the major share of increased power coal demand will be met most from these coalfields.

The pit head power stations have been observed to be more cost effective on the basis of various techno-economic studies. From the point of view of various operational issues, pit head power stations have definite edge over load centre power stations. However, as the growth in local power demand of the pit head power stations is restricted, there is need for fast growth in the transmission network for evacuation of power to different regions.

For meeting the regional power demand many of the power stations are also likely to come up in the load centres in the future years which will require rail transport of coal to these power stations. There is need for augmentation of rail capacity from the major potential coalfields such as Korba, Ib, Talcher and North Karanpura.

It is necessary to make in-depth study of cost benefit as well as other operational issues before coming to any conclusion for selection of site for power stations. From the experience of the past, it is necessary to remove regional bias for such selection of site. It is essential to make the most apropriate choice for such site selection for power station to ensure growth in power generation capacity which is an initial input for economic development of the country.

REFERENCES

Annual Report on the working of State Electricity Boards and Electricity Departments - Power and Energy Division of Planning Commission.

National Paper on Energy Scenario in India - Indian Member Committee of the World Energy Council, The Institution of Engineers (India).

DIVISION - III

ENERGY CONSERVATION AND ENERGY EFFICIENCY

ENERGY CONSERVATAION: STRATEGIC OPTION FOR REDUCING ENERGY SHORTAGES

40

BHASKAR NATARAJAN
Director
Energy Management Centre, New Delhi, (INDIA)

ABSTRACT

The role of Energy Management Centre in Planning for the power sector is described. The situation leading to shortages due to administrative load shedding is dealt with. The role of Energy Management Centre is to coordinate initiatives for all end use sectors with a view to providing a thrust for energy efficiency activities in the country.

The Indian Power Sector

The Indian power sector is expected to add 100,000 MW in the next 10 years. Add to this the present installled capacity of over 83,000 MW, the Indian power sector will be one of the largest in terms of installed capacity, in the world , next only to USA,UK, JAPAN and perhaps Canada. Inspite of such large installed capacity, the per capita consumption of electricity will be still a samall fraction of per capita consumption of electricity in the developed countries such as France, Denmark, USA, etc. Even though, all the villags would be electrified under the various plan programmes, it is expected that there would still be a percentage of the population would not have access to electricity.

The energy consumption in the utility sector has been growing at 8-9 % over the last two decades. This rate of growth is expected to continue well into the Tenth Plan. Some of the sectors such as domestic, commercial and agriculture have grown at a much faster rate and this varies from 8-12% across the States in the country.

Increasing financial resource crunch coupled with the fact that there is a global pressure to limit environmental impacts, sustaining this level of energy consumption is not feasible. Energy efficiency in energy production, transmission and efficient end-use is but one of the elements in the entire starategy of meeting rapid increase in the demand for electricity.

The Fourteenth Electric Power Survey had estimated a need based additions to capacity of about 38,000 MW estimated at Rs. 120,000 crores. The target for the Eighth Plan was set at 30,538 MW. However, due to funancial constraints, the resource allocation for the power sector was placed at Rs. 79,500 crores at 1991-92 costs. Reviews indicate that a capacity addition of the order of 20,000 MW is likely in the Eighth Five Year Plan.

The Indian economy has embarked upon a series of measures aimed at liberalizing the economy and freeing economic activity from government control. Markets forces will increasingly decide demand and supply and to that extent price will be set by the market. Electricity generation and supply was till in the early 90's fully controlled by the government.

Amendments were made to the Electricity Supply Act to enable private investors both domestic and foreign, to invest in power generation facilities in the country. These private power stations would sell all the generated electricity in bulk to the State electricity board. Series of financial incentives were given with a view to attracting Indian and foreign investment into the power sector. While a large number of MOUs have been signed, considering the status of even some of the fast track power projects, it is unlikely there will be any significant capacity addition from the private sector during the Eighth Plan. Benefits from this are likely to be realised only from the Ninth Plan onwards.

While there was very little incentive for cosumers to use energy efficiently in the 60s and 70s, this was primarily due to low energy prices and the fact that most of the key sectors in the economy such as iron and steel, cement, sugar, etc., were all under government control and prices were set up on a cost plus basis. Also, there was a need to find new sources of cosumption since capacity addition took place at a fairly rapid rate. However, all this has changed with the opening up of the economy, decontrol of prices in most key sectors and with the enormous resource crunch for adding to capacity. The entry of the private sector is currently providing a signal of high costs on electricity generation which means a substantial increase in tariffs over the next 2-5 years time frame.

The State Electricity Boards (SEBs) are responsible for electricity generation and distribution to consumers at the State level in India. In addition, there are central sector power generating companies which sell power in bulk to the SEBs. In a situation of shortages that have prevailed since late seventies, administrative load shedding, both planned and unplanned are resorted to, to manage the demand-supply-gap. In addition, staggering of weekly holidays in industrial areas, rostering of agricultural pumpsets have been adopted to flattern the load curve.

More than energy shortage, the peaking shortages assumes significane in India, for the domestic and commercial consumers must be given electricity during the eveining hours and it is invariably the industries who are affected by way of reduction in supply.

Nodal Ministry and Agency Initiatives

The Ministry of Power (MOP) is the nodal ministry for coordinating efficiency and conservation activities and initatives in the country. The Energy Management Centre (EMC) is the executive agency of MOP which implements the programmes and policies as well as provides technical advice and support to MOP on matters relating to energy efficiency and conservation. Efficiency related to electricity generation is undertaken by the generating companies and the SEBs, and the activities are coordinated by the Central Electricity Authority. Electricity transmission and distribution is again the activity of the SEBs and the distribution utilities and they do undertake initatives in the area of reducing the T&D losses in their area of supply. Energy efficiency in end-use is to be undertaken by the end-users and the EMC coordinates the initiatives in this regard, addition to the other areas mentioned above. The primary role of EMC is to coordinated initiatives for all end-use sectors, across energy sources with a view to providing a thrust for energy efficiency activities in the country.

Activities in the area of energy efficiency are undertaken by the consumers in order to reduce the cost of their operations. Since, there exists no law on conservation at present, the activities of EMC are purely promotional and catlytic with a view to increasing the awareness on energy conservation in the country and to provide an enabling environment for energy efficiency investments in the country. The activities currently ongoing are

(a) policy and sub-sectoral studies on measures, equipment, technologies etc.
(b) training programmes, workshops at the industry level and also for specific end-uses
(c) promotional campaign for increasing awareness
(d) implementing international energy efficiency cooperation programmes
(e) implementing demonstration projects
(f) school, college and university based programmes
(g) information dissemination

To statrt, with 11 lead agencies have been established in the country, to provide a major thrust to energy audits during the period 1990-93. All these agencies, which received support during the early stages, are now rendering energy efficiency consultancy services to the industries purely on a commercial basis. In addition, 15-20 new agencies are in the process of going through training and field experience process to provide energy consulting to the industry. These agencies have also been given initial training on energy service companies (ESCO) with a view to encouraging ESCOs in the country.

MOP/EMC have supported a large number of studies in various end-uses and specific sub-sectors to identify needs of industry, technologies and barriers to adoption of energy efficiency. These reports are widely disseminated to the concerned target groups through the training programmes and workshops that are organised regularly. International experts also address these meetings, through the different international cooperation programmes that are ongoing.

Information Dissemination and Awareness

Information campaigns and awareness can play a very strong role in providing information to consumers on the costs and benefits of energy efficiency, on the benefits to the economy and last but certainly not the least, benefit by way of cleaner environment. Energy efficiency is a strategy where there is a win, win and win situation as far as the consumer, manufacturer and the economy are concerned. Presently in the entire market efforts undertaken by the various manufacturers of energy efficient equipment, energy efficiency receives little or no importance. The words "energy efficient" may be found in a full page advertisement without any infomation on the levels on energy efficiency and on the benefits on energy efficiency. Also in some cases, consumerism has diverted the subject of energy efficiency and the demand for products presently goes in a direction that is not conducive for energy efficiency. One of the leading examples in this area is the introduction of refrigerators in the coloured models which were priced at 10-15% higher than the conventional white model when they were introduced several years ago. The consumer was willing to pay a higher price for a different colour of the product but the same consumer is not willing to pay an initially higher cost for energy efficiency. There is certainly something wrong in the market mechanism.

It is here that the Government of India sees an important social responsibility for itself to promote energy conservation. They Energy Management Centre handles the awareness campaign programme of the Ministry of Power and will be launching a substantial information dissemination drive to increase the conumer awareness on energy efficiency. This will be done through a multi-media campaign on press, T.V. and through any other channel that may be considered relevant in this regard. The ultimate aim is to ensure that the consumer demands energy efficiency and that the manufacturers are forced to provide what the consumer demands.

DEVELOPMENT OF ENERGY EFFICIENCY AND DSM SERVICES—EXPERIENCE OF U.S. UTILITIES

41

WILLIAM S. FLEMING
Science Applications International Corporation
U.S.A.

ABSTRACT

Demand Side Management (DSM) has not made energy demand entirely predictable. The demand side management options have been covered in the paper. The DSM in commercial building, in industrial processes and issues, and the reasons for high cost of investments in DSM are dealt with. In particular, the Indian energy efficiency and DSM and the items for consideration are covered.

INTRODUCTION

At ever increasing levels, the North American utility industry is operating in a competitive business environment. Competition is present from natural gas, independent power producers, cogeneration and technology developments. The North American utility industry must also operate in a regulatory environment while maintaining sound financial business practices. Increased competition is a major reason for recent utility industry efforts to increase customer satisfaction. No longer can the utility industry succeed by just providing energy as a product. More and more it must provide value-added services to its offerings to meet the competition.

In response to this business environment, "Energy Efficiency and Demand-Side Management (DSM)" continues to be a major component of a utility company's corporate strategy. Because of marketing strategy, the term "DSM" has been downplayed by some utilities. Similar utility service is provided with the terms "consumer satisfaction/retention," "cost-effective energy service," etc. This paper will use the term "DSM" for "energy efficiency," "DSM" and the above items.

Demand-side management (DSM) has not made energy demand entirely predictable. However, with financially sound planning and communication between the electric power supplier, commercial/industrial consumers and construction industry professionals, demand-side management service and technology is a valuable North American resource. Such a resource can have reasonable economic value to building owners, building and industrial management, building and industrial occupants, operating professionals and of most importance--an electric utility.

North American utilities benefit from DSM by inducing changes in the time pattern and magnitude of electric demand. The change is intended to maximize productivity and cost-effectiveness of a utility's resources. The commercial and industrial customer benefits by being able to control energy use and lower operating costs.

DEMAND-SIDE MANAGEMENT (DSM) OPTIONS

North American commercial/industrial consumers and construction industry professionals determine which demand-side management options and parameters best meet their needs. Options include customer or utility load control, strategic energy efficiency, and energy storage. Parameters include financial, technical and effect. Financial parameters are initial cost, operating cost, payback and return on investment. Technical parameters are equipment life, past experience and operational knowledge. Effect parameters are comfort to building occupants, work habits, function, industrial production rate, product quality and cost/benefit. As DSM services and products were introduced and marketed to commercial/industrial consumers and construction industry professionals, it became most important that both the North American utility and its consumers understood technical and economic considerations of each DSM option.

Customer or utility load control can be categorized in two ways. First, customers may control their load by voluntarily altering equipment use in response to price signals. Second, utilities may control customers' loads either remotely or at point-of-use.

If customer or utility load control is considered, the commercial/industrial owner, manager or construction industry professional must have closely scrutinized a contract with the electric utility company. Price quotations must not be allowed to vary during contractual periods. The technology must be beneficial to the consumer and the electric utility.

Strategic energy efficiency applies to an effort by a utility to reduce monthly utility costs to programme participants by reducing the customers' energy consumption. Benefits to a North American utility are consumer retention and load change(s). Benefit to a North American commercial/industrial consumer are lower operational costs.

A commercial/industrial building owner, manager, ator or construction industry professional must ensure technologies that reduce peak demand or energy utilization have been proven and used on similar commercial buildings, industrial process, functions, and consumers. Technology and/or service offered for energy efficiency must not infringe on the commercial/industrial consumer's primary business functions.

Energy storage is an option that involves equipment designed to use off-peak energy supplies while maintaining a required energy delivery schedule throughout the day.

When considering DSM commercial/industrial renovation, Thermal Energy Storage (TES) system size depends on a building or industrial plant's design, cooling/heating load and operational profile. All are factors of building or plant function, occupancy, and existing or current markets. Commercial/industrial owners, managers, operators and construction industry professionals require a utility to provide technical support and training. North American utilities also must provide training to building/industrial plant staff to ensure energy storage systems are properly operated and savings are realized.

Demand-Side Management (DSM) — Commercial Buildings in Detail

Four major factors contribute to energy savings within the commercial building sector: (1) building shell, (2) lighting, (3) mechanical systems, and (4) load management.

— **Building Shell:** A building's shell consists of its exterior walls, roof, foundation, doors, windows, skylights, dampers, and other openings. Energy-efficiency improvements to the shell typically include the addition of insulation to walls, floors, attics and/or ceilings;

window upgrades or treatments; and shell "tightening" measures to reduce air infiltration and exfiltration. Shell improvements are most critical for those buildings that have large exterior surface areas relative to their internal volumes.

TABLE 1 : **Common Energy Efficiency Measures in Retrofitted Commercial Buildings (Percentage of Retrofitted Buildings)**

Lighting	(%)	Mechanical Systems	(%)
High-Efficiency Fluorescent Ballasts	62	Variable Air Volume System	78
High-Efficiency Fluorescent Fixtures	50	Air Economizer	59
High-Efficiency Fluorescent Lamps	47	High-Efficiency Motors	47
Programmable Exterior Lighting	44	Point-of-Use Water Heaters	35
Task Lighting	32	Optimized Motor Sizing	32
PL/SL-Type Fluorescent Lamps	29		
Metal Halide Lamps	29		
Shell	**(%)**	**Load Management**	**(%)**
Vestibules	71	EMCS Programmable Thermostats	50
Thermal Break Windows	68	Curtailable-Load Wiring	29
Increased Wall Insulation	41		
Increased Roof insulation	38		
Low-Emittance Glazing	26		

Source: American Consulting Engineers Council (ACEC, 1990)

— **Lighting:** There are several important points to remember when working to improve lighting energy efficiency and/or demand-side management (DSM): (1) do not over-illuminate; (2) use efficient fixtures, lamps, and ballasts; (3) control lighting efficiently and keep fixtures and lamps clean; and (4) tailor lighting levels to the task or function being performed at an illuminated space.

With use of efficient lamps and ballasts, adequate office lighting can typically be obtained at an energy level of less than 1.5 W/ft^2. Generally, building spaces with appropriate lighting levels can reduce their fluorescent lighting energy use by around 25% with conversion to higher-efficiency lamps and electronic ballasts. In spaces highly overlit (3 W/ft^2 or more), reductions exceeding 60% can often be obtained.

TABLE 2 : **Average Payback Periods for Lighting Upgrades**

Lighting Upgrade	Simple Payback (years)
Compact Fluorescent Replacing Incandescent	1.1 - 2.3
High-Efficiency Fluorescent Lamp Replacing Standard 40W	0.5 - 1.2
High-Efficiency Magnetic Ballast Replacing Standard Magnetic	1.4 - 2.8
Electronic Ballast Replacing Standard Magnetic	3.9 - 8.0
Daylighting Controls	1.6 - 9.6
Occupancy Sensors	1.3 - 5.4

Note: Based on 3,500 hrs/yr operation, electricity at $0.07/KwH.
Source: Commercial Building Equipment Efficiency: A State-of-the-Art Review (Geller, 1988)

— **Mechanical Systems:** Opportunities for mechanical system retrofits to heating, ventilating, and air conditioning (HVAC) systems are numerous and varied due to the wide assortment of HVAC systems and equipment in buildings. If the efficiency or performance of an existing system can be reliably determined, efficiency gains from retrofit or system replacement can often be accurately estimated.

Common retrofits for maximizing mechanical system demand control and energy efficiencies include: (1) operation and maintenance (O&M); (2) control system improvements; (3) ventilation and distribution system improvements; (4) replacement of existing equipment with higher-efficiency equipment; and (5) improvements to existing equipment.

1. *Operation and Maintenance.* O&M resources are no-cost, low-cost items that must be implemented in existing commercial buildings. O&M-type resources are popular because opportunities are abundant and these measures offer substantial reductions in operating costs, often for very little capital. Some examples of O&M-type resources would be periodic maintenance procedures, quality control programs, staging of multiple heating or cooling systems to improve part-load performance and minimize operating cost, and use of manual cutoffs, time clocks or setback thermostats to reduce runtimes.

2. *Control System Improvements.* HVAC control system retrofits are also common. The popularity of these measures is based on their potential for rapid payback. As shown in Table 3, the Texas Loan STAR Program found that the median paybacks for many control measures were well under five years (Saman, 1995).

TABLE 3 : **Payback of Control Measures in the Texas Loan STAR Programme**

Controls Retrofit	Number of Projects	Median Payback (years)
Chiller/Chilled-Water Controls	8	3
Replace/Install EMS	24	3
HVAC Controls	60	2
Motor Controls	77	4
Pump Controls	11	1
Steam Controls	9	4

Source: Proceedings of the 1995 ASME/JSME/JSES International Solar Energy Conference (Saman, 1995)

What are Commercial Building DSM Issues?

It is a priority to improve a utility's and consumer's understanding of barriers to efficiency, and thus the utility's ability to devise effective marketing and policy strategies to overcome the barriers. For example:

— Why are most energy efficiency investments so high? In commercial purchase decisions, payback periods must be one to two years or less. Industrial energy investments typically require a two- to three-year payback. Although the question is phrased in economic terminology, the reason for long payback periods is likely to be found in non-economic areas such as poor access and/or proof of reduced operating costs.

— How do consumers make energy investment decisions? In the commercial sector, the interaction of owner, architect and tenant/occupant becomes crucial. This DSM action is one obvious opportunity for utilities to work with private sector allies in the building component or appliance industries.

— Understanding the distribution infrastructure and market intermediaries for efficient technologies. In marketing to building sectors, 70% of resistance is the "hassle factor". Distributors and equipment specifiers do not have a gut feeling for new technologies.

— How do utility customers use their energy bills to make decisions? How can energy bills be modified to induce greater energy efficiency?

Retail Stores	It seems like the utility always has an excuse. It is the same thing in a state's heat pump program. We went to the state and used heat pumps they specified. We went to get our rebate, "No we have to see the original drawings at the beginning of the concept. Therefore, we can't give the money to you." They always have an excuse (incorrect paperwork, etc.).
National Retail	That is exactly right. You just throw your hands up and say, "I'm not participating in anything, anymore."
Offices	It is the same thing with high-efficiency lighting. We had to show the originals, the rewires, and had to go back to the utility more than five times. Finally, we just said, "I don't need the money." The incentive programmes are too bureaucratic and take too much time. Time costs money.

Demand-Side Management (DSM) — Industrial Facilities and Process in Detail

In a recent Asia Pacific Economic Cooperation (APEC) DSM Workshop, industrial DSM programmes that provided large savings to North American industry fell into the following categories: lighting, compressed air, chillers, motor drives, process heating, self-generation, and improved operation and maintenance. A list of the technology group, technologies and application is provided in Table 4.

When excluding fuels used as non-energy feedstock, but including energy used to produce and deliver electricity, the North American industrial sector accounted for 36% of all primary energy consumption in the United States.

When "industry" includes not only manufacturing and processing, but also mining and agriculture, energy consumption by manufacturers is approximately 80% of the industrial sector. Seventy percent (70%) of non-feedstock energy use is attributable to four energy-intensive North American industrial sectors: petroleum and coal products, chemicals, primary metals, and pulp and paper.

The industrial sector becomes more energy efficient as older equipment and facilities are replaced. During the last 15 years, increased energy costs have accelerated this turnover and retooling. Newer equipment and facilities have given North American industry additional fuel-flexibility and improved emissions control. To the extent that manufacturers can switch from one fuel to another in response to price differentials, fuel-flexibility in industrial processes adds to cost-effectiveness and provides DSM.

TABLE 4 : **Industrial DSM Technologies**

Technology Group	Technologies (Energy Consumption Index) *	Applications
Lighting	Incandescent Lighting (1.00) Standard Fluorescent Lighting (0.38) High-Efficiency Fluorescent Lighting (0.30) Standard HID Lighting (e.g., metal halide) (0.18)	Lighting
Compressed Air	Conventional Compressed Air System (1.00) Efficient Compressed Air System (0.80) - Leaks fixed and monitored - Manual system shutdown High-Efficiency Compressed Air System (0.65) - Maintenance of compressor controls - Efficient compressor controls - Selected task compressor installations State-of-the-Art Compressed Air System (0.50) - Distribution system redesign - Correct system sizing - Efficient dryers - Automatic system shutdown	Compressed Air
Chillers	Conventional Chilling System (1.00) High-Efficiency Motors w/o ASD's (0.75 - 0.85) - Efficient motor, compressor and heat exchanger	Space Cooling
Motor Drive	Standard Motors w/o ASD's (1.00) High-Efficiency Motors w/o ASD's (0.93 - 0.97) Standard Motors w/ASD's (0.55 - 0.85) High-Efficiency Motors w/ASD's (0.50 - 0.82) Fabrication/Materials Processing	Process Cooling Liquid Pumping Ventilation Materials Handling Fabrication/Materials Processing
Process Heating	Industrial Process Heat Pumps (0.10 - 0.30) Passive Heat Recovery (0.05 - 0.20) Membranes (0.40 - 0.70) Freeze Concentration (0.40 - 0.70)	Preheating/Heating Water Removal Separations Washing, De-Sizing, Scouring Dyeing Water Removal - Bulk
Self-Generation	N/A	Electric Power Supply

* The Energy Consumption Index for a specific technology reflects the relative energy input needed when compared to the energy input requirement associated with a conventional technology providing the same level of output. The conventional technology will have an index equal to 1.00, while the high-efficiency technology will have an index less than 1.00. The index for a given technology may be reported as a range that reflects different end-use applications and/or industry applications. In the case of process heating, the conventional technology for the applications listed is often direct-gas heating or fossil-fueled boiler systems.

What are Industrial Facility and Process DSM Issues?

The most important stems from two related problems confronting North American utilities. First, many utilities cannot convince customers they are aware of an industrial need or attitude. Second, and partly as a result of the first problem many customers don't trust utility information and advice concerning technologies. Effective DSM programmes have helped establish better customer relations.

Another industrial issue is how a utility may target their industrial DSM efforts. Rather than targeting "high growth/major electric" market segments, utilities may target industrial customers that are not growing because they represent a large constituency or load. However well-intentioned, there is a risk that these efforts to retain existing load do not always succeed and some industrial consumers will fail for reasons beyond their energy costs (e.g., high labour costs, changing markets or bad management). When this happens, a DSM investment is lost.

An issue that discouraged DSM by some of North America's largest industrial firms is that they enjoyed inexpensive electricity and the economic incentives to be more efficient, or shift loads off-peak are limited.

Applied Energy Efficiency and DSM Programme Experience

DSM action should explicitly tie to specific technologies and implementation efforts.

— Preliminary action on market acceptance and price resistance should proceed in parallel with technology transfer. What skill levels will be required to specify, install, and maintain new types of equipment? Potential problems in marketing could be fed back to design decisions.

— Develop DSM marketing strategies that work. Utility involvement would be appropriate for problems and opportunities that are not associated with particular energy products, but are generic to selling many types of demand and energy efficiency measures. For example, utilities could provide means to market energy-efficient products based on their cost-effective performance (energy service company). There is also a need to develop effective strategies that stimulate better maintenance of mechanical equipment and systems.

— Operation and maintenance of commercial buildings, industrial facilities, and industrial process. A key component of DSM programme action must be the capabilities of maintenance staff to operate controls and fit into a system of rewards and social interaction for efficient building operation. For example, a common fate of highly sophisticated controls is that they are bypassed or disabled so that operations' staff can perform direct manual control. Interaction and conversation is needed to determine whether such barriers can be solved by job incentives, retraining or redesign of the controls themselves.

National Retail	I really think we need a clearing house that avoids people with the black boxes. People peddling black boxes seem to get the name of a CEO. The CEO's send the letters down and they expect a response. Many times the letters and salesperson will come up with a statement that says, "We can save you 50%." Well, to a CEO, 50% is a lot of money! When you tell the salesperson to do a survey, he'll come back and say, "You know what? You already delamped a four-lamp fixture to two-lamp." I say, "Great, put that in a memo and send it to me." The salesperson says, "No, we can't do that." It becomes a tremendous problem.

Finally, one building type where innovation in building operations is very high is the hotel/motel industry where building performance is literally sold on a daily basis (analogous to a high-vacancy rate office space market in miniature). The American Hotel/Motel Association (AHMA) has found that interest remains high among their members (AHMA publishes a newsletter *The Energy Tip of the Month*). Significant energy savings have taken place in this industry over the past few years.

Indian Energy Efficiency and DSM—Items for Consideration

What recommendations can be made for the Indian electric industry faced with high-energy demand, good economic growth, and customers which may not be familiar with DSM? An initial item would be to consider a price of electricity that reflects the real marginal costs of production and distribution. Without sufficient price signals, little progress may be possible in the adoption of energy efficiency and DSM. This need not be done all at once, but direction of price changes and benefits of efficient and stable growth should be clearly communicated.

Environmental awareness will inevitably grow. DSM efforts should not ignore the environmental attributes of new electric technologies in the world market.

Cogeneration can be encouraged in a manner that benefits the utility and consumer. This could also expand the utility generation and distribution system (Figure 1). Cogeneration projects may also attract foreign investment and allow domestic capital to be applied to other investment needs.

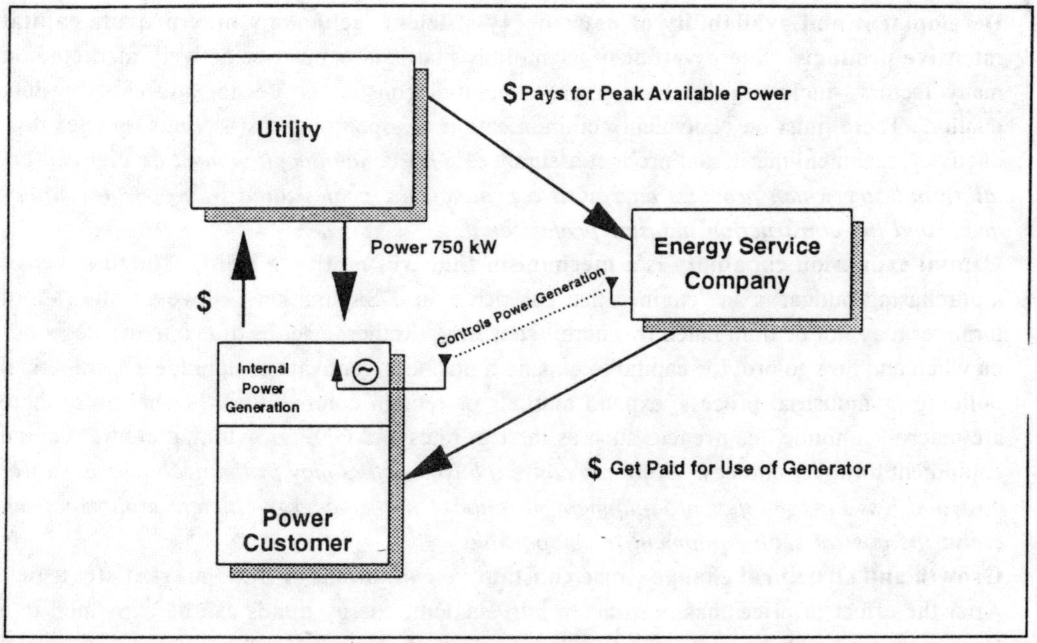

FIG 1.

When circumstances make renewables attractive, such as the unavailability of gas or electric infrastructure, capacity constraints on distribution, or local availability of biogas, wind, and solar, the development of indigenous resources can be conducted through DSM incentives. For example, new high-efficiency pulse combustion boilers are now available that operate on biogas fuels.

Consider load control, energy efficiency, and electric power dispatching within and between major consumer sectors (agricultural, residential, commercial, industrial) at peak load times. Develop better relations with existing and new customers. By developing good relations with customers, utilities will earn their trust and successfully promote DSM programmes that benefit both the utility and its customers.

Indian Energy Efficiency and DSM — Items to Remember

When a DSM programme is considered by a North American utility and its commercial/industrial consumers, the programme success may be based on five items.

1. **Energy prices strongly, but not singularly, affect energy trends.** In the commercial building and industrial process market, increasing energy prices that affected operating costs did not always provide an actual shift in energy use. Improvements in energy efficiency of the building, its equipment or industrial process was directly related to the percent of energy cost to total expense. *However, just as high prices stimulated changes in energy utilization when prices became competitive, growth of energy-efficient technology and its marketplace slowed.*
2. **Development and availability of new energy-efficient technology may provide capital intensive products.** Energy-efficient technology that is new must be actively marketed on many factors—including purchase price. Potential consumers are not aware of product quality. There must be equivalent communication on operating cost savings, greater productivity, technical merit, and product assurance. *This communication must be between the utility industry which provides energy as a product, the professional who operates equipment, and the construction industry professional.*
3. **Capital expansion capability is a mechanism that will motivate DSM.** The turnover of a purchasing budget is an "engine" that can drive the DSM market. However, the rate of turnover may not be dominated by energy concerns. Rather, a series of economic decisions on when and how to provide capital to expand a business, renovate or upgrade a commercial building or industrial process, expand markets or remain competitive. Furthermore, there are macro-economic occurrences such as interest rates and GNP growth that control capital equipment turnover choices. *DSM can cause a business that may instinctively choose lower cost and less energy-efficient equipment, to choose higher energy-efficient equipment because the cost of such equipment is comparable.*
4. **Growth and structural change cause constant re-evaluation of DSM market strategies.** After the effect of prices has been taken into account, energy trends can be explained in a large measure by changes in level and composition of activity. Commercial and industrial economic change(s) can accelerate, decelerate or stop renovation and retrofit which, together with its energy use, is only predictable in the near-term. *Therefore, DSM programmes should be re-evaluated for any change in consumer benefit priorities.*
5. **Customer behaviour has been responsive to changes in energy prices but must be nurtured.** While prices provide the stimulus to DSM, individual choices can either reinforce or diminish DSM gains as a new, renovated or retrofit building becomes occupied or industrial process is expanded. Just as higher prices may cause less energy use, comparative lower prices in future years may encourage higher energy use. *Communication and inter-*

action between the utility providing DSM and the commercial/industrial owner and construction industry professional must include a realistic long-term forecast of building energy costs and utility demand.

SUMMARY

Commercial/industrial customers are not interested in DSM activities and services that simply assist the utility's corporate goals for demand reduction. Nor are the majority of commercial/ industrial renovation customers interested in future savings at the expense of near-term real or imagined costs. A proper DSM programme allows the commercial/industrial consumer to choose a technology and/or service that provides a multitude of benefits--including lower operating cost. Proper DSM provides an attractive choice for sustained and cooperative action between utilities, government, construction industry professionals and customers.

Proper implementation of effective DSM programmes requires a commitment by utilities. Utility professionals must understand customer's needs, desires and actions. Utility and consumers must communicate. Utility, commercial/industrial consumers, construction industry professionals, and government can create successful DSM programmes. The success will allow a good probability that energy and environment near- and long-term goals may be met. Successful DSM will provide financial and social benefits to all.

ENERGY CONSERVATION OPPORTUNITIES IN COAL WASHERIES OF COAL INDIA LIMITED

42

M. P. RAKESH, M. N. SINGH and P. K. Mishra
Central Mine Planning & Design Institute Ltd.,
India

ABSTRACT

Due to growing demand of energy in day to day life, conservation of energy has become the crying need of the day in our country as well as all over the world. "Energy saved is energy produced", has become the universal slogan. The higher consumption of energy in any coal washery plays a vital role in enhancing the beneficiation charges and thus making the washed product costlier for core sector end users like steel, power and cement etc. This has also adverse cumulative effect on growth/development of the nation. Washery is a power intensive unit, employing a huge number of equipment, large as well as small, requiring considerable amount of electrical energy. In view of the above, all efforts are required to reduce the energy consumption in coal washeries by imparting awareness among the users regarding conservation of energy and by the use of improved and efficient energy system. Presently there are 15 coal washeries in Coal India Ltd. (CIL). Besides this, two coking coal and two non-coking coal washeries are under construction stage and six more washeries are expected to come up in Coal India Ltd in near future. On the basis of study of energy consumption pattern of the coal washeries of CIL, it transpired that the power consumption in respect of a few washeries were very high as compared to standard norms as indicated by reputed designing firms for coal washeries. With a view to analyse the reasons for higher power consumption and to suggest corrective measures to reduce the power consumption, cost of power and resultant cost of operation as well as cost of washed products, two washeries say Washery A and Washery B have been identified for detailed study. For this study, following major areas have been identified:

1. **Study of past power consumption rate of washeries.**
2. **Study of power feeding system.**
3. **Study of Electrical installation system of the washery.**
4. **Study of actual operating hours and the input rate of the washery.**
5. **Study of general operating conditions of the plant and machineries.**

This study reveals that there is a scope for considerable energy saving in the coal washeries by proper motivation/awareness of the employees regarding energy conservation, by maintaining desired level of power factor in the electrical system and operating the washery at normal or near normal input rate. There should be a cell at Washery level to study and identify Energy conservation opportunities, plan action for long term and short term measures and implement the same. A small Cell at Company Headquarters level should also be set up to closely monitor the actions taken in different washeries in this respect.

The views expressed in this paper are that of the authors and not of the organisation they serve.

INTRODUCTION

Due to growing demand of energy in day to day life, conservation of energy has become the crying need of the day, in our country as well as all over the world. "Energy saved is energy produced," has become the universal slogan. The higher consumption of energy in any coal washery plays a vital role in enhancing the beneficiation charges and thus making the washed product costlier for core sector end users like steel, power and cement etc. This has also adverse cumulative effect on growth/development of the nation. Washery is a power intensive unit, employing a huge number of equipment, large as well as small, such as crushers, screens, feeders, reclaimers, stackers, pumps, blowers, compressors, belt conveyors, main beneficiation equipment, centrifuges, filters and other related equipment requiring considerable amount of electrical energy. In view of the above, all efforts are required to reduce the energy consumption in coal washeries by imparting awareness among the users regarding conservation of energy and by the use of improved and efficient energy system.

PRESENT SCENARIO IN COAL WASHING

Existing

Presently there are 15 coal washeries under Coal India Ltd. (a Public Sector Unit of Government of India) with a total raw coal input capacity of 27.84 million tonnes per annum (mtpa). Besides this, there are 7 washeries functioning in other companies (SAIL, TISCO, DPL) with total raw coal input capacity of 9.94 mtpa, thus making the overall total capacity of the existing washeries (under PSU and TISCO) as 37.78 mtpa of raw coal. Companywise number and capacities of these washeries are given in Table-A1.

TABLE - A1 : Existing washeries under PSU & TISCO

Sl. No.	Company	No. of plants	Annual raw coal input capacity (mt)
1	BCCL	9	14.33
2	CCL	5	12.31
3	WCL	1	1.20
	Total (CIL)	15	27.84
4	SAIL, TISCO & DPL	7	9.94
	Total (PSU & TISCO)	22	37.78

Washeries under Construction

In CIL, there are two coking coal washeries under construction stage, namely, Madhuband (2.50 mtpa) in BCCL and Kedla (2.60 mtpa) in CCL, which are expected to be commissioned during 1996-97. The total raw coal input capacity of these two washeries under construction comes to 5.10 mtpa.

There are two non-coking coal washeries under construction, namely, Bina (4.5 mtpa) in Northern Coalfields Ltd. (NCL) and Piparwar (6.5 mtpa) in Central Coalfields Ltd. (CCL), which are expected to be commissioned during 1996-97.

Washeries to be set-up under "Build-Own-Operate" Scheme

To meet the growing demand of beneficiated coal, washeries have been identified to be set up under "Build-Own-Operate" (BOO) scheme for supply of beneficiated coal to Power, Sponge Iron and Cement Sectors:

Under the above scheme, the following coal washeries are to be set up which are under different stages of Tender/contract finalisation. The details of washeries with their raw coal input capacities are given in Table-A2.

Table - A2 : Washeries be set up under "BOO" Scheme

Sl. No.	Name of the washery/ Company		Annual Raw coal input capacity (mt)
1	Ananta-Bharatpur	(MCL)	5.2
2	Kalinga	(MCL)	8.0
3	Dipka	(SECL)	6.0
4	Sasti	(WCL)	1.5
5	Raigarh	(SECL)	0.6
6	Tikak	(NEC)	0.6
	Total	(CIL)	21.9

Besides the above, a number of washeries are under different stages of identification/planning, to be set up under "BOO" scheme in near future to cater the needs of beneficiated coal of the different consuming sectors.

CASE STUDY FOR CONSERVATION OF ELECTRICAL POWER

Energy consumption pattern in different coal washeries of BCCL OF CIL

Electric power consumed per tonne of raw coal input in KwH for different washeries of BCCL for the year 1994-95 and 1995-96 (Till Dec.1995) are given in Table-A3.

Table-A3 : Electric power consumed per tonne of raw coal input in different washeries of BCCL of CIL

Sl. No.	Name of washery and installed capacity	(mt/yr.)	Electric power consumption per tonne of raw coal input (KwH)	
			1994-95	1995-96 (Till Dec1995)
1	Dugda-I & II	(4.80)	14.19	17.52
2	Bhojudih	(2.00)	8.87	8.91
3	Patherdih	(2.00)	10.68	11.15
4	Sudamdih	(2.00)	17.22	28.17
5	Moonidih	(2.00)	17.31	21.89
6	Lodna	(0.48)	19.61	20.73
7	Mahuda	(0.63)	11.92	12.50
8	Barora	(0.42)	18.64	20.46

From the figures of the above table, it transpires that the power consumption of Dugda I & II, Sudamdih, Moonidih, Lodna and Barora are very high as compared to Bhojudih, Patherdih and Mahuda washery as well as from the standard norms as indicated by reputed designing firms for coal washeries (varying generally from 4 to 8 KwH/tonne of raw coal input). Out of the above, the power consumption in case of Sudamdih, Moonidih, Lodna and Barora were observed to be considerably higher.

Identification of Washery for Case Study

With a view to analyse the reasons for higher power consumption and to suggest corrective measures to reduce the power consumption, costs of power and resultant cost of operation as well as cost of washed products, two washeries, say washery A and Washery B have been identified for detailed study. The raw coal input capacity of washery A is 2.0 million tonnes per annum and Washery B is 0.42 million tonnes per annum and corresponding hourly normal raw coal input rate are 500 tonnes and 100 tonnes respectively.

Major areas of study

The following major areas have been identified for study of power consumption of the washeries under study . Each item is separately elaborated hereafter.

Study of past power consumption rate of the washeries

The monthwise power consumption, raw coal input and specific power consumption (i.e. power consumption per tonne of raw coal input) for the washery A and Washery B for the year 1994-95 are given in Table-A4.

TABLE-A4 : Monthwise power consumption for Washery A & B for the year 1994-95

Month	Washery A			Washery B		
	Consumption (KwH)	Raw coal input (Te)	Specific Power consumption (KwH/Te)	Consumption (KwH)	Raw coal input (Te)	Specific Power consumption (KwH/Te)
April '94	1466667	94000	15.60	502000	24800	20.29
May '94	1466667	83000	17.67	494830	27300	18.13
June '94	1466667	85700	17.11	434720	27700	15.69
July '94	1466667	80600	18.20	490300	25000	19.61
August '94	1466667	79000	18.57	550100	27500	20.00
Sept '94	1466667	75500	19.43	536690	28000	19.17
October '94	1466667	76000	19.30	578870	24000	23.99
Nov '94	1466667	83300	17.61	511310	24000	21.30
Dec '94	1466667	91000	16.12	496000	26500	18.72
Jan '95	1466667	92000	15.94	557350	35000	15.92
February '95	1466667	9100	16.12	4932.00	27500	17.93
March '95	1466667	91200	16.08	493200	31800	15.52
Total/Avg.	**1,76,00,000**	**10,22,300**	**17.22**	**61,35,570**	**3,29,100**	**18.64**

From the Table-A4, it can be seen that the power consumption per tonne of raw coal input in case of washery A varied from 16-20 KwH and in case of Washery B varied from 16-24 KwH during the year 1994-95.

Study of power feeding system

For Washery A, power is received from Damodar Valley Corporation (DVC) through a sub-station at 33 KV by two parallel feeders. The details of transformers of Washery A are mentioned as follows :

(a) In the main washery 1750 KVA - 6 Nos. (out of which 2 are standby).
(b) In the CHP 500 KVA - 1 no.
(c) In secondary crusher house 250 KVA - 1 no.

For Washery B also, power is received from DVC through another sub-station by two feeders. The capacity of the main transformer is 1750 KVA. There are two other lighting transformers of 100 KVA and 50 KVA.

Study of Electrical installation system of the washery

In both the washeries A and B, the capacitor banks were not in operating condition, causing low power factor and thus higher power consumption. The power factor in washery A was 0.77, whereas in Washery B it was 0.80 as against the desired level of 0.92 (minimum).

The connected load and the operational load of the equipment of Washery A and Washery B have been examined for the purpose of this study. Equipment and their connected as well operational loads for Washery A and Washery B are given in Tables-A5 and A6 respectively.

Table - A5 : Connected and operational load of the equipment of washery A

Unit No.	Item	Connected load (HP)	Unit No.	Item	Connected load (HP)
4	Feeders	60.0	111	Dilute Media Pumps	300.0
6	Dump Hopper Reclaim Conveyor	125.0	113	Double Drive Separator	40.0
9	Rotary Breaker	100.0	116	Clarifying O/F Pumps	600.0
11	Product Conveyor	75.0	117	Tailings Thickener (Drive & Lift)	21.0
12	Transfer Conveyor	20.0	118	Thickener U/F pumps	120.0
13	Reject Conveyor	20.0	120	Clarified Water Pumps	490.0
15	Conveyor	125.0	121	Disc Filter Agitator	15.0
16	Conveyor	125.0	124	Vacuum Pumps	400.0
-	Duplex Feeders	200.0	125	Filterate Pumps	50.0
28	Conveyor	50.0	126	Blowers	100.0
30	Conveyor	75.0	127	Seal Water Pumps	40.0
32	Conveyor	100.0	128	Sump Pumps	200.0
35	Trippper Conveyor	30.0	145	Ball Mills	250.0
40	Crusher	545.0	150	Fresh Magnetic Pumps	40.0
42	Vibrating Screen	80.0	151	Clean Coal Conveyor	175.0
45	Return Belt	50.0	154	Clean Coal Filter Cake Conveyor	75.0
46	Crushed Product Belt	75.0			
47	Washery Feed Conveyor	200.0	158	Clean Coal Tripper Conveyor	30.0
51	Scrapper Conveyor	75.0			

Table 5A - *Contd.*

Unit No.	Item	Connected load (HP)	Unit No.	Item	Connected load (HP)
53	Desliming Screens	100.0	159	Middling Conveyor	75.0
57	Clean Coal Depulping Screens	100.0	172	Middling Loadout Conv	50.0
62	Primary HM Pumps	1500.0	177	Clean Coal Loadout Conveyor	100.0
68	Middling Rejects Depulping screens	60.0	188	Fresh Water Pumps	50.0
76	Middlings Depulping Screens	40.0	205	Compressors	200.0
88	Cyclone Feed Pumps	375.0	217	Pressurising Fan	15.0
93	Fine Coal Circuit Flotation Cell	300.0		- Workshop	30.0
94	Froth Paddlers	16.0		- Laboratory	10.0
98	By-Pass Scrapper Conveyor	40.0		- Welding Machines	115.0
102	Clean Coal Conveyor No.1	50.0		- Lighting Load including Plant & Colony	400.0
107	Middling Conveyor No. 1	20.0			
108	Middling Conveyor No. 2	20.0			
110	Dilute Media Clarifier (Drive and Lift)	10.5			
	Total connected load (HP)				8,752.5
	Total operational load (HP)				7702.5

TABLE - A6 : Connected and operational load of the equipment of washery B

Unit No.	Item	Connected load (HP)	Unit No.	Item	Connected load (HP)
4	Conveyor	25.0	63	Centrifuge*	75.0
7	Conveyor	5.0		Vibrator*	7.5
10	Single Roll Crusher	60.0		Lub*	0.5
13	Conveyor	7.5	71	Pump (2)	200.0
15	Conveyor	25.0	72	Magnetic separator (4)	20.0
20	Conveyor	25.0	77	Pump	100.0
21	Jug	15.0	84	Thickener (2)	10.0
	Elevator (2)	10.0	85	Thickener U/F Pump	20.0
28	Screen	20.0	87	Thickener O/F Pump (2)	150.0
32	Jig Blower	60.0	88	Disc Filter*	3.0
34	Pump	60.0		Agitator*	5.0
35	Double Roll Crusher	120.0	90	Pump*	7.5
37	Conveyor	20.0	91	Pump*	215.0
39	Conveyor	15.0	92	Filter Blower*	20.0
43	Screen (2)	40.0	93	Conveyor	20.0
47	Froth Flotation Cell (3)*	90.0	95	Conveyor	15.0
	Paddle*	4.0	97	Conveyor	20.0
52	Screen (2)	40.0	99	Conveyor	15.0
55	Centrifuge*	75.0	101	Motorised Gate*	2.0
	Vibrator*	7.5	104	Conveyor*	150.0
	Lub.*	0.5	196	Feeder (10)*	100.0
60	Pump	5.0	114	Sump Pump (4)	30.0
61	Screen (2)	40.0	115	Booster Pump	10.0
				New tailing dewatering system	20.0
	Total connected load (HP)				1985
	Total Operating load (HP)				1222

* Item not in constant operation/By-passed/Removed.

Study of actual operating hours and the input rate of the washery

Monthwise raw coal input and actual operational hours for Washery A and Washery B for the year 1994-95 have been studied in order to compare the actual operational hours with the desired operational hours (on the basis of normal feed rate of the washery) and thus to calculate extra operational hours in case of both these washeries.

These data are given in Table-A7.

TABLE-A7 : Monthwise raw coal input and actual operational hours for the year 1994-95

Washery	Month	Raw coal input (te)	Actual operational hours (hrs.)	Feed rate (tph)	Desired operational hrs. as per normal feed rate (hrs.)	Extra operational hours (hrs.)
Washery A	April '94	94000	339:30	276.9	188:00	151:30
	May '94	83000	307:30	269.9	166:00	141:30
	June '94	85700	325:00	263.7	171:00	154:00
	July '94	80600	316:30	254.7	161:00	155:30
	Aug. '94	79000	320:00	246.9	158:00	162:00
	Sep. '94	75500	277:30	272.1	151:00	126:30
	Oct. '94	76000	349:00	217.8	152:00	197:00
	Nov. '94	83300	334:30	249.0	167:00	167:30
	Dec. '94	91000	362:30	251.0	182:00	180:30
	Jan. '95	92000	363:00	253.4	184:00	179:00
	Feb. '95	91000	372:00	244.6	182:00	190:00
	March '95	91200	359:00	254.00	182:00	177:00
Total		1022300	4026:00	254.5	2044:00	1982:00
Washery B	April '94	24800	309:00	80.0	248:00	61:00
	May '94	27300	322:00	85.0	273:00	49:00
	June '94	27700	343:00	81.0	277:00	66:00
	July '94	25000	334:00	75.0	250:00	84:00
	Aug. '94	27500	328:00	84.0	275:00	53:00
	Sep. '94	28000	353:00	79.0	280:00	73:00
	Oct. '94	24000	309:00	78.0	240:00	69:00
	Nov. '94	24000	335:00	72.0	240:00	95:00
	Dec. '94	26500	226:00	117.0	265:00	39:00
	Jan. '94	35000	373:00	94.0	350:00	23:00
	Feb. '95	27500	334:00	80.0	275:00	69:00
	March '95	31800	356:00	89.0	318:00	38:00
Total		329100	3932:00	84.0	3291:00	641:00

The close examination of the above data show that the actual average hourly input rate for Washery A and Washery B during the year 1994-95 were 254.5t and 84t as against normal input rate of 500t and 100t respectively.

Due to lower input rate, Washery A had to run for extra hours to the extent of 1982 and Washery B to the extent of 641 hours, thus consuming extra power. It is clear from the above that if these washeries could be operated at normal input rate or near normal input rate, considerable electrical energy could be saved.

The reasons for lower actual input rate is the frequent power failures, breakdowns, bunkers full due to shortage of wagons and other operational problems. Once the washery is shut, it takes considerable time to start the same and stabilise and till such time plant has to be operated at reduced load affecting the actual input rate of the washeries.

Study of general operating conditions of the plant and machineries

(i) It is generally observed that lightings are "ON" even during day time in Raw Coal Receipt and Storage Section, Product Despatch Section, other sheds and streets.

(ii) For the purpose of study of general operating condition of the plant and machineries, the reasons for the loss of operational hours in respect of these washeries have been examined. These figures are given in Table-A8.

TABLE-A8 : Reasons for loss of operational hours in 1994-95

Name of the Washery	Reasons for loss of production	Duration Hours	Min
Washery-A	(i) Power shortage	124	00
	(ii) Raw coal shortage	270	00
	(iii) Breakdown	380	00
	Total	774	00
Washery-B	(i) Power shortage	168	30
	(ii) Raw coal shortage	92	45
	(iii) Clean coal bunker full	01	00
	Total	262	15

Note: These data have been obtained from BCCL

From the above figures, it is amply clear that considerable operational hours are lost due to power shortage, raw coal shortage, wagon shortage and breakdown and operational problems.

ANALYSIS OF DATA AND ESTIMATION OF SAVINGS DUE TO IMPLEMENTATION OF CORRECTIVE MEASURES

By implementing corrective measures, savings in terms of energy and money can be realised which are estimated hereafter.

Savings due to improvement of power factor

By installing a suitable capacitor bank the power factor of the system can be improved to 0.92 (minimum). The resultant savings are indicated below :

Washery-A

Let us consider the operational load of Washery A.

= 7702.5 HP = 5746 Kw (Refer Table-A5)

At 0.77 Power factor, current drawn will be

$P = \sqrt{3}.V\ I_1\ \cos\phi 1$

$$I_1 = \frac{P}{\sqrt{3}\times V\times 0.77} = \frac{5746\times 10^3}{\sqrt{3}\times 415\times 0.77} = 10382 \text{ Amp.}$$

At 0.92 power factor, current drawn will be

$$I_2 = \frac{P}{\sqrt{3}.V.\cos\phi_2} = \frac{5746\times 10^3}{\sqrt{3}\times 415\times 0.92} = 8689 \text{ Amp.}$$

Thus, less current drawn due to high power factor = $I_1 - I_2$

= 10382–8689 = 1693 Amp.

Therefore excess power consumed at 0.77 power factor with respect to 0.92 power factor = $\sqrt{3}$ x 415 × 1693 × 0.77 = 937009 Watts = 937 Kw

Thus the energy saved per annum due to improved power factor = 937 × 13 × 300 = 3654300 KwH

Washery-B

Let us consider the operational Load of Washery B.

= 1222 HP = 912 Kw (Refer Table - A6)

At 0.80 Power factor, current drawn will be

$P = \sqrt{3}.V\ I_1\ \cos\phi 1$

$$I_1 = \frac{P}{\sqrt{3}\times V\times 0.80} = \frac{912\times 10^3}{\sqrt{3}\times 415\times 0.80} = 1586 \text{ Amp.}$$

At 0.92 power factor, current drawn will be

$$I_2 = \frac{P}{\sqrt{3}.V.\cos\phi 2} = \frac{912\times 10^3}{\sqrt{3}\times 415\times 0.92} = 1379 \text{ Amp.}$$

Thus, less current drawn due to high power factor = $I_1 - I_2$ = 1586 – 1379 = 207 Amp.

Therefore excess power consumed at 0.80 power factor with respect to 0.92 power factor = $\sqrt{3}$ × 415 × 207 × 0.80 = 119030 Watts = 119 Kw

Thus the energy saved per annum = 119×13×300 = 464100 KwH

Where, 13 is the total effective working hours in a day and 300 is the total working days in a year.

Savings due to operation of the washery at normal input rate

From para 4.3.4, (Table - A7) it is obvious that by operating the washery at normal input level, 1982 hours of extra operation can be saved in Washery A and 641 hours of extra operation in case of Washery B.

The resultant saving in energy per annum will be as follows :

For Washery-A

$$\text{Energy saved per annum} = \text{Total energy consumed in the year} \times \frac{\text{Extra opern. hrs.}}{\text{Actual opern. hrs.}}$$

$$= 1{,}76{,}00{,}000 \times 1982 \,/\, 4026 \;=\; 86{,}64{,}480 \text{ KwH}$$

For Washery-B

$$\text{Energy saved per annum} = \text{Total energy consumed in the year} \times \frac{\text{Extra opern. hrs.}}{\text{Actual opern. hrs.}}$$

$$= 61{,}35{,}570 \times 641 \,/\, 3932 = 10{,}00{,}229 \text{ KwH}$$

Savings by switching off 50% of lighting load during day time

By "switching off" 50% lighting loads during day time (from 7 AM to 5 PM i.e. 10 hrs.), electrical energy saved for both the washeries will be as follows:

As per the data collected, the lighting load of Washery A and Washery B are 300 Kw and 20 Kw respectively.

For Washery-A

Energy saved per annum = 50% of lighting load (in Kw) × Hrs. × Days = 150 × 10 × 365 = 547500 KwH

For Washery B

Energy saved per annum =10 × 10 × 365 = 36500 KwH

CONCLUSIONS

This study reveals that there is a scope of considerable energy saving in the Washery by proper motivation/awareness of the employees regarding "Energy Conservation", by use of natural light during day time, putting off light when it is really not required, maintaining desired level of power factor by providing suitable capacitor bank and operating the washery at normal or near normal input rate.

This study shows an estimated saving of energy to the extent of 1,28,66,280 KwH and 15,00,829 KwH per annum for Washery A and Washery B respectively. In terms of money the same works out to Rs.1,92,99,420 and Rs.22,51,244 for Washery A and Washery B respectively. In actual practice, if even 50% of estimated savings are achieved, this will be a quite considerable amount and it will have positive effect on overall economics of the Coal Washeries.

The break-up of the quantified savings for the two washeries are given hereafter:

TABLE - A9 : Estimated savings per year

Washery	Item	Likely savings per year (KwH)	Likely savings per year (Rs.)
Washery-A	1. Installation of capacitor bank	36,54,300	54,81,450.00
	2. Operation of plant at normal capacity	86,64,480	1,29,96,720.00
	3. Switching off 50% lighting during day time.	5,47,500	8,21,250.00
	Total	1,28,66,280	1,92,99,420.00
Washery-B	1. Installation of capacitor bank.	4,64,100	6,96,150.00
	2. Operation of plant at normal capacity.	10,00,229	15,00,343.50
	3. Switching "off" 50% lighting during day time.	36,500	54,750.00
	Total	15,00,829	22,51,243.50

Note: Taking power tariff rate of Rs.1.50 per KwH.

RECOMMENDATIONS

— For energy conservation, proper motivation/awareness of the employees like use of natural light during day time, "switching off" electrical lighting when it is really not required, are to be created for better and sustained result.

— Desired level of power factor should be maintained by providing suitable capacitor banks in the electrical system.

— The general condition of the plant and machineries of washeries should be improved by proper and preventive maintenance to ensure troublefree/smooth operation of the plant with a view to operate the washeries at normal or near normal input level. This will reduce the energy consumption, reduce the operating cost and improve the overall economics of the washeries in the greater interest of the coal company and the nation.

— In every washery there should be a small cell for looking after energy conservation aspects. They should study the plant in greater depth, identify the ECO and prepare short term and long term measures for conservation of energy. This cell will also ensure implementation and regular follow up of the corrective measures.

— There should be a cell at Company Headquarters level to closely monitor the study for energy conservation and implementation of suggested measures.

ABBREVIATIONS USED

BCCL	-	Bharat Coking Coal Limited
CCL	-	Central Coalfields Limited
DPL	-	Durgapur Projects Limited
ECO	-	Energy Conservation Opportunities
MCL	-	Mahanadi Coalfields Limited
NCL	-	Northern Coalfields Limited
NEC	-	North Eastern Coalfields
PSU	-	Public Sector Undertaking
SAIL	-	Steel Authority of India Limited
SECL	-	South Eastern Coalfields Limited
TISCO	-	Tata Iron & Steel Company Limited
WCL	-	Western Coalfields Limited

REFERENCES

Reading materials given at EMCAT - IDBI - USAID Auditor Training Course in India in March 1995 on "Fundamentals of Industrial Energy Auditing".

Operational data collected from the concerned washeries/coal company.

43

CLEAN COAL AND HIGH ASH COAL UTILISATION TECHNOLOGY

B.C. MISRA
Chairman-cum-Managing Director
Central Mine Planning and Design Institute Ltd, Ranchi (INDIA)

ABSTRACT

India is endowed with nearly 202 billion tonnes of coal compared to 765 million tonnes of oil and 707 billion cu.mtr of natural gas. At the present rate of consumption, coal will last for 200 years compared to 24 years in case of oil and 23 years in case of natural gas. With environmental restriction on use of nuclear power, India will have to depend on coal as a primary source of energy for its sustainable growth even beyond 2020.

Indian coals, in general, are high in ash and already all the coking coals going to steel sector are washed (direct feed coal is limited to 0.5 million tonnes/year), cement and power sectors are also aware of the benefits of utilising washed coal. A good amount oft spadework has been started to convince the customers, particularly cement and power sectors to go in for washed or deshaled coal use in their installations.

The main properties of coal which influence power plant performance are : specific energy, ash content, ash fusion temperature, size distribution, moisture content, sulphur content, volatile matter, ash resistivity, HGI etc. Indian power coals are, in general, high ash with difficult cleaning characteristics. Consequently, such coals are having comparatively low specific energy and low HGI which adversely affect the power plant performance. However, they have some positive features such as high ash fusion temperature, low chlorine and sulphur, low toxic trace elements etc.

Similarly, use of weakly coking and low volatile coking coals, which do not normally find use in steel sector, has been discussed alongwith the technological options for its use in steel sector.

This paper deals with the prospect of utilising indigenous coal with due consideration to environment friendliness.

Introduction

India is endowed with nearly 202 billion tonnes of coal (as on 01.01.1996 upto a depth of 1200 metres and coal seams of thickness 0.9 mtr and above) which is likely to last for more than 200 years. India's coal reserve position is shown in Table-I. As against the above reserve position of coal, the reserve of crude oil is 765 million tonnes and that of natural gas is 707 billion cu. mtr. At the present rate of production, both crude oil and natural gas may last for another 23 years.

TABLE I : Total Coal Resource of India as on 01.01.1996

(In million tonnes)

Type	Proved	Indicated	Inferred	Total
Coking				
Prime	4236.45	1063.55	-	5300
Medium/blendable	10987.64	12049.91	1474.84	24512.39
Sub. Total	15224.09	13113.46	1474.84	29812.39
Non-coking	54,867.02	76,530.24	39,857.13	171,254.39
NE Region (Tertiary)	352.02	106.16	428.84	887.02
Grand Total	70,443.13	89,749.86	41,760.81	201,953.8

Source : *GSI*

It is evident from the Table-1 that, of the total coal reserve, only about 15% is of coking variety and the rest 85% is of non-coking variety. Of these resources, the estimated reserve down to a depth of 600 m. (considered workable) is around 90%. Out of this, around 85% is non-coking coal having ash upto 50%. If this non-coking coal is categorised into low ash (less than 24%), intermediate ash (24% to 35%) and high ash (35% to 55%) quality, it is observed that only about 13% of the workable reserve is of superior quality (low ash).

This is reflected in the qualitywise production programme of Coal India Ltd. as shown in Table 2.

TABLE 2 : Qualitywise Production Programme of Coal India Ltd.

(In million tonnes)

Quality	1996-97	1997-98	1998-99	1999-00	2000-01	2001-02
Direct Feed coking	0.5	0.5	0.52	0.53	0.53	0.56
Washery grade coking	21.76	23.37	23.91	24.57	25.9	26.05
Non-metallurgical coking	17.54	17.41	17.83	17.39	15.9	15.75
Non-coking superior gr. (A & B)	28.5	29.1	30.71	33.9	32.36	32.75
Non-coking Intermediate gr. (C&D)	76.77	79.89	84.76	86.11	91.02	94.45
Non-coking high ash. (E, F & G)	107.38	112.97	117.27	120.57	126.67	135.44
Grand Total	252	263.24	275	283.07	292.38	305

Source : Working Group on Coal & Lignite for Ninth Plan

This Table indicates that by the turn of the century, major portion of the coal utilised will be having high ash. Only about 60% of the coking coal produced will be of washery grade and the balance will be consumed by the non-metallurgical sector and more than 50% of the non-coking coal produced will be of high ash. Keeping in view the massive growth in energy demand and with due consideration to environmental aspects, ways and means have to be evolved for clean and efficient use of this high ash coal.

Sectorwise coal demand projection in India is shown in Table - 3.

TABLE - 3 : Sectorwise Coal Demand (Projected)

Sector	VIII Plan end (1996-97)		IX Plan end (2001-02)		X Plan end (2006-07)	
	in mill. te.	%	in mill. te.	%	in mill.te.	%
Steel (including private cokeries)	40	13.4	55	12.9	59.1	10.9
Power (Utilities)	185.5	62	279	65.4	379	70.2
Cement	17.1	5.7	25	5.9	30	5.6
Fertiliser	4.4	1.4	4.7	1.1	4.7	0.9
Others	52.4	17.5	62.7	14.7	67.2	12.4
Total	299.4	100	426.4	100	540	100

Source : CMPDI's own assessement

From this table, it may be seen that the power sector will lead the sectoral demand followed by steel. Consumption of nearly 70% of the coal production in power sector emphasises the need for putting serious consideration for clean and efficient use of coal for power generation. In view of the scarcity of good grade coking coal, utilisation of more and more inferior coking coal for metallurgical purpose requires to be emphasised. In other sectors (like cement , domestic etc.) also, utilisation of inferior quality coals becomes imperative.

Power Sector

The main properties of coal which influence power plant performance are :

— Specific energy
— Ash content
— Fusion temperature of the ash
— Size distribution
— Moisture content
— Hardgrove Grindability Index (HGI)
— Sulphur content
— Volatile content
— Ash resistivity etc.

In this context, it may be noted that Indian power coals are generally high ash with difficult cleaning characteristics due to intermixed mineral impurities. Consequently, such coals are having comparatively low specific energy and low HGI which adversely affect the power plant performance. However, such shortcomings notwithstanding, Indian coals have some positive features also. Some favourable characteristics of Indian coals are :

— In general the ash fusion temperature is high and the ash is of refractory nature
— Silica alumina ratio is favourable (about 2:1) from slag viscosity point of view, in case of blast furnace
— Low chlorine content (< 0.1%)
— Low sulphur content (< 0.5%)
— Low toxic trace elements (compared to coals of many of the developed countries).

These inherent characteristics of Indian coals make their use less detrimental to the environment. But keeping in view the magnitude of coal use for power generation, measures have to be taken for making their use more environment friendly, which in other words, calls for adoption of clean coal technologies.

Various means exist for making use of coal environmentally acceptable which may be categorised as follows :

— Pre combustion treatment
— Control during combustion
— Post combustion treatment
— Conversion to more acceptable fuel.

In Indian context, the pre-combustion treatment i.e. beneficiation of feed coal has been tried and encouraged with result, few non-coking coal washeries for beneficiation of power coals are being planned. But in other modes of environment friendly measures, little effort has so far been put in. With massive growth in coal use and increasing awareness on environmental aspects, it is recommended that these measures should also be adopted in new power plants or the old ones being revamped. It involves substantial cost, but for the sake of cleaner use of coal, the additional investment and operating expenditure has to be kept in view..

Precombustion Treatment

The shortcomings of Indian power coals, as discussed above, may mostly be overcome by beneficiation, even by limited beneficiation. The cleaning characteristics of Indian coals are not very good. As such, even by beneficiation, the ash content cannot be brought down to a level at par with power coals used abroad. Still, limited beneficiation is desirable for bringing in consistency in ash content, calorific value and size of the coal supplied to power plants. Power plants designed for the use of coal with certain ash level can operate smoothly if the quality of coal supply is consistent. But with wide fluctuation in coal quality, performance of the power plant is badly affected.

On application of appropriate beneficiation technology, most of the extraneous inorganic impurities can be removed from the coal which in turn helps in reducing release of particulates including toxic trace elements during the combustion process. Ten out of the eleven toxic trace elements (identified by US Clean Air Act Amendment 1990) are arrested in the fly ash. Only mercury escapes with the flue gas as vapour.

Beneficiation of power coal also helps improve the power plant performance and economy of power generation (for distantly located plants). Reduction of sulphur to some extent may also be achieved through beneficiation (if the sulphur is present in pyratic form), however, this is not of much relevance in Indian context since Indian coals are already having low sulphur. But for a location having cluster of power plants, desulphurisation of flue gas has to be kept in view to maintain the SOx emission within reasonable limit.

The gains/advantages of beneficiation of power coals (even limited Beneficiation from 40/42% ash to 34% ash) may be assessed from the outcome of actual R&D trial of beneficiated coal at Satpura power station in a 210MW unit for a period of about one month, as listed below:

— Improvement in plant utilisation factor from 73% to 96%
— Reduction in specific coal consumption from 0.77 to 0.553 Kg/KWH
— Reduction in specific auxiliary power consumption by around 1.5%
— Reduction in outage/down time of coal mills
— Reduction /elimination of fuel support
— Reduction in boiler slagging
— Improvement in boiler efficiency by about 2%
— Reduced smoke and dust emission
— Reduced maintenance of boiler and its auxiliary.

The net benefits of pre-combustion treatment are reliability of operation, improvement in economy besides reduction in particulate and SOx emission.

Control during Combustion

The major pollutants viz. oxides of sulphur and nitrogen can be removed during combustion of coal either through adoption of appropriate combustion technology or through modification in existing coal combustion system. In most of the conventional power plants, pulverised coal firing method is adopted. In recent years, for small capacity captive power plants (and also for steam generation), fluidised bed combustion system is also being widely adopted for utilisation of high ash coals and washery rejects (otherwise finding no use).

In fluidised bed combustion system, instead of burning pulverised coal, crushed coal (mixed with limestone or dolomite in case of high sulphur coal) is burnt in a bubbling bed. The SOx formed during combustion is fixed by the additives in the bed material and thus the release of SOx into the atmosphere is restricted. The added advantage of fluidised bed combustion is that it operates at a low temperature restricting the formation of NOx, as formation of NOx depends on factors like fuel nitrogen of combustion percentage excess air, compared to PC system. Development FBC technology for large scale technology is essential.

In pulverised coal firing system, *in-situ* sulphur capture is difficult and hence SOx generated is arrested downstream. However, formation of NOx can be restricted through use of low NOx burners. Such burners are designed to delay the mixing of the fuel and air in the burner zone to keep the primary combustion zone deficient in oxygen resulting in reduction in NOx formation by 50%. Another approach for reducing NOx formation in combustor is to provide air ports in the furnace wall above the top row of burners to create fuel deficient low NOx combustion zone. Such system can reduce NOx formation by 15-30%. In other system, a small amount of natural gas is injected above the normal combustion zone to form an oxygen deficient zone. Air inlet above the natural gas ports at appropriate locations ensure complete combustion. Such system may control NOx formation by about 60%. Injection of dehydrated lime over the overfire air port can reduce SOx by 50%. Reduced excess air in the primary combustion zone is essential for control of NOx formation.

Post Combustion Treatment

Pollutants generated during combustion of coal are mainly particulates , SOx and NOx . The particulate pollutants can be arrested by electrostatic precipitators and fabric filters. Recovery to the extent of 99.9% may be achieved through such systems.

Post combustion SOx removal may be achieved either through in-duct sorbent injection or through scrubbers. In-duct sorbent injection works inside the duct connecting the boiler with the stack. Absorbents like lime are sprayed into the duct and the humidity is carefully controlled. In this system, 50-70% of the SOx present in the flue gas can be removed. The scrubber is installed outside the main plant where the flue gas is treated for removal of SOx and NOx . Selective catalytic reaction is extensively adopted for control of NOx from the products of combustion . In this technology, ammonia is mixed with flue gas and is passed through a separate reaction chamber before the scrubber where NOx and ammonia react to produce nitrogen and water. The system may reduce NOx emission by 50-80%. Another method of NOx reduction is a non-catalytic process where urea solution is sprayed into the convective zone at carefully controlled temperature when urea combines with NOx to form nitrogen and water. This system can reduce NOx emission by about 30%.

Conversion to more acceptable Fuel

To take advantage of the liquid and gaseous fuels, coal can be converted to these convenient forms at a cost. Technologies exist for gasification of high ash coals to produce fuel gas as well as synthesis gas. While the fuel gas may be used as a fuel, the synthesis gas may be converted into the liquid fuels. The fuel gas derived out of coal may be utilised for generation of power through combined cycle which helps achieving higher efficiency and environment friendliness.

Steel Sector

Steel sector uses metallurgical coke which is produced through carbonisation of coking coal. In view of scarcity of good grade coking coal, beneficiation of coking coal is widely practised to bring down the ash content within the acceptable limit. For widening the resource base, attempts are being made to include low volatile medium coking coals also in the coke oven feed coal blend. Beneficiation of coking coal to an ash level of almost 19% generates sufficient quantity of rejects, the disposal of which in a environmentally friendly manner is essential. For the purpose of FBC, power plants of smaller capacity are being installed in the country.

Increasing emphasis on environmental aspects on modern coke production has led to renewed interest in non-recovery coke ovens for the production of metallurgical coke. . The by-product plant is a significant source of air and water pollution. Non-recovery ovens have a distinct environmental advantage over conventional ovens in that they are operated under suction condition which reduces the potential for emission to very low level. In addition, non-recovery plant does not require coke-oven gas processing facilities. The prominent examples of development in this area are the Jewel-Thompson oven (USA) producing power as by-product utilising its waste heat and the Illawara Non-recovery oven (Australia). Under development is a highly productive Jumbo Coking Reactor (Germany) where each reactor is capable of producing 100 tonnes of coke daily.

Additional technologies are being adopted to maximise use of inferior coking coals and even non-coking coals for metallurgical purpose. The technologies in this direction are :

— Partial briquetting of coal charge
— Stamp charging
— Selective crushing
— Coal dust injection etc.

Partial Briquetting of Coal Charge

A part of the coal charge is converted into briquettes which are mixed with the balance quantity of feed coal. The aim is higher densification of charge coal in the coke oven. It is claimed that even non-coking coal to the extent of 20% can be utilised in making the briquette, without affecting the coke quality. presently in Italy, this technology is used for improving the coke quality only.

Stamp Charging

This is another way of densification of coal charge having potential for improving the coke quality. This technology helps in getting better coke strength using the same coal charge. The coke strength can be maintained accommodating inferior coking/semi-coking in the charge.

Selective Crushing

Coal is subjected to controlled crushing. The reactive components are subjected to milder crushing while the harder materials to intense crushing. This is achieved in two stage crushing and results in improved coke quality.

Coal Dust Injection

Pulverised/dust coal is injected through tuyer with an aim to reduce consumption of coke in almost direct proportion, thus reducing requirement of coking coal . However, low ash (14 -15%) non-coking coal is needed for this purpose.

Cement Sector

Adoption of energy efficient dry process utilising pre-calcination is preferred option these days in India. For the purpose of pre-calcination, coals with relatively high ash can be used separately for calcination of calcium carbonate. Due to deterioration in quality of coal, cement plants in India are switching over to this technology. It is expected that the specific coal consumption may come down to 0.16 Kg/Kg of clinker from the present level of around 0.25 Kg/Kg. For meeting the coal quality specification for the cement industry, beneficiation of non-coking coal for some of the cement plants are now under consideration.

Domestic Sector

Low grade coals are generally used for the production of soft coke for household purpose. Technology has been developed for production of improved quality domestic coke (commercially known as Special Smokeless Fuel or SSF) from low grade coal in an environmentally acceptable manner. This technology has been adopted by private entrepreneurs to produce SSF which will reduce consumption of non-commercial fuels like firewood, cow-dung , agro-wastes etc. to reduce ecological imbalance.

Global Warming Issue

Swedish chemist Arrhanious used the term "Greenhouse Effect" to describe his prediction that large scale combustion of fossil fuels would increase the CO_2 level and thereby cause global warming. Greenhouse gases generated during coal combustion process are NOx and CO_2. Another polluting gas that may be emitted (during combustion of high sulphur coal) is SOx which can be controlled either through the use of low sulphur coals or through adoption of SOx arresting devices commercially available. While one of the greenhouse gases, viz. NOx, can be reduced through the use of improved combustion technology or use of post-combustion treatment of the flue gas, reduction of CO_2 (as is allegedly responsible for greenhouse effect) is possible either through choice of fuels like natural gas which generate less CO_2 or through improvement in energy conversion efficiency to reduce the total coal consumption for same energy output.

The concept of global warming due to CO_2 emission, however, needs a relook. In 1992, more than 500 leading scientists from all over the world (including more than 60 Nobel Lauriates) sharply protested in the so called "Hydelberg Appeal" against this concept which is supported by pseudo-scientific arguments or false and non-relevent data. The Director of Environment Unit of London's Institute of Economic Affairs, Roger Bate is of the view that global warming is a political issue and decisions in this regard are subject to the pressure of international diplomacy and the democratic process.

In fact, contrary to the concept, CO_2 is not a pollutant and is rather vital and useful for life system. Moreover, the proportion of entire man-made CO_2 is only about 4%, the remaining 96% being supplied by nature itself. As such, reduction in man-made CO_2 emission does not significantly affect the global warming even if at all it is caused by CO_2.

Regarding global warming, the finding of the office of Climatology of the Arizona State University, USA, may be referred to, in terms of which, in Germany and other countries of medium latitudes, there was no increase of temperature in the last 200 years even though the combustion of fossil fuels in those countries was very high. In fact, the computer models predicting the steady increase of temperature are faulty.

Some facts about CO_2 are :

(i) It is not a pollutant, but vital for life on our planet.

(ii) Increased level of CO_2 would stimulate plants growth resulting in more availability of fuels and food for mankind.

(iii) The share of man-made CO_2 is about 4%. As such, its reduction is not going to affect our climate significantly.

(iv) Rise of temperature due to increase in CO_2 content, if at all, may be beneficial to the mankind by way of reduced fuel consumption.

(v) The massive consumption of fossil fuels started in 1850 with the industrial revolution. However, in the history of the earth, there have been much warmer period than today.

(vi) Restricting release of CO_2 to the level of 1990, as recommended by Climate Change Convention would halt the development activity, particularly in the developing and under-developed countries.

Conclusion and Recommendations

Coal is and will continue to be the major source of energy in India. With gradual deterioration in coal quality and increased awareness on environmental protection, advanced technologies have to be adopted in various coal consuming sectors. The recommendations made in this paper are:

— Beneficiation of power coal for maintaining consistency in the quality of coal supply and reducing pollutants emission.

— Adoption of fluidised bed combustion on a larger scale

— Adoption of technology for controlling SOx and NOx emission during-combustion and post-combustion.
— Conversion of coal to cleaner fuels
— Adoption of larger capacity improved non-recovery type coke ovens
— Adoption of advanced technologies of coke making for increased use of inferior coking/non-coking coals and improve the coke quality.
— Adoption of pre-combustion and dry process for cement manufacture.
— Production of coal based domestic fuels for reducing use of non-commercial fuels like firewood, dungcake and agro-wastes.
— Use of rejects from the coking coal washeries for the generation of power through FBC route

ENERGY CONSERVATION THROUGH EFFICIENT PROCESS TECHNOLOGIES IN PAPER MILLS

44

A. G. KULKARNI
Director.
Scientist 'F' & Head.
ABHA GUPTA
Scientist
B. P. THAPLIYAL & N. J. RAO
Chemical Recovery, Energy Management & Effluent Treatment Division,
Central Pulp & Paper Research Institute, P.O.Box.174, Saharanpur - 247001 U.P. (INDIA).

ABSTRACT

Paper industry ranks sixth amongst the energy intensive industries and the cost of energy has become one of the primary factors which determines the cost of production. The energy consumption figures are still much higher than those in developed countries. Obsolescence of technology, scale of operation, lack of adequate process control systems and non-implementation of energy efficient technologies are some of the reasons which have resulted in poor energy efficiency in our paper mills. Mills abroad are dependent only on 30-40% of external energy resources, but the mills here are still purchasing nearly 70% of their total energy requirement from external sources. Studies conducted by various agencies clearly indicate that there is a net saving potential of atleast 20-25%.

Energy efficient technologies could improve the energy situation in Indian paper industry. The article with the help of few case studies discusses the key areas where the adequate modernisation through adoption of energy efficient process technologies is possible for long term energy saving programmes. The paper also highlights the efforts made by Central Pulp & Paper Research Institute (CPPRI) as R&D institute in propagating such programmes.

Introduction

Energy has become the third largest cost sharing component behind fibre and labour in producing a tonne of paper . Due to the rising trend in the cost of energy,it becomes imperative to consider the ways and means to conserve energy. Considerable energy savings is possible through process optimisation, advanced control systems, conversion to lower cost fuels, increased use of renewable energy sources and energy auditing on routine basis. Modernization of mills by adoption of energy efficient process technologies is the immediate step towards energy conservation. The present paper highlights the attempts made by CPPRI energy team in identifying the various areas, where energy could be conserved.

DISCUSSION

The average energy cost in paper production in India is about 22%. The major energy consuming plants are the chippers (for wood based), digesters, refiners, paper machines and evaporation plants. The specific steam consumption in Indian mills varies from 10.0-16.0 tonnes/tonne of paper and power from 1300-1950 Kwh /tonne of paper . Indian mills on an average are consuming 1.5-2.0 times higher energy for their unit paper production in comparison to the paper mills in developed countries. The major factors responsible for higher energy consumption in most of the Indian paper mills are due to[1,2].

- Low scale of operation with lower capacity utilization.
- Old and technically inefficient machines.
- Lack of modernisation and instrumentation for better process control.
- Over sized motors, pumps and fans.
- Lower boiler efficiency.
- Live steam parameters are low but process steam parameters are high.
- Poor waste heat recovery.
- Frequent power cuts, load sheddings lead to wasteful use of energy.
- Inefficient cogeneration system, .
- Besides, the new technologies are not fully known .

Developed countries have conserved appreciable amounts of energy by adoption of such technologies like use of continuous digesters, adoption of Oxygen delignification, alkaline sizing, effective way of waste heat recovery, etc. Also, there is lack of cooperation between the industry, power utilities/equipment manufactures and R&D institutions.

Scope of Energy Conservation Potential in different Sections is as described below.

(i) Raw Material Handling and Storage

Adoption of proper raw material preparation, storage methods and cleaning techniques is the first step which reduces fibre losses, chemical requirement and restricts entry of non-fibrous silicaceous matter, pith, rice husk, etc. especially in case of agro-based raw materials, which causes problems during various paper making operations.

Chippers

Indian paper industry employs wide range of fibrous raw materials for paper manufacture and requires different methods of preparation. Most common chippers in Indian Paper Industry are the disc chippers. Recently, Palman drum type Chippers are becoming common for bamboo chipping. The drum chippers are of higher capacity and energy efficient and it is possible to reduce energy consumption from 35 KwH/t to as low as 7KwH/t of dry chips. It is already being practiced in some of the mills. Further,with inefficient chipping system, the dust formation is as much as 5-8% which goes as a waste, if not utilized[3].

(ii) Pulping

(a) Installation of Continuous Digesters

The predominant practice is the use of batch digesters. Two large mills have Kamyr Continuous digesters. Some of the agro based mills use Pandia Continuous Digester.

The advantages of continuous digesters are :

— Uniform cooking and better pulp quality
— low specific energy consumption
— less alkali charge
— shorter cooking cycle
— higher liquor concentration going to recovery.

TABLE -1 : ENERGY CONSUMPTION IN BATCH and CONTINUOUS DIGESTERS[4]

Type of Digester	Cycle (hrs.)	Steam t/tp
Spherical (Small Mills)	8-10	4-5
Stationary (direct heat)	4	1.5-2.0
Pandia (Continuous)	3	1.5-2.0
Kamyr (Continuous)	2	1.5

The constraints for not using continuous digesters are -

- Higher outlays for machinery replacement in the existing units;
- Lack of flexibility of the continuous digester system to accept varying raw material mix.
- The relatively smaller size of the mills which makes it uneconomical to use continuous digesters more efficiently.

(b) Modified Batch Cooking

The present trend is towards low energy batch cooking. The modified batch system is an extension of conventional batch cooking process in stationary digesters. The salient features of the process are : Lower steam consumption by about 25%, Reduced cooking cycle, Improved pulp washing, High viscosity of pulps

For a large integrated mill, the energy savings with modified batch process is estimated to be 52800 t/year of steam equivalent to Rs. 24 million, out of total annual savings of nearly

Rs 37.0 million accrued on account of reduced operational costs, increased productivity and reduced chlorine consumption. Moreover, mills going for pulping of 2 to 3 different raw materials, modified batch cooking is preferred over continuous cooking.

(c) Blow Heat Recovery

Although blow heat recovery is recognized and practiced in Indian mills, but its utilization is not exploited to the fullest extent. The system is advantageous as it recycles the excess quantity of steam consumed in batch digesters. As much as 0.7 t/tp, steam can be recovered in the form of hot water or flash steam and the investment is very less approximately to the tune of Rs. 20 lakhs. The main limitation of blow heat recovery in the Indian paper industry is poor hot water system integration in the existing mills as the requirement of hot water is less than what is generated[5].

(d) Pulping of Agricultural Residues : Vapour Phase Pulping

Wide range of agricultural residues constitute the second major fibre resource after forest based raw materials.

TABLE -2 : VAPOUR PHASE PULPING OF RICE STRAW[7]

Particulars	Conventional liquid phase	Vapour Phase
NaOH,% w/w	13.7	13.6
Straw : liquor ratio	1:5	1:3.1
Black liquor solids, % w/w	9.6	18.4
Steam consumption, t/t pulp	4.8	1.7

Nearly 36% of the country's installed capacity is fulfilled by small mills based on agricultural residues in comparison to 38%, in case of forest based mills[6]. Most of these small mills are depending entirely on purchased energy. Batch digester or tumbling digesters are commonly used. Higher bath ratio (raw material to cooking liquor ratio) are required to be maintained, which consumes more steam. Pilot plant studies at CPPRI have shown that over 60% steam saving can be achieved in vapour phase process when compared to conventional liquid phase pulping.

(iii) Washing

(a) Installation of Efficient Washing System

The chemical recovery and environmental impact starts with pulp washing stage. Washing of pulps is an important operation from the view point of chemical losses and steam energy requirement in subsequent stages of evaporation of black liquor (Spent pulping liquor). Table-3 shows the impact of washing on energy and environment.

TABLE-3 : Impact of Washing on Energy and Environment

ENERGY IMPACT		ENVIRONMENTAL IMPACT
Initial Conc.	Steam demand	
14% 16%	213 t/day 178 t/day	200 kgs. day of additional Cl_2 demand with 0.5% B.L. carryover
Saving in steam for 2% rise in concentration.	35 t/day	

Recently the energy efficient washers have been introduced to achieve high liquor recovery efficiency with reduced investment and operating cost viz.[8]

— Pressure diffuser washer.

— Fourdrinier type (Flat) washers such as Black Clawsons Chemiwasher and Sunds ultrawasher, both with 5 to 6 washing stage forming a complete wash plant with efficiency over 98% and dilution factor of 1.0 t/tp.

— Twin wire belt filters such as Andritz Twin-belt filter where the energy demand is hardly 1/3rd of the vacuum drum washer system.

— Drum displacer washers,where all the washing stages are in-built in a single drum and Presses.

Horizontal belt washer have proved to be an efficient washing system for washing of agro-residue pulps. The advantage of horizontal belt washers are summarised in Table 4, which are based on pilot-cum-commercial scale trial being carried out at CPPRI.

TABLE - 4 : COMPARISON OF CONVENTIONAL VACUUM WASHER and HORIZONTAL BELT WASHER (Basis 30 tpd mill)[9]

Washer type → Particulars	Horizontal Belt Washer	3-stage vacuum Washer
Dilution factor,	~2	> 3
Washing efficiency,%	90-95%	~ 80.0
Elect. energy, KwH	1200	3600
% w/w dissd. solids going to recovery	> 12.0	~ 6.0
Steam for evaporation to achieve 40% cocn., tonnes	263	450

(iv) Bleaching of Pulps

The developments in bleaching processes during the last 10-15 years include;

— Displacement bleaching, improved conventional bleaching through equipment modification and new process configuration, short sequence bleaching, Oxygen bleaching as first bleaching stage. Oxygen addition to extraction stage.

(a) Oxygen Pre-bleaching

Substitution of CE stages of bleaching by oxygen delignification has several advantages in environmental, economical and energy related considerations. Energy and environment related benefits include[10,11]:

— Recovery of heat from burning of the organic material removed from the pulp in oxygen stage. It is estimated that for a 120 tpd mill about 21 tonnes of extra steam/day can be generated.
— Lower refining energy requirement for oxygen bleached pulp.
— Manufacture of oxygen requires only about 1/3 of energy (0.5 KwH/kg.) required to make chemically equivalent amount of chlorine. (1. KwH/kg).
— Reduced formation of toxic dioxins.

The following figures in Table -5 show the potential of Oxygen delignification in terms of energy savings and reduced cost of chemicals.

TABLE 5 : ENERGY CONSERVATION AND COST SAVINGS THROUGH OXYGEN DELIGNIFICATION

Bleaching sequence		CEHED			OCE(O)HD	
Kappa No.		30			15	
Chemicals,→	kg/tp	Cost Rs./tp	Energy KwH/tp	kg/tp	Cost Rs./tp	Energy KwH/tp
Chlorine	75	225	127.5	37	110	63
Sodium Hydroxide	30	450	120	20	300	80
Hypochlorite	40	140	-	15	52.5	-
Chlorine Dioxide	15	180	75	10	120	50
Oxygen	-	-	-	30	75	15
TOTAL	-	995	-	-	657.5	-

(b) Other Improvements in Bleaching

When a bleach sequence is shortened for instance, from 5 to 3, both power and steam consumption are reduced. The reduction in power demand is 15 to 20 KwH /tonne/stage less than any increase in power use, because of more efficient mixers. The heat savings are very significant since, two hot stages are typically removed. With oxygen in the extraction stage reduced Cl_2 demand by 0.8 to 1.5 Kg ClO_2/kg O_2 is achieved. The steam consumption is slightly reduced (by 0.10 to 0.15 MJ/tp. depending on ClO_2 solution and tower temperature). The power demand may increase by about 5 KwH/tp due to the additional mixer.

Most of the Indian mills today use chlorine based bleach chemicals with barometric drop leg or displacement type washers. Common bleaching sequence are CEHH, CEH of which CEHH is most common . Some large mills have started using ClO_2 and H_2O_2 e.g. sequence CE(p)HH. It is expected that in near future, the mills would gradually reduce chlorine by introducing more dioxide and oxygen in the extraction stage[12].

(v) Stock Preparation

Stock preparation is a major power consuming area in paper making process. The energy requirement in refining is influenced by the fibre properties. Larger the fibre more will be the refining energy. The Indian mills use beaters, conical and /or wide angle refiners which are not energy efficient. Therefore, these mills should go for energy efficient double disc refiners. Table 6 shows the efficiency of different refining systems.

Installation of energy efficient refining equipment can save substantial portion of energy required for stock preparation, besides upgradation of quality of fibres. It has been noted that although not much refining is needed for pulp from nonwoody fibrous raw materials, still most of the Indian mills are using refining energy as high as 200-300 KwH/t paper. There is wide scope to conserve energy by installation of disc refiners or double disc refiners.

TABLE 6 : Energy Requirement with Different Types of Refining Equipment

Beating/Refining equipment	Energy KwH/°SR/t	Energy KwH/t to raise 17°SR to 35°SR
Conical refiner	17	300
Wide angle refiner	14	254
Double disc refiner	9	162

°SR - Canadian Standard of freeness.[13]

(vi) Use of Double Dilution System

Approximately 80% water and as high as 40% fibre and fillers leaving the slice are drawn into the white water trays. A huge quantity of water and stock is forced through the centricleaners at a considerable head, resulting in considerable energy consumption. The energy consumption can be reduced by by-passing a part of the already cleaned stock through the centricleaners. In this system the overflow from the machine head box, the secondary screen accepts and a part of the wire tray water, which have already cleaned, by-pass the centricleaner system. Adoption of double dilution system would enable a mill to save 20-25% in power demand compared to single dilution system.

(vii) New Pressing Technologies

Over 97% of the total water is removed in the forming section, about 2% in the press section and the remaining 0.5-1% in the dryer section. Yet the forming section is responsible for 10-14% of the total dewatering costs, while the press section accounts for 18-25% and the dryer section for the remaining 61-72%, so it is more economical to remove water in press section than dryer section. A typical 3-5% increase in dryness in the last press will provide a 10-15% reduction in steam consumption in dryer section. Today extended nips and hot pressing can be effectively used to save energy or to increase the dryness in press section. The amount of water removed in press section is directly proportional to the viscosity of the water in the nip. In this process, by increasing the temperature of web in press section, viscosity of water reduces considerably and high water removal takes place and results in high off press dryness.

(viii) Pumping

Nearly 40-50% of the total energy requirement in a pulp and paper mill is spent in pumping dilute suspension of fibres. The pulp suspension are usually at low consistency in the range of 3-5%. If pumping can be carried out at higher consistency,i.e. 8-10% for which pumps are available today, a lot of energy can be saved.

(ix) Alkaline Sizing

Alkaline paper making has been developed recently and adopted in many countries. It is considered as an energy efficient sizing technique. The alkaline system permits increased filler levels which has a known effect in reducing refiner power consumption. The cellulose fibres in alkaline conditions swell and wet at a faster rate and makes the refining process more efficient. Also, the steam consumption for drying in case of alkaline sizing varies from 3.5 to 4.0 t/t of paper as against figures of 3.5 to 4.5 tonnes in case of acid sizing.

(x) Paper Machine

Mills abroad are using twin wire formers, whereas, the mills in India use conventional fourdrinier system with few exceptions like Hindustan Newsprint Ltd, using twin wire formers for newsprint, Tamil Nadu Newsprint Ltd, and West Coast for writing and printing grades of papers. The advantages of twin wire formers are ;

- Effective removal of water from the pulp stock coming out of head box which reduces the load on the press and dryers and this helps in reducing the steam consumption.
- High productivity
- Less space
- Can take up the weaker pulps also.

(xi) Chemical Recovery

(a) Evaporators

Black liquor evaporation is an important operation and the efficiency of the evaporation units will determine the steam economy. Black liquor properties of different raw materials influence the evaporation and combustion efficiency in the chemical recovery. The type of evaporators being used in the Indian Paper Industry has been long/short tube evaporators, which currently are being replaced by the 7-effect free falling film type evaporators. The 7-effect reduces the steam consumption by 25-30% in the evaporator section. Increasing the number of effects of falling film evaporators, a steam economy as high as 6.0 can be achieved[3].

(b) High Solids Evaporation through Thermal Depolymerisation

The energy in terms of steam that can be generated/ton of black liquor solids is dependent on the the concentration of black liquor that is fired to recovery boiler. The high viscosity of black liquor from agricultural based raw materials make it difficult to concentrate and handle in evaporators[13].

The heat treatment technology developed at CPPRI for black liquors is based on liquor thermal treatment at temperatures higher than pulping temperatures,which reduces the viscosity of black liquor tremendously thereby making it possible to evaporate the black liquor to more than 65% solids. The advantages of black liquor heat treatment are following;

- Lower pumping energy required to pump less viscous black liquor obtained after thermal treatment.
- Case studies worked out at CPPRI have shown that by firing black liquor of 65% solids concentration instead of 55% solids, it is possible to generate about 60 tons of additional steam per day for a 100 tpd bagasse based mill.

— The net thermal efficiency is likely to reach about 70% in case of 65% solids concentration instead of 50-55% thermal efficiency in 55% solids black liquor.
— Elimination of direct contact evaporator used in conventional evaporation thereby reducing gaseous pollution.

(c) Desilication of black liquors : Lime Reburning

Bamboo and agricultural black liquors are characterized by high silica content. The presence of silica in black liquor has a detrimental effect on smooth operation of recovery system.

TABLE - 7 : Energy Savings with Black Liquor Desilication[14]

	With Desilication	Without Desilication
ENERGY IMPACTS :		
Steam economy in evaporators	4.0	3.5
Saving in steam, t/d	100	-
Lime sludge dryness %	65	40
Fuel oil requirement in lime kiln Kg/t of lime	150	200
ENVIRONMENTAL IMPACTS :		
Solid waste generation, kgs/tp.	130	900
Environmental hazards	Low ground water contamination	Severe ground water contamination

The mills based on silica rich raw materials do face such problems. Apart from scaling problems, the presence of silica makes the lime sludge burning difficult , as a result the mills have to use fresh lime for causticization . The desilication of black liquor will facilitate in reducing silica in lime sludge making it suitable for reburning and reuse. There is an improved energy efficiency in evaporators due to reduced scaling and also the fuel oil consumption will be less for lime kilns due to increased dryness of lime sludge after desilication.

(d) Energy Cogeneration through Boiler Modifications

Lot of improvement in configurations of recovery boiler with respect to liquor firing, efficient three zone air distribution, auxiliary fuel, furnace construction corrosion, protection and steam pressure are required. Earlier boilers were giving hardly a steam pressure of about 20 Kg/cm^2 and now mills are able to produce steam pressure of about 40-60 Kg/cm^2. Today boilers are designed to produce high pressure steam of 100 Kg/cm^2. West Coast Paper Mills are now going for recovery with steam pressure of 80 Kg/cm^2 by incorporating the above said improvements.

(xii) Recycling of Secondary Fibres – Input of Energy

During the recent years, the use of secondary fibres has gained momentum particularly in industrialized countries. The driving force for use of secondary fibre has traditionally been economic which has now changed to a new concept i.e.

— Lower energy requirement in pulping and conversion process.
— Lower waste disposal costs.
— Environmental implications.

In the developed countries, the amount of recycled paper available is increasing rapidly.

The use of recycled pulp does save energy depending on the pulping method from 10 to 1000 KwH/t of pulp.

TABLE -8 : Energy Savings on 100 TPD Capacity Substituting Virgin Fibres with Recycled Fibre

Virgin:Recycled Fibres	Energy Savings G.Calories/100 ton.
100:0	500
80:20	428
60:40	356
40:60	284

Substituting the virgin fibres with recycled fibres, has a great impact on not only energy conservation but also on natural resources. Table 8 shows energy saving pattern on the basis of 100 tpd, almost 57% reduction in energy saving is achieved in a fibre furnish of 60% substituted virgin fibre with recycled fibres.

(xiii) Process Control and Instrumentation

It has generated its own importance in pulp and paper industry for a well defined improvement of the product and cost reduction by achieving close tolerance in quality and reducing the down time in losses. Above certain levels of plant capacity, complexity and machine speed, controls are only possible by fast acting electronic microprocessor and intricate control loops. The important process variables which need to be controlled and measured are : flow, level, pressure, temperature , consistency, speed, moisture, basis weight and tension etc.

(xiv) Increased Co-generation through Utilization of Non-Conventional Energy Resources

Cogeneration of energy using internal resources can help to reduce the extent of purchased fuel in paper mills. In developed countries the integrated mills, where self sufficiency in fuels for steam generation may not have been achievable without energy conservation, the conversion to low cost fuels is an attractive investment for reducing energy cost. The wastes that earlier had to be destroyed with the aid of fossil fuels can today be pressed and dried to provide a net energy input for increased co-generation. A developing country like Brazil, despite of its poor economy, has made significant progress in the area of energy conservation by effectively utilising the black liquor and other waste biomass as fuel.

Case study carried out has shown that nearly 75% removal of pith from whole bagasse results in 105 tonnes steam production/day for a 70 tonnes/day bagasse based mill, besides improved product quality at low cooking chemical charge.

(xv) Energy Auditing

The primary objective of energy audit is to determine ways to reduce energy consumption/ unit of product output. It attempts to balance the total energy inputs with its use. Industrial energy audit is an effective tool in defining and pursuing comprehensive energy management programmes. Preliminary audit is performed in a limited span of time accounting for atleast 70% of total energy requirements. The detailed audit goes beyond quantitative estimates to costs in savings. It focusses on[2].

(i) Current energy consumption
(ii) Assessment of energy savings potential
(iii) Evaluation of energy savings options by various means e.g. improved maintenance of equipments, recycling of waste material, improved energy use management, integrated process design and improved insulation.
(iv) Recommendations of energy conservation measures and their financial implications.
(v) Recommendation of any additional equipment or measuring instruments for future energy audits.

Conclusion

Major segment of the industry still continues to operate on obsolete technology,which is much more energy intensive. Modernisation and keeping pace with energy efficient process technologies needs to be taken up in all important key areas. The adoption of energy efficient technologies to a large extent should reduce the burden on external fuel sources,while cutting down the energy bill to the tune of 200 crores of rupees. The excess energy utilised today can become spare energy for additional capacity coming up in the next 5-10 years period.

REFERENCES

1. Energy Conservation in pulp and paper industry in India, Panda A., International Seminar on Energy Conservation in Pulp and Paper Industry, IPPTA,18-19th March, 1985, New Delhi.
2. Study on Energy Conservation - Audits carried out in Indian Pulp and Paper mills, Jyrki Halme, Paperex-93.
3. Report submitted to Industrial Development Services for ADB-GOI Industrial Energy and Environment Improvement Studies. Vol. 1, Kulkarni A.G., CPPRI, 1993.

4. Cost effective pulping and papermaking through improved energy management practice., Garg U.K., Thapliyal B.P., Kulkarni A.G and Pant R., 2nd International Non-wood Fiber Pulping and Papermaking Conference, Shanghai, P.R. China, 1992.
5. Report submitted to M/s Nath Paper Mills by CPPRI under UNDP assisted project on "Assistance to Non-wood Industries", 1995.
6. Forest based fibre and Indian Paper industry, Singhania, H.P., Paperex-95.
7. Mathur, R.M., Bist,V., Naithani, S. Kulkarni, A.G and Pant , R., IPPTA Convention Issue: (101), 1983.
8. Report on "Pollution Control and Cleaner Technology", submitted to Development Council for Paper, Pulp and Allied Industries by CPPRI, 1996.
9. Application of Double Wire Belt Washer concept for efficient washing of agricultural residue pulps. Bhaskar, S., Jain, R.K., Bohidar, P.R., and Mohindru V.K.., IPPTA Silver Jubilee International Seminar and Workshop on Appropriate Technologies for pulp and Paper Manufacture in Developing Countries, New Delhi, Nov. 1989.
10. AF-IPK Energy Compendium, Nygaard Johan, 1991.
11. Report submitted to Industrial Development Services for ADB-GOI, Industrial Energy and Environment Improvement Studies. Vol-II, Kulkarni A.G., CPPRI, 1993.
12. Total chlorine elements in Newsprint manufacture in Indian context, Fernandes J.D., Raghuvullu V.V. and Bhargava S.C., Paperex-93.
13. Process modification and process innovation - A prelude for future energy management in pulp and paper industry. Jain R.K., Gupta Abha, Kulkarni A.G. and Pant, R., IPPTA .March, 1993.
14. Calculations made at CPPRI during pilot scale desilication trials. 1993.

ENERGY CONSERVATION EFFORTS IN STEEL AUTHORITY OF INDIA LIMITED

45

U.K. MITTAL
Director (Operations),
Steel Authority of India
(INDIA)

ABSTRACT

Steel Making and Finishing is highly energy intensive process. Steel industry accounts for 35% of the total energy consumed by industry in India. High level of energy consumption in this vital infrastructure industry not only is important from the cost point of view but also is a continuous addition to environmental degradation from mining stage to the end of the energy cycle when the fuels are converted to carbon dioxide and other pollutants. Efforts to conserve energy in steel making, thus benefit not only the producers of steel but nation as a whole.

Steel Authoirty of India Limited, one of the largest industrial enterprise of India and 13th largest steel producer of world has successfully reduced energy consumption for last 10 consecutive years. Energy consumption in SAIL has come down by 23% over this period and Bhilai Steel Plant of SAIL is the lowest energy consuming steel plant in India at 7.76 Gcal/tcs. Energy conservation in SAIL started getting major thrust about a decade ago when the company evolved a three tier structure to monitor the energy conservation efforts. The Plant Level Task Force on Energy coordinates the energy conservation at shop floor level, increases the energy awareness and is responsible for implementation of the energy conservation schemes. The Corporate Level Task Force (CLTF) of energy was constituted to monitor the energy performance, provide the macro energy indices for budget making exercise of the company. CLTF also had a separate budget was created to enable fast track approval for small energy conservation schemes which have low gestation period and high pay-back. In all about 141 schemes involving an expenditure of 42 crores have been implemented so far. At the highest level in the Organisation a Steering Group on energy policies, headed by the CEO of company himself provides overall direction to energy drive and recommends the technological upgradations towards cutting down energy consumption and costs.

Besides the energy conservation schemes of very specific nature, SAIL has also taken up modernisation and technological upgradation of the steel plants. These modernisation schemes aim at phasing out obsolete, energy inefficient technologies like Open Hearth Furnaces and ushering in state of art technologies like Bell Less Top Charging in Blast Furnaces, Basic Oxygen Furnace Process of Steel making and continuous casting of steel. Modernisation of Durgapur Steel Plant has been completed while those of Rourkela Steel Plant and Bokaro Steel Plant will be completed during 1996-97 and 1997-98 respectively. With these the energy consumption levels in SAIL are expected to come down further.

Introduction

Energy is a vital input for the growth of industrial sector, which is the largest consumer of commercial energy in the country. In India, this sector is responsible for about 50% of the total commercial energy consumption. The iron and steel industry is the largest consumer of energy in the industrial sector. Energy cost constitutes 30% of the production expenditure of iron and steel.

Steel production in India is namely through (i) the Integrated Steel Plants – based on the blast furnace – open hearth / basic oxygen furnace route and (ii) the Mini Steel Plants – based on scrap or sponge iron – Electric Arc furnace route.

At present, there are seven integrated steel plants in India with an installed capacity of 16.35 million tonnes of crude steel. Five of these seven units are under Steel Authority of India Limited and are located at Bhilai, Durgapur, Rourkela, Bokaro and Burnpur. The other two are Tata Iron and Steel company at Jamshedpur and Visakhapatnam steel plant at Vizag. Apart from this, India also has about 269 Electric Arc furnaces with an annual steel making capacity of 9.3 million tonnes. With the recent changes in the economic policies in India, many new steel plants are also coming up fast.

Energy Consumption Pattern in SAIL

Steel Industry by its very nature is highly energy intensive with its many heating, cooling, melting and solidification cycles. The major energy inputs to the integrated steel unit of SAIL are Coking coal, Boiler coal, Purchased power and Petrofuel. Coking coal is the prime fuel for production of blast furnace coke and accounts for about 77% of the total energy input in the steel plant. Boiler coal which is required for steam and power generation in the plant accounts for 13% , Petrofuels meant to compensate for the fuel deficit or special process requirement account for 2% and Purchased power meant for supplementing the captive power generation accounts for 8% of the total energy input to the plant. Figure 1 gives the stagewise energy consumption of integrated steel plants in SAIL.

Purchased electrical power is one of the costliest form of input energy. Utility power supply situation has been unsatisfactory both in terms of reliability and quality. SAIL plants had suffered huge production loss on account of inconsistent poor supply from the utility. Therefore, SAIL decided to augment its captive generating facilities. At present SAIL is able to meet about 56 of the total demand from captive generation. Captive generation is not only reliable but also significantly cheaper as compared to the utility power. SAIL is in the process of installing Supervisory Load Control Centres for optimising its captive generation vis-a-vis import from utility for its plants located in the eastern region.

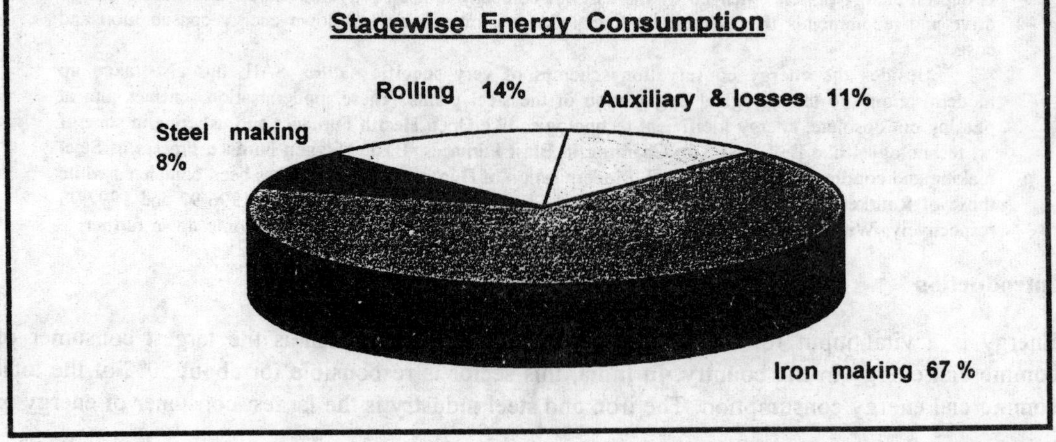

The specific energy consumption in an integrated steel plant is computed as per the standardized methodology prescribed by the International Iron and Steel Institute, Brussels. Typically, the specific energy consumption of integrated steel units in SAIL ranges from 7.5 Gcal/tcs to 13 Gcal/tcs. Amongst them , Bhilai Steel Plant has been recording the lowest Specific Energy Consumption for the last few years.

Very often it is stated that the energy consumption level in Indian steel industry is double compared to that in the advanced countries (Table 1.0).

TABLE 1.0

Unit: - Gcal/tcs

Country	Specific Energy Consumption
United Kingdom	4.7
U.S.A.	5.1
Germany	4.8
Japan	4.2
Italy	4.5
Spain	3.8
Sweden	4.7
India	7.5-13

However, such comparisons are often misleading as the figures presented for foreign countries are averages which include electric arc furnace route of steel making also (Table 1.1). The scrap-based EAF method of steel production requires considerably less energy than the integrated route of steel making.

TABLE 1.1

Unit: Percentage

Country	Steel Making Route		
	OH	*BOF*	*EAF*
U.K.	-	74.2	25.8
U.S.A.	-	60.6	39.4
Germany	1.5	76.9	21.6
Japan	-	68.8	29.7
Italy	-	42.5	57.4
Spain	-	42.3	57.7
Sweden	-	64.0	36.0
India	25.8	47.4	26.6

Moreover, in advanced countries, the steel plants have completely switched over from the Open hearth route of steel making to more energy efficient BOF route of steel making. Similarly, Ingot casting route has been replaced by continuous casting technology (Table 1.2). However, Indian integrated steel units are still continuing with energy inefficient Open Hearth (OH) steel making process where the energy consumption is about ten times compared to the energy consumption in the LD process.

TABLE 1.2

Unit : Percentage

Country	BOF	Concast Technology
U.K.	74.2	86.1
U.S.A.	60.6	85.7
Germany	76.9	93.9
Japan	68.8	95.7
Italy	42.5	96.5
Spain	42.3	95.9
Sweden	64.0	88.7
India	47.4	43.5

Energy Conservation in SAIL

The importance of energy conservation as a precondition for economic viability was recognized by SAIL way back in the eighties. SAIL identified Energy Conservation as one of the major corporate thrust areas in 1987. Consequently guidelines for energy conservation were laid under the Corporate Plan of the company. The Energy conservation plan envisages a 30% reduction in the specific energy consumption level by the year 2004-5 AD. The three-fold strategy formulated to achieve this ambitious goal involves :

1. Restructuring of the Organisational Framework,
2. Formulation of Energy Conservation Plan, and
3. Implementation of energy conservation measures.

Restructuring of the Organisational Framework

Organisational structure both at corporate level as well as plant level was restructured in order to take up energy conservation activities in the real earnest. A three tier organisation for planning, implementation and monitoring of energy conservation measures was evolved.

1. **Steering group** — The steering group on energy headed by the Chairman, SAIL with Managing Directors, Director (Finance) and Director(Operations) as members is the highest body which formulates the policy on energy conservation and approves medium term schemes involving relatively larger capital investment.

2. **Corporate Level Task Force on Energy** — The Corporate Level Task Force headed by Executive Director (Operations) has representatives from different steel plants and RDCIS. It regularly reviews the performance of energy parameters and also implementation of various energy conservation measures.

3. **Plant Level Task Force on Energy** — The Plant Level Task Force on Energy under the

chairmanship of head of works is responsible for preparation and implementation of the energy conservation plan for each steel plant. It monitors the energy performance of the steel plants on monthly basis.

Apart from this, Energy Management Department was set up in each plant for energy auditing, maintaining the energy facilities, and monitoring of energy flow in the plant.

The Energy Conservation Plan

The close nexus between energy and economics of a process necessitated the formulation of energy conservation plan.

The Energy Conservation Plan of SAIL lays out targets to be achieved by systematic and sustained efforts in the following broad areas :

— Operational improvements
— Equipment improvements
— Modernisation measures

Briefly these measures revolve around :

1. Reduction in coal to Blast Furnace coke ratio,
2. Reduction of coke rate in blast furnaces,
3. Reduction in hot metal to liquid steel ratio,
4. Improvement in metallic yield in steel making,
5. Suppressed gas recovery system for LD converters,
6. Reduction in liquid steel to crude steel ratio,
7. Reduction in petrofuel consumption through increased recovery of bye-product gases namely coke oven gas,
8. Decrease in specific heat consumption of soaking pits and reheating furnaces,
9. Introduction of waste heat recovery systems in different areas, and
10. Elimination of wastages and leakages of steam, compressed air and oxygen.

The plan also quantifies the energy consumption level to be achieved by each unit of SAIL as a result of implementation of the energy conservation measures (Table 2.0).

TABLE 2.0

Unit : Gcal/tcs

SAIL steel unit	1997-98
Bhilai Steel plant	6.5
Durgapur Steel Plant	7.75
Rourkela Steel Plant	8.0
Bokaro Steel Unit	7.0
Indian Iron & Steel Company (Burnpur)	11.0

Implementation of Energy Conservation Programme

SAIL had commissioned the expert team from British Steel Corporation under UNDP / UNIDO assisted project for carrying out a detailed energy audit of Bhilai, Bokaro and Rourkela units. Similarly, SAIL had also commissioned expert teams from NKK, Japan under a SAIL-NKK contract for its Durgapur, Rourkela and Burnpur units. The recommendations made by the two teams helped in identifying the potential energy saving areas in SAIL. Thereafter, based on the nature of the scheme, the payback period and the cost involved, the energy conservation schemes to be implemented were determined and classified under

- Short term measures,
- Medium term measures, and
- Long term measures.

Short Term Measures

The short term measures incur minimal investments and are taken up on a regular basis. These include -

- Elimination of steam, oil and compressed air leakages
- Improved house-keeping
- Maintenance of proper air-fuel ratio in reheating furnaces
- Adherence to optimal thermal regimes etc.

Medium Term Measures

The medium term measures include among others installation of microprocessor controlled combustion systems, introduction of variable speed drives, hydrojet cleaning of oven doors. Annexure 1 gives the medium term measures that have already been implemented in the integrated steel units, the list is indicative but not exhaustive.

Long Term Measures

The long term measures include the schemes taken up as a part of technology upgradation and modernization programme such as Partial Briquetting of Coal charge, Stamp charging in coke ovens, Coal dust injection in blast furnaces, Replacement of open-hearth furnaces by basic oxygen furnaces, Introduction of bell-less top in blast furnaces Installation of combined cycle power plant. Some of these measures are already being implemented across the company.

SAIL has also introduced a system of setting up shopwise targets of energy parameters for different shops in each steel plant alongwith the target for overall energy consumption. The energy management personnel closely interact with operational departments to determine the reasons for variance from targets. This analysis helps in evolving the corrective measures to be taken for achieving the targets.

Apart from the technical aspects, the culture of an organisation has a marked bearing on the energy conservation in a steel plant. Realising the significance of human factor, SAIL in its pursuit to conserve energy has made concerted efforts in educating and training its employees through workshops, classroom lectures, banners, posters etc. This has immensely helped in generating energy consciousness in the entire workforce of SAIL.

It is the result of implementation of the aforementioned energy conservation measures that the energy consumption level in SAIL has decreased by 23% from 1986-87 to 1995-96.

Future Prospects

There are a number of new and improved energy saving technologies which are technically established but often display low levels of investment efficiency. However, these can be examined for their technoeconomic feasibility w.r.t SAIL steel units. Some of these technologies are :

Coke Dry Quenching

The coke extracted from the ovens is at very high temperature. In the conventional system, it is cooled down by spraying water on it. Dry quenching of coke helps in extracting the sensible heat of the high temperature coke and converts it to high pressure steam which can be further used to generate power.

Top Pressure Recovery Turbine

The modern blast furnaces are high capacity furnaces. These furnaces operate at high top pressure. The high top pressure of the gas is exploited in the Gas expansion turbines to generate power.

Natural Gas Injection

Use of natural gas in blast furnaces alongwith oxygen is a proven technology for maximising the productivity of the furnace and decreasing its coke consumption.

Coal Bed Methane Gas

The possibility of recovering coal bed methane gas from the eastern coal fields is being examined at the national level. In case the project materialises SAIL can consider using coal bed methane gas as a fuel injectant in the blast furnaces to cut down on coke consumption of the furnaces.

Alternate Routes of Iron-Making

The need for an alternative iron-making technology arises out of the demand to conserve the depleting resources of good quality coking coal. In view of the success of Corex technology in the process of iron making, SAIL is considering the development of an indigenous concept of adoption of the imported technology.

SAIL has also signed a memorandum of understanding with AMET for imbibing Russian technology in the Romelt process of iron making. It is a simple, one-step process for direct iron smelting. This process utilizes simple coal fines as both the fuel and the reductant. It is also capable of smelting wide range of ores and metallurgical waste materials.

Combined Cycle Power Plants

Following modernisation of SAIL plants large amount of surplus gas would be available. This gas could be utilised for augmentation of captive generating capacity through installation of energy efficient low CV gas fired combined cycle power plants.

Conclusion

Although SAIL has achieved a remarkable reduction in its energy consumption level by adopting energy saving measures, but it still has a long way to go. Through continued efforts for energy conservation in the existing facilities as well as through modernisation of facilities, it is planned to further accelerate the energy conservation drive in SAIL.

ANNEXURE 1

ENERGY CONSERVATION SCHEMES

Some of the energy conservation schemes that have been taken up in SAIL steel plants are :

1. Monitoring of maximum demand of electrical power : BSP
2. Installation of modified water cooled skids in reheating furnace No. 1 and 3 of Merchant Mill : BSP
3. Modified Ladle Heating Stand for SMS-1 : BSP
4. Insulation of cold blast main for blast furnaces : BSP
5. Time switches for switching off street lights : BSP
6. Micro-processor based controller in reheating furnaces of Wire Rod Mill : BSP
7. Soft start and pulsar AC variable speed drive for hot saw mill motor of Rail and Structural Mill : BSP
8. Use of ceramic fibre lining in Annealing Furnaces of CRM-1 : BSL
9. Automatic street lighting on-off devices in township : BSL
10. Conversion of high pressure mercury vapour lamps to high pressure sodium vapour lamps: BSL
11. Oxygen enrichment of cold blast, BF4 : BSL
12. On line steam leakage plugging : BSL
13. Double skid insulation in reheating furnaces : RSP
14. Thermal insulation of fuel oil tanks : RSP
15. Ceramic fibre sealing in hood annealing furnaces : RSP
16. Insulation of fibre glass transluscent sheets for roof tops in mills area : RSP
17. High emissivity coating in electric sheet mill furnace : RSP
18. HYSIL brick lining in soaking pits : RSP
19. High emissivity coating in sleeper plant reheating furnace : DSP
20. High emissivity application in soating pit No.3 : DSP
21. Electrical tape heating in SMS area : DSP
22. Installation of booster for coke oven gas at sinter plant : DSP
23. Micro-processor based control system for soaking pits : IISCO
24. Ladle drying covers : IISCO
25. Dynamic reactive power compensation for Arc Furnace operation : ASP
26. Reactive Power Compensation : DSP
27. Reactive Power Compensation : RSP
28. 220 KV power supply for DSP and ASP
29. Supervisory Load Control Centres at DSP, BSL, RSP and Calcutta for optimisation of captive power vis-a-vis import from utility.

Key

BSP - Bhilai Steel Plant
DSP - Durgapur Steel Plant
RSP - Rourkela Steel Plant
BSL - Bokaro Steel Plant
IISCO - Indian Iron & Steel Company, Burnpur
ASP - Alloy Steels Plant

ENERGY CONSERVATION IN IRRIGATION PUMPING SYSTEM

46

A. K. SUNDARESHAN
Superintending Engineer (Ele) KEB, Appeals I
K.R. Circle; Bangalore-560001

ABSTRACT

Energy is very precious. It is the duty of every individual to preserve it and save it. Energy saving is indirectly generating more electricity. There is very big constraint of electricity in India in general, and more acute in Karnataka in particular. The load growth of Karnataka State per annum, is abot 600 MWs whereas, the increase in generation is almost nil. This has resulted in severe shortage of electricity in respect of peak demand and energy.

This paper deals mainly on the prospects of energy conservation in respect to Irrigation pumpsets. There are about 10 lakh Irrigation pumpsets working in Karnataka as on 31.3.1995. Out of these there are about 6.25 lakh open well pumpsets and 3.75 lakh bore wells. The water table differs from district to district. It is about 200-450 ft in Kolar district and varies from 100-250 ft in other districts. Day by day the water table is going deeper and deeper. A pilot study has revealed that an average of 3000-4000 KwH* of energy is used in respect of open wells and about an average of 8,000 - 10,000 KwH/ annum* in respect of bore wells.

Open wells

Assuming about 30% of open wells as defunct and 5% of bore wells as defunct, the energy consumption of irrigation pump sets in Karnataka works out to 4800 MU per annum at the consumer point. Taking a line loss of atleast 22%, the energy consumption of open well IP sets at the generating point is 4000-MU + 880 MU = 4880MU. Assuming an average cost of Rs. 1.00 per KwH cost of IP energy at generating point is 488.00 crores/annum.

Bore wells

The energy consumption per anum is 356250 × 10000 = 3562 MU. At generating point = 3562 + 783.64 = 4345 MU. Assuming an average cost of Rs. 1.00 per KwH, cost of IP energy at generating point is 434.5 crores.

The total cost of energy from IP sector = 488+434.5

= 922.5 Crores.

* Data from KEB statistics department.

INTRODUCTION

Basic Information relating to IP Sets

India is a very big country having an area of 30,65,027 square kilometers, with a huge population of 838.584 million as per 1991 census. The greatest problem before the planners is to provide food for all with the available resources of land, man power and electric power. It was a challenge to adopt most modern methods of agriculture for growing more food in minimum area using high yield seedlings, providing the most needed fertilisers with judicious feeding of the available scarce water resources to cater to the needs of the crops and getting maximum yields to the 840 million hungry mouths. It was a revolution in the field of irrigation, when we were importing wheat and other foodgrains from abroad in the 1960's and today even with the present explosion of population we are proud that we are self sufficient in food production. We are in a position of exporting foodgrains to other countries.

Out of the total area of 30,65,027 sqKms, land that is fit for irrigation is only about 20%. This has led to the exploitation of minor irrigation through irrigation pumping system. The ground water resources, water available in rivers, streams, canals, tanks, open wells and bore wells are tapped and used for agricultural production.

1.2 Description of Irrigation Wells in the State

Karnataka State is a small State but, a jewel in the crown of India. It has an area of 1,91,791 sqKms and a population of 44.977 million as per 1991 census. State of Karnataka is only 5% of India in both area and population. In Karnataka people have utilised the limited water resources to a maximum extent by adopting irrigation pumping system. There are about 10,00,000 irrigation pump sets in Karnataka, out of which there are 6.25 lakh open wells and 3.75 lakh bore wells. The average HP of each pump set is 5 HP. The approximate load of irrigation pumping system is 50,00,000 HP or 3730 MWs. We are irrigating 5 million acres of land using minor irrigation through IP sets. That is about 2 million hectares of land is being irrigated using IP sets. About 6 million tonnes of food grains are grown from the IP sector. Water availability is the only constraint. Wherever there is less water our farmers grow crops like, ragi, jowar, pulses and vegetables that need less water.

Earlier in the 1960s and 1970s, farmers were using fully open wells were used to pump water for irrigation. The water table of these open wells was varying from 25 ft to 50 ft. Even today we can see some of the good open wells, about 50 - 75 ft diameter fully lined with stone masonry for the entire inside which looks like mini swimming pool. Farmers were using hand operated *Yetas* and Bullock operated *Yetas* which used to render about 15-50 Lts of water per operation. This operation continued for hours together till water was available in the well.

As the availability of water became less and less rich farmers thought of the philosophy of bore well. This came into being after 1980. In the early 80's they were getting good water yield ranging from 2000 - 6000 gallons per hour within 100-150 ft deep bore wells. This was quite handy for the farmers for agricultural operations. In fact some farmers were giving water to the adjoining lands, as long as their water source permitted them to do so. The average efficiency of an irrigation pumping system in the field is 20% only. 80% of the energy is wasted as friction to the flow of water and other losses like, transmission of power, inefficient foot valves, high friction pipes, inefficient pumps and low power factor of IP set motors etc. It is quite posible to improve the efficiency of irrigation pumpsets to at least 50-60%, thereby saving a big chunk of energy which is being wasted today as losses.

The I.P. sector consumption today in Karnataka is about 44% of th total consumption of the State. That is out of 65MU per day, IP consumption is 28.6 MU per day. The free power supply to this vital sector has impaired the finances of Electricity Boards which are mostly running under loss. An honest attempt is being made here to save energy which is being wasted without giving any benefit to anybody. The saving in energy can be made use of for the improvement of the working of electricity boards, by giving good quality power supply with better voltage, frequency and lesser interruptions to its consumers.

In a centrifugal pump the suction head* plays a very important part. For best results, the suction head in a centrifugal pump will have to be maintained between 3-4 meters. Sometimes in summer, the water level in the well goes down to 25-30 ft, when the farmer will have to construct 2-3 platforms inside the well. They started to shift their pumps to the ready-made platforms already made in the well such that, the suction head was maintained at 3-4 meters**.

When the suction head* of open wells increased above 25 ft, the water pumped diminished to a very low value and became zero when it touched 30 ft, depending upon the elevation of the place from sea level.

The depth of the bore wells, went on icreasing after 1985 due to the water table going down and now we can find bore wells, which are as deep as 400 ft in Kolar and Tumkur districts. After 1985 the philosophy of bore wells went on increasing and finally, now out of every 100 pump sets serviced, about 90-95 are bore wells only. We can also find 3-4 bore wells in each village for drinking water purposes. Most of them are hand operated and one or two bore wells pump water directly to the taps in the village, and in some other cases, they pump water to the overhead water tank and distribute to the various outlets by gravity flow. Now we also find about 5-10% of bore wells, which have become dry and defunct and have been abandoned.

People were suffering without drinking water, whenever the bores got dried up. Wherever there were deeper agricultural bore wells in the vicinity of drinking water bore wells, people suffered without drinking water. The Government of Karnataka, after studying this problem, has made it mandatory to leave at least 805 ft from the drinking water bore well to other bore wells. The farmers will have to also bring NOC from the Tahsildar of the taluk to get electricity to their bore wells.

1.3 Irrigation Pumpsets on Wells and Lift Irrigation from Rivers and Streams

Farmers who were having lands by the side of rivers and streams where, perennial source of water was available started digging open wells near the rivers and streams where water was available

* Suction head is the depth from the centre line of the pump to the water level in the well.

** For maximum efficiency of pumping in open well the suction head should be ranged from 3-5 meters.

in abundance. They got high H.P. water pumps operating to irrigate rich crops like sugarcane and paddy, whose water requirement is very high. This was more prevalent on the banks of Ghataprabha, Malaprabha, Krishna, Bheema and Cauvery basins, where regular canal irrigation facilities are not available. After studying the benefits of such irrigation Karnataka Government, has started lift irrigation schemes by establishing high H.P. Pumps. This was achieved by lifting the water and taking water to a high level storage sump tanks of larger capacity. From these tanks, water was drawn to the various beneficiaries for agricultural operations to grow rich crops by gravity flow. After establishing the scheme, the Government formed co-operative societies with the co-operation of local leaders and handed over these schemes to them for maintenance.

Lift irrigation through electricity is a great success particularly in Belgaum district, where KEB has drawn 11 KV sub-transmission lines along the river beds which facilitate farmers to take new high HP pumpsets to draw water and use it for agricultural purposes. This has helped the farmers in a very big way.

1.4 Historical Trend in the Number of IP Sets - District-wise

In Karnataka out of the 20 districts, the watertable is quite good in some districts and scant in the other districts. Belgaum tops the list with 1,10,300 pumpsets followed by Kolar, Bijapur. Tumkur and Bangalore districts.

The Government of Karnataka after deep thought, about the benefits of irrigation pumping systems has magnanimously come forward to give free electricity to the irrigation pumpsets which caused a severe revenue loss of about 1845 crores per annum*** due to which the finances of KEB have suffered a lot.

1.5 Discernible Trends in Open Wells, Bore Wells and Lift Irrigation Pumps in Watertable Levels

As already commented, the water table in Karnataka has gone down. Today in Kolar district there are bore wells which are 400-450 ft deep. This means that the power required for lifting of water is more. For lifting the same quantity of water, we require more HP IP sets with multi stage pumps. This is narrated clearly in the technical paras of this paper. This will cause a heavier burden on KEB and in addition there may not be sufficient water for the crops. If the same trend continues by 2010 AD we may not be left with any water to pump. The agricultural produce goes down drastically which creates scarcity of essential commodities, which ultimately leads to heavy inflation.

---

* This data is taken from the Agricultural Department of Karnataka Government.

** This data is taken from the KEB statistics department.

*** This is as per KEB's figures of 44% consumption from IP sector of the total consumption. Taking an average cost of energy at Rs. 2.00 per KwH.

ELECTRICITY CONSUMPTION BY I.P. SETS

2.1 Consumption Patterns of Electricity, Sector-wise in Karnataka since 1950s (Industrial Domestic, Commercial, Agricultue etc.,)

Karnataka has the pride of being the first in Asia to have had the highest generation of 4.3 MWs from Shivasamudram in 1902 and the longest 33 KV transmission line in the world at this time, from Shivasamudram to K.G.F. for the goldmines. It slowly grew up by improving its generation capacity after commissioning of M.G.H.E power house during 1935 to 45 MWs. There was a time when KEB was requesting industrialists to set up major industries in Karnataka. One off-shoot of this is the Indian Aluminium Co., Belgaum during 1972. The giant hydro electric project namely, Sharavathi Valley project, started its operation during 1970. During the 1970's, KEB infact sent appeals to all the domestic lighting consumers to avail the facility of `All Eletric Home' and to use more electricity when all the 10 generators of the Sharavathi project started functioning. We had really surplus electricity at that time.

The load gradually increased year after year, and has now exceeded our generation. It was a necessity to introduce power cut in Karnataka in 1982. From then onwards, we are never having a good time. Power cut continued every year and became more and more, year after year. For a balanced electrical system there should be a mix of Hydel and Thermal generation, with the thermal generation taking the base load continuously with a high plant load factor and Hydel generation to assist the system during peak load periods. This universally accepted theory was thought of and Karnataka started working on Raichur Thermal Power Station. Today Raichur Thermal power station is a every big complex with 4 × 210 MW units. Since transportation of coal is a problem, it is not in a position to generate electricity from all the 4 units every day.

During the 1980s, there was a very big load growth in the irrigation pumping sector and the same trend continues even today. From 1990-91 and onwards every year we are energising about 60,000 pumpsets per annum. Following table gives the details*

No. of IP sets serviced from KEB from 1979-1995 :

Sl. No.	Year	IP sets serviced	Sl. No.	Year	IP sets serviced
1	1979-80	12826	10	1988-89	45630
2	1980-81	18455	11	1989-90	41786
3	1981-82	23653	12	1990-91	70138
4	1982-83	25697	13	1991-92	72879
5	1983-84	37688	14	1992-93	52537
6	1984-85	45413	15	1993-94	47865
7	1985-86	48377	16	1994-95	55962
8	1986-87	47857	17	1995-96	
9	1987-88	49083			

* Figures are taken from KEB Statistics section.

The consumption pattern of Domestic, Industrial, Commercial and Agricultural consumers are changing year after year and the pattern of consumption is given in table below :

Categorywise energy consumption in Karnataka

Sl.No.	Year	Domes. Light.	All Elec. Home	Com. Light.	Indus. Power Lt	Indus. Power HT	Public Light.	IP sets	Licensees & others
1	1979-80	254.21	342.62	121.43	460.12	3211.39	69.48	360.91	38.07
2	1980-81	278.54	417.80	132.71	489.38	3374.14	70.99	384.36	40.56
3	1981-82	280.22	508.60	141.39	504.93	4050.17	77.02	425.53	41.76
4	1982-83	331.31	649.57	161.69	620.71	3698.67	86.62	575.62	42.61
5	1983-84	422.00	761.00	175.00	629.00	3691.00	96.00	593.00	28.00
6	1984-85	378.67	804.13	166.60	554.60	3655.20	124.61	1233.42	36.50
7	1985-86	457.21	883.64	159.66	606.38	3022.74	128.24	1803.45	50.23
8	1986-87	556.25	967.11	180.38	614.53	3065.11	135.28	2225.10	-
9	1987-88	602.00	854.00	197.00	556.00	3316.00	140.22	2477.00	55.43
10	1988-89	641.69	878.89	197.86	692.19	3974.86	143.27	2848.58	43.00
11	1989-90	711.89	973.30	225.30	773.02	4051.95	145.55	3535.76	69.98
12	1990-91	769.77	1032.93	244.31	722.75	4706.35	148.24	4485.75	71.03
13	1991-92	819.45	1109.49	280.35	799.61	4727.31	150.57	4584.99	67.69
14	1992-93	828.38	1155.66	294.72	791.84	4117.25	152.87	5473.99	73.40
15	1993-94	1064.79	1196.94	373.93	861.40	4117.43	154.93	6213.14	61.74
	%age	7.5%	8.52%	2.66%	6.13%	29.32%	1.10%	44.24%	0.45%

All energy consumption figures are in million units.

Figures are taken from KEB Statistics section.

From a review of the table it could be seen that the agricultural sector has increased its share of consumption from about 5% during 1970 to 44% during 1994-95. From this we can clearly see the tremendous interest shown from our farmers for irrigating their land by IP irrigation. Assuming that about 7.93 lakh live pumpsets are working, an average pumpset rating of 5 HP per pumpset and that 1 HP can irrigate about 1 acre land, in Karnataka State we are irigating 14 lakh hectares of land by irrigation pumps alone.

2.2 Reasons for Increase in Share of Electricity Consumption by IP Sets

The statistics in tables above show, that year after year the energisation programme of IP sets increased, and what were earlier open wells gradually became bore wells and those bore wells which were 100-150 ft deep became deeper and deeper due to water table going down. Earlier KEB had an annual target of energising 20,000 pumpsets per annum and this tempo increased to 60,000 pumpsets per annum and even today the same exists.

Perhaps the main reason for the tremendous rate of increase of IP consumption is, that year after year the number of IP sets increased and more and more bore wells got serviced. The energy required to lift water from deeper bore wells is more.

Work done is = Q × H KW

where `Q' is the quantity of water lifted in Kgs and `H' is the total head to which water is lifted in meters.

i.e., Energy utilised for pumping of water - KWs x Hours of working

Naturally, KwH energy required to pump water from the bore wells increased day by day, as the watertable went on going down. The consumption per pumpset gradually, increased from 5% during 1970 to 44% during 1994. The graph enclosed gives the pattern of increase in consumption year after year. The share of IP consumption is increasing steadily.

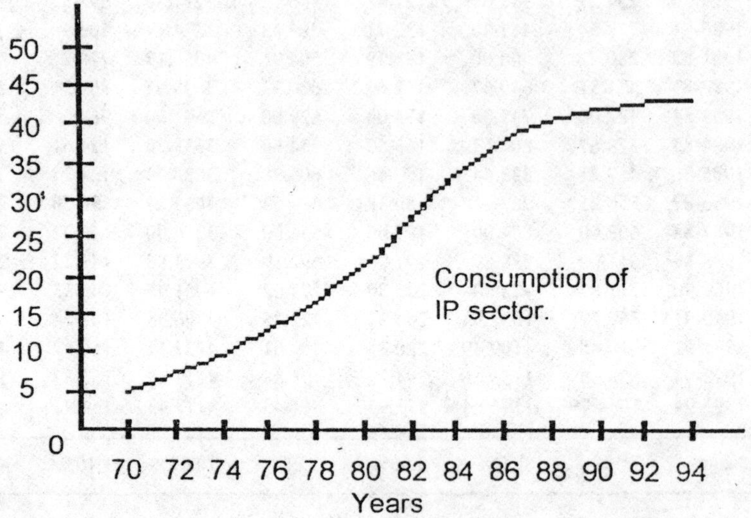

It will reach saturation perhaps after another 5-8 years. This can be clearly seen from the graph.

Another important factor that can be attributed to this increase is that, the Government of Karnataka in the year 1980 started demetering of IP sets. Then the IP sector consumption was only 5% the tariff for IP sets was Rs. 50.00 per HP per annum. This was not a big burden on the Electricity Board and it was affordable. Further, every year the issue was politicised and in the year 1992 IP sets were made fully free. Today the impact of this facility is costing the Electricity Board every year by 1845** crores which by any standards is unbearable for any organisation. The Karnataka Government is not even subsidising this great loss. For a gross revenue of Rs. 1850 crores per annum, Rs. 822* crores is 44.4%. Due to this reason the economy of KEB has crippled. It is quite encouraging that the present government is thinking of re-introducing some tariff to the IP sector during the next tariff revision.

2.3 Average Consumption of Electricity in Irrigation Pumping System

(a) Open Well Pumpsets

The maximum suction head of this pumpset cannot be more than 30 ft at sea level. The maximum efficiency of an I.P. Set achievable in the field is 65%. The average consumption of an IP set is a most complex one. It depends on —

* Taking the average cost of energy at Rs. 2.00 per KwH.

** Revenue of IP sector consumption at Rs. 1.00 per KwH.

(i) Availability of water
(ii) Horse Power of the pumpset
(iii) Extent of land irrigated
(iv) Nature of Crop
(v) Depth of water in the well
(vi) Seasonal variations
(vii) The overall efficiency of the irrigation pumping system.

Availability of water in the well is the first and foremost consideration. If a good water source is available, then water pumping can be done as long as there is electricity. Farmers can go for commercial crops like sugarcane and paddy which requires more water to get better benefits.

If the land to be cultivated is more, naturally the H.P. of the pumpset will have to be more and the consumption will naturally be higher. As a thumb rule, 1 H.P. power is required to irrigate 1 acre of land in case of open wells. The depth of water in the well is also an important parameter while determining the average consumption of an IP set. Since the power ie., $(Q \times H)$ willl be more when the energy $Q \times H \times t$ is also more, as long as `H' goes on increasing (keeping other parameters as constant).

The consumption pattern of IP Sets vary with season also. During rainy season, the water requirement of crop is less and the farmer does not pump more water. Further during rainy season the depth of water from the ground level is low and hence energy consumed to lift the water is also less.

During summer, the water below the ground level is deeper. The availability of water is less whereas the requirement of water to the crops is more. The energy required to lift water from that depth is more. Hence the energy consumption is higher.

Another important factor is the efficiency of the pumpset itself. The average efficiency of pumpsets in the field as per pilot study* conducted by KEB research station is only 20%. It means that 80% of the input energy is a waste. Only 20% of the energy is used for pumping of water. The wastage of energy is in the form of

1. Heat losses in the windings of the motor
2. Friction losses
3. Transmission of power from the motor drive to the pump
4. Loss due to velocity head V2/2g etc.
5. Head loss due to bends in the piping.

It is possible to achieve atleast 50% efficiency in the irrigation pumping system in the field. Naturally, the farmers use the cheapest pumpset. G.I. Pipe, foot valves etc., available in the market, which have very high friction coefficient resulting in poor efficiency.

There are two methods of assessing the average consumption of IP Sets.

1. By fixing energy meters to the IP sets, calibrating the meters for their acuracy, recording the monthly regular consumption for one full year and then arriving at the average consumption.

2. To select one rural transformer centre, feeding purely to IP sets. To provide an L.T. 3 Phase meter with L.T.C.T's of the appropriate rating, recording the actual consumption every

---

* KEB Research Station conducted a pilot study on 100 pumpsets during 1984-85.

month on a fixed date and then to arrive at the average consumption. Here again the HP's of bore wells differ from one another according to their requirement. It is better to arrive at the average electrical consumption per HP which will be more realistic.

3. Assuming in case of open wells that,

(a) IP sets run at an average of 6 hours per day working for 200 days in an year
(b) The average HP of each pumpset is 4 H.P.

On this basis the average consumption is worked out to 3000 units per annum.
In case of bore wells that,
(a) IP sets run for 8 hours a day working for 300 days in an year
(b) The average HP of each pumpset is 6 H.P.

On this basis the average consumption is worked out to 10000 units per annum.

International Energy Initiative studied intensively in coordination with KEB and has arrived at similar figures.

2.4 Discontinuance of Energy Meters on IP Sets and Reasons

As already commented earlier, this important subject has been politicised. This issue has been used as an effective tool for capturing the vote banks. During 1980 KEB brought an order demetering all the IP sets. They based their argument, that the entire IP consumption of the IP sector is only 5-6% which is negligible and that, from the demetering of IP sector KEB gets back 3 phase energy meters which could be used for other sectors by stopping/adjusting or buying of 3 phase energy meters. The manhours saved by not reading IP meters, the stationary saved in this process and the manhours saved for billing were all taken into consideration. KEB on that day was getting 7.5 crores per annum (for 350,000 pumpsets), charging a nominal tariff of Rs. 50.00/HP/Annum. Ultimately the overall loss was so less that they decided to demeter the IP sets and charge them nominally at Rs. 50.00/HP per annum.

Whatever said and done this was an incentive given by the Karnataka Government and Electricity Board to the farmers to grow more food which was absolutely essential to improve food production to minimise the food deficit facing the State at that time. Many farmers have made use of this facility and have come up in life in addition to making the country self sufficient in food production inspite of this alarming rate of population growth.

2.5 Method of Ascertaining Average Consumption of Electricity in IP Sets

This aspect has already been discussed in clause 2.3 K.E.B. has taken up study of IP consumption in many parts of the State. The consumption recorded for a period of 12 months, reveal that the average consumption of open/dug well is 3000 KwH per annum, bore well IP set as 10,000 KwH per annum and the river bed pumpset as 8000 KwH per annum.

Taking this average consumption if we compute the total energy consumed in the IP sector as follows:

Assuming 10 lakh pumpsets with 6.25 lakh open wells and 3.75 lakh bore wells about 30% of open wells as defunct and 5% bore wells as defunct

No. of working open wells = 4,37,500 Nos.
No. of working bore wells = 3,56,250 Nos.
Energy consumption in open wells = 4,20,000 × 3000 = 1312.5 MU per annum.
Energy consumption in bore wells = 3,80,000 × 10,000 = 3562.5 MU per annum.
Total energy consumption from IP sector at consumer point = 4875 M.U. per annum
Energy consumption at the generating station per annum from IP sector = 4875 M.U. + 22% line losses = 5947.5 M.U.
Taking an average cost of energy at Rs. 2.00 per KwH of energy
Total revenue from IP sector per annum = Rs. 1189.5 crores.

If a tariff of Rs. 300 per H.P. per annum is levied, taking average of 6HP per IP set, then for 10,00,000 pumps sets. It is 60,00,000 HP.

KEB revenue from IP sets = 10,00,000 pumpsets × 6 HP × Rs. 3000.00
= Rs. 180 crores.

This study has been conducted in the taluks and districts of Karnataka by KEB research station.

2.6 Relationship between Depth of Well, and Electricity consumed

As already discussed the power required to lift water increases as the depth of the bore well increases.

Work done (power input) = (QH / 101.9306) KW's.

where Q is the quantity of water pumped in lits/sec and where H is the static head from the water level in the bore well to the point of delivery of water in meters.

Power is directly proportional to the static head where `Q' remains as constant. The product of this power and the duration of pumping gives us the energy consumed.

Energy = P × T (where T is duration of pumping in hours)

Keeping the quantity of water to be pumped as constant, power/energy required is directly proportional to `H' i.e., the greater the depth of well/bore well the more power/energy is required for pumping of water.

In the case of open wells the suction head could be a maximum of 30 ft at sea level beyond which the pump does not work.

PROJECTS IS KARNATAKA STUDYING THE CONSUMPTION OF ELECTRICITY AND CONSERVATION OF ENERGY IN IP SETS

3.1 Year in which Studies were Conducted and Results

In the year 1984-85, KEB research station has taken up the possible energy conservation in irrigation pumpsets. In this direction a pilot project was taken up to study the open wells first. 100 typical open wells around Bangalore were taken up at Channapatna, Belandur in Sarjapur Road, Konaiahana Agrahara limits in Hosur Road etc.,

Before we take up the analysis of this study let us have a fair idea of where exactly energy conservation could be achieved.

An irrigation pumping system consists of a motor, a water lift pump coupled to the motor or monoblock pump, suctionpipe line, delivery pipeline and a foot valve or a reflex valve. All these put together in an irrigation pumping system. All the components of the pumping system contribute to the wastage of electrical energy.

(a) Suction and Delivery Pipes

Whenever water or any liquid flows through a pipe there will be friction losses. During 1984 farmers were using G.I. Pipe for both suction and delivery pipes. G.I. Pipes are susceptible to be rusted within one year.

The friction loss of head to the flow of water can be expressed as
$hf = f\, l\, v^2 / 2gd$.

where,

`f' is the friction coefficient.
`l' is the length of the pipe
$v^2/2g$ is the loss of head due to the velocity of flow of water
`d' is the diameter of the pipe and
`g' is the acceleration due to gravity of earth.

Due to rusting of the internal walls of the G.I. Pipe, encrustation will be formed which it restricts the free flow of water. About 20%-50% of losses are due to friction in pipes due to wrong selection. Instead of G.I. Pipes suitable RPVC or HDPE pipes are used. The frictional losses in suction and delivery pipes can be reduced to 31.8% as their inner surface is smooth. The friction losses in G.I. Pipes go on increasing as time lapses. The smooth inner surface of the PVC pipes will not deteriorate as time elapses.

The Hazon Williams formula for determining friction loss in pipes is

$$Q = 0.2343 \times C \times D^{2.63} \times S^{0.54}$$

where,

`Q' is the rate of discharge of water in gallons/minute
`C' Hazon Williams constant (c = 100 for GI pipes & 150 for PVC pipes)
`D' Diameter of the pipe in inches.
`S' Slope of Hydraulic gradient (usually 8/1000)
when worked out

(i) for a 2.5" Dia GI Pipe 5.05 lits/sec
(ii) for a 3" Dia GI Pipe 8.16 lits/sec
(iii) for a 4" Dia GI Pipe 17.39 lits/sec
(iv) for a 5" Dia GI Pipe 31.27 lits/sec
(v) for a 6" Dia GI Pipe 50.51 lits/sec

`hf' head loss due to friction in RPVC pipes is 47% of the head loss in G.I. Pipes for the same internal diameter of pipe.

hf for PVC pipe $= 15 \times Q\ 1.852 / D^{4.87}$
hf for GI pipe $= 32 \times Q\ 1.852 / D^{4.87}$

Thus it can be seen that friction in pipes contribute to high energy losses. By replacing the existing G.I. Pipes by HDPE on RPVC Pipes we can reduce the losses, improve discharge and save pumping time to achieve energy conservation for the discharge of the same quntity of water.

(b) Foot valves

A foot valve is a non return value always submerged in water for all centrifugal pumps used for open wells.

In earlier days the foot valve due to its shape and restricted opening (neck) was contributing for loss of energy. Foot valves or reflex valves are fixed in the suction line of centrifugal pumps. They create a loss of head which is measured in terms of K-value. The `K-value' of old type of foot valve varies from 2-13. But M/s. Kirloskar & Sujala have designed a less friction foot valve whose `K-value' is 0.5. If the new type of foot valve is used an energy saving of 8% can be achieved*. It costs about Rs. 300.00. Every month it saves energy by 51 units and 612 units per annum. The pay back period is only 6 months.

(c) Motor and Pumpsets

This plays a major role in the working of the irrigation pumping system. The maximum efficiency of a motor and a pump, which can be achieved in the field is 65%. If proper components are not used in the manufacture of pumpsets, the overall efficiency of the pumpset can only be 30-35%.

Overall efficiency of the pumpset

η = Motor $\eta \times$ pump η

Max - $90 \times 70 = 63\%$

Min - $75 \times 45 = 33.75\%$

Care must be taken to select a good ISI marked pumpset while selecting the pumpset.

iv) Mismatching of the Capacity of Pumpset:

Before selecting the pumset we will have to take into account the availability of water in the well, the area of land to be irrigated and the crop they are going to grow. Based on all these factors the HP of the pump will have to be determined. When a high HP pumpset works on low load, the efficiency of the pumping system is low. Ultimately it is the availability of water which is the deciding factor.

HP of electric motor = $Q \times H \times 1.1^* / 75 \times$ pump $\eta \times$ Transmission $\times$
(50-60%) (85-95%)

KW rating of electric motor = QH / 101.9376 $\times$ pump $\eta \times$ Transmission η

where,

Q in Lts/sec

H in metres

(v) Selection of Suction and Delivery Pipe Diameters

For efficient operation of the pumpset, the selection of the suction and delivery pipe diameters is very important. the velocity of flow of water inside the pipe should not be more than 1.5 - 2 meters/sec.

$V = Q / A$

where,

* This is as per the findings of the pilot studies conducted by KEB Research Station during 1984.

Q in lts/sec

A cross-sectional area of pipe

If the velocity of flow is more, there will be constraint for flow of water and the friction to the flow of water inside the pipe will be more and will result in lower efficiency of the pumping system.

(vi) Mechanical & Transmission Losses

These losses also play vital role whenever there is slippage in transmission of mechanical power from the motor to the pump. The efficiency will be low. The bearings of motor and pump will have to be of less friction type. The pump impellers shall have correct shape and smooth surface for pumping water efficiently. Now a days polycarbonate moulded smooth impellers are used to improve pumping efficiency.

The results of study conducted by KEB research station

(a) open wells

Study conducted on 16 open wells - Replacing old foot valves by new less friction foot valves

(i)	KW Demand before conversion	42.21
	KW Demand after conversion	43.38
	Increase in KW Demand	1.17
	% increase in KW Demand	2.77
(ii)	Energy requirement per annum (KwH) before conversion	33,768
	Energy requirement per annum after conversion KwH	29,755
	Savings in energy consumption per annum KwH	4,012
	%age savings in energy per annum	11.88
(iii)	Discharge in lits per second before conversion	155.78
	Discharge in lits per second after conversion	181.69
	Improvement in discharge	25.91
	%age improvement in discharge	16.63

Replacing the existing G.I. Pipes by less friction HDPE pipes of the same diameter (for 10 cases)

(i)	KW Demand before conversion	29.48
	KW Demand after conversion	29.96
	Increase in KW Demand	0.48
	%age increase in Demand	1.63
(ii)	Energy requirement per annum KwH before conversion	23,584.00
	Energy requirement per annum KwH after conversion	16,731.59
	Savings in energy per annum	6,852.41
	%age savings per annum	29.06
(iii)	Discharge in liters/sec before conversion	100.74
	Discharge in liters/sec after conversion	144.31

* Provision for low voltage of EB.

Improvement in Discharge	43.57
%age improvement in Discharge	43.25

Results after changing both the old foot valves and G.I. Pipes by less friction HDPE Pipes of the same internal diameter.
(for 10 cases)

(i)	KW Demand before changing	28.38
	KW Demand after conversion	29.96
	Increase in KW Demand	1.58
	%age increase in KW Demand	5.57
(ii)	Energy requirement per annum before change (KwH)	22704.00
	Energy requirement per annum after change (KwH)	15175.00
	Savings in energy (KwH) per annum	7528.64
	%age savings in energy	33.16
(iii)	Discharge in liters/sec before change	91.37
	Discharge in liters/sec after change	144.31
	Improvement in Discharge	52.94
	%age improvement in Discharge	57.94

(b) Bore wells

In bore wells there are no foot valves. The study was conducted on 10 submersible pumps.

(i)	KW Demand before conversion	37.50
	KW Demand after Conversion	36.62
	Decrease in KW Demand	00.88
	%age Decrease in KW demand	2.35
(ii)	Energy requirement per annum (KwH) with G.I. Pipe (before conversion)	1,50,000.00
	Energy requirement per annum (KwH) with same dia HDPE Pipe (After conversion)	92,425.00
	Savings in energy requirement/annum (KwH)	57,575.00
	%age savings in energy per annum	38.38
(iii)	Discharge in Lts/sec before conversion	29.61
	Discharge in Lts/sec after conversion	46.61
	Improvement in Discharge Lts/sec	17.00
	%age improvement in Discharge	36.47

From the studies, it is extremely clear that there is a good lot of energy savings if the existing G.I. Pipe are replaced by HDPE pipes of the same internal diameter for delivering the same water discharge.

Similarly there will be savings in energy consumption if the old foot valves with high K-value are replaced by less friction foot valves.

The effect of changing of G.I. Pipes by HDPE pipes or old foot valves by new less friction foot valves has only marginal effecting on the input power demand (KW).

The effect of changing G.I. Pipes and foot valves has very little effect on the power factor.

The improvement of efficiency of submersible pumpsets increased from 16% to 29% by this process.

3.2 Sample Size and Geographical Spread

As already discussed, a pilot study on the various aspects of irrigation pumping system involving open wells was conducted on 100 pumpsets. The results of the study are given. Today we are concerned more about bore wells. KEB Research Station studied the behaviour and the possible energy conservation on bore wells in its second pilot study. The paper giving the result of the study is given.

It can be seen that the bore well culture has spread to Kolar, Bangalore, Tumkur, Belgaum, Bijapur, Hassan, Chitradurga and Bidar Districts. The IP Set energisation programme of KEB is covering 50,000 - 60,000 pumpsets every year. This is being continued for another 6-8 years when it may reach a saturation stage. Further the ground water potential is also going down day by day which may be the ultimate limiting factor for the programme.

The crop pattern of IP sets irrigated areas differ. We grow water rich crops in Belgaum and river bed areas such as sugarcane and paddy, whereas, they grow Mulberry, Ragi, Jawar, Maize, groundnut, sun flower pulses, vegetables etc., in low water potential areas.

The bore well culture has spread all over Karnataka as the Jilla panchayats, taluk panchayats etc., are going on with a planned programme for arranging loans to prospective farmers for establishing bore well IP Sets in a very big way. This will gain momentum and more farmers will get the benefit in the years to come to grow more food for the nation.

3.3 Studies conducted by KEB Research Station on Open Well and Bore Wells

KEB research station has conducted 2 pilot studies on

(a) 100 open wells and
(b) on 10 bore wells

To find out the possible savings in energy per annum and pay back period etc. The results of the study are furnished in para 3.1

While conducting pilot studies KEB research station has envisaged lot of problems. KEB had demetered IP sets and introduced a nominal tariff of Rs. 50.00 per HP per annum. Then it was made free of cost. The farmers had very strong apprehensions that KEB may reintroduce metering for IP sets again and they will have to pay heavily. Many farmers did not allow our testing team to conduct studies. Some farmers, critically and verbally accused KEB for the very poor voltage consequent to burning of their motors. This is one of the major reasons for their woes. since their motors burn often due to low voltage they will have to spend at least Rs. 2500-3000 to get their motor rewound.

The objective of this study is to exactly know the extent of energy savings and the pay back period due to the replacement of foot valves and G.I. Pipes, by less friction foot valves and PVC Pipes. It was also contemplated to prepare a project report for changing the foot valves and G.I. Pipes for 1000 IP installations in Kanakapura and Channapattana taluks with the loan assistance from REC (Rural Electric Corporation), New Delhi. When REC Ltd. after studying the various energy saving figures of the project report, came forward to give grant to Electricity Boards for taking up of such projects.

3.4 Follow-up on the Studies :

REC, New Delhi, came forward after having convinced of the tremendous savings from this project and took up mass conversions of 1000 IP sets by fixing low friction foot valves and replacing the existing G.I. Pipes by PVC Pipes. It was also decided to study the benefits from this mass rectification by selecting a purely 11 KV rural feeder predominantly feeding IP sets in Kanakapura taluk. The energy meter fixed to the feeder gave the desired results of energy savings. Some of the greedy farmers sold out the PVC pipes and again fixed G.I. Pipes for the sake of petty gains. However the project was a great success. If such projects are taken up for implementation there will be very good energy conservation. The government gets the full benefit of it after 2 years which is the pay back period to save the full amount spent on this project. The economics of replacement of pipes, foot valves etc., will be discussed later.

4. IMPROVEMENTS (RETROFIT) ON IP SETS

4.1 and 4.2 Improvements Possible on Open Wells and Bore Wells

As already discussed, improvements to the existing open and bore wells are possible in the following avenues:

(i) Changing the old high K-value foot valves by, the new K-value foot valves, can give 8-10% energy saving. By spending about Rs. 300.00 farmer can save about 240 (units) KwH per annum (taking the average consumption of one IP set at 250 KwH per month and that an IP set works for 240 days in a year). The pay back period is about 6 months only after which all the savings is bonus to the farmer and the EB (taking Rs. 2.00 per KwH at the IP set point).

(ii) Changing of the existing G.I. Pipe for suction and delivery pipes by PVC pipes of the open well can save about 29% energy. There can be saving of 870 units per annum. The approximate cost of replacement of 20 meters of GI Pipe of 50 mm HDPE pipe of 48 mm internal diameter is about Rs. 3800.00. The pay back period is about 2 years 4 months taking Rs. 2.00 per KwH.

(iii) In respect of a bore well there are several apprehensions to our farmers whether HDPE pipe can carry the weight of the motor and pump? This has been amply clarified and shown to the farmers at Hosakote area. The labour for erection of a HDPE Pipe into the bore well is about 60% of the cost of labour for hoisting of G.I. Pipe. The average saving in energy is 38% and the pay back period is one year and 10 months only.

(iv) Energy Efficient Monoblock Pumps:

Whenever there is a motor and a pump, the motion will have to be transmitted from the motor to the pump. This can be done in many ways such as, Belt Drive, Gear Drive, Flange Drive etc. In any of these motion transmissions, there will be loss of mechanical energy. Now-a-days energy efficient monoblock pumpsets with ISI markings have come and when they operate at their duty point we can get maximum efficiency. In submersible pumpsets, the stages of the pumps will have to be decided precisely by gauging the draw down level of water in the bore well. The pump will have to be operated at its duty point for best results.

If we select an ISI marked monoblock pump and when it is operated at its duty point then we can get 5% savings in energy. This has been the findings of the pilot study conducted by KEB Research station.

(v) Capacitors

The average power factor of an irrigation pumping system is about 0.7 or even less. Low pf will amount to loss of energy by way of wattless power by heating unnecessarily. With all the above measures a possible 35-40% energy can be saved.

4.3 Cost of Improvements

(a) Less Friction Foot Valves

The shape of foot valve and its opening arrangement plays a significant role in the overall efficiency and energy savings in the irrigation pumping system. Foot valves/reflex valves are used in centrifugal pumps where the suction head does not exceed 9 meters. The friction/opening resistance of a foot valve, is stated as K-value of the foot valve. In old food valves the K-value was varying from 3-12 whereas, the K-value of a less friction foot valve designed by M/s. Kirloskar, M/s. Sujala Corporation is of the order of 0.5 friction foot valve.

It has already been made clear that the new foot valve costs around Rs. 300.00 for a 2 1/2" (65 mm) dia foot valve.

In the new less friction foot valve it can be seen that the operating flap opens fully for the flow of water whereas in the old foot valve the flap opens only by 50%. Thereby the flow of water is restricted resulting in loss of energy. The saving in energy by the less friction foot valve is about 8-10%. This has been established already by the pilot study

Costing

(a) Open wells

Assumption (cost of energy Rs. 2.00 / KwH)*

i)	The cost of less friction foot valve is	Rs. 300.00
	Labour for replacement of old foot valve by new foot valve (2 persons for 1 day)	Rs. 100.00
	Total	Rs. 400.00

Taking average consumption of open well as 250.00 KwH/month.

Changing the old foot valve by new foot valve saves energy by 8% — Rs. 40/month.

It can be seen that the pay back peiod is about 10 months. For any electricity board or Government this is a welcome measure and this can be adopted in a big way to achieve the results of energy conservation.

(b) Changing of suction and delivery pipes (GI) by RPVC/HDPE pipes in open wells.

From the studies made it is clear that by replacing the exsting GI Pipes by new RPVC/HDPE pipes of the same diameter, the energy saving is 29% per annum. The result of the pilot study have given the above result.

The average requirement of PVC pipe for one pump set for replacement of both the suction and delivery pipes is 20 meters.

The cost of 20 Mt. of 4 kgf 65 mm dia HDPE Pipe (Assuming cost/Mt of 65 mm pipe @ Rs. 57.00 Mt)	20 × 57 = Rs. 1140.00
Labour charges for replacing GI Pipe by HDPE pipe (Assuming 4 persons can complete 1 IP set / Day (Rs. 75.00 as wages/person/day)	Rs. 300.00
Cost of 2 joints foot valve + suction flange @ Rs. 200.00/joint	Rs. 400.00
Total expenditure for rectification	Rs. 1840.00

Savings in energy per annum = 3000 units × 29 / 100 = 870 units/annum

Pay back period is only 2 years and one month.

(b)*   *Bore wells

The requirement of 53 mm OD HDPE pipe 10 Kgf rating for changing the existing GI Pipe is about 65 meters (average value)

* Average cost of energy can be taken as per actuals of EB's

Cost of 53 mm OD HDPE Pipe = 233 ×65 = Rs. 15145.00
(for 65 meters.)
(i) Expenditure for 2 joints Rs. 800.00
(ii) Expenses for labour four persons/day Rs. 300.00

Total cost Rs. 16245.00

Annual consumption of each bore well 10,000 KwH
Savings is 3838 units at Rs. 2.00 / KwH or Rs. 7676.00/annum.

Pay back period = 16245 / 7676
= 2 years 2 months.

It can be seen that the pay back peiod for recovery of the cost of investment is only about 2 years and 2 months. Savings in energy by using HDPE pipes is 38.38% (as per pilot study)

(c) Monoblock Pumpsets and Usage of Static Capacitors

By using monoblock pumpsets the losses in transmission of mechanical power by gearing, belts etc., can be saved to an extent of 5% in case of open wells.

It is also true that by fixing static capacitors (LT of appropriate rating another 5% energy can be saved in case of centrifugal or submersible pumps.

Costing

(i) open wells

Cost of 3 KVAR LT capacitor Rs. 800.00
Average energy consumed/pumpset 150 KwH/month
Energy saving is 5% = 12.5 KwH/month = Rs. 25.00/month.
Pay back period is about 2 years and 8 months which is a good proposition.

(ii) Bore wells

Cost of 3 KVAR capacitor Rs. 800.00
Energy consumption/annum Rs. 1000.00
Energy Savings = 10000 × 5/100 = 500 units
Cost at Rs. 2.00 per KwH = Rs. 1000.00
The pay back period in this case is only 10 months which is very good.

4.4 Savings Potential from Improvements in KwH and in Rupees at Current L.T. and H.T.

As already stated there are about 10,00,000 irrigation pumpset in Karanataka out of which there are 6.25 lakh bore wells and 3.75 lakh open wells. Till mid 1980's there were more open wells. After that the number of bore wells increased in a very big way. About 90% of the new IP sets serviced were bore wells.

In accordance with the study conducted by KEB research station 8-10% energy savings from replacement of old foot valves by less friction foot valves 29% energy savings from replacement of both suction and delivery G.I. Pipes by less friction PVC pipes and 33% energy savings in respect of open wells (centrifugal pumps) where foot valves and suction and delivery GI pipes are replaced by PVC pipes and 38% energy savings in respect of submersible pumps where GI pipes are replaced by HDPE pipes of the same diameter.

The average consumption of IP sets per month
Energy savings wise
(i) In open wells - 3000 KwH/annum
(ii) In bore wells - 10000 KwH/annum

Results :

(i) By changing foot valves 8-10% savings amounting to 300 KwH/annum can be achieved from each pumpset
(ii) By changing suction and delivery GI Pipes by PVC pipes 29% energy savings amounting to 870 KwH/annum can be achieved from each pumpset.
(iii) By changing both the high friction foot valves by low friction foot valves and G.I. pipes by PVC pipes energy savings amount to 33%
(iv) By changing of GI pipes of a submersible pump by HDPE Pipe of the same internal diameter energy savings amounting to 3888 KwH/annum can be achieved from each pumpset.

Total savings achieved can be worked out as follows :

Total No. of open wells in the State	6,25,000 Nos.
Open wells where there is no water about 30%	1,87,500 Nos.
Total No. of live open wells in the State	4,37,500 Nos.
Total No. of bore Well IP sets in the State	3,75,000 Nos.
Assuming 5% bore wells do not have any water	18,750 Nos.
Live bore wells in Karnataka	3,56,250 Nos.
Total No. of IP sets both open and bore wells	7,93,750 Nos.

(1) Open Wells

If we replace all the old foot valves by less friction foot valves and replace the suction and delivery GI pipes by HDPE/RPVC pipes savings in energy is 33%.

Total energy consumption of all the live open well IP sets $= 437500 \times 3000$ KwH
$= 1,31,25,00,000$

=1312.5 MU

Therefore savings per annum = 1312.5 × 33/100 = 433.125 MU

Assuming average energy cost at Rs. 2.00 / KwH.

Monetary value = 433.125 × 2 1000000 = Rs. 8662500000

= Rs. 86.625 Crores .... (A)

(b) Bore wells :

Replacing all the GI pipes of submersible pumps by HDPE pipes in bore wells we can save energy as follows :

Total energy consumption of all the live bore well IP sets = 3,56,250 × 10,000 KwH

= 3562500000

= 3562.5 MU

Energy savings if the GI pipe is replaced by HDPE pipe = 3562.5 × 38 / 100

= 1353.75 MU

Assuming average energy cost as Rs. 2.00/KwH

Monetary value = 1353.75 × Rs. 2.00 = 270750 = 270.75 Crores ...(B)

Total savings achievable from both open wells and bore wells per annum

(A) + (B) = 86.625 + 270.75

= Rs. 357.357 Crores.

If we can change the less friction foot valves of all the open wells and change the suction and delivery GI pipes by HDPE/RPVC pipes of the same ID and change the GI pipes of all submersible pumps by HDPE pipes of the same ID there we can save Rs. 357.375 crores per annum.

4.5 Cost Benefit Analysis

(a) Open Wells :

In open wells the possible energy conservation measures are

(i) Changing of the old foot valve by the less friction modern foot valve.

Changing of old foot valve costs (labour @ 20%)

Rs. 300.00 + Rs. 60.00 (labour) = Rs. 360.00 ... (A)

(ii) Changing of the old GI pipe by new RPVC/HDPE pipes of the same internal diameter. Assuming,

(a) Each pump requires 20 meters pipe for both the suction and delivery pipe put together.

(b) Cost of 6 kgf HDPE pipe to cost Rs. 157.00 per meter.

Rs. 3140.00 + Labour Rs. 628.00 = Rs. 3768.00

+ 2 Joints at Rs. 200 each = Rs. 400.00

Total = Rs. 4168.00 ... (B)

(iii) Changing of the old centrifugal pump + motor by an efficient new monoblock pumpset of same HP rating (5 HP pumpset)

labour at 20%

Rs. 7500.00 + Rs. 1500.00 (labour) = Rs. 9000.00... (C)

(iv) Fixing of adequate capacity LT static capacitor

Static capacitor 3 RKVA

Rs. 300.00 + Rs. 50.00 = Rs. 350.00 .... (D)

Total cost of full rectification of pumpset = A+B+C+D

Rs. 360.00 + Rs. 4168.00 + Rs. 9000.00 + Rs. 350.00 = Rs. 13878.00

Credit old motor + pumpset + GI Pipe (Scrap value) (–) = Rs. 500.00

Net rectification cost/pumpset = Rs. 13378.00

Energy savings per annum/pumpset with all the above 4 conservation methods

Pay back period = Rs. 13378/1740

= 7 years and 7 months

Out of this, one can select judiciously and economically where one can get maximum benefit from minimum investment.

Possible energy savings in open wells

Energy consumed in open wells = 1312.5 MU/Annum

(i) Energy savings by changing the high friction foot valves by less friction foot valve and the old GI Pipes for suction and delivery = 433.15 MU ... (a)

by PVC pipes - 33%

(ii) possible energy savings by replacing low efficiency centrifugal coupled pumps by high efficient Monoblock pumps 5%

= 1312.5 × 5/100 = 65.625 MU ... (b)

(iii) Possible energy savings by fixing adequate static capacitor... 5%

= 1312.5 × 5/100 = 65.625 MU ... (c)

Total energy savings by adopting all these methods = (a) + (b) + (c)

= 564.365 MU

Monitory value by taking average cost of energy at Rs. 2.00 / KwH

= 564.365 × 2 = Rs. 1128.73 million or 112.873 crores/annum.

Per pumpset per annum it works out to Rs. 2580.00

Expenditure for rectification of all the above four components

Foot valve + Replacement of suction and Delivery GI pipes + Replacement of old pumpset by monoblock pumpset + Providing static Capacitor

Rs. 360.00 + Rs. 4168.00 + Rs. 9000 + Rs. 350.00 = Rs. 13878.00

Less cost of scrap materials (–) = Rs. 500.00

Cost that will have to be incurred on each open well pump set for best results Rs. 13378.00

Pay back period = Rs. 13378.00 / Rs. 2580.00 = 5.185 years

If we adopt all the energy conservation measures it will take about 5.185 years to get back the investment.

By option the electricity boards can take up mass rectification programmes by loking at the savings vis-a-vis investment. For example if we take up rectification of item (i) and (iii) by leaving the replacement of old pumpset by monoblock pumpset.

Then energy savings per annum = 433.115 + 65.625 = 498.740 MU

Taking average energy cost as Rs. 200/KwH = 498.74 × 2
= Rs. 99.75 Crores.

or Savings per pumpset/annum = Rs. 99.75 Crores/4,37,500
= Rs. 2280.00

Cost of replacing high friction foot valve, old GI pipes and capacitor per open well pumpset=Rs. 4878.00

Pay back period = Rs. 4878.00 / Rs. 2280.00 = 2.14 years.

This is a very reasonable proposition and it can be taken up by electricity boards and get the benefit.

Possible Energy Savings in Bore Wells

Here the starter wire, panel capacitor and the pumpset (submersible) are all the same. Only we have to work out the cost of replacing the B. Class GI pipe by HDPE pipe.

Assumptions

(i) Quantity of water discharge is a variable, but the internal dia of both the GI pipe and HDPE pipe are almost the same.
(ii) As per design - let us use a 5 HP, 10 stage standard (ISI) submersible pump for a head of 80 meters inside a 150 mm bore well.

(iii) Cost of energy is Rs. 2.00 per KwH and Rs. 2.49 assuming 22% live losses at generating point.
(iv) Usage of 65 mm OD, 48 mm ID HDPE pipe of 10 kgf rating for replacement of 50 mm bore GI pipe.
(v) Cost of 65 mm OD, HDPE pipe — 10 kgf rating is Rs. 233.00/meter
(vi) Cost of labour is assumed at 10% of cost of material in case of HDPE and 20% in case of GI pipe.
(vii) Cost of new class B pipe 50 mm ID is Rs. 150.00/meter
(viii) Cost of 2 Nos. of HDPE terminations Rs. 200.00/each end=Rs. 400.00

Cost of 80 meter GI pipe = 150 × 80/meter	= Rs. 12,000.00
Cost of 80 meter HDPE pipe (233 × 80 + Rs. 400)	= Rs. 19,040.00
Cost of releasing GI pipe – 20% (labour)	= Rs. 2,400.00
Therefore total cost of rectification—	
Rs. 19,040.00 + Rs. 2400.00 + Rs. 2000.00	=Rs. 23,440.00
Savings in energy by using HDPE pipe (study result)	= 38%

Therefore savings in energy per annum
(Taking energy consumption of 10000 KwH/annum/bore well IP set)
10,000 × 3800 KwH

Assuming a cost of Rs. 200/KwH at bore Well point		
Monetory savings per annum	= Rs. 7600.00	...(i)
At the generation point assuming 22% T&D losses		
Monitory savings = 7600 × 1.22	= Rs. 9272.00	...(ii)
Pay back period = Expenses for rectification/Monetory savings per annum		
= 23440 / 9272	= 2.52 years	

It is quite encouraging for anybody to do this job on top most priority.

Cost of Installation of a New Submersible Pump

1.	The cost of a 10 stage 5 HP submersible (ISI mark) pump	... Rs. 16,500.00
2.	Cost of starter 3 core PVC cable & control panel	... Rs. 3,500.00
3.	Cost of 2 KVAR LT capacitor	... Rs. 300.00
4.	Cost of 10 kgf, 65 mm OD, 48 mm ID. HDPE pipe 80 meter long Rs. 233.00/Mt.	... Rs. 18,640.00
	Total material cost	... Rs. 38,940.00
3.	Cost of Labour for installation 10 for HDPE pipe	... Rs. 3,894.00
	Total Cost	... Rs. 42,834.00

Total cost of installation of a 5 HP ISI submersible pump with 10 kg HDPE pipe 80 meters + cable + capacitor + control panel say Rs. 43,000.00

4.6 *Cost of Digging Open wells, Bore wells, Lift Irrigation Schemes - Agencies for doing the work.*

(i) open wells

The farmers in the early 70's and 80's were going in for open wells when the water table was low. Now percentage of open wells is very low. Say a maximum of 10%. The digging of open well depends mainly on the nature of soil in the region. Usually a good open well may be about 30-50 ft dia and about 40-60 ft deep. The well will be riveted using granite stone lining with joints well bounded by rich cement mortar. Such wells cost about Rs. 75,000.00 - Rs. 1,00,000.00 today and is rather expensive. If there is rock below the earth the cost of digging an open well will be more. However an average cost of open well with about 20 ft dia and 40-60 ft deep if soft earth is available, is around Rs. 8000.00 There is no surety of getting water in summer. The granite stone lining is given to retain the excavated earthen sides from caving in. However the open well philosophy is very less but for those places where there is a perennial water source like a river, *nalah*, etc.

(ii) Bore wells

From the late eighties onwards bore wells have come into the fore. The philosophy of bore wells is very well understood by farmers. Here, the selection of site is a very important job. First the ground water potential should be studied by a geologist. Then in the land, the most water yielding point will have to be identified through scientific instruments and then the bore well will have to be drilled. The topography of the area will have to be studied by looking at the formation of rocks in the area, etc., by geologists. The water inside the bore can be identified by electro sounding. It can also be identified by measuring the earth resistance and reading it on the graph. The water table differs in Karnataka from district to district. For example the water table has gone down to 300-450 ft in Kolar district, from 250-300 ft in Tumkur district and from 200-250 ft in Bangalore district. In the northern part of Karnataka the water table varies from 150 ft - 350 ft.

Thus the water level is going down and down day by day. This is due to reduction in rainfall and due to reduction of greenery.

Digging of Bore Wells :

The drilling operation of a bore well itself is a big subject. The type of drilling of a bore well also differs from one type of soil to the other. There are mainly 4 types of drilling rigs, namely

(i) Calyx rig,
(ii) Percussion rig,
(iii) Pneumatic rig, and
(iv) Hydraulic rig.

The drill bit tips of the drill rigs are made of tungsten carbide which is a very hard material. The drill bits however hard will wear out after use and their drilling diameter reduces in accordance with their service.

The process of drilling is to first drill 8½" bore till the soft earth is drilled and to drive the casing pipe to see that the soft earth from the drilled portion does not cave in. The provision of casing pipe is the costly portion of drilling. Usually the casing pipe is made up of No. 6 SWG MS class B. pipe 8" outer dia. It weighs 19-20 kgs per meter and supplied in standard lengths of 6 meters each. The rate is about Rs. 12.00/kg. Each pipe costs Rs. 1440.00 for 6 meters long.

The bore well drilling process is continued. Usually the standard diameter of bore well is 6" or 150 mm. The drilling process is continued till water jet comes out. The water that flows out will have to be measured accurately for issuing of bore well water yield certificate which is measured using a 90°V-notch method. The discharge of water is allowed to flow through the V-notch duly forcing compressed air to the bore well and displacing water. This will also clear the bore inner surface.

Usually the drilling process is continued for about 25 ft even after getting water source to ensure flow of water. When the diameter of the drilling comes below 146 mm it will have to be discarded.

The farmers are taken for a ride by some private drillers using 146 mm worn out drillbits and claiming money for a 150 mm bore well.

In Karnataka the Department of Mines and Geology will undertake the job of drilling bore wells at a rate of Rs. 175.00 per meter (Rs. 54.00 per ft).

There are private bore well drilling agencies who undertake drilling of bore wells and issuing water yield certificate. Some of them drill bore wells at Rs. 42.00 to Rs. 45.00 per foot. But the diameter of the bore is not guaranteed.

If a 80 meter deep bore well is drilled the cost would be

(i) if the soft earth portion is 6 M × 2	- 12 meters.
Casing pipe cost for 12 meters is	Rs. 2880.00
(ii) Cost of drilling 80 meters bore well	Rs. 175 × 80
	= Rs. 14000.00
Total expenditure	= Rs. 16880.00
Say Rs. 17000.00	
Cost of submersible pump + motor control panel, capacitor, 3 core cable along with erection	= Rs. 43,000.00
Total	Rs. 60,000.00

Thus it can be seen that the total cost for drilling of a 150 mm bore well and providing a 5 HP ISI mark, 10 stage submersible pump with necessary pipe, etc., will cost Rs. 60,000.00

ENERGY SAVINGS IN LIGHTING — THE FASTEST AND PRACTICAL WAY FOR DSM

47

K. SESHADRI,
Vice President
(Energy and Environment Strategies),
Philips Lighting Asia Pacific Group, Singapore

ABSTRACT

The power demand in the whole of Asia is expected to rise tremendously in the next two decades. If the Region's economic growth is to continue, the Asian Development Bank has estimated that an additional 300,000 MW will be required already by the end of this Century (which is only four and half years away)! By the year 2020, the power demand in India alone is estimated at 385,000 MW, requiring about US$380 Billion in investment! Faced with acute financial limitations, and the long delays involved for power plants, countries are trying to tackle the problem by encouraging Independent Power Plants, revamping and refurbishing existing plants, increasing Plant Load Factor, better Demand Side Management (DSM) etc.

The focus of DSM is to influence the end-user's peak load-demand at any given time. When we take the case of lighting in homes, this is mainly switched on after sunset and all the load comes on simultaneously at about 7 pm to stay on till about 10 pm, affecting peak load demand severely.

The Incandescent lamp which is used in most households is fairly inefficient in Lumens/Watt and takes high Watt loading. They can be easily changed to Compact Fluorescent Lamps (CFLs) to cut the lighting load by as much as 80%. But there are some techno commercial and logistic issues to be tackled first. This Paper describes DSM in lighting, specifically through energy-saving CFLs and the follow-up actions required through :

— Role of Government (Ministry of Energy, Ministry of Environment, Ministry of Finance etc).
— Role of Electric Supply Companies (Utilities)
— Manufactures of (reliable) Lamps, Luminaires and Gear
— Retailers; and
— The Public

The tremendous environment pollution saved in the process, will be described. The paper will also cover the technical characteristics of selected Lamps and Gears and the optimum ways to get load-relief by Governments and Utility companies.

INTRODUCTION

With the investments required to increase Electric generation capacity in Asian countries, it is doubtful :

— whether such huge investment funds could be raised at all!
— even if raised and implemented, whether it will be enough to meet the growing demand by then!
— whether there will be tolerance for Ecological disturbance (for new hydroelectric dams) or pollution (from Thermal power stations).

Thus, load-reducing and energy saving devices and systems are needed badly and immediately. The widely accepted term here is Demand Side Management (DSM), to help keep the demand within planned capacity.

Within DSM, Lighting application-areas stand out as most practical because:

— it can be implemented immediately
— it is cost effective to end-users, Utilities and society
— it can be implemented via different scenarios of action—to suit countries, provinces or even small cities

For this paper, the focus is on Compact Fluorescent Lamps as a flagship instrument of DSM.

THE COMPACT FLUORESCENT LAMP FAMILY

Compact Fluorescent Lamps (CFLs) are developed for use in those applications that were traditionally the province of incandescent lamps. At present, the following common CFLs can be distinguished:

(1) SL Lamps

A small-bore fluorescent tube folded into a compact form and placed in a glass or plastic outer envelope. Integral with the lamp are a starter and a conventional ballast, and the whole is fitted with a Screw or Bayonet cap (Fig. 1). The SL lamps belong to this category.

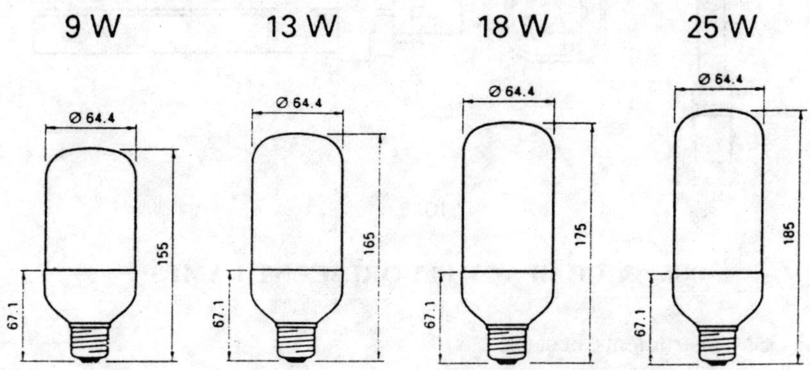

FIG. 1

(2) PLEC and PLET Lamps

A single small-bore tube folded into a flat compact form, or two or more parallel tubes interconnected in such a way as to offer a continuous pathway for the electric discharge. The lamp is provided with an electronic ballast, and a screw cap (Fig. 2).

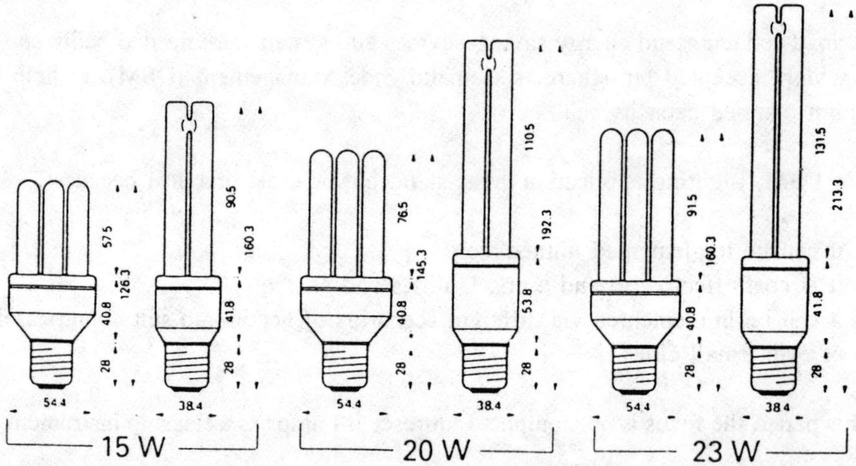

FIG. 2

(3) PL-S/PLC Lamps

Two or more parallel small-bore tubes, interconnected near the ends in such a way as to offer a continuous pathway for the electric discharge (Fig. 3). The lamp is fitted with a special cap that mostly contains a starter. The ballast must be connected externally, somewhere on the luminaire or nearby.

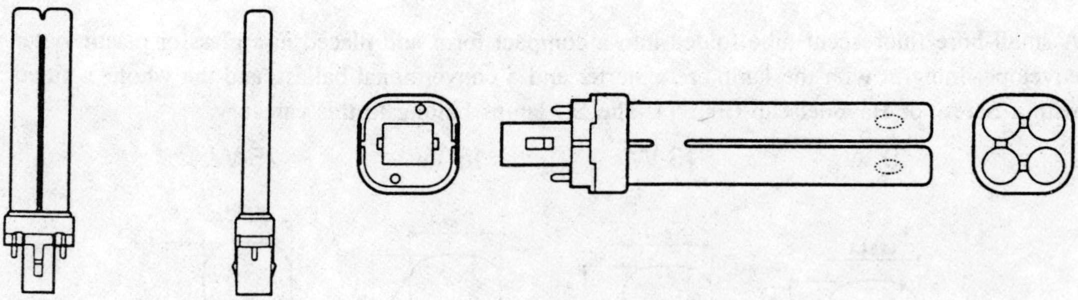

FIG. 3

DSM & COMPACT FLUORESCENT LAMPS

CFLs are ideal DSM instruments because :

— Compared to other possible devices, they represent relatively a small investment per unit

— Possibility to achieve rapid penetration because they retrofit with incandescent lamps which are replaced frequently and hence CFL can take its place any time.
— They consume 75-80% less energy than incandescent lamps with similar light output and lifetime 8-10 times longer, as seen below.

GLS Incandescent Equivalent	40W	60W	75W	100W
PLEC/PLET (internal Electronic ballast)	9W	11W	15W	20W
SL (internal Magnetic ballast)	9W	13W	18W	25W
PL-S (external ballast)	7W	9W	11W	–
PLC (external ballast)	8W	10W	13W	18W

Consider the change over of one GLS 75W incandescent lamp per household with one SL18W (for 10 Million homes)

(1)	Load demand wth GLS 75W × 10 Mln/(Power Factor 1.0)	=	750MVA
(2)	Load demand with SL 18W × 10 Mln/(Power Factor 0.5)	=	360MVA
(3)	Demand reduction	=	390MVA
(4)	Cost of Constructing a 390MVA Thermal Station (approx)	=	US$390 Mln
(5)	Price of 10 Mln SL 18W lamps @ US$ 10.70 = Rs 375 each)	=	US$107 Mln
(6)	Investment Saving	=	US$283 Mln
(7)	Demand Reduction on Power Grid	=	Immediate!
(8)	Pollution Reduction of Environment	=	Immediate!

If an investment of US$107 Mln in SL lamp saves 390 MVA in Demand, that works out to about US$275 per KVA reduced. In a home with about 3 hours of use per day, the lamp-usage is 100 hours per year. A good CFL (from a reputable manufacturer) can last 8000 hours or 8 years of life. The amortisation of a Power plant is usually 15 years. During this period, each CFL point at home would have taken up 2 lamps. In other words, the CFL would really cost US$550 to save one KVA of demand for 15 years. But the cost of constructing a Power Station together with transmission and distribution lines, in developing countries is US$ 1500 to US$4000/KVA. The US$550 for CLF, can be regarded as a maximum based on the assumption that the utility pays full cost of the lamps. In reality, it is less, since the cost of CFLs is partly paid by the Endusers also!

The effectiveness of CFLs has been studied more in depth (in a feasibility stydy called ILUMEX, by the International Institute for Energy Conservation, (December 1992) prior to the start of a 1.7 Million CFL campaign by CFE, the Utility in Mexico. The study compares the costs of CFLs for the Utility, Society and End-users with their respective benefits, as follows:

TABLE 1 : Costs and Benefits of CFLs

	Utility	Society	End-users
Costs :			
- CLF price*	✓	✓	✓
- project and promotion	✓	✓	✓
- loss of electricity sales	✓		
Benefits :			
- Less investment in new electricity generation	✓	✓	
- Less pollution to environment	✓	✓	✓
- Less electricity bills			✓

* The costs of the lamps were shared by the Utility and the End-user (consumer).

The ILUMEX study showed the following Net Present Value per CFL (= benefit minus cost in today's Dollar) as follows :-

	For Utility	For Society	For End-users
Net Present Value of CFL	US$32	US$37	US$6.70

In the ILUMEX program, 1.7 Million CFLs postponed the investment for about 80MW. Note: Postponing or preventing investments in new power plants is difficult to measure. But what has been measured is the reduction of peak load. The PLN (state utility in Indonesia) did a pilot study in 650 homes in 1993. In the so called R1 homes (mostly low income homes, representing 80% of households in Indonesia), incandescent lamps with 3 or more burning hours per day were replaced by CFLs. The reduction in peak load alone was 30% !

Consumer and the Compact Fluorescent Lamp

The obvious question then arises : If CFLs are that cost effective for Utilities, Society and End-users, why doesn't everyone use CFLs? The answer lies in the (price) perception of these lamps. In a recent study (by the Research Institute of Danish Utilities, June 1996) about the domestic use of CFLs in Europe, the barriers for not buying CFLs were investigated. The outcome is presented in the Table 2.

The conclusion is clear : many consumers find the CFLs too epensive to consider buying. This attitude apparently has not improved over the years. The reason could be that lamps are a low 'personal involvement' product. People are not very interested in them, as can be seen from the table where "no interest" has a high score. Buying lamps is not regarded as a pleasant and exuberant idea of shopping, since people are simply annoyed when their old lamp has burned out. "Low involvement" is a relative term and must be compared with `high involvement' products like cars, Television, stereo equipment, etc.

Another, (equally important) reason for the `perceived' high price of CFL is lack of knowledge.

TABLE 2 : **Reasons for not buying CFLs in % (multiple answers possible). Danish ELSAM Survey**

	December 1991 %	April 1992 %	March 1994 %
Price	40	41	48
Poor durability	4	5	5
Design	7	7	8
Does not fit into existing fixture	12	10	17
Lack of knowledge	9	5	1
Dissuaded	2	2	1
No interest	21	21	5
Habits	25	9	18
Other	9	10	18

Consumers know about them (99% awareness in Denmark) but they do not know or believe in the benefits. This is illustrated in Table 3 :

TABLE 3 : **"No real saving is achieved by buying CFLs, compared to the purchase of ordinary bulbs?" (Figures do not add up to 100% because of some no response) Danish ELSAM survey.**

	December 1991	April 1992	March 1994
I totally disagree	23	25	8
I disagree	8	12	28
I agree	7	5	19
I totally agreè	30	25	6
I do not know	16	10	22

(Note that the percentage of "non-believers" is about the same as the percentage consumers who perceive the CFL price as too high, in the previous Table)

The Price Perception of CFL will be a Bigger Problem in Asian Countries

It is also due to the fact that the most lamps available in the market are of low quality, in terms of lifetime and light output. Hence, promoting high quality CFL from reputable manufacturers (with a proven track record) is one sure way to remove such doubts on the benefits of CFL and justification for its higher price. But the end-user will still be reluctant to buy CFL unless the cost is shared by the other beneficiaries of Demand reduction.

Lead by the Utility Company

Number of CFLs in use in household (1995) in some Europeans countries :

Country	Average number of CFLs/household
Netherlands	2.3
Denmark	2.0
Ireland	1.5
Germany	1.3
Finland	1.0
Sweden	0.3
Spain	0.2
Italy	0.2
Norway	0.1
Total	1.0

The top three countries are Netherlands, Denmark and Ireland. What they have in common is the fact that, unlike the other countries, they have (had) nationwide DSM campaigns to promote the use of CFLs for several years (3-5 years). In these campaigns the Utilities have taken the lead actively by both promoting and financing CFLs for consumers. This was done together with manufacturers and retailers. A description of the different kind of campaigns, often referred to as 'Utility CFL campaign' can be found further.

Why are utility CFL campaigns more effective in persuading consumers to install CFLs? Because the Utility or the Energy Authority automatically lend their image and credibility to the benefits of CFL. They are more effective than manufacturers or retailers because they are regarded as objective institutions, i.e. People trust their message better. This can be seen from a research by the Dutch Association, as given in Table 4.

TABLE 4 : "Information about energy saving : whom do you trust most?" (1991)

	Before Utility CFL campaign June 1991	After December 1991
Utility		
Consumer guide	45%	52%
Municipality	30%	28%
Shop	9%	8%
NOVEM*	5%	4%
Other	7%	8%
	4%	

*NOVEM = Dutch organisation for stimulation of energy efficiency

Therefore Utility CFL campaigns have a high impact and result in a significant increase in the use of these lamps worldwide. A few examples :

— European CFL campaign conducted until mid 1991 were targeted at 7.2 Million households and resulted in 2.2 Million lamps installed;

— EGAT, Electricity Generating Authority of Thailand during (1993 - 1994), joined with manufacturers in campaigns to replace the standard fluorescent tubes (TL 12 which is 38mm diameter) by more slim tubes (TL8 with 26mm diameter) at an energy saving of 10% per lamp. Today, 95% of fluorescent lamps sold in Thailand are TL8 type.

— In Mexico, CFE, the Utility invested in a CFL campaign which resulted in installation of 1.7 Million lamps in homes in 1995.

Utility CFL Campaign—The Scenarios

We can divide these campaigns into 3 types.

(a) Joint effort by utility, Energy Authority, Manufacturers and Retailers
(b) Utility alone
(c) Utility, subcontracting to Manufacturers

All the scenarios have one thing in common : The Leader of the campaign is the Utility company. As mentioned earlier, they are regarded by consumers as objective and impartial, resulting in higher credibility of the message. This credibility will be further enhanced if the Utility practices what it preaches, i.e. combine its promotion with a financial incentive for consumers. How this can be done is described below.

(a) Joint Effort and the Role of Participants

Utility

1. They must promote the campaign under their name, which means they are seen as the messenger. This is done via mass media (Newspaper, Radio, TV) and also via direct mailings, which is more expensive but has a higher response.
2. Offer a financial incentive for consumers. This will greatly increase the response, sometimes up to more than 20%. (Average response rate in Europe is 15%). This can be done by allowng consumers to pay for their lamps via their monthly electricity bill. This is realised by offering consumers a coupon which must be signed and handed over at a distribution outlet. At this outlet, the consumers receive the lamps. Payment is done by monthly instalment via their electricity bill. In case payment via the bill is administratively not possible, the financial incentive can also be by offering consumers a coupon which gives them the right to buy at a discount at selected outlets.

Energy Authority

This is usually the Ministry of Environment or the Ministry of Energy. They take charge and pay for the National awareness campaign for energy saving and acts as a catalyst of publicity.

Manufacturers

1. Manufacturers should ensure sufficient supply of lamps at the Utility campaign's distribution outlets in the geographic area of the campaign. This is vital for the success of the capaign.
2. Arrange promotion material (displays, brochures, posters etc) to be present at the distribution outlets. This is called 'below the line' promotion.

Retailers

(i) Keep enough stock of lamp
(ii) Display promotional materials
(iii) Act as Utility's distribution point to sell CFL at a discount. Collect coupons which allow that discount. The retailers will hand over the coupons to the Utility, which will pay them the discount (with a small handling profit)

If all aspects of the campaign are executed correctly by the parties, it could have a high response rate. The earlier mentioned study of 51 European utility campaigns mentions an average response rate of 15% with an average of 2.3 CFL lamps per buyer.

(b) Campaign by Utility Alone

Sometimes the utility prefers to carry out all aspects of the campaign by itself, as in the ILUMEX campaign in Mexico. This is possible if they have their own distribution outlets, or if the lamps can be distributed door to door via the collectors of electricity bills. Also in this scenario the response rates can be high if the offer to the consumers includes a financial incentive by means of a discount coupon and payment of the balance charge for CFL via the Electricity bill in 3 or 4 instalments.

The disadvantage of this scenario is the fact that all the work and cost of carrying out the financing, distribution and promotion will be on the shoulders of the Utility.

(c) Utility Campaign, Subcontracting to Manufacturers

First practised by Edison Electric, the US Utility, it is also called "manufacturers' rebates". Here, the utility calls for a tender from manufacturers. The tender includes not only CFL lamps and prices, but also promotion and distribution of the lamps. The winning manufacturer signs a contract with the utility, in which the Utility commits to pay a subsidy directly to the manufacturer and the manufacturer commits to :

(i) deduct the subsidy from their selling price to distributors (and to the ultimate customer)
(ii) distribute the lamps in the campaign area
(iii) make promotion for lamps and energy saving theme

This scenario has been carried out in Poland in a CFL campaign which was subsidized by the Utility GEF and coordinated by the IFCA (World Bank). The IFC took also care of generic promotion of CFLs in mass media (TV, radio, newspapers). The campaign was very successful; consumer sales increased by 100% during the campaign period. As in Netherlands, such a campaign has to run for 2-3 years to achieve maximum penetration of energy-saving lamps.

The advantage of this scenario for the Utility is that more work is passed to manufacturers and retailers. Disadvantage: more paperwork and administration to prove that the manufacturer(s) have used the subsidy as described in their contract.

CONCLUSION

Any of the Lighting DSM programmes, if executed well, can result in a high penetration of energy saving lamps and huge reduction in the Peak load-demand on existing (strained) Power system. Suddenly it can become a manageable (and pleasant!) power situation that contributes to higher standards of living and stimulates further economic activities and growth all around.

ENERGY EFFICIENCY IMPROVEMENT - DECISION MODELS FOR SELECTION OF ALTERNATES AND COST EFFECTIVE SOLUTIONS

48

G.G. Rajan
Chief Manager (Energy & Environment)
Cochin Refineries Ltd., Cochin, (India)

ABSTRACT

For achieving higher industrial productivity, energy efficiency is one of the major parameters that has to be considered in depth. Efficiency monitoring is very imperative for achieving higher GNP which has direct impact on the nation's economy. In International Competition, this will be the 'key parameter' for the survival of all energy—intensive process industries and power generation units as well. It is likely that many units may close-down due to high operating costs and low profit margin compared to energy efficient ones. Optimal overall energy efficiency could be achieved by fuel mix, better energy dissipation systems (e.g. better burner efficiency, better combustion efficiency, better conversion efficiency of thermal energy into mechanical energy to electrical energy (e.g. high efficiency turbine and turbo generator), efficient power transmission and efficient prime mover equipments at user's end.

Improving the energy efficiency of the existing system by retrofits / revamps is observed to be more cost-effective and could contribute to achieve higher output at low energy input (i.e.increase in productivity). Dynamic programming approach could be used to get the best solutions, even under non-optimal conditions. This paper highlights some of the mathematical/statistical models that could be used to solve this complex problem with ease. Impact of various parameters on economy could also be predicted. Various alternates could be analysed to meet the energy demand at the least cost using these models. A number of strategies are discussed in this paper to achieve the highest efficiency at minimum cost .

1. Introduction

Energy system and sub-system analysis could be used effectively to monitor the energy performance. A number of parameters such as load factor, energy quality, type and mix, energy distribution system efficiency, ageing of the system etc affect the overall energy efficiency of the total system.

Under these conditions, even the most experienced operations manager of a Utility Plant may find it difficult to pin point the problem area for higher energy consumption. In such situations, mathematical models could be used to identify the problems easily than one could visualize.

1.1 (a) Benefits of Performance Models

— Performance models are based on actual process variables which are dynamic in nature and fluctuate due to exogenous or endogenous factors. Since the effect of these manipulable variables form the inputs to the model , the outcomes will naturally be far more precise and accurate.

— Performance models may be used for carrying out the most needed sensitivity analysis of the dynamic system to identify the output trend for a set of new variables and / or operating conditions.

— Performance models may also be used to determine and monitor the system efficiency as a whole and also the sub-systems contributing them. Hence, corrective actions may be taken more precisely.

— Performance models are real life operating models and combines the expertise of the operations group and the system behaviour to arrive at the right decisions compatible with the system.

— Performance models are deterministic in nature as the variables and their inter-relationships are evaluated by observed values and performance rather than theoretical or design mode models.

— Performance models are system specific and cannot be generalized to all similar situations. For developing a generalized model all variables constituting the total macro system will have to be considered.

1.2 (b) Decision Model Methodology

A decision flow diagram for an Energy Efficiency Monitoring System using Performance Models is given in Fig 1. This involves a series of steps as given in the figure.

Step 1
Involves systematic collection of data of the system, sub-systems, elements and exogenous endogenous controllable, non-controllable parameters, energy input, output etc.

Step 2
Energy efficiency of each system, sub-system and/or elements is evaluated from the collected data. Simultaneously observed losses controllable and uncontrollable are identified .Parameters controlling these losses are also ascertained.

Step 3
Using these observed parameters , performance models are developed for each sub-system and total system. Then these Models are tested by incorporating the parameters. Models capable of predicting the behavioural pattern of the system with reasonable accuracy are selected and validated. These activities are given by a decision flow diagram in Fig1 which is self explanatory. This concept could be applied to any system at micro / macro level without exception. The analysis covered in this paper refers to a power generation industry at micro level and covers *steam - power - transmission - user - conversion cycle.*

1.3 System

For efficiency performance monitoring by models, a power industry is considered which comprises of

1. High Pressure Steam Boiler (to generate steam)
2. A steam turbine (to convert thermal energy into mechanical energy)
3. An alternator (to convert mechanical energy into electrical energy)
4. An electrical sub-station (to increase the voltage to HTV and transmit the same to user's end)
5. User Equipment (to convert back the electrical energy into mechanical energy to produce finished products.)

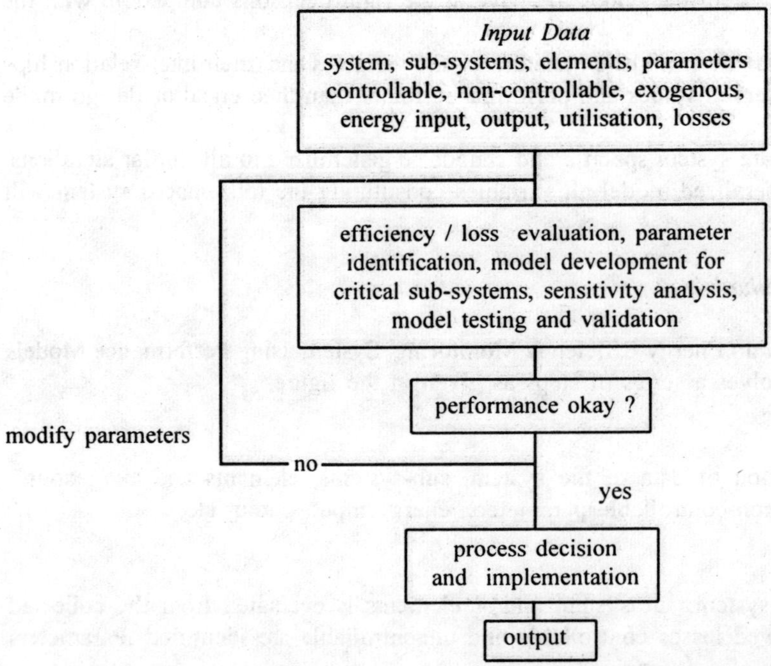

FIG 1 : *Decision Flow Diagram for Energy Efficiency Monitoring*

A number of intermediate voltage-booster stations may also be there in real life situation which is not considered in the productivity analysis.

Starting from the power boiler to the user's end, it may be noted that efficiency plays a major role at each stage. In the case of boiler for example, the efficiency of boiler is a function of load factor, tube condition, burner efficiency, fuel type, quality, ageing etc. A list of parameters that affect the efficiency at each stage is given in Fig 2.

From Fig 2 it is obvious that the power demand at consumer's end and the energy efficiency of sub system at each stage has direct impact on energy inputs to the system. For sustaining the energy efficiency of the overall system, it is imperative to monitor efficiency of each sub system. This involves generation of a suitable data base for each system, recording the data and analyzing the data. *This exercise itself could reveal a number of parameters responsible for poor energy efficiency.* An innovative approach of applying theory to practice is imperative to tackle the problems using the data base logically .

If E1, E2, E3, E4 & E5 are the efficiencies at each stage respectively the overall efficiency of the system could be represented by

$$E = E1 * E2 * E3 * E4 * E5$$

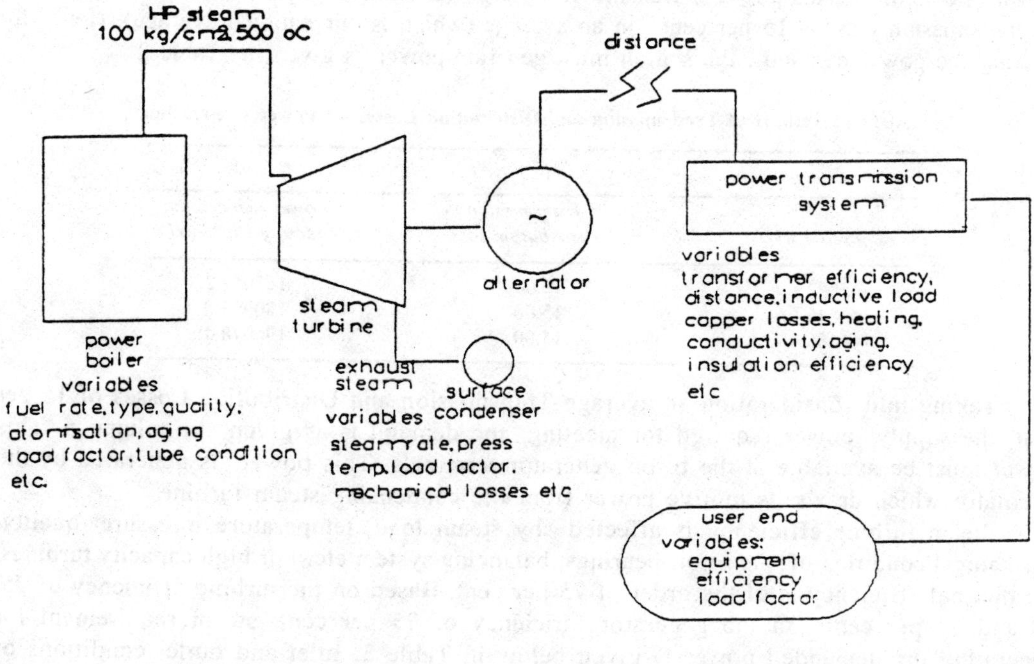

FIG. 2 : *Energy Performance Monitoring Cycle for Power Generation Plant*

where E is the overall efficiency of the system under consideration. Improvement of stage efficiencies is very cost effective and could be generally achieved by certain operational changes and / or with minor revamps / retrofits of the system.

1.4 Stagewise Productivity Analysis

An industrial sector having an average electrical demand of 100 mw/h is selected for the study. These industries have a number of power intensive equipments like pumps, compressors, fans, blowers, electric arc furnaces, etc. The impact of energy efficiency at consumer end affects the required electrical input to meet the demand is given in Table 1.

TABLE 1 : **Impact of Equipment Efficiency on Power Consumption.**

(1) Power required KwH	(2) Efficiency of elec eqpt %	(3) Power input required in KwH
100000	70.00	142857.1
100000	65.00	153846.1
100000	60.00	166666.6

For meeting the total power requirement of the industry, the power supplied must be as shown in column.3. This power is transmitted by a power station located at a far off place and the transmission loss is 15 per cent on an average (which is our national average). Hence for meeting the power demand , the station must generate power as given in Table 2.

TABLE 2. **Impact of Transmission and Distribution Losses on Power Generation.**

(4) Power required KwH	(5) % Transmission & Distribution loss	(6) Power Generation required in KwH
142857.1	15.00	168067.2
153846.1	15.00	180995.5
166666.6	15.00	196078.4

Taking into consideration an average Transmission and Distribution Losses of 15 per cent, the supply power required for meeting the demand is as given in column 6. This power must be available at the turbo generator terminals. This power is generated by the alternator which draws its motive power from the condensing steam turbine

Steam turbine efficiency is affected by steam load, temperature, pressure, quality, mechanical condition of the rotor, bearings, balancing system etc. For high capacity turbines, the thermal efficiency is of the order of 75 per cent. Based on the turbine efficiency of 75, 70 and 60 per cent and the generator efficiency of 95 per cent, steam requirement for generating the demanded power is given below in Table 3. Inlet and outlet conditions of steam are as given below.

Steam Inlet and Outlet Conditions

Steam @	Pressure Kg/cm²	Temperature °C	Enthalpy kcal/kgm
(a) inlet	151.0	540.0	817.7
(b) outlet	0.08	41.2	485.1
(c) dryness fraction	0.77		
(d) entropy	1..5525	0.1402	1.9664
(e) delta enthalpy used	332.6		
(f) corrected enthalpy			333.0

Based on the steam enthalpy of 333 kcal / kgm, steam requirement for generating the power is given in Table 3, Column 6.

TABLE 3 : **Steam Input Data to Turbo Generator**

(4) Power generation (KwH)	*(5)* Combined efficiency of turbo generator %	*(6)* Steam input for power generation t/h
168067.2	71.25	557.98
180995.5	66.50	643.82
196078.4	61.75	751.12

Taking into consideration 10 per cent of steam for internal consumption (which is the normal value), the fuel requirement (as liquid fuel having *gross calorific value of 10,000 kcal / kgm*) for generating the required power has been estimated as shown in Table 4.

TABLE 4 : **Steam Load Data for Power Generation**

(7)	*(8)*	*(9)*	*(10)*	*(11)*
Steam input to turbine (t/hr)	*Internal consumption (t/hr)*	*Total Steam generation (t/hr)*	*Boiler efficiency %*	*Fuel used in t/hr*
557.98	55.80	613.78	91.0	53.14
643.82	64.38	708.20	91.0	61.32
751.12	75.11	826.23	91.0	71.54

Based on the power equivalent of the fuel used , the overall efficiency of the system has been arrived at as given in Table 5 and column 15 gives the overall efficiency of the system.

TABLE 5 : **Overall Efficiency of the system**

(12)	*(13)*	*(14)*	*(15)*
Fuel Consumed in t/hr	*Power equivalent in KwH*	*Power available KwH*	*Overall efficiency %*
53.14	618007.6	100000	16.1810
61.32	713085.7	100000	14.0235
71.54	831933.3	100000	12.0202

1.5 Observations

1. The maximum system efficiency available is very low which indicates that about 90% is lost in the process of conversion, transmission and distribution.
2. Efficiency at each stage of conversion / transmission / distribution has a direct impact on the overall efficiency.

1.6 Stage Efficiency Impact Model

Based on the above methodology, a stage efficiency impact model has been developed by a combination of efficiencies which could be manipulated by improved equipments and/or technologies /methods. Basic data used in the modelling program is given in Table 6.

TABLE 6 : **Input Data for stage efficiency impact model for a SGP**

User equipment efficiency	*Transmission & distribution efficiency*	*Turbo-generator efficiency*	*Internal consumption in %*	*Steam generator efficiency*	*Overal! efficiency*
70.00	85.00	71.25	10.00	91.00	16.18103
65.00	85.00	66.50	10.00	91.00	14.02355
60.00	85.00	61.75	10.00	91.00	12.02019
70.00	90.00	71.25	10.00	91.00	17.13285
65.00	90.00	66.50	10.00	91.00	14.84847
60.00	90.00	61.75	10.00	91.00	12.72726
70.00	85.00	71.25	10.00	92.00	16.18103
65.00	85.00	66.50	10.00	92.00	14.02355
60.00	85.00	61.75	10.00	92.00	12.02019
70.00	90.00	71.25	10.00	92.00	17.13285
65.00	90.00	66.50	10.00	92.00	14.84847
60.00	90.00	61.75	10.00	92.00	12.72726
70.00	85.00	71.25	10.00	93.00	16.18103
65.00	85.00	66.50	10.00	93.00	14.02355
60.00	85.00	61.75	10.00	93.00	12.02019
70.00	90.00	71.25	10.00	93.00	17.13285
65.00	90.00	66.50	10.00	93.00	14.84847
60.00	90.00	61.75	10.00	93.00	12.72726

Variables used in the model are ......

Variable 1 is eff4%
Variable 2 is eff3%
Variable 3 is eff2%
Variable 4 is intl%
Variable 5 is eff1%
Variable 6 is ovreff%

Overall efficiency of the system could be predicted by the following model.

$$\text{overall eff \%} = 2.160613\text{E-}05 * (\text{eff4})^{.69742} * (\text{eff3})^{1.01874} * (\text{eff2})^{1.34877} * (\text{intl})^{-.31198} * (\text{eff1})^{0.22199}$$

Standard Error of the model = 0.11989

1.7 Validity of the model

Validity of the model for the system has been tested by comparing the model output value with that of observed overall efficiency and carrying out deviation analysis as given below.

1.8 Stage Efficiency Impact Model for a Conventional Power Generation Unit

Observed overall efficiency %	*Simulated Eff % by Sci. Model*	*Error term*
16.18103	16.18110	– 0.00007
14.02355	14.01830	0.00525
12.02019	12.02789	– 0.00770
17.13285	17.11518	0.01767
14.84847	14.82753	0.02094
12.72726	12.72222	0.00504
16.18103	16.18848	– 0.00745
14.02355	14.02469	– 0.00114
12.02019	12.03338	– 0.01319
17.13285	17.12298	0.00987
14.84847	14.83429	0.01418
12.72726	12.72802	– 0.00076
16.18103	16.19577	– 0.01474
14.02355	14.03101	– 0.00746
12.02019	12.03880	– 0.01861
17.13285	17.13069	0.00216
14.84847	14.84097	0.00750
12.72726	12.72802	– 0.00076

Standard Error of the estimate = 0.11989

Since the standard error of estimate is very small, the model is accepted as a valid model and is used for performance evaluation.

1.9 Maximum Achievable Overall efficiency

Based on this model , best overall efficiency achievable for the system under consideration is given below under the following conditions.

1. Boiler efficiency	93.5%
2. Turbo generator efficiency	75.0%
3. Transmission & distribution efficiency	92.0%
4. Equipment efficiency of user	75.0%

Maximum overall efficiency for long distance transmission	19.83395 % (by model)

1.10 Analysis of Alternates

One of the alternates for improving the overall efficiency of the system is to set up power plants of the required capacity in industrial belts (i.e. decentralisation) which could *reduce the transmission losses to around 4 per cent as these losses are proportional to the distance, conductor resistance etc.* In addition to this the power factor of the system could also be maintained high. This improves the efficiency of the electrical systems also by about 2 per cent. The above model has been used for identifying the impact of relocation of power generation unit as given below.

Boiler efficiency	93.5%
Turbo generator efficiency	75.0%
Transmission & distribution efficiency	96.0%
Equipment efficiency of user	78.0%
Estimated Overall efficiency of the system	21.28719 % (by model)

1.11 Where All the Energy has gone?

It must be noted that even with the best efficiency at each stage, the achievable overall efficiency is as low as 19.85 per cent. From the data analysis, it may be noted that substantial energy is lost in the cooling water which is used to condense the input steam to turbine. Percentage of Energy lost is of the order of 70 per cent and from thermo dynamic angle, the system chosen cannot give better efficiency than this. Hence for achieving higher system efficiency, a better technology/ system retrofit/revamp is imperative for the same inputs .

1.12 Other Alternates

A critical micro analysis of the sub system (industry) will reveal that most of the electrical power is used basically in mechanical drives such as pumps, compressors, blowers etc and only a very low percentage for lighting purposes.

This gives rise to the following options on energy utilisation.

(a) Combined heat power cycle (extraction steam to process)
(b) Maximise Conversion of electrical drives to steam drives
(c) Conversion of existing system to gas-turbine mode with co-generation using the existing facility
(d) Incorporation of Organic Rankine Cycle to the existing power generation system.

Option (c) could be used for enhancing the efficiency of the overall system as given in Fig 3.

In this system, coal may be used for auxiliary firing and the existing steam turbine could be used for cogeneration. Combined efficiency of such a system has been found to be much higher.

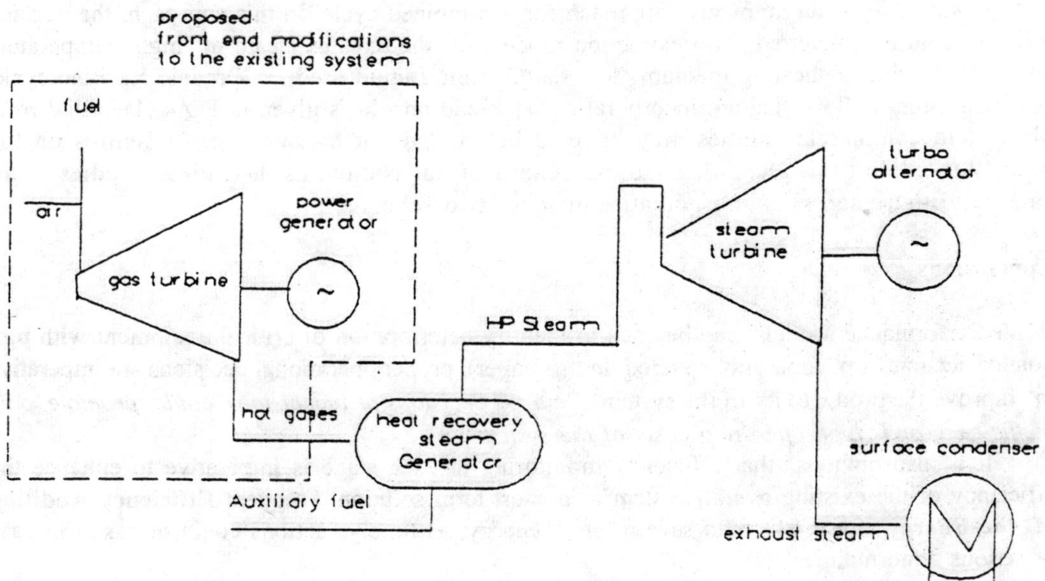

FIG. 3 : *Retrofit of Existing System or Achieving High Overall System Efficiency.*

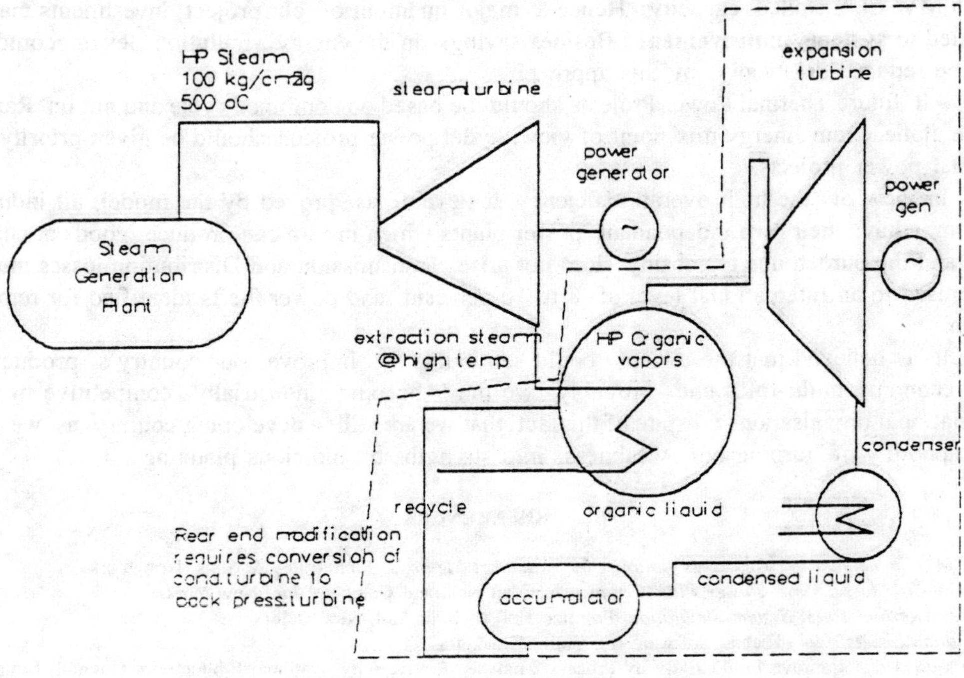

FIG. 4 : *Combining Organic Rankine Cycle to the existing system for achieving Higher Overall Efficiency*

Option (d) is an innovative approach for a combined cycle. In this approach, the existing turbine could be converted into extraction mode and the exhaust steam at high temperature could be used as a heating medium for the Organic Liquid used in Organic Rankine cycle. A typical process flow diagram incorporating back-end retrofit is given in Fig 4. Detailed pilot plant and commercial studies may have to be carried out by our R & D centres on top priority to exploit this technology to the benefit of our Nation, as theoritical studies indicate very high energy saving potential in these two schemes .

Conclusions

While Performance Models may be used to identify deterioration of critical equipment with pin-pointed accuracy (examples not covered in this paper), proper operational decisions are imperative to improve the productivity of the system. *Delayed decisions or indecisions could promote only inefficiency and high operating costs of the system.*

It is also obvious that efficiency monitoring at each stage is imperative to enhance the efficiency of the existing overall system as a short term solution. Frequent **Efficiency Auditing of the Energy Systems** could save a lot of energy, as timely actions could be taken in case of serious abnormalities .

From energy efficiency point of view , system retrofits / revamps are essential to bring down the colossal energy loss which constitutes a major chunk of our national expenditure. It is economical to improve the system efficiency and meet the energy demand than going in for new projects and increasing inefficiency further . Typically, one MW saved is equivalent to 1.5 MW of installed capacity. Hence a major quantum of our project investments may be diverted to **system improvement** . Besides savings in the energy , pollution levels could also also be reduced drastically by this approach

All future Thermal Power Projects should be based on combined cycle and not on Rankine Cycle alone. From Energy mix point of view, hydel power projects should be given priority over thermal power projects.

In view of the high overall efficiency achievable as proved by the model, all industrial belts may have their own independant power plants which in turn can produce goods at cheaper rates and the burden due to subsidy does not arise. Transmission and Distribution losses must be minimised to an International level of 8 to 10 per cent and power thefts identified for remedial action.

It is believed that these steps could undoubtedly improve our country's productivity and economy multi-folds and probably we may become industrially competitive to other international organisations in spite of the fact that we are still a developing country as we have the capability of turning our weakness into strength by judicious planning .

REFERENCES

David M. Himmelblau, 1970, *Process Analysis by Statistical Methods,* John Wiley & Sons, New York.

Frederick. T. Morse, 1965, *Power Plant Engineering,*Van Nostrand Company Inc, New York.

Geoffrey Gordon, 1993, *System Simulation,* Prentice Hall of India Ltd, New Delhi

*Information Sheets,*1994, Techno Software International, Madras

*Rajan. G.G.,*1989, Improve Productivity by Process Analysis, Productivity, National Productivity Council, India.

Smith. J. M & Van Ness. H.C., 1965, *Introduction to Chemical Engg Thermo Dynamics,* McGraw-Hill Book Company, New York.

ENERGY USE EFFICIENCY IN INDUSTRIAL, DOMESTIC, AGRICULTURE AND TRANSPORT SECTOR

49

V. B. MAHENDRA
ENCON Thermal Engineers (P) Ltd.
New Delhi (INDIA)

ABSTRACT

Our experiences in energy efficiency and saving in the secondary steel sector have been illustrated as case studies. Recuperative Ladle preheating energy consumption has been reduced several folds. Similarly, with open heater wall, heat shield with gap, chimney with ceramic fibre and billet reheating furnace, energy saving has been achieved.

INTRODUCTION

The industrial growth in India—qualitative as well as quantitative—had been phenomenal from the time of liberalisation with the growth, the energy sector had to gear up with :

1. More production
2. Bold energy saving methods

We had tried to take some bold steps in innovative concept engineering which had proven as substantial energy savers in the respective fields.

Highlights of resultant factors can be enumerated as under (and elaborated in different case studies):

1. 70–95% savings in electric power
2. 10–50% savings in oil consumption
3. Increased life of the consumable materials like crucible, heaters, skids and trays etc.

To share our experience, we are illustrating few case studies in the later part of the paper.

Apart from the above, there are certain fields, where still a lot has to be explored in India for example :

(a) Use of Oxy-fuel burners in Steel melting through Arc Furnace process.
(b) Use of Monolithic lining for repairing and other applications.
(c) Use of Low-Excess Air film burners developed by IIP and PCRA.
(d) Use of Recuperators.
(e) Use of Low mass lining.
(f) Intelligent use of alternate energy.

So on and so forth.

To start with, we are pleased to share our experience with the energy savings achieved in the secondary steel manufacturing industry.

STEEL PLANT

1. ENERGY CONSERVATION SYSTEMS AND THEIR PRACTICAL APPLICATIONS IN THE SECONDARY STEEL MANUFACTURING INDUSTRY

A. LADLES

Recuperative-Ladle Preheating Station

This application is universal in all mini steel plants. The average fuel consumption for heating a Ladle had been 10-12 Ltrs/Ton or more. Today with our efforts and stepwise modifications, we have come down to the levels of 1.5-2 Ltrs/Ton, of the ladle capacity heated. Our race to higher efficiency levels route through the following efficiency (Fig. 1) which I would like to elaborate stagewise.

As we observe from Fig. 2, the conventional ladle preheating station consists of a wall lined with refractory bricks or castables mounted on a trolley or simply a plain wall provided with a blower, burner, oil pipeline and a pumping unit to provide oil (with a drive mechanism in case of trolley unit). The ladle was placed maintaining the same centre line as that of the system or Ladle preheating station and the system was driven closer to the Ladle as per Fig. 2.

Due to conventional method, the system had to be provided with a gap for escaping of combustibles and to provide room for dross formation on the ladle rim to avoid any damage to the heating system.

Open Heater Wall with 0.5 M Distance

We observe from the efficiency chart (Fig. 1) that with a difference of 0.5 mtr. distances, the efficiency of the total system is just 10% (According to my observation, this system is still prevailing in 60% plants of the country) as per Fig. 2.

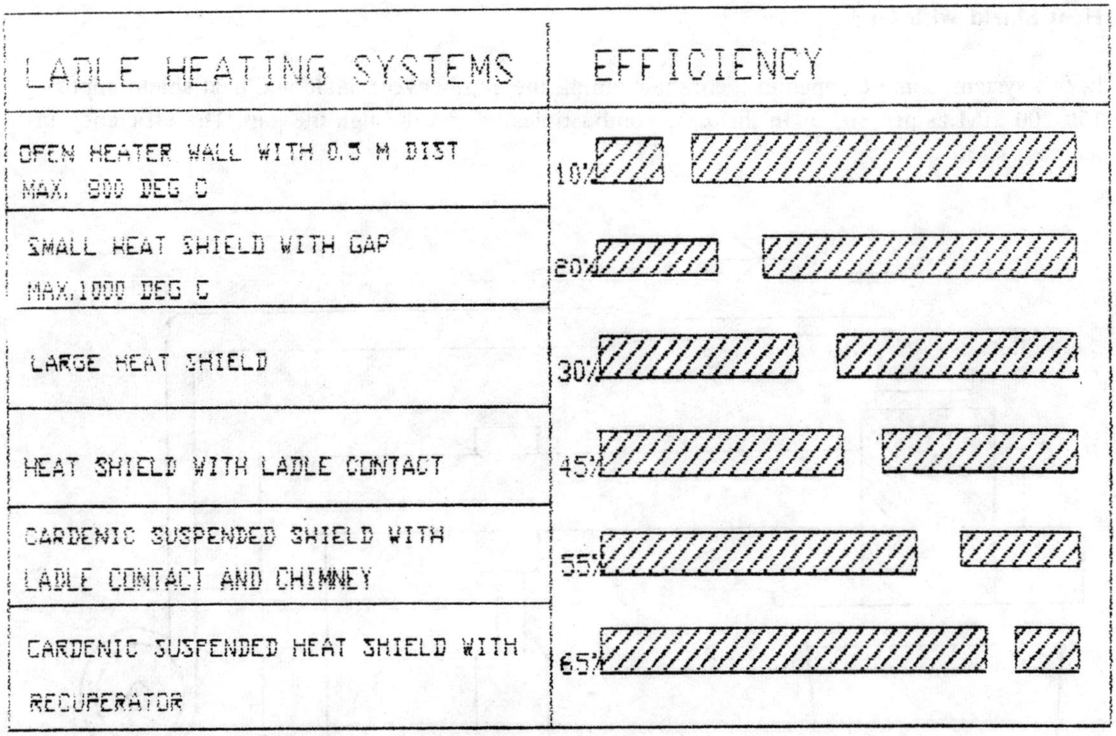

FIG. 1 : *Efficiency of Various Preheating Concepts Heat Retention of A 200 TON Ladle at 1100 Deg C.*

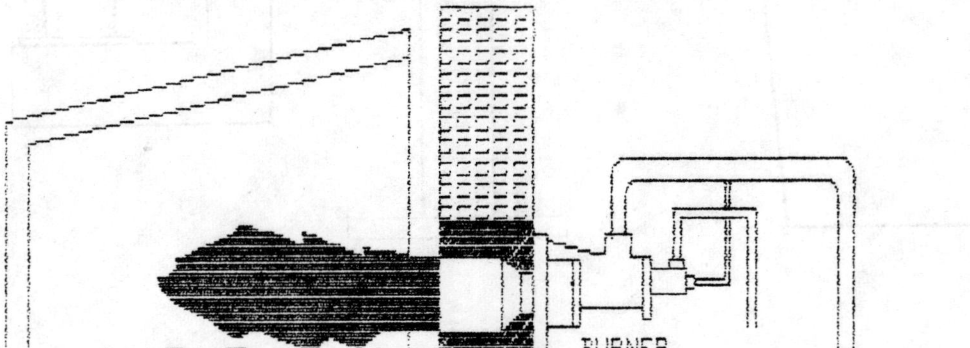

FIG. 2 : *Horizontal Ladle Heating Station*

Heat Shield with Gap

In this system, some companies were maintaining the gap between ladle and heat shield approx. 150-200 MM as per Fig. 3. In this case combustibles passed through the gap. The efficiency in the case is 30%.

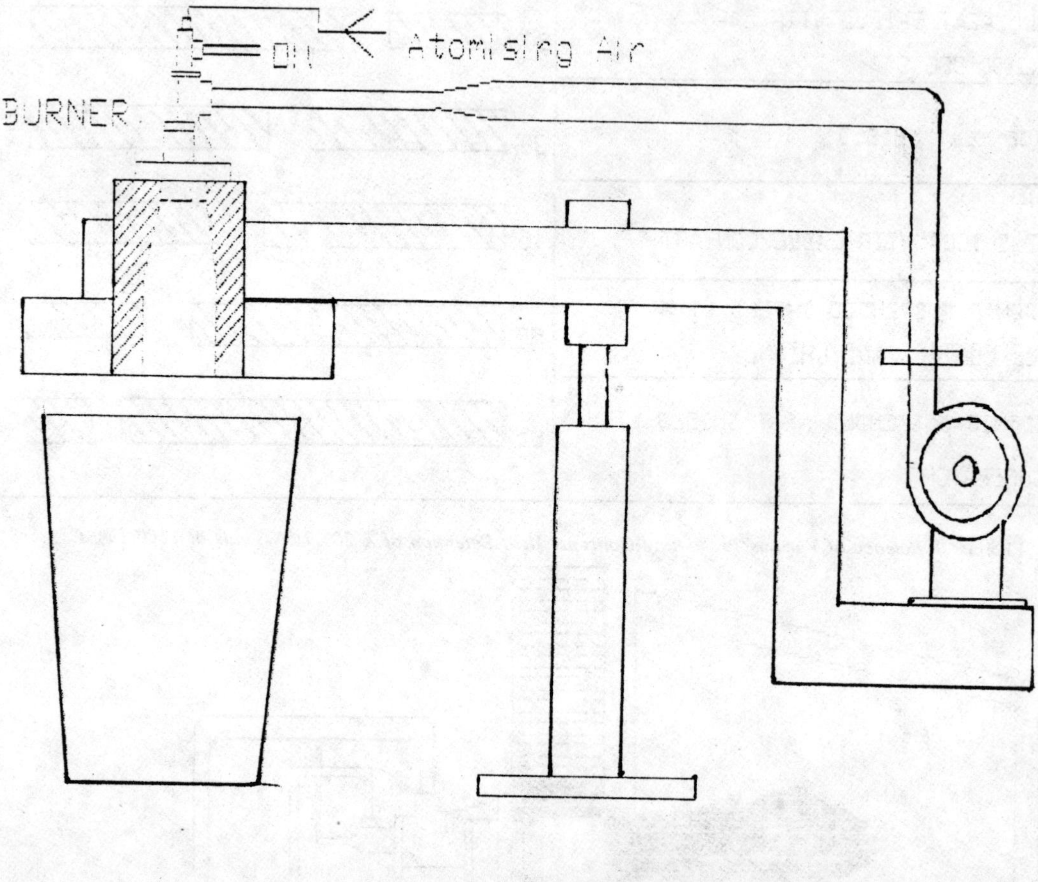

FIG. 3 : *Vertical Ladle Preheating Station*

Heat Shield with Contact and Chimney with Ceramic Fibre

In this system, there will be no gap between ladle and heat shield. The combustibles will pass through the chimney, provided in the heat shield. This system will be lifting/lowering type upto 85-90 deg through hydraulically or mechanically . The efficiency in this case is 55% as per Fig. 4.

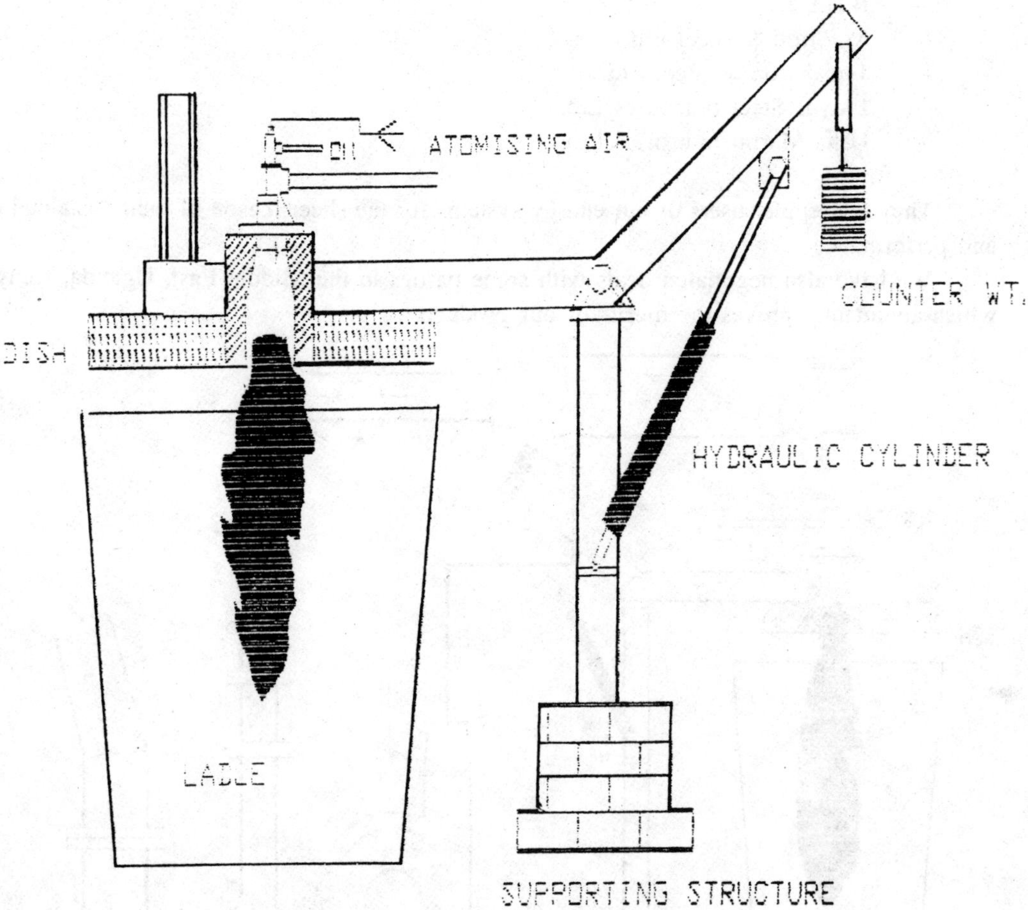

FIG. 4 : *Vertical Ladle Preheating Station*

Ceramic Fibre Lined Heat Shield with Recuperator

We have proved our worth in WASTE HEAT RECOVERY system as well and our Recuperators are further saving 12-15% in fuel costs in a number of prestigious companies.

Our ladle preheating systems are another ace up our sleeve. A sophisticated design using the most revolutionary ceramic fibre lining and the durability of the system coupled with our fuel efficient combustion equipment has made us one of the most reputed manufacturers of ladle preheating systems as per Fig. 5.

Some of our prestigious clients are :

— Mukand Ltd.
— Special Steel Ltd.

— B.H.E.L.
— M.P.Iron & Steel Ltd.
— Jindal Iron & Steel Ltd.
— Lloyds Steel Industries Ltd.
— Usha Martin Industries Ltd.

They are regular users of our energy systems for the sheer reason of their sustained quality and performance.

We have also negotiated deals with some parties in the Middle East, Uganda, Kenya etc., which abundantly proves the merits of our products.

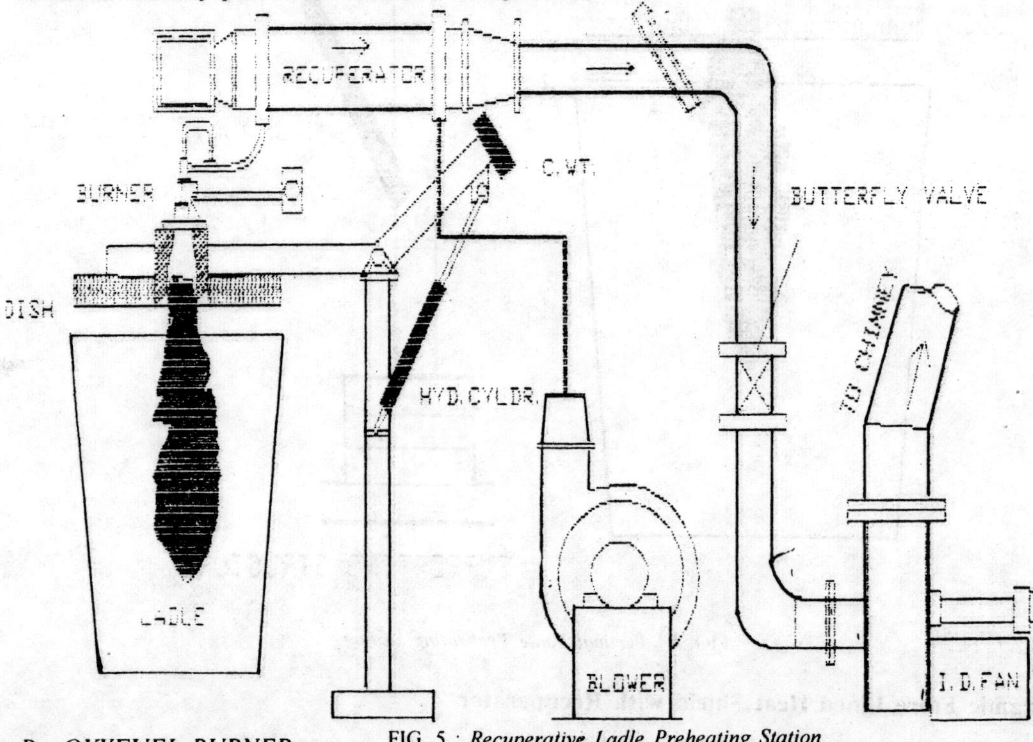

FIG. 5 : *Recuperative Ladle Preheating Station*

B. OXYFUEL BURNER

In the steel plants, the oxyfuel burners are used to preheat the scrap to a temperature of 900 deg C. This preheating saves the electric consumption of induction and arc furnaces by 30%.

The 30% saving of electricity through the preheating of the scrap effects the steel prices to enormous levels.

This technology of oxyfuel burners is finding new horizons in steel plants.

C. BILLET REHEATING FURNACE

A Rolling mill having 4 Tons/Hr billet reheating furnace was studied, and with the co-operation of the owners the step-wise modifications were carried out resulting in 50 percent fuel economy in totality as per Fig. 6 and 6A.

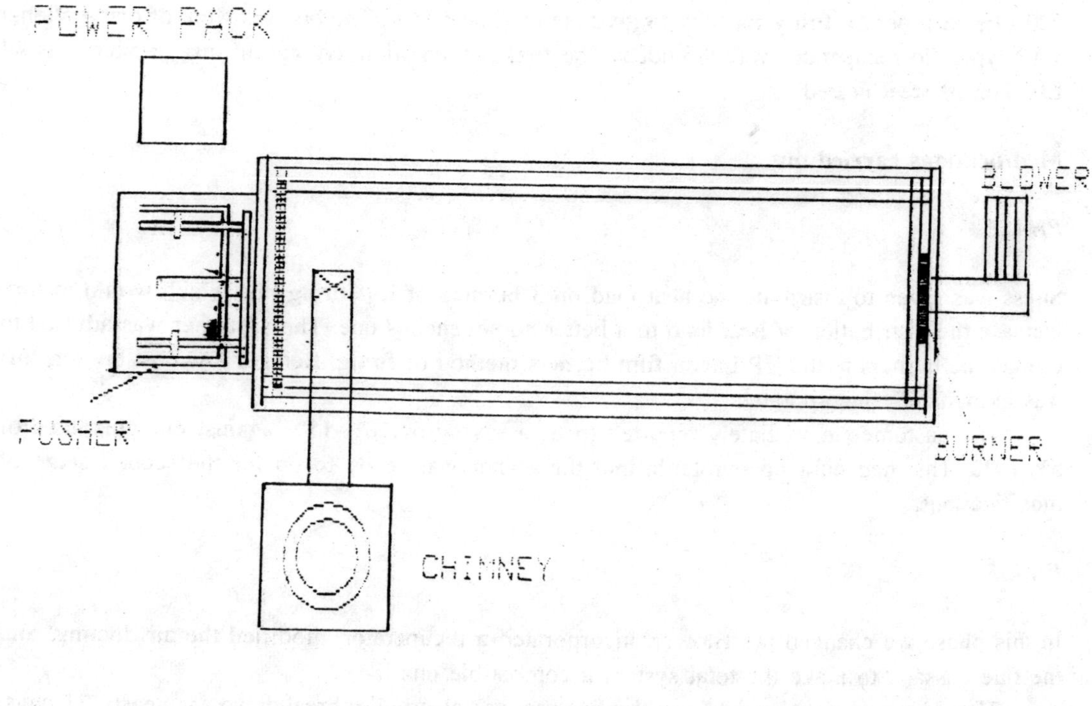

FIG. 6 & 6A : *Billet Reheating Furnace*

The furnace was lined with refractory bricks in a conventional method, two nos. burners of 120 Ltrs each per hr. firing capacity to give the total heat load. The burners were of conventional LAP type. No recuperator was provided. The fuel consumption average in this furnace was 84 Ltrs/Ton of steel heated.

Modifications carried out

PHASE - I

Stress was given to distribute the heat load on 3 burners of less firing rate which would in turn increase the distribution of heat load to a better/homogeneous one. The customer was advised to change the burners to the IIP Encon film burners method of firing. Being a low cost burner, this was immediately incorporated.

The customer immediately reported to us a saving of 10%-15% against our guarantee of 5%-10%. This had built up our faith, and the owner was ready to go for the second stage of modifications.

PHASE - II

In this phase we changed the Blower, incorporated a recuperator, modified the air ducting, and the flue passage to make the total system a compatible one.

This involved a major cost and the furnace was also under breakdown for nearly 21 days. A bypass system was incorporated to facilitate the cleaning of the recuperator as and when required.

The owner marked a saving of nearly 15-20% by making all the changes. The fuel consumption came to nearly 55 Ltrs/Ton from 84 Ltrs/Ton.

By now the customer was totally sold to our ideas and was ready to incorporate anything that could have resulted in the saving of fuels.

Having run the equipment for over 2 years we again suggested a major modification in the furnace lining. The total preheating zone roof was changed from conventional refractory lining to ceramic fibre lining.

Through the modifications carried out here are under process but as per our observation. We anticipate the saving of not less than 15% by incorporating this latest design of ceramic fibre.

PHASE - III

Instrumentation and controls

The next phase for achieving further fuel economy will be incorporating automatic controls for the furnace and also monitoring the combustion efficiency.

(*NOTE* : Though these systems are already incorporated in bigger furnaces in India, it is much more desirable to incorporate the same in smaller furnaces also.)

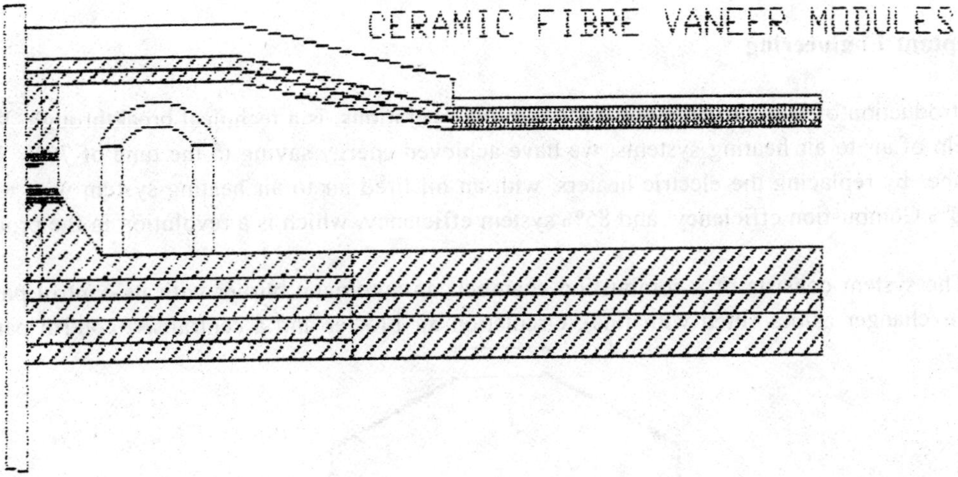

FIG. : *Billet Reheating Furnace*

OUR AIM

One of the best run plants is averaging 28 Ltrs/Ton of the steel heated on a production of 40 T/Hr. We would definitely like to achieve a figure of 33-36 Ltrs/ Ton in the range of 40-45 T/Hr.

D. MORGANITE LINING

Remote gunning process is a technical breakthrough in the field of relining of a blast furnace. The remote gunning process of relining is much easier than the other methods of relining.

In the remote gunning process, the neck of the blast furnace is punctured and a rack carrying high pressure jet nozzle and camera is lowered into the furnace.

The camera is mounted on a gantry looking vertically upward on its centre line.

The water coming out of the high pressure nozzle removes the loose refractory and dust from the surface. These dust particles are carried out of the furnace through the tuyeres.

Once the surface is clear, the module is lowered to the area to be relined and material is applied to the stack by rotating the lance through 360 deg. and gradually raising the module until the new lining is completed.

With the help of remote gunning, the time required for the relining is reduced to 3 days from 90 days which brings major savings.

In steel plants, the single day production is around 2000 ton, which means a sum of Rs. 1.2 cr., so if we calculate the loss of 87 days then it will be around 1,74,000 Tons.

2. PROCESS INDUSTRY

A. HOT AIR GENERATOR

Conceptant Engineering

The introduction of Hot air generator, in the drying operations, is a technical breakthrough. With the help of air to air heating systems, we have achieved energy saving to the tune of 70%. This was done by replacing the electric heaters, with an oil fired air to air heating system with more than 90% Combustion efficiency and 85% system efficiency, which is a revolution in the process industry.

The system consists of a combustion chamber, an automatic burner with its control panel, a heat exchanger (tubes fitted with fins), circulation air blower and a centralised control panel.

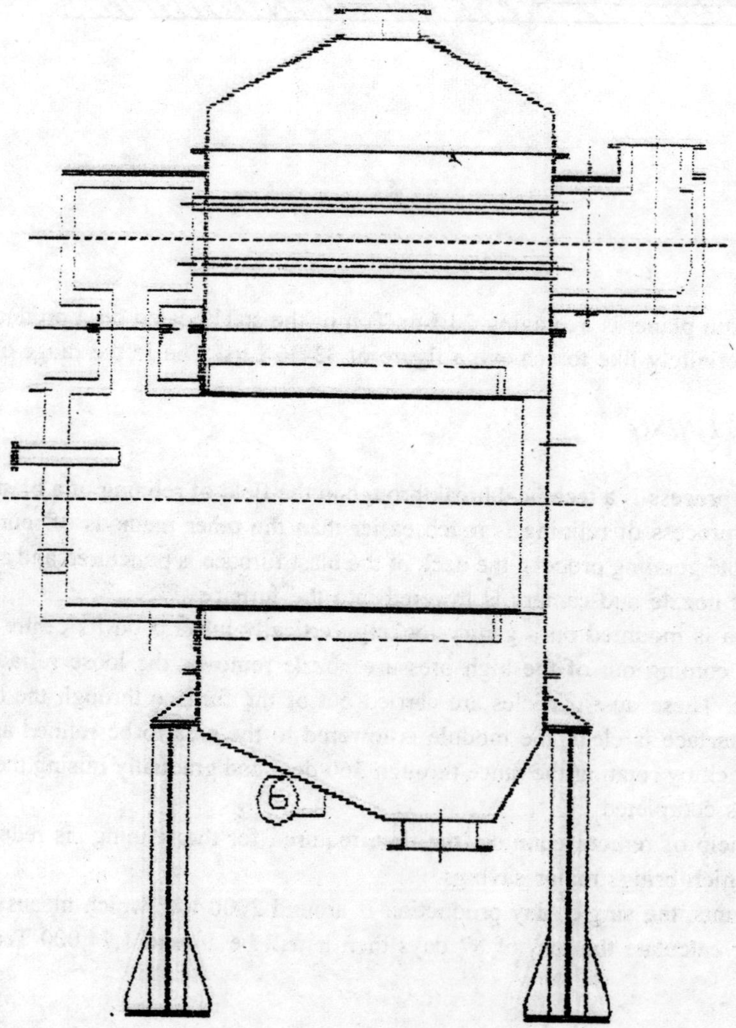

Working Principle

1. Preheating Zone

In this zone, the fresh air delivered by the circulation air blower, is heated to a certain temperature in the indirect way.

The zone is fitted with a combustion chamber, consisting of an automatic burner. The salient features of the burner are as follows :

- Sequence controller – which is the heart of the burner and controls the whole sequence of the burner.
- Auto ignition
- Flame failure sensor
- Inbuilt blower
- Pumping system with pressure regulation and in-built by-pass system.

The fins are provided on the combustion chamber to make the system more efficient and compact.

2. Heating Zone

The flue gases coming out of the combustion chamber are passed through the heat exchanger for further heating of the preheated air and to get the desired temperature.

The heat exchanger comprises of tubes fitted with fins to increase the heat transfer coefficient and for higher efficiency.

The flue gases are delivered to the atmosphere after passing through the heat exchanger.

The hot air coming out of the generator, after conditioning and checking, is fed to the desired place.

Advantages and Salient Features of the System

1. With the help of air to air heating system, the boiler, economisers and thermic fluid are deleted.
2. The system is fully automatic and it is very easy to handle as compared to boiler systems.
3. No preheating is required in the system hence the system is much more instantaneous.
4. The oil pumping system and oil leakage problems are eliminated (as in the case of thermic fluid heater).
5. The higher burner efficiency result is a pollutionless and noiseless system.
6. The power consumption will be reduced by 70% (approx.), in case the present heating is by electricity power.

Case Study

With the use of hot air generators in the drying operations, M/s. Hyderabad Industries Ltd., one of the companies of Birla Group, has earned a saving of almost 70%.

The oil consumption has come down from 570 ltrs to 540 ltrs/day while production has gone up by 1 Ton per day.

By circulating the hot air uniformly throughout the drying oven, drying operation becomes more efficient and faster. After drying the material, hot air coming out of the chamber is sucked and again heated and fed to the chamber.

WIRE INDUSTRY

CASE STUDY OF A LEAD BATH

By modifying the Lead bath furnace of M/s. Special Steels Ltd.,Bombay, we have achieved a saving of 22%, by incorporating the following :

1. BURNER

A burner which needs low excess air energy to prepare air and fuel for a desired combustion and which produces such flame and gases as to bring out high rates of heat transfer. With the help of such a burner, it is possible to affect fuel conservation.

With the help of IIP burner, 5–7% saving was observed during the case study.

2. RECUPERATOR

To reduce the stack losses of the furnace, a waste heat recovery system was installed on the flue passage in order to extract the heat carried out by the flue gases to the atmosphere.

After getting the preheated combustion air at a temp. of 200 deg C from the recuperator, air was fed to burner. A saving of around 10% was observed during the case study, as the preheated air made the combustion complete and more efficient.

3. AUTOMATIC TEMPERATURE CONTROLLING SYSTEM

After incorporating the automatic temperature controlling of the furnace, a further 5% savings were observed.

In order to maintain the constant temperature throughout the furnace, the air and oil flow rate were interlocked with the temperature of the furnace through the programmable controller and sensor.

The flow rate keeps on varying according to the variation in the temperature of the furnace, which helps in maintaining the uniformity of the temperature of the furnace,resulting in increased pot life.

ALUMINIUM INDUSTRY

We share our findings and observations with a new lining and automatic burner management system in the Aluminium melting concept. In Crucible type aluminium furnaces, we have achieved 94.5% electric and 34% oil saving.

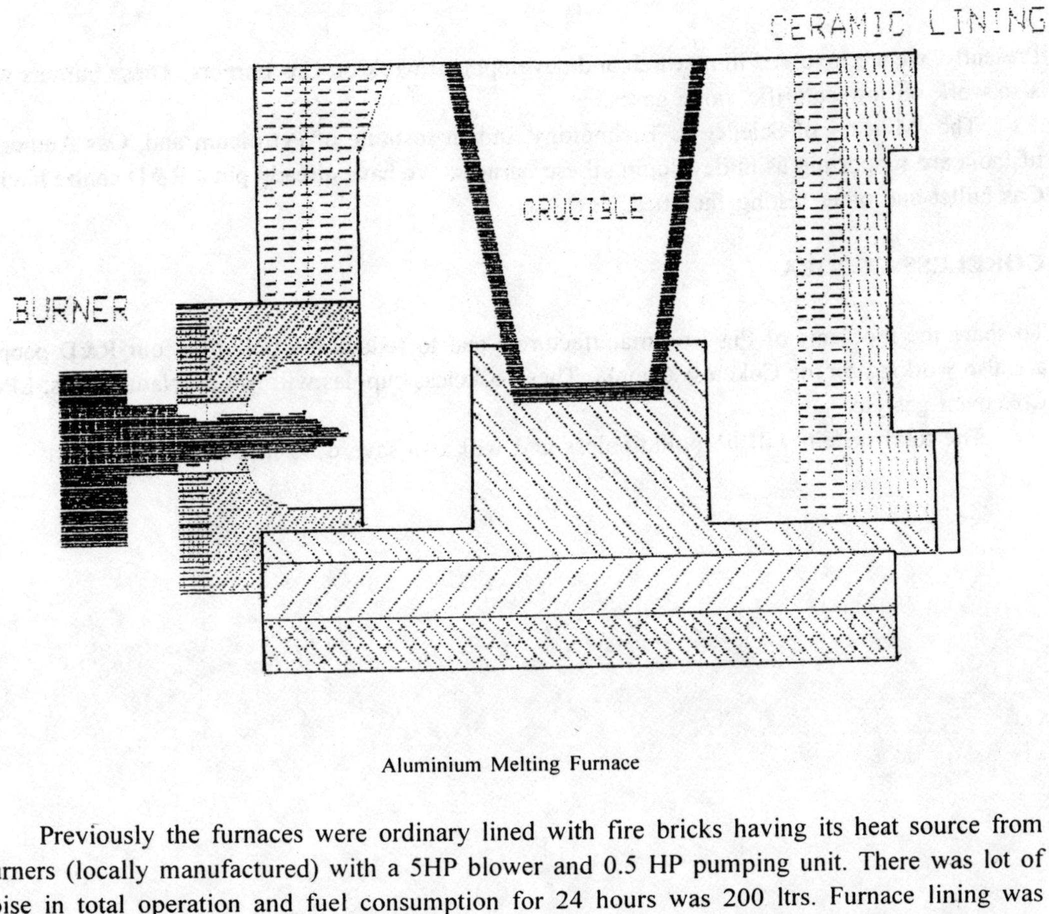

Aluminium Melting Furnace

Previously the furnaces were ordinary lined with fire bricks having its heat source from burners (locally manufactured) with a 5HP blower and 0.5 HP pumping unit. There was lot of noise in total operation and fuel consumption for 24 hours was 200 ltrs. Furnace lining was changed to high density ceramic fibre lining. The burners were changed with monoblock type auto ignition burners with built-in oil pumping system. The burners are automatically controlled on On/Off basis, regulated by thermocouple provided in the molten bath.

On achieving the said temperature, the firing shuts off and restarts on its own, when the temperature goes below the said point. The total system is automatic.

Savings Observed

1. Electric consumption came down from 5.5 HP to 0.25 HP for Blower and Heating and Pumping unit.

2. Oil consumption came down from 6000 ltrs/month to 4000 ltrs/month.
3. Increased crucible life (Almost double)

R & D PROJECT

Gas Burners

Presently, we are involved in research and development work on Gas burners. These burners will also work on low calorific value gases.

The Ministry of Science & Technology, Indian Institute of Petroleum and, Gas Authority of India are supporting us in developing these burners. We have already put a R&D centre having Gas bullet and other testing facilities.

COKELESS CUPOLA

To share the problems of Pig Iron manufacturers and to reduce the pollution, our R&D people are also working on the Cokeless Cupola. These cokeless cupolas will run on Natural Gas, LPG, Cokeoven gas etc.

The total system will be pollutionless and will also save 67% in energy costs.

ARTIFICIAL NEURAL NETWORK BASED HIGH EFFICIENCY DRIVE OF A SEPARATELY EXCITED DC MOTOR

50

E. G. JANARDANAN
N.S.S. College of Engg., Palakkad, (INDIA)

F. GAJENDRAN
Regional Engg. College Calicut (INDIA)

ABSTRACT

A method for controlling a thyristor fed separately excited DC motor ensuring high efficiency at low load factors is presented. The control system employs two artificial neural networks (ANN) : one for adjusting the field current for achieving maximum efficiency and the other for adjusting the speed controller parameters for achieving fast speed response. Both ANNs are simulated using back propagation algorithm. A PI controller is used in the field to control loop to ensure very quick field current response. The scheme is illustrated with its application to a specific motor using simulation studies.

INTRODUCTION

Separately excited DC motors are widely used in industrial fields, where high accurate speed control over wide range is necessary. In most cases the control is achieved by adjusting the armature impressed voltage with constant field excitation. This scheme suffers from the disadvantage of decreased efficiency at low load factor. The majority of installed motors operates at two-thirds or less their rated load. So with regard to energy saving, it is important to improve the efficiency of motors at light loads. The possibility of improving the efficiency by simultaneous control of armature and field volatages has been reported[4] The notable works on the control of DC drives with optimum efficiency have been reported by Petrov[1], Slezhanovskii[2],Cathy[3], A.Kusko[4], T. Egami *et. al.*[5], Nickols Margaris[6] , and S.C.Hong[7]. Most of the work adopts the loss model controller (LMC) approach.TheLMC approach requires either the direct calculation or look up table method for optimal excitation value for a given speed and armature current.The look up table method seems to be more effective. But the method requires large memory for considering the complete range of operating conditions.The use of ANN is attractive as it requires very less memory and it offers high computation ability.

The proposed scheme employs ANNs for achieving fast speed response ensuring high efficiency at any operating points characterised by torque and speed. The optimum field current is selected by an ANN by observing the load torque and steady state speed. A PI controller is used in the speed loop to ensure fast speed response. The parameters of this controller are selected by an ANN that observes the reference field current for ensuring high efficiency. A PI controller is employed in the field loop to achieve fast field current response. The scheme is illustrated with the aid of simulation studies.

PRINCIPLE OF LOSS MINIMIZATION

In this section, the loss minimization condition for high efficiency motor is investigated. The performance of a DC motor is governed by the three important variables; the speed (ω_s), the armature current (I_a), and the air gap flux (ϕ). For an operating point characterised by the load torque and speed there exists an optimal combination of armature current and field current that gives maximum efficiency. To derive the condition for maximum efficiency the following loss components are considered:

1. Armature copper loss(P_a) given by $I_a^2 R_a$, where R_a is the armature resistance including brush contact and I_a is the armature current.
2. Field copper loss (P_f), characterised as $I_f^2 R_f$, where I_f is the field current and R_f is the field resistance.
3. Armature iron loss (P_i), characterised as $(k_1\omega + k_2\omega^2) I_f^2$, where k_1 and k_2 are the coefficients of hysterisis and eddy current losses respectively, and ω is the speed.
4. Friction and windage losses(P_m), a function of ω.
5. Stray load loss (P_{st})

The total loss (P_t) is given by,

$$P_t = P_a + P_f + P_i + P_m + P_{st} \tag{1}$$

Assuming linear magnetic circuit and negligible armature reaction the controllable loss (P_c) is given by;

$$P_c = P_a + P_f + P_i \tag{2}$$

The loss is minimum for an operating point (T_L, ω) when,

$$\left.\frac{\partial P_c}{\partial I_f}\right|_{T_L,\,\omega} = 0$$

The armature current, I_a in terms of torque is given by

$$I_a = (T_L + T_{nL}) / (K_t\, I_f) \tag{3}$$

where T_L = load torque

T_{nL} = no load torque

K_t = torque constant

Performing the partial differentiation the condition for minimum loss is given by;

$$I_f = (R_a / (R_f + k_1\omega + k_2\omega^2))^{1/4} ((T_L + T_{nL}) / k_t)^{1/2} \tag{4}$$

ANN IMPLEMENTATION OF LOSS MODEL CONTROLLER

To realize the principle, the field current and armature voltage should be separately controlled. A trained neural network is proposed for implementing the loss model controller. A three layer network, with two input neurons for receiving the signals proportional to speed and load torque and one output node to suggest appropriate field current command is used. The training data are generated using equation (4)

SYSTEM DESCRIPTION AND ARCHITECTURE OF THE PROPOSED SCHEME

A separately excited DC motor is considered.The dynamic equations of the motor are given below :

$$d\omega(t) / dt = -\omega(t) / \Gamma m + k_t I_f(t) I_a(t) / J - T_L(t) / J \quad (5)$$

$$dI_a(t) / dt = - I_a(t) / \Gamma_a - k_b I_f\omega(t) / L_f + V_a(t) / L_a \quad (6)$$

$$dI_f(t) / dt = - I_f(t) / G_f + V_f(t) / L_f \quad (7)$$

where

$\omega(t)$	=	rotational speed, rad/sec.
$I_a(t)$	=	armature current, Amp.
$I_f(t)$	=	field current, Amp.
$T_L(t)$	=	load torque, N-m
$V_a(t)$	=	armature supply voltage,Volts
$V_f(t)$	=	field supply voltage,Volts
Γ_m	=	mechanical time constant
Γ_a	=	electrical time constant
K_t	=	torque constant
K_b	=	back e.m.f. constant
L_a	=	armature inductance, Henries
Γ_f	=	field circuit time constant

The DC motor is represented in the Fig.1.

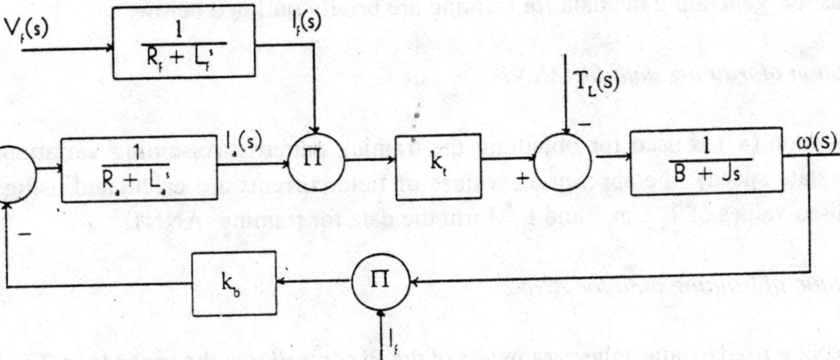

FIG. 1 : *Block diagram of the DC motor*

The field and armature are supplied from three phase bridge converters. Both field and armature supply voltages are controlled for ensuring maximum efficiency and fast speed and field current responses.The block diagram of the proposed scheme is shown in Fig. 2.

The control system employs two ANNs. ANN1 is used to provide reference field current signal for achieving maximum efficiency at any operating point. ANN2 is used to generate signals for adjusting the coefficients of the PI controller (K_{ps} and K_{is}) provided in the speed loop. The PI controller,with coefficient K_{pf} and K_{if}, in the field current circuit is used to obtain very fast field current response.

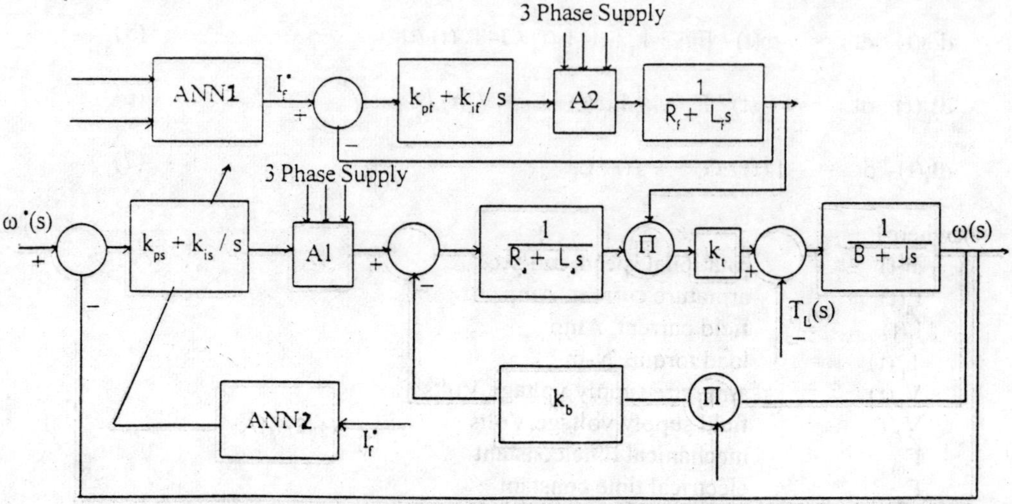

FIG. 2 : *Block Diagram of the Proposed Scheme*

SIMULATION OF ANNs

The ANN1 and ANN2 are simulated using back propagation algorithm with adaptive learning and momentum.For using these networks in the actual system they are to be trained with known data.The methods for generating the data for training are briefly outlined below.

Generation of training data for ANN1

The equation (4) is used for obtaining the training patterns. Assuming variations in T_L and ω_{ss} (steady state speed) the appropriate values of field currents are calculated using (4).The sets of normalised values of T_L , ω_{ss} and I_f^* form the data for training ANN1.

Generation of training data for ANN2

The ANN2 is used to adjust the parameters of the PI controller in the speed loop.The input for ANN2 is the reference field current. The outputs correspond to the signals for adjusting the coefficients of PI controller. For generating the training data the controllers for known field current are to be designed.The design procedure of the controller is described below.

Design of Controller[8]

For a given operating point characterised by the steady state speed and load torque the speed control loop can be linearised.The linearised speed loop is shown in Fig. 3.

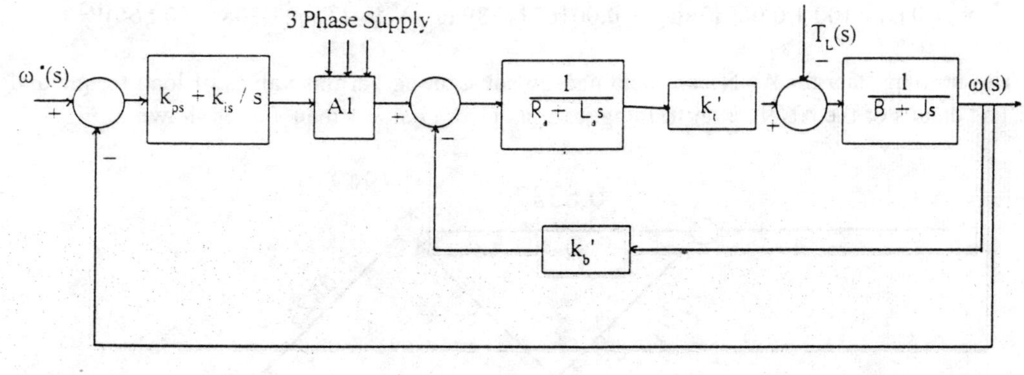

FIG. 3 : *Block diagram of the Sheed Loop*

In Fig.3. $k_t' = I_f^{\,*} k_t$, $k_b' = I_f^{\,*} k_b$, I_f^* is the field current satisfying the loss minimization condition for the specified operating point (T_L,ω_{ss}). The block diagram of Fig.3 can be reduced to that shown in Fig. 4 (the load torque is assumed as zero for design purpose) [9].

$$\omega^*(s) \rightarrow \boxed{\frac{A1\,(k_{ps} + k_{is})}{JL_a s^3 + (J R_a + BL_a)\, s^2 + (k_t' k_b' + BR_a + {}^*A1k_{ps})s + A1k_i}} \rightarrow \omega(s)$$

FIG. 4. *Reduced Block diagram for design purpose*

The design is based on the pole placement technique. Let the desired locations of the closed loop poles be at s = –a. Then the controller coefficients are obtained by using the following equations (with the assumption of negligible armature inductance):

$$k_{ps} = (2a - (BR_a + k'^2) / JR_a) \;/\; A1\, k_t\, I_f^* \qquad (8)$$

$$k_{is} = a^2 / A1\, k_t\, I_f^* \qquad (9)$$

The normalised values of I_f^* forms the inputs and corresponding normalised values of k_{ps} and k_{is} form the outputs for training the ANN2.

ILLUSTRATION

A DC motor drive with following ratings and parameters is considered:

1.5 Kw , 1500 rpm , 100 V

rated torque = 9.058 N-m , rated armature current = 18.5 A, R_a = 0.6 ohm, L_a = 0.003 H , R_f =100 ohm, L_f= 11.3 H, B= 0.009 , J=0.098, k_t= 0.566, k_b = 0.566 and A1 =A2=25.0

Rated losses with full excitation (i.e. I_f = 1.0 A) are:

Field copper loss	=	100 W
Eddy current loss	=	14 W
Hysterisis loss	=	40 W
Armature copper loss	=	205 W
Mechanical loss	=	25 W

The field current for minimum loss condition are obtained by using the following equation:

$$I_f^* = (0.6 / (100 + 0.089126\omega_{ss} + 0.0016211389\ \omega_{ss}^{2})^{1/4}\ ((T_L + 1{,}3089) / 0.566)^{1/2}$$

The training data for ANN1 are obtained by substituting various values of load torque and speed.The details of the ANN1 after training it for an error goal less than .001 is shown in Fig 5.

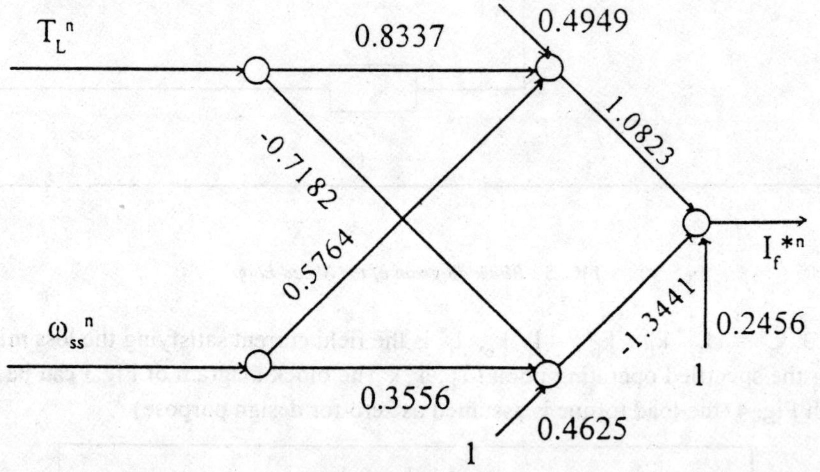

FIG 5 : *Details of ANN1*

The training data for ANN2 are generated using the technique outlined in section 5.2. The equations for obtaining K_{ps} and K_{is} for the placement of closed loop poles at s = –10 are given in (10) and (11)..The details of the trained network is shown in Fig.6

$k_{ps} = (20 - ((0.0054 + 0.566\ I_f^{*2}) / 0.0588) /14.15\ I_f^*$ (10)

$k_{is} = 100 / 14.15\ I_f^*$ (11)

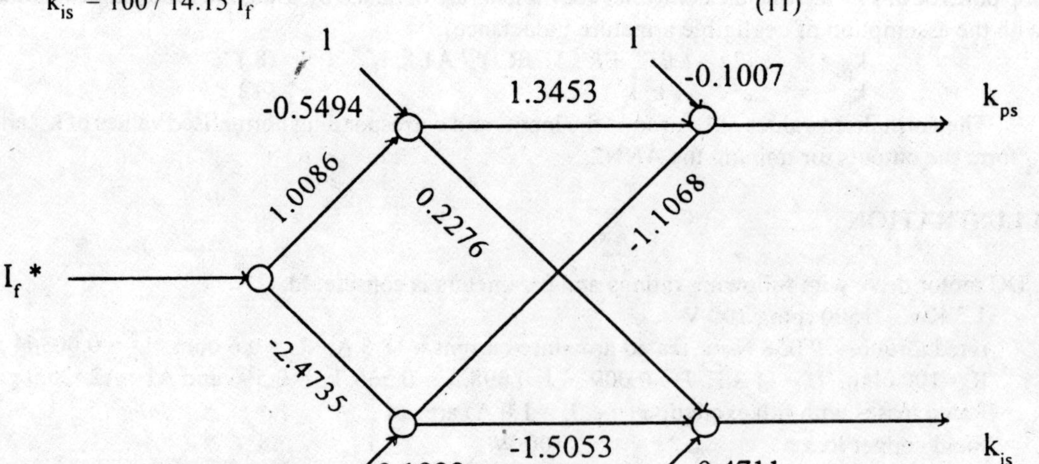

FIG. 6 : *Details of ANN2*

For achieving fast field current response PI controller with following coefficients are selected for field current loop as :

$$k_{pf} = 113.0 \qquad (12)$$
$$k_{if} = 1000.0 \qquad (13)$$

With this controller the unit step response of the field circuit is obtained as shown in Fig 7. The response indicates that the current reaches its steady state value within 0.02 seconds without any steady state error and overshoot.

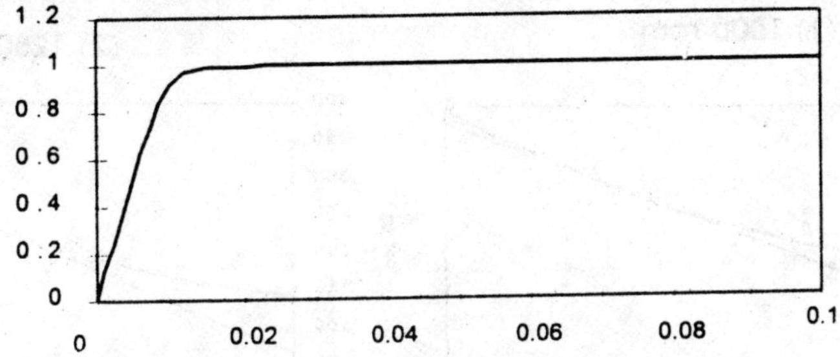

FIG.7 : *Unit Step Field Response*

The speed response, with T_L = 3.0 N-m and for a unit step input is shown in Fig 8. From this figure it can be seen that the speed response is fast, overshoot is small and steady state is reached within 0.9 seconds without any steady state error.

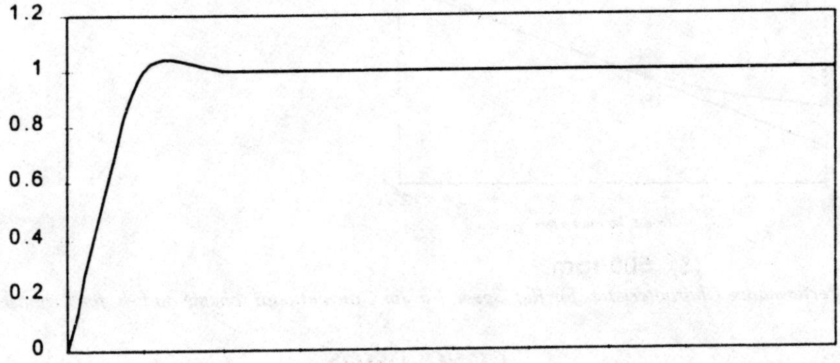

FIG. 8 : *Unit Step Speed Response with T_L=3.0 Nm*

For the comparison of losses with conventional and proposed schemes both schemes are simulated for various operating points. The results are shown in Fig. 9. The figures indicate that the loss is much reduced by employing the proposed scheme at low speeds and low load factors.

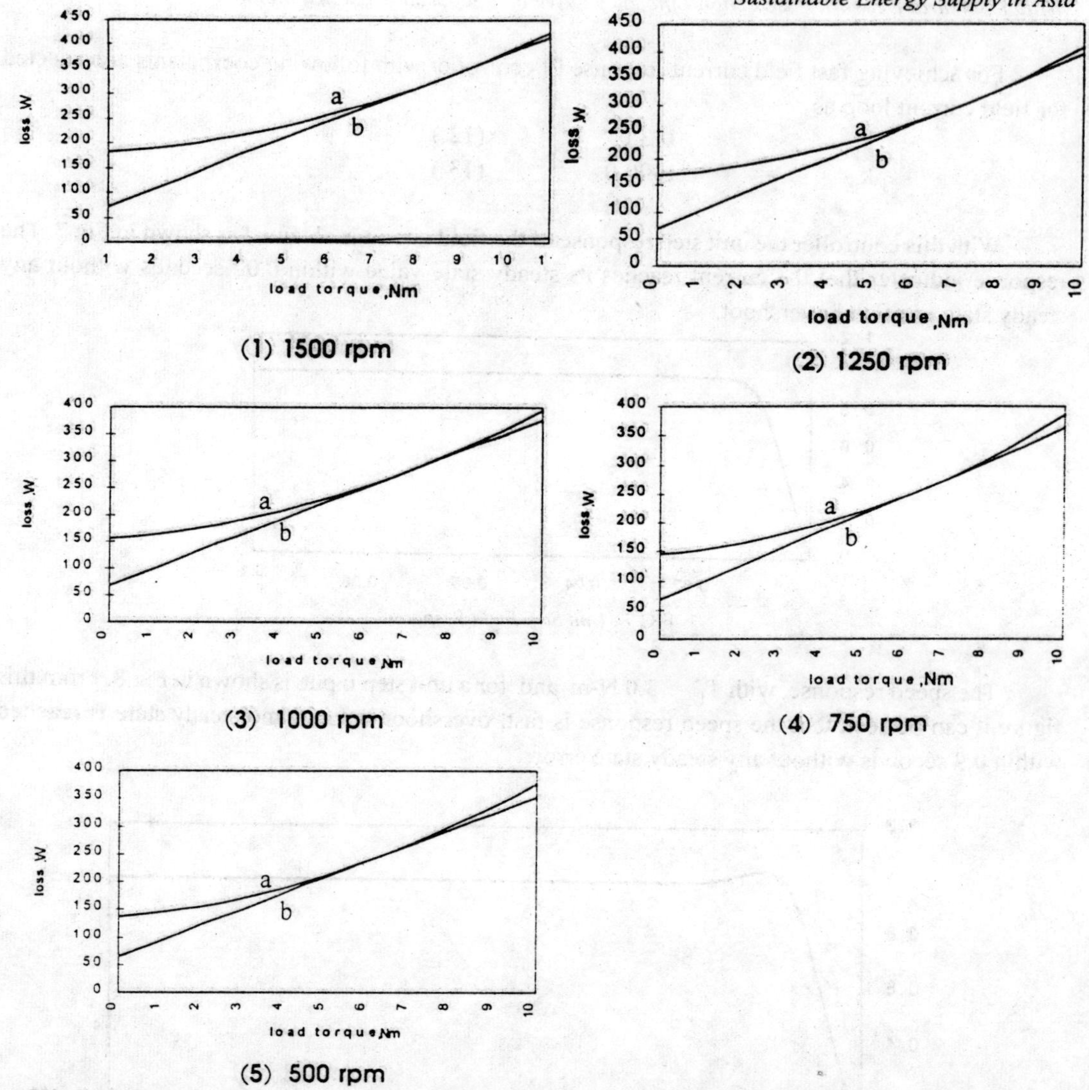

FIG. 9 : *Performance Characteristics for Ref. Speed ('a' for Conventional scheme and 'b' for Proposed scheme)*

CONCLUSION

In this chapter a control scheme for DC motor that ensures high efficiency at any operating point is presented.The control means for fast response for armature speed and field current are also presented.The simulation results show the effectiveness of the proposed scheme.

REFERENCES

1. Petrov, "Optimal laws for electric drive control," Automatic remote control,Vol 2, July 1959, pp 860-867.
2. O.V. "Slezhanovskii," Optimum variations of basic parameters and optimal structure of control systems for electric drives", Electrical Technology,USSR,Vol.2, May 1960, pp 149-171.
3. J.J. Cathy, "Reduction of DC traction motor armature copper loss through optimal control," Electro mech, 1979, pp 269-283.

4 A. Kusko and D. Galler, "Control means for minimization of losses in AC and DC motor drives," IEEE Trans. on Industry Appl., Vol. IA 19, No. 4, July/Aug. 1983, pp 561-570.

5. T. Egami, J. Wang and T. Tsuchiya, "Efficiency optimized speed control system synthesis method based on improved optimal regulator theory - application to separately excited DC motor system," IEEE Trans. on Industry Appl., Vol. IE 32, No. 4, Nov. 1985, pp 372-380.

6. N. Margaris, T. Goutas, Z. Doulgeri and A. Paschali, "Loss minimization in DC drives," IEEE Trans. on Industry Electronics., Vol. 38, No. 5, Oct. 1991, pp 328-334.

7. S.C. Hong and M. H. Park, "Microprocessor based high efficiency drive of a DC motor," IEEE Trans. on Industry Electronics., Vol. IE 34, No. 4, Nov. 1987, pp 433-440.

8. E.G. Janardanan and F. Gajendran, "A quick response high performance adaptation scheme for DC motor control using artificial neural network," Proc. of National systems conference, Dec. 1995, pp 83-87.

9. K.B. Naik, G. K. Dubey and V. K. Jain, "Design, modelling, and stability analysis of closed loop controlled chopper fed sparately excited DC motor," Journal of Institution of Engineers (INDIA), Vol. 72, PT. EL 6, Feb.1992, pp 248-259.

ENERGY USE EFFICIENCY IN INDUSTRIES, DOMESTIC, AGRICULTURAL AND TRANSPORT SECTORS

51

A. G. MATANI
Lecturer-Mech. Engg.,
Govt. Polytechnic, Amravati (INDIA)

ABSTRACT

Energy shortage is a global problem, this crisis is sought to be overcome through energy conservation i.e. without either impairing the economic growth or the quality of lighting. Energy saving occur either by improving the energy conversion processes, by recycling the waste energy or by resuming the waste materials.It is believed that less than 50 % of all the energy consumed throughout the world is used effectively, the rest is wasted in conversion losses, heat radiation and in other ways. The National Energy Efficiency Programme of Govt. of India has targetted savings of about 5000 MW to be realised by the end of Eighth Plan through both demand-side (2750 MW) and supply-side (2250 MW) efficiency improvements.

INTRODUCTION

It is ironical that the developed world which can afford the cost of energy is involved seriously in the energy conservation, while the developing world,which needs to stretch their existing electricity distribution capacities with energy conservation, finds that all energy-saving propositions have not registered a sizeable difference. Electricity generation is not only trailing way behind demand , but in most cases there is wasteful management with a PLF of about 55 % and distribution loss of about 30 %.

MANAGING INDUSTRIAL ENERGY CONSERVATION

Industry uses various forms of energy say coal, lignite, petroleum fuels, natural gas, electricity, steam, compressed air, and of late, solar energy, wind energy, bio-gas, etc. The various energies are in turn derived in the industry from certain primary energy sources like coal, petroleum,electricity bought from outside the industry or produced in the industry. The energy conservation management involves all the facets of usage of energy in the industry.

Identification of Opportunities

The first step is to conduct an energy survey. The objectives of such a survey are :

- to zero in on equipment processes that operate inefficiently and pin point specific energy leaks
- to identify energy saving opportunities in the form of projects designed to correct each energy wasting situation
- to calculate the specific cost of savings to be derived from implementing each project
- to assist priorities to the various projects after reviewing not only the cost saving posed by each, but also other considerations such as time required to implement

Planning

- decide what to accomplish
- decide how to accomplish
- decide how to apply skills and assign responsibilities
- decide the timeframe and production planning for non-capital and capital projects

Organising

- establish a structure
- arranging and co-relating tasks
- describing a framework of jobs

Implementation

- differentiate non-capital projects and capital projects
- implement non-capital projects at the earliest
- implement capital projects taking into consideration production planning

Controlling

Energy conservation programmes exercise continous and warning supervisory controls:

Maintenance and Surveillance

- keep periodic inspection to uncover energy leakages to reduce energy costs
- make additional preventive maintanance practices, if requred
- monitor periodic operation
- maintain instrumentation
- maintain energy usage graphs plant/production-wise
- review and reconduct the whole programme every year

SUITABLE ENERGY CONSERVATION/ EFFICIENCY TECHNIQUES

The following suitable measures can be implemented for effective energy conservation techniques:

— regeneration or recuperation, where the exhaust energy is recycled back as input energy conservation in the same process is a vital technique for energy conservation in the form of heat recovery linked with retrofit energy efficient process technology.

— installation of shunt capacitors, addition of new lines and substations, optimum conductor loading and stepping up of voltage level has led to reduction of T & D losses by 1 %.

— promoting design of subsystems of energy efficient techniques such as integrated coal gasification combined cycle (IGCC).

— through renovation and modernisation (R & M), additional power can be made available faster and at a cost far lower than by setting up of new capacities.

— advising State Electricity Boards (SEBs) to switch from outdated Aluminium Conductor Steel Reinforced (ACSR) modes of transmission (having 21-22 % losses) to All Aluminium Alloy Conductors which save power by about 16 %.

— invention of more efficient devices, more efficient processes and the institution of a methodology for waste heat management in plants.

— the Government has to design and adopt appropriate energy conservation laws and regulations to promote energy efficiency and demand side management, like requiring industry to progressively reduce energy consumption per unit of goods produced.

— good housekeeping practices alone can help to achieve as much as 10 % savings in energy inputs to industries. Technology related medium and long term measures can enhance savings in energy by an additional 10-15 %.

— energy labeling has been successfully used in many developed and developing countries to encourage consumers to buy energy efficient products.

— for the agricultural sector, it makes sense for the electric utility companies to finance demand side management measures. The utilities can sell the power saved in the agricultural sector (consuming power at highly subsidized rates) to the industrial sector by paying a premium on the power supplied. The payback for the portions of retrofilling the existing pumpsets is about 1.65 years, while for installing new energy efficient pumpsets will be just over 3.5 years.

CONCLUSIONS

Energy saved is energy generated. There need to be incentives for efficiency and disincentives for inefficiency. Cogeneration is a potential area for promoting energy conservation. The energy sector consumes 50 % of final commercial energy. Enforcing energy audits and encouraging energy efficient technologies by offering financial assistance, tax-incentives, etc, would promote energy savings and consequently reduce cost.

So far, our energy policies have been confined to supply side management. It is high time that emphasis is given on genuine demand side management measures. Enhancing energy conservation measures will not only help in solving energy supply constraints, but also help in tackling environmental problems.

REFERENCES

Mantripragada C. S. M. Deepti Energy Consultants, Ankaleshwar.
Mohanty S, Kedia A. Power Shortages : The Negawatt Options, Tata Energy Research Institute, New Delhi.
Roy S. N, Peaking Problems after 1997, *The Economic Times*.

PERFORMANCE ANALYSIS AND RETROFITTING OF INTEGRATED GASIFICATION COMBINED CYCLE POWER PLANTS

52

Dr. D.N. REDDY,
E. NANDA KISHORE.
M. V. RAMANA MURTHY,
University College of Engineering,
Osmania University, (INDIA)

ABSTRACT

Energy is the most basic and critical input for rapid industrialisation of the country. Energy recovery from various sources of industry is essential to meet the gap in power needs. Co-generation and combined cycle power plants are advances made in the energy recovery systems.In the present study existing combined cycle power plant of NTPC, AURIYA plant is retrofitted with integrated gasification combined cycle power plant by suitably designing the parameters. Retaining existing capacity of gas turbine power plant (224 MW) consisting of 2 GTs and steam turbine power plant (110 MW) consisting of 1 steam turbine, gasification plant is designed for supply of coal gas for gas turbine combustion chamber, replacing existing natural gas supply. The gasification plant delivers 91.1 Kg/sec of coal gas to each gas turbine. Gas turbine output is thereby increased marginally (2 MW) to 114 MW from 112 MW. Steam turbine plant parameters are kept unaltered. Various sources of heat recovery from gasification plant are explored (like bleed air cooler, ash coolers, heat recovery from hot coal gas) and additional steam turbine developing 56 MW is proposed using steam generated by heat recovery boiler. Modifications required for existing gas turbine, heat recovery steam generator and power plant cycle piping/equipment are suggested. The proposed project cost is estimated to be Rs. 650 crores (this includes cost of existing project also).

INTRODUCTION

India's requirements of commercial energy are primarily met through fossil fuels i.e., coal, lignite, oil and gas. Among them coal and lignite offer major contribution to energy sector not only because of their abundant availability but also scarcity of other fuels.

Coal as fuel plays dominating role in power generation. About 65% of total installed capacity is coal based. The coals used for power generation are sub-bituminous, having low calorific value and high ash content. Most of the coal based power plants are pulverised coal fired plants with conventional steam cycle. High ash content in coal not only created nuisance of handling huge volumes of air (for transportation of Pulverised Coal) and flue gas; but also leads to severe erosion of boiler components resulting in reduction of life and availability. Moreover power plants installed earlier as well as recently are under severe pressure by statutory authorities to reduce particulate matter emission through chimney. Efficiency that could be achieved is also less which is around 36%.

Next to coal is natural gas which is also used for power generation. Natural gas is used as fuel in Combined Cycle Power Plants. This power plant contains a gas turbine plant operating on Joule's Brayton cycle. Exhaust gas from gas turbine power plant is fed to steam turbine plant which works on Rankine cycle. Combination of these two cycles give better efficiency than any single cycle and it works out to 45 to 50%. But this type of power plants also suffer from setback owing to limited availability of gas in India. Present gas reserves of India amount to 700 biilion cum which may last for next 23 years at the rate of present consumption.

Continuous effort to take advantage of high efficiency of combined cycle power plants and availability of coal has led to development of integrated gasification combined cycle systems. These are expected to give plant efficiency of 40% (which is expected to be improved in future). Besides high efficiency the integrated gasification combined cycle (IGCC) have a number of other advantages over PC (Pulverised cool.) plants such as:

1. Excellent environmental performance – i.e., low CO_2, NO_x and SO_x emission.
2. Low water consumption.
3. Smaller quantity of solid residue as compared to PC plants.
4. Comparable cost of generation with PC plants.
5. Possibility to retrofit of such systems in combined cycle plants
6. It can handle high ash content coals without reduction in efficiency.

Integrated gasification combined cycle (IGCC) power plant constitutes a gasifier which gasifies coal and this coal gas is supplied to gas turbine power plant as fuel. The exhaust of gas turbine is led to HRSG which supplies steam to steam turbine. Coal can be gasified by introduction of steam and air into coal bed and resulting chemical reaction gives coal gas as output. This coal bed can be fixed, entrained, fluidised, moving fluidised and pressurised fluidised depending upon capacity and type of coal. For Indian (high ash content) coals pressurised fluidised bed gasification (PFBG) is most suitable process.

The advantage of PFBG include high gasification rates, feasiblity of operation over a wide range of outputs and flexibility to handle a wide variety of coals while preventing production of tars and oils as by products.

As the availability of natural gas is expected to be short lived (compared to coal) or to divert the scarcely available gas to most efficient systems (using natural gas) retrofitting of existing combined cycles power plants with IGCC is gathering momentum. Present work highlights this idea of retrofitting existing gas based combined cycle power plant with IGCC. All the performance calculations are made keeping retrofitting and minimum modification required in view.

PRESENT STATUS OF GASIFICATION/IGCC TECHNOLOGY IN INDIA

Eventhough coal gasification is not new phenomenon for India (being used in fertiliser plants for Ammonia production) higher capacity gasification plants suitable for power generation are not yet developed. A 1300 TPD gasification plant was designed by BHEL for large scale power stations. IICT, Hyderabad developed and erected a pilot plant of low pressure fluidised bed gasifier and it is being used for its internal gas requirements.

With the above background a conceptual IGCC plant is designed and it is retrofitted to NTPC, Auriya project. Gasifier design procedure adopted by BHEL R&D is considered in the study.

IGCC Power Plants

For many decades conventional coal fired power plants have reliably supplied electric power to industry and domestic customers. For the past decade a number of new technologies and modifications to conventional concept have been proposed. Some of them have even been tested in pilot plants. Combined cycle plants, due to rapid progress in coal gasification technology have surpassed highly efficient, well designed conventional coal fired power plants, for their low emmission and ability to handle high ash content coals. (Indian coals)

Coal Gasification

Coal gasification is a process in which coal's potential heat is converted into fuel gas by reacting with a gasification agent.

$$\underset{\text{(Dry and hot - Heat added)}}{\text{coal (2C)}} + \underset{\text{(Liquid water - Heat added)}}{\text{steam } (2H_2O)} \xrightarrow{\text{Heat Energy Added}} \underset{\text{(coal gas)}}{CH_4 + CO_2}$$

Depending upon the medium of gasification, reaction temperature and pressure, product gases with different composition and calorific value can be obtained. These fall in the following categories.

1. Low calorific value of gas (H_2, CO, CO_2, N_2)
   CV = 1000 to 1700 Kcal/NM_3

2. Medium calorific value of gas (CH_2, H_2, CO, CO_2)
   CV = 2500 to 5000 Kcal/NM_3

3. High calorific value of gas (mainly CH_4)
   CV = 8000 to 10000 Kcal/NM_3

Obtained from Hydrogen and Oxygen blown gasifiers.

Coal Conversion Technologies

The low CV, medium CV or high CV gases can be produced by anyone of the following coal gasification technologies according to the method of coal contacting with gasification medium within the reactor.

1. Fixed bed coal gasification
2. Fluidised bed coal gasification
3. Entrained bed coal gasification
4. Molten bed coal gasification.

The fixed bed, fluidised bed and entrained bed technologies have been developed to commercial status while molten bed coal gasification technology is still in the developmental stage.

In the present case study, a pressurised fluidised bed is used for production of coal gas and open circuit-closed cycle combined cycle.

Design of IGCC Power Plants

The existing combined cycle power plant, with gasification consists of 2 Nos. of gas turbines developing 224 MW (total) and a steam turbine developing 110.4 MW. The gasifier produces raw coal gas at 950° C. The raw coal gas coming out of gasifier is cleaned in a train of three cyclones and passed through two heat recovery boilers (HRB1 and HRB2). In cyclones dust level is reduced to about 30 gms/cu.m. Steam produced in HRB1 is utilised for power generation, where as HRB2 steam is diverted to gasifier requirement and gas heating. Gas coming out of HRB2 will be at 200° C. Further this raw coal gas is passed through the quench column and ventury scrubbers to reduce dust level further to the gasifier requirements. The clean gas coming out of gas cleaning system is at around 105° C and it is heated to 220° C to suit combustion chamber requirement by using extraction steam. A part of this clean coal gas is used for recycling fine particles into gasifies after boosting its pressure. Air is bled from GT compressor for gasifier requirement and is admitted to gasifier after compressing it to 30 Kg/cm^2. The pneumatic transport of coal and its pressurisation in the coal lock is done utilising the bleed air of GT compressor. A steam turbine generating @ 56 MW using import steam is proposed as existing ST cannot accommodate import steam.

Methodology

1. The Gas turbine set and HRSG of Auriya Power project of NTPC are considered for analysis.
2. Gasifier design concept is taken from BHEL, Corporated R&D division.
3. The coal gas requirement for developing same power by gas turbine is calculated. Gas turbine outlet temperature for the given flow is designed to suit existing HRSG steam flows. Then gas turbine inlet temperature for same power output of GT is designed. Steam turbine inlet conditions are kept undisturbed.
4. In gasifier part heat given away by raw coal gas is used for generating import steam in HRB1 and process steam in HRB2. The heat in bleed air is utilised for condensate heating.
5. Specific heat of flue gas in HRSG at different temperatures is calculated from gas table as

per its composition. The value of Coal gas Cρ at required temperatures is taken from IGCC analysis form the available data from BHEL Corporate R&D division.

Retrofitting of existing CC plant with IGCC power plant

The demand for natural gas is growing faster than its supply and availability. As it is compulsory to allocate natural gas for its specific purposes like fertiliser, Ammonia production, its availability to power plants will be limited. Hence approval of future CC power plants using natural gas becomes difficult and it is also possible that gas supply to existing power plants may be cut off before their life is over.

Above constraints have led to choose alternate fuel to existing CC plants. One such choice is coal gas which replace natural gas for power generation purpose. Integrated Gasification combined cycle exploits the advantages of high cycle efficiency of CC plants with the use of abundantly available fuel i.e., coal. Advantages of IGCC power plants are already illustrated earlier.

Retrofitting of CC plant with IGCC involves erection of additional equipment and modification of existing equipment. CC power plant situated at Auriya (UP) owned by NTPC is chosen for retrofitting.

For retrofitting above plant with IGCC following additional equipment are needed.

1. Gasifier and its Auxiliaries
    a. Cyclone separators
    b. Hot gas cleanup arrangement
    c. Gas heating equipment.
    d. Clean gas compressor
    e. Coal handling system.
    f. Bleed air compressor
    g. Associated electrical and control equipment.
2. Import steam turbine and its auxiliaries
3. Generator and auxiliaries for import steam turbine.

Peformance at "Off Loads"

As the turbines became larger, many installation required that they operate at less than full load. It became obvious then to able to predict the performance of a specific heat recovery system at various gas turbines outputs and at various ambient air temperatures. Hence, it is necessary to design the system for the normal or base load and then calculate the performance of that system at the various alternate loads. In addition, the change in gas temperature effects the LP steam flow adversely so that an increase in gas temperature actually results in a decrease in the production of LP steam. Steam drums and steam lines must be designed to accommodate these changes experienced during OFF peak loads in operation. Pinch point change is proportional to the change in gas flow and gas temperature.

This factor results in unstable operation due to low pinch points at low loads. Hence the effect of pinch points must be considered if the heat recovery system is to be operated at low loads for extended periods.

Power Calculations

Gas Turbine Output :

GT turbine inlet temperature = 1185 deg. C
Enthalpy of gas at 1185 deg. C = 1364.99 kj/kg.
Gas flow = 360.25 kg/sec.
Amount of cooling air = 67.61 kg/sec.
Temperature = 387 deg.C
Enthalpy = 413.32 kj/kg
Gas flow at turbine outlet = 427.86 kg/sec
Temperature = 532.8 deg.C
Enthalpy = 582.34 kj/kg
Compressor flow = 405.57 kg/sec
Inlet temperature = 27 deg.C
Enthalpy = 27.38 kj/kg
Outlet temperature = 387 deg.C
Enthalpy = 413.32 kj/kg.
Power developed by gas turbine = W_{Gt}

$$W_{Gt} = (360.25 \times 1364.99 + 67.611 \times 413.32 - 427.86 \times 582.34) - (405.57 \times 413.32 - 405.57 \times 27.38) = 114000 \text{ KWe.}$$

Steam turbine output :
As steam turbine inlet and outlet conditions remain same power developed by steam turbine remain same and it is given by

110.404 MW (gross)

Import steam turbine output :
Flow = 27.76 kg/sec.
Steam flow from HRB = 55.52 kg/sec
HP turbine inlet steam pressure = 60 kg/sq.cm.
Temperature = 493 deg.C
Enthalpy = 812.1 kcal/kg.
Condensor pressure = 0.1 kg/sq.cm (abs.)
LP turbine inlet pressure = 6 kg/sq.cm.

Temperature = 220 deg. C
Enthalpy = 688.53 kcal/kg.
Efficiency of HP turbine = 84.8% (same as existing steam turbine)
Efficiency of LP turbine = 77.78% (same as existing steam turbine)
Enthalpy drop in HP turbine = ($H_1 - H_2$) = 84.8 (812.10 - 666.38)/100
= 123.57 kcal/kg.

LP turbine inlet steam enthalpy = 688.53 kcal/kg
LP turbine efficiency = 77.78%

Enthalpy drop in LP turbine = ($H_2 - H_3$) = 77.78(688.53 – 533.82)/100
= 120.33 kcal/kg
Enthalpy of steam at LP turbine outlet = 568.23 kcal/kg
Dryness fraction = 91.5%

Gross Power developed by steam turbine = 55.52(123.57 + 120.33) × 4.1868
= 56694.84 KWe.
= 56.7 MW

Cycle efficiency :

Total power developed by plant = 2×112.86 + 109.3 + 56.13
(at generator terminals) = 391.15 MW
Coal input = 35.37×2 = 70.74 kg/sec
LHV = 3647.2 kcal/kg
Cycle efficiency = 391149/(70.74 ×3647.2)×4.1868
= 36.21%

Estimated cost of the project

Approved cost of AURIYA project	Rs.Cr. 326.00
Cost of gasifier plant	Rs.Cr. 246.74
Cost of modification of GT	Rs. Cr.@8.00
Cost of modification of HRSG	Rs. Cr. 1.00
Cost of modification of DEAERATOR	Rs. Cr. 1.00
Cost of import ST	Rs. Cr. 60.00

CONCLUSIONS

Retrofitting of integrated Gasification Combined Cycle using low calorific value coal gas to the existing combined cycle plant using natural gas is possible. Even though fuel flow is to be increased by 9 to 10 times the gas flow across the gas turbine increases by 7 to 10 percent. This is because low calorific value coal gas produced lower combustion temperatures in combustion chambers and requires less amount of air for cooling combustibles to turbine inlet conditions. Rest of the air is diverted to gasifier through bleed air cooler to extract available heat energy.

Heat is extracted from gasifier plant from all possible ways through ash coolers, heat recovery boilers to increase energy conversion rates of gasifier plant. In entrained bed the ash is in fusion state and ash coolers cannot be employed and energy conversion rates are low. Hence, it is not suitable to high ash content Indian coals. Additional power of 56 MW can be generated by retrofitting IGCC plant to CC plant using steam generated in the coal gasification plant.

Efficiency of IGCC plane developed in this study is 36.2%. This can be improved to around 40% by :

1. Optimising heat recovery system
2. Using oxygen enriched air blown gasifiers.
3. Checking whether flue gas temperature leaving stack can be reduced as H_2S content in clean gas is minimal.
4. Adopting reheat in import steam turbine. This is possible if flue gas temperature leaving stack is reduced.

The efficiency of CC plant using coal gas keeping gasifier out of bettery limits would be close to CC plant efficiency using natural gas. Efficiency of new IGCC plant with multi-high pressure HRSG would be more than that of retrofitted one. In the proposed IGCC plant efficiency of approximately 0.9% to 1% is lost due to retrofitting to take care of excess cost factor in modifying HRSG.

The cost of retrofitting of existing CC power plant is less than installing a new PC plant in case of shortage of natural gas. The cost of generation of proposed IGCC plant after retrofitting is in between cost of generation of CC plant NTPC, Auriya and that of a new PC plant with 2.75 cr./MW investment.

Particulate matter and flue gas emissions like NOx, SOx are very less in IGCC plant compared to PC plant. Hence IGCC is a promising clean coal future technology. Increase in overall efficiency and reduction in cost can be expected with future technologies.

RETROFITTING

Auriya power plant's existing combined cycle parameters, proposed IGCC parameters are shown in table given below. Difference to be accommodated for retrofitting IGCC to existing CC plant is also shown.

SL. NO.	DESCRIPTION	CC PLANT	IGCC	COMMENTS
1.	GASIFIER	—	6 × 1050 TPD	
2.	HEAT REC. BOILER-I	—	99.94 T/Hr. Steam at 65 Ata. Sat.	Additional quantity of steam is produced Additional Power is generated.
3.	HEAT REC. BOILER-II	—	1. 21.53 T/Hr. Steam at 30 Ata 380 deg. C	For process requirement. No. Addl. power is generated

SL. NO.	DESCRIPTION	CC PLANT	IGCC	COMMENTS
			2. 18.94 T/Hr. Steam at 30Ata. Sat.	
4.	GAS TURBINE			
A.	COMPRESSOR:			
	FLOW (KG/Sec.)	405.57	405.57	—
	r.	13	13	—
B.	COMBUSTION CHAMBER			
	AIR (Kg/Sec.)	337.96	269.16	BURNER IS TO BE MODIFIED TO ACCOMMODATE EXCESS FLOW
	FUEL (Kg/Sec.)	8.045 (NG)	91.1 (Coal Gas)	
	LHV (KCal/Kg.)	10118	934	
C.	GAS TURBINE INLET			
	FLOW (Kg/Sec.)	346	360.25	14.25
	COOLING AIR	67.61	67.61	Suitable Blade Coatings are to be applied to withstand increased turbine inlet temperature.
	INLET TEMP. DEG.C	1154	1185	
	OUTLET FLOW (Kg/Sec.)	413.61	427.86	
	TEMP. deg.C	513	532.8	
5.	HRSG			
A.	HP STEAM FLOW (Kg/Sec)	46.55	46.55	
	TEMP. deg. C.	495	495	—
	PRESSURE (Ata)	65	65	—
B.	LP STEAM			
	flow(Kg/Sec.)	12.25	14.13	Steam for deaerator is produced within LP section. Condensate preheating is also done.
	PRESSURE (Ata)	7	7	
	TEMP. deg.C	164	164	
C	HEAT EXCHANGER SURFACE AREA Sq.M			
	1. HP SUPER HEATER	8730	SUFFICIENT	
	2. IMP. STEAM SH	—	4239.76	
	3. HP EVAPORATOR	20800	SUFFICIENT	SUITABLE MODIFICATION IS TO BE DONE ACCORDINGLY
	4. HP ECONOMISER-II	13860	SUFFICIENT	
	5. LP Evaporator	13860	Additional area required	
	6. LP ECONOMISER	870	Aditional area required	
	7. HP ECONOMISER-I	5200	SUFFICIENT	
	8. CPH	10400	To be reduced	
6.	STEAM TURBINE			
	A. HP STEAM FLOW (T/Hr.)	335.2	335.2	—
	INLET TEMP. deg.C	493	493	—
	PRESSURE (Ata)	60	60	—

SL. NO.	DESCRIPTION	CC PLANT	IGCC	COMMENTS
	B. LP STEAM FLOW (T/Hr.)	415	415	—
	INLET TEMP. deg. C.	200	200	—
	PRESSURE (Ata.)	5.5	5.5	—
	C. CONDENSOR PRESSURE (Ata.)	0.0952	0.0952	—
	D. POWER DEVELOPED	110.4	110.4	—
7.	DEAERATOR			
	STEAM FLOW (Kg/Sec)	1.56	5.32	
	COND. FLOW (Kg/Sec)	116.055	120.7	
	PRESSURE (Kg/CM²)	4	4	
8	IMPORT STEAM TURBINE			
	STEAM FLOW (T/Hr.)	—	216.58	A new Steam turbine is to be installed which would generate 56 MW.
	PRESSURE (Ata)	—	60	
	TEMP. deg. C	—	493	
	POWER DEV. (MW)	—	61.4	
9.	TOTAL POWER DEVELOPED (MW)	335	391	56

REFERENCES

1. J.H. Horlock. Combined Power Plants, Pergamon Press.
2. BHEL, Heat Recovery Steam Generators, BHEL.
3. V Ganapathy, HRSG Design Method Optimises Power Plant Efficiency, Power Engg., May 1991
4. Report, Coal Gas with Nuclear Energy-Prospect for Unproblematical Burning of Coal *URJA*, July 1984.
5. H. E. Vierrath, Electrical Coal, *URJA*
   P. K. Herbert. The British Gas/Lury, October, 1986.
   B. H. Thompson; C. F. Griel, Gasification for C. C. Power Generation.
6. J.P. Tessier, Waste Heat Boiler Design, for Gas Turbine C. C. Station, *URJA*, October 1986.
7. Data Consultancy Services, Feasibility Report on NTPC, Auriya Project, NTPC Corp. Div.
8. RSTPS, Performance Budget, NTPC, RSTPS.
9. Report, First International Conference on C. C. Power Generation, Jan. 1994.
10. J. S. Rao; K. S. bandyopadyay; K. Basu; System Parameter Selection, Criteria for GT Based Cogen. Plants, Seminar on Co-generation
11. K. Basu, BHEL Developmental Strategy for Coal Based C. C. Plants, Power Directory
12. Steven Collins, Advanced Gas Turbines, Power, Nov. 93.
13. Steven Collins, Special Report on Gas, Turbine Power Plants, Power Jun. 94.

CONTROLLED SERIES COMPENSATION AND AUXILIARY CONTROLLED SVS FOR FLEXIBLE AC TRANSMISSION SYSTEMS

53

DR. NARENDRA KUMAR
Asstt. Professor
Deptt. of Elect. Engg.
C. R. State College of Engineering,
Murthal (Sonepat) -131039

DR. M. P. DAVE
Professor
Deptt. of Elect. Engg.
University of Roorkee,
Roorkee-247 667

ABSTRACT

In this paper an SVS controller, known as the combined reactive power and frequency (C.R.P.F.) auxiliary controller, has been developed and incorporated in the SVS control system located at the middle of the transmission line. A digital computer simulation study, using a nonlinear system model, has been performed to illustrate the effectiveness of the coordinated application of CSC and the proposed SVS auxiliary controller developed.

Introduction

In recent years SVS has been employed to an increasing extent in the modern power systems due to its capability to work as VAR generation and absorption systems[1,2]. On the other hand the controlled series compensation (CSC) is the novel technique proposed under Flexible AC Transmission Systems (FACTS) concepts. But investigations on how to control these devices for improving stability of power systems are still in preliminary stage[3].

In the present paper, an attempt has been made to employ the SVS auxiliary controllers in coordination with the controlled series compensation (CSC) in a long series compensated transmission line. A continuous control has been provided by the midpoint located SVS with the bang-bang form of control of the CSC. A detailed system model has been developed using a network model represented by single lumped parameter π-circuit on either side of the SVS to evaluate the dynamic and transient performance of the system.

The controlled series compensation having bang-bang control characteristics has been provided in the series compensated transmission line for improving the system performance under large disturbance conditions and to test the transient stability of the system.

System Modeling

The study system (Fig. 1) consists of a generator supplying power to an infinite bus over a 400 KV, 600 Km. long series compensated line. The series compensation has been provided towards the line side of the transformer at the sending end of the line. A fixed capacitor-thyrister controlled reactor type SVS is considered located at the middle of the line.

Generator Model

In the detailed machine model[4] used here, the stator is represented by a dependent current source in parallel with the inductance (Fig. 2). To have a common axis of representation with the network and SVS, these flux linkages are transformed to the synchronously rotating D-Q frame of reference. The generator model includes the field winding and a damper winding along d-axis and two damper windings on q-axis. The linearized state and output equations are as follows :

$$\begin{aligned} X_R &= (A_R)\,X_R + (B_{R1})\,U_{R1} + (B_{R2})U_{R2} + (B_{R3})U_{R3} \\ Y_{R1} &= (C_{R1})\,X_R + (D_{R1})U_{R1} \\ Y_{R2} &= (C_{R2})\,X_R + (D_{R2})\,U_{R1} + (D_{R3})U_{R2} + (D_{R4})U_{R3} \end{aligned} \qquad \text{...(1)}$$

Where,

$$X_R = (\Delta\psi_f\ \ \Delta\psi_h\ \Delta\psi_g\ \Delta\psi_k)^t,\ Y_{R1} = (\Delta I_D\ \Delta I_Q)^t,\ Y_{R2} = (\Delta\dot{I}_D\ \ \Delta\dot{I}_Q)^t$$
$$U_{R1} = (\Delta\delta\ \ \Delta\omega)^t,\ U_{R2} = (\Delta V_F),\ U_{R3} = (\Delta i_D\ \ \Delta i_Q)^t$$

Mechanical system

The state and output equations of the mechanical subsystem are derived as :

$$\begin{aligned} \dot{X}_M &= (A_M)\,X_M + (B_{M1})\,U_{M1} + (B_{M2})\,U_{M2} \\ Y_M &= (C_M)\,X_M \end{aligned} \qquad \text{...(2)}$$

Where, $X_M = (\Delta\delta\ \ \Delta\omega)^t$, $Y_M = (\Delta\delta\ \Delta\omega)^t$, $U_{M1} = (\Delta I_D\ \ \Delta I_Q)^t$, $U_{M2} = (\Delta i_D\ \Delta_Q)^t$

Excitation System

The state and output equations of the linearized IEEE type-1 excitation system model are derived as :

$$\begin{aligned} X_E &= (A_E)\,X_E + (B_E)\,U_E \\ Y_E &= (C_E)\,X_E \end{aligned} \qquad \text{...(3)}$$

Where, $X_E = (\Delta V_f\ \ \Delta V_s\ \ \Delta V_r)^t$; $Y_E = (\Delta V_f)$, $U_E = (\Delta V_g)$

Network

The transmission line is represented by a single lumped parameter π-circuit on either side of the SVS. The network has been represented by its α-axis equivalent circuit which is identical

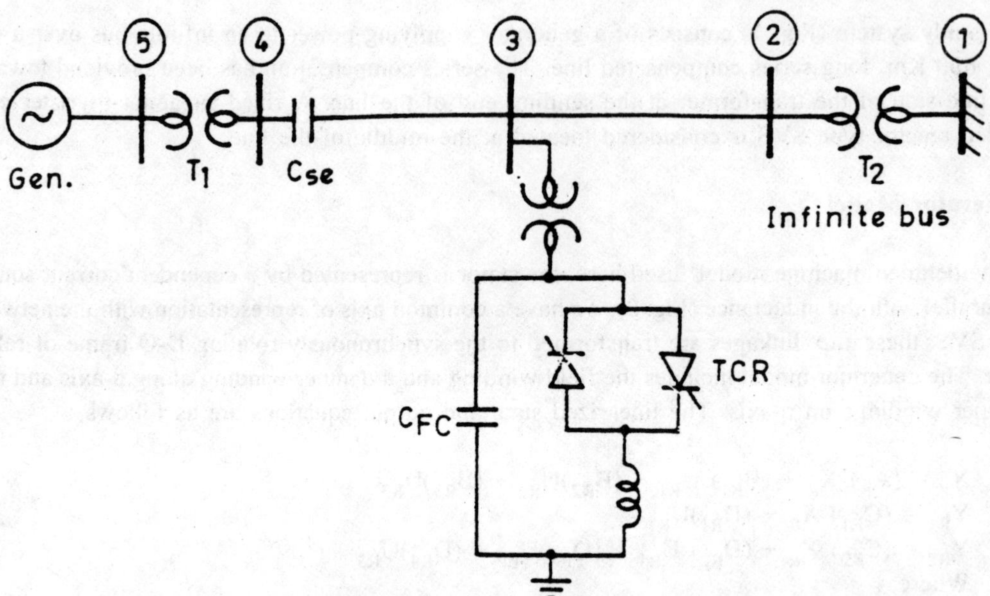

FIG. 1 : *Study System*

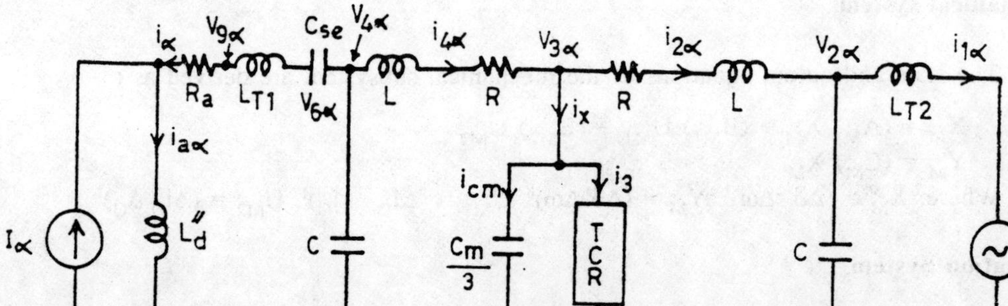

FIG. 1 : α - *Axis Representation of the Network*

with the positive sequence network (Fig. 2). The linearlized state and equations of the network are as follows :

$$\begin{aligned} \dot{X}_N &= (A_N)\, X_N + (B_{N1})\, U_{N1} + (B_{N2})\, U_{M2} + (B_{N3})\, U_{N3} \\ Y_{N1} &= (C_{N1})\, X_N + (D_{N1})\, U_{N1} + (D_{N2})\, U_{N2} + (D_{N3})\, U_{N3} \\ Y_{N2} &= (C_{N2})\, X_N + Y_{N3} = (C_{N3})\, X_N \end{aligned} \quad \text{...(4)}$$

Where,

$X_N = (\Delta i_{1D}\ \ \Delta i_D\ \Delta V_{2D}\ \Delta V_{4D}\ \Delta i_{1Q}\ \Delta i_{1Q}\ \Delta i_Q\ \ \Delta V_{2Q}\ \ \Delta V_{4Q})^t$

$U_{N1} = (\Delta i_{2D}\ \ \Delta i_{2Q})^t,\ U_{N2}\ (\Delta I_D\ \Delta I_Q)^t,\ Y_{N3} = (\Delta \dot{I}_D\ \Delta \dot{I}_Q)^t$

Static Var System

The block diagram of a general SVS control system [2] is shown in fig. 3. The linearized state and output equations of SVS model are obtained as :

$$\begin{aligned} \dot{X}_S &= (A_S)\, X_S + (B_{S1})\, U_{S1} + (B_{S2})\, U_{S2} + (B_{S3})\, U_{S3} \\ Y_S &= (C_S)\, X_S + (D_S)\, U_{S1} \end{aligned} \quad \text{...(5)}$$

Where, $X_S = (\Delta i_{2D}\ \ \Delta i_{2Q}\ Z_1\ Z_2\ Z_3\ \Delta B)^t,\ Y_S = (\Delta i_{3D}\ \ \Delta i_{3Q})^t$
$U_{S1} = (\Delta V_{2D}\ \ \Delta V_{2Q})^t,\ U_{S2}\ (\Delta V_{ref}),\ Y_{S3} = (\Delta V_F)$

Auxiliary Control of SVS

To effect the SVS control system, any general auxiliary signal U_C is implemented through the controller G(S) as shown in Fig. 4. The state and output equations are given by

$$\begin{aligned} \dot{X}_C &= (A_C)\, X_C + (B_C)\, U_C \\ Y_C &= (C_C)\, X_C + (D_C)\, U_C \end{aligned} \quad \text{...(6)}$$

Where, $X_C = (Z_5)$ and $Y_C = (\Delta V_F)$

Line Reactive power auxiliary controller

The line reaction power entering SVS bus is expressed as :

$$Q_2 = V_{2D}\, i_Q - V_{2Q} i_D \quad \text{...(7)}$$

Linearizing equation (7), gives the deviation in line reactive power ΔQ_2 which is selected as auxiliary control signal. This is expressed in matrix notation as :

$$U_C = (F_{CR})X_R + (F_{CM})X_M + (F_{CE})X_E + (F_{CN})X_N + (F_{CS})X_S \quad \text{...(8)}$$

Where, $U_C = (\Delta Q_2)$ and $(F_{CR}) = (F_{CM}) = (F_{CE}) = (F_{CS}) = 0$

or $U_C = (F_{CN})\, X_N$ ...(9)

Where, $(F_{CN}) = (0 - V_{2Q0}\ i_{Q0}\ O\ O\ V_{2D0} - i_{D0}\ 0)$

The suffix '0' indicates the corresponding operating point values.

Bus Frequency Auxiliary Controller

The svs bus frequency is given by

$$f_{SVS} = \frac{d\theta_2}{dt} = \frac{d}{dt}\left(\tan^{-1}\frac{V_{2Q}}{V_{2D}}\right) \quad ...(10)$$

After linearizing equation (10) Δf_{SVS} is selected as auxiliary control signal

$$\Delta f_{SVS} = \frac{V_{2D0}}{V_{20}^2}\Delta V_{2Q} - \frac{V_{2Q}}{V_{20}^2}\Delta V_{2D} \quad ...(11)$$

The equation (11) can be written in matrix form as :

$$U_C = (F_{CR})X_R + (F_{CM})X_M + (F_{CE})X_E + (F_{CN})X_N + (F_{CS})X_S \quad ...(12)$$

Where, $U_C = (\Delta f_{SVS})$ and $F_{CR} = F_{CM} = F_{CE} = 0$

or $U_C = (F_{CN})\,X_N + (F_{CS})\,X_S$ ... (13)

Where,

$$(F_{CN}) = \left(\frac{K_4}{C_n} - \frac{K_4}{C_N}K_{3\varpi_o}\ 0 - \frac{K_3}{C_n} - \frac{K_3}{C_n}K_{4\omega_o}\ 0\right)$$

$$(F_{CS}) = \left(\frac{K_4}{C_N} - \frac{K_3}{C_N}\text{oooooo}\right)$$

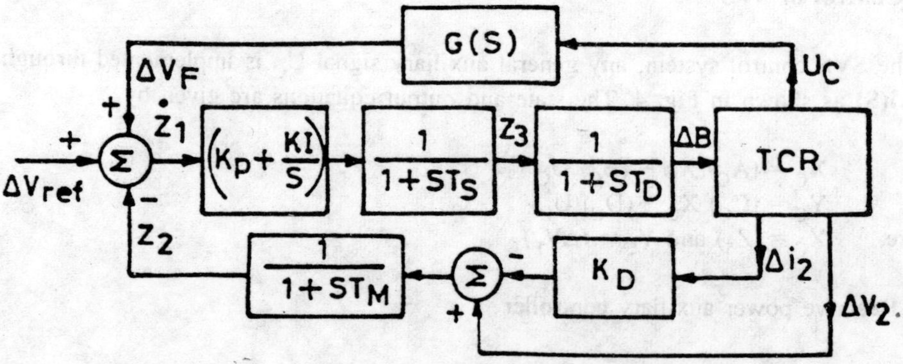

FIG. 3 : *SVS Control System with Auxiliary Feedback*

Where, $K_3 = \frac{V_{2Do}}{V_{2o}^{2}}$ and $K_4 \frac{V_{2Q0}}{V_{2o}^2}$

The suffix 'o' indicates the operating point values.

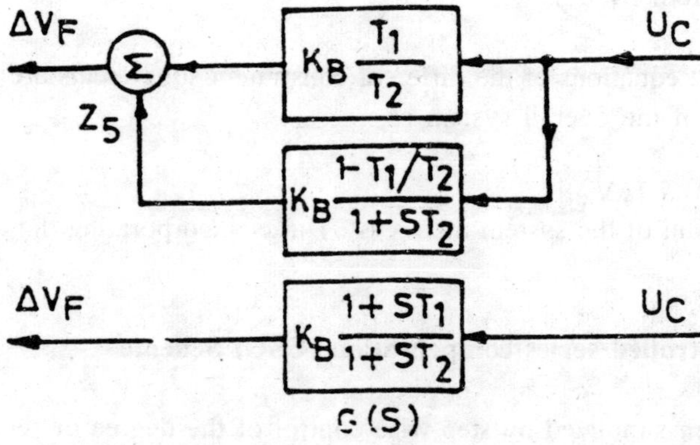

FIG. 4 : *General First Order Auxiliary Controller*

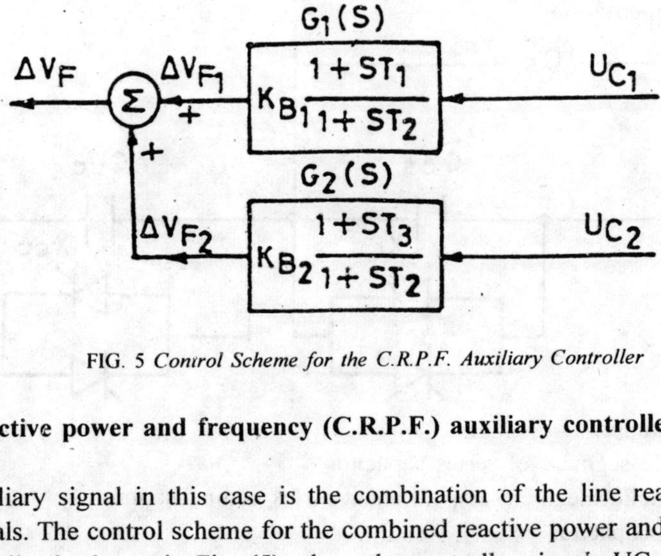

FIG. 5 *Control Scheme for the C.R.P.F. Auxiliary Controller*

Combined reactive power and frequency (C.R.P.F.) auxiliary controller

The auxiliary signal in this case is the combination of the line reactive power and bus frequency signals. The control scheme for the combined reactive power and frequency (C.R.P.F.) auxiliary controller is shown in Fig. (5) where the controller signals UC_1 and UC_2 correspond respectively to the line reactive power and bus frequency auxiliary controllers. The state and output equations of the C.R.P.F. auxiliary controller can be written as follows.

$$\begin{bmatrix} \dot{XC}_1 \\ \dot{XC}_2 \end{bmatrix} = \begin{bmatrix} AC_1 & o \\ o & AC_2 \end{bmatrix} \begin{bmatrix} XC_1 \\ XC_2 \end{bmatrix} + \begin{bmatrix} BC_1 & 0 \\ o & BC_2 \end{bmatrix} \begin{bmatrix} UC_1 \\ UC_2 \end{bmatrix}$$

$$[Y_C] = [CC_1 CC_2] \begin{bmatrix} XC_1 \\ XC_2 \end{bmatrix} + [DC_1 DC_2] \begin{bmatrix} UC_1 \\ UC_2 \end{bmatrix} \qquad ...(14)$$

Where, AC_1, BC_1, CC_1, and DC_1 are the matrices of the reactive power auxiliary controller and AC_2 BC_2, CC_2, and DC_2 are the matrices of the bus frequency auxiliary controller.

Development of the System Model

The state and output equations of the different constituent subsystems are combined resulting in the state equation of the overall system as :

$$\dot{X} = [A]\ X + [B_{S2}]\ [\Delta V_{ref}] \tag{15}$$

The overall dimension of the system matrix is 31 after incorporating the auxiliary controllers.

Development of the controlled series compensation (CSC) Scheme

The CSC scheme has been simulated by step-wise control of the degree of series compensation by means of sections with thyristor by-pass switches. The series capacitor Cs of the line is split into to two segments Cse and Ccs with capacitive reactances Xce and Xcs respectively. Cse is left as permanent series compensation

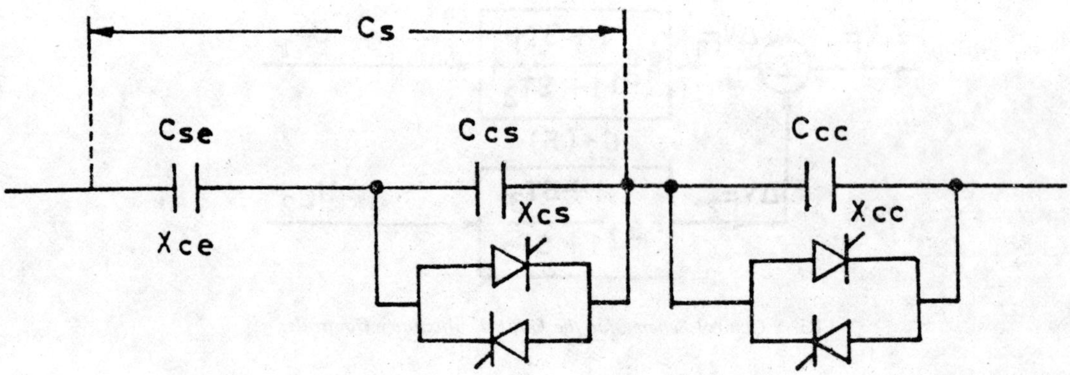

Cs = Series capacitor
Ccs = Controllable segment of series capacitor
Cse = Series capacitor segment as permanent series compensation
Ccc = Controlled capacitor
Xcc = Xcs = Controllable capacitive reactance
Xce = Capacitive reactance of permanent series capacitor segment.

FIG. 6 : *Controlled Ceries Compensation Arrangement*

while Ccs is made controllable segment. Another capacitor segment Ccc equal in magnitude that of Ccs, and having capacitive reactance Xcc is also added in series with the line (Fig 6). During steady state condition this capacitor segment remains by-passed. The controlled capacitive reactance can be defined as :

$$Xccap = \begin{cases} -Xcc & \text{when Ccc and Ccs are inserted} \\ o & \text{when Ccs inserted and Ccc is by-passed} \\ +Xcc & \text{when Ccc and Ccs are by-passed} \end{cases}$$

The control algorithm used senses the angular speed deviations from the steady state value and inreases or decreases the power transmission by inserting or by-passing the controlled series capacitors when angular speed deviation ($\Delta\omega$) reaches a threshold value of +0.25 or –0.25 rad./sec. respectively. Otherwise, the controlled series capacitor Ccc remains by-passed. In such a way, the full bang-bang control characteristics of the CSC is obtained.

Dynamic Performance

The load flow study is carried out (Table 1) for varying generator power with sending end, midpoint and receiving end voltage being maintained at 1.05, 1.05 and 1.0 p.u. respectively. The SVS rating for the line has been chosen to be 200 MVAR inductive to 300 MVAR capacitive. The series compensation has been provided by locating a capacitor bank equivalent to 200 MVAR rating towards the line side of the transformer at the sending end of the line. About 25% series compensation is achieved in such a manner.

The eigenvalues have been computed for the system with the and without auxiliary controllers incorporated in the SVS control system. Table-2 shows the eigenvalues without any auxiliary controller, with the reactive power, the bus frequency, and the C.R.P.F. auxiliary controller for 800MW generator power (Pg). When no auxiliary controller is incorporated in the SVS control system, the rotor mechanical mode having frequency 4.9966 rad./sec is unstable. The line reactive power, the bus frequency, and the C.R.P.F. auxiliary controllers, when applied, shift the unstable mechanical mode to the stable region. The damping of the mechanical mode is slightly higher in case of reactive power auxiliary controller than the damping in case of the bus frequency auxiliary controller. But the damping is highest when the C.R.P.F. auxiliary controller is used.

The auxiliary controller parameters, based on the root locus study, are taken as KB = –0.04, T1=0.39, T2=0.2 in the case of the reactive power auxiliary controller, and KB = –0.12, T1=0.002, T2=0.1 in the case of the bus frequency auxiliary controller.

TABLE 1 : Steady State Performance of the Transmission System

	Generator Bus				SVS Bus				Infinite Bus	
S.No.	Pg (MW)	Qg (MVA)	Vg (pu)	δ (DEG)	Pg (MW)	Qs (MW)	Vs (pu)	θs (DEG)	P (MW)	Q (MW)
2-	200	–156.12	1.05	12.589	0	–68.78	1.05.	7.792	–194.98	–111.43
5-	500	–156.18	1.05	31.305	0	61.58	1.05	19.551	–470.74	–19.18
8-	800	–115.09	1.05	50.298	0	286.92	1.05	31.524	–725.90	131.04

TABLE 2 : System Eigenvalues

	Without Aux. Controller	Reactive Power Aux. Controller	Bus Frequency Aux. Controller	C.R.P.F. Aux. Controller
	1786± j 4.896	– .5975±j5.6372	–.4657±j5.1505	– 1.6115±j5.695
	–38.0213	– 33.0788	– 43.189	– 7.037+j1.918
Gen.	–3.0346	–2.9791	–3.043	–2.97
	–.8802 + j.9177	–.5396+j.8518	–.8923+j.9241	–.5983+j.8512
	–33.2483	–33.0788	–30.84	–32.336

AVR	-.8802 -j.9177	-.5396-j.8518	-.8923-j.9241	-.5983-j.8512
	-25.75 ±j24.11	-26.62±j24.636	-26.56±j23.68	-25.75±j24.05
	-3.24±j3497.45	-3.26±j3498.4	-3.27 ±j3499.1	-3.27±3499.11
NET.	-3.24±j2869.45	-3.26±j2870.38	-3.47±j2871.10	-3.27±j2871.11
	-13.12±j2489.8	-13.22±j2493.0	-13.23±j2495.5	-13.24±j2495.5
	-14.92±j1861.8	-14.91±j1865.0	-14.90±j1867.5	-14.89±j1867.5
	-12.74±j1101.6	-12.73±j1129.7	-14.88±j1147.5	-15.11±j1147.0
	-20.51±j473.33	-19.53±j507.35	-8.20±j518.23	-8.75±j521.731
	-12.74±j444.08	-0.34±j446.10	-14.49±j444.58	-12.57±j445.93
	-10.293±j191.9	-14.19±j199.92	-6.14±194.84	-10.78±j201.65
SVS	-5.12±j311.33	-5.72±j312.61	-4.98±j309.25	-4.88±j310.30
	-47.15±81.25	-124.349	-57.99±66.10	-10.83±j32.29
	-546.98±j77.7	-546.02±j79.59	-546.70±j72.93	-547.05±j76.97
		-1.758j29.92	-8.478	-127.946
				-7.037-j1.918

Time Domain Simulations

To illustrate the comparative effectiveness of the controlled series compensation and the C.R.P.F. auxiliary controller under disturbance conditions, a digital computer simulation study using the non-linear system model, has been carried out for the system having a series compensated transmission line. The disturbance is simulated by applying a pulsed torque disturbance of 40% for 0.1 second. the simulation study has been carried out at PG=800MW.

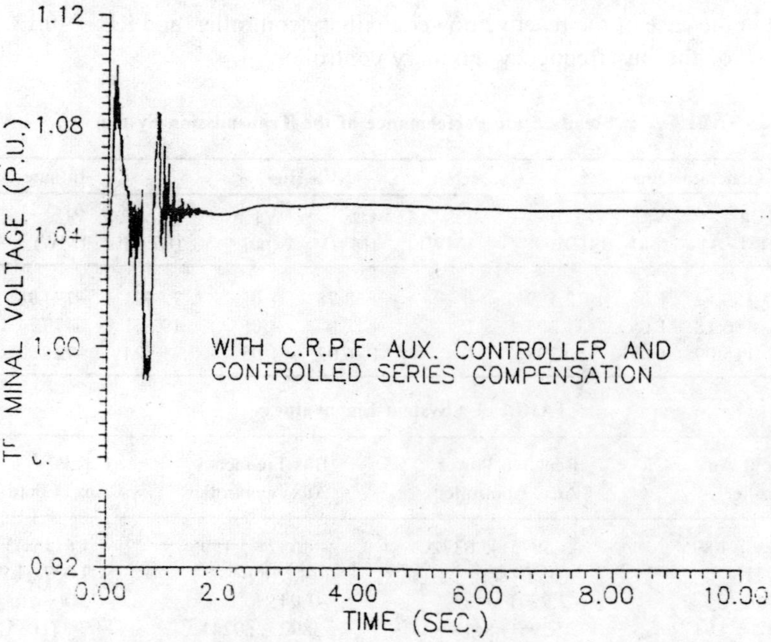

FIG. 7 (a) : *Pg=800 MW, Variation of Terminal Voltage due to 40% increase in Tmech for 01. Second (PI-Circuit Model)*

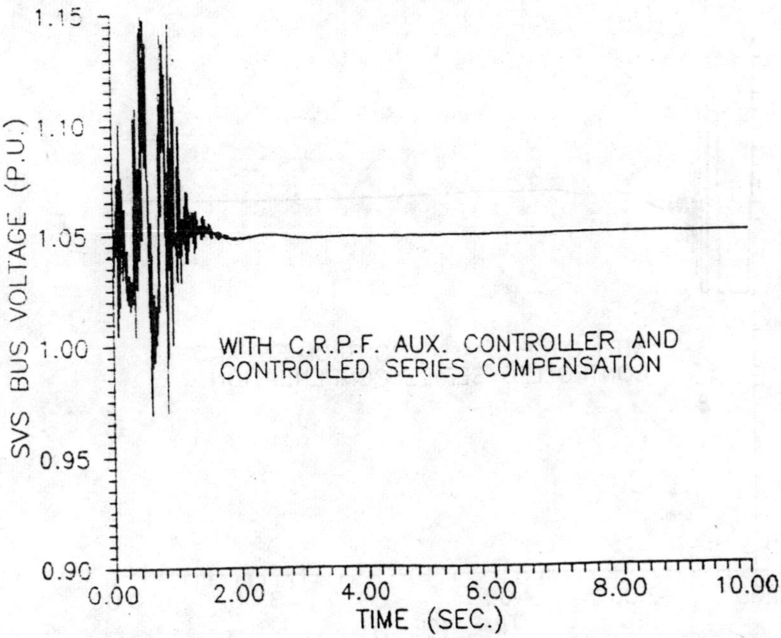

FIG. 7 (b) : *Pg=800 MW, Variation of SVS Bus Voltage due to 40% Increase in Tmech for 0.1 Second (PI-Circuit Model)*

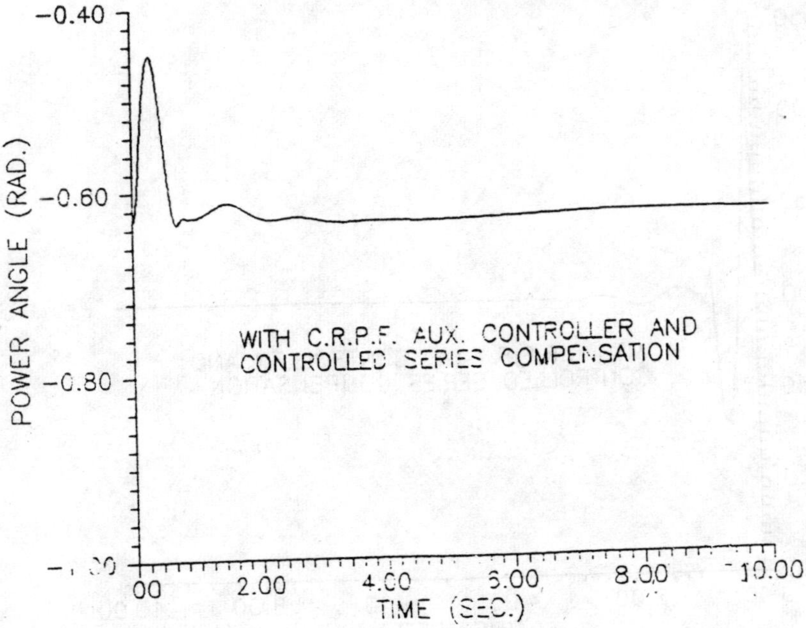

FIG. 7 (c) : *Pg=800 MW, Variation of Power Angle (Rad.) with 40% Increase in Tmech for 0.1 Second (PI-Circuit Model)*

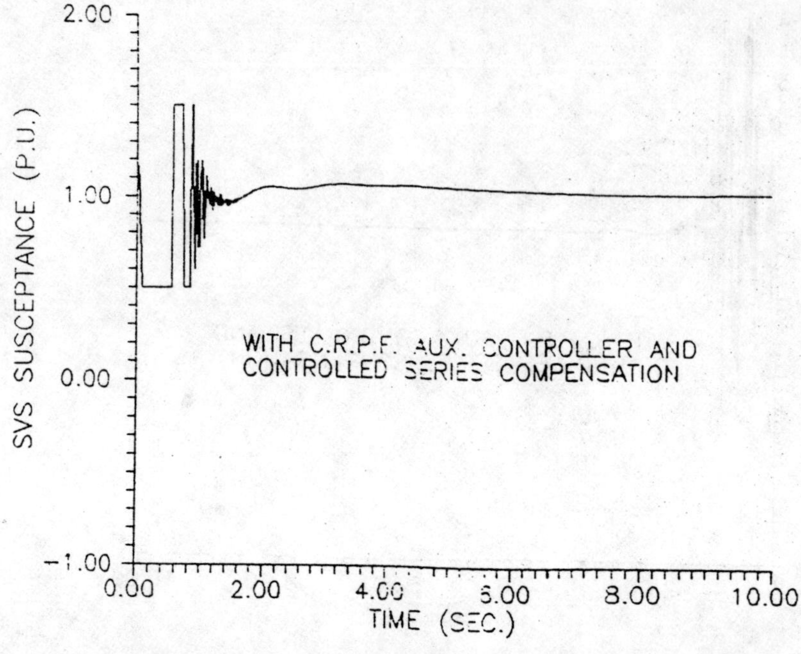

FIG. 7 (d) : *Pg=800 MW, Variation of SVS Susceptance (P.U.) due to 40% Increase in Tmech for 0.1 Second (PI-Circuit Model)*

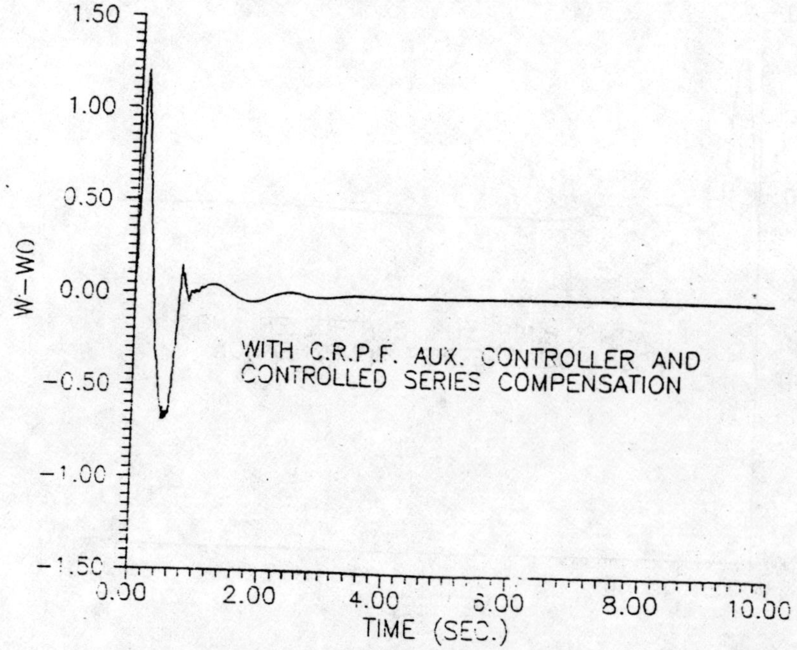

FIG. 7 (e) : *Pg=800 MW, Deviation in Angular Velocity due to 40% Increase in Tmech for 0.1 Second (PI-Circuit Model)*

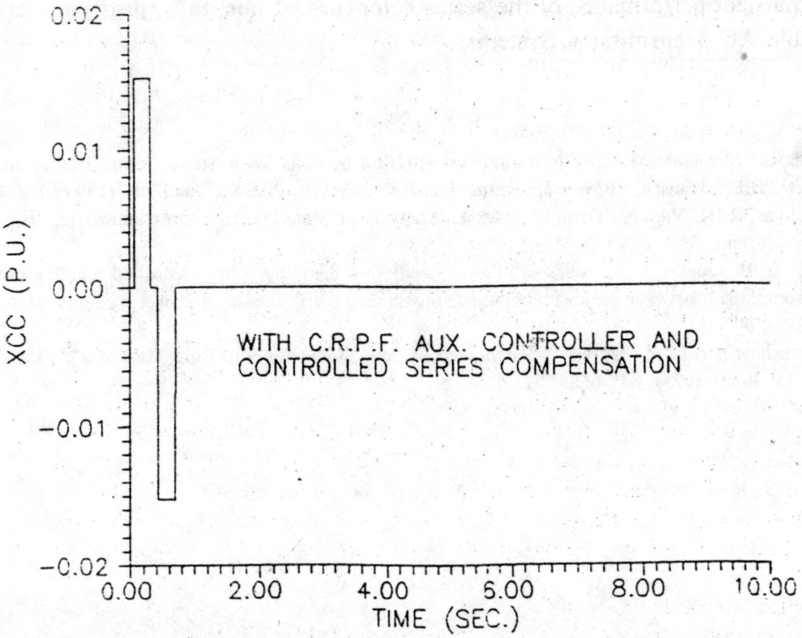

FIG. 7 : (f) *Pg=800 MW, Variation of controlled series compensation due to 40% Increase in Tmech for 0.1 Second (PI-Circuit Model)*

Figures 7a to 7f show the response curves with the C.R.P.F. auxiliary controller in coordination with the CSC. It is seen that the coordinated application of the C.R.P.F. auxiliary controller and CSC smoothens both voltage and power angle oscillations very effectively and better dynamic and transient performance is achieved. The settling time is considerably reduced and efficient damping of the voltage and power angle oscillations is obtained. However, the bang-bang control characteristics of the CSC gives rise to the voltage transients. These voltage transients are controlled by closely restricting the reactive power limits of the SVS (from 0.5 to 1.5 p.u. in the present case) and avoiding the frequent switching of the controlled capacitor.

CONCLUSION

The following conclusions can be drawn from the study performed.

1. The unstable system modes are stabilized by the bus frequency, the line reactive power, and the C.R.P.F. auxiliary controllers developed for the SVS located at the middle of the transmission line. But the largest damping is achieved by the C.R.P.F. auxiliary controller.

2. The coordinated application of the CSC and the C.R.P.F. auxiliary controller developed for the SVS located at the middle of the transmission line greatly improves the

transient performance of the series compensated line, thus, justifying its use in Flexible AC Transmission Systems.

REFERENCES

1. S.C. Kapoor, "Dynamic Stability of Long Transmission systems With Static Compensators and Synchronous Machines, "IEEE Trans. on Power apparatus and Systems, Vol. PAS-98, Jan./Feb. (1979) 124-134.
2. K.R. Padiyar, R. K. Varma, "Damping Torque Analysis of Static Var System Controllers, "IEEE Transactions on Power Systems Vol. 6 No. 2 May (1991) 458-465.
3. L. Angquist, B. Lundin, J. Samuelssos, "Power oscillation damping using controlled reactive power compensation–A comparison between series and shunt approaches", IEEE Trans. on Power Systems, Vol. 8, No. 2, May (1993) 687-700.
4. R. S. Ramshaw and K. R. Padiyar, "Generalised System Model for Slip Ring Machines", Proc. IEE, Pt C, Vol 120, No. 6, June (1973), 647-658.

ENERGY MANAGEMENT BY CONSERVATION MEASURES IN TEXTILE INDUSTRY

54

K. R. CHANDRAN, R. PRAKASAM and P. MUTHUKUMARASWAMY
The South India Textile Research Association (SITRA)
India

ABSTRACT

The Textile Industry consisting of the organised mill sector and the unorganised handloom, powerloom and knitwear sectors is the second largest industry in India. The economic well-being of a country is closely linked to its level of energy consumption. Textile Industries consume about 9-10% of the total energy available in the country. Energy management based on systematic and regular energy auditing has helped the Textile Mills to reduce the energy consumption without affecting production. SITRA is regularly conducting comprehensive energy audit in Textile Mills. Studies have revealed that it is possible to obtain as much as 15 - 20% savings in energy consumption with improvement in equipments and processes. SITRA has evolved Energy Norms, Power Consumption Index and Units per kilogram of yarn or fabric production and has implemented several energy conservation measures. Energy efficient components have been developed/identified and introduced in the regular process after thorough evaluation. The emergence of new generation—high productivity textile machines and their impact on energy consumption is also discussed. There is a compelling need to devise a training and re-orientation programme for the personnel engaged in the industry. Energy conservation, either by its more efficient use in existing processes and products or by the development of improved ones, is ultimately the key to future growth.

INTRODUCTION

The Textile Industry consisting of the organised mill sector and the unorganised handloom, powerloom and knitwear sectors is the second largest industry in India. With 1400 units in the organised mill sector and cluster of tiny units in the decentralized sector, the industry produces 2.09 billion Kg. of yarn, 27.07 billion sq.metres of fabrics and exports Rs.30,000 crores worth of textiles and clothing every year. It is among the highest net foreign exchange earners, with limited import intensity and accounts for 20% of industrial production and 18 million direct employment. As the clothier of the nation, the textile industry's fortunes are closely linked with the country's economy. The rate of economic growth can be greatly facilitated through the increased availability of energy at reasonable prices. Hence, the management of energy supply and demand becomes the primary concern for all developed and developing countries. The auditing of the use of energy to identify the possible areas for efficient utilisation and conservation plays a major role in fulfilling the need for rapid economic development coupled with the aspiration for a better qualify of life.

ENERGY SCENE IN INDIA

India has come a long way since independence in industrializing the country and there have been commendable achievements in the production and consumption of energy too. Still the per capita energy consumption in India is very low with higher energy requirements comparing to the available resources. A comparison of energy consumption in Indian industries with that of developed countries reveals that the energy consumption per unit of product is much higher in Indian industries. Moreover, the planning efforts in India hitherto have been directed towards increasing the generation of electricity rather than improving the efficiency of its use. The transmission and distribution losses are inordinately high for the available power system. Also, the declared weekly 'Power - Holidays' by state electricity boards have far-reaching repercussions on the economy – endangering economic growth and employment opportunities and giving rise to social unrest.

The industrial sector in India accounts for about 60% of consumption and of this, more than 88% is consumed in energy-intensive industries like Iron and Steel, Chemicals, Fertilizers, Aluminium, Textiles, Cement, Paper and Collieries. It is estimated that the conservation potential in industrial sector is 20 - 25%. Numerous attempts have been made to achieve energy conservation either by harnessing energy from renewable sources or by improving the operating efficiency of equipments used in generation, transmission and utilisation of energy.

ENERGY MANAGEMENT IN TEXTILE MILLS

Energy availability and its cost plays a significant role in the production cost. The share of energy cost in the total manufacturing cost is increasing day-by-day and it is ranging from 5 - 17% for different yarn counts. At present, the energy has turned out to be the most expensive item next to raw materials and labour cost in textile industry. Textile industries consume about 9-10% of total energy available in the country. Studies have revealed that it is possible to obtain as much as 15-20% savings in textile industrial energy consumption with improvement in energy equipments and processes. Energy management based on systematic and regular energy auditing will help the textile mills to reduce the energy consumption without affecting production. Loss of energy can be minimised by adopting good operational and maintenance practices, besides continuous updating of technology.

Energy Audit

Energy Audit is the starting point of an energy management plan which can give a positive maintenance and quality control programmes vital for production and utilisation activities. Also, it is an important process for taking stock of the available system, so that it can highlight various issues for improvement. SITRA is regularly conducting comprehensive Energy Audit in Textile Mills and has so far covered 113 mills both within the country and abroad – Philippines, Thailand and Malaysia. The energy consumption pattern of a composite textile mill which consists of spinning, weaving and wet-processing is given in Fig.1. Similarly, the energy consumption pattern of a modern textile spinning mill is depicted in Fig.2. The monthly consumption of energy for a spinning mill with a capacity of 25000 spindles is 7 lakh units (Kwh). According to the prevailing power cost, the monthly energy expense is Rs.17.0 lakhs.

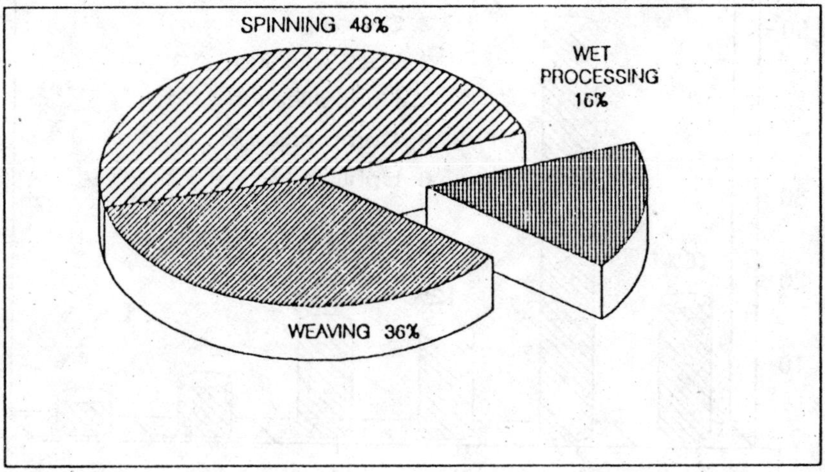

FIG. 1 : *Energy Consumption Pattern in A Composite Textile Mill*

SITRA has evolved Energy Norms, Power Consumption Index (PCI) and units per kilogram (Ukg) of yarn or fabric production for textile industry. The PCI is the ratio of actual energy to the expected energy based on SITRA energy norms. This rates the performance of the mill from the industrial average and 100 is the normal PCI. But anything higher than that indicates the level of deviation on the higher side from the average. The PCI of a typical modern textile spinning mill with spindle capacity of 15,000 is given in Table-1. Based on the production data provided by the mill under consideration and actual measurements made during the audit, the Ukg of yarn production for 40s converted has been computed. This has been compared with the SITRA Energy Norm and the details are tabulated in Table-2.

Based on the computed PCI, Ukg. and Energy Norm, further in-depth analysis has been made to find out and pin point the excess energy consumption in a particular machine. This, further analysis is based on energy consumption of main drives, auxiliaries, machinery parameters, processing parameters, electrical parameters and electrical system in the textile mills. The PCI of a modern mill equipped with new generation — high productivity textile machines is about 2.5 times as that of an equal capacity mill with conventional machineries. The excess consumption is mainly attributed to the higher consumption in auxiliaries, like dust extraction plants, common pneumafil suction, compressors, fully automatic air-conditioning plants etc., and high level of automation. This auxiliaries consumption accounts for about 45% of total consumption. This has to be compensated by way of improved quality, higher earning through exports and labour complements.

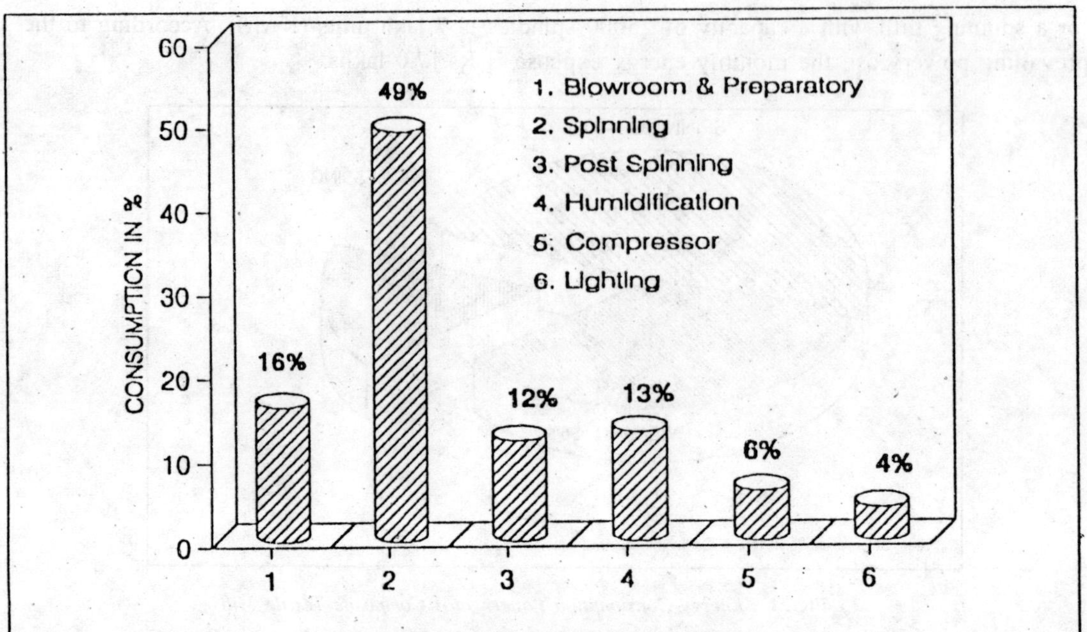

Fig.2 ENERGY CONSUMPTION PATTERN IN A MODERN SPINNING MILL

TABLE-1 : **Power Consumption Index**

Month	W. Days	Units	Count	Speed	PCI
MAR '95	31	727376	36s	13512	185
APR '95	29	653273	36s	12926	160
MAY '95	30	686621	36s	12815	170
JUN '95	30	717378	36s	14395	183
JUL '95	30	721392	36s	15180	166
AUG '95	30	745640	39s	14541	149
SEP '95	30	693577	38s	14496	157
OCT '95	28	599543	37s	13637	158
NOV '95	30	607048	38s	14773	140
DEC '95	31	604945	37s	14685	141
JAN '96	29	585443	36s	14728	110
FEB '96	29	612846	36s	14368	133
TOTAL	357	7955082	37s	14171	154

TABLE-2 : Comparison of Ukg

Dept	Production (40s Conv.) in kg.	Units Consumed in Kwh	Conv. 40s Ukg	SITRA Norm 40s Ukg
Main Processing Machineries				
Blowroom	110261	18012	0.16	0.16
Carding	103048	39210	0.38	0.22
Drawframe	98591	15390	0.16	0.10
Flyframe	98296	14787	0.15	0.12
Ringframe	94973	241416	2.54	2.32
Autoconer	94500	50040	0.53	0.05
Sub Total		378855	3.92	2.97
Auxiliaries				
Humidification	94500	50800	0.54	0.40
Compressor	94500	25400	0.27	—
Lighting	94500	20100	0.21	0.15
Others	94500	2600	0.03	0.04
Sub Total		98900	1.05	0.59
Total		477755	4.97	3.56

Energy Conservation Measures

Energy consumption is not distributed evenly over the many operations of a textile mill. The energy conservation measures undertaken by SITRA in various departments of a textile mill viz., spinning, preparatory, post-spinning, humidification, compressors, lighting and power system are enumerated with suitable energy efficient components.

Ring Spinning

In a textile spinning mill about 50% of energy is consumed by ring spinning department alone. SITRA has conducted exhaustive energy conservation studies on ringframe, spindles, spindle tapes, oils, bobbins, belts, motors and package size. The energy consumption is increased as a square of the spindle speed. Significant amount of saving is achieved in ringframe energy consumption but still the ringframe spindle and bolsters together with spindle driving tapes account for 30-40% of total energy consumption and provides more scope to conserve energy. Spindle power is very much influencd by (a) Weight and wharve diameter of spindle, (b) Type and quantity of spindle oil used in bolster and (c) Type and weight of spindle tape.

SITRA energy saving spindles with 205 g weight and 20.2 mm wharve diameter consume 13% lesser energy than the conventional spindles with 325 g weight and 25.0 mm wharve diameter. The payback period is only 2 years. Similarly spindles with 18.8 mm wharve diameter saves 3.6% of the total ringframe energy consumption when compared to spindles with 20.2 mm wharve diameter. Spindle power in ring spinning frame is also very sensitive to the quantity of oil used and its viscosity. The energy saving in synthetic spindle oil over the conventional mineral spindle oil is about 2-4%. The minimum saving in using synthetic spindle oil is about Rs. 5.00/spindle/year. Spindle tape is another important component of spindle power as it accounts for about 17-20% of total ringframe power. SITRALISED energy saving tape is lighter than synthetic tape and consumes 1.4% less power. The saving in using SITRALISED Energy Saving Tape is Rs. 4.00/spindle/year.

Energy consumption is influenced significantly by the mass of objects which are rotating at higher speeds. The lower weight of plastic bobbins with 22-25 gm instead of 30-35 gm weight bobbins save 3.6% energy consumption. The lower weight plastic bobbins save Rs.15.00 spindle/year. Package size in a ringframe is determined by the spindle lift and ring diameter. At present, 7" lift spindles and 42 mm diameter rings are predominantly used in ring spinning frames for most of the yarn counts. Smaller package size, apart from saving significant amount of power, also leads to reduction in end-breakages because of lower tension acting on the yarn balloon which would enable the mill to go in for higher spindle speeds. Therefore, for yarn counts 60s and above 6" × 38 mm size can be used to optimize ringframe energy consumption.

Motor plays a dominant role in power transfer and high efficiency motor saves about 1-3% of the total ringframe power because of its superior design and construction. The five characteristics viz., light-weight, thin, strong abrasion resistant and elongation free are important for any belts. The Flat belts when compared to V-belts save energy to the tune of 3-4% of total consumption of a ring spinning frame. Tin-roller consumes about 10% of total ringframe power. The lower weight individual ABS pulleys instead of tin-rollers save considerable energy in ring spinning frames.

Preparatory

The textile machineries like, blow room, carding, drawing, combing and flyframe constitute the preparatory section. The energy consumption of these machines is about 16% of the total energy consumption of the mill. Substantial saving is achieved in using one pneumafil box for two cards instead of individual pneumafil motor for each card. Generally, motors in the drawing and flyframe sections are loaded below 50% even after full doff position. Fluid coupling drive or electronic soft-start drive saves energy. Flyframe bobbins made of polypropylene instead of high density polyethylene save energy to the tune of 5% and the payback period is just 2 years.

Post-Spinning

The energy consumption in post-spinning section is about 12% of the total consumption. This section constitutes winding, doubling, two-for-one twisters, singeing and other machines depending upon the end-product of that particular textile mill. In recent years, the energy consumption of the post-spinning machines have increased tremendously. Compared to Rotoconer winders, Auto-coners consume 5-6 times more energy because of the high level of automation and yarn quality achieved. The conservation potential in auto coners is very minimum. Most of the measures taken in ring spinning frame can be directly applied in doubling frames.

Humidification

Humidification system provides required relative humidity and temperature for textile processing and better environment inside the departments. It consumes about 13% of the total energy consumption of a mill and it consists of supply, exhaust and pumping systems. In multi-fan system, the non-working supply fan should be isolated. Otherwise the stationery fan will rotate in the opposite direction and act as a supply fan to the working fan and then create a local circulation rather than supplying humidified air to the department. Pneumafil air must be collected in a separate duct and a positive exhaust must be provided for the same to avoid back pressure in the pneumafil system. The capacity of the exhaust and pumping system should be 80% and 45% of the supply system respectively to avoid excess consumption. Provision of Rotary filters for exhaust systems, FRP fan instead of aluminum fan, V-type filters for supply systems, PVC clamp-on type water nozzles, attic area ventilation, separate energy meters for plants are some of the other measures for conservation of energy. Above all, about 5% of energy can be saved only through proper plant operation and maintenance.

Compressors

Most of the modern textile machines are equipped with compressed air handling systems. On an average, about 6% of the total power bill of the mill accounts for the generation of compressed air. Reciprocating compressors consume more power compared to screw compressors. As the screw compressors are designed for higher capacities, they can be selected for the capacities above 100 cfm compared with the reciprocating compressors. Refrigerated air dryers are preferable when compared to heatless air dryers. The energy loss in heatless air dryers is about 5-15% due to purging of compressed air. Centralization of compressors saves about 5% of energy consumption. High pressure compressed air at about 7-8 Kg/cm^2 should not be used for cleaning purposes and air leakages should be arrested completely and monitored periodically.

Lighting

Illumination level required in textile industry is around 75 lux in spinning and in all other departments around 50 lux. The energy consumption of lighting is about 4% of the total mill consumption. Installation of servo controlled voltage stabiliser at power house for lighting circuit saves energy. Installation of electronic ballast in place of conventional ballast saves energy. Replacement of 40 W tube fitting to 36 W slim tube fitting conserves energy. Low pressure sodium vapour lamps are better for outside lighting.

Electrical Systems

The transformer yard should be as close to the power house — MV panel room as possible to avoid wastage of energy. As far as possible, multiple transformation i.e. installation of intermediate transformers should be avoided. The location of the power house should be near to the load centre of a textile mill i.e. Ring spinning department to save substantial amount of energy.

Acquisition and periodic monitoring of energy data by proper instrumentation help to save energy in the shop floor level. Proper selection of cables, motors, installation of power capacitors, separate energy meters for different departments are some of the other measures. Apart from this, so many other valuable Energy Conservation tips are available at SITRA, for both electrical and thermal energy utilisation in textile industry.

Training and Awareness Programme

SITRA is advocating to form a separate Energy Cell in each and every textile mills through its training and awareness programmes. The Energy Cell consists of Chief Executive / decision maker and Personnel from different departments including costing, personnel and utility sections. The Energy Conservation Culture has been created now in textile industry to save all forms of precious energy.

CONCLUSIONS

The growth of the textile industry worldwide will come from increasing population and per capita income in the developing countries. India has got all the advantages necessary to be a leading player in world trade in textiles particularly in the post GATT era. In spite of the many-fold productivity and quality improvements from the latest generation of textile machinery, power consumption per unit of production is generally higher than the older conventional machineries. With the conservation measures initiated by SITRA, an overall saving of about 10% has been achieved by the textile mills. Energy conservation Programme is not an one-time affair and it is a continuous process. Both Central and State Governments should declare that the Energy Audit is a statutory obligation for every industry. It may be noted that investment needed to save one unit of energy is substantially less than the investment needed to supply an incremental unit of energy at the point of use. There is a compelling need to devise a training and re-orientation programme for the personnel engaged in the textile industry. The challenge for the coming 21st century therefore lies in motivating the inefficient mills to adopt efficient energy practices and improvised technology. Energy conservation either by its more efficient use in existing processes and products or by the development of improved ones, is ultimately the key to future growth.

ENERGY AUDIT TO MARSHALL ENERGY RESOURCES

55

P. C. THOMAS
Kirloskar Electric Co Ltd
(INDIA)

ABSTRACT

This is the age of energy shortages. It is generic to almost every industry. This deficit can be partially offset at three levels.

(A) the energy policy level
(B) the macro level
(C) the consumer level

The first-two approaches involve the industry associations, entail capital expenditure and also promise large energy savings. This paper focuses on the energy audit method at the third level. It is inexpensive and totally within the control of the user. More importantly, energy management at the consumer level is a must for the exchequer.

The energy level configuration consists of the material input, energy input and the manpower input to the value addition process, leading to the finished goods output. Thus, the consumer has to contend with a minimum quantum of finished goods output and a maximum allowable quantum of energy input. Hence, the user is appraised of a specific energy consumption (energy consumption per unit of production) which is crucial to his profitability. The energy inputs can assume several primary forms as well as its derivative secondary energy forms which together feed the value addition process.

Hence it is imperative that all energy inputs (in either primary or secondary form) be tracked, identified and quantified in its route to the final value addition point. The basic aim of the energy audit is to minimize the specific energy consumption. .This paper is a precursor to the energy audit methodology.

INTRODUCTION

'In the present scheme of things, energy shortages will be a recurring feature for some time to come. This can be tackled at three levels

(A) The policy level
(B) The macro level
(C) The consumer level

It is the objective of this paper to remedy the situation at the third level by employing the energy audit. Essentially, this powerful tool identifies the energy consumption per unit of production at major production centres. This figure is also known as the SEC (specific energy consumption). Once the SEC is determined, heat/energy balances are drawn up for every production centre and for every major process. This helps in properly identifying the areas where energy losses can be contained.

THE APPROACH

Generally, the energy audit is conducted on the following lines.

(A) Energy survey
(B) Quantification of production in major centres
(C) Ensuring a minimum of energy metering
(D) Study of load patterns
(E) Quantification of energy usage in major centres
(F) Detailed energy studies of various energy forms
(G) Methods study
(H) Determination of SEC
(I) Recommendations

Energy Survey

At the consumer level, the energy configuration is depicted in figure-A 1. In this value addition process, the user is constrained by

(A) Maximum permissible energy allowed (by regulatory and market conditions)
(B) Minimum amount of production (for break-even operation)

The user is appraised of a SEC which is crucial to profitability. At the elementary level, a modified SEC, involving the percentage of total energy costs to the total value of finished goods, can be worked out. If this value is higher than about 10 % , the commissioning of the energy audit is justified.

Quantification of Production

Imperative that a common production figure-A 1 can be expanded to include several zones of value addition. This is shown in figure-A 2. Thus, the finished goods output of one department will form the material input for its customer department. Hence, if the SEC of each department is to be determined, the corresponding production of that department is to be known. A problem is that of a variety of production items. Thus, the commonality of the production unit is lost. Therefore this unit can be fixed, say, on the basis of material weight, which can be computed easily.

Energy Metering

A minimum of energy metering should be rigidly complied with, for an effective energy audit. Generally, there is adequate power metering at the substation level, power house, compressor room and for some major process equipment. Correspondingly, at the tanker entry point, fuel is also metered satisfactorily. However, at downstream points, metering is generally inadequate. If the survey indicates a gross lack of metering systems, the energy audit should be postponed until adequate metering is installed.

Study of Load Patterns

The study of the load pattern at the consumer level is a very important one. This study consists of monitoring the load pattern on an hourly basis for 24 hours, spread over a minimum of two weeks, especially covering the last week of the month. It consists of the electrical and fuel load pattern study. The electrical part determines the approximate instance when the maximum demand KVA occurs. Since the maximum demand attracts a penalty, a proper analysis of the connected load will help in reducing the demand. The fuel pattern study is of significance to consumers who operate arc furnaces, heat treatment furnaces, induction furnaces, tunnel kilns and other heat chambers. In these cases, uniform material loading spread out over the available shifts help in containing heat losses and prevent cold start-ups.

Quantification of Energy Usage

Every production centre involved in value addition, uses many kinds of energy inputs. These comprise of the primary energy forms such as

(A) Power
(B) Fuel – LSHS, FO, LDO, MS, Kerosene etc.
(C) Fossil fuels - coal, coke, lignite, peat, fines etc.
(D) Others

These may also use secondary energy forms. These are derivatives of the primary forms and may consist of

(A) Compressed Air
(B) Hot Thermic Fluid
(C) Hot Drying Air
(D) Hot Water
(E) Water Gas
(F) Producer Gas

Most often, the value addition process is fed by a mixture of both energy forms. For this purpose, a commonality of energy units is a must. Hence, the units KwH, BTU, Kcal (whichever is applicable) are to be reduced to any one unit. This commonality has got an intrinsic safeguard, in that, savings in one kind of fuel cannot be offset by excessive use of another kind. The total energy is thus monitored effectively.

Detailed energy Study

Compressed Air

Compressed air is an important utility. Through the medium of the unloader, the compressor motor runs at load and no-load alternatively. Hence, energy conservation is already built in. However, air leaks in the oil traps, filters and nozzles account for about 90 % of the leaks and about 20 % of the energy losses. These form potential power savings. The extent of losses can be assessed by determining the time taken for a full cylinder to drop from rated pressure to about 15 psi.

Soft Starters*

Energy savings during motor starting were introduced about a decade back. By the artifice of employing soft starters. These invertor drives reduce the voltage impressed on the motor terminals during starting and gradually increase it on attaining the rated speed. Substantial energy savings have been credited to these drives. Care should however be taken to ensure that the starting torque of the motor is not sacrificed.

Power Control of Processes

Many processes are run by equipment supplied by Oem's. These processes have single or multistage operation(s). The power levels for the various stages are generally specified by the Oem's. However, in practice, the working conditions and material feed characteristics require changes in power levels. The correct levels can be determined if a heat balance is drawn up for the process. A case in point is the alloy steel route through the arc furnace by scrap feed. Here, the power levels required can be determined by working out the heat balances for the melting, reduction and oxidation stages separately.

Optimal Motor Sizing

Motors generally operate at higher efficiencies when loaded at about the rated value. This feature translates into a higher operating power factor, reduced losses and lower power input. Energy savings can be realized by replacing very lightly loaded motors by motors of smaller ratings. This is generally inadvisable in the case of equipment supplied by an Oem. For example Machine tools. However, utility motors can accept such changes , even to the extent of downsizing the frame size itself. This is subject to changing the mounting dimensions of the motor. Consumers having large repair shops and sufficient stocks of a variety of medium range motors are in a position to adopt this technique.

Drawing Heat Balances

The heat balance is an elementary, but very versatile technique of integrating an energy input into its component losses. This can be done on an electrical or thermal basis. The latter one is of more interest to the consumer, especially when furnaces, kilns and ovens are used. The electrical balance is of little use to the consumer. However, the thermal heat balance can only be calculated by the user. Once the constituent losses are known, the benefits of any proposed improvement(s) can be reckoned easily.

Preheating of Charge

Preheating of charge, by itself is of relevance to only the foundry. However, the technique can be used by others too. In any production department operating continually for more than two shifts, the cycle time for a process is reduced by means of performing the preliminary process on an auxiliary equipment. The auxiliary equipment is much smaller than the main one . It acts as a feeder to the main equipment.

With specific reference to the jobbing foundry and the investment casting industry, baled scrap was preheated in the heat treatment furnace and then fed as main charge to the electric arc furnace and induction furnace. A combined heat balance for both processes determines the exact quantum of power savings. Faster throughput is an extra advantage.

Improvement of Combustion Efficiency

This is applicable to consumers of furnaces, kilns and ovens where burners are widely used. Maintaining a CO_2 percentage of about 17 % in the flue results in good combustion efficiency. This can be monitored by the use of a spirax analyzer at frequent intervals by the operator, with accompanying corrections of the blower air and fuel inlet settings.

Study of air-conditioning plants

Air conditioned zones form a large area to certain consumers. The compressor motors constitute a sizeable load in such cases. Since these operate on load and no-load alternatively, the scope of energy savings lie in increasing the no-load operation period. This is achieved by reducing the losses in the air-conditioned zone. The first step lies in drawing up a heat balance for the zone, taking into account air-locks, wall radiation, occupancy, cladding and equipment losses. Subsequently, insulation schemes, rerouting of personnel movement and other methods can be employed to reduce the losses.

Optimal Loading of Heat Chambers

Heat chambers pertain to furnaces, kilns and ovens. Optimal loading material to be processed in such chambers, yield energy savings to the extent that the job absorbs more of the useful heat than the furniture and fixtures within the chambers. This is a scheduling exercise and has got a relationship with the fuel demand pattern study.

Methods Study

Some of the detailed energy studies discussed demonstrate their affinity to the methods used in the processes. In fact, the methods study is of more importance, when one considers that semi-automatic processes outnumber the automatic ones. The methods study, traditionally done by the industrial engineering group, thus needs to be totally integrated with the energy study for comprehensive benefits .

DETERMINATION OF THE SEC

After having quantified the production units and the energy units in every major production department and important process, the SEC can be determined individually. This is the starting point for any improvements leading to energy savings. It also serves as a reference benchmark for evaluating the cost-worth analysis of a new energy-efficient equipment. Processes can thus be defined and streamlined. This is also where the work starts in earnest.

Incidentally, the gross SEC that is quoted by many industries, is actually a misnomer. These figures refer only to the final value addition process. Ideally, it should cover all the processes right from the receipt of material upto shipment.

CONCLUSION

Energy conservation is a small step towards alleviating the energy crisis in the country. However, it is significant when one considers that a multitude of users can contribute collective savings by private enterprise. In fact, it is encouraged to a great extent by the State energy development agencies. An energy survey will reveal the extent of proper management of the energy resources available to the consumer. If found feasible and necessary, the full scale energy audit can be carried out to assess specific energy consumption levels and patterns on a detailed scale. On a case-to-case basis, engineering recommendations, methods improvements, process modifications and introduction of energy efficient equipment are made, with accompanying cost-worth analyses and pay-back periods. The adage of an unit saved is worth two generated is worth thinking about.

LIST OF APPENDICES

FIG . A1. *The energy configuration*

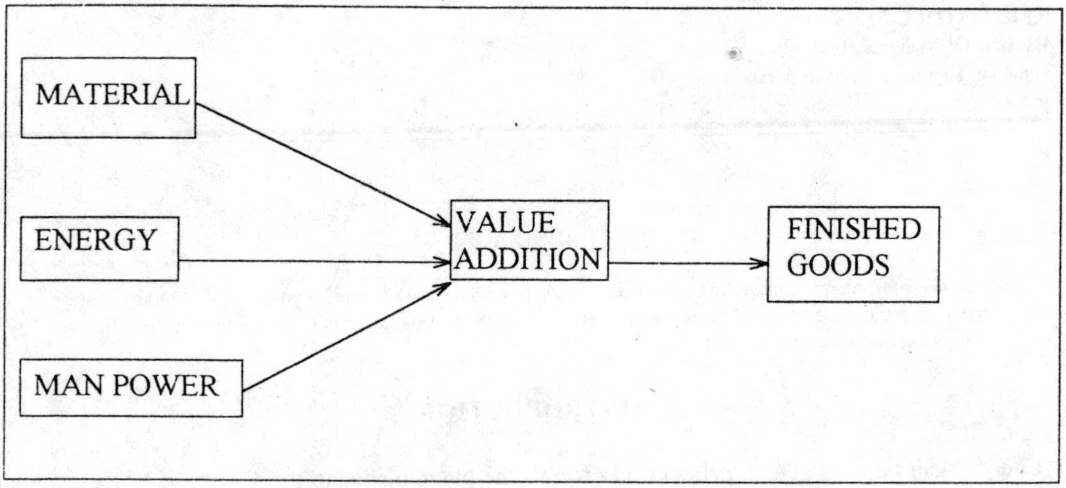

FIG. A2 : *The Production Center Configuration*

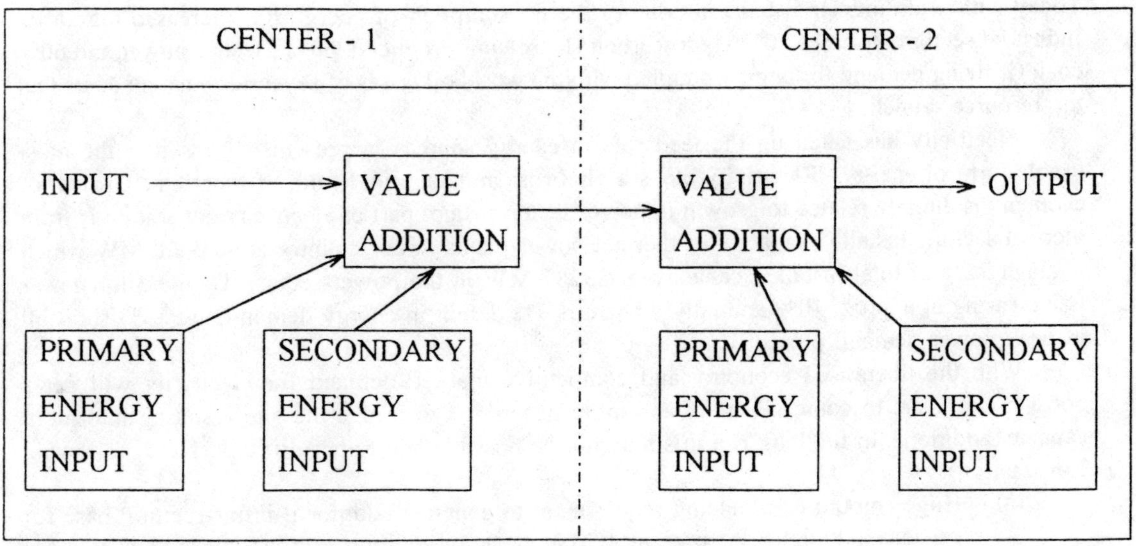

ROLE OF NTPC IN ENERGY CONSERVATION

56

B. N. OJHA, Executive Director (OS)
National Thermal Power Corporation
A. K. MAITRA,
Dy General Manager (EC)
National Thermal power corporation,
India

ABSTRACT

The need for energy conservation in power sector is emphasised. A brief renovation and modernisation schemes are described for various power sources. The critical parameters to be monitored for effective energy conservation are presented.

INTRODUCTION

"Energy" can be termed as a primemover for all the phenomena in the Universe. Advancement of human civilization corresponds to ever increasing demand of energy. Modernisation is, therefore, synonymous to the growing consumption of energy.

Since the fifties, our country has been witnessing increase in industrial production capacity. Today with wide industrial base in country the consumption of energy has increased manifold. Industrial sector consumes 20% Hydrocarbon (Petroleum Products), 60% Electric power and 60% coal. Growing demand for energy coupled with the world oil crises of seventies have made us feel the resource crunch.

Electricity has taken up the lead role in energy sources in present day, as it is the most usable form of energy. Electricity plays a vital role in economic health of a nation. Growth of economy is directly related to growth in power sector. Major part of electricity generation is from thermal sector. Installed capacity of thermal power plants in our country is 60398.8 MW which is about 72% of total installed capacity of 83287 MW in the power sector. Demand for power is increasing at approx. 10% annually. There is 7% deficit in energy demand and 15% shortfall in peak power demand presently.

With the liberalised economy and competitive market, demand for electricity will get a boost in the years to come i.e. more and more generation to bridge the gap resulting into more capacity addition. In the long run this may not be possible and rather difficult due to following constraints:

(a) Huge amount of financing requirement to generate additional infrastructural base for capacity addition. At present 1 MW capacity addition implies investment of more than rupees 3.5 crores. Further, this amount will be much higher with the incorporation of various environmental protection devices like FGD etc.

(b) Stringent environmental protection laws and increased amount of Co, CO_2 and NOx level in environment further poses restriction against coming up of new fossil fuel fired thermal power plants. Nuclear power plants also have different type of environmental hazards. Already developed countries' power utilities have been under severe restriction in this regard.

(c) Depleting natural resources further signifies the necessity to use energy in an efficient way. The emerging difficult future power scenario leaves us with no option other than to seriously think of following course of action:

(a) Energy Conservation; and

(b) Energy substitution (by non-conventional source of energy i.e. wind, hydel, solar etc.)

'Energy Conservation' is simply a "Demand side management" instead of conventional "Supply (Capacity addition) side management". i.e. to reduce the demand of consumer without affecting/curtailing his due power requirement which means reduction of energy wastages" and maximisation of efficiency of energy consumption and energy production.

Of late, the attention of our planners have been drawn towards energy conservation and thus conserve environment also. This is evident from the fact that in VIII plan, capacity addition of approx. 5000 MW is envisaged through energy conservation alone. The seriousness of it can be gauged from the fact that Ministry of Power has initiated a separate wing i.e. Energy Management Centre to help promote and identify schemes of energy conservation, be it generation or distribution and adopt policies for rational and efficient use of energy in end use applications.

ENERGY CONSERVATION IN POWER SECTOR

NTPC is today the biggest power utility in the country contributing about 20.2 % in total installed capacity, 27.8 % in thermal installed capacity, 31.2 % in thermal generation and 24 % in total electrical generation in the country. Its performance parameters like PLF, Availability, Fuel Oil Consumption, Auxiliary Power Consumption, Heat Rate etc. are amongst best in the country. This could be possible through better design selection, O&M practices and dedicated efforts towards Energy Conservation. Awareness Campaigns, Workshop, Seminars etc. are being held on regular basis to spread the message of Energy Conservation amongst all sections of society in NTPC. Energy Audits and Quality Circle Meetings have become regular features for identification of areas of Energy Conservation. Besides, NTPC is also concerned about Conservation of Conventional Sources of Energy by using sustainable and Non-Conventional Sources of Energy like Solar Power, Wind Power, Mini Hydels, Bio-Mass etc. A number of renovation and modernisation schemes are in the pipe line for updation of technology for improvement in plant performance and quality of environment under its sustainable energy development programme.

In case of power sector energy conservation is important due to following reasons:

(i) As per national average, 10% Aux. Power is being consumed by power industry itself. As per 1995-96 figures the country generated approx. 397700 MU (thermal) i.e. 39770 MU had been lost as Aux. Power. 1% improvement in Aux. Power means a saving of the order of 3977 MU i.e. $ 20 million approx.

(ii) On the same lines reduced "Specific Oil Consumption" and "Sp. Coal Consumption" will not only result in efficient generation of power but will also save precious foreign exchange and natural resources of the country. It may be an indicator to the nation that Specific Oil Consumption in NTPC is being maintained around 1 ml/Kwh as against a national average of about 5 ml/KwH.

(iii) Energy Conservation efforts in power sector will ultimately become a wide base for industry and manufacturers to bring in new, innovative energy efficient devices.

(iv) It will give due impetus to demand side management of grid/consumers as huge energy conservation margins are left to be tapped with Agricultural Pump sets, Compressors, Blowers etc.

(v) With reduced power consumption at consumer end Corresponding margins will be available at transmission losses which in turn will result in cascading of savings due to energy conservation efforts.

ROLE OF NTPC IN ENERGY CONSERVATION

As a pioneer in the field of electricity generation, NTPC had taken lead in the field of Energy Conservation. NTPC philosophy of Energy Conservation is broadly based on the following themes:

1. Efficient use of existing energy resources.
2. Examination of Energy use pattern to improve energy efficiency and curtail waste.
3. Energy Conservation to maximise power generation.
4. Energy Conservation as an essential element of environment protection.
5. Energy Conservation as a measure to improve nation's economic health.

In order to implement the above philosophy into action, NTPC has created Energy Conservation Cell at Corporate Office and at each of its projects. Further, energy circles consisting of experts from various concerned departments in the power plant, like operations, maintenance etc. have been formed for each stream/system. These energy circles deliberate on various measures to improve the energy efficiency of equipments and system. Finally the implementation of the energy conservation measures are carried out by operation and maintenance division. The energy conservation cells also organise regular interaction amongst different category of working personnel, awareness campaigns to inculcate deep sense of commitment for energy conservation in the employees without whose active involvement any tangible success in conservation efforts is simply not possible. The activities of the energy conservation cells at these units are closely monitored by corporate energy conservation cell.

MONITORING CRITICAL PARAMETERS EFFECTIVE TOOL OF ENERGY CONSERVATION

The term Energy Conservation is of significance in the context of Coal fired thermal power stations. The main emphasis being on reduction of inputs required for generation of each unit of electricity on one hand and increasing the output with the above referred inputs on the other. The salient areas that needs to be attacked in chronological order are as under:

1. Increase in the availability of the station in turn leading to higher PLF.
2. Reduction in partial loss by increased availability of auxiliaries by carrying out scheduled preventive maintenance programmes.

3. Reduction in heat rate, fuel oil consumption, auxiliary power consumption and DM water consumption.
4. Improvement in condenser vacuum, control over excess air and unburnt carbon loss.
5. Optimisation of start up time and procedure.
6. Effective and efficient lubrication of all equipments.
7. Periodic calibration of energy meters.
8. Bringing awareness among the masses for the need for energy conservation.

Energy Conservation Areas

1. Heat Rate

For reducing heat rate following areas were taken up:

(a) Reduction in Coal Consumption.
(b) Reduction in Furnace Oil Consumption
(c) Reduction in D.M. water consumption.

(a) Coal Consumption

The units were operated near to their designed parameters like pressure and temperature etc. and areas of heat losses were identified and attended.

Condenser Vacuum: Efforts were made to keep it at design value.

Heaters: Preventive maintenance was done in a way that heaters remained in service for maximum duration.

Air Ingress: Air ingress was controlled for attaining maximum boiler efficiency.

Coal Fineness: Classifiers of coal mills were maintained and adjusted to keep coal fineness to within design limits thereby keeping combustible at bottom ash and fly ash within design limits.

(b) Furnace Oil Consumption

Furnace oil consumption was controlled by monitoring oil consumption in every shift. The oil support is being taken only when it is technically required. Every shift, analysis for oil consumption is being done and discussed.

Start up: A systematic start up procedure have been adopted to minimise start up time and wastage.

Fuel Oil Header Pressure: Fuel oil header pressure was controlled in a way that oil consumption per hour at each burner was optimised.

(c) D.M. Water Consumption

DM water quality like presence of Silica etc. was removed to reduce blow down in boiler. Steam leakages and DM water leakages were attended on line. DM water being wasted were recycled.

2. Auxiliary Power Consumption

Auxiliary Loading

Main plant auxiliaries running hours, their power consumption etc. were calculated and compared with readings on energy meter. Similarly power consumed by different auxiliaries in off site areas with respect to their running hours were calculated and compared with advances on energy meters. The areas identified for higher consumption were analysed. In order to optimise coal handling and minimise breakdown in coal handling area, exercises for improving coal quality were also taken up with Coal India Authorities. Their help minimised receipt of big stones and foreign materials leading to reduction in auxiliary loading.

Close Monitoring of MDBFP Operation:

Partial Loading: The partial loading was minimised by carrying out scheduled maintenance of equipments, arresting air ingress, leakages etc. This optimised ID Fan, PA Fan and other auxiliary loadings resulting in minimising partial load at the station.

Calibration and Testing of Meters: In order to have accurate metering within permissible limit of error metering circuit and meters were timely tested and calibrated.

The above actions taken with dedicated team approach have improved the human as well as machine productivity by reducing the consumption of coal, oil, auxiliary power and spares etc. In addition, this has resulted in maximum capacity utilisation improving plant load factor and stability in generation as also brought about improvement in Energy Conservation areas like Heat Rate, Specific Oil Consumption, Aux. Power Consumption and plant outages etc.

SPECIFIC MEASURES TAKEN TOWARDS ENERGY CONSERVATION

In NTPC Energy Conservation activities have been taken up in institutional way and with zeal and enthusiasm. Apart from conducting "ENERGY AUDITS" and close monitoring of equipment/ systems, following specific measures have been undertaken:—

(i) Split Coal Nozzles

(a) To reduce the minimum load on the unit without the use of oil support.
(b) To reduce emission of Nitrogen Oxide.

(ii) *Energy Conservation Through Condenser Care:*
On load tube cleaning system have been commissioned with the philosophy that clean condenser must remain clean. The results are quite encouraging.

(iii) *Energy Conservation by Automatic intermittent evacuation of electrostatic precipitator hoppers*
This modification reduces the water consumption significantly and saves lot of pumping power (both ash water and ash slurry) contributing in a big way to Energy Conservation.

(iv) *Energy Conservation by Pulse Energy Station of Electrostatic Precipitator.*
The technique has been adopted for proper precipitator of highly resistive dust. This not only reduces stack emission level but also contributing in a big way in conservation of energy.

(v) *Ultra-To-Visible Flame Scanner Conservation Kit - an effective means of Energy Conservation*
The alteration not only resulted in terms of saving of huge amount of HSD (Earlier used in ignitors) but also brought safety of boiler and working personnel.

(vi) *Densoveyor System of Mill Rejects*
This new system of mill rejects brought sea change in environment and conservation of hot air (210°C) through mill reject and thereby improved mill/boiler efficiency.

(vii) *Variable Frequency Drive Motors*
These are quite efficient particularly at low load operations.

(viii) *Heat of Compression Type Driers*
Instead of conventional driers used in instrument air compressors, these type of driers have been used resulting in good amount of energy saving.

(ix) *Modification in C.W blow down scheme at BALCO T.P.S.*

(x) *Energy Conservation Through Lighting*
In NTPC modification in various lighting system has been taken up in a systematic way. Results are self encouraging and there is substantial saving by this activity alone.

NTPC has come a long way in establishing itself as one of the most efficient thermal power utility in the country. This has been achieved primarily due to concentrated efforts by NTPC O&M personnel towards energy conservation. NTPC has not only improved its own efficiency but also improved performance of other utilities by imparting them expertise in the field of energy management. NTPC's sustained high levels of performance is self explanatory as viewed in annexures A, B, C & D.

ANNEXURE A

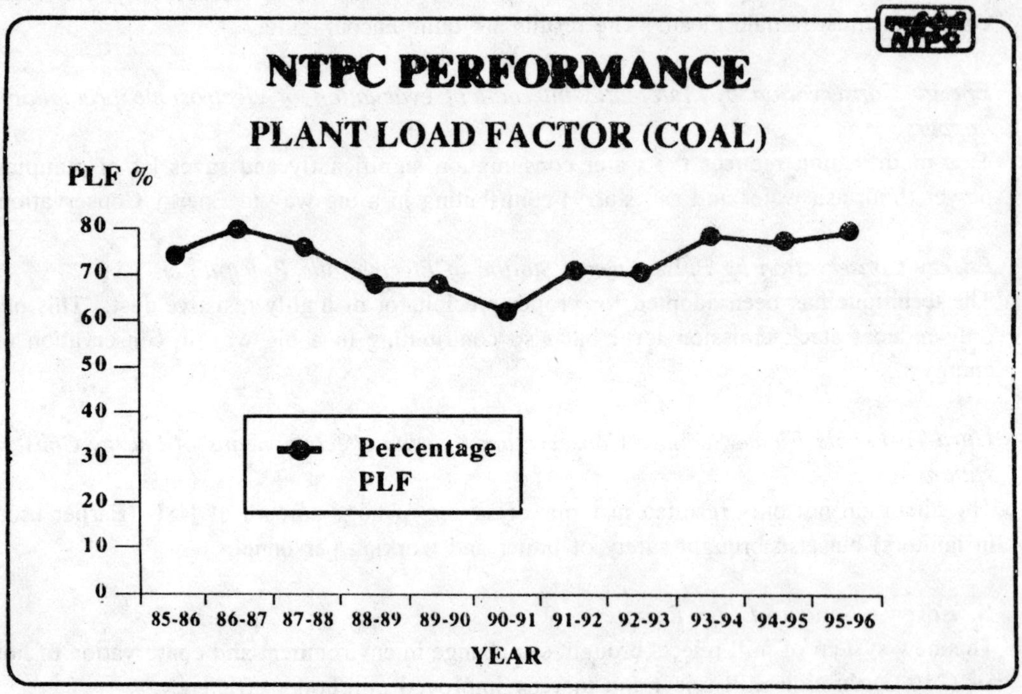

ANNEXURE B

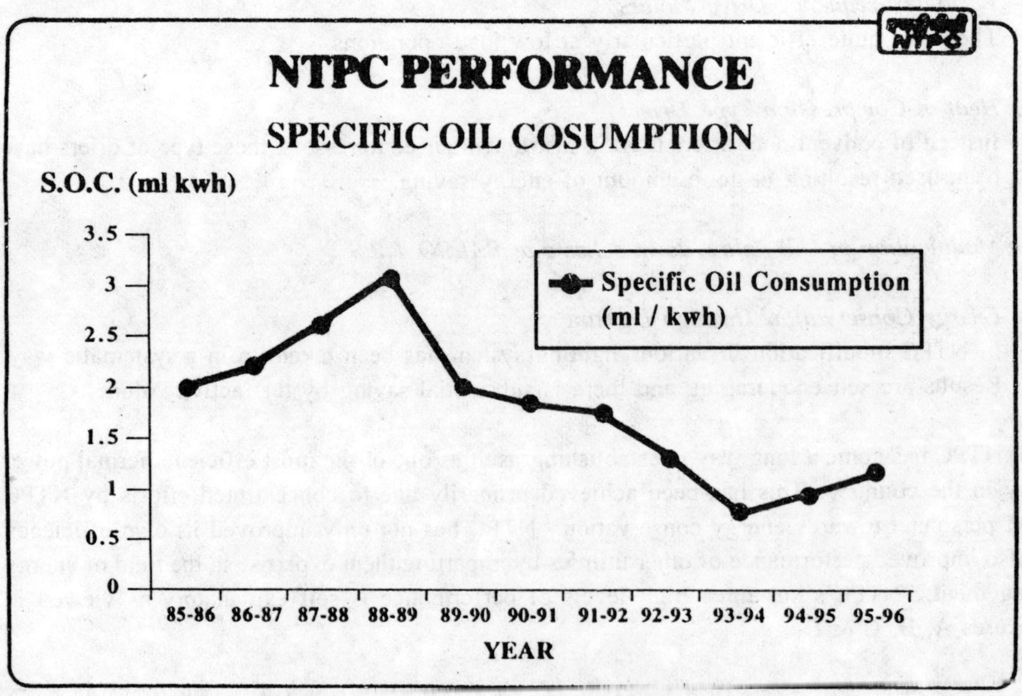

ANNEXURE C

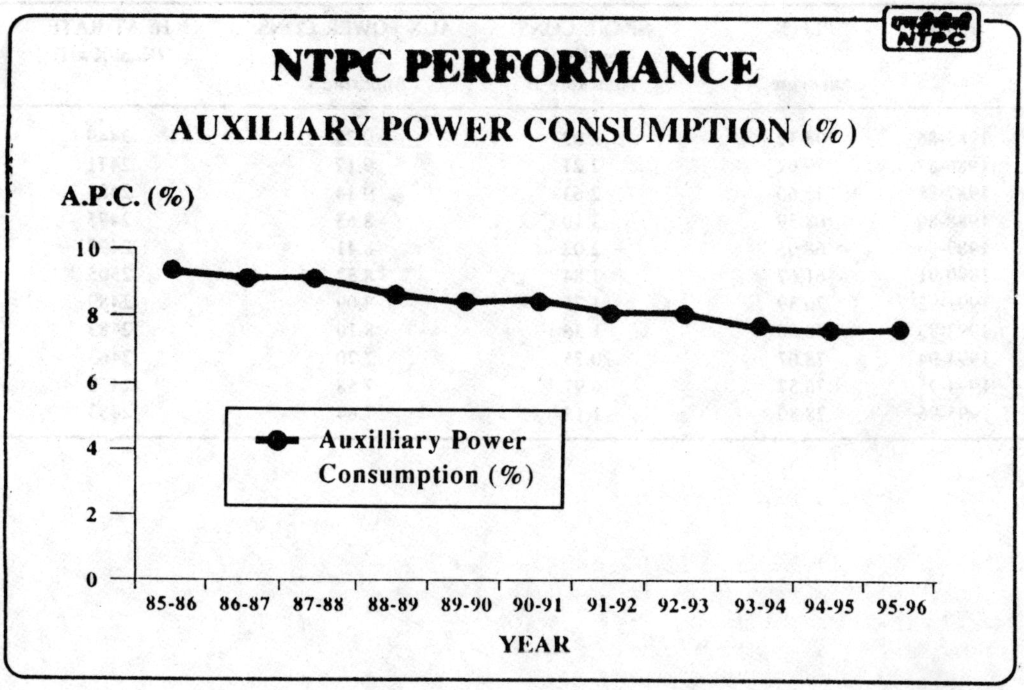

ANNEXURE D

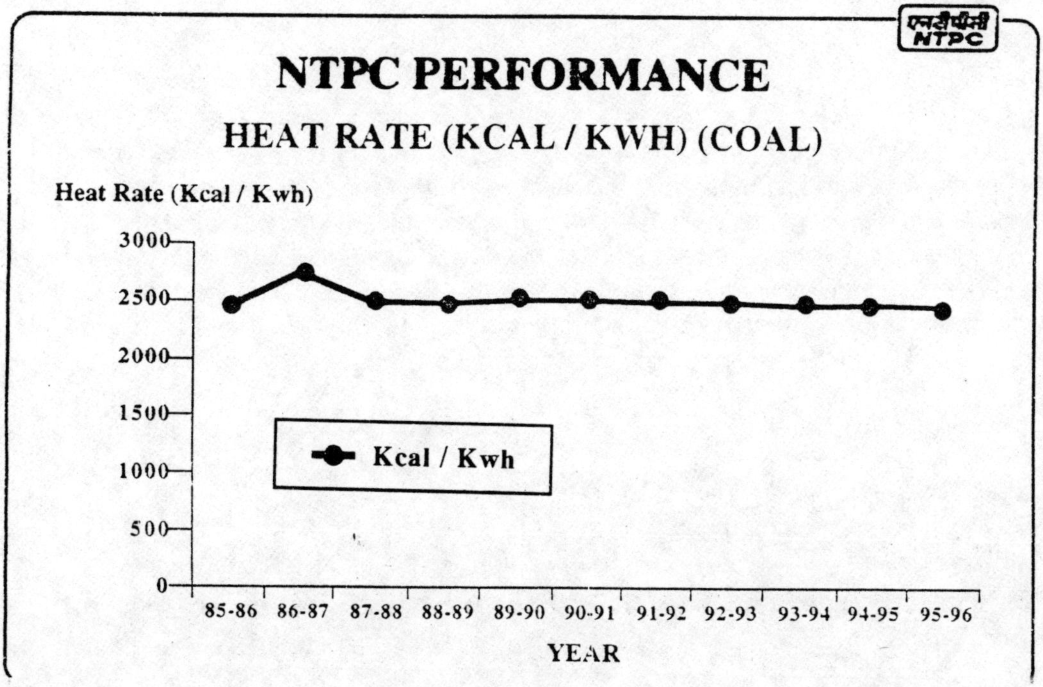

DATA USED IN PREPARING ANNEXURES A, B, C and D

YEAR	PLF(%) Annexure A	SP.OIL CONS. (ml/KwH) Annexure B	AUX POWER CONS. (%) Annexure C	HEAT RATE (Kcal/KwH)
1985-86	74.10	2.02	9.33	2449
1986-87	79.62	2.21	9.17	2471
1987-88	75.60	2.63	9.14	2483
1988-89	68.39	3.10	8.63	2475
1989-90	68.05	2.02	8.41	2498
1990-91	61.67	1.84	8.52	2505
1991-92	70.59	1.75	8.09	2489
1992-93	70.00	1.36	8.10	2483
1993-94	78.07	0.75	7.70	2463
1994-95	76.57	0.97	7.58	2472
1995-96	78.80	1.15	7.64	2453

USE OF PRIMARY ENERGY FOR ENERGY CONSERVATION AND EFFICIENCY

57

Dr. MANOHAR PRASAD
Professor
Refrigeration and Airconditioning
Department of Mechanical Engineering
Indian Institute of Technology, Kanpur - 208 016 (INDIA)

ABSTRACT

This paper emphasises on the utilisation of primary energy with a view to fulfil different requirements of industries as well as public sector undertaking. Because several industrial organisations require power and refrigeration in process heating. A few systems have been described and their salient features are presented by giving calculated data. Some relations are also presented for calculations of quantities of interest for typical applications. The usefulness of the present systems is highlighted through the calculations carried out for systems III and IV. A percentage saving of 27% has been found for both fuel consumption and decrease in release of carbon dioxide. Hence, the present approach would be highly beneficial for large sized plants/industrial organisations.

Introduction

These days demand for energy is increasing very fast, conventional fuel reserves are depleting rapidly, creating enormous environmental problems and financial involvement. Hence, there are concerted efforts by all concerned to arouse awareness among users and publicise for development energy efficient systems. The direct use of primary energy has tremendous scope for energy conservation. This approach also provides better use of energy. As an example primary/waste heat driven absorption system for airconditioning. Advantages of this approach are reflected from elimination of the following energy conversion processes:

i. Thermal to mechanical
ii. Mechanical to electrical
iii. Electrical step-up
iv. Electrical step-down
v. Electrical to Mechanical
vi. Other Mechanical Losses.

Hence, direct use of primary energy provides much more efficient method of achieving the objectives. A few systems are described over here to demonstrate the distinguishing features as how the same renders better proposition:

I. System I: Directly coupled refrigerant compressor to prime mover
II. System II: Gas turbine power plant with refrigeration
III. System III: Gas turbine with process heating and refrigeration
IV. System IV: Steam power with process heating and refrigeration.

System I. Prime Mover Directly Coupled to Refrigerant Compressor

In this arrangement a gas turbine or a prime mover drives the compressor directly as shown in Fig. 1. Here compressor drives the centrifugal compressor at high speed thereby advantages presented above are utilised. Of

$$- T_1 (1 - \varepsilon) [1 + (Pr^{(\gamma-1)/2\gamma} - 1)/\eta_c]\}/ |\Delta H_s| \quad (4)$$

where ΔH_s as heating value of the fuel, Pr = Pressure ratio.

$$\eta_{th} = \frac{P}{m_f |\Delta H_s|}$$

Thus, for a refrigeration system of TR ton capacity (1 ton is equivalent to 3.5 KW of refrigeration/cooling) power is found from:

$$TEUR = \frac{3.5.TR}{Q} = \frac{3.5.TR.\eta\ th}{(2 m_a\ c_p)[\eta_t T_6 (1-1/X) - T_1 (X-1)]} \quad (6)$$

TEUR for 1000 ton capacity, $T_6 = T_8 = 1200$ K, $T_1 = 330$ K Pr = 6, $\varepsilon = 0.8$, $\eta_c = 0.85$, $\eta_t = 0.87$ is found to be 1.82. This is much higher than the electricity driven system. If mechanical to electrical, step-up and step-down, transmission and electrical to mechanical efficiencies are considered, then TEUR becomes 1.32 which is just 72.5% of the prime mover driven system.

II. Gas Turbine Power Plant with Refrigeration

Here primary energy is first used to generate power and the waste heat is directly used for process heating and refrigeration. This arrangement again eliminates various energy conversion processes. Figure 2 shows the schematic diagram having a two-stage gas turbine coupled to a two-stage compressor. The intermediate pressure is the same for the turbines and compressors (geometric mean of the upper and lower operating pressure limits due to the minimum work of compression and the maximum work output). The products of combustion pass through a steam generator for a vapour-absorption system. By this arrangement the electric power becomes available for external use as absorption system operates on the waste heat. One may derive expressions for various quantities for the present system. The net power output corresponding to minimum work input to compressor with perfect intercooling and maximum work output with perfect reheating is obtained from:

$$P = 2\dot{m}_a c_p [\eta_t T_6 (1 - 1/X) - T_1 (X - 1)] \quad (7)$$

where m_a is mass flow rate of air and $X = Pr^{(\gamma-1)/(2\gamma)}$. T_1 and T_6 are the minimum and maximum temperature limits, respectively.

The fuel consumption is found from:

$$\dot{m}_f = \dot{m}_a c_p \Big\{ T_6[(1-\varepsilon)+\eta_t\varepsilon(1-Pr)^{-(\gamma-1)/(2\gamma)})] + T_6\eta_t[1-Pr^{-(\gamma-1)/(2\gamma)}]$$

$$-T_1(1-\varepsilon)[1+(Pr^{(\gamma-1)/(2\gamma)}-1)/\eta_c]\Big\}/|\Delta_s| \quad (8)$$

where $T_6 = T_8$ assumed, and ΔH_s as heating value of the fuel, Pr = Pressure ratio.
Energy available for refrigeration is given by: (9)

$$Q_g = \dot{m}_a c_p (T_{10} - T_{11})$$

course, this is suitable for large capacity refrigeration system due to economic consideration. This type of system has been fabricated by Mitshubisi up to 10,000 ton capacity. This gives very high energy efficiency ratio.

The thermodynamic analysis is carried out for the system comprising refrigeration system being driven by a prime mover with a view to energy efficiency. The same is derived based on the thermal efficiency of a power generator operating under allowed temperature limits. Foregoing derivations, the net power output corresponding to minimum work input to compressor with perfect intercooling and maximum work output with perfect reheating is obtained from:

$$P = 2\dot{m}_a c_p [\eta_t T_6 (1 - 1/X) - T_1(X - 1)] \quad (1)$$

where $\dot{m}_a$ is mass flow rate of air and $X = Pr^{(\gamma-1)/(2\gamma)}$. T_1 and $T_6 = T_8$ are the minimum and maximum temperature limits, respectively.

Power requirement for the refrigeration system can be related to refrigeration capacity though COP as

$$P = 3.5\ TR/COP \quad (2)$$

where COP of the system can be calculated from the operating condition of the vapour-compression refrigeration system. For airconditioning it may be 4.0.

From Eqs, (1 & 2) one gets:

$$\dot{m}_a = 3.5[TR/(2c_p.COP)]/[\eta_t T_6(1 - 1/X) - T_1 (X - 1)] \quad (3)$$

The fuel consumption is found from:

$$\dot{m}_f = \dot{m}_a c_p \{T_6[(1 - \varepsilon) + \eta_t \varepsilon (1 - Pr)^{-(\gamma-1)/2\gamma)})] + T_6\eta_t[1 - Pr^{-(\gamma-1)/(2\gamma)}]$$

Then, we define a term called Thermal Energy Utilization Ratio (TEUR) as:

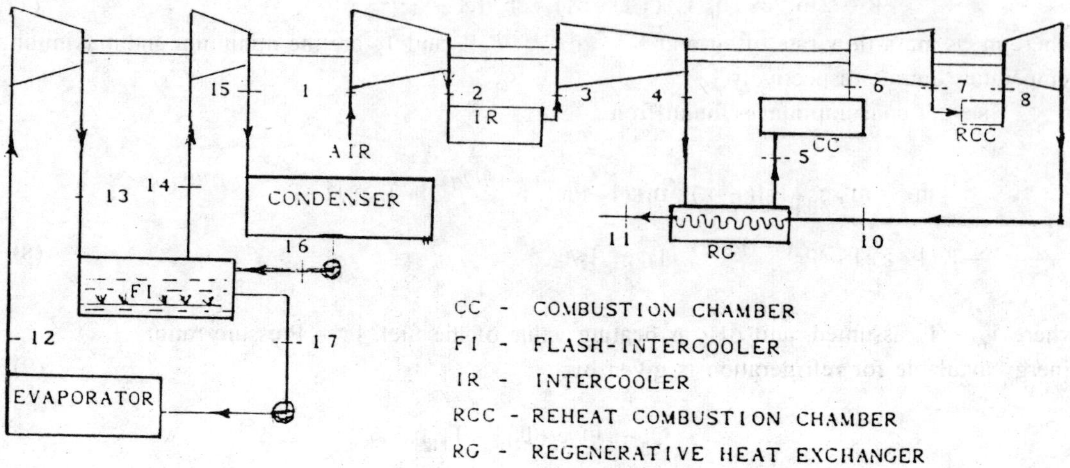

FIG. 1 : *SYSTEM 1: Prime mover driven refrigeration system.*

$$\text{TEUR} = \frac{P_{output} + Q_{ref}}{Q_{supplied}} \tag{10}$$

$Q_{supplied} = m_{fuel} |\Delta H_s|$, asuming that reactants enter at 298°K temperature at which enthalpy of combustion is given.

Calculations were carried out for process heating = 4000 KW, refrigeration load = 1000 ton, operating temperature limits = 1500 K and 313 K, COP = 0.7, η_t = 0.86, η_c = 0.84, Pr = 6, e = 0.86. It gave power output 14,853 KW and TEUR = 0.65. So one should get exact picture if the investment for the system is considered. The expected investment for the power side would be Rs. 3.2 × 2 = Rs. 6.4 crores. This investment can be compensated in about 4 years. Thereafter, the user will be saving around Rs. 3 crores/year, rendering better utilisation of fuel.

System III. Gas Turbine with Process Heating and Refrigeration

In this arrangement a gas turbine drives two compressors, Fig. 3. The exhaust from the first turbine is reheated up to the allowed maximum temperature and then expands through the second stage up to the ambient pressure. The exhaust gas passes through a regenerator where the compressed air from the second-stage compressor is heated to utilize a part of the waste heat. Thereafter, the rest heat of exhaust is used to generate steam for the process heating for industrial purpose and generator heating for the vapour-absorption refrigeration system. Finally, the exhaust gas is discharged into atmosphere. A brief description is presented for various quantities as given below:

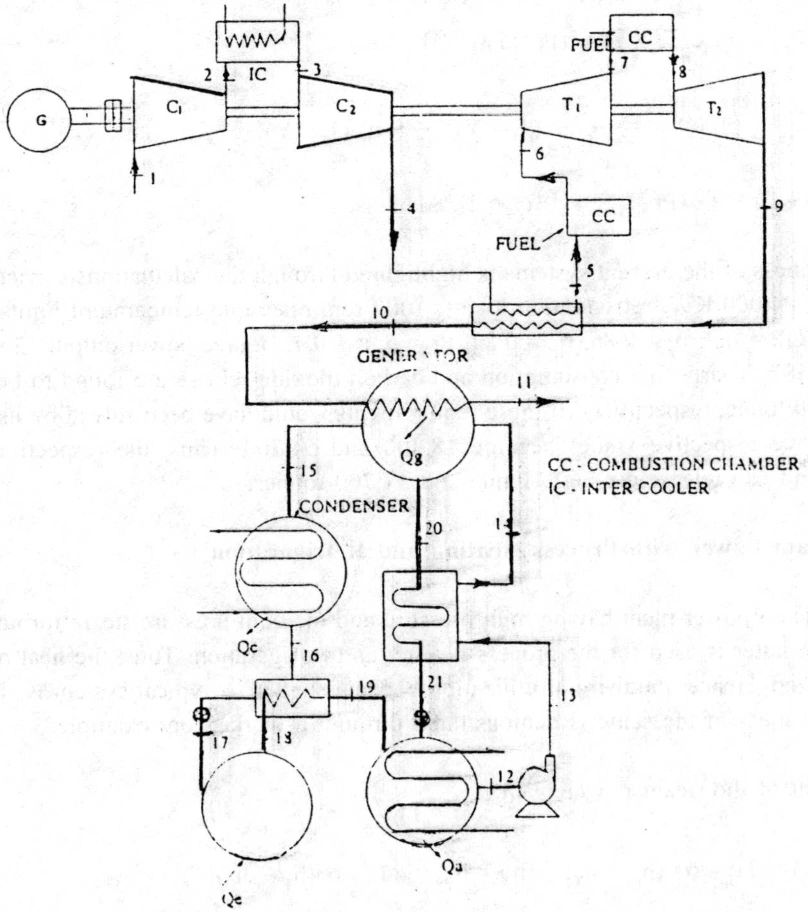

FIG. 2 : *System II: Generation of power and production of refrigeration*

Minimum power supplied to compressors is found from:

$$P_c = 2\,m_a\,c_p T_1[Pr^{(\gamma-1)/(2\gamma)} - 1]/\eta_c \tag{11}$$

The exit pressure for the first turbine is found from: $P_{t1} = P_c$

i.e.,
$$\eta_t.m_a\,c_p T_6[1 - Pr\,t^{-(\gamma-1)/\gamma}] = 2\,m_a\,c_p T_1[Pr^{(\gamma-1)/\gamma} - 1]/\eta_c \tag{12}$$

where Prt is the expansion ratio for the first turbine. Then,

$$Pr\,t = \left[1 - \frac{2T_1[Pr^{(\gamma-1)/(2\gamma)} - 1]}{T_6\eta_t\eta_c}\right]^{\gamma(\gamma-1)} \tag{13}$$

The power output of the turbine with $Y = (Prt/Pr)^{(\gamma-1)/\gamma}$ is:

$$P_{net} = P_{t2} = \dot{m}_a c_p T_6 \eta_t \, [1 - (Prt/Pr)^{(\gamma-1)/r}]$$

$$\dot{m}_f = \dot{m}_a c_p \, \{T_6[(1 - \varepsilon) + \eta_t \varepsilon \, (1 - Y)] + T_6 \eta_t \, (1 - Y)$$

$$-T_1 \, (! - \varepsilon) \, [1 + (Pr^{\,(\gamma-1)/2\gamma)} - 1) \, \eta_c]\}/|\Delta H_s| \tag{14}$$

The usefulness of the present systems is highlighted through the calculations carried out for process heating = 4000 KW, refrigeration load = 1000 ton, operating temperature limits = 1500 K and 313 K, COP = 0.7, η_t = 0.86, η_c = 0.84, Pr = 6, ε = 0.7. It gave power output 12,966 KW and TEUR = 0.687. Yearly fuel consumption and carbon dioxide release are found to be 14,400 tonne and 43,000 tonne, respectively. If these requirements would have been solved by individual manner, the above respective values become 18,300 and 54,700. Thus, the respective yearly saving in fuel and CO release are 3600 tonne and 11,700 tonne[2].

System IV. Steam Power with Process Heating and Refrigeration

It comprises a steam power plant having high pressure and medium pressure steam turbines. The exhaust from the latter is used for the process heating and refrigeration. Thus, the heat rejection is fruitfully utilized. Hence, the thermal utilization becomes better. A typical system is shown in Fig. 4. The usefulness of the same is demonstrated through a worked out example.

Net power output of the steam power plant:

$$P = \dot{m}_5[(h_5 - h_6) + (1 - \alpha)\,(h_6 - h_7) - (h_3 - h_2) + (1 - \alpha)(h_1 - h_0)] \tag{15}$$

where α is the fraction of steam bled for the process heating and refrigeration.

The steam consumption for the process heating is found from:

$$\dot{m}_6' = Q_{process}/(h_6 - h_2) \tag{16}$$

Amount of steam needed for the refrigeration purpose is obtained from:

$$\dot{m}_{refn.} = 3.5TR/[(h_6 - h_2)COP] \tag{17}$$

where TR is tonnage of refrigeration system and COP for the refrigeration is based on heat requirement in the generator.

Also from mass balance among the masses of the steam for process heating, refrigeration and bled steam from HP turbine.

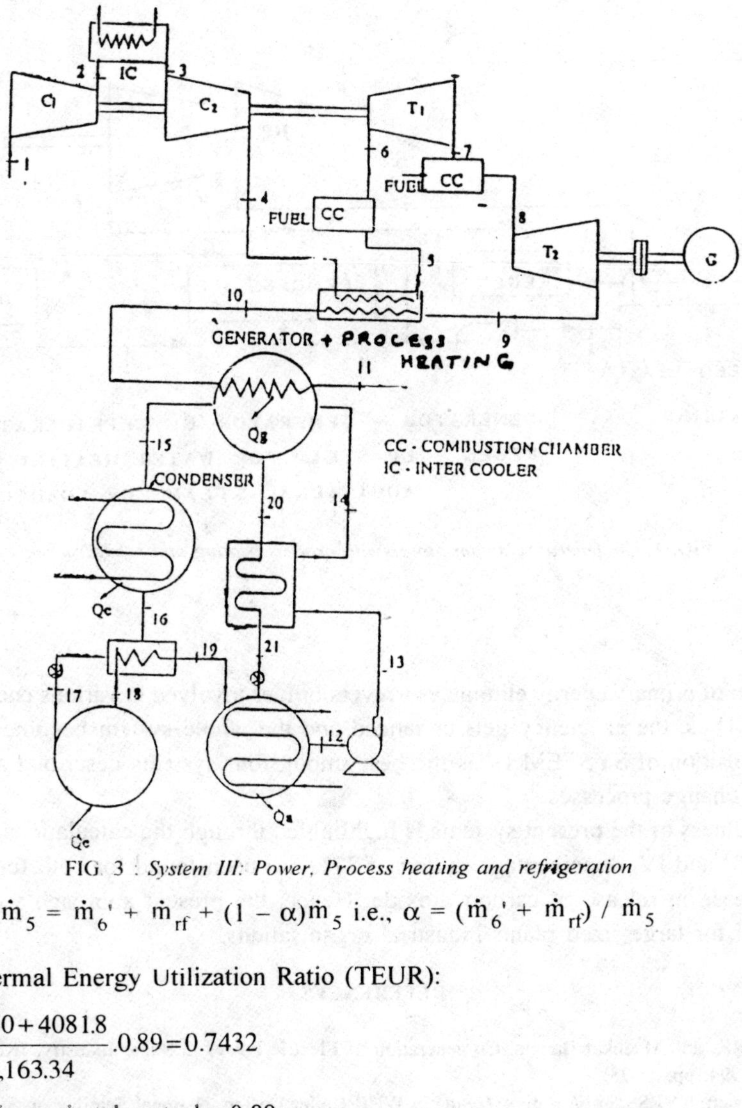

FIG. 3 : *System III: Power, Process heating and refrigeration*

$$\dot{m}_5 = \dot{m}_6' + \dot{m}_{rf} + (1 - \alpha)\dot{m}_5 \text{ i.e., } \alpha = (\dot{m}_6' + \dot{m}_{rf}) / \dot{m}_5 \quad (18)$$

Overall Thermal Energy Utilization Ratio (TEUR):

$$\frac{Q}{Q_5} = \frac{10{,}250 + 4081.8}{17{,}163.34} 0.89 = 0.7432$$

when boiler efficiency is taken to be 0.89.

Calculations were carried out for process heating = 5000 KW, refrigeration load = 1500 ton, operating temperature limits = 1500 K and 313 K, COP = 0.65, η_{boiler} = 0.86, η_t = 0.86, steam at 100 bar and 773 K, bleeding at 40 bar, process heating at 5 bar fed pump efficiency = 0.8. It gave power output 4,000 KW and TEUR = 0.65. Yearly fuel consumption and carbon dioxide release are found to be 19,200 tonne and 28,000 tonne, respectively. If these requirements would have been solved by individual manner, the above respective values become 25,560 and 37,500. Thus, the respective yearly saving in fuel and carbon dioxide release ar 6360 tonne and 9500 tonne.

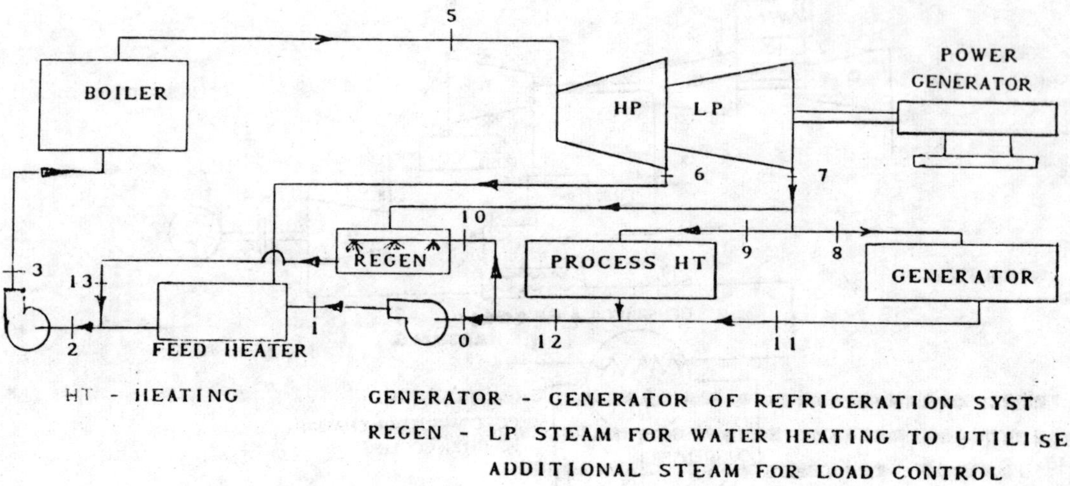

FIG. 4. *Co-generative steam power with process heating and refrigeration*

Conclusions

1. Utilisation of primary energy eliminates irreversibilities involved in various energy conversion systems. Thus, the efficiency gets enhanced and the whole system becomes compact.
2. The proposition of SYSTEM IV is the best among four systems described above. Because of phasé change processes.
3. The usefulness of the present systems is highlighted through the calculations carried out for systems III and IV. A percentage saving of 27% has been found for both fuel consumption and decrease in release of carbon dioxide. Hence, the present approach would be highly beneficial for large sized plants/industrial organisations.

REFERENCES

1. Sugandhi, RK, and Madhkar Bangla, Co-generation of Electric Power in Sugar Industry, IREDA NEWS, Vol. 5, No. 4, 1994, pp. 21-25.
2. Subrahmanyam, NVRSS, *et al*, Future Trends in WHR Boiler Design, National Seminar on Steam & Gas Based Co-generation Systems, Indian Institution of Plant Engineers (AP Chapter), 1993, pp. II-2-1 to 13.
3. Swamy, CN, Gas Turbines for Combined Cycle Applications, Same as [2], pp. 1-3-1 to 5.4 TERI ENERGY DATA DIRECTORY & YEAR BOOK, 1995/96, pp. 39-97.

DIVISION - IV

FUNDING MECHANISM AND SOURCES

BRIDGING THE ENERGY GAP IN INDIA (1997 TO 2012) : AN ESTIMATION OF THE FINANCIAL CONSTRAINT AND RELATED POLICY OPTIONS

58

SANJEEV S AHLUWALIA
Senior Fellow, Tata Energy and Resources Institute
India

This paper quantifies the availability of finance for energy investments over the period 1997 to 2012 in India. It examines the relationship between growth in India, growth in the rest of the World, financial liberalization, global financial integration and capital supply. It assesses investment demand in the energy sector in India under a three separate economic scenario identified by differing assumptions of growth rates and examines under each scenario the impact of liberalized trade in energy goods and services. Rapid export growth is crucial in making available optimum economic choices regarding sourcing of capital and products. The challenge of meeting the energy gap increases in scale and dimension as the country pushes for higher levels of growth. Bridging the gap involves the making of a set of correct choices between the role of the State and market mechanisms in determining the use of foreign capital, inter-sector flow of capital, share of the government in available commercial capital, formulation of energy use standards and commercialization of energy suppliers.

The Financial Crunch

2. Investments in power remained fairly steady at 2.4% of GDP during 1981 to 1991 but declined in the Eighth Plan (1993-97) to 2% of GDP. Investments in coal similarly declined from 0.5% of GDP in the Sixth Plan to 0.4% of GDP in the Seventh Plan and subsequently to 0.3% of GDP post 1992. Conversely investments in oil and gas exploration, oil refining and gas supply increased from 0.7% of GDP in the Sixth Plan to 1% of GDP in the Seventh Plan and subsequently to 1.2% of GDP post 1992. The energy shortages in the electric power and coal sectors are therefore more intense. Investments in the energy sector have been almost exclusively reserved for the Public Sector. In power, except distribution and generation by licensees in Calcutta, Bombay, Ahmedabad and Surat and captive generation by industries, the sector has been the preserve of the Central and State Public Sector which till 1995 controlled 90% of the capacity and a similar share in investment. In the hydro carbon sector, crude oil production is still exclusively with the Public Sector though private companies have been active in exploration and are expected

to start production in 1996/97. In the refining sector the public sector, with thirteen of the fourteen refineries, has a majority share. In Coal the entire capacity is with the public sector though recently captive mines have been permitted for power and steel plants.

3. The near exclusive reliance on public sector finance has inevitably resulted in a constraint on investment due to the increasing commitment of the government to downsize its expenditures and to relocate its focus on the social sector. Public investment in general has declined from a high point of 11.6% of GDP in 1986/87 to 8.7% in 1994/95. Energy investments have consequently suffered. Public investments in energy[1] rose from a level of 3.2% of GDP in the period 1981-85 to a high of 4.3% of GDP in 1987 before declining to 3.5 % in subsequent years. Figure 1 charts the trend over the period 1980 to 1996.

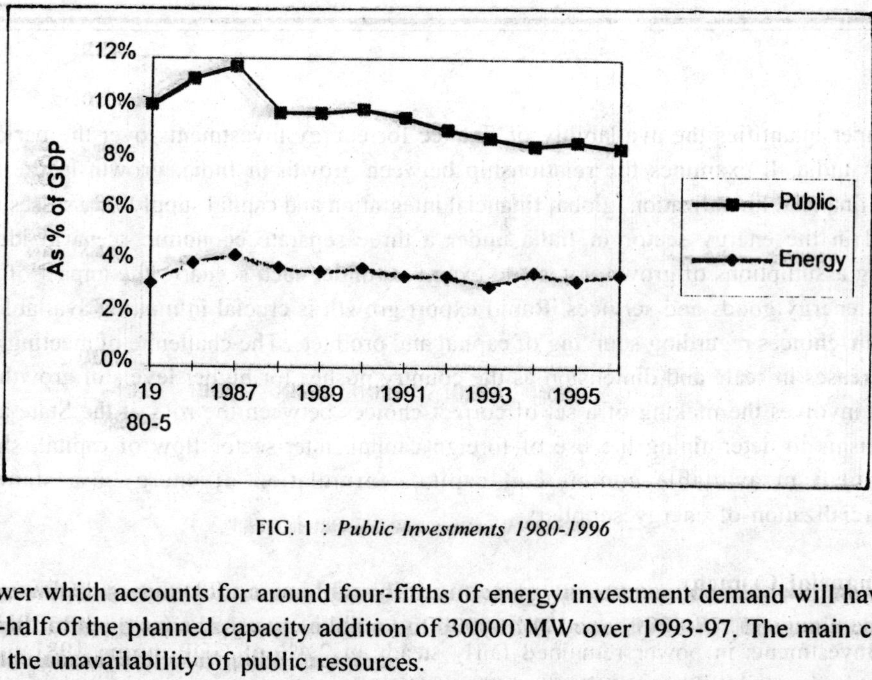

FIG. 1 : *Public Investments 1980-1996*

Power which accounts for around four-fifths of energy investment demand will have added only one-half of the planned capacity addition of 30000 MW over 1993-97. The main constraint has been the unavailability of public resources.

Investment Demand

4. While the productivity of energy is not very well known in India, perceived energy shortages have reached levels where they constrain economic output; hence the concern to bridge the energy gap. What is the dimension of the anticipated investment in Energy over the period 1997 to 2012 to bridge the energy gap currently estimated at around 15% in electric power? Table 1 provides some estimates formulated on a business-as-usual basis. The *inter se* sub-sector financial requirements do not have a strict validation. The sub sector investment needs are dependent on interfuel pricing, technology options and import versus production decisions which are difficult to predict. Second, in the event of market based energy product prices there is likely to be considerable

variation in the energy demand pattern with a possible overall reduction in energy demand. A similar effect can be anticipated with formulation of more rigid energy use standards and minimum energy efficiency levels. Energy demand can also be compressed by providing fiscal incentives for the use of preferred technology, as for example fuel efficient vehicles or electric traction mass transport systems.

TABLE 1 : **Investment Demand 1997 to 2012**

$ billion

Energy sub-sector	Low Case	Base Case	High Case
GDP growth annual	5.6%	7%	8.4%
Comm. Energy Elasticity	0.71	0.77	0.8
Coal Mining	12	19	27
Oil & Gas Exploration & Prod.	64	64	64
Oil refining	13	13	13
Power	220	322	452
Total	308	417	556
% of GDP	4.3%	5.1%	5.9%

Note : 1994/95 constant prices

Inter-fuel consumer prices are held constant thus ignoring the possibility of inter-fuel substitution. The secondary energy demand has been estimated using income elasticity of demand of 0.6 for coal, 0.9 for oil and 1.1 for electric power, under three scenarios of low, base case and high GDP growth. A low elasticity of electric power demand is assumed as compared to the more conventionally used income elasticity of 1.5 on account of anticipated price elasticity effects emanating from imminent price rationalization in this sector. The primary energy demand for electric power has been estimated assuming a share of 62% for coal thermal, 10% for gas, 5% for other hydrocarbons and 23% for hydro, nuclear and renewables by 2012 with energy conversion efficiencies of 35% for coal and 45% for hydrocarbon fuels. Table 2 details the physical capacities and estimated unit costs of investment.

TABLE 2: **Energy Capacity and Capital Costs (Base Case)**

	Coal MT	Oil & Gas million tonnes	Refining million MT	Power MW
Installed capacity in 2012	618 million	44, 20	70	257000
Incremental capacity 1997/2012	366 million	12, 14	134	155000
Unit cost $ million	51	**	200	2

** A lump sum of $58 billion assumed.

The unit costs for power at 1994/95 prices are the system cost, including investments required in power transmission and distribution per MW of installed capacity. A lump sum amount has been taken for oil and gas production as per the estimates in the Sundarajan Committee Report 1995.

Capital Supply

5. Capital supply is capped by constraints at two levels. At the national level, the constraint relates to the overall availability of capital in India. In the domestic sector, the constraint is the low savings rate, especially in the public sector. Public sector savings peaked at 4.9% of GDP in 1976/77 but declined to 0.5% in 1993/94 and were 1.7 % of GDP in 1994/95. The low level of public savings give rise to the second constraint. The continuing need for the government to borrow heavily in domestic markets, to meet its fiscal deficit "crowds out" private investment. Fiscal deficits are currently lower than the peak of 8.3% of GDP in 1990/91 though still in the range of 6% of GDP.

6. India has a very nominal share in the international flow of capital. Net foreign capital inflows have averaged 2.3% of GDP over 1989 to 1994. Despite this nominal share, increasing the inflow of foreign capital will be constrained by the opportunities available to investors in other countries, many of which are competitively liberalizing foreign capital controls. It is tempting to assume that the level and mix of private foreign capital available in India correctly indexes the appetite of foreign investors for Indian assets. However information gaps and artificial capital controls are time consuming to remove and hence it is possible for proactive, investor friendly policies to vastly improve private foreign capital inflows on a sustainable basis. Clearly, though, there will be limits to the inflow of private foreign capital, which need to be quantified for forecasting purposes.

7 Our concern is with capital flows to the energy sector. This imposes the third level constraint of the intersectoral share of energy in available capital flows. This paper does not attempt to estimate the share which will be a function of the pace of reforms in the energy sector, risk weighted returns available to investment in the sector and competing opportunities in other sectors. Since no statistically significant determinants of capital flows to the energy sector have been identified so far, this paper stops at presenting the demand and aggregate capital supply estimates.

Parameters of Capital Supply

8. The incremental supply of private capital in India over the period 1997 to 2012 has been modelled on the following assumptions:

Foreign Capital

The stock of investible private capital in Industrialized countries is assumed to grow at the same rate as World GDP. The share of such capital invested in emerging markets increases in proportion to the rising share of the GDP of developing countries in World GDP. Increasing levels of financial integration, the dilution of capital controls in developing countries and higher returns in these countries supplement this trend. A higher rate of economic growth in India as compared to the World pulls private foreign capital into India. The constraints on the supply of foreign capital is that it is capped by the share of India's GDP in World GDP and India's export growth. No incremental inflows of multilateral debt have been assumed on account of the large outstanding stock of such debt. Inflows are expected to match repayments.

Domestic Capital

In the domestic market, financial liberalization increases the assets of domestic institutional investors like Insurance companies, pension funds and mutual funds, however the share of the private sector is limited to 25% of such assets with the rest invested in Government securities. Consistent growth increases the investible capital of domestic promoters, whose profitability increases outpace GDP growth. Reform of the Public Sector industrial enterprises generates investible capital through a positive return on their assets. No direct budgetary support for capital investment from the government to energy producers has been assumed.

9. How much of the estimated capital supply will be available for energy investments? The share of energy in public outlays has varied between 27% to 30% over the period 1980 to 1995. Public investments in electric power alone had a share of around 25% in total public investments. Will private investors have similar investment priorities? This paper does not attempt an estimation of capital flows to energy. It is assumed that so long as energy investments are a reasonable ratio of the GDP risk return criterion will guide private capital to energy. Table 3 provides estimations of the availability of private domestic and foreign capital during the period 1997-2012 in India under the base case scenario of 7% per annum GDP growth in India compared to 3% per annum growth in the World.

TABLE 3 **Supply of Private Investment Capital in India 1997 to 2012**

$ billion

Inter-source share %	*% of GDP*	*Amount $ billion*	*Source* <<<<	>>>> *Source*	*Amount $ billion*	*% of GDP*	*Inter source share %*
34	4.3	347	Foreign Capital Total	Equity Total	656	7.98	66
9		90	Bonds	Foreign Portfolio equity investment	23		2
15		153	Export credit	Direct Foreign Investment	82		8
2		23	Foreign Portfolio equity investment	Promoters equity	148		15
8		82	Direct Foreign Investment	Domestic Portfolio Equity investors	205		20
66	8.15	671	Domestic Capital Total	Public Sector Surplus	198		19
12		120	Domestic Long Term Debt	Debt Total	363	4.4	34
15		148	Promoters equity	Bonds	90		9
20		205	Domestic Portfolio Equity investors	Export credit	153		15
19		198	Public Sector Surplus	Domestic Long Term Debt	120		12
100	12.4	1018	Total Capital Supply	Total Capital Supply	1018	12.4	100

1994/95 constant prices

As compared to the base case scenario, under the low growth scenario the capital availability falls to $ 882 billion while under the high growth scenario it increases to $ 1201 billion. At 12.4% of GDP, under the base case scenario, the supply of commercial investment capital is higher than the past trend. Total private investment in all sectors has varied between 11.5% of GDP to 15.5% in the last decade. Consequently our estimate of commercial capital which caters only to the large industrial and infrastructure sectors represents a considerable increase in the supply of commercial capital to this sector. The second major diversion from the past trend is the increase in volume of foreign capital. As compared to the trend share of around 2.2% of GDP for net foreign inflows, under our estimates, the ratio of net foreign capital to GDP doubles to a level of 4.3%.

10. The supply of foreign capital has to be viewed from the following two perspectives

Financing of Capital Imports

(i) Is the supply of foreign capital sufficient to meet the incremental foreign exchange investment demands? The share of foreign capital is just over one-third of the total eligible capital flow. This is somewhat higher than the average estimated capital requirements for foreign exchange in energy projects, which would more typically be around 25% to 30% of total capital cost. Since energy projects do not generate net foreign exchange earnings, (except export oriented projects as are being conceived now in the oil refining sector) any exposure to foreign capital exposes the project to exchange rate depreciation risk. Consequently it is reasonable to expect that investors would prefer to limit their exposure to foreign capital servicing and repayment risk. Assuming that the entire investment demand of $417 billion under the base case scenario is met, the foreign exchange requirement, assumed at 30% of total demand would be $ 125 billion. This comes to 36% of the available foreign capital of $347 billion under this scenario (Refer Table 3). A share of 36% in foreign capital flows for energy is slightly higher than past allocations channelled through the public budget but not unreasonable. Hence we can surmise that there is no foreign exchange availability gap for financing energy imports. In any case with foreign exchange liberalization, importers can purchase foreign exchange directly from the market and need not compulsorily cover the capital import cost from foreign loans as in the past.

Is the supply of foreign capital sustainable?

(ii) What impact will foreign capital inflows of this magnitude have on the external account? Concerns of this nature are voiced mainly in the context of debt inflows. Equity inflows are assumed to be more productively employed since they flow to corporates unlike debt which can be used by Governments for consumption purposes. Consequently no commonly acceptable indicators of sustainable equity flows exist. We will therefore focus exclusively on the issue of debt sustainability and use the commonly accepted World Bank criterion for classification of the debtor status of developing countries.[2] Two export scenarios are generated for each of the three GDP growth scenarios in Table 4.

TABLE 4: **Export and GDP Growth Scenarios**

	Low Case- GDP growth 5.6%	Base Case- GDP growth 7%	High Case-GDP growth 8.4%
Low export growth % per year	7.9	9.3	10.7
High export growth % per year	9.8	11.3	12.7

Under the high export growth scenario exports increase from a level of 10% of GDP in 1996 to a level of 20% of GDP by 2012. Under the low export growth scenario the ratio of Exports to GDP in 2012 is capped at 15%. Table 5 below compares the levels of sustainable incremental debt calculated under the World Bank guidelines for remaining within the "less indebted" category, with the estimated availability of incremental debt under the three GDP growth and two Export growth scenarios.

TABLE 5: **Normative and Estimated Incremental External Debt levels**

$ billions

Norm	Low Case- GDP growth 5.6%	Base Case-GDP growth 7%	High Case-GDP growth 8.4%
Debt/GDP@30%	111	157	214
Debt/Exp.@165%. Low export growth.	101	148	195
Debt/Exp.@165%.High export growth.	165	216	278
Availability of incremental debt 1994/95 constant prices	204	243	292

The Debt to GDP norm permits a higher level of LDOD since the share of exports in India is relatively small. Since a low export to GDP ratio is a structural constraint, we prefer to adopt the Debt to Export norm as the proper comparator. The most striking feature is that the estimated availability of incremental debt exceeds the outer bounds prescribed under all scenarios. However the gap between the two decreases as we go from the low GDP growth, low export growth scenario to the high GDP growth and high export growth case. From being 200% of the norm ($204 billion against the norm of $101 billion), the estimates of foreign debt supply become a more reasonable 150% of the norm ($292 billion against $195 billion) under higher growth assumptions.

11. Does this undermine the projection of available foreign commercial debt as an overestimate? We think not, except for the low growth scenarios which become less feasible. Under the World Bank criterion, India had an unsustainable level of External debt (LDOD) in 1996. LDOD to GDP was 34% while LDOD to exports was 319%. Simultaneously however India had an investment grade rating from Moody's of Baa3 while S&P revised its rating in mid 1996 upwards to BB+ What is the discrepancy between the two sets of credit quality indicators? Investors consider future trends along with current status. Since the trend in India is positive, investors are willing to dilute the conventional World Bank criterion. Consequently the foreign capital inflow estimates are not vitiated due to their resulting in the debt norms being exceeded.

However minimum GDP annual growth rates of 7% (which is the base case scenario) and annual export growth rates of at least 11% become critical to preserve investor interest. In the absence of these criterion being fulfilled foreign capital flows would be lower than those estimated.

Debt Equity Profile of the Available Capital

12. The availability of capital is biased in favour of equity, more particularly in the domestic sector. The current norms target a debt equity ratio of around 67:33. The supply of foreign capital under the base case scenario is in conformity with this norm with a debt equity ratio of 73:27. However in domestic capital debt is in short supply with an inverted debt equity ratio of 30:70.

13. How sustainable is an inverted capital structure? The excessive use of equity is an expensive and wasteful use of capital. Equity funds need to be leveraged for maximum advantage. Not to do so, reduces the total availability of capital and increases the cost of funds since, unlike debt, the servicing of equity is not tax deductible. An example will help. Let us assume an assured post tax rate of return on equity of 16%, a higher nominal interest rate of 20% in keeping with current interest levels and a marginal tax rate of 43% on corporate profits. For an investment of $ 417 billion, the cost of interest and equity servicing increases under the equity intensive capital structure option of 33:67 as compared to the more leveraged option of 67:33 by $ 23 billion per year or $ 230 billion over a ten-year period. In a cost plus tariff fixation system the utility just passes on the cost to the consumer who thereby end up subsidizing the inefficient use of capital. Consumers would therefore actually be better off by incurring an up front cash outlay in subscribing to a utility debt issue and earning interest income thereafter rather than by paying more through higher purchase cost of power.[3] The second adverse impact is the related contraction in outlay available for investment. Assuming a normative debt equity ratio of 67:33, the available equity, under the base case could finance $1988 billion of investments or nearly twice as much investment as is possible under the base case. This highlights the need to adopt debt enhancing policies.

Sensitivity Analysis of Capital Demand and Supply

14. Both capital demand and supply have been related to GDP growth in our model. Table 6 provides the sensitivity of demand and supply of capital to change GDP growth under three scenarios. The base case is of GDP growth of 7% pa. The low case is of a 20% lower growth rate of 5.6% and the high case is of a 20% higher growth rate of 8.4%.

TABLE 6 : **Sensitivity Analysis of Demand for Energy and Supply of Capital in India (1997 to 2012)**

Low Case: GDP Growth 20% lower –5.6% pa	Base Case: GDP Growth in India 7% pa	High Case: GDP Growth 20% higher –8.4% pa
– 26%	Investment Demand in Energy	+ 33%
– 13%	Total Capital Supply	+ 18%

As modelled by us investment demand is closely correlated to GDP growth. Due to capped import volumes (to ensure external account sustainability) higher growth increases the pressure on domestic production of petroleum products and coal resulting in a proportionately higher investment demand. Capital supply is less sensitive to change in GDP growth since much of the incremental capital is assumed to become available through a restructuring of the investment profile of foreign and domestic Institutional Investors along with growth in the aggregate pool of resources. An increasing share for India in foreign capital and similarly increasing shares for private investment in domestic institutional investors are examples of such restructuring.

15. Table 7 below provides absolute values for the demand for capital in the energy sector and the estimated supply of capital under alternative scenarios.

TABLE 7 : **Energy Investment Demand and the Supply of Capital in India (1997 to 2012)**

Parameter		Low Case 5.6% pa GDP growth	Base Case 7% pa GDP growth	High Case 8.4% pa GDP growth
Demand for Capital in Energy	$ billion	308	417	556
Demand/Cumulative GDP	%	4.3	5.1	5.9
Supply of Incremental Comm. Capital in India	$ billion	882	1018	1202
Supply/Cumulative GDP	%	12.2	12.4	12.8
Energy capital demand / Total capital supply	%	35%	41%	46%

Note 1994/95 constant prices.

The supply of private capital varies around 12.4 % of GDP. Meeting the demand for investments in energy would imply a share for energy, in available investible capital of between 35% (low case) to 46% (high case scenario). The share of energy investments in total plan allocations has varied between a low of 12% in the Second Plan to a high of 28% during the Sixth and Seventh Plans. While the determinants of capital flow to the energy sector are not known, past experience shows that the share of energy would be capped at 30% of available capital. Investments in energy have never exceeded 4.3% of GDP in any one year (1987) and 3.9% of GDP in any plan (Seventh) since 1980. Consequently it would be imprudent to target a share for investment in energy at higher than 4% to 4.5% of GDP defines the outer bound.

Energy Investments, GDP Growth and External Balance

16. At a maximum targeted level of 4 to 4.5% of GDP for investments in energy, the only feasible area is the low GDP growth case of 5.6% per year. Higher levels of growth, unless accompanied by energy demand optimizing and hence investment reducing measures, would be unsustainable in terms of the increasing draft that investments in energy would make on total capital supply. However the problem with lower levels of GDP and export growth is that the

estimated incremental debt becomes unsustainable. A minimum GDP growth rate of 7% and export growth rate of 11% optimizes both the foreign debt servicing and the investment demand constraint.

17. Consequently aggressive energy conservation measures are the only feasible option to reduce the demand for incremental capacity. The World Energy Council (WEC) has classified an energy efficiency improvement level of 1.5% per year till 2025 under a high energy conservation scenario. A more achievable energy conservation norm for India would be a target of 1% per annum for the next sixteen years. Through appropriate supply side measures a 15% increase in energy efficiency can be achieved in India. This will reduce the investment requirements under the base case scenario to $ 354 billion bringing it to an achievable level of 4.3% of GDP.

18. Even at this reduced investment level of $ 354 billion the requirement of debt under the target debt equity ratio of 67:33 would be $ 237 billion. This implies a share of 65% for energy in the actual availability of debt of $ 363 billion. The apportionment of such a substantial share of investible long term debt funds by energy is infeasible. It is not also immediately clear that the energy sector is advantageously placed vis-a-vis other sectors to be able to attract capital away from the other sectors in a market determined situation. Meeting the financing gap will therefore require proactive intervention from the government.

The Role of the State in Bridging the Financing Gap

Managing the Transition

19. Till now we have looked only at the aggregate investment demand and availability of capital. The relative shortage of capital perceived in aggregate terms becomes even more pronounced when viewed on an annual basis. Investment demand is front loaded while capital supply is back loaded over the period 1997-2012. Gestation periods of 3 to 10 years in energy projects makes the commitment of investment immediately necessary if energy demand gaps are to covered by the end of the period. Capital supply contrarily will be an outcome of the multiple restructuring and reforms suggested earlier. Consequently there is likely to be a severe capital constraint in the transition period till 2002. The Government needs to take account of this differential time phasing of capital demand and supply. Consequently continued Government support during the transition period of the Ninth Plan (1997 to 2002) will be necessary. Government support need not be in the form of direct expenditures on energy projects. Financial support to domestic Developmental Financial Institutions (DFI)[4] through the provision of Sovereign Guarantees, equity capital and preferential tax regimes would be the preferred option.

Domestic Developmental Financial Institutions

20. It is anticipated that the role of the DFI's will become less critical with development of capital markets and financial liberalization. However the financial sector is likely to be

restructured sequentially later than the real sectors of the economy. Consequently DFI's need to be prepared to meet the capital requirements of energy projects in the transition period. Over the Ninth Plan (1997 to 2002) these are likely to be around $ 175 billion. Over the period 1971-84 nearly 55% of the project cost of new Indian companies was met from DFI's and bank loans. This ratio has decreased in the last decade to around 31%. Even at this reduced ratio the DFI and commercial banks in India would need to provide an outlay of $ 52 billion for energy investments alone. In comparison the total assets of all the DFI's in India were only around $ 19 billion in end 1994/95. Assuming an accelerated target share for energy investments of 30% in the total assets of DFI's the total volume of assets would need to increase to $ 175 billion by 2002 implying a ratio to GDP of 42% in that year and an improbable annual growth rate of 37%. More important even if funds of this magnitude are channelled through the DFI's to tide over the period of transition will not that create the inertia typical of public financial giants? The gradual fading away of financial intermediation for industrial credit and the direct exposure of industrial borrowers to the discipline of the capital markets via equity and debt based securities is a part of the process of financial development. Consequently, it is imperative that financial intermediation by DFI's, in the transition period, is in the nature of investment banking and guarantee operations rather than via fund based intervention.

21. DFI's are second level financial intermediaries. They are not predominantly original sources of long term capital. They source their funds from Institutional Investors. Hence even if the regulatory controls on investment by Institutional Investors, which favour investments in government securities, are relaxed, adaptation to a market based investment pattern will take time. During this period DFI guarantees and Government counter-guarantees will provide considerable comfort to new Institutional Investors venturing into investment non-Government securities and loans. Adequate capitalization of DFI's to enable them to provide guarantees would therefore be a key initiative for the Government. It seems necessary from this viewpoint not to dilute the 'public' character of DFI's. This is contrary to the current trend where the Government has proposed to set up a Infrastructure Development Finance Corporation with minority participation of the Government.

Promote the Growth of Domestic Institutional Investors

22. Within the available volume of capital long term debt is in short supply. What is needed is enlargement of the resource base of natural long term investors like Pension Funds, Life Insurance Companies and Mutual Funds even beyond the fairly vigorous growth rate of 12% per annum which we have assumed in our base case growth model. The World Bank has estimated an annual rate of growth of 15% for the assets of institutional investors in East Asia. GDP growth in East Asia is admittedly higher. The total volume of Institutional assets in India is only around 43% of the GDP in 1995. The ratio of Institutional assets to GDP is more than one in developed financial markets. Consequently there is ample scope for encouraging a faster growth rate for institutional assets to achieve a volume equal to GDP by 2012. Growth of these assets will provide an ideal pool for long term debt.

Reduce the Fiscal Deficit

23. Currently over 85% of the assets of Institutional Investors (Insurance companies and Provident Funds) are invested in Government or Quasi-Government securities. Consequently we have conservatively assumed that only 25% of these funds would be available in future for non government borrowers. If the share of the government is reduced to 50%, the availability of debt approaches a debt equity structure of 50:50. Second, the overall availability of private capital increases by $141 billion to $ 1159 billion or an increase equal to 1.7 % of cumulative GDP. At this higher volume of capital availability (14% of cumulative GDP) the investment demand in energy of $ 354 billion, under the high energy efficiency improvement scenario, is a reasonable 31% of the available capital. Lowering the borrowing requirement of the government, by reducing the fiscal deficit by at least 2% of GDP, will release long term debt for meeting the energy financing gap.

Shift Energy Investment Overseas

24. Capital investments in energy can be reduced by shifting expenditures from the capital to the current account by importing energy goods and services, under "Open Economy" assumptions, rather than producing them. What are the limits of such investment shifts? Table 8 details the possible reductions in investment capital demand. The options examined are limiting oil imports to the minimum level for feeding the installed refinery capacity of 60 million MT per year (domestic oil production is at 30 million MT and expected to peak at 44 million MT) and importing the incremental petroleum products. Import of electric power from 3000 MW of generating capacity to be installed in Nepal. Limiting the domestic production of coal to 530 million MT instead of 616 million MT in the "Less Open Economy" scenario.

TABLE 8: **Reductions in Investment in the 'Open Economy' Scenario**

$ billion

GDP growth		Low case	Base case	High case
Reductions in investment $ billion		34	34	34
Energy imports. - Open Economy	(As % of high exports case)	31%	33%	35%
Energy imports Less Open Eco	(As % of high export case)	22%	25%	28%

The reductions in energy investment outlay are nominal and range from 6% to 11% of the outlay under the 'Less Open Economy" scenarios. However, even assuming moderate real increase in import price and high export growth rates, the share of energy imports in export earnings increases to unsustainable levels. Real import price increases beyond 2% per annum would further increase the import burden. In comparison the import burden under the "Less Open Economy" scenario is more reasonable and sustainable.[5] Energy security considerations and a shallower international market for petroleum products, as compared to crude, further limit the potential of this alternative. Consequently there does not appear too much scope for reducing the investment outlay through this route in the long run.

Shift Investment to Non-energy Sectors

25. Demand side management provides a less capital intensive, lower variable cost and environmentally superior option to supply enhancement in bridging the energy gap. In addition it has the advantage that it 'externalizes' investments from the energy sector to the end use sectors. Table 9 provides comparative capital and annualized costs for supply enhancement and demand side management measures in electric power.

TABLE 9 : Cost Comparison of Supply Side and Demand Side Options (1995 costs)

	Coal Fired Plant	Co-generation	Demand Side Management in Agriculture	Wind Power
Capital Cost of a 15 mW plant or equivalent capacity Rs millions	765	276	128	652
Energy Supply Cost Rs per KwH	3.01	1.59	1.51 to 1.75	6.04

Source : TERI estimates.

Demand side management is the cheapest alternative to bridge the energy gap, followed by co-generation. Both these alternatives are cheaper both in terms of cost of power generated as well as capital intensity as compared to a conventional coal fired power plant. A "Negawatt" generated costs less than a "Megawatt" even if the avoided environmental costs are ignored.

26. Increasing the efficiency of electric power consumption in the domestic sector requires investment in housing design and construction, improving the efficiency of end use equipment like consumer durables and lighting appliances. Similarly increasing energy efficiency in agriculture, requires investments in better water pumping equipment, motors and in the better utilization of water. Often better management and maintenance are enough, and the investment levels required to bring about significant energy use efficiencies are nominal. Similar efficiencies are possible in the use of petroleum products through better efficiency in the use of water pumps, tractors and vehicles. Price increases are often a prime motivator to induce better efficiencies in use. However where the price motive is absent due to social considerations formulation of energy use standards for end use equipment is a must. Since there are large negative externalities associated with the inefficient use of energy, Government intervention through regulation, fiscal signals, creation of property rights and the removal of informational asymmetries will be required.

Privatization

28. The experience in Latin America has established that privatization has the potential to pull in large if 'lumpy' investments. The net assets of the Industrial Public Sector[6] are around $ 247 billion of which the net assets of energy related public sector entities would be around $140 billion. However since many public sector entities have been incurring losses their net worth

would be lower. Privatization of energy utilities would also distance the government from the temptation of financial irresponsibility via provision of un-targeted heavily subsidized energy services and products. Privatization increases the revenue of the government through sale of assets and reduces the revenue expenditure through transfer of utility losses to the new private owners. It also infuses capital into balancing equipment, modernization of the existing assets which is a more cost effective alternative for enhancing supply. It is estimated that the costs of modernization are less than one-half of the costs of new capacity. Lastly in a situation where the supply of equity exceeds the supply of debt, accelerated privatization is an ideal solution to absorb the higher availability of equity. The capital structure of public sector utilities becomes highly leveraged once their balance sheets are cleaned up by writing off losses against equity. The blend of fresh equity and the existing outstanding debts in such enterprises is an efficient solution for absorbing the excess supply of equity as compared to debt.

Utility Reform and Restructuring

28. This paper has assumed that reform of the industrial public sector would generate internal resources[7] of 5% of their net assets thereby adding to the revenue inflow of the Government. This is a modest target. The internal resource generation in the top 500 private corporates in India during 1994/95 has been estimated at over 15% of Net Assets. Reform in the Electric Power Sector has commenced. The Coal Sector is still to initiate a similar move though recent liberalization in the sale price fixation method is anticipated to add to the bottom line of Coal India. The Petroleum sector is expected to be reformed over the Ninth Plan period. These are all positive signs. However the speed of reform and consequently the rate of increase in the generation of internal resources by the industrial public sector as a whole will be critical in either constraining or facilitating the requisite investments by up to 2% of GDP.

Conclusion

29. High GDP growth rates provide the necessary objective conditions for enhancing the supply of capital, especially from foreign sources. However energy investment demand increases rapidly with higher levels of economic growth. Consequently energy supply side options to bridge the energy gap becomes unsustainable. Aggressive demand management measures are needed to constrain energy demand within sustainable limits. Investments in demand side measures not only reduce the total investment demand but they also 'externalize' these demands to energy end use sectors. Investment levels of between 4 to 4.5% of GDP required to meet the energy gap are possible given the adoption of proactive and supportive Government initiatives to enhance internal resource generation from the public sector, capitalize domestic Developmental Finance Institutions, enhance the long term resources of Institutional Investors in India and reduce its draft on the assets of the banking system and the domestic capital market.

30. The role of the Government will be critical in managing the transition period, till 2002 during which, the capital supply gap is especially significant due to the front loading of investments and the back loading of capital supply. Resource constraints have traditionally provided the impetus for technological change. These in turn generate incomes, which fuel higher levels

of consumption. In this sense, "the fuelish gap" can never be bridged. However at a practical level, energy constraints need to be removed not only to achieve targeted GDP growth but also to meet the objectives of social equity. Energy gaps and low growth affect the poor the most. In India the energy gap is a child of government policy. Bridging the gap requires some parental guidance and the imposition of considerable discipline. There is no alternative to a proactive government policy, effective independent regulation and a market determined economy outlook.

NOTES

1. Investments in energy are defined as investments in crude oil and gas exploration and production, oil refining, coal mining and electric power generation, transmission and distribution.
2. The World Bank classifies "Less indebted" developing countries as those which have a long term external debt (LDOD) to exports ratio below 165%, LDOD to GDP below 30%, Debt servicing to Exports below 18% and Interest to Exports to below 12%.
3. A company can reduce the effective rate of servicing of equity by letting share folders benefit through bonus shares rather than pay out of dividend since capital gains are taxed at a lower rate of 20%.
4. Industrial Development Bank of India, Industrial Credit and Investment Corporation of India, Industrial Finance Corporation of India Small Industries Development Bank of India are the multi sector industrial finance agencies. Rural Electrification Corporation, Power Finance Corporation and Indian Renewable Energy Development Agency are the specialized power finance corporations.
5. Crude and petroleum product imports were equal to 25% of export earnings in 1995/96. The ratio is expected to be higher in 1996/97.
6. The Industrial Public Sector has been defined as all public net capital stock in mining, manufacturing, electricity, gas and water supply estimated at $247 billion in 1992/93. Capital stock in transportation has not been included.
7. Internal Resources are defined as Profit After Tax plus Depreciation. Net Assets are Gross Assets less depreciation.

REFERENCES

Ahluwalia, S. Investment Demand and Capital Supply for the Electric Power Sector in S. E. Asia 1994 Background paper, International Trade Div World Bank.

Ahluwalia, S. Financing the Power Sector in India 1995 Background paper for the Committee on Commercialization of Infrastructure, GOI.

GOI, Central Statistical Organisation, National Account Statistics 1994

GOI, Ministry of Finance, Economic Survey 1995/96

GOI, Planning Commission, Annual Report on SEBs and EDs, 1995

GOI, Planning Commission, Eighth Five Year Plan, 1992

Gooptu, Sudershan, Portfolio Investment Flows to Emerging Markets 1993 World Bank

Reserve Bank of India, Report on Currency and Finance 1994/95

Samuel, Cherian, The Stock Market as a Source of Finance, 1996 World Bank

Singh, Harinder & Kwang W Jun, Some New Evidence on Determinants of FDI in Developing Countries 1995 World Bank

TERI, Demand Side Management in Agriculture in UP, 1996

TERI, Economics of Wind Power, 1996

TERI, The Impact of Investments in the Power Sector in India on Public Finance 1995

United Nations, World Investment Report, 1996

WEC, Global Energy Perspective to 2000 and beyond, 1995

World Bank, World Debt Tables 1995

World Bank Capital Expenditures for Developing Countries in the 1990's, Feb.1990

World Bank, The East Asian Miracle 1993

World Bank, World Development Report, 1994, 1995

World Bank, The World Bank's Role in the Electric Power Sector, 1993

World Bank, The Emerging Asian Bond Market 1995 World Bank

GROWTH AND INVESTMENT IN THE ENERGY SECTOR OF INDIA TILL 2020[1]

59

J. PARIKH and **N. S. MURTHY**
Indira Gandhi Institute of Development Research (IGIDR)
Mumbai - 400 065, (INDIA)

ABSTRACT

The growth of energy in the country is described both in rural and urban areas. The future growth is projected. The investments in various secctors of energy production and use have been covered. Mathematical models for the estimation of investments have been proposed.

Introduction

The growth potential of the indian economy is perceived to be high for the next 25 years. At the same time this can be realized only with a corresponding development and growth in the energy sectors. Capacity addition in energy sectors requires sustained high levels of financial investment as construction of new electric power plants, opening of new coal mines and exploration and extraction of petroleum are capital intensive activities and in many cases involve imports of equipment. Investment in the energy sectors are constrained by the availability of foreign exchange and investible funds. This paper forms gross estimates of capital expenditure required over the next 20 years to meet the projected levels of energy demand upto the year 2020.

We begin with a brief overview of the broad developments in the Indian energy sector in recent four decades. The total energy supplies, commercial and non-commercial together, in India increased from 82.7 mtoe (million tonnes of oil equivalent) in 1950-51 to about 291 mtoe in 1990-91. This represents an increase by a factor of 3.5 which is equivalent to an annual average growth rate of about 3.2%. A sizeable share of energy requirements, especially those of the rural household sector, is met from non-commercial sources like fuelwood, crop residue, animal waste and animal draught power. Fuelwood alone accounts for 65% of the total non-commercial energy consumed in the country. It is an important source of energy for rural and to some extent, urban households. The efficiency at which useful energy is presently being realized from the resources is very low at about 10-15%. The share of non-commercial fuels in total energy supply has, however, declined from 74% in 1950-51 to 41% in 1990-91 due to substitution by commercial forms like coal, oil and electricity which provide energy of a much higher quality and with greater efficiency, though at a higher cost. Because of this substitution process, the growth rate of commercial energy has been higher than that of total energy and has varied between 4.2% to 5.5% in the last four decades. Elasticity of consumption of coal, oil and electricity with respect to real GDP has been about 1.3, 1.2 and 1.8 respectively during 1971 to 1991.[2]

The trends in consumption of primary commercial energy are given in Table 1 and Figure

1. While Table 1 gives the information on use of different forms of energy in their individual physical units. Figure 1 gives it in comparable energy units. The relative share of oil and gas has increased between 1950-51 and 1990-91, whereas the share of coal has declined. Coal continues to be the main source of not only primary commercial energy use in industry but also for secondary energy use through power generation. However, there has been a significant step up in the production of oil and natural gas. In 1990-91, India imported about 16% of the petroleum products it consumed. In addition, about 40% of the crude processed domestically was also imported. Natural gas is likely to play an increasingly more important role in providing energy in the coming years. Availability of hydro-electricity has also increased steadily over the years. There have been modest additions to nuclear power generation capacity.

TABLE 1: Trends in Consumption of Primary Commercial Energy *

Energy Form	Unit	1950-51	1960-61	1970-71	1980-81	1990-91
Coal	mt	32	55	73	114	212
Lignite	mt	-	0.1	3.4	5.1	13.8
Petroleum pr.**	mt	3.3	7.7	17.9	30.9	55.0
Natural Gas	MCM@	-	-	1445	2358	17998
Hydro Power	TWh$	2.5	7.8	5.2	46.5	71.7
Nuclear Power	TWh	-	-	2.4	3.0	6.1
Others	TWh	1.5	3.2	5.4	8.4	25.1

Source Ministry of Finance (1994)
* The consumption figures for energy forms other than oil are based on respective production figures.
** Includes imports of petroleum products. Part of the domestic production is by using imported crude oil.
@ Million cubic metres.
$ 1 Twh = 10^9 Watt-hours

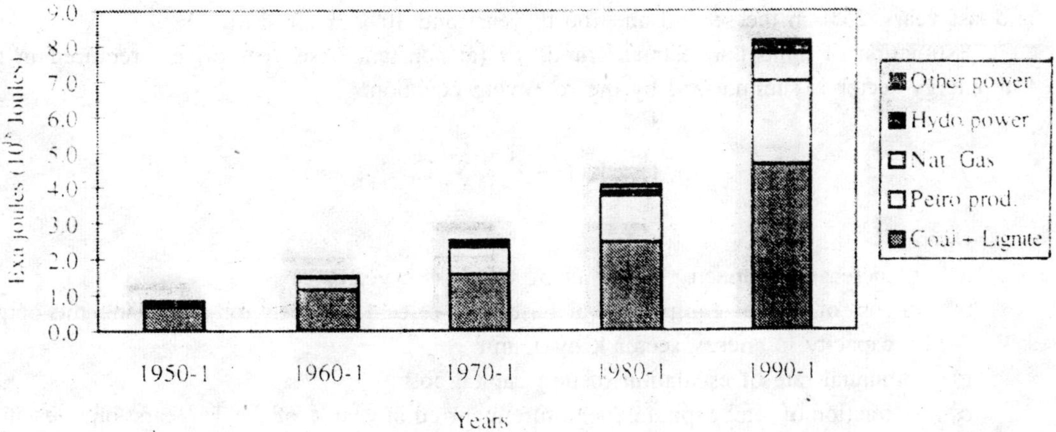

FIG. 1: *Commercial energy supply by category of source, 1950-1990, India*
Source Murthy, Panda and Parikh (1996)

Methodology for Estimating Investment Requirements in the Energy Sector

The capital requirements arise in three major sectors: electricity, coal and oil and gas. In the case of electricity, an imperative investment is for augmenting generating capacity for which a capital cost of Rs. 20,000 per KW of capacity added is assumed at 1989-90 prices. There will be additional investment expenditure associated with expansion, modernization and maintenance of transmission lines and also for organizing the increased distribution of power especially in the rural areas but we do not account for it in this paper.

In the case of coal the capital costs arise in mining and transporting it. At 1989-90 prices this cost is estimated at Rs. 2,000 per tonne of coal and is assumed to be rising at the rate of 3% per annum on account of the need for mining deeper deposits. Costs involved in the exploration and development of new coal fields is not considered in this paper.

Oil and gas sectors require two kinds of investments: exploration, development and operation of oil and gas fields and the transport of crude by pipelines to the refinery locations. We have taken a figure of Rs. 6,000 per million tonne of crude oil output to represent the capital cost of crude oil and gas production. In addition, there are sunk costs involved in the exploratory drilling for new fields and their development into a regularly operating well. While these costs are not included in this analysis they should also be counted among the capital costs for supplying crude oil.[3] Crude oil (domestically produced as well as imported) but not natural gas is processed further in refineries for which a capital cost of Rs. 600 per tonne of oil products. Both these figures are taken at 1989-90 prices. The capital coefficients used in this paper are summarized in Table 2 below.

The required gestation lag for investment to mature into new production capacity is taken as 5 years in each sector with the necessary investments being phased over this duration of time. The distribution of capital expenditure over the gestation period is assumed to be 10% in the first and last years, 25% in the second and fourth years and 30% in the third year.

Estimation of annual investment funds $I^k t$ (at constant base year prices) required in the k-th energy sector is summarized by the following equation:

$$I_t^k = \beta_0^k (1+g^k)^t \sum_{\tau=1}^{T^k} \alpha_\tau^k N_{t+\tau}^k$$

where

$N^k{}_t$ = increase in capacity of output of sector k in year t

$\beta^k{}_0$ = cost of capital equipment (at base year prices) required for increasing the output capacity in energy sector k by 1 unit

g^k = annual rate of escalation of unit capital costs

$\alpha^k{}_\tau$ = fraction of total capital expenditure incurred in year τ of the T^k-years long gestation period for capital investment to mature into new output capacity.

The values of various parameters used for different sectors in the above set of equations are given in Table 2 below.

TABLE 2: **Equation Parameters**

k	β^k_0 (at 1989-90 prices)	g^k	α^k_1	α^k_2	α^k_3	α^k_4	α^k_5
Coal	Rs. 2,000 per tonne	0.03	0.10	0.25	0.30	0.25	0.10
Crude oil & gas	Rs. 6000 per tonne	0.00	0.10	0.25	0.30	0.25	0.10
Oil refinery	Rs. 600 per tonne	0.00	0.10	0.25	0.30	0.25	0.10
Electricity generation	Rs. 20,000 per KW	0.00	0.10	0.25	0.30	0.25	0.10

Source: Majumdar and Parikh (1996) for estimates of β^k_0, g^{coal} and α^{elec}_τ

Estimates of Capacity Expansion

The capacity expansion variables appearing in the previous section, N^k_t are computed on the basis of an input-output (I-O) model[4] of the Indian economy from 1990 through to 2020. Macroeconomically consistent projections of gross domestic product (GDP) and the different final demand categories—private consumption expenditure, government consumption expenditure, gross fixed capital formation (investment demand), change in stocks and inventories, exports and imports (oil and non-oil are projected separately) are produced by the SIMA model. These macro-aggregates are then used to scale the I-O model results giving the total energy demand as the sum of sectoral energy demands by each category of energy source viz., coal, oil and gas and electricity. The I-O model has the advantage of maintaining inter-sectoral consistency which is important as the bulk of the energy consumed in the Indian economy is at various intermediate stages in different production sectors. At the same time the detailed representation of the production activities at 40-sector level in the I-O model permits simulations of technological changes in some energy intensive sectors and evaluating the impact on the energy demand. The projected requirements of output from the energy sectors are as given in Table 3 below.

TABLE 3: **Gross Output Requirements from the Energy Sector, India**

Year	Coal (10^6 t)	Crude oil & gas (10^6 t)	Petro. prod. (10^6 t)	Electricity (10^9 KwH)
1990	219	48.5	45	282
1995	282	53	53	358
2000	362	54	72	455
2005	479	54	102	593
2010	634	54	142	774
2015	859	54	200	1033
2020	1164	54	278	1380

Source: Computed by authors, IGIDR.

This paper assumes that enough coal reserves are available to meet the desired levels of coal demand. On the other hand, the position of oil and gas reserves in India is not at all comfortable. It is assumed that the current production trends—35 million tonnes (mt) of crude petroleum and 19 million tonnes of oil equivalent (mtoe) of natural gas — will be maintained upto 2020. Needless to say, this in itself is a stiff target to meet and will involve sustained exploration and development activity in this sector. As far as the electric power sector is concerned the only constraint on capacity expansion considered in this analysis is the availability of finance for building new power plants. Additional electricity requirements are assumed to be met by continuing with the current supply mix and capacity utilization of power plants.

To facilitate the computational exercise, it is assumed that out of the estimated demand of oil and gas products the requirements of oil products are met entirely by the output of the domestic refineries. Capacity expansion in the refinery sector is therefore computed at rates which keep pace with the increase in the consumption of non-gas oil products.

Depreciation of capital equipment in each energy sector is assumed at the rates indicated by the consumption of fixed capital in the corresponding sectors notified by CSO (1996) in its yearly presentation of National Accounts Statistics. In the case of coal and crude oil, the corresponding sector is "Mining and Quarrying" while for the oil refinery it is the "Manufacturing (Registered)" sector. Electricity corresponds to the "Electricity, Gas and Water Supply" sector in the National Accounts Statistics. The annual rates of depreciation are 8.33% for coal and crude oil sectors, 6.79% for oil refineries and 5.07% for the electricity sector. Allowing for depreciation implies that the part of the existing capacity which is depreciated must be replaced by fresh investment just to maintain the capacity in each sector. We assume that the capital equipment in the mining sectors, coal and crude oil and gas, can be shifted from one mined site to another so that the depreciation of the machinery is independent of the life of the mines/oil and gas fields.

The estimates of required capacity expansion are shown in Table 4 below.

TABLE 4: **Capacity Additions* (in output units) Required in the Energy Sector, India**

Year	Coal (10^6 t)	Crude oil & gas (10^6 t)	Petro. prod. (10^6 t)	Electricity (10^6 KW,
1991	30	5	5	7
1995	36	5	5	9
2000	46	4	9	11
2005	64	4	13	15
2010	85	4	18	20
2015	118	4	26	27
2020	160	4	35	36

Source. Computed by authors, IGIDR

* Inclusive of replacement for depreciated capital equipment.

Investments Funds Needed for Meeting Increased Energy Needs upto 2020

Using the results from the previous two sections, we estimate the annual requirement of investment over the next 20 years. The results are summarized in the Figure 2. The annual capital expenditure increases from nearly 250 billion Rupees in 1990 to a requirement of more than 1350 billion Rupees by 2015.

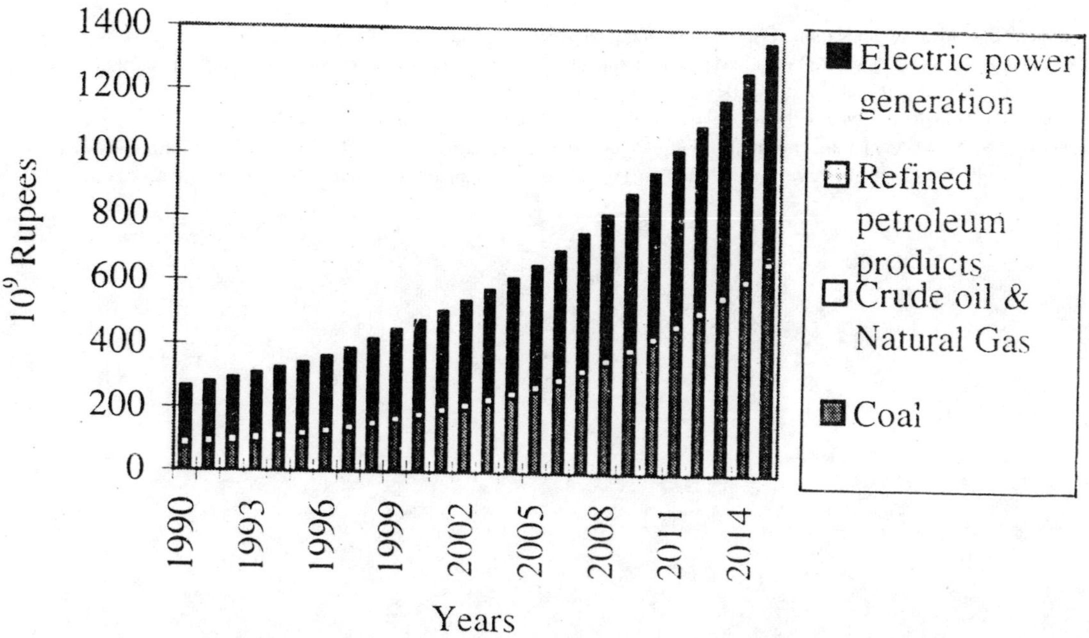

FIG. 2: *Annual investment requirements upto the year 2015 to meet additional demand upto the year 2020 given by the I-O Model (Murthy, Panda and Parikh, 1996)*

Discussion

It is seen that the bulk of the capital requirements come from capacity additions in the electric power sector. Given its reliance on adding coal based power plants this translates into high rates of investment into the coal sector as well. The capital expenditure in oil sector is seen to be quite low over the years which is due to two reasons: firstly the share of oil in meeting the total energy demand does not increase over the years. Secondly the expenditure on exploration and development of oil and gas fields not counted as capital expenditure here should correctly be treated as investment as it is essentially a sunk cost.

NOTES

1. This paper draws substantially on energy modelling work done at IGIDR: by Majumdar and Parikh (1996) using the SIMA model and Murthy, Panda and Parikh (1996) using the Input-Output model for the SDC project.
2. If we exclude the coal consumed for secondary energy production (electric power generation), then the elasticity for coal drops to 0.67.
3. Considering the distribution of unit cost crude oil and gas production over the three activities–exploration, developmen and operation — roughly half of it falls under the category of development costs.
4. The input-output model is calibrated based on the Planning Commission's I-O Table for 1989-90 and is described in Murthy, N.S., M. Panda and J. Parikh (1996).

REFERENCES

Central Statistical Organization: CSO (1996). National Accounts Statistics 1996, Government of India, New Delhi.

Majumdar, S. and J. Parikh (1996), "Macroeconomically Consistent Energy Scenarios for India", in Parikh., J. (eds) Energy Models for 2000 and Beyond, Tata McGraw-Hill, New Delhi [forthcoming].

Ministry of Finance (1994). Economic Survey 1993-94, Government of India, New Delhi.

Murthy, N. S., M. Panda and J. Parikh (1996), "Consumption Patterns, Energy Demand and CO_2 Emissions in India: 1990-2020", in Structural Transformation Processes — Sustainability, Project No. xxx, INFRAS, Zürich.

FINANCING OF POWER PROJECTS IN INDIA

60

B.S. SACHDEV
Senior Manager, Power Finance Corporation Ltd.
Chandralok, 36, Janpath, New Delhi - 110 001.

ABSTRACT

None of the lender likes to risk their investment in having financial arrangement with loss making borrower However, certain compulsions of the borrower in the form of social needs/obligations are contributing to the loss, although commercial operation of the utilities is recommended by different agencies. On the other hand, major concern is preconditions of international funding agencies like ADB and World Bank as well as priorities of private financing agencies for a lucrative return in other fields with shorter pay back periods.

Since power is the back bone for any development therefore needs to the funded by alternative sources. Relaxations in preconditions by the lenders, joint ventures, leasing of equipments, private participation and top management support for commercial operation of the system. Load management is of vital importance when the demand surpasses the supply. Further high level demand in the region is obviously a big attraction for multinationals provided conducive environment is created.

Certainly there is no future without power. Even distribution of shares to employees and public to the tune of 10 - 15 per cent will attract considerable amount to complete ongoing projects after retaining 51% equity with the Government. On diagnostic study, parameters identified for inefficiently needs to be implemented with all seriousness.

INTRODUCTION

Indeed a direrct correlation exists between the level of economic development and the consumption of energy. Recognising the critical importance of energy as principal element of growth and treating power as priority, most of the countries has earmarked considerable outlay for this sector. Inspite of heavy investments, the gap between demand and supply is increasing. The exercise carried out by the various groups revealed that requirements for the energy by the year 2020 will pose a challenging task in terms of arranging for adequate energy supplies and investments that would be required. Even the recent study by ADB reviewed that Asian countries needs to spend $ 280 per year on infrastructure.

Now when we look at the resources, we find persistent fiscal deficit of most of the south Asian contries are indicative of their financial health. Financial constraints on account of balance of payments affected adversely the efforts to supplement additional capacity and forced the policy makers to search for the alternatives. Further political uncertainties had slowed down the developmental activities. Inter-rivalry wars added fuel to the fire.

Keeping in view the highly capital intensive nature of the power supply industry and fast rising trends in this sector, it would be becoming more and more difficult to find adequate financial resources for implementation of additional generation capacity and transmission facilities programmes. Bringing additionality of funds through various sources has it down limitation in terms of the volume of funds that needs to be mobilised. Therefore, one of the most serious constraints faced by these countries is the lack of adequate resources, infrastructure and capacity to undertake large scale programmes in power sector to meet the present shortfall as well as future requirements.

One key reform considered is the opening up to multinational companies to attract private investment in general and foreign investments in particular. Sectors such as power, private participation be now encouraged as a matter of policy although priorities differ. In India broad policy changes have been effected to attract private investor but the quantum of investment actually flowing in has been far below expectations. The main reason, is the absence of commercial orientation with existing Public Sector — for instance, the State Electricity Boards which are run as extension of State governments. This has meant poor recovery of revenue, massive losses and paucity of investible resources. This is the reason that most of SEBs are not in a postition to achieve minimum 3 per cent rate of return, a precondition for ADB and World Bank loans. Further funding is restricted to major equipments only, therefore balance funds has to be arranged by the State Electricity Boards.

The latest Indian Economic Survey estimates the annual bill for providing subsidised power for agricultural and domestic purpose at Rs. 15,000 crores. This recovery means a substantial investment in new generation and transmission and distribution system. As regards the areas of concern, the priority must go to system improvement schemes having quick returns and second priority for ongoing projects. There are a number of project financing agencies in the public sector which extend assistance by appraising reports. Power Finance Corporation is the only agency which stresses on institutional development apart from funding power projects. PFCs concern is not only the efficient utilisation of funds provided but the efficient operation of the power utility as a whole. The agency is further making all out efforts to create awareness about the commercial operations and the manner in which these should be achieved.

It is prudent to mention that experience of private participation in power sector even in some of the western countries are not encouraging. Therefore, we must concentrate on recovery of realistic revenue and technology upgradation apart from arranging funds for new projects by various methods:

(a) Approaching international funding agencies for development assistance and soft term loan.

(b) Invitation to multinational companies for joint venture.

(c) Participation by public/private companies and employees.

(d) Leasing of equipments.

(e) Institutional development.

CONCLUSION

The introduction of privatisation have set the scene for major cultural change. From being an engineering dominated culture, the electricity supply industry is rapidly changing to commercially dominated one. Experience in different countries have shown that there is very little economic difference between well managed nationalised and privatised utilities, yet privatisation is sweeping the electricity utility industry in many countries. Inspite of numerous examples, how do we justify the liberal operations in privatisation by replacing the concept of national interest in looking after the long term goals in power industry.

INDIA'S PERSPECTIVE ON PRIVATE INVESTMENT IN THE OIL AND GAS SECTOR

61

DR. A. BHATTACHARYA
Economic Adviser
Ministry of Petroleum & Natural Gas
(Contract cell)
Government of India
New Delhi

Backdrop

India is one of the few countries in the developing world that made a significant attempt to develop a domestic energy sources before the first oil crisis of 1973. Starting from a meagre production of 0.25 million metric tonnes per annum (MMTPA) of oil in 1947, the country today produces more than 30 MMTPA. Refining capacity has also increased from 0.25 MMTPA to over 60 MMTPA during the same period. National oil companies like ONGC and OIL produced over 0.5 billion tonnes (cumulative) for the last four decades. To begin with, most of the petroleum sector projects received budgetary support from the Government of India. Subsequently, they began to generate sufficient internal resources which were ploughed back into new projects. Also, petroleum sector PSUs have resorted to various types of internal and external borrowing, including project-tied loans from international development banks, bond issues abroad, buyers' and suppliers' credits, multi-lateral credits and syndicated loans. ONGC and IOC had successfully tapped the international bond markets to raise financial resources and new issues in European market to meet part of their foreign exchange requirement.

Role of Government

As India is a net importer of petroleum products, the downward adjustment in the exchange rate of Indian rupee against other currencies not only increased the cost of debt services, but led to an overall increase the cost of exploration, development and exploitation of hydrocarbons, refining capacity addition and infrastructure development (e.g. port upgradation, pipeline network, crude and product tankages and other marketing/distribution facilities). However, the silver lining came with the Government's announcement of various policy reforms which ushered in an era of liberalisation of the Indian economy and opened it up to greater private and foreign investment. The Government decided to reduce its role in managing PSUs and currently in the process of disinvesting and diluting its equity as part of its stake in various petroleum PSUs to private investors. At the same time, the Government chose to cut down budgetary support to oil PSUs. Out of a total plan outlay of Rs.122.52 billion for the entire petroleum sector, the Government's contribution was only Rs.1359 million. For the last 3 years i.e. 1993-94 to 1995-96, due to the need for substantial investment to meet the growing demand for petroleum products, internal resources of the PSUs had been able to finance only 55 to 66% of total expenditure. It is estimated that around US $ 20-25 billion will have to be invested in upstream and downstream petroleum projects before end of this decade, out of which US $ 10 billion will have to be invested to provide additional refining capacity of atleast 40 MMTPA. Over US $ 5 billion worth of projects are underway in the western offshore region of ONGC alone. In addition, investment is required for upgrading processing facilities in existing refineries to meet new product quality and environmental specifications. Investment in crude product pipeline and storage will be required to bring new refinery projects online. Furthermore, infrastructure in terms of upgrading port facilities in the East and West coast, new product tankages at the various tap-off points of the new product pipelines, other marketing and distribution facilities are to be created to cater to an estimated demand for petroleum products of 113 MMTPA by the year 2001-02.

AVENUES FOR RAISING FUNDS

In the case of domestic oil producers viz. ONGC and OIL, the debt-equity ratio is high since they had to often borrow from overseas markets to meet foreign exchange requirements of their projects. This coupled with the colossal exchange looses incurred on revaluation of external commercial borrowings in the wake of persistent depreciation of the rupee, has adversely affected the debt equity ratio of the oil producers. Major oil companies like ONGC and IOC are raising financial resources from international money market at very competitive and low rates, due to their very sound financial health, economies of scale and high degree of acceptability with international investing community. These oil companies are also presently providing a bench mark rate for the country. Some of oil companies (e.g., ONGC & IOC) are carrying some high cost debts mainly taken from international financial institutions.

NEED FOR FINANCIAL RESTRUCTURING

One of the pre-requisite for public sector oil companies to be internationally competitive is to improve their financial competitiveness. Till now a need for a closer look of the financial structure of the oil companies was not felt, because all the oil companies were under administered price mechanism, and they were largely meeting their fund requirements for expansion programmes through internal resources, where budgetary support from Government of India was not required and the gap between the internal resources generation and the fund requirements were met through debt route only. Further, option of raising finance through equity were not available. The options in the debt route were also limited due to lack of adequate financial autonomy and in certain cases, strategic national considerations, overriding the effective cost evaluation of foreign currency debts. These factors have attributed to present financial structure of oil companies. The need for financial restructuring is also felt due to substantial fund requirements in the next five years.

One of the main objectives for financial restructuring is to evolve ideal debt-equity structure. Though no tailor-made equity-debt structure can be prescribed for each PSUs' oil company, nevertheless, it will be appropriate that long range policy guidelines are prescribed to all oil PSUs, with the overall objective of maximising cost effectiveness of financial resources. The debt-equity blend may be assessed by each company based on cost of debt, asset composition, capital servicing capability, period of resource requirement and also macro-economic indicators, risk factors and falling interest rates in the country.

THE EQUITY ROUTE

In the present context, it may be recommended that all the oil companies shall first be allowed to meet their major portion of the fund requirement through equity route due to the following reasons:

- equity base of all oil companies is very low.
- the capacity to service the equity is substantial.
- equity reserve ratios are very high.
- shares of the oil companies which are presently traded in the stock exchanges attract high PE multiplies.
- high equity premium can be charged by most of the companies.
- the substantial growth in earnings of oil companies is expected in the near future in view of the ensuing deregulation and capacity expansion.
- the cost of domestic borrowings is very high. Even globally, the interest rates are rising and foreign currency loan carries the risk of rupee devaluation.

It may be noted that a number of oil companies particularly, major oil companies have so far not made any public offer of its equity shares. Therefore, it may be recommended that such offer should be allowed to issue equity in the domestic/international market upto 20% of their existing paid-up capital. In respect of companies, who have already made public offers, a further issue of equity in domestic/international market may be allowed upto 10% of their paid-up capital. However, it may be desirable that such blanket approval should be subject to certain broad parameters, which are enumerated below :

- Government of India's holding should not go down below 51%.
- Within three years from the public offer, the earnings per share (EPS) shall reflect a minimum compounded annual growth of 3% above the prevailing rate of inflation.
- There shall not be any drop in the quantum of retained earnings.
- The rate of dividend and pay out ratios shall not decrease significantly.

THE DEBT ROUTE

Notwithstanding the cost effectiveness and other advantages to raise resources through equity route as enumerated above, it would be prudent for the oil companies to source a part of their fund requirement through debt route due to the following reasons :

- Substantial enlargement of equity to raise required resources may lead to implications associated with servicing the large equity, particularly in the context to the Indian investors, who expect substantial growth in investments made through equity route.
- Debt route provides flexibility for adjusting level of funds deployment since it can be discharged, which equity route creates permanent obligation of serving.
- The enlargement of equity may adversely affect the market value of equity owned by Government of India.

DILUTION OF ENLARGED EQUITY

In order to bring down the Government holdings upto 51%, Oil companies should be allowed to dilute their shares by issue to fresh equity rather than disinvestment by Government as is presently carried out. The disinvestment made so far is in the form of issue of shares to financial institutions/ mutual funds etc. resulted into very small floating stock in the market. Consequently, market prices of shares of these companies are not truly representative due to lack of adequate market capitalisation. Further, the disinvestment process has enabled the Government to generate additional funds, but no funds were made available to respective Oil companies. Dilution through fresh issue of capital would, however, help the oil PSUs to raise funds required for their expansion/modernisation programmes, apart from creating enough floating stock in the market. The PSUs will also be able to retain the cost-free premium available on the issue of these shares.

At present, a number of Oil companies are carrying costly foreign exchange debts, which have adversely affected their debt-equity ratio; debt-servicing ratio etc. Therefore, it is suggested that Government may consider allowing the Oil companies to prematurely repay the costlier loans out of surplus funds generated by them. Possibilities for refinancing some of the costlier loans with carrying lower interest rates also need to be explored by these companies.

NEED FOR FINANCIAL AUTONOMY

The public sector oil companies suffers from the limitation to inadequate financial autonomy. The financial restructuring shall be incomplete and ineffective without providing a greater financial autonomy to public sector oil companies, with a view to ensure that public sector oil companies have enough financial independence to compete nationally and internationally. The areas where immediate attention is required as under :

- Capital investment decisions.
- Freedom for raising finances.
- Diversification
- Autonomy in other areas of financial management.

The financial sector worldwide in general and India in particular is rapidly changing which requires immediate and quick response to ever-changing external environment. Under the present system of approvals and associated procedures for raising the debt or equity by PSU oil companies, the response time is very slow. Hence, in most of the cases the opportunity of raising finance through least cost option could not be fully exploited. Therefore, there is a need for providing greater autonomy in regard to pricing, sourcing, type of instruments, hedging, forward deals and the other related areas which should be left to the decisions of the Board of Directors of the oil sectors PSUs.

Most of the oil companies world over are integrated companies with about 60 to 70% of their investments in exploration and production activities (E&P) and the balance in refining, marketing petrochemicals etc. In the international scenario, one rarely finds a pure E&P company or refining/ marketing company as we have in India. With the current trend towards globalisation, the oil sector PSUs should also be allowed to diversify their activities into other related areas. The refining and markeitng companies could take up exploration and production of hydrocarbons preferably as joint ventures. Likewise, the oil producing companies could also take up the related areas of refining, marketing and petrochemicals. In the case of a pure E&P organisation, if there is no success in discovery of hydrocarbons (which could happen, keeping in view the very nature of business) profitability would be severely affected. Perhaps, it is because of this risky nature of this business that international oil majors are not engaged in pure E&P activities alone.

ROLE OF PRIVATE SECTOR

Regardless of the Government's ability to support upstream and downstream expansion in the petroleum sector, PSUs themselves may seek private participation which offers numerous benefits, including improved capitalisation, greater managerial efficiency and advanced innovations in technology. The involvement of the private sector in the petroleum industry has increased greatly following the liberalisation of the Indian economy. Government had offered acreages for exploration to private companies since 1980. To date, 9 rounds of exploration bidding alongwith two rounds of discovered oil/gas fields and one round of seismic survey have been announced. The contracts take the form of production sharing with no financial obligation on the part of ONGC or the Govt. during the exploration phase.

CONCLUDING REMARKS

Thus, petroleum sector will continue to rely to a certain extent on commercial borrowings in India and abroad, because of the opening up of several other capital-intensive sectors to private investment and domestic market, probably, will not be capable of raising sufficient debt financing. The magnitude of this required investment implies that the petroleum sector must continue to source foreign long term debt to its expansion programme. With the entry of foreign institutional investors, there will be enhanced appetite in profit making PSU stock. The global depository receipt and convertable bond issue of Indian corporations in the international capital market have been well received. The petroleum sector PSUs like ONGC, IOC, BPC are watching this source closely and they are expected to tap the Indian and international capital market very soon. They are also exploring the possibility of forming joint venture with other private parties to part-finance, development projects and embitious accelerate programme on exploration which has been invested to augment India's hydrocarbon resources by the turn of the century.

DIVISION - V

INDIA ENERGY SEMINAR

COAL : INDIA'S MAIN ENERGY RESOURCE AND LONG-TERM POLICY ISSUES

62

N.N. GAUTAM
Director (Technical), Ministry of Coal. New Delhi.

ABSTRACT

Coal, currently is, and will remain the primary source of commercial energy in India for several decades. Although, current recoverable reserves of over 75 billion tonnes are adequate to last for more than a century even at the enhanced rate of 653 MT per annum, massive investments are needed for achieving the incremental production of about 365 MT by 2006-07. The decline in the Govt. budgetary support to PSUs has led to greater reliance on internal generation and market borrowings for financing their new coal projects so crucial to Indian economy. Notwithstanding these efforts, the need to open up the coal mining sector for private participation, over and above the captive use policy already in vogue, seems to be inevitable. Removing the transport bottlenecks especially in certain fast growing coal fields also requires immediate attention. Development of lignite resources and introduction of clean coal technologies for generating power will go a long way in not only conserving coal but also in improving the environment and reducing CO_2 emissions, a potential green house gas.

1 Introduction

1.1 Coal is the major source of commercial energy in India, currently meeting about 54% of the country's requirement and contributing to nearly 70% towards electricity generation in the country. The per capita consumption of commercial energy, one of the accepted indicators of the level of development of a country, is only 236 Kgoe in India. This is projected to grow to 447 Kgoe by 2010 leading to an increased share of commercial energy in the total energy use in the country. Given the country's endowment of primary energy resources, it is certain that coal will continue to be the main source for meeting the rapidly growing requirement of commercial energy in the coming decades as well.

1.2 The consumption of commercial energy in the developing countries of Asia has grown at a faster rate than in the developed OECD countries. According to World Energy Outlook, 1996, the East Asia, South Asia, China and India will continue to grow at a much faster rate than the world average. The table below depicts the percentage growth in energy consumption for the period 1971-93 and 1993-2010 :

Percentage Growth in Energy Consumption

Country/ Region	1971-93	1993-2010
World	2.2	2.2
OECD	1.5	1.4
East Asia	6.8	4.6
South Asia	5.8	5.8
China	5.3	4.2
India	7.2	7-8

1.3 In India, the domestic sources of commercial energy supply have not kept pace with growing demand resulting in increasing imports. Commercial energy imports as percentage of total imports currently stands at 25%. With the increasing share of commercial energy in the total energy supply in India in future and with almost doubling of the per capita energy consumption in next 15 years, Indian coal industry has not only to play a greater role in meeting the growing requirement of commercial energy but also in containing imports reasonably.

2. Coal Resources

2.1 Geological reserves of coal have been assessed at about 202 billion tonnes in India. Fifteen percent of the Indian resources are of the coking variety and 85% of the non-coking coal variety. Qualitative break-up of coal resources is as follows :

	Resources (bt)	% Share
Coking coal		
Prime coking	5.3	2.6
Medium coking	22.9	11.4
Semi-coking	1.6	0.8
Sub total (Coking)	29.8	14.8
Non-coking Coal		
Grade A to C (Ash level upto 24%)	24.8	12.3
Grade D (Ash level 24-29%)	22.5	11.1
Grade E, F & G (Ash level 29-50%)	84.2	41.7
Inferred reserves	39.8	19.7
High Sulphur low ash coals of North-East	0.8	0.4
Sub total (Non-coking)	172.1	85.2
Total (Coking + Non-coking)	201.9	100

The recoverable resources upto a depth of 600 m are, however, estimated to be of the order of 75 billion tonnes. Nearly 56% of these reserves are amenable to opencast mining.
2.2 Although, Indian coals are high in ash content, these have some very positive features like low sulphur and low toxic trace elements. While coal reserves are abundant in the eastern and the central parts of the country, the south and the west are endowed with substantial lignite resources.

Coal resources of India are concentrated mainly in the States of Bihar (32%), Orissa (23%), Madhya Pradesh (20%), West Bengal (13%), Andhra Pradesh (6%) and Maharashtra (3%).

A large number of potential blocks have been identified in coalfields like North Karanpura and Rajmahal (Bihar), Talcher and Ib valley (Orissa) and Korba coalfield (Madhya Pradesh) for future development.

3. Production and Consumption Pattern

3.1 During the last two decades (1973-96), since the industry was nationalised, India's coal production has risen from 78 million tonnes to 288 million tonnes. India now ranks third amongst the coal producing countries in the World. 74% production comes from open-pit mines and 26 % comes from underground mines.

3.2 The consumption of coal during 1995-96 was about 283 mt of which 66% was consumed for generating electricity, 13% for production of steel and the balance 21% in thousands of other industries like cement, textiles, fertilizers, refractories, brick kilns etc.

Past consumption trend has been as follows :

	1991-92 (mt)	1995-96 (mt)	Annual Growth (%)
Steel	32.05	35.63	2.7
Power utilities	137.05	186.87	8.1
* Railways	4.42	0.27	(–)50.3
Cement	9.95	11.00	2.5
Others	48.51	48.85	0.2
Total	231.98	282.62	5.01

* Railways are switching over to dieselisation and electrification.

3.3 Growth of Power Utility Sector

The power sector in India has emerged as the single largest consumer of coal accounting for 66% of total coal consumption. There has been substantial growth of the coal based electricity generation capacity from a level of 8,000 MW in 1973, when the coal sector was brought under the State control to about 50,000 MW at present. Central Electricity Authority in India has estimated the demand for electricity in the coming years as under:

	1995-96 (Actual)	2001-02 (Projected	2006-07 (Projected)
Gross generation (BU)	380	570	836
Total Capacity (MW)	81,000	1,46,000	1,81,870
Coal Based Capacity (MW)	50,000	87,170	1,16,400
Coal Based Generation(BU)	250	410	566

With the share of coal based generating capacity remaining at around 60-65%, the power sector will continue to remain the largest coal consumer in India.

3.4 Growth of the Steel Sector

According to projections made by the Ministry of Steel, India's production of finished steel is expected to reach 37mt by 2001-02 and 42 mt by 2006-07. Presently, 60% of the steel making capacity is in the primary sector and the remaining 40% in the secondary sector. In future, more and more steel is likely to be produced in the secondary sector thus reducing the share of steel from the integrated steel plants route. This would mean less of steel, in percentage terms, would be produced through the use of coking coals and more from the use of non-coking coals. Direct Reduction Technology is already well established in the country and requires non-coking coals of less than 24% ash. Incidentally, India is presently the world leader in coal based sponge iron production.

With the availability of new steel making technologies dependent on non-coking coal, this sector which was traditionally using coking coal, would be looking for good quality non-coking coal as well and remain the second largest consumer of coal. The future coal demand for this sector, therefore, has to provide for the conventional coking coal for producing steel through the conventional BF-BOF route as well as non-coking coal for new liquid iron making technologies and DRI and also for pulverised coal dust injection into the blast furnaces.

4. Projected Growth of Coal Demand

4.1 According to the recent projections by the Working Group on Coal and Lignite the coal consumption pattern will be as under:

Sector	1996-97	2001-02	2006-07
Steel	40.50	49.60	64.60
(Growth %)		(4.1)	(5.2)
Electricity	215.00	300.00	477.00
(Growth %)		(6.9)	(8.3)
Cement	17.50	19.90	30.00
(Growth %)		(2.6)	(8.5)
Others	60.00	70.50	112.00
(Growth %)		(3.3)	(9.7)
Total	333.00	440.00	653.00
(Growth %)		(5.7)	(8.2)

4.2 The Working Group has also estimated indigenous coal availability as 348 mt by 2001-02 and 473 mt in 2006-07. The projected gap in the domestic coal availability and demand is as under :

Domestic Coal Availability and Projected Gap

(Million tonnes) Demand/ Gap	2001-02	2006-07
Coal demand	440	653
Coal availability (From public sector coal companies and present captive sources)	348	473
Gap	92	180

5. Possible Options for Bridging the Gap

The gap as projected earlier will need to be bridged and following actions need to be considered.

5.1 Use of Other Fuels

While coal will continue to be the fuel of choice for India, the options for using other non-coal energy sources need to be pursued vigorously. The country is already dependent on import of petro products. Hydel power has its own limitation, of long gestation as well as substantial impact in terms of displacement of local population. However, there is good scope of harnessing non-conventional energy like solar and wind power. Emphasis should also be laid in tapping potential of Coal Bed Methane.

5.2 *Management of Coal Demand*

5.2.1 Use of coal more efficiently need no emphasis. Since power sector is the largest consumer of coal, significant improvements can be achieved by more efficient use of coal in this sector. We have a substantial number of power houses with a specific fuel consumption close to 1 Kg per unit of electricity generated. Even a modest 10% reduction in specific coal consumption by power sector over the current levels can substantially reduce the gap in the coming years.

5.2.2 Low price of coal does not encourage efficient use and conservation of coal. Further, the thermal efficiency of the existing coal based power plants is 25% or less. Even the present guidelines for IPPs stipulate an efficiency of 33% as against the world's norm of 38%. This, therefore, calls for renovation, modernisation and retrofitting of old power houses for improving efficiency of utilisation and adoption of clean and efficient combustion technologies.

5.2.3 With the progress made in combustion technology, it has become possible to achieve a thermal efficiency much higher than usually attainable with conventional pulverized fired coal boilers. In this context it is necessary to switch over to more efficient, as well as environmentally friendly coal combustion technologies like supercritical and ultrasupercritical pulverized fired technology, Pressurised Fluidised Bed Combustion (PFBC) and Integrated Gasification Combined Cycle (IGCC) etc. There is also a need to go for beneficiation of power grade coal. These technologies can be adopted by the new power plants to be set up in India by the private sector to achieve on one hand efficient utilisation of coal and, on the other, the cleaner environment and lower GHG (CO_2) emission per unit of electricity generated. This would, however, need a policy direction which motivates the IPPs to opt for such technologies.

5.3 Import of Coal

The import of coal is on OGL. While the import duty on coking coal is 5%, import tariff on steam coal has been reduced from 85% to 20%. Although, inadequate port handling capacities, constraints in rail movement and huge outflow of foreign exchange will inhibit large scale imports, nevertheless, this is likely to grow with the development of infrastructural facilities. The alternative of domestic production is superior to the option of imports with obvious advantages of local employment and development of ancillary industries. It is in this context our coal companies have to improve its operational efficiency and produce coal at competitive prices. We also need to encourage increasing participation of private sector in the Indian coal mining industry.

6. Increase Domestic Production

6.1 The production potential of the indigenous coal industry can be significantly achieved if necessary policy initiatives are taken in this regard. The operational and financial initiatives have been outlined in the succeeding paragraph.

6.2 Transfer of Forest Land

At present the major constraint in expanding coal production is the inordinately long time taken in obtaining permission for mining in forest lands. Even detailed explorations in forest areas are suffering affecting future projectisation of regionally explored reserves. The process is getting delayed due to :

(i) Delay in identification of land for compensatory afforestation by States.
(ii) Delay in mutation of compensatory land in favour of Forest Department.
(iii) Demand for compensation for environmental losses.

Clearances for coal mining projects should be given on payment of cost for afforestation of degraded forest lands upto twice the extent of forest land required. Responsibilities for mutation should vest on State Governments.

6.3 Strengthening of Coal Evacuation Infrastructure

At present 56% of total coal despatches in India are moved by rail. Many rail coal routes are nearing saturation. Evacuation constraints presently limit production in potential coalfields like Korba, Talcher, Ib valley, North Karanpura and Rajmahal. Large scale investments are, therefore, necessary for development of rail infrastructure to maximise coal availability from the potential coalfields. Higher budgetary allocation and increase of the excise duty levied under Coal Conservation and Development Act would be required for creating necessary transport infrastructure. Coal is a bulky commodity. As such, road transportation is not feasible on a large scale. Also this mode of transportation is expensive.

6.4 Exploitation of Lignite Reserves

Another important aspect which needs to be addressed is development of lignite deposits located in Tamilnadu, Rajasthan and Gujarat. Large lignite based power generation capacity can be created in these States which can sustain a power generation capacity of over 20,000MW against the present capacity of 2040 MW. The potential for lignite based pithead power stations needs to be tapped immediately as this will reduce need to move coal over long distances.

6.5 Encourage Pit Head Generation

Several power stations are being proposed which will require substantial coal movement by rail. However, maintenance of supplies by rail will be difficult as large investments will be needed in creating rail infrastructure. Recent studies indicate pit-head generation alongwith transmission of power to be cheaper than generation at load centres with long distance rail transportation and that pollution levels in Singrauli and Korba, having large pit-head generation capacity, are well within the norms. In the emerging scenario possibility of optimal utilisation of coal in large pit-head power stations merits urgent and serious consideration.

6.6 Encourage Private Participation in Coal Mining

6.6.1 In order to enable private sector participation in coal mining the Coal Mines (nationalisation) Act, 1973 was amended in June, 1993. This limited or partial opening-up enables power generating companies, steel producing companies and cement producing companies to undertake mining of coal for captive use. It is noteworthy that captive mining can be taken up by a separate company, in which the coal user company or the consortium of user companies must hold at all times not less than 26% voting equity and ensure compliance with the requirement that coal is used only for the given (captive) purpose. Coal washing in the private sector is freely permitted and setting up of washeries near the coal mines is encouraged.

6.6.2 Forty-seven mining blocks with an ultimate production capacity of 150 mt have been earmarked for allotment to parties interested in captive mining. Of these, till now 15 blocks have been identified for utilisation in power plants and 5 for utilisation in the steel sector. Three parties who propose to setup power plants have taken steps to actualise coal mining and coal production is likely to commence towards the beginning of the next century. The response to captive mining of coal has not been very encouraging due to :

- Low administered price of coal.
- Development of infrastructure at captive mines in new coalfields is time consuming.
- Restriction on sale of coal to other parties.

6.6.3 The liberalisation measures in the coal sector shall not only have to be continued but further strengthened by widening up of the scope of private participation because of :

(i) Widening gap between demand and domestic supply and steeply increasing energy demand.
(ii) Only captive mining may not be able to fully bridge the gap.
(iii) Port and railway infrastructure is inadequate to handle imported coal.
(iv) Shortage of financial resources with public sector coal companies.

6.7 Fund for Investment

6.7.1 The Government budgetary support to public sector coal companies for investments was meeting almost the total requirements till 1985-86. This has now come down sharply and at present there is practically no budgetary support. Public sector coal companies have, therefore, to rely solely on their internal resources and borrowings for sustaining their developmental programmes.

6.7.2 However, the present Administered Pricing Structure of inferior grades of coal, which constitutes over 60% of the overall coal production, has not been remunerative enough to generate adequate investible surplus. Further, there has been no revision of coal prices of the inferior grades for last 2-1/2 years. This has affected investments in new projects. In the regime of Administered pricing, the private sector consuming industries find it more economical to obtain coal from the public sector coal companies rather than going for captive mining.

6.7.3 Since the financial health of many of the public sector coal companies is not sound, borrowings from market has become very difficult. To improve the situation it is necessary to have remunerative price by total decontrol of coal prices, time bound action plan to liquidate past arrears from coal consumers and strict enforcement for advance payments for coal supply.

6.7.4 Other aspects in improving the investment scenario would be :

(i) Duty on imports of mining equipment including spares should be reduced and should atleast be lower than the duty on import of coal.

(ii) World bank loan with Japan Exim Bank co-financing of about US $ 1 billion is under discussion for rehabilitation of the Indian coal sector. However, this loan is tied up to more policy reforms in the coal sector which include :

   (i) A time-frame for reduction in manpower.
   (ii) The sector to be further opened up to private investments.
   (iii) A time-frame for deregulating of prices and distribution of other grades of coal.

Recent initiatives taken by the Govt. of India and coal industry clearly demonstrate that these issues are being already addressed and some progress has been made in this regard. However, it may be noted that the changes will take some time and cannot be made in one go. Availability of this fund without any further delay will help in the growth of the Indian coal industry significantly.

7. Beneficiation of Coal for Improving the Quality

7.1 Indian coals are generally high in ash content. Further deterioration in quality of coal takes place due to large scale opencast mining. In power generating sector poor quality of coal leads to lower availability, higher O&M costs and high pollution at the power plant end apart from higher cost of transportation of coal. Similar problems are faced in other coal consuming sectors due to degradation of coal quality.

7.1.1 Beneficiation of Coal for Steel Sector

In the area of coking coal, almost all the coals are being washed before despatch to steel plants. The present ash content of 20-22% in the washed coal against the desired 17±0.5% results in low productivity and higher specific energy consumption. Modernisation of existing washeries is to be completed soon for improving quality of washed coal.

Another area which needs immediate attention is the utilisation of the LVC coals for steel making at the earliest. Following results of the studies conducted at CFRI and RDCIS, these coals have been successfully tested at Bokaro Steel Plant. These LVC coals exhibit excellent coking properties which are even better than other indigenous as well as imported coking coal if washed down to 17-18% ash level. Almost 50% of the coking coal being produced currently are of this variety and can make available an additional 5-6 mt washed coal if suitable washing facilities are put in place at the earliest.

7.1.2 Need for Beneficiation of Non-coking Coal

Use of high ash coal (40-45%) in the power sector places a great burden on the railways who have to move high ash coal to distant power stations. Further, it not only results in higher coal consumption per unit of electricity generation, it also affects the power plant performance. According to a rough estimate, by 2006-07, about 31% of the coal for the power sector will be consumed in pit-head power houses, 14% in power houses located within a distance of say 500 kms and 55% in those power houses which are located beyond 500 kms. A recent study indicates that if all the power coal transported over 500 km is washed to reduce ash from over 40% to 34%, there will be a saving in transportation to the extent of about 30 mt — an additional availability of railway capacity at no extra investment, besides substantial reduction in flyash as well as bottom ash.

Distance-wise Requirement of Coal for Power Sector

Distance	1993-94	2001-02	2006-07
Pit-head	61	109	143
< 500 Km	39	51	61
>500 <1000 Km	22	30	53
>1000 Km	43	95	190
TOTAL	165	285	447

The Ministry of Environment and Forests has also initiated measures to push the use of washed coal in all new power stations and in a phased manner in existing power plants. This will generate the need to set up a large number of washeries in the next few years. If all the coal for power sector required in 2006-07 is to be washed, the total investment required would be of very high order of Rs. 14,000 crores. For other sectors, beneficiation will depend upon consumers' willingness to upgrade their coal quality.

7.1.3 Progress in Non-coking Coal Beneficiation

A beginning in beneficiation of non-coking coal has already been made and two washeries for a total capacity of 11 mt for upgrading power grade coal are under construction. It has been decided that all the future washery plants for non-coking coal will either be constructed under B-O-O arrangement by private/ public parties or by consumers themselves. Construction of four more washeries of a combined capacity of 20mt are to be taken up in the private sector, under a "build-own-operate" arrangement for which agreements with private parties, both Indian and foreign, are being discussed. Some consumers have come forward for setting up of their own washeries at the pit-head and then transporting the beneficiated coal to their required destination. Coal companies are providing the necessary facilities like land, water etc. Such initiatives needs encouragement.

8. Conclusion

Although there is no finality about long term projection of supply and demand of energy, it is atleast certain that coal will continue to dominate the energy scene in India in the foreseeable future. Cost-effective exploitation of indigenous coal resources and its utilisation in an environmentally benign manner are of paramount importance if we are to provide a dependable, economical and sustained source of energy to ensure fast-track growth of Indian economy that has opened up with the recent liberalisation. Keeping in view the quantity of coal requirement, import can, at best, be of peripheral substitution. The growth of indigenous coal industry and the associated infrastructure in the coming years will need massive investment. Improvement in operational efficiencies of the nationalised coal sector and increasing investment of private capital would be vital to ensure the growth plan of the coal industry in the coming years.

INDIAN ELECTRICITY SUPPLY
Issues and Options for Tomorrow

63

R.N. SRIVASTAVA
Member (Thermal) Central Electricity Authority & Ex-Officio
Additional Secretary to Govt. of India.

ABSTRACT

'Economic reforms undertaken over the last five years have led to revival of national economy. The process of liberalisation has also resulted in gradual integration of Indian and the global economy. The growth rate achieved, however, needs to be sustained and further improved. Development of adequate, cost effective, and reliable power supply infrastructure in the country will be vital in years to come. The paper attempts to look into a few of the important issues involved and options available for optimal development of the power sector in the country'

1. INTRODUCTION

The reforms undertaken over the last five years have led to a revival of national economic growth. Overall economic growth accelerated significantly in 1994-95 to attain 6.3 per cent and touched 7 per cent in 1995-96. The average growth over the first 4 years of the Eighth Plan at 5.9 per cent is higher than the plan target of 5.6 per cent . In this background the development of adequate and reliable infrastructure services at reasonable cost and with sustainable financing and pricing policies play an important role. It would be difficult to sustain the momentum of growth in industry, agriculture and export unless there are major improvements in the infrastructure area. In this context the cost effective development of country's power sector is of vital importance. The paper attempts, to project the future needs of the sector, identifies issues and analyses available options for formulation of required strategies to bridge the demand-supply gap most optimally.

2. POWER SCENARIO

The present installed capacity of power plant in the country is 83,288 MW (As on March,1996) comprising 60087 MW thermal, 20976 MW hydel and 2225 MW nuclear. Present all India maximum peak demand is of the order of 60,000 MW out of which 49,000 MW could be met resulting in the peaking shortage of about 19 per cent. Corresponding energy shortage is around

NOTE: The views expressed in the paper are those of the author and not of the organisation he belongs to.

10 per cent. The current year,1996-97 is the terminal year of the Eight Five Year Plan and it is anticipated that the all India installed capacity would be raised to about 86,225 MW comprising around 22,000 MW Hydro, 62,000 MW thermal and 2,225 MW Nuclear. The likely achievement in terms of new capacity addition during the Eighth Plan is 18,000 MW as against the target of over 30,000 MW. The present peaking and energy shortages are expected to continue at the same level by the end of the Eighth Plan.

The Central Electricity Authority undertook a study in September 1991 to project a perspective of National Power Development upto the year 2006-07 (End of Tenth Plan). The electricity demand projections as contained in the 14th Power Survey Committee Report formed the basis of working out the generation capacity expansion plan upto the year 2006-07AD. The first year of the study was the first year of the Eighth Plan (1992-93). These studies indicated that to be able to meet the peak demand of 144,000 MW and energy requirement of 824 billion units by the year 2006-07 at the reliability level of 2 per cent (Loss of Load Probability) and energy supply restrictions not exceeding 0.15 per cent, new capacity additions of the order of 142,000 MW alongwith matching transmission and distribution networks would be required in that time frame of 15 years *i.e.* covering the period of Eighth, Ninth and Tenth Five Year Plans. Studies also indicated that this additional generation capacity need to come from hydel (52,000 MW), thermal (82,000 MW) and nuclear (8,000 MW) supply options. During the Eighth Plan new capacity of 18,000 MW is likely to be added against the required capacity of over 30,000 MW which would result in low level of power availability and in turn the demand growth would also be slowed down. However, taking into account the liberalisation of the national economy it is assumed that the demand would grow faster and new capacity of about 125,000 MW would still be required during Ninth and Tenth Plans periods.

Main recommendations of the study on the Supply and the Demand sides were:

3. SUPPLY SIDE RECOMMENDATIONS

— Double the present pace of hydro power development to add 52,000 MW of hydel capacity (1992-2007).
— Accelerate nuclear power development to create at least 10,000 MW by 2006-07 AD.
— Reduce T&D losses to at least 18 per cent from the present level of 23 per cent.
— Continue Renovation and Modernisation programme of thermal and hydro units.
— Ensure formulation and operation of National Power Grid by the end of Tenth Plan (Benefits - 10,000 MW).
— Ensure Benefits of atleast 15,000 MW from non-conventional sources of Energy namely Wind, Solar, Biomass, Micro hydel by the end of Tenth Plan.

4. DEMAND SIDE RECOMMENDATIONS

— Flatten regional power system load curves to achieve system load factors above 70 per cent atleast.

— Achieve energy conservation target of atleast 10 percent, 10 percent and 15 percent in the domestic, agriculture and industrial sectors respectively (The above would result in capacity benefit of about 30,000 MW)

5. POWER GENERATION RESOURCES

In the context of the requirement of the new capacity of about 125,000 MW during Ninth and Tenth Plans, it would be important to review the power generation resources availability. The present status of these resources is discussed in the following paras:

The primary energy resources for power generation available in the country are hydro power, fossil fuels namely coal and lignite, oil and natural gas and nuclear power. Some other non-conventional and renewable sources of energy such as fuelwood, biomass, tidal, solar, wind and geo-thermal energy are also available but they are in preliminary stages of development. These resources are not evenly distributed over various Regions/States in the country. It is estimated that hydro power and fossil fuels i.e. coal and lignite will remain main source for power generation in the country in the foreseeable future. Share of nuclear power is also expected to increase appreciably in future. The details regarding various resources available for power generation are as follows:

1. HYDRO (Conventional)	—	84000 MW at 60% L.F. (untapped potential 80%)
2. PUMPED STORAGE HYDRO	—	94000 MW
3. COAL RESERVE	—	202 Billion Tonnes
4. LIGNITE	—	9186 Million Tonnes
5. CRUDE OIL	—	779 Million Tonnes
6. NATURAL GAS	—	735 Billion Cu.M.
7. URANIUM	—	78000 Tonnes
8. THORIUM	—	363000 Tonnes
9. NON-CONVENTIONAL SOURCES	—	BIOMASS (6000 MW)
	—	WIND (20000 MW)
	—	TIDAL (9000 MW)

As coal is expected to play the dominant role in country's future power development it is important to go into details of its availability. As on 1.1.1996, the coal resources of India (down to a depth of 1200 metres) have been estimated by the GSI at 201953.70 million tonnes. The State-wise distribution is as follows:

State	Proven	Indicated	Inferred	Total
Andhra Pradesh	6582.47	3502.36	2935.67	13020.503
Arunachal Pradesh	31.23	11.04	47.96	90.233
Assam	228.37	26.83	65.01	320.213
Bihar	30927.17	28352.38	5880.82	65205.373
Madhya Pradesh	10097.53	21979.22	8982.84	41059.593
Maharashtra	3524.53	1448.23	1663.64	6636.403
Meghalaya	88.89	69.73	300.71	459.433
Nagaland	3.43	1.35	15.16	19.943
Orissa	6869.74	22296.83	17553.09	46721.663
Uttar Pradesh	662.21	400.00	-	1062.213
West Bengal*	11382.46	11659.76	4315.91	27358.163
Total	70443.13	89749.76	41760.81	201953.703

There is significant potential for generation of power from non-conventional energy sources such as wind power (20,000 MW), small hydro power (10,000 MW), bagasse cogeneration (3,500 MW) etc. Under the new Strategy and Action plan of the Ministry, special emphasis has been laid on generation of grid quality power from non-conventional energy sources. The targets for the Eighth Plan have been significantly enhanced and the revised targets envisage generation of 2000 MW of power, through wind (500 MW), small hydro (600 MW), bio-energy (800 MW), solar (55 MW) etc. The programmes are being given a market orientation to make them commercially viable and sustainable. While the budgetary allocations are being utilised for selected demonstration projects, the programmes are mainly being implemented through private sector projects with mobilisation of resources from financial institutions, external assistance and private investment.

As is seen from the above, India is endowed richly with various resources for power generation. The new generation capacity additions from these resources would, however, be difficult to achieve mainly on account of the present financial crunch. At present costs capital investment required to support additions of new capacity of the order of 125,000 MW alongwith matching transmission and distribution would be in the range of Rs. 800,000 crores to Rs.1100,000 crores based on present day price level. Mobilisation of finances of this order for country's power sector alone may not be possible. It is therefore essential that other options available for mitigating optimally the demand-supply gap be explored both on the supply as well as the demand side. A few of the important strategies in this respect are discussed in the following paras.

6. SUPPLY SIDE ISSUES AND OPTIONS

6.1 Private Sector Participation

Since the liberalisation of country's economy, the private sector both domestic and foreign is being encouraged to participate in the power development process. The doors of power sector were opened for private participation 5-6 years back, the contribution from such participation however has been practically negligible. The progress has been slow and the hurdles to be crossed in working out the mutually acceptable Power Purchase Agreement have been formidable.

The private sector, particularly the foreign investors, would be entering the country for the first time and are, therefore, looking for due coverage for their anticipated investment risks involved apart from maximising the profits. The government on the other hand has its obligation of protecting consumer interests. A few projects have taken off the ground in the private sector and it is expected that mutually acceptable terms would be found in the near future.

6.2 Mobilisation of Finances

In order to mobilise financial resources the Government have mooted the proposal for setting up of Infrastructure Development Fund (IDF). For funding the infrastructure sector including the power sector, the IDF is proposed to have contributions by the Central and State Governments, industrial users and assistance from financial institutions including multilateral agencies like World Bank and the Asian Development Bank.

6.3 Hydel Development

There has been an alarming declining trend in the share of hydro power in the total system resulting in sub-optimal utilisation of existing power plant capacity in the country and in the absence of adequate hydro power plant capacity, the peaking power is being provided by thermal power plants particularly, in the predominantly thermal eastern and western regions. This often results in breakdown of thermal capacity during off-peak hours. Therefore, there is an urgent need for accelerating the present pace of hydro power development. Paucity of funds, environmental problems and difficulty in land acquisition are some of the problems in development of hydro stations. Essential pre-requisites for accelerating the pace of hydro development therefore are allocation of sufficient funds, timely cash flow, public awareness of the benefits, expeditious acquisition of land, early resolution of problems relating to environment, rehabilitation, inter-state disputes etc. The Central Government may have to play a greater role in this regard. Private Sector also needs to be encouraged for investment in hydro development.

6.4 R & M Of Power Plant

6.4.1 Thermal Power Plants

In order to improve the performance of the existing thermal power stations, a massive renovation and modernisation (R&M) programme was launched all over the country in the Seventh Plan. Programme covered 163 thermal units (13570.50 MW) in 34 selected stations. Total sanctioned cost of various renovation schemes was Rs.1165 crores. An additional generation of more than 10,000 MU per annum have been achieved after the R&M as against expected benefit of 7000 MU per annum.

Keeping in view the encouraging results achieved from the implementation of the above R&M programme of selected old thermal units during Seventh Plan, second programme for R&M of 46 nos. of thermal power stations was taken up in the year of 1990-91 by the Government of

India for implementation during Eighth Plan. Under this programme, 44 nos of thermal power stations consisting of 198 nos thermal units aggregating to a total capacity of 20869 MW are covered. The total cost of all the schemes covered was estimated at about Rs. 2,500 crores. The expected benefits are:—

(1) Additional generation of about 8000 MU/Yr;
(2) Increase of 100 MW peaking capacity;
(3) Life extension of 24 nos of thermal units by 10-15 years; and
(4) Environmental improvement of 93 units (13,000 MW) were envisaged after the completion of phase. II programme. This programme was originally scheduled to be completed within a period of 3-4 years.

The pace of progress of this programme has been very slow mainly due to paucity of funds with many utilities specifically of the Eastern and Northern regions. Power Finance Corporation provided funds for part financing of 23 nos of schemes out of total 44 nos of schemes now under implementation as some of the utilities did not fulfil the eligibility criteria laid down by the PFC. The same therefore is required to be looked into as the R&M schemes identified are highly cost effective.

6.4.1.1 Future Strategies

The above two R&M programmes for Seventh and Eighth Plans have covered all the smaller size thermal units above 20 MW capacity and below 200 MW units except few units of 200/210 MW units . Some of these units below 200 MW size would be approaching towards the end of their economic life span during the Ninth Plan period. Hence, it is required to carry out life assessment studies (LAS) on such units.

6.4.1.2 Private Sector Participation

Keeping in view the financial resource crunch in the country, Government of India thought of encouraging private sector participation in R&M and life extension of old thermal stations and issued certain guidelines for private sector participation in these works. However, the response has not been upto the expectations. This needs to be encouraged.

6.4.2 Hydro Power Plant

A National Committee was set up in 1987 to identify the hydro projects in the country that would need renovation, modernisation and uprating. Based on subsequent reviews a programme for renovation, modernisation and uprating of hydro power stations was formulated in which over 50 schemes were identified with an aggregate capacity of about 10,000 MW. These schemes are estimated to cost about Rs. 1,500 crores and give benefits of the order of 2500 MW.

6.5 Technology Upgradation

The analysis of country's perspective of power development indicates that the coal would remain the major energy resource for power generation in years to come, although alternative fuels such as natural gas, low grade liquid fuels and other non-conventional fuels are also being exploited. Even in the post liberalisation scenario, proposals received from Independent Power Producers mainly concentrate on use of coal (Indigenous and Imported) using conventional technologies and liquid fuels through combustion turbines.

Indian non-coking coal available for power generation is generally of low quality with high ash content of upto 40 to 50 per cent. The situation is further aggravated by increase in non-coaly materials due to open cast mining. This makes the task of using this coal for power generation somewhat difficult as well as expensive.

Notwithstanding the fact that conventional technology of pulverised coal firing has taken long strides over the year and Indian power sector has long experience in using such low grade coals, the difficulties experienced with low grade coal and greater environmental concerns make it imperative to have a look at the new coal technology options as available today. The main problems reported by the power station authorities relating to the quality of coal which adversely affect their performance are inconsistency in coal quality, presence of non-coaly matter such as shales, metallic pieces etc. Besides being responsible for jamming of the coal conveyors, breaking of hammers and high rate of wear and tear of the coal handling plant, bad coal also causes erosion of boiler tubes, mills etc which result in low availability of plant equipment.

With the growing concern for environment and also to achieve better efficiencies and improved performance from the plants, it has been the endeavour to identify new technologies/ processes such as fluidised bed combustion, coal gasification and advanced emission control methods. Power sector being the largest user of coal, it is essential that this sector keeps itself fully responsive to adoption of latest technologies for cleaner use of coal, each of these technologies having their own merit for specific areas of application.

6.5.1 Integrated Coal Gasification Combined Cycle (IGCC)

Integrated Coal Gasification Combined Cycle (IGCC) technology is currently attracting attention as a mode of power generation. The IGCC process involves gasification of coal which is then burnt in a combustion gas turbine to produce power. Main advantage with the process is its high thermal efficiency which is of the order of 40-50 per cent as against 35-38 per cent for conventional coal based plants. In addition, the technology promises environmental superiority in terms of reduced emission of particulate matter, SOx, NOx, less water and land requirements. At this stage of development of coal gasification technology, there is, however, no commercially operating plant which has used high ash (40%) and high ash fusion temperatures with Indian coals.

6.5.2 Fluidised Bed Combustion (FBC)

Fluidised Bed Combustion (FBC) technology was primarily developed to reduce SOx and NOx emission but in the Indian context it is more relevant in burning low grade high ash coals having large quality variations. FBC boilers reduce NOx emissions by virtue of the lower and uniformly distributed furnace temperature and also have the advantage of sulphur capture in the furnace itself by limestone injection. Of the three main variants of FBC technology, Atmospheric Bubbling Fluidised Bed Combustion (AFBC) and Circulating Fluidised Bed Combustion (CFBC) are now well established with many small size units already operating in the country. CFBC boilers are more suited for higher capacity installations as compared to bubbling bed FBC. Choice of FBC technology would largely be guided by the cost economics as these boilers are likely to be costlier than conventional pulverised coal fired boilers. CFBC boilers find specific use for burning high sulphur lignite where sulphur capture can be achieved in the furnace itself.

6.5.3 Coal Beneficiation

Present policies envisage use of washed coal for power generation. This may not only help environmentally by way of reduced emission of particulate matter, reduce disposal of ash etc, but may also give rise to operational benefits such as higher availability, improvement in efficiency, reduced maintenance etc. Beneficiated coal could also uprate the old thermal power stations which were originally designed for low ash coals but have now been derated due to coal degradation. The need of the hour is to have simple and inexpensive coal washing technology so that the accrued benefits outweigh the cost of beneficiation.

The coal beneficiation at the same time would result in large quantity of washery rejects, which contain significant amount of heat value (1350-1600 KCal/kg). The coal washery rejects, if not utilized would need to be either dumped in abandoned mines or in nearby areas resulting in ecological degradation. With the progress made in fluidized bed combustion technology these washery rejects can effectively be utilized for power generation.

Availability of low grade coal for power generation, adoption of large unit sizes for economic reasons, constraints of transportation sector and increased concern for environmental compatibility have forced to integrate new technologies of coal production, coal washing and power generation.

6.5.4. Control and Instrumentation

Control and Instrumentation system play a very crucial role in safe, reliable, efficient and economic operation of modern capital intensive thermal power plants. With the fast technological developments in the field of electronics, the capabilities of C&I systems have increased manifold in past few decades. As a consequence, today's C&I systems are far more complex, sophisticated and reliable. C&I systems have come a long way from the days when local operator used to control the handwheel of the valve looking at the gauge mounted nearby on the piping. Today, complete process is accessible through single window (CRT/keyboard) from where we can control as well as monitor. Micro-processor based distributed digital control systems are the order of the day. In India, these systems were first introduced in Vijayawada and Trombay. The confidence level has since increased and all the new installations are going in for such systems.

One of the tasks expected of C&I systems is the optimisation so that the plant runs under optimum conditions so as to give best efficiency. Optimisation in control means setting the best operating points of the plant and set points of controllers under varying operating regimes. This required resorting to dynamic modelling of the process. In power plant so far, optimisation in this context has been used in limited area by way of providing model based predictive controls/state variable controls for superheater/reheater steam temperature controls. Packages comprising of process system models and control system models are also available which guide operator in optimising the operating conditions of boiler and turbine to minimise the controllable losses.

From a near monopolistic situation in 1960s when there was only one C&I manufacturer in the country, today we have a number of C&I vendors each having collaboration with reputed manufacturer abroad. Almost all the foreign renowned manufacturers of C&I system have presence in India viz. Siemens, ABB, Honeywell, Hitachi, Yokogawa, H&B, Taylor, Westinghouse, Fischer-Resemount etc. So far in India, we have been adopting the technologies developed abroad. Now with good manufacturing base already developed, manpower and software skills available in the country and access to latest technology abroad, Indian C&I industry should set its sights rather far and think in terms of technological upgradation and innovations through indigenous R&D. They should not only market their capabilities to the foreign investors who are setting up power projects in India but also exploit the export potential of C&I equipment.

6.6 Clearance of Power Projects from Environment and Forest (Conservation) Act 1980 angles

The projects involving diversion of forest land are required to be cleared from Forest (Conservation) Act 1980 angle. Subsequently the Environmental Impact Assessment Notification, 1994, *inter alia*, provided that expansion or modernisation of a new power project shall not be undertaken unless accorded environmental clearance by the Central Government. This, however, does not apply if the investment cost is less than Rs. 50 crores.

The EIA notification 1994 sets out procedure for clearance of projects. For site specific projects such as hydro electric projects and pit head thermal power stations, site clearance is to be obtained from MOEF for initiating any investigation and surveys. Therefore, for such projects and also for other (not site specific) projects EIA reports are to be submitted to MOEF for clearance which may be considered by the Committee of Experts. As per the notification, the assessment shall be completed within 90 days from the receipt of requisite information and data from project authorities and completion of public hearing, wherever required, and decision conveyed within 30 days thereafter.

During the period January 1996 to August, 1996 over 40 thermal schemes with aggregate installation of about 30,000 MW were considered by the Committee of Experts and about 30 schemes with total installation of 15,000 MW have been recommended for clearance from environmental angle. These thermal schemes do not involve forest land and clearance from forest (Conservation) Act angle was not required.

As regards hydro electric projects over 40 schemes with total installation of about 10,000 MW are awaiting clearance from environmental/forest angle. In addition around 50 transmission schemes are awaiting clearance from Forest (Conservation) Act 1980 angle.

6.6.1 Some Important Issues

The data and studies required to be presented in the EIA report are well defined in the guidelines for thermal stations issued by MOEF. The various environmental parameters to be complied with are also available in various notifications issued from time to time. The time taken for clearance of the projects depends on the quality of data collected and studies presented in the report. The EIA reports are generally prepared by the consultants. The EIA reports prepared adequately have been cleared by the Experts Committee in one meeting only. It is, therefore, important that the required EIA studies are completed ensuring full compliance with the relevant provisions of the Act.

In case of hydro electric project which may involve forest land and rehabilitation problems, the delays sometimes are considerable. In the process of the Environmental/forest clearance to the power projects various environmental management plans such as under are insisted upon by the MOEF:

- — Rehabilitation Master Plan for the Project affected people. Flora, Fauna, Forests,Wildlife Protection and Rehabilitation
- — Catchment Area Treatment Plan.
- — Rehabilitation and relocation of Archaeological, Religious and Historical Monuments
- — Compensatory Afforestation Plan.
- — Disaster Management Plan.

Though the above plans are drawn and monetary provisions are made in the estimates of the projects, difficulties of identification of non-forest land for compensatory afforestation takes a considerable time. If the land for compensatory afforestation in the state is not available, a certificate is to be obtained from the Chief Secretary of the State regarding non-availability of non-forest land for compensatory afforestation. In such cases degraded forest has to be identified for compensatory afforestation. The whole procedure sometimes takes few years. Such delays could be minimised if the concept for "Forest Bank" to offset for the diversion of forest land, is followed.

6.7 Recommended Options

6.7.1 Raising th monetary limits for project requiring clearance

As per the Environmental Impact Assessment Notification 1994 the implementation of power projects with investment less than Rs. 50 crores does not require environmental clearance by the Central Government. Consideration may be given to raise this limit.

6.7.2 Delegation of powers to the State Government

The procedure for the environmental clearance could be streamlined/simplified and in this connection consideration could be given to delegate power to the State Governments for according clearance to power projects. The necessary guidelines/norms could be evolved for this purpose. It may be mentioned that the Ministry of Environment and Forest has already constituted a Working Group for evolving norms for delegation of greater powers to the State Government(s) for environmental clearance of Thermal Power Projects which would submit its report on 15th October,1996

6.7.3 Inter Ministerial Coordination Committee

It may be desirable to concentrate efforts on the clearance of projects which are in an advanced state of appraisal. An Inter-Ministerial group comprising officers of MOP/CEA and MOEF could meet at regular intervals to identify and prioritise projects for clearances on the basis of their stage of appraisal so that efforts of the two Ministries are concentrated simultaneously on the same projects enabling expeditious clearance from the Central Government.

8. DEMAND SIDE ISSUES AND OPTIONS

Widening gap between demand and supply of power and environmental considerations have compelled us to look for other ways for meeting energy requirement in addition to installation of new capacity. Energy management, energy conservation and end use efficiency have been identified as areas with potential of enormous saving in energy consumption. It has thus become imperative to improve energy efficiency in the fields of power generation, transmission, distribution and end use consumption.

Ministry of Power and Central Electricity Authority have taken major initiative for promotion of energy conservation in all the possible areas. Exhaustive Energy audits in energy generation industry and major consumer industries like aluminium, chemicals, petro-chemicals, steel, fertilisers, pulp and paper, cement and textile etc are being carried out in order to identify the various controllable losses. Special tax incentives and subsidies for carrying out energy audits and energy conservation measures are being provided. Awareness campaigns, national awards for productivity and energy conservation, renovation and modernisation are some of the other important measures undertaken.

In the rural sector, programme of rectification of about 1,20,000 agricultural pump sets in various parts of the country has been implemented with financial assistance provided by Central Government. It has been established that electricity consumption can be reduced by 30 per cent by carrying out suitable rectifications. Detailed guidelines in regard to implementation of energy conservation in agricultural pump sets have already been issued.

In India T&D losses are about 23 per cent as compared to 10-12 per cent in advanced countries and there is urgent need to reduce the losses. In this context the Government has initiated many measures like installation of capacitors and fail safe transformers, transmission at higher voltages, installation of LT switch gear and amorphous core type transformers, development of distribution automation etc. Other measures in this regard are saving energy in evening peak through efficient lighting system, load management schemes, energy audits in power system and resorting to TOU tariffs for selected consumers.

Thermal plants have been operating at much less than design efficiencies. This leads to costly power, increased pollution and wastage of scarce energy sources like coal and oil. CEA has undertaken a scheme approved by Ministry of Power for carrying out detailed energy audits at 30 thermal power stations, covering different sizes of the units.

Energy audits aim at pinpointing the areas where operating efficiency have deteriorated from design values, to analyze the reasons and specify the steps for restoring original efficiencies. All sub-systems like regenerative systems, main turbines, condenser system are subjected to detailed testing to get to the root cause of the decline in efficiency of generation. Of course, boilers and turbines as a whole are also tested for heat rates, efficiency, specific fuel consumption, auxiliary power consumption and DM water consumption etc.

9. CONCLUSION

Country's need for creation of additional generation, transmission and distribution facilities is beyond any debate. Meeting this need through large-scale government budgetary support does not appear possible. At this point six alternatives seem feasible which need to be followed for next several decades;

(i) Autonomy for the utility (Board) to operate a viable tariff structure.

(ii) Mobilisation of resources by encouraging private power producers.

(iii) Formulation of a viable coal policy, as coal is going to be the main resource for power generation.

(iv) Efficiency improvements on the supply side including reduction in T&D losses, better hydro thermal mix and development of national power grid.

(v) Efficiency improvements on the demand side by better energy management and optimal end use consumption.

(vi) Energy conservation both on the supply as well as demand side.

India will have to base its future power development on these strategies to be able to avoid energy constraints on its development.

SKILL AND SUPERVISORY DEVELOPMENT IN ENERGY SECTOR THROUGH DISTANCE MODE – A PLANNING CONSTRUCT

64

Prof. J. JHA
Pro-Vice-Chancellor
Indira Gandhi National Open University
Maidan Garhi, New Delhi - 110 068 (India)
and
Prof. VIJAY V. MANDKE
Director
School of Engineering & Technology
Indira Gandhi National Open University
198. Asian Village Complex, New Delhi - 110 049 (India)

ABSTRACT

Rapid industrialization of countries has placed hitherto unknown demands of production in Energy sectors worldover. These demands are in terms of technology modernization needs in conventional energy sector, thereby calling for retraining of existing workers, as also in terms of preparing work force for new skills for use of non-conventional energy sources such as solar, wind energy, bio-mass as source of energy, ocean thrmal energy, etc. This has thrown before educational planners. requirement to plan for continuing education and further training of work force at skill and supervisory levels. This paper develops a planning construct for responding to this manpower development need in energy sector by proposing a flexible and modular Distance Education design for skill and supervisory training. Specifically, keeping in mind the training needs in conventional and non-conventional energy sector, the paper presents an approach to learner specific programme planning, designing multi-media, self-instructional material using print and audio-video techniques, planning for learner support services with employer support and learner assessment at learner's work place; thereby increasing the reach of Energy sector worker training framework. The paper concludes by identifying innovational requirements — educational as also implementational — in achieving the Distance Learning model proposed.

1. Introduction : Energy Needs for National Development

Significance of Energy Sector, for nations engaged in tasks of establishing and maintaining their urban centres. transport systems and manufacturing industries, needs no emphasizing. Thus most nations — developed as also developing — are heavily dependent on commercial energy sources of petrol and heating oil, coal, gas and electricity. Speaking about India alone, in just 19 years its coal consumption has increased 2.98 times, while from 1970-93 in 23 years, its consumption of crude oil has increased 2.9 times and that of petroleum products 3.3 times. Further. in same time duration, to meet its energy needs, India has increased its natural gas production 24.9 times and Electricity generation 5.8 times.

As India has abundance of coal and lignite, it is self-sufficient in its coal consumption requirements. However, steep rise in commercial energy use and generation as above, has necessitated India, during the period 1970-93, to increase its crude oil imports 2.6 times and imports of petroleum products 10.5 times.

2. Energy Policy Options — Possibilities and Issues

2.1 Increased Domestic Production of Primary Energy

A conventional solution to the problem posted by increasing oil imports is to increase domestic production of primary energy in order to reduce the level of oil imports.

Several options are available. As already mentioned, India already produces coal in significant quantities. However, it is costly fuel to recover and transport, and coal mining also has substantial requirements of modernization in areas of process equipment, automation, instrumentation and control, communcation, material handling systems, etc., which in turn have engineering design, development, manufacturing, operation and maintenance requirements.

There are other technologies available for electricity generation. Instead of oil-fired power stations, developing countries may switch to coal fired generation, nuclear plants, or geothermal or hydroelectric generation.

However, nuclear energy technology is highly capital intensive and has requirements of modern technology in areas mentioned above and extending further to areas of materials science, transfer processes, structural analysis and design reactor analysis and design etc.

Similarly, investment costs per unit of hydroelectricity are also relatively high resulting in concentration of capital resources by large hydroelectric projects and creating regional imbalances in respect of benefits accruing therefrom.

Further production of primary energy as above inveriably poses issues in respect of waste management and environmental engineering, which in turn then have their own technological requirements and operational cost consequences.

Thus primary energy production is capital intensive and is characterized by technology modernization, in trun calling for highly skilled manpower for energy sector technology design, development, manufacturing, operation and maintenance.

2.2 Energy Policy for Agricultural and Rural Development

With exception of scattered enclaves of mechanical and chemical intensive farming associated with green revolution, the peasant farms and small enterprises in rural sector of developing countries like India depend for their energy supply on traditional non-commercial fuels—wood. animal dung and variety of agricultural wastes, particularly for cooking and other domestic purposes. Further, animal power and human labour provide the direct mechanical energy required in cultivation of small farms. Kerosene and electricity have entered rural economy, but their progress is critically dependent on ever increasing constraints in oil consumption.

It is within this framework, country like India has been working towards "renewable energy technologies" for agricultural and rural development using : bioenergy, fuel-wood, agricultural residues (crop residues, agro-industrial by-products, animal and poultry wastes), animal and human waste, other organic wastes and residues, draught animal power, solar energy, wind power and other renewable energy sources, namely, tidal power, geothermal energy and chemical sources of energy including batteries and hydrogen. In what follows, the paper briefly presents issues concerning renewable energy technologies so pursued.

Biogas is perhaps the best known example of the new generation of non-conventional technologies. It offers tangible benefits to households—especially womenfold who might otherwise spend time and effort in the collection of fire wood—in the form of methane gas for cooking, lighting and space heating, with, as a valuable by-product, a rich slurry which can be used as fertilizer.

However, Indian experience reports that, though meant for rural population, main beneficiary of this technology are middle class families owning more than 3 cattle, the minimum required to make a plant.

Another technology pursued is "improved chulhas". Total number of improved chulhas installed since initiation of the project by Governmnet of India is over 21 lakhs. Further, India has also developed community stoves.

However both for "improved chulhas" as also for "biogas" plants, studies indicate that there is much attrition in numbers of chulhas/plants in actual working condition, within a period of 6 months from installation. Further, even is case of operation stoves/plants, the improvements in efficiency, reduction in fuel consumption and/or gains in other benefits are in many cases much less than claimed or expected on the basis of laboratory tests.

Coming to solar technology it covers a range of devices and systems : Solar driers to process foods in a simple way, solar stills to purify brackish water, flat plate collectors to provide hot water for domestic use and for space heating, cooking and washing, solar irrigation pumps, photovoltaic cell arrays combined with battery storage.

Techno-economic viability of these solar technologies is yet to be conclusively established, though a market for flat plate collectors has opened up in India, Solar-powered irrigation pumps are much more sophisticated in their design and have started finding use in India. Large scale use of solar cookers is yet to take off. Finally, photovoltaic cell arrays hold considerable promise if price of cells can be reduced rendering the electricity produced by such arrays cost-competitive with electricity supplied to rural areas by grid extension.

However, all these solar energy capture devices may well be more useful in towns. where manufacturing and repair and maintenance facilities are available; in turn calling for trained work force availability for the purpose.

Like solar energy, another renewable energy source available is wind power. This source leads to windmill technology which is available in variety of designs. shapes and sizes. Power captured by a windmill can be harnessed directly to drive irrigation pumps, or converted by a dynamo into electricity and subsequently stored in batteries. In this context mention may also be made of R&D in India to setup a wind farm. In such a farm, a number of windmills are setup in a cluster to generate and supply electricity to a combined grid system.

However, while windmill operating costs are low, initial costs are relatively high, especially taking into consideration the likelihood of low levels of capacity utilization because of daily and seasonal variations in wind speed. Indeed, the major difficulty about windmills is the sensitivity of their design to specific geographical locations. This is turn then calls for location specific skills for dsign, manufacturing, operation and maintenance of windmills and related energy systems, as trined personnel for grid connected non-conventional energy sources or stand along energy sources are in short supply.

India has also been experimenting with tidal power, micro hydal power, hydrogen energy project, chemical sources of energy in the field of storage batteries, bio-conversion of solar energy into electricity and hydrogen energy, ocean thermal energy, geothermal energy, battery powered vehicles, etc. However, these systems have a long way to go before becoming operational.

Finally, there is also conventional rural energy option. At one end this option, as presented through use of animal power has requirements of better designed bullock carts for improved rural transport for agricultural productivity, and at the other end, it has requirements of efficient implementation of rural electrification systems. Trained manpower who can locally improve rural transport as above and who can ensure efficient transmission and distribution of rural energy with minimum losses is also in short sapply, though under Rural Electrification Corporation Limited there are training efforts in the context.

2.3 Energy Conservation Option : Towards an Integrative View of Energy Policy

An alternative to simply increasing domestic production of primary energy so as to meet spiralling energy needs is to make more efficient use of energy, i.e., by "conserving" the energy. This option will also supplement the "renewable energy technologies", in turn leading to an integrated and comprehensive Energy Policy Option.

3. Manpower Training Needs to Implement Energy Policy Options

To implement Energy Policy Options as above so as to meet the Energy demands of a modern nation, one then has to meet the demands of (a) technology modernization in primary energy sector, (b) of innovation for renewable energy technologies and (c) of planning designing and managing energy system integration characterized by conservationist principles. This certainly is a quality Energy Engineering objective and calls for :

(a) Technical and Management manpower at Planning, design and development level in the area of Energy Engineering, and

(b) Skill and Supervisory Manpower at Manufacturing, Operation and Maintenance levels in the area.

Starting since 1947, India has developed a massive technical education system comprising about 500 Engineering Colleges and 1070 Polytechnics accommodating over 70,200 students at degree education and 1,22,400 students at diploma education levels. This has understandably provided an adequate technical manpower support at planning, design and development levels in different sectors of engineering activities, but much is desired in respect of vocationally trained manpower available for quality manufacturing, operation and maintenance of modern engineering systems. This is true of Energy sector, too.

3.1 Supervisory and Skill Development Requirements

Specifically, as can be seen from issues facing various Energy policy options discussed above, there is definite need for vocational training of manpower for:

(a) Modernization of primary energy systems and their operation and maintenance.
(b) Construction for local adoption and repair and maintenance of renewable energy systems and devices, and
(c) Implementing and maintaining energy systems designed for energy conservation.

As can he seen, these Energy engineering activities would mainly encompass manufacturing, processing, servicing, mining construction and transport sectors, and would require traning at Skill Development and at Supervisory Development levels to perform a range of functional tasks as follows:

- **For a given Engineering Sector, at Skill Development level, training to cover following range of Energy Engineering responsibilities :**
  - Construct Plant/device
  - Perform plant/process/device/service operations
  - Use resources machine/material/financial-optimally.
  - Detect change(s) in machine/material/work/device/systems.
  - Work in team.
  - Demonstrate positive approach to work.

- **For a given Engineering Sector, at Supervisory Development, level training to cover following range of Engineering responsibilities :**
  - Maintain and improve plant/process/device/service operations.
  - Recommend, monitor and control use of machine/material/human/financial resources.
  - Identify cause, find solutions and implement them in case of change(s) in machine/material/work/device/systems.
  - Contribute to implementation of change(s) in process/device/systems.
  - Develop teams, individuals and self.
  - Plan, allocate, monitor and evaluate work carried out by teams and individuals.
  - Maintain and develop a positive working environment.
  - Collect and evaluate information relating to personnel, departmental and organizational objectives.
  - Exchange information to analyse problems and make decisions.
  - Obtain information, co-operation and support through establishing and maintaining relationships.

3.2 Need for Traning those at Work

Certainly training areas as above, defined in terms of learning of work functions to be carried out at Skill and at Supervisory levels, are at variance with what conventional educational system with emphasis on "knowledge" based learning would teach. This is because, firstly, conventional technical education system, be at degree, polytechnic or ITI level, aims at first time learners, while training needs in Energy Engineering as above would additionally call for continuing education, retraining and further training for Supervisory and Skill Development of those at work employed in the Energy Sector, henceforth also termed as Energy Employment Sector. Secondly, in implementing any skill and Supervisory Development training as required, curriculum and its learning will have to be "practice" based and will have to be done in collaborative participation with those multi-skill tradesman, expert supervisors and employers who manufacture, implement, operate and maintain the technologies under Energy Sector. This calls for innovation in instructional method.

3.3 Need to Train Large Numbers

There is another aspect of the task at hand and that relates to sheer number of traineess, Specifically, industrial sectors of manufacturing, processing, servicing, mining, construction and transport, covering the bulk of energy intensive industries, employ, as per 1991 statistics, 43.3 Mn, i.e., 15.5% of main workers out of the total number of 278.9 Mn. man workers. If one considers Sixth and Seventh Five year Plan figures along with 1981 census figures, one finds stock of technical diploma holders who are economically active being estimated at varying levels from as high as 0.86 million to 0.33 million. If it is assumed that industrial category-wise percentage distribution profiles of main workers and of employed technical diploma holders are of same order, then estimated stock of technical diploma holders in energy intensive industries assumes a number between 1.33 lakh to 0.51 lakh. Certainly, this is a very large variation, but the figures give an idea of the order of magnitude one is dealing with.

It is this number of technical diploma manpower averaging 0.92 Mn which would be mainly contributing at Supervisory level responsibilities, while the number of workers concerned with skilled tasks will be accounting for the bulk of the remaining estimated 43.3 Mn main work force in energy intensive industries; in turn making the total main work force (in need of training at Supervisory and Skill Development levels in energy intensive industries) run into multiple of lakhs, if not more, and considering that 1980 estimate of out-turn of S&T personnel under entire technical education system in the country is only 1.69 lakh, any such demand for manpower training and upgradation responsibility at Supervisory and Skill levels in energyintensive industry is something that far exceeds any national training supply capability in the context.

It is these two issues, one pertaining to need for instructional innovations built around work, employer and user needs and other pertaining to requirement for training a very large learner group, dispersed across the length and breadth of the country, that then call for an innovative approach to the task at hand. Approach to teaching of engineering and technology through Distance Mode (which can train 10-20 learners per one learner in formal system) as implemented by Indira Gandhi National Open University, offers a viable teaching learning system for this purpose at hand.

4. Skill and Supervisory Development through Distance Mode—A Viable Mechanism

Presently the world is experiencing the electronic and information technology revolution, which in turn has offered distance learning as a new educational and cost effective technology for reaching learners in large numbers and for flexible, open learning systems which are consistent with requirements of education and training at work.

While teaching of humanities courses through distance mode with heavy emphasis on print texts and interaction in writing and on recordings of spoken word, radio and TV, Video Recordings, Telephone and Face-to-Face communication, has a more ready acceptance, when it comes to Science and Engineering and Technology courses, those who regard face-to-face teachnig as an integral part of education find it difficult to accept distance mode of learning.

However educational researchers report the while cognitive objectives and those affective objectives critical to learning Science and Engineering which are concerned with academic socialization (unprejudiced search, etc.) and some psycho-motor objectives (skills in drawing, calculating) can even be well attained through non-contiguous components under distance mode, the contiguous face-to-face interaction, as under distance mode, could be meaningfully utilized for the learning requirements like practicing skills in laboratories and in similar situations, verbal skills through personal communication, training in co-operation, etc.

Thus open learning models world over have emphasized relevant approaches through education—work linkages to designing curriculum, preparation of self-instructional material and student support services, so as to offer viable mechanisms for teaching of Science and Engineering at professional, technician and skill levels through distance mode.

5. Distance Education Programmes in Engineering at IGNOU

The Indira Gandhi National Open University (IGNOU) was set up in 1985 by an Act of Parliament with the objectives of, not only democratizing education but also bridging the gap between the programmes of the conventional institutions and the emerging demands of education, In a short span of eight years, the University, adhering to the principles and methodology of Distance Education and Open System of learning, has started as many as 57 programmes—with an annual student intake in 1994 of over 80,000 and an enrollment strength of 220,000—comprising 13 certificates, 17 diplomas, 24 bachelor's degrees and 3 master's degrees in disciplines of Humanities, Social Sciences, Applied Sciences and Engineering and Technology.

Specifically, from 1994 IGNOU has launched its Employment related Engineering programmes aiming at continuing and further education and retraining for those at work. In the first instance IGNOU has identified employment sectors of "Construction" and "Water Resources Engineering" and launched post-diploma programmes leading to B. Tech. Civil degree with specialization options in "Construction Management" and in "Water Resources Engineering". These programmes have modular structure characterized by multiple entry and multiple exit points, in turn offering in-service training and continuing education packages for learners at intermediate levels like Advanced Diplomas in specialization areas, while at the same time offering opportunities to proceed upto degree level.

IGNOU has developed these employment related programmes in participation with industry and employer groups, Specifically, "Construction" and "Water Resource" sector industries/organisations and employer groups have participated with IGNOU in delineating functional, field-oriented course content under the programmes, thereby reflecting in the syllabus appropriate in-service training and continuing education needs as perceived by the work places. Further, professionals of industries have also participated with IGNOU in preparing self-instructional print material for the functional, field-oriented courses so identified.

As regards to the course delivery through distance mode, IGNOU recognizes "specialized" needs of student support services for learning engineering courses which, apart from "knowledge", emphasize learning of "practice" or "problem solving" capabilities. Thus IGNOU has planned for "Programme Specific Study Centres" implemented in collaboration with engineering institutions for face-to-face counselling in respect of academic course and in respect of "laboratory" courses, while for face-to-face counselling for functional, field-oriented courses IGNOU has planned for "Work Centres" at industries/employer groups participating with IGNOU. These work centres will be attached to Programme specific Study Centres in the region of their location. Finally, all counsellor-student interactions at these Programme Specific Study Centres and at these Work Centres will be implemented during off-time or vacation time of participating institutions/industries.

6. Supervisory Development Programme for Energy Sector through Distance Mode - A Design Basis

It is based on the Distance Learning model as developed by IGNOU for its Engineering programmes, that a cost-effective Supervisory Development Programmes model through Distance mode can be conceived to meet the manpower development task at hand for learners at work place in large numbers.

Specifically, Table 1 presents Functional-Supervisory-Area vs. Energy-Employment-Sector matrix for determining Energy Sector Supervisory Development Programme curriculum. As can be seen, columns give various Energy Employment Sector organizations, while rows indicate Functional Supervisory activities. What is visualized is that a survey questionnaire may be administered to industries/institutions/ organizations/departments/NOGs from Energy Employment Sector categories to seek their perception of supervisory level knowledge and problem solving learning requirements for the work force with them. Needless to say, the questionnaire should identify these curricular requirements for each of the functional Supervisory activities.

It is in terms of so identified curricular requirements that the self-instructional learning package can then be developed, as also specific Supervisory level work projects can be identified basically reflecting work outcome requirements from the functional supervisory activities. It is this self-learning package the learners could then learn and undertake to complete work projects in assistance with participating industries/organization/NOGs and learner's employer. Figure 1 gives a conceptual schematic giving systems representation for learner support system to deliver Supervisory level training through distance mode in participation with Energy Sector industries, departments, organizations and NGOs.

Open learning institutions could then undertake responsibility to implement these training programmes, with IGNOU having additional responsibility for quality monitoring.

TABLE 1 : **Functional-Supervisory-Area Vs. Energy Employment Sector Matrix for Determining Energy Sector Supervisory Theory and Practice Learning Requirements**

Functional Supervisory Y↓ / Energy Employment Sector → X	Primary Energy Sector	Renewable Energy Sector	Energy Conservation Sector
Areas/Activities			
Maintain and improve plant/process/device/service operations			
Recommend, monitor and control use of machine/material/human/financial resources			
Identify causes, find solutions and implement them in case of change(s) in machine/material/work/device/systems		XiYj Through Survey Questionnaire administration, identify learning requirements as perceived by Energy employment sector in terms of : – Academic courses : Knowledge., practical, problem solving – Functional field based courses: supplementary knowledge for work; work methods, technologies and techniques: work practice	
Contribute to implementation of change(s) in process/device/systems			
Develop teams, individuals and self			
Plan, allocate, monitor and evaluate work carried out by teams and individuals			
Maintain and develop a positive working environment			
Collect and evaluate information relating to personnel, departmental and organizational objectives			
Exchange information to analyse a problem and make decisions			
Obtain information, co-operation and support through establishing and maintaining relationships			

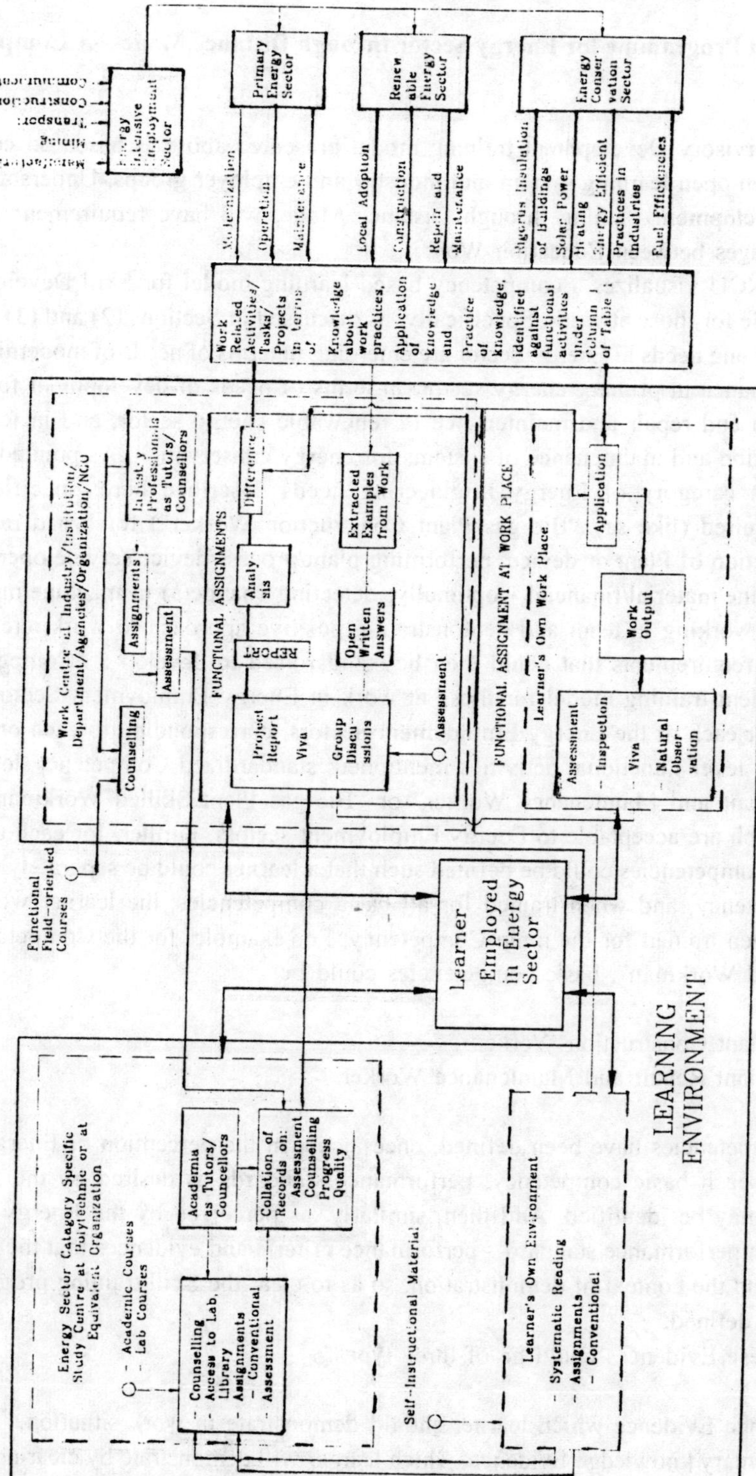

FIGURE 1 *Conceptual Schematic Presenting Energy Sector Supervisory Development Programme Delivery System Through Distance Mode*

7. Skill Development Programme for Energy Sector through Distance Mode—A Competency Based Approach

As can be seen, Supervisory Development training model presented above is based on collaborative linkages between open learning system and industry and employer groups. Understandably offering of Skill Development training through Distance Mode, will have requirement for yet more integrative linkages between Education Work.

As a result. IGNOU visualizes a competency based learning model for Skill Development through Distance Mode for those at work. Specifically, as discussed in Sections (2) and (3) above, Skill Development training needs in Energy Sector are emerging in terms of needs of modernization, operation and maintenance in primary energy sector, in terms of needs of development for local adoption, construction and repair and maintenance in renewable energy sector, and in terms of application and operation and maintenance of systems for energy conservation. As pointed out in Section (3), for each category of Energy Engineering needs described here, specific Skill developments are required (like say "Bio-gas Plant Construction Worker") for broad range of activities of : construction of Plant or device; performing plant/process/device/service operations; using resources machine/material/financial - optimally; detecting change(s) in machine/material/work/device/systems; working in team and demonstrating positive approach to work; etc.

It is these skill requirements that could then be standardised to develop a "Competency" based Skill Development training model for those at work in Energy Employment Sector.

To elaborate, for each of the Energy Employment Sectors, corresponding to each or group of Skill Development level Functional activities mentioned, standardized Competency levels as say "Solar Pump Repair and Maintenance Worker" or "Bio-gas Plant Skilled Workman", etc. could be defined, which are acceptable to Energy Employment sectors, Further, for each of such Competencies, basic competencies could be defined such that a learner could be separately trained for each basic competency, and when trained for all basic competencies, the learner would be considered to have been trained for the main Competency. For example, for the Competency of "Bio-gas Plant Skilled Workman", basic competencies could be :

— Bio-gas plant Construction Worker
— Bio-gas Plant Repair and Maintenance Worker

Once basic competencies have been defined, once again, in the perception of Energy Employment Sector, for each basic competency, performance standards as desired by the Energy Sector professionals may be identified. And then, similarly, as perceived by the Energy Sector professionals, for each performance standard—performance criteria and evidences that the learner should demonstrate and the context of demonstration, so as to clear the Skill training programme successfully, may be defined.

Specifically, these Evidences could be of three types:

— Performance Evidence which learner should demonstrate in work situation,
— Supplementary knowledge Evidence, which learner will demonstrate by clearing work-related knowledge courses, and

— Prior achievement requirement, if any, as desired by Energy Employment Sector, like say previous work experience.

It is these Evidences when generated by the learner under various channels through distance mode, that could then be assessed by IGNOU for successful Skill Development training programme completion by the learner.

Thus as can be seen, Skill Development programme module visualized is highly modular in the sense a learner can collect basic competencies separately and as per work need, as also at his/her pace. Further, all these basic competencies can be obtained from workplace itself, as they are strictly about work.

IGNOU perceives that such offering of Skill Development programmes will bind world of learning and world of work together in an integrative and dynamically evolving manner, in the sense, while teaching-learning systems will get built around work methods and work outputs and become relevant, the work requirements by themselves will also get standardized and, therefore, become more amenable for developing new technologies and, therefore, new knowledge.

8. Conclusion

Energy sector has manpower training needs at Skill and Supervisory work force level, so as to facilitate increased stategic efforts in primary energy sector for modernization and for better maintenance, in renewable energy sector for its acceptance, and in energy conservation sector for application of energy conservation technologies. However, developing such training programmes calls for innovation in instructional methodology by application of distance education as, in addition to instructional methods, need is also to train learners in large numbers. Further, this also calls for imaginative Education-Work linkages, as neeed is to build programmes around needs of work contributions at Skill and Supervisory levels as perceived by Energy Employment Sector. in concrete terms this would mean participation by Energy Employment Sector with Open Learning System in defining curriculum at Supervisory level and competencies at Skill level and in preparing self-instructional multi-media material as also in its delivery by offering Work Centre facilities at energy sector industries, institutions, departments, agencies, and NGOs. This certainly is a challenging task requiring networking of pedagogic resources available from every possible end and is going to be mind-bogling task. Responding to such task by Energy Sector and Open Learning System together, however, would have all ingredients of educational innovators, work practitioners and technologies coming together for a very purposive technological action.

REFERENCES

"Energy for Tomorrow's World", World Energy Council, Kogan Poage, 1993.

Ghosh, P.K., "Energy Policies and Third World Development, international Development Resource Books, No. 4, Greenwood Press, 1994.

Gusain, P.P.S., "Renewable Energy in India", under the auspices of Society for Development Alternatives, Har-Anand Publications, 1990.

Handbook, India Society for Technical Education, 1994-1995.

Holmberg, Borje, "Growth and Structure of Distance Education", Croom Helon, UK, 1986.

Mandke, V.V. and Mishra, A.K., "The Challenge of Vocational Education through Distance and Open Learning - Strategies (A planning construct), Paper Presented at SAARC Workshop on Distance Education Organised by MHRD, GOI at New Delhi from 11 - 13 April, 1991.

Mandke, V.V., "Face-to-face components in engineering education through distance channel-issues and possibilities", Paper presented at the Asian Association of Open Universities Annual Conference, Colombo, Sri Lanka, 1991.

Mandke, V.V., "Learning of Science and Technology through the Distance Mode : Opportunity for Continuing Education at the Work Place", Special issue - Engineering Education in India of The International Journal of Engineering Education, Vol. 10, No. 1, pp 61-69, 1994.

Rumble Greville and Harry Keith (editors), "The Distance Teaching Universities", Croom, London, 1982

Statistical outline of India 1992-93, Department of Economics & Statistics, Tata Services Limited.

POLICY OPTIONS FOR MEETING LONG TERM ENERGY NEEDS IN ASIA

65

RAJENDRA SINGH
Chairman & Managing Director
National Thermal Power Corporation Ltd.

While outlining policy options for meting long term energy needs in Asia under the overarching concerns for sustainable energy supply, we need to appreciate the issues concerning the continent which is going to be the epicentre of world's economic growth in the decades to come. The 21st century has been predicted to be the Asian Century. The focus will shift from the Atlantic to the Pacific/Indian Oceans. According to an estimate, by the middle of the next century the industrialised nations would account for about 12% of the world economy, in contrast to over 70% at present and the share of the developing countries of Asia will rise from under 10% to over 55%. India and China are likely to account for 10% and 15% respectively of world's income while USA, for only 5%. India and China together have more than one-third of the world's population in less than one-tenth of its area and both have a burgeoning middle class throbbing with talent, ambition, resources and enterprise.

Availability of adequate low cost energy, especially electrical energy is critical to the growth of most of the Asian economies which, it must be noted, are at varying stages of development and present a highly differentiated picture in terms of economic growth, energy/power development and quality of life indices, Japan and the Asian Tigers are at par with many OECD countries while countries of South Asia with their very low per capita consumption of power e.g. Nepal and Bangladesh (below 100 KwH), India (283 KwH) and Pakistan (416 KwH) are way behind. Even China, with its tremendous growth explosion in gross terms is far from having a good per capita power consumption level. The correlation between greater eletrification and high growth in GDP is very direct as the economies of Asia would bear out. Indonesia is registering a GDP growth of 6% per annum and is increasing its power generation capacity by 12% per annum Malaysia is having an annual GDP growth of 8% with annual electricity growth of 14%. China, Thailand and Korea, too have impressive statistics, especially Korea which ranks among world's most vibrant economies. All these 'Asian miracles' have been backed by sound and intensive power development efforts in terms of both quality and quantity. The rapid economic growth of Asian economies and the energy-intensity (especially power intensity) of such growth is indicated by the following projection regarding electricity demands in Eastern and Southern Asia:

— Upto 15,000 MW per year over the next decade in China
— About 7,000 MW every year during the remainder of the nineties and probably over 10,000 MW/year thereafter.
— 25,000 MW in the remainder of the nineties and 42,000 MW in the first decade of the 21st century in the ASEAN countries.

The enormous need for power in Asia indicates that out of the US $ 30 Trillion of the finance required by the energy sector worldwide between 1990-2020 (which incidentally is half as big as the total 1990 world GDP and is very difficult to raise), a large chunk would be required for Asia. Financing this power growth shall be the biggest constraint in evolving realistic and realisable long term energy options in Asia. And this requires an urgent and concerted efforts by the Governments, companies and the funding agencies of the entire world in this era of ever increasing inter-dependence.

Increased private participation, trans-national investment, restructuring of the existing power industry, redrawing the legislative administrative framework, or in short, deregulation substantial privatisation and re-regulation are required to be done in several Asian economies which are going through macro economic structural adjustment programmes at varying speed and intensity.

Availability or development of appropriate, affordable, technologically amenable and environmentally benign fuel is another major area, after finance, which will determine the long term policy options for Asia.

Asian fuel scenario is also highly diversified which is natural for the largest continent of the world. While the 'middle east' (which is, in fact, West of the continent) is oil-rich, Japan imports 90% of its fuels. India and China have vast coal reserves and these two Asian giants are to play a dominant role in the future 'Asian Drama'. These countries will have to (and they are already taking steps towards it) encourage clean-coal technologies and overall efficiency of conversion. All things considered, oil and coal shall provide the mainstay to the energy supplies and their use by Asian economies should not be condemned as short-termism. Some studies and opinion-building-exercises from the industrialised countries have already started raising the scare of India and China contributing to high levels of green-house gas emissions in the future. Before rebutting this projection and the arguments based thereupon, I must underline a few clear propositions. Development of economies or energy or power must always be sustainable. Sustainability requires that the cost of today's growth should not be transfrerred to the future generation. But equally emphatically, sustainability also requires that the cost of growth of the rich should not be transferred on to the poor. Sustainability is a dynamic concept. In the long term there need not be tension between the drive for sustainability and economic growth, as sustainability is about the degree and efficiency of use of resources and adopting sustainability principles should promote economic well-being. Sustainability, however, has different connotation for different countries and regions depending upon their overall priorities and levels of socio-economic development. This is of increasing importance as countries with rapid economic growth strive to meet their increasing energy needs. The environmentally prudent developing countries should not be made to pay for the environmental profligacy indulged in so far by the developed countries. The 'survival emissions' of the developing people should not be equated with the 'luxury emissions' of those in the industrialised countries. In fact, according to a rather revealing World Bank estimate, emissions trading can transfer US $ 70 Billion each year from the North to the South in the form of automatic transfer, which is more than the total official development aid today. And it must be clearly underlined that poverty is a potent threat to ecology and environment. Poverty alleviation through intensive economic growth is an unavoidable compulsion. The fundamental right of growth of the under developed regions of Asia and of the world cannot be denied in the name of a homogenized recipe for sustainability. Coming back to the `greenhouse emissions', it is recorded that while USA and Russia have 10% of world's population, they emit 40% of net CO_2, Likewise USA and Canada emit over two-thirds of methane. The following table is rather enlightening:

Carbon Di-oxide Emissions from Industrial Processes for Selected Countries of the World: 1991

Countries	Per capita carbon di-oxide emissions (tonnes)
Africa	1.0
Nigeria	0.8
South Africa	7.2
China	2.2
Japan	8.8
Canada	15.2
United States	19.5
Argentina	3.6
Brazil	1.4
Germany	12.1
United Kingdom	10.0

Source : World Resources 1994-95: A Report by the World Resource Institute in collaboration with UNEP and UNDP.

In all fairness, the responsibility towards preservation of environment will naturally have to be in proportion to the level of emissions mentioned above.

This is not to under-emphasise the need for concerted action on the part of the resurgent Asian economies to be environmentally conscious in designing their growth. In fact, China and India have been very alert to the environmental aspects of their power-development efforts and have shown the commitment to conservation and preservation which is so characteristic of societies which have nurtured uninterrupted march of civilization and culture through millennia.

However, India and China can take a leaf out of Japan's Book. Japan has shown the highest order of efficiency in utilizing its largely imported energy resources. In most Asian countries, the efficiency in conversion of energy is relatively much lower. India and China, with their vast coal resources could further develop their technologies to be adopted for future which would be conversion efficient and environmentally benign. International co-operation and pooling of mental and material resources can be effectively utilized for developing such technologies—An example of such co-operation is the Centre for Power Efficiency and Environmental Protection (CENPEEP), set up in 1994 by NTPC with the assistance of USAID and with active involvement of US Deptt. of Energy. NTPC, the largest power utility in India, has been identified as the agency to implement the Greenhouse Gas Pollution Prevention Project (GEP) of the US Government in India's power sector through CENPEEP. The project aims at reduction in emission of greenhouse gases per unit of electricity generated by increasing energy productivity and promoting more efficient and advanced coal conversion for power generation.

Countries with less abundant or more expensive fossil fuels will have to strive to achieve low production cost through inventiveness. Everyone concerned shall have to develop means of using associated gas which gets wasted in large quantities. Gas will be transported either by pipeline or on LNG over increasingly long distances. Utilisation of gas will require close mutual regional interaction over-ruling political or ideological or national prejudices. The Eurasian experience of Siberian gas being supplied to West Europe by the erstwhile USSR is an example of energy issues overcoming geopolitical barriers. Likewise North-east Asia, especially northern China and far-eastern Russia can go for gas and other fuel supplies mutually. Not only for gas but for tapping and optimally utilising the energy resources generally, all the countries (e.g. the SAARC nations) should develop integrated power generation and distribution plans to achieve greater generation capacity/efficiency and to have better voltage profile. Outside of SAARC also, Burma, China, India, Nepal, Bangladesh, Pakistan etc, can have plans for utilising resources by overcoming politically driven xenophobia. India could be the pivot for the power development efforts in South Asia while China, in the Far East and North of the continent. India, Pakistan, Nepal and Bangladesh could go in for power trading to have better peak demand management and to have better voltage profile. They can also plan inter-country projects to maximise utilization of natural reserves, topography and economies of scale, wherever appropriate.

The mutual regional cooperation among the Asian countries becomes all the more necessary for developing technologies which would help utilise their natural resources and geographical features. For instance, developing solar photovoltaics at viable costs can be very useful for the tropical countries like India. But reducing costs to viable and competitive level calls for high priority, intensive research and high investments. Solar energy cannot be so high a priority for the countries of higher latitudes. So the `hot' Asian countries should, in right earnest, pool their resources to quickly develop cost-effective conversion technologies for tapping solar energy. The same approach should be extended to developing other new/renewable sources like wind, biomass, OTEC etc. Although most energy demands during the short and medium run will be met by established fossil fuels only, there is a strong case for developing non-fossil fuel based technologies, so that by the middle of the next century there start taking over a larger share of total energy sources. This would help sustainability of our planet as a whole.

Besides having a regional perspective of energy development to encourage optimality, the Asian countries must utilise their own unique topographical and geological features and resources to the fullest. For example, India can develop a large number of mini and micro hydel power plants which would be ideal peak-load stations with no threats to the biosphere, in addition to exploiting its large hydro potential and achieving a good thermal-hydro mix.

There is also a strong case for re-evaluating the appropriateness and efficacy of nuclear energy. Getting out of the shadows of Chernobyl, the countries should reconsider the option in view of the happy experience of countries like Japan and France, Intensive R&D should be done to find appropriate answers to the nuclear fuel cycle including waste-disposal/storage. Technological advancement to optimise costs and minimise hazards and opinion building to encourage social and political acceptability may help the cause of nuclear power development as a long term energy solution.

The movement towards long-term energy options should be optimally calibrated so that, at every stage, the transition is smooth and cost effective (including social - environmental costs). The immediate and short run imperatives must be met effectively before we move on to the medium long term. For instance the immediate compulsion for economies like India is to go for low-cost and high-return approaches to augment their power supply situation. In India, these approaches would include:

— Renovation and Modernisation of existing capacities — 10% improvement in the PLF of existing thermal capacity equals 8300 MW of new capacity addition at about one-third cost.
— Reduction of Transmission and Distribution losses — the high loss of about 21%, if brought down to around 10%, would be like adding 9200 MW in much shorter time and at a traction of the expenses required for capacity addition through administrative-technical initiatives.
— Energy Conservation — the industry and agriculture sectors can, through efficiency and conservation drives, be made to save power amounting to 20,000 MW at roughly one-third of the capacity addition cost.
— Demand Side Management (DSM) — through peak demand management, power dispatching and pricing, time of the day metering, inter-regional wheeling of power etc.

In addition to these, India must go for quick and large volume capacity addition to meet the projected requirement of over 140,000 MW capacity addition by the year 2007. The time horizon of upto 2020 would naturally see manifold increase in demand.

Dealing with the energy crunch facing the capital scarce economies like India requires a skillful blending of the conventional and the new approaches. India is on the threshold of an economic explosion. And if infrastructure sector, especially power sector, provides the right thrust to the economy, India's gains will have a healthy impact on the overall growth of the entire region.

THE ENERGY SCENARIO AND THE OIL AND GAS SUPPLY OPTIONS

66

R.C. GOURH,
Director (Technical), Oil & Natural Gas Corporation

ABSTRACT

The global energy scenario is briefly described. The per capita energy consumption in India is among the lowest in the world. However, an exponential growth in natural gas consumption has taken place. The work being done at ONGC is described.

Introduction

Energy is the driving force behind the growth of a modern economy. The per capita energy consumption of any nation is a fair index for benchmarking the progress made by it. The economies of developed countries of Western Europe and America have made galloping progress and ensured a high quality of life for its people. Some of the Asian nations have also made phenomenal progress and economies of Japan and South Korea are the best examples. Other Asian nations are also making concerted efforts for development of their economies. The growth potential of this region has been recognised by the World Community and the international focus has now shifted to this part of the globe.

Realising the factors such as opportune time, change in perception of developed nations, need to become an integral part of global economy for accelerated growth, the major economies such as India, China, Taiwan, Vietnam etc have started economic reforms to act as a catalyst to this whole process change. Though the Asian Region is attracting foreign direct investment, there is a specific aim underlying the rapid industrial growth witnessed here, which is not on agenda of developing countries. In this part of the world, the basic challenge remains to be that of the improvement in standard of living of the people and uplifting the economically weaker sections of the society and this forms the centre theme for economic growth. This aspect becomes more pronounced in developing economies such as India.

It is this background that makes this international conference, with its objective of "identifying the options for ensuring sustainable energy supply for fulfilling the socio-economic needs of Asian region", very timely and important.

1. World Energy Consumption and GDP Growth

The global energy consumption has been steadily increasing over the years. From a level of 5,913 million tonnes of oil equivalent (MMTOE) in 1973, it has increased to 7804 MMTOE in 1993 and is expected to reach a level of 12169 MMTOE by the year 2010. The annualised average growth for 1973-83 was 0.84 per cent, which grew to 1.95 per cent during 1983-93 and is expected to grow at the rate of 2.6 per cent during 1993-2020. The world energy consumption, by fuel type, when analysed shows that as a percentage of annual rate of increase over the last two decades, the growth of natural gas has been very impressive at 38 per cent followed by coal at 25 per cent and oil at 17 per cent. It has been estimated that the consumption of coal as a percentage of total energy consumption will marginallly increase in the coming years with the share of natural gas remaining steady and the consumption of oil showing a declining trend. However, oil, gas and coal are going to account for 90 per cent of the total increase in consumption. Another interesting feature in world energy scene is the growth LNG trade. During 1993 it was accounting for around 4 per cent of the world's gas demand. During 1983-93, LNG trade has grown at an annualised rate of 6.31 per cent. In fact, Japan accounts for about 64% of the total traded LNG in 1993. With increasing investments in infrastructure and technology, LNG would grow to become a much more tradable commodity than it is today.

The global production of natural gas is expected to increase at an average rate of 1% even in a situation where oil prices remain low. The rate is expected to go up to a high of 3% should the oil prices increase in real terms. Even at this high rate of growth, depletion of world's conventional gas reserves by 2010 would be only a little over 40 per cent. At the current production levels, the proven oil and gas reserves are expected to last for another 43 and 65 years respectively. Here again the gas scores over oil and is expected to last another two decades more.

As the energy consumption rises on one hand, the world GDP has been declining from 4 per cent in the seventies to 3.2 per cent in eighties and is estimated to be 2.9 per cent for the period 1993-2010. Thus, more energy than in the past is required to achieve the same GDP growth. This has an important bearing on the developing economies as major GPD growth would be witnessed here. Higher efficiency in use of energy through better technologies and other conservation measures would be needed to convert the large energy sources to real wealth in terms of increase in GDP.

2. The Indian Scene

The economic development of a country manipulates to a larger extent the energy demand. In a developing country like India, the rate of growth in energy consumption is higher than the rate of growth in GDP. Our energy consumption grew from 4.6 per cent to 6 per cent against the GDP growth of 3 per cent to 5.6 per cent in the seventies and eighties. But the per capita energy consumption of India is about 226 kgs. of oil equivalent which is among the lowest in the world.

The Indian energy scene was dominated by non-commercial energy in early years after independence. In 1953, it accounted for nearly 75 per cent of the total energy consumed. In the nineties, it has dropped to 47 per cent. The displacement of non-commercial energy by commercial ones raises the standards of living of the people apart from helping to maintain the ecological balance. By 2010 and beyond, we expect that more than 25 per cent of energy needs would be met by commercial energy sources.

An analysis of the primary energy consumption in India reveals that the annualised growth of coal and oil are more or less stagnant at around 6 per cent, since the seventies through the nineties, whereas an exponential growth rate has been witnessed in natural gas.

A typical breakup of the energy consumed by different sectors in India shows that roughly 50 per cent of energy is consumed by industry; 25 per cent by transport sector; 14 per cent by household, 9 per cent by agriculture and the rest by others. The energy planners need to keep this aspect in mind when the energy options are explored in the Indian context.

3. The Energy Options

The profile of future energy demand for coming years shows that hydrocarbons and coal have the lion's share with hydroelectricity and neclear contributing to merely about 12 per cent of the total demand.

This is going to be the situation all over the world. Growing safety and environmental concerns are the main reasons for the shift away form these sources in the world. However, in Indian context smaller hydroelectric capabilities may be a suitable option for the hilly regions purely from a point of view of social needs rather than commercial considerations. The active participation of NGOs and the fiscal incentives offered by the Government can usher in perceptiable generation capabilities from this avenue.

Renewable energy sources constitute an important alternative for meeting large demands of energy in our country in the 21st century. Out of the estimated 20,000 MW potential for wind energy, a capacity of about 375 MW has already been established. The tax concessions given by the Government is instrumental in generation of capabilities in this sector. Solar energy represents a very important source in our country where 4 to 6 KwH solar radiation per square metre is received for 200-300 days in a year. Government is giving great thrust to the development of renewable energy Sources not only to augment energy supply but also to provide environmentally benign energy options for the country. But all these options are aimed at reducing to the extent possible the dependency on the two main options of coal and hydrocarbons.

Although coal is available in plenty, there is a serious problem as regards its quality. It is estimated that India has mineable reserves of about 100 billion tonnes which are not yet projected as mines and this can ensure coal supply for over 234 years at the current rate of production of approximately 267 million tonnes per annum. In view of the favourable geological configuration like seam depth, thickness, incline etc. it is possible for the coal industry to supply one of the cheapest fuels to the economy closely competing with natural gas. It is expected that the consumption of coal could exceed 600 million tonnes by the year 2010. But steps are required to augment the present production on a short term basis. Large investments, both Indian and foreign, are required to be brought in for achieving this, Government may have to think of policy framework to enthuse meaningful response to attract these investments. Better technology to mitigate the associated environmental hazards, recourse to pithead power generation and its transmission have to be thought of.

The production of coal bed methane has been getting the focus of attention in the energy sector globally. Oil majors such as AMOCO have been extensively working on this option and have been commercially producing coal bed methane in US for over a decade. ONGC is also striving to tap this resource at Durgapur depression of Raniguni Coal fields and in Gujarat.

4. Hydrocarbon Sector

Out of India's total area of 3.2 million sq. kms. its 26 sedimentary basins cover an area of 1.78 million sq. kms. Of this, 1.46 million sq. kms. is onshore and 0.32 million sq. kms. is offshore, in waters upto 200 metres in depth. Presence of hydrocarbons has been established in nine of these

26 basins. Six of the basins have been put on commercial production.

The total discovered geological reserves in these basins is about 6 billion tonnes amounting to only 29% of estimated (prognosticated) reserves. The offshore basins of the country in water depths exceeding 200 Mts. is expected to have a large potential.

Indian production of crude oil in 1995-96 is 35.15 MMT and that of gas is around 22.64 billion cubic metres. It is important to understand that the offshore production of crude oil is as high as 2/3rd of total indigenous production, with western offshore accounting for almost the total offshore production. The production estimates for 1996-97 is 35.87 MMT of oil and 23.3 BCM of gas. But these figures are simply dwarfed by demand projection of 78 MMT of oil in 1996-97 growing to 164 MMT by 2010-11.

Government of India has initiated several measures to give a new thrust to the hydrocarbon sector. ONGC has been converted from a commission into a Corporation. This gives freedom to raise external funds from within and outside the country. For the five year period of 1992-97, ONGC launched an ambitious programme to enhance oil and gas production from existing and newly discovered fields located in western offshore of India. This involved implementation of seven major offshore development schemes with an estimated investment of US $ 5 billion to derive a cumulative production of 150 MMT of oil and 65 BCM of gas upto the year 2011.

ONGC Videsh Ltd. (OVL), a wholly owned subsidiary of ONGC, is actively pursuing several exploration and development projects in different parts of the world. It is active in Vietnam where it has discovered gas in association with BP and Statoil, in Egypt and Yemen with British Gas and in Tunisia with Command Petroleum

Changing from a strict monopoly, now the petroleum sector is open to foreign and private participation. 125 exploration blocks have been offered to private sector and multinationals since 1991. Contracts have been awarded for 4 blocks and negotiations are in progress for 10 blocks. In addition to the 5 medium sized fields under JVC operation, 13 small fields have been given to private parties under Production Sharing Contracts. It is expected that development of discovered oil and gas fields in joint venture will yield production of 4 MMTPA during the Ninth Plan.

It is expected that crude oil production will pick up and reach a plateau level of around 45 million tonnes per annum at the turn of the century but the oil demand is estimated to be 164 MMT during that time. To meet this demand we need 2.3 to 3 billion tonnes of recoverable oil reserves against India's prognosticated reserves of 21 billion tonnes. This calls for investments to the tune of US $ 60 -100 billion. ONGC on its part is giving a major thrust to exploration activity through the Accelerated Programme of Exploration (APEX). This envisages an outlay of about US $ 1 billion during Eighth plan. Plans are afoot to invest more than US $ 1 billion in a number of projects in pipeline to get a cumulative production of around 11 MMT and 9.8 BCM of gas.

The indigenous gas production in the Ninth Plan has been projected as about 88 million cubic meters per day out of which about 68 million cubic meters would be available considering the internal usage and shrinkage etc. The demand supply gap of natural gas in India has been estimated as about 78 MMSCMD in 1999-2000 which would grow to 120 and 215 MMSCMD respectively in the succeeding ten years, in steps of five years. Considering the profile of energy consumption in India, in which industry and transport are on the top of the list, the use of natural gas automatically occupies an important place.

Power generation capacity of 55,000 MW is planned to be added in the Ninth Plan. When one considers the "imputed economic values" of the fuel alternates available such as imported coal, naphtha and natural gas, it can be found that gas can comfortably edge out the other fuels upto a price tag of about US $ 3.8 per Million British Thermal Units.

The next question is that when would the gas be available for use freely? In addition to the efforts of ONGC to augment the indigenous production of the hydrocarbons, Government of India signed MOUs with Government of Islamic Republic of Iran and Sultanate of Oman in 1993, for gas imports to India. The Oman-India pipeline is aimed at transporting 56 million cubic meters of gas per day to India and Iran-India pipeline is envisaged to contribute another 50-70 million cubic meters of gas per day. But there are certain difficulties that arise form technical and political issues. Apart from these options, many multinational companies are trying to explore the gas transportation through pipelines form Myanmar, Bangladesh to India, dovetailing the proven gas reserves at Tripura. however, these attempts are in an infancy stage only.

All these factors point towards import of LVG into India as a practical option. Let us consider certain interesting facts. We are ideally situated between the biggest gas reserves of Middle East on one side in countries such as Oman, Qatar, Saudi Arabia, UAE, Iran etc. and the sizeable gas reserves of the South East of the countries like Indonesia, Malaysia etc on the other. Also West Asia is emerging as another major exporter of LNG after Asia-Pacific. In the next 3-5 years, Quatar, Oman and Abu Dhabi are planning supply large quantities of LNG. The close proximity of Indian market cannot be overlooked by these LNG developers. Thus the timing is most suitable for LNG projects in India and this would help us to secure most attractive contractual terms for LNG purchase. But attempting an LNG project involves huge front end financial commitments. For example, to import about 20 million cubic meters of gas per day in the form of LNG, would involve a project entailing an investment of around Rs. 18,000 crores of which nearly half the amount would go for field development and liquefaction facilities in the exporting country.

As the LNG supply contract are typically long term running to about 20 years with Buy or Take commitments, the supplier needs to be convinced of the financial strength of the buyer. This aspect of LNG trade necessitates that the buyer has to have concrete commitments of the consumer. If you look at the European market such a problem may not be there since there exists a mature gas market. On the other hand the example of Japan, which is the major importer of LNG in the world, shows that imported LNG is directly consumed by the importer's power plants located near the coast. In our case, either the importer has to have such a captive facility of his own or he should tie up with power producers. For such a chain to develop, Government should think of giving some of sort of guarantee to the LNG importer regarding the collection of his dues from the power producer who in turn depends on the payment from SEBs for the power sold to them. Moreover, the import of LNG would be feasible for the Southern part of our country as the transportation cost of even the future piped gas arriving at the western coast to Southern region would make the LNG competitive with this gas. The leading PSUs in the oil sectors such as ONGC, IOC and GAIL are working towards this option. Once the chain is established, the reality of LNG availability at an affordable price will act as the catalyst for the gas market to establish and grow to maturity.

While concluding another important factor of energy management – the conservation needs to be stressed. With the growing GDP on one hand and the growing population on the other hand, the energy demand would be galloping. The world energy council in one of its studies points out that developing countries even with the existing machinery can increase their energy efficiency by 25-30%. Increase to the extent of even 60% would be possible through the adoption of energy efficient technologies and well thought out processes of manufacture. It has been estimated that we can reduce as much as 100 MMTOE of energy through conservation measures by the year 2010.

COAL INDUSTRY

A. K. SAHAY
Director (Technical)
and
U. P. SINGH
New Technology Divn., Coal India Ltd. (INDIA)

1. PREAMBLE

Energy is the engine of growth in any society and therefore it is the prime obligation in part of the Energy Sector to rise upto the expectation of the end users and meet the challenges posed by the growing economy of the country/society. Fossil fuels still dominate the energy supplies in India. The primary source of energy available in India are coal, oil, natural gas, hydroelectricity and nuclear power. India is having a coal reserve of 202 Billion Tonnes upto a depth of 1200 Mts., out of which proven reserve is 70 Billion Tonnes, in 56 known coal basins, which is approximately 6.73% of world proven coal reserves.

The most crucial energy resource sector of the country i.e. coal was given major thrust to expedite its development and modernisation immediately after the independence when the level of production was only around 30 MT. It was considered that private sector alone would not be able to achieve the desired growth. This led to formation of NCDC under public sector in 1956 and was given the task to develop outlying coalfields. However, the level of coal production stagnated around 73 MT in the early 70's i.e. an increase of only 43 MT over a period of about 25 years.

The task of stepping up coal production within the time horizon envisaged, was considered far beyond the financial capabilities and organizational competence of the private coal mining industry in the country. Therefore, the ownership of coking coal mines and non-coking coal mines were taken over in Oct. '71 and Jan. '73 with the objective of establishment of a single authority to whom task of massive development of country's coal reserves could be entrusted and accordingly Coal India Ltd. took birth in 1975. Since then there is a quantum jump in the coal production in the country, i.e. 237.25 MT for CIL and 270.05 MT for the country as a whole, in the year 1995-96. The nationalisation of the coal sector in the early 1970s marked the beginning of a comprehensive program for the development of coal mines in India.

1.1 Indian Coal Reserves

Indian coal resources are likely to last for more than 250 years at the current rate of consumption and is and will remain the predominant commercial energy source in India. Though there is some decline in its share in total commercial energy, however, still more than 2/3 rd of commercial energy consumption is from coal and coal will continue to be the main source of primary commercial energy, not only for direct energy use in industry, but also for indirect energy use through power generation in future years.

At present, coal & lignite has 53.4% share of commercial energy in the country compared to 16.4% for indigenous petroleum, 17.7% for imported petroleum, 2.7% for hydel, 8.9% for natural gas and 0.9% for nuclear energy.

COAL RESOURCES IN INDIA
(AS ON 1.1.1996)

(BILLION TONNES)

COAL	PROVED	TOTAL	% OF TOTAL
COKING	15.22	29.81	14.8
NON-COKING	55.22	172.14	85.2
TOTAL	70.44	201.95	100.0

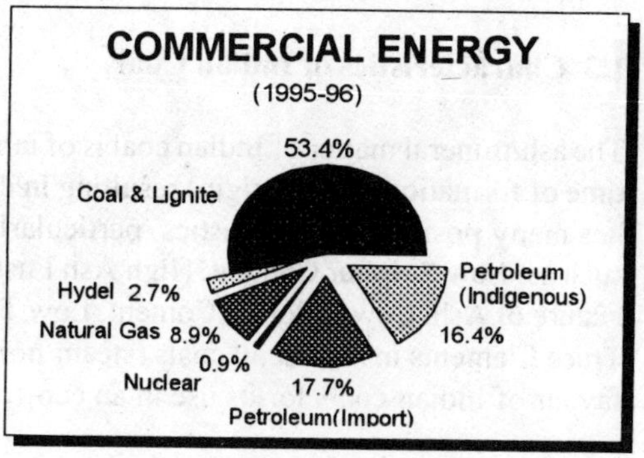

The country ranks tenth among the coal bearing countries and third among the coal producing countries of the world. India's total reserves comprise coking 29.81 Billion Tonnes and 172.14 BillionTonnes non-coking category, out of which proved reserves comprise 15.22 BillionTonnes coking and 55.22 Billion Tonnes of non-coking coal.

1.2 Global Energy Scenario

It is worthwhile to have a look at the Global Energy scenario in order to examine the role of coal in the energy development. It has been noticed that there is a wide gap in the rate of per capita energy consumption between the low income economies and higher income economies. There is also a trend of slowing down of growth rate of energy consumption by the higher income economies, which is mainly due to effective energy efficiency and conservation measures and partly due to attainment of near saturation level.

WORLD COAL RESOURCES

RESERVES	WORLD	INDIA	% OF WORLD RESOURCES
TOTAL RESERVES	24460	202	0.83
PROVED RESERVES	1040	70	6.73

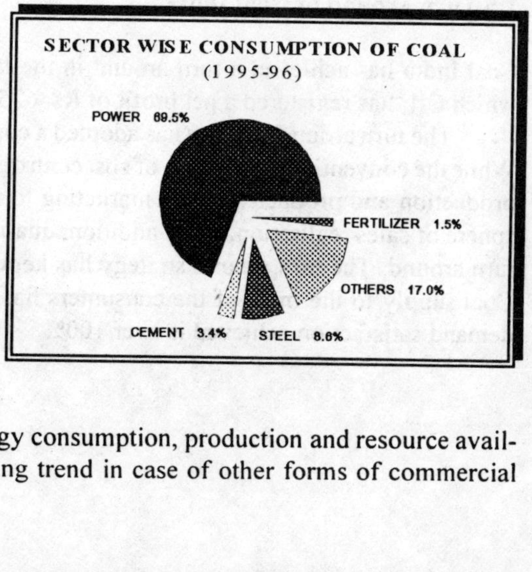

While examining the World commercial energy consumption, production and resource availability trend, it is observed that contrary to declining trend in case of other forms of commercial energy, there is an increase in growth rate in the case of solid fuels (i.e. coal & lignite), which is more pronounced in case of low & middle income economies. This reflects the increased importance of coal towards growth of commercial energy consumption world over. The world coal reserves of about 1040 Billion Tonnes is expected to last for more than 225 years at the current rate of consumption.

1.3 Characteristics of Indian Coal

The ash/mineral matter of Indian coal is of inherent nature i.e. intimately mixed with coal mass at the time of formation (Drift Origin) resulting in difficult cleaning characteristics. However, Indian Coal has many positive characteristics, particularly with respect to environmental aspects and end-use, such as: Low Sulphur Content, High Ash Fusion Temperature, Low Iron Content in Ash, Refractory Nature of Ash, Low Chlorine Content, Low Toxic Trace Elements, etc. Low Sulphur and Low Toxic Trace Elements in almost all coals (steam/non-coking and coking) are the biggest possible factors in favour of Indian coals for its use in an eco-friendly manner in almost all spheres of end use.

1.4 Indian Coal Industry

Coal Industry has pursued its main objective of providing energy support to India by satisfying the needs of the coal consuming sectors, such as Power, Steel, Cement, Fertiliser and others.

Coal production in India is largely in the Government sector except a few captive mines of Tata Iron & Steel Co. (TISCO). Coal India Limited is the largest public sector entity having produced 237.25 MT of coal during the year 1995-96, others being Singreni Collieries (SCCL), Indian Iron and Steel Co. Ltd. (IISCO) and Damodar Valley Corporation (DVC). Presently 98% coal is produced by public sector coal companies, out of which Coal India Ltd. and its subsidiaries produce 88% whereas Singreni Coal Co. produces 10% and others 2%.

Coal Industry has been able to meet the demands of the Industries for the last two decades and has registered phenomenal growth in production. Coal has literally powered the Industrial growth of the country with 100% demand satisfaction of the power sector and the coking coal supplies to the steel plants have also been to the envisaged levels.

1.5 Turn around of Coal India

Coal India has achieved a turn around in the last four years. This is the fifth consecutive year in which CIL has registered a net profit of Rs. 625.25 crores.

The turn around strategy has adopted a combination of conventional and innovative approach. While the conventional approach of cost control, effective and higher utilisation of equipment, higher production and productivity and marketing has paid its dividends, the innovative approach in the sphere of sales realisation, value addition, quality control and dispatch have also contributed to this turn around. The turn around strategy has kept the consumer satisfaction as the most valued goal. Coal supply to the most of the consumers has met their demand satisfactorily. For power sector, demand satisfaction achieved is over 100%.

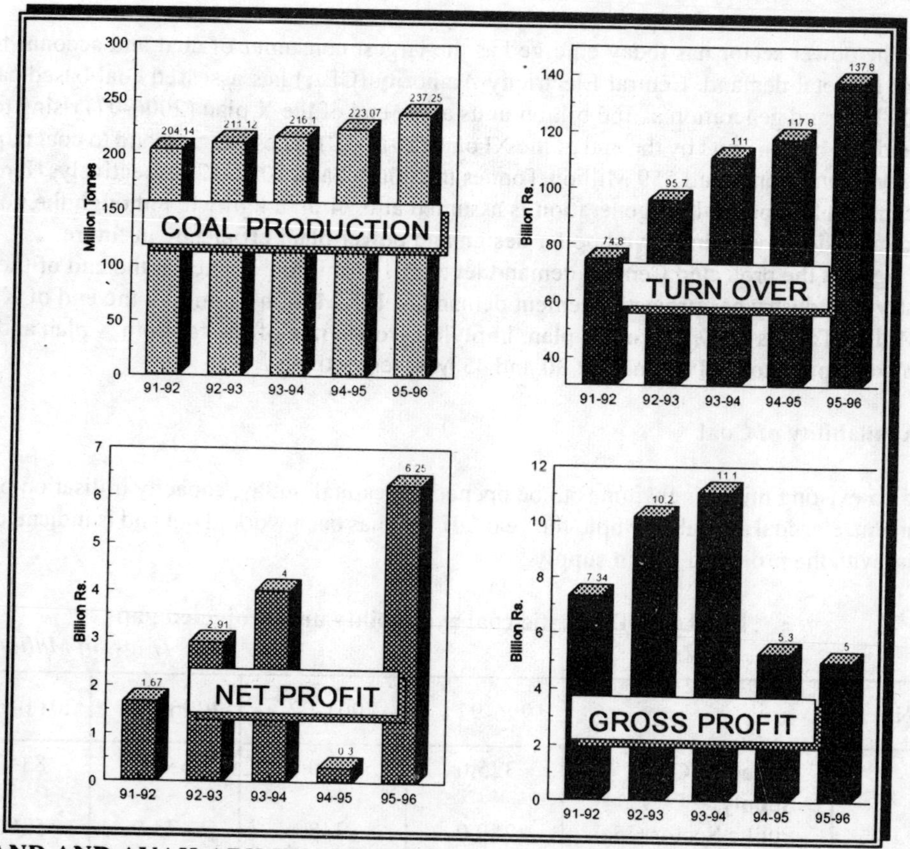

2. DEMAND AND AVAILABILITY OF COAL

2.1 Demand

The sector wise consumption trend of coal indicates that power sector consumes 69.5 % of coal, whereas steel consumes 8.6%, cement - 3.4%, fertiliser -1.5% and others -17.0% of total coal produced in the country (1995-96). The demand of coal based on indications received from various consumers upto year 2011-12, is shown in table 1.

TABLE 1. Sectorwise Coal Demand

(Figs. in Million Tonnes)

Sl. No.	Sector	Demand			
	1995-96	2001-02 (IX Plan)	2006-7 (X Plan)	2011-12 (XI Plan)	
1.	Power	215.00	325.00	447.00	559.00
2.	Steel	40.50	50.00	64.00	78.00
3.	Cement	17.50	22.00	30.00	45.00
4.	Others	52.00	78.00	112.00	153.00
5.	**Total**	**325.00**	**460.00**	**653.00**	**835.00**

The power sector has today emerged as the largest consumer of coal and account for nearly 70% of the total demand. Central Electricity Authority (CEA) has assessed coal based capacity at 116400 MW and generation at 566 billion units at the end of the X plan (2006-07) rising to 138300 MW and 690 billion units by the end of the XI plan (2011-12). These correspond to coal requirement of 447 Million Tonnes and 559 Million Tonnes in 2006-07 and 2011-12 respectively. Here specific coal consumption per unit of generation is assumed at 0.74/0.75 kg/kwh, although the norm is expected to go down with improved boiler design and power plant efficiency in future.

Against the projected Cement demand level of 107 million Tonnes at the end of the IX plan, Ministry of Industry has projected cement demand at 158 Million Tonnes at the end of X plan and 233 Million Tonnes at the end of XI plan, implying growth rate of 8% for both X plan and XI plan, which correspond to coal demand of 30 and 45 MT, respectively.

2.2 Availability of Coal

Based on existing mines, new mines to be opened up, capital outlay, capacity utilisation of HEMM equipment, etc. coal availability upto the year 2011-12 has been worked out and is indicated in Table 2 along with the projected gap in supply.

TABLE 2. Domestic coal availability and projected gap

(Figs. in Million Tonnes)

Sl. No.		1996-97	2001-02	2006-07	2011-12
1.	Expected Coal	325.0	460.0	653.0	835.0
2.	Supply				
i.	Public Sector Coal Companies	289.0	348.0	473.0	530.0
ii.	Captive mining	NIL	35.0	45.0	60.0
iii.	Expected Import	NIL	15.0	20.0	25.0
	Sub Total of Supply	289.0	398.0	538.0	615.0
3.	Gap	36.0	62.0	115.0	220.0

It is observed that wide gap is expected in coal demand and its availability in the country and, therefore, the prime job of coal companies is to bridge the gap of demand and supply. Reduction in demand will certainly jeopardize the industrial growth of the country, which no country can afford and, therefore, the only alternative to bridge this gap is to increase the supply position of coal. Thus coal industry must insure that this challenge of meeting the demand of coal in coming years for industrial growth of the country would be met both quantitatively as well as qualitatively.

3. CHALLENGES AHEAD

3.1 The major challenge ahead with the coal industry is to meet the demand of industries in the coming years. Coal industry has to play a pre-dominated role here. This obviously calls for heavy investment to develop:

— new capacity;
— improve the productivity of the existing capacity; and
— improve the quality of coal to meet the industry requirement.

13.2 Despite the need for growth and investment in the coal sector, the recent economic reforms has also forced a formidable challenges for coal industry. Within a short period, the budgetary support to finance the plan outlay, which used to be nearby 100% earlier and even 95% during the 7th plan period, has now been reduced drastically, to less than 12% in the year 1995-96 and is heading towards complete withdrawal. The coal industry thus have to depend mostly on the followings for meeting its growth and investment plan:

- — internal generation of resources;
- — external borrowings for funding plan outlay;
- — inviting foreign investments; and
- — promotion of joint ventures.

3.2 As Govt. has not only removed restrictions on import of coal but has also reduced the import tariff on non-coking coal from a level of 85% to only 35%, where as the import duty on coking coal is already at a nominal 5%, thus the coal industry has now been exposed within a short time of 2/3 years to achieve self-sufficiency, not only in financing investment and growth, but as to face global competition in terms of quality and price of coal.

3.3 In the recent years, with the advancement of deep drilling technologies and its application for Coalbed Methane (CBM) development, a new resource of energy has been identified, which is associated with coal i.e. CBM, and coal Industry puts major thrust in exploitation of this resource, which necessitates further generation of Internal resources.

3.4 Even for improvement of quality of coal to be supplied to the industry coal may have to be imported for proper blending, which should be of supplementary characteristic to Indian coal. This further needs resources for initial investment.

3.5 With the liberalization of economy and reduction in the custom duties on import of coal, a level playing field has been created in which domestic coal production has, of necessity, to be competitive with the imported coal. Competitively priced coal from South Africa, Indonesia and Australia is now the in the realm of possibility, especially for the coastal consumers.

3.6 Apart from this, captive coal mines in the private sector, especially for power generation, have already been permitted, which may put further competition for coal produced in state sector.

In view of these challenges, it is imperative that coal production in the public sector be competitive.

4. CONSTRAINTS IN OPERATION

Though, the coal industry is to gear up to meet the new challenges mentioned above, by way of cost reduction, improvement in efficiency, etc., it has to face the various difficult situations under which it has to operate, such as:

4.1 Coal industry is inheriting a large manpower strength right from the time of nationalization, the total employees strength at CIL now stands at 6.34 lakhs. The wages still continues to form the major cost of operation (more than 43%). Operating in 8 different states, which are being controlled by several different political parties, the coal industry has to face various socio-economic-political influences which are not common in other industries. This create major hindrance in reducing the cost of operation and hence cost of production.

4.2 While the coal industry is now to depend more on internal generation of resources, the continuance of administered price restrict such scope unless there is gradual improvement in efficiency of operation. Situation deteriorates further due to large outstanding dues from consumers, specially the bulk consumers, remain unrealised for quite a long period.

4.3 Gradual shrinkage in production capacity of the taken-over mines necessitates the call for higher growth in production from the new projects. The availability of land is the single biggest constraint for coal mining operations. Land requirement for coal mining has two unique features:

— The requirement is very large compared to other sector; and
— The requirement is site specific.

4.4 Moreover, the new projects are not coming up due to some of the following reasons:

— Opening of new mines is faced with non availability of land, especially forest land, for coal projects.
— Legal constraints inhibit even exploration operations in forest areas.
— Severe restriction on allowing coal mining over forest land areas have seriously affected development of many new coal projects as majority of new coal deposits are available in areas covered by forest land.

4.5 The new coal projects are not financially viable at current administered prices and are not likely to be sanctioned. Presently coal mining is being done upto a depth of 300 meters. Average depth of operation is 150 meters. The maximum production which can be achieved at these depths is 400-500 MT. Mining of coal at greater depth shall not be economical at today's price of coal. The prices of Indian coal shall have to go up and become more comparable with Global prices in case the demand is to be met from indigenous sources.

4.6 Opening of new mines in the outlying coalfields is also faced with problem of lack of infrastructure such as transports, power supply, civic amenities, etc.

4.7 Old uneconomic mines need to be closed. However, such closure of mines are not allowed from the conservation point of view as well as present socio-economic-political scenario makes it further difficult. Even after mines are being closed due to exhaustion of reserves, redeployment of unutilised labour is invariably faced with stiff opposition.

4.8 New underground mines which are being taken up for working underlying virgin seams in areas where upper coal seams have been extracted in the past, are faced with the problems of old subsidence, water-logged workings, fire etc. Exploitation of deeper reserves has also to face more difficult geo-mining conditions.

14.9 Thrust to develop new outlying coalfields for meeting the demand in different regions, has resulted in availability of inferior quality coal due to lack of superior coal in such coalfields.

4.10 Contract labour Act is not favourable for development of coal industry.

4.11 Unrealistically low prices of coal are leading to inefficient use of coal energy.

4.12 Relative technological backwardness in underground coal mining, due to limited success to longwall support technology in this country, has been largely responsible for its low level of productivity and profitability.

4.13 Efficiency in operation of opencast mines depend to a large extent on efficient maintenance of Heavy Earth Moving Machinery (HEMM) which constitutes the major part of the investments. Inadequate support from manufacturers in supply of suitable spares in time and development of servicing facilities are serious handicap in such operation.

4.14 The constraints, in increasing underground coal production (to get better quality of coal) from CIL command areas (including TISCO/IISCO/DVC) can be appreciated if the coalfieldwise break-up of the recoverable coal reserves amendable to opencast and underground mining is considered as given in Table 3. From the study of the table following is inferred:

— The entire coal production from Rajmahal, Ramgarh and Singrauli coalfields will be from opencast mining because, in the present pricing system, it will be uneconomical to exploit the coal reserves amendable to underground mining.

TABLE 3. Coalfieldwise Technology Mix of Coal Reserves and Production

Sl No.	Coalfield	Recoverable Reserves (Upto 300m)		Coal production (2002-02)	
		OC%	UG%	OC%	UG%
1	Raniganj	54.5	45.5	30.9	69.1
2	Rajmahal	79.8	20.2	100.0	-
3	Jharia	65.4	34.6	55.1	44.9
4	East Bokaro	78.0	22.0	87.7	12.3
5	West Bokaro	42.2	57.8	83.2	16.8
6	Ramgarh	91.8	8.2	100.0	-
7	S Karanpura	46.3	53.7	69.7	30.3
8	N Karanpura	85.4	14.6	94.9	5.1
9	Singrauli	64.0	36.0	100.0	-
10	Wardha	44.2	55.8	84.9	15.1
11	Kamptee	11.9	88.1	26.3	73.7
12	Pathekhera	-	100.0	-	100.0
13	Pench-Kanhan	5.6	94.4	2.5	97.5
14	Korba	64.2	35.8	93.5	6.5
15	Korea Rewa	22.4	77.6	28.7	71.3
16	Ib Valley	69.6	30.4	89.9	10.1
17	Talcher	77.1	22.9	98.1	1.9

— For similar reason, the share of opencast production in North Karanpura, Korba, Talcher and Ib Valley coalfields are higher than the share of coal reserves amendable to opencast mining.
— In East Bokaro and West Bokaro coalfields, the underground coal reserves can not be exploited for technological reasons, before the bulk of the opencast coal reserves are exploited.

4.15 The infrastructure facilities directed to effective growth of coal sector are railways, road, transportation, ports and MGR system. Inadequate infrastructure for coal sectors poses another constraints for coal industry, especially inadequate coal evacuation facility in Talcher, Ib Valley, Korba and North Karanpura coalfields. Coal production can increase rapidly in these coalfields if the evacuation facilities are enhanced by way of:

— Doubling of Talcher Paradip link;
— Doubling of Champa Gevera Road section;
— New line connecting Korba coalfield with Katni;
— Adding line between Champa and Durg.

MODE-WISE MOVEMENT OF COAL

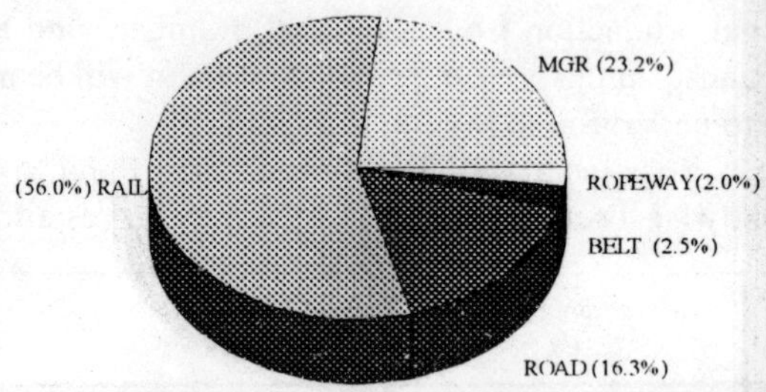

4.16 Movement of coal by rail connecting producing centers to consuming centers has been one of the most critical issues in fulfilling the demand needs of different consumers. Coastal transport and movement by road, only supplement the said movement of coal. Consumers at and around the pitheads are normally served by MGR system. The need for linking by rail has been stressed time and again but a number of lacunae still exist and large scale improvement in evacuation facilities by l rail in terms of major availability and loading arrangement are necessary.
4.17 Coal sectors have been facing the problem of power shortage for a long time, especially in eastern region, causing reduction in coal production.

5. STRATEGIES

Under the changed environment, in which coal industry has to operate is well understood. Time has come for coal industry to operate with well conceived strategic plan to meet the above challenges. Cost reduction in all the operation has been identified as a key thrust area.

5.1 As mentioned earlier, gainful utilisation of existing manpower is very essential to reduce the wage cost and to generate the internal resources. The socio-economic-political situation does not allow in reduction of manpower, therefore, only alternative left is not to consider this larger manpower as burden but as a resource and utilise the existing manpower for alternative jobs, such as:

5.1.1 Large scale tree plantation in the evacuated mining areas, which will serve the cause of environment protection, utilisation of excess manpower gainfully and providing timber for future use in underground mines. This activity has to be carried out on commercial basis with the help of Govt. agencies engaged in promotion of afforestation with requisite funds and technical knowhow available with them. This is a labour intensive job and the existing excess large manpower resource of coal industry can be utilised here gainfully.

5.1.2 As regards opportunities for alternative deployment, with some training, it is possible to engage them for following activities: Trainee overman, trainee surveyor, trainee foreman, technical trainees, etc.; ward attendant in hospital; security guard in the depot and in other places; cane manufacturing jobs; motor winding; lamp room cabins; fan operator; sampling mazdoor; lab mazdoor, etc.

5.1.3 Encouragement of VRS, specially in white collar jobs has to be practiced.

5.2 Coal industry has to work towards complete deregulation of coal prices. Though Government of India, has deregulated prices and control of distribution of A, B & C grades of non-coking coal and all coking coal, it is required that prices should be deregulated for all the grades of coal. This will give self sustaining strength to coal industry.

5.3 Coal Price

The present gradewise distribution of coal price is based on Useful Heat Value (UHV) basis. The grade pricing does not provide an incentive to the coal producers as The heat value range for a particular grade is too wide. Not only that, UHV concept was put forward for pricing in 1964 based on the then existing combustion technologies. Now with the improvement in combustion technologies, such as Pulverized Fuel firing in major thermal power station and popularization of Fluidised-Bed Combustion of high ash coal, therefore, it is high time to switch over to coal grading based on Gross Calorific Value (GCV), which is more relevant and scientific with present combustion technologies. Moreover, world wide coal grading is based on GCV.

It will provide incentive to coal producer also in terms of realisation of better price for the coal supplied.

5.3.1 A scheme of this nature would provide incentive to the management at the mine level to improve quality of coal produced. Obviously this would fetch a better price for the coal produced, but at the same time this would not put any additional burden on the consumers. The extra price paid by him by way of bonus would be off-set by the higher calorific content of coal. Changeover from UHV scheme to GCV scheme for determining the coal price, will help coal produce to get better return on its investment, and generation of additional revenue.

5.3.2 Further, growing importance of protecting ecology and environment as well as reclamation and restoration of damaged land due to mining operation, particularly in opencast mining put further burden on coal producing companies which can be met by putting levy of cess @ 1% in coal price for environmental protection.

5.4 Technological upgradation and modernization in different phases of operation in coal industry, right from exploration, planning of extraction technology and its end use, is called for overall improvement in the coal industry. This necessitates a through restructuring of the coal industry and its working culture. The latest management techniques, such as Business Process Reengineering (BPR), etc., should be applied to bring out overall working cultural revolution in the organisation to make it more vibrant to withstand the new challenges posed for the coal industry.

5.5 The imbalance in the OC-UG technology mine in CIL needs to be corrected as only about 31 per cent of coal reserves in CIL commands area (down to 600 m depth) are amendable to opencast mining and with the present emphasis on opencast mining, quarrible coal reserves are exhausting at a rapid rate. It is difficult to specify any technology mix of opencast and underground coal production as optimal. This varies from coalfield to coalfield and depends upon the geo-mining conditions, depth of coal reserves, coal to overburden ratio, the quality of coal, the cost of coal production and price of coal, etc.

Generally speaking, underground mining is uneconomical in the present pricing system, if the coal mixed is lower than Grade 'C'. Thus ultimately, market forces play the dominant role in deciding the technology to be adopted for coal production in future.

5.6 One of the most important areas for achieving better operational efficiency is to improve the equipment utilisation which constitutes over 70% of total investment.

Measures have been taken to reduce idle time of equipment by ensuring continuous availability of operators at the site by providing suitable facilities i.e. mobile service van, mobile canteen, transport vehicle etc.

In addition computerised truck despatch system has been introduced to optimise the use of dumper fleet. Computerised condition monitoring maintenance and materials management are in use to help in better equipment availability. Suitable workshop facilities at the unit level as well as centrally in area and also at headquarters level are being developed for improvement in proper maintenance of HEMM. Optimisation of production of digging equipment is to be ensured through optimisation in design of drilling and blasting pattern.

The major areas in which development should be carried out, for improvement of efficiency, are : Upgradation in equipment size; Use of advanced electronics and computer control system for equipment reliability; Improvement in the operational efficiency of the equipment and system; Alternatives to wheeled transport for coal and overburden i.e. mobile in pit crushing and conveying of coal and overburden; Steep angle conveying of coal; Selective winning of thin bands using special equipment; Continuous Mining method, etc.

5.7 Almost 82% of the total underground coal production still comes from Bord & Pillar mining. Mechanised loading by the introduction of Side Discharge Loaders (SDL) and Load Haul Dumpers (LHD) gaining momentum in underground coal mines. Unfortunately, the geo-mining conditions in India is not favourable for large scale introduction of mechanised longwall mining which contributed only 4.38 MT in 1995-96. It is necessary to overcome the existing constraints to facilitate its larger application so as to obtain at least about 11 MT coal from such faces in 2001-02. Stress has to be given for technology development for working deeper deposits.

5.8 Development of ports and linked facility now assumed paramount importance in the context of imported coal supplies. Port development projects is a major component of infrastructure sector to improve availability through import of coal both coking and non-coking. Over capacity utilisation of ports is already causing problem and drawing disadvantage to Indian business units and it is very much required to develop the following facilities: setting up of container terminal; setting up of storage facilities; operation and maintenance of cargo handling terminals; dredging; maintenance of various port crafts and other equipment; leasing of equipment; etc.

5.9 The following issues may need consideration for appropriate balancing of coal production movement and consumption: removal of all bottlenecks in the existing rail linkage movement; development of railway line evacuation facility in new coalfields; to promote coast shipment of coal with adequate inland rail movement facilities.

5.10 Since import of coal have become necessity for maintaining targeted generation of power sector, movement of coal from ports to proximate power stations by slurry pipelines may be given due consideration.

5.11 Benificiation of non-coking coal, particularly for the coal to be used in the distant power stations, by setting up washeries at loading points. This will ease all ready constrained wagons availability. Moreover, such coal will go a long way to reduce specific consumption of coal.

5.12 All the suggested changes above need money and therefore the thrust area for the coal industry is to become fully self supporting. Apart from deregulation of coal prices, this would be possible by : Waste management in each and every activities and bringing it to zero level; arresting the low production and productivity in underground mines together with improving the overall productivity of machinery and manpower; evolving and implementing an action plan for the control of coal quality, which should also provide for beneficiation of non-coking coal; improving the marketing set up so as to ensure customer satisfactions and timely realisation of the full value of coal sold; import of coals from such sources as to have supplementary characteristics to the indigenous coal and after proper blending provide better customer satisfaction and realisation of better price.

5.13 Export of Coal

There is a potential for export of coal to neighboring and distant countries and joint venture participation in this sphere should be welcomed. There is export potential for Indian coal, especially in two areas.

(i) To neighboring land-locked countries or to parts of neighboring countries away from the port and connected with rail and road to India. Here, the comparatively better quality of coal of 20-25% ash can find market, being more competitive in terms of cost. Limited benificiation of even 25 to 30% ash coal can be competitive in such locations.

(ii) In the International market, served by sea-route, the eco-friendly and combustion friendly characteristics of Indian non-coking/steam coal, particularly with reference to low-sulphur high-ash fusion temperature, low Iron content refractory nature of ash, low chlorine content and low toxic trace elements make Indian coal an attractive commodity, inspite of its high ash, for blending with high-sulphur high toxic coal for reasons of environmental and combustion considerations/advantages. The high ash of coal can be brought down to suit the requirement by limited beneficiation or total beneficiation as required.

The approach of blending low heat value but eco-frindly coal is already being adopted in certain countries where high moisture sub-bituminous coal (with 40 to 42% moisture) is being moved over 100 to 800 miles over land and blended with high sulphur coals for this purpose. This new aspect, coming up in International coal trading, must be tapped for getting better market for our kind of coal.

5.14 A major thrust area is maintenance of coal quality. This would go a long way in avoiding quality dispute with the consumers and thus improving the financial viability of the industry. The aim is to reduce the ash content but more emphasis is on coal with consistent quality.

All coal supplied to core sector consumers should be after beneficiation and the pricing policy should take into account the cost of beneficiation.

One of the main problems relating to quality of coal identified in the Eighth Plan was the high ash content in the coking coal supplies to steel plants. The average ash content of washed prime coking coal had been varying around 20.22% during the last several years, against the stipulated 17% for use in steel plants. Apart form this, supply of oversized coal and presence of extraneous matter in supplies to power plants and variations in quality, are other main problems relating to quality.

It is suggested that all coal would pass through coal handling plants, having crushing, screening and sizing facilities. A beginning has been made in the beneficiation of non-coking coal with construction of Piparwar Coal preparation plant and decision to set up Ananta-Bharatpur and Sasti coal preparation plants under the Build-own-Operate (BOO) concept.

5.15 Eastern region has been facing the problem of power shortage for a long time. To overcome this problem, captive power stations have been taken up in some of the selected areas where rejects from coal washeries are being gainfully utilised. This will not only solve power problem but also make available power at reduced cost.

Segregation of feeders for coal mines has also been taken up to ensure direct supply of power from generating agencies.

5.16 It is also possible to reduce cost of operation by off-loading some of the activities which are not of regular nature to outside agencies at competitive rates. Besides civil construction, road transport, etc.; such jobs which can be given to outside agencies include underground drifting, construction of ventilation stopping, construction of haul roads, etc.

5.17 Care for Environment

Care, concern and protection of nature and environment with sustainable development is the central theme of Coal India. Some of the environmental protection measures in vogue in Coal India are:

* Air pollution	:	Green Belt Barriers, Ground Bunker Shielding, Dust suppression through hydrojet and use of chemical additives, etc.
* Water pollution	:	Closed Circuit operation of washeries, Provision of affluent treatment plants, Acid mine drainage treatment, Sewerage treatment, etc.
* Land Reclamation	:	Bio-reclaimation of degraded and subsidence affected areas, Planting of more than 60 million saplings during the last eight years, etc.

Over the years, technical capabilities have been developed to deal with multi-disciplinary dimensions of environmental complexities in coal and other sectors. A multi-disciplinary team of experts from mining, environmental engineering and science, forestry, geology, hydrogeology, remote sensing and landscaping deals with environment related matters and forms the core group of "Institute of Environment" (IOE), a separate establishment under CMPDI, Ranchi for monitoring environmental quality and suggesting/evolving environmental upgradation measures.

6. PRIVATE SECTOR PARTICIPATION

With continuos high growth achieved in the past and anticipated high growth of coal industry in future also with turn around over the last five years, coal industry is poised to meet the challenges and avail the opportunities as mentioned earlier. With liberalisation of Indian Economy, Coal industry has opened its door for participation by entrepreneurs and business organisation in its activities in close association.

Despite the over all efficiency improvement and strategic business planning, Public Sector Coal industry may find it difficult to meet the projected demand of coal alone and , therefore, it is required to supplement 112 MT and 180 MT deficit of coal by the year 2001-02 and 2006-07, respectively, by way of inviting Private Sector participation in coal mining.

6.1 Government of India has already ammended the Coal Mines (Nationalisation) Act, 1973 w.e.f. 9.6.1993 to allow captive mining by companies engaged in generation of power, in addition to companies engaged in production of Iron and steel, which is allowed from 1976 onwards. The amendment also enables the Government to notify other end users for which captive coal mining would be permitted. It has been decided now to extend the facility of captive mining to the cement sector also.

Apart from this, any separate company in which coal user company has 26% voting equity can also take up captive mining.

Government is also considering permitting private sector participation in Joint Sector coal mining with public sector coal companies, allowing up to 49 % equity participation by the Private Party.

6.2 47 coal mining blocks with an estimated reserve of 14150 Million Tonnes of coal, representing 6.94% of the Nation's reserves as on 1.1.1995, have so far been identified for captive mining. The companywise details and the quantity of estimated reserves are shown in Table 4.

TABLE 4 : Captive Mining Blocks

Sl. No.	Coal Companies	No. of Bocks identified	Estimated Reserves (million Tonnes)
1	Estern Coalfields Ltd.	0.9	4,290.0
2	Central Coalfields Ltd.	9.0	3,758.0
3	Western Coalfields Ltd.	14.0	1,233.0
4	Northern Coalfields Ltd.	1.0	242.0
5	South Eastern Coalfields Ltd.	5.0	1,700.0
6	Mahanadi Coalfields Ltd.	9.0	2,927.0
	Total	**47.0**	**14,150.0**

So far, 16 captive mining blocks have been offered/alloted to various parties proposing to set up thermal power stations. In addition, one block has been offered to a party setting up a Sponge Iron Plant and 4 coking coal blocks have been offered for captive mining by Steel Authority of India Ltd (SAIL).

6.3 However, the response has been lukewarm. Not much progress has been made in the development of mining blocks, offered for captive development. The response to captive mining has not been very encouraging, is due to the following reasons: Temptation to low administered price of coal; Outside sale from the captive mine is not permitted; Development of infrastructure at captive mines will take time; Blocks identified for captive mining are comparatively difficult to mine; Lenders are not willing to provide capital funds to IPPs for captive mining.

6.4 This necessitates the thorough examination of whether the private sector could be permitted to take up coal mining operations not only for captive requirement, but also with the freedom to sell the coal produced by them to any consumer. Some of the following steps are desired to encourage private sector participation in coal mining: Govt. should permit majority participation by private sector in Joint Sector Coal mining; The private operator should be free to choose method of operation within the framework of laws pertaining to safety, conservation and environment; Special consideration should be given to entrepreneurs interested in developing coking coal reserves. Tax holiday should be allowed for all coking coal mines for some period after commencement of commercial production;

However, the private sector operator should be required to make an initial up front payment for acquiring mining rights, followed by royalty payments for production. Blocks with good reserves should be auctioned or given out as per international practice. Failure to meet commitments to a time bound exploration and production progress should result in cancellation of the concession.

6.5 Coal industry's door is not only open for captive mining for power, steel and cement producers, but is also open for participation in captive power plants based on washery rejects, to be constructed on "Build -oven-operate" (BOO) basis or washing of coal on BOO principle, supply of HEMM and other equipment on gain/risk sharing basis, etc. Close association in the sphere of mechanised longwall, short wall, bord and pillars mining, continuous mining and hydraulic mining, etc. are also required. In the sphere of beneficiation, the "difficult to wash coal of Gondwana formation or washing of fine coals need special washing circuits to be developed.

Coal washing in private sector is freely permitted and setting up of washeries near the coal mines is encouraged. 4 coal washeries have been finalised and are in initial stages of being set up. However, there is reluctance on the part of consumers to pay the washing costs.

7. COAL UTILISATION

The short term prospects of coal are optimistic but the long term prospects solely depends on how closely the producing company and user industry work together. It depends not only upon the capabilities of coal industry to provide clean coal with consistency in quality depends on, but also how efficiently coal industry promote the cause of introduction of better combustion technologies i. e. clean coal technologies in the country, which are more eco-freindly and have better conversion efficiency as well as quite compatible with indigenous coal.

7.1 Initially adoption of these technologies may appear to be costly but if we look into the initial investment with an integrated approach and add the cost of environment protection, better utilisation of national resource for conversion into other form of energy, and of course the conversion cost; it works out to be cheaper and ultimately society as a whole would be gainer. The time is very ripe that user industry come forward and work in close association with producing companies to achieve this national objective.

7.2 It has been found that about 40 MT of lower grades of coking coal and A, B & C grades of non-coking coal are still being used in the older power stations, due to adoption of poor coal utilisation techniques, especially in West Bengal and Bihar. This is the shear wastage of the national resource. No one can deny this fact that more efficient use of coal is need of the hour.

7.3 STEEL SECTOR

Coal is going to meet the major share of energy in the years to come. The present existing coal utilisation technologies always ask for better quality of coal, which is scare, planned utilisation of coal has to be ascertained for the country. To achieve this, development are required in some of the following areas:

(i) It is necessary for the coking coal industry to improve coal preparation techniques both for optimising clean coal recovery and up grading quality parameters of coke. Concurrently, some efforts have to be made in part of user industry or at Govt. level to adopt such technologies so as to utilise some deficient coking coals, by blending them with suitable matching coking coals, instead of using high grade coking coal.

(ii) Suitable R & D work may be taken up to design coke ovens so that poor cooking and non-coking coals may substitute (partly) prime coking coal. Special technologies like using (a) Taller Coke Ovens, (b) Partial Briquetting of the Coal Charge (P.B.C.C.), (c) Stamp charging, (d) Selective crushing, (e) Dry quenching, etc. should be tried out at commercial level.

(iii) Reductions in coke rate in the blast furnaces should be achieved through the injection of coal dust in the blast furnace.

(iv) Simultaneously, it is necessary to restrict/reduce the use of coking coals by using low ash non-coking coals in direct injection to replace the coking coals in blast furnaces as much as possible. Corex process is the most developed of the new processes for this purpose.

7.4 POWER SECTOR

Coal is the only fuel abundantly available which can be exploited safely and economically for meeting the rapidly accelerating growth in the national/international demand for energy, especially electricity.

The main problems associated with the use of coal for power generation are the low conversion efficiency and emission of greenhouse gases. Conventional coal fired thermal power plants generally operate at an efficiency level of around 33% (Coal energy to Electrical energy). This generation efficiency is sought to be increased by way of utilising better combustion technologies which will reduce the emission of greenhouse gases, simultaneously. The development in the area of controlling emission of pollutants such as So_x, No_x, Co_2 and particulate have to be intensified. Some of the advanced combustion technologies to be put in practice, which are:

(i) Pulverized fuel system with co-generation, (efficiency upto 45%), Fluidised Bed Combustion (thermal efficiency 42-44%) (FBCs - Atmospheric and Pressurised), etc. should be used extensively as combustion technologies, which have better thermal conversion efficiency and are more eco-friendly. With increasing adoption of coal beneficiation, large amounts of washery rejects will be generated which can be gainfully utilised in FBCs.

(ii) Integrated Coal Gasification Combined Cycle (IGCC) and Hybrid Combined Cycle (HCC) for power generation which have net efficiencies of approximately 43% higher than the efficiencies achieved with the most advanced pulverised fuel plants.

(iii) Commercialisation of coal fueled heat engines, fuel cells and Magnetohydrodynamics (MHD) system, etc.

7.5 Other Clean Coal Technologies

(i) Scope of using coal, which have been discovered in the course of exploration for oil, and which are too deep to be mined commercially with present technologies, for producer gas/synthesis gas should be investigated.

(ii) Tapping of coal resources for methane recovery which could offer a major alternative and environmentally compatible energy resource. According to a conservative estimate, Indian coal has approximately 2 Trillion cubic Meter of Coalbed Methane (CBM).

(iii) Possible commercialisation of new Coal Liquefaction process to produce gasoline and diesel fuel by way of gasifications of coal to produce a gas which can then be converted into a liquid (indirect liquefaction) or by adopting technologies producing a range of products suitable for standard refinery process (direct liquefaction). Both the techniques have already been applied in Germany on commercial basis between 1927 and 1945. It may be mentioned that the production of oil from coal has for long been commercially exploited by South Africa Coal Oil and Gas Corporation (SASOL) which presently consumes some 40.4 MT of coal annually to produce gasoline fuel and numerous by-products.

(iv) Use of non-recovery coke ovens which have a distinct environmental advantage over conventional oven in that they are operated under suction condition which reduces the potential for emissions to very low levels. In addition, it does not require coke oven gas/by product processing facilities.

8. ISSUES TO BE RESOLVED BY GOVERNMENT

Though coal industry has already tighten its belt for meeting the challenges ahead, but it can only be possible if Government (Central as well as State) is resolving some of the following issues:

— Extending the de-regulation of prices of all grades and varieties of coal;

— Import duty on non-coking coal is such that the domestic coal industry is neither put to disadvantage nor given undue protection;

— Removal of Import duty on plant and machinery deployed in coal mining so that the best equipment is available to indigenous coal producers at competitive prices and the Indian coal can be enabled to compete with imported coal on equal footing;

— Subsidising activities such as shaft sinking, waiving import duty on winding and haulage machinery, mechanical ventilators, etc. and by reducing royalty on underground coal production;

— Relaxing Contract Labour Abolition and Regulation Act provisions under the changed environment. Jobs such as mechanised coal loading/unloading, mechanised overburden removal, drivage of stone drifts and stone cutting, transport of coal, middling and sand, removal of slurry, maintenance management, township services, etc. could better be performed more efficiently and economically on contractual basis;

— The public sector coal producers to be encouraged to form joint ventures with international operators without any limit on equity participation by foreign parties in such joint ventures;

— Govt. funding of activities such as detailed drilling, master planning etc. and to recover the cost from individual coal producers. Govt. to take initiative for development of infrastructure facilities in the virgin coal fields and virgin area of operating coalfields;

— Permission to the coal industry to utilise HEMM on contract basis. In order to increase production. Such step would enable the rapid increase of production in certain selected mines. This should be permitted to be practiced preferably on gain/risk sharing basis;

— As per the fuel supply agreement, power sector desires that risk on fuel supply should be entirely borne by coal companies. Power producers are assured for 16% return on 62.5% PLF. In case PLF falls below the level, due to inadequate fuel supply, the financial losses to the power producer is proposed to be passed on to coal companies in the form of liquidated damages. It is estimated that the penalty per Tonne of short supply will be of the order of Rs. 2700 compared to the price of coal at the rate of Rs. 400/Tonne. This level of penalty cannot be borne by coal companies. At the most coal companies can borne the difference in price of alternate fuel and it should not exceed the value of goods not supplied;

— More efficient use of coal is the need of the day, however, in India, we continue to plan power stations based on traditional designs. Even the policy guidelines of the Ministry of Power issued in 1992 to attract IPPs stipulate the rate of 2500 k.cal. per unit of electricity generated (about 33% conversion efficiency). However, world over no power stations are being built with a conversion efficiency of less than 38% (heat rate of about 2,250 K.Cal./KG). This poor conversion efficiency needs to be raised;

— Establishment of power stations at pit heads with transportation by captive transporting systems . Pit head generation is not only cheaper but at the same time pollution is and will remain well within the norms;

— Abolishment of Coal India Ltd. as a holding company to generate fair competition among each subsidiary companies;

— Facilitating land acquisition and rehabilitation of land oustees by changing the legislation as per requirement.

9. CONCLUSION

In a developing country like India, the process of economic development has a close connection with commercial energy consumption. Coal occupies and will continue to occupy a premier status among various forms of energy available. Till date, coal production in India is largely in the State sector and has become a vibrant part of national economy. The State sector has performed admirably in the past.

Now, coal industry is passing through a challenging phase under the liberalised economy in the country and is poised to take up the challenges of the future.

The need of the day calls for strategic advance planning to take care of all the uncertainties under which it has to operate. It is imperative that planning has to be integrated, in fine tuning with private sector response for participation in coal mining as well as with Government's directive in resolving the pertaining issues, not individual, and then only coal industry would be able to protect the interest of various stake-holders, such as consumers, coal producers, general public, employees, environmental agencies, etc.

The major thrust area is to generate enough internal resources to get self reliance in production for meeting the country's demand with consistent coal quality. This calls for not only the waste management to zero level but also for improvement in performance and efficiency in all sphere of coal activities; starting from men, material, money, management and the total approach towards the business, in part of coal producers as well as their associates in the field of beneficiation, supply of equipment, etc. It has to be fully supplemented by Govt.'s effort and its policies towards price, infrastructure, availability of land, contract labour, private participation, import of coal etc. However, certainly, it is the coal users who are having final say in success of coal industry, especially in the long run. Their willingness alone, to adopt to better coal utilisation technologies, can ensure sustained future growth of coal industry.

If all of us are waking up together and have a closely knit strategic planning, coal industry is bound to meet, not only the present challenges, but even all challenges of future, also. Time has come to look into the entire activities commercially: consumers must get what they want; coal companies must provide and be willing to provide what consumers want. But someone has to pay for this enhanced product quality and services; and certainly it has to be the coal consumers.

Import of coal may be a short term alternative and time gap arrangement to meet the challenges but prolonged and large import will be a strain on economy.

Coal companies can fulfill this need of the consumers and country with an open fair competition, not only with the imported coal or with private entrepreneurs, but also among the subsidiary companies of Coal India as well.

Sustainable Energy Supply in Asia

SUNSTAINABLE ENERGY SUPPLY IN ASIA

(Proceedings of the International Conference, ASIA ENERGY VISION 2020, organised by The Indian Member Committee – World Energy Council, under The Institution of Engineers (India), during November 15-17, 1996 at New Delhi)

VOLUME TWO

Edited by
PRADEEP CHATURVEDI

World Energy Council
CONSEIL MONDIAL DE L'ENERGIE

For
Indian Member Committee-World Energy Council
The Institution of Engineers (India)

Published by
CONCEPT PUBLISHING COMPANY, NEW DELHI 110059

First-Published 1997

ISBN 81–7022 – 632–5 (Vol. II)
ISBN 81–7022 – 633–3 (Set)

Printed and Published by
Ashok Kumar Mittal
Concept Publishing Company
A/15-16, Commercial Block, Mohan Garden,
New Delhi - 110059 (INDIA)

Phone : 5554042, 5504042
Fax : 091-11-5598898

FOREWORD

The Indian Member Committee of the World Energy Council, IMC-WEC organised an International Conference on Energy - 'Asia Energy Vision 2020' with the theme Sustainable Energy Supply in Asia in New Delhi during November 15-17, 1996. Experts from the Asia region as well as other parts of the world took part in the conference.

This conference provided an opportunity for interaction amongst policy framers, industrialists and researchers to meet on a platform and discuss and synthesise on issues crucial for energy sector development in Asia.

The Papers were of a very high quality. All participants thought that the same should be published and made available to wider readership. The Institution of Engineers (India) is fulfilling its obligation by publishing the proceedings in two volumes.

I would like to compliment M/s Concept Publishing Company for publication of the Two Volumes.

I am sure the proceedings will be found valuable.

N.C. VAISH

Chairman
Indian Member Committee of the World Energy Council
and
President
The Institution of Engineers (India)

November 23, 1996

INTRODUCTION

The Indian Member Committee of the World Energy Council organised 'Asia Energy Vision 2020', an International Conferene on Energy, with the focus on Sustainable Energy Supply in the South Asia Region. This was a major initiative preparatory to the 17th Congress of the World Energy Council planned for Houston, USA, in September, 1998. The significance of the Conference can be realised from the fact that it was inaugurated by the Hon'ble Prime Minister of India, Mr. H.D. Deve Gowda. Its National Advisory Committee was chaired by the Secretary to the Government of India, Ministry of Power, and had representatives from all ministries of the government, and major energy industries from the private and public sectors had supported the Conference. The Ministry of Power played the key role as the sponsoring agency.

The interest aroused in the delegates and the participants from India and other parts of the world has helped in getting papers on policy, international funding, regional cooperation and technical matters. The experts hailed this Conference as having technical inputs of a very high order.

The Technical Committee conducted the technical sessions in the following four divisions:

Division 1: Sustainable Energy Supply Options
Division 2: Energy Sector Developments in South Asia
Division 3: Energy Conservation and Energy Efficiency
Division 4: Funding Mechanism and Sources

The Conference also had three important Discussion Panels as follows:

Panel I : Role of Nuclear Power in attaining Sustainable Energy Supply in Asia.
Panel II : Policy Options for Meeting Long Term Energy Needs in Asia.
Panel III : Strategies for Energy Sector Financing in Asia.

In keeping with the tradition laid down by the World Energy Council, at the 16th Congress at Tokyo, India Energy Seminar was organised to focus on energy requirements in India and the possibility of its fulfillment through international funding and technology transfer. In addition, Overview Presentations on the global, regional and national levels were also organised.

All these papers have been accordingly segregated in different technical divisions and as per their policy relevance. Some of the policy papers were presented by the distinguished experts at the Conference and not submitted in advance so that they are able to update the information to the latest stage. Whereas the First Volume consisting of all the papers received in advance was published earlier and circulated to the participants at the Conference to support the discussions, papers received at the time of the Conference have been published as Volume II of the Proceedings. Volume II of the Proceedings also consists the New Delhi Declaration consisting of Conclusions and Recommendations. These two volumes are complimentary to one another, giving deep insight into the present status, internationally, regionally and nationally.

Wishing everyone a useful information input,

PRADEEP CHATURVEDI
Chairman
Technical Committee
Asia Energy Vision 2020

New Delhi
November 23, 1996

CONTENTS

INAUGURAL ADDRESS

1

SHRI H.D. DEVE GOWDA
Hon'ble Prime Minister of India

Minister of State for Power, Shri Venugopalachari, Secretary General of the World Energy Council Mr I.D. Lindsay, Chairman of the World Energy Council Mr John Baker, distinguished guests, ladies and gentlemen,

1. It is my great pleasure to be with you today.

2. Humanity is passing through a crucial phase where innovative actions have to evolve to deal with the challenging tasks of fulfilling the basic minimum needs of the poor on the one hand and preserving the global environmental balance on the other. Energy is essential to economic and social development, and to improving the equality of life. Therefore, it has appropriately been on the top of the international agenda. Adequate and guaranteed supply of energy has always been a matter of concern, despite the apparent abundance of oil in the market since the late 1980s. The recent stable energy prices should not make us complacent about the long–term availability of commercial sources of energy. This could well be a temporary, during which new strategies and aproaches for future sustained energy supply should be developed. If this is not done now, I am afraid, our future generations will be put to even greater difficulty.

3. The centre of concern has shifted from the fear of resource depletion to that of the impact of energy production and use on the environment, as seen in The Agenda 21 adopted at the United Nations Conference on Environment and Development in June 1992. Though there are differences in perception between the developing and the industrialised countries, it is evident that the energy needs of the developing world will grow at a faster pace than in the past and there is need to find low cost solutions for making energy consumption patterns more environment friendly.

4. It is important for the members of the World Energy Council to realise that unless a match is found between the growing needs of the society, mainly in the developing Asian countries, and the policies of the international funding agencies, multinational corporations and governments, the global energy situation can assume a serious dimension. It is necessary that all concerned should start thinking in terms of the essential needs of user agencies and try to fulfil them at a low or reasonable cost. For sustainable energy supply the strategy which is chosen has to be in keeping with the capabilities of individual countries and their socio economic structures.

5. In South Asia, the needs of the energy sector are growing rapidly. Dictated by the international funding mechanism these countries are looking for efficient technologies and funding from

foreign and private sources. Recent efforts of countries of this region have met with a certain amount of success, but not to the desired level.

6. The funding mechanism for energy projects cannot be left entirely to market processes. In the initial phases Government will have to intervene both in terms of funding pattern as well as administering energy prices. With almost 75% people living in rural areas, with low purchasing and payment capacities, governments in these countries cannot rely only on the market mechanism, at least initially. The structural changes reuired in the energy sector will also have to suit the needs of each country.

7. Secondly, energy supply in this region has to be sustained and will not always be accompanied by introduction of most efficient technologies. This will take time, and happen in phases, when more efficient and high cost conversion systems and processes are introduced. There is an apprehension that emissions due to energy consumption in the developing world will increase over the next 20 to 25 years. These projections are based on present technologies. However, there have been significant advances in the South Asian region in the agricultural and industrial sectors which will have positive impact in near future. The energy scenario in terms of the environmental impact for this region has to be developed independently of the global energy scene.

8. There are certain issues that will have a bearing on policies for attaining a long–term sustainable energy supply. Very often it is said that political will is required. It is not only the political will or the Government's action that is important, but it is the collaborative effort of the Government, private sector and user agencies that will ensure sustainable energy supply in the future. This has to be done on appropriate management solutions, rather than technical solutions alone.

9. The next important issue is availability of the energy resource and its acceptability to the user country. Many countries are fortunate to have their own indigenous resources, whereas others are dependent on supplyfrom overseas. Accessibility to energy supply has to be ensured within the capabilities of importing countries, which is not easy. In many developed countries, the real cost of energy has come down by almost 30 per cent in the last 30 years. Developing countries have not really benefited, and have in fact been subjected to higher financial burden in the local currency. Developing countries have, therefore, to focus on identifying long-term energy sources.

10. In India, coal is the primary source of energy and is expected to be so in the near future. Renewable sources of energy, mainly biomass, and the recent advancements in wind, mini and micro hydel and solar energy have considerable potential, which needs to be expolited through consistent R&D effort. This is perhaps true for other South Asian countries as well.

11. To encourage greater investments by the private sector, enable State Utilities to become financially viable entities and to meet the power needs efficiently, a number of initiatives have been taken in India. The legislation governing the electricity sector was amended in October, 1991 to provide for a more liberal administrative, legal and financial environment for private licensee companies than was so far available. As State Utilities are at the cutting edge of the policy, we have taken initiatives to facilitate their reforms and restructuring. The Government is increasingly simplifying clearance procedures. The limit of capital expenditure of schemes requiring concurrence of the Central Electricity Authority has been raised from Rs 400 crores to Rs 1000 crores. The Central Government is closely cooperating with the States. Our objective is to devolve greater powers to them so that decisions and projects are not delayed. State Governments are being encouraged to set up independent regulatory authorities.

12. Given the similarity of problems and circumstances in South Asia, regional cooperation in the energy sector can be of immense benefit to all countries. Such cooperation by involving governments, industry, and research institutions can lead to cost effective solutions for ensuring sustainable energy supply. The World Energy Council should take upon itself the responsibility of evolving a mechanism whereby regional cooperation is encouraged and region specific solutions are evolved. The Indian Member Committee of the World Energy Council under the Institution of Engineers (India) could take the initiative.

13. I am glad that this conference is being hosted in India where a number of initiatives in the energy sector have been launched. I can assure you that we are seriously concerned about energy and environment issues in our region and will be willing to learn from the experience of others, and share our own experiences.

14. I wish the conference all success. The Government of India will carefully examine the Recommendations and the Plan of Action that emerges from your deliberations.

Jai Hind !

PRESIDENTIAL ADDRESS

2

DR. S. VENUGOPALACHARI
Hon'ble Minister of State for Power
and Non-conventional Energy Sources

It is a privilege for me to address the Opening Ceremony of the Asia Energy Vision 2020 having such a distinguished audience. The presence of the Chairman of the World Energy Council Executive Assembly and the Secretary–General of the World Energy Council adds to the significance of the seminar. Realizing that this is an initiative preparatory to the 17th Congress of the World Energy Council, scheduled to be held at Houston, in September 1998, the significance of this conference takes the top priority on the energy agenda. This conference is quite appropriately held in India after the 16th Congress has identified the two major challenges to be responded, with urgency and determination to the plight of over two billion people mainly in the lower income developing countries without adequate access to electricity and commercial energy and the need for achieving a path of sustainable development. I would like to complement the World Energy Council and the Indian Member Committee of the WEC for taking this initiative.

Energy is at the very heart of the socio-economic development. The globe has two different scenarios, one of high intensity use, where even substantial increase do not make any perceptible change in the quality of life and the other of the poor countries, where even incremental increases in energy availability make a tremendous impact on the quality of life. It is, therefore, necessary for countries of the South Asian Region to focus on wide ranging issues including small energy devices to large energy infrastructure. The energy situation, globally seems to be stable at the moment. But this situation should not lead to any complacency amongst the global actors in the energy scene. Such a situation has, however, not resulted in any advantage to the developing world, where primary energy is supplied in small quantities per capita; and the cost of acquiring the same in relation to the income per capita is substantial. The energy availability in the Asia region has to meet the minimum needs on one hand and also plan for the growing needs of the expanding economies on the other.

A number of innovative examples regarding energy management in the Asian region are available, and the experience can be shared by various countries in this region. The World Energy Council with its very nature of structure, where governments, industries and the non-government institutions have been networking on the same platform, is a mechanism that can be used much more effectively. Realizing this, the Ministry of Power of the Government of India, sponsored this event

so that the recent advancements in improving the quality and quantity of energy is discussed with the wider cross section of energy experts from South Asia and from other parts of the world. The Ministry of Power has opened the whole sector and has been responding to the demands for deregulation and sincerely look forward to active partnership from the private sector to meet the targets in the energy sector.

May I on behalf on the Ministry of Power extend you a warm welcome to India and wish you all success.

FUTURE ENERGY MARKETS

3

JOHN BAKER
Chairman, National Power PLC
and Chairman, WEC Executive Assembly

We are living through a revolution in the supply of energy. Revolutionary times are exciting : they provide huge opportunities for new ideas to flourish and new ways of doing things to emerge. They release a rush of human energy. At the same time, the outcome is often different from what the initiators expected – remember Bonaparte and remember Stalin. And the winners and losers cannot always be predicted with accuracy. So what Id like to do today is to give a brief overview of our state of revolution, and try to derive a few lessons.

But let me first sketch the background. From the perspective of the industrialised or western economies, energy is abundant, prices are relatively low, and there seem to be ample reserves of fossil fuels to cater for at least the next 50 years subject only to modest technical development and the mobilisation of plentifully available capital. So in the West we are reasonably relaxed. The oil market is working well; gas supplies are increasing year by year with significant environmental as well as economic benefits; and the world traded coal market is flexible enough to meet customers' needs and growing demand without any great upheaval. We have discovered this decade that market principles can be extended to cover the gas and electricity sectors as well.

But there is another perspective that covers the economies in transition and the developing countries. This is a world where 2bn people have no access to commercial energy and where there is huge latent demand, a world where potentially fast economic growth is held back by lack of power supplies, a world where concepts of plenty and abundance are not known. Given that world population and world energy demand will both broadly double over the next 50 years, the stresses to which this picture could give rise are obviously concerning. There are geopolitical issues. There are environmental concerns. There are moral issues. It is not my intention today to dwell on these major and serious issues – since I take it for granted that we must seek global solutions to issues of sustainability– the reconciliation of economic growth with environmental protection and regard for the needs of our children and our childrens' children, recognising that it takes decades to change our energy systems and to build new energy infrastructures. I want rather to concentrate on how new energy markets can facilitate the changes that must be made. These are global themes, and it is, therefore, appropriate that the revolution about which I'm talking today is that of the continuing globalisation of energy markets. First, energy markets today are being served by global companies not just in oil, but in coal, electricity and gas. Second, governments are increasingly looking outside their own borders for resources to develop their national energy systems. And, third, technology and investment funds are flowing to those energy markets that are best placed to receive them.

So let me review what is happening in the different sectors of the world today as energy markets go through the various changes of liberalisation, restructuring, deregulation and even privatisation.

Developing Countries

Privatisation and deregulation in the developing countries seems to be driven off their population growth and industrialisation, which can lead to rapid economic growth of 7 or 8% per annum or more. This, in turn, is producing massive infrastructure requirements in energy, transport and public services, both to meet latent demand and also new incremental demand, especially for industry. Crucial to that infrastructure provision is the mobilisation of capital, recognising that the old multilateral lending agencies can no longer come remotely near the provision of adequate finance to meet all the needs. In particular, many of these countries traditionally have a high savings ratio, but underdeveloped financial markets and savings instruments to unlock those indigenous savings for investment in the local economy. Hence a first requirement is to develop the markets and the savings instruments that will attract local savings into them.

Secondly, Developing Countries are in many cases in need of inward investment and with it, technology transfer. To secure both the mobilization of local capital and inward investment, the key is economic liberalisation; investors, whether they are local or inward investors, need a financial system which is transparent, and subject to appropriate financial regulation, including guarantees about the free flow of capital and the repatriation of dividends. Equally, economic liberalisation normally demands some form of pluralism in order to ensure that economic liberalisation is not going to be turned off at some stage by political dictate.

Against this background, we can consider the range of activity now apparent in some leading developing economies, distinguishing broadly three types of approach; encouragement of investment in independent Power Projects (IPPs); restructuring including corporatisation of existing state-owned energy organisations; and privatisation of currently state-owned organisations.

In these economies, it is unusual to see competition introduced into the electricity or gas supply markets. Whilst it is possible to take advantage of the benefits of competition in terms of bidding for the right to build an IPP and to operate it, or for the opportunities to buy into existing assets, without the maturity of markets that exist in the developed economies such as Nortn America and Western Europe, it is unrealistic to look for the establishment of new commodity markets in gas or electricity. In particular, there would be major problems in looking for,competitive retail or even wholesale markets in many of the Asian countries in that the customer frequently pays less than a full market price for the energy he consumes. For example, large amounts of the electricity consumed may be stolen, or those who do pay, may be paying artificial subsidised prices bearing scant relation to costs of production or distribution.

Market systems cannot be run satisfactorily on artificial prices that lack a sound pricing base, or where the customers – whether retail or intermediaries in the form of distributors – are uncreditworthy.

In South America the picture is a little different. Chile, along with the UK, was one of the first countries to get seriously involved in restructuring existing assets in an effort to rid the government of responsibility for the management and financial burdens of owning and operating infrastructure enterprises. In South America there has been less activity than in Asia in terms of encouraging IPPs, but a certain amount of restructuring has been going on separating generation, transmission and distribution and privatization is proceeding or planned as part of programmes to reduce the public finance burden in several countries.

Economies in Transition

For the Economies in Transition, the situation again is very different. Here, the new governments have inherited, in the main, highly inefficient plant, serve environmental impacts, and a very uncertain legal framework. Often, the markets for the goods the Central European countries were producing have disappeared. Ninety per cent of the output of the Czech Republic, for example, previously went to the USSR, while today it is less than ten per cent. These countries have needed to turn to Western Europe, and then have faced formidable problems due to the lack of competitiveness of their industries.

A strong feature of these economies was extremes of artificial pricing. At one time, for example, in the city of Moscow, electricity was being sold at around one mil per kilowatt hour, and people regulated heat in their houses by opening or shutting the windows. When energy is provided that cheaply you do not need sophisticated means for controlling its use. Clearly with such a price structure, it is impossible to attract new capital investment, because it can not be remunerated.

Faced with this position, several countries have seen the need for economic reform and there has been an emphasis on privatising existing assets. In Poland, for example, a framework to encourage inward investment in the form of IPPs, has been set up, and they are restructuring their existing institutions, and privatizing in a well-considered and orderly manner. As regards the Russian Federation, there has been little useful activity in the power sector reflecting the severity of the changes required to the legal and commercial infrastructure. In the oil and gas sector, however, there has been significant inward investment reflecting the ability of the Russian Federation to export hydrocarbons for hard currency.

Mature Economies

For the mature economies, the picture is very different again. Here the drivers are the existing financial burdens of infrastructure ownership falling on government accounts, and the fact that there are sophisticated trading markets both in finance and commodities on which to build. There is an increasing general recognition that deregulation is a desirable thing across the whole range of enterprises in order to encourage efficient management and a developing understanding that security of supply can come through diversity rather than through central planning.

Additionally, the costs and rigidities of central planning are being increasingly recognised. Customers are demanding to be enfranchised: they no longer accept the utilities that say, "We are giving you good service, and isn't it lovely". Rather customers are saying "Let us choose, and we will tell whether it's lovely or not"

Restructuring of existing enterprises is, therefore, to the fore in the mature economies. Much of the approach in these countries is based on promoting competition and creating markets were customers are increasingly free to choose. Privatisation, which is seen as a way of locking in efficiency gains and insulating them from political change, is being pursued in a variety of patterns; and enormous ferment of ideas and systems is being tried.

Leaming the Lessons

From this overview, it is clear that different countries are pursuing different means and have different priorities as they work their way to new energy markets. That is wholly good: in my experience there are few models that translate readily from one country to another in any aspect of life. But there are, I believe, some common themes which hold true for the new emerging markets and which are, therefore, fundamental to meeting local, regional and global needs for more and more energy :

(i) Market pricing. There is a need for convergence of energy prices to market prices, for without market prices as the basis of what the customer pays, energy will not be produced or consumed efficiently and new capital will not flow into investment in necessary energy infrastructure projects, be they new power plants or the exploitation of new coal oil and gas reserves. Given the gross departure from market prices in many developing countries and economies in transition I have already described, and the widespread use of subsidies, it will obviously take many years for convergence on market prices to be achieved if unacceptable social and economic disruption is to be avoided. But there is no useful alternative.

(ii) Investments must be bankable. That is to say that in order to attract funds at the scale required, particularly where there are overseas investors involved, there must be the rule of law, laws of property and contract that are enforceable and free of political caprice, and financial systems that deal properly with risk, be that in the form of fully open and transparent financial markets where capital and revenue can cross frontiers unhindered, or the provision of appropriate government or quasi–government guarantees where that is not the case.

(iii) Regulation. Whether regulation of the new energy markets is primarily a matter for competition as in the case of the mature economies or for government or government appointed regulatory bodies where mature markets don't yet exist, what is required in all circumstances is a coherent and consistent system that is stable and transparent (so as to provide confidence to all the parties) and well-administered. Obviously these qualities need to apply to environmental as well as economic regulation.

Provided these three principles hold, the exact format of the new energy markets is very much a secondary issue. Fully competitive markets open to all comers may suit mature western markets. Privatisation and restructuring may well be most relevant to economies in transition. And for developing countries, the priority may be the economic liberalisation and establishment of a body of enforceable contract and property law that attracts inward investment, eg. g. into IPPs, but for whom wholesale and retail competition would be a huge and unproductive distraction.

At this point, I would like to address a few remarks to my Indian hosts to use them as an illustration of how these new markets can look through the other end of the telescope. India is an attractive market to many because of its obvious need for electricity, the demand growth, the well established legal system and administrative system and its sophisticated bureaucratic Infrastructure. In addition. I believe English companies in particular are attracted by the ties that exist and the knowledge and understanding between the two countries. However, what attracts us also frustrates us, in particular the endless and seemingly purposeless bureaucracy which seems to lack direction but pursues detail apparently for its own sake. The outsider observes rival and competitive bureaucracies crawling over large projects which need approval, both at the State level and Government level, with each level, and each department at each level, apparently engaged in an endless game of seeking to prove who has the most sophisticated way of saying "no" or "come back tomorrow"

The process is very technical and very detailed and is, therefore, resource intensive for the promoters. In this level of technical detail really necessary? Other countries have found other ways, such as competitive tariff bidding, so that there is evidence that the best project has been chosen, but on a reasonable timescale and without the endless second-guessing that currently takes place.

If India is to get inward investment, then India must recognise that project promoters will be relying on international project finance. Therefore, a degree of certainty has to come into the project income streams.

I recognise, of course, that if in India – or in any other country – you decide it is more important to play the game than secure a result, that is the nation's privillege. But remember that India is also in another game, the competitive game of persuading project promoters and international finance to come to India rather than go elsewhere in Asia or elsewhere in the world. In the case of National Power, since my company decided to go international in 1992, we have been able to establish businesses in China. Turkey, Pakistan, Australia, Spain and the USA, but here in India, where we started first, four years on we are still entangled in the bureaucratic web. We have taken our opportunities elsewhere but I believe all these delays are a loss to India not only in shortage of power but also in reputation. In short, in the new energy markets, capital and with it the enthusiasm of management will flow where it is made most welcome, not necessarily where it is most needed, and so Governments need also to understand this aspect of globalisation.

So let me conclude. Energy markets are in global revolution. This revolution can bring great benefits – in state of the art technology transfer. In the flow of investment resource, and, through the power of markets, in securing competitively priced energy supplies. This globalisation is a powerful engine, therefore, to the more efficient and hence less environmentally damaging exploitation of energy resources. It is a great engine for economic development. There are many approaches that can be followed, but in this market, countries are as much in competition with each other for re-

sources as companies are for projects. What must unite us is a common view of the need to apply market principles consistently and efficiently, best adapted to the local needs, in order to bring energy to the countries that need it – for that is our common goal, a goal that will help us all alleviate the communal energy deprivation that still affects 2bn out of the 5bn people in the world today.

A GLOBAL ENERGY OVERVIEW

4

IAN LINDSAY,
WEC Secretary–General

The 16th Congress of the World Energy Council took place in Tokyo in October 1995. This Congress analysed, discussed and commented on the international energy scene, its current issues and its probable longer term development, and, I will, therefore, base my views of a global energy overview largely on the Conclusions and Recommendations of that Congress.

The WEC, as most of you know, was founded in 1923 and has over the years become the world's leading non-governmental energy policy forum. It now has Member Committees, including the dynamic and increasingly prominent Indian Member Committee, in over 100 countries covering some 92% of the world's energy consumption, and this gives it three unique characteristics:

1. it has the spread of view and range of opinion to be authoritative in what it does and says;
2. it has the ability to work "bottom up" from the market-place which gives great credibility to its work; and
3. by blending government, academic, industry and consumer views in its reports and action, its work achieves a balance which renders it closer to being definitive than any other non-governmental energy organisation.

In addition, the WEC has, after achieving a mind and a will of its own, taken on the role of "facilitator" to assist in "getting things done". Examples are: the holding of the first ever African energy ministers conference in Tunis last year. This has assisted in setting up the African Energy Commission whose task will be to coordinate African energy development. My second example is the request from the Republic of Namibia to identify an integrated energy policy with which to enter the next millennium.

I will present the 16th Congress Conclusions and Recommendations in three parts:

1. the Context,
2. the Issues, and
3. the Conclusions and Recommendations.

1. The Context

First let us remind ourselves of the basic analysis derived from the WEC Commission report "Energy for Tomorrow's World" with its horizon year of 2020. Global population by that year will rise from today's 5.4 billion people to 8.0 billion. Global economic development is likely to run between now and 2020 at between 1.6% and 2.4% annual growth. All the traditional major variants required to examine future energy perspectives have been built into the Commission's four scenarios. These variants have been built up from regional data.

Three of the scenarios used in the Commission produce increased energy demand from 1990 to 2020 of between 50% and a staggering 95%. The fourth, the environmentally led scenario, is unlikely to be achieved.

What will this huge energy growth mean? It will mean that by 2020 more than 90m barrels of oil a day will be consumed, an increase of some 27m B/d, or the equivalent of the whole of today's OPEC production.

It means that coal output will double to about 7 billion tonnes per annum, more than twice the UK's or Canada's known total reserves.

It means that annual gas demand will more than double, to reach some 4 trillion cubic metres, almost as much as the USA's total current gas reserves.

It means that more electricity generating capacity will be built over the next 25 years than in the previous century.

And by our analysis it means that 90% of this energy growth could take place in the currently defined developing countries particularly here in Asia and in Latin America, providing these markets are able to develop satisfactorily with the support and assistance of their authorities.

It means that the developing countries which today consume 30% of the world's total energy will consume 50% by 2020 and probably 70% by 2100.

It also means that by 2020 those same developing countries will emit CO_2 from fossil fuel burning equivalent to the total CO_2 worldwide emissions from the industrialised countries in 1990.

2. The Issues

Against this contextual background, today's general global scene is one of considerable demand growth, currently met by more than adequate energy supplies at relatively low prices. But this of course masks the fact that two people in every five in the world today have no access to commercial energy supplies at all. I say low prices because oil, for example, sells today at approximately 40% only of its price in real terms of 15 years ago. I do not of course refer to the importation cost to many developing countries as a percentage of their import bills or of its impact on their GDPs. In more mature markets in the industrialised countries of North America and in Western Europe, development has been slow, and disappointing. Disappointing for three reasons:

— little structural change had occurred (with the exception of the move to deregulation in some countries), despite calls for more attention to be paid to energy efficiency and conservation;

— little tangible advance appeared in the scientific research associated with possible global climate change and practically no action has been taken following the 1992 Rio Earth Summit to implement "Precautionary Policies" to mitigate possible results of any global climate change

— little was done to improve the flow of capital or technology from North to South, although the private banking sector has considerable resources available for project financing, if enough viable projects can be found. Technology advances have made great strides, particularly in the oil industry in the North. Could this mean that the processes involved in attracting both capital and technology to the South should be speeded up and enhanced? "Yes" is the answer, and this is a major point which many would do well to consider.

The development of more liberal energy markets in E. Europe and the CIS has inevitably been slow. This has been particularly so in the Russian Federation where oil consumption 1994/1993, for example, declined by some 20%. These markets are taking much longer to adapt and are much slower to attract inward capital investment than was thought likely when their political systems changed so radically six years ago.

The pace of international energy cooperation and coordination has been slow, and given the magnitude of the problems likely to confront the energy and energy-related sectors over the next 30 years, it appears to the WEC that this is an issue which requires considerable attention and development.

Oil with 44 years of reserves, gas with 70, and coal with some 250 years of reserves, at present rates of consumption, give the impression of plenty, but mask the fact that supply lines are lengthening and becoming more vulnerable, both technically and due to political pressures.

Uranium supplies are plentiful, but it is in the domain of the renewable forms of energy, particularly solar and wind, that current energy costs and lack of government support have combined to slow their development and installation – particularly in the developing countries. Total new renewable energy supplies excluding major hydro and traditional biomass currently only represents some 2.0% of current world energy consumption. By our estimates, they are unlikely to rise above 2.5% to 3.0% of the much greater world energy consumption by the year 2020 unless they receive support.

There is today considerable complacency in the industrialised world among both governments and consumers alike about the current availability of energy and the way it is used. This sits ill alongside the fact that:

— 2 billion people of the total world population of 5½ billion remain in the poverty trap and have no access to commercial energy at all, and

— the world's population is expected to double during the next century, which together with estimated global economic development is likely to more than double the world's demand for energy, with all its consequences.

This huge requirement for new energy supplies cannot – in the interests of sustainability – be allowed to occur simply by relying on fossil fuels.

A sustainable route to improved living standards is required so that tomorrow's generations can satisfy their own requirements for economic and environmental wellbeing.

Energy systems, however, can only be changed gradually. Energy projects have notoriously long lead times and so global action for change, particularly for the development and installation of new renewable forms of energy (solar, biomass, wind, etc.), for revitalised and safe nuclear development and for improved environmental performance of fossil fuel use, must start now.

This seems to us to be the core longer term issue associated with today's global energy sector scene. It is, we believe, up to governments and industry alike to recognise the problem (which does not only have huge humanitarian implications, but major long–term global geopolitical ones as well), and to take action to mitigate the problem now.

Overall, however, the 1995 Congress felt that the dominant use of fossil fuels is likely to remain for several decades to come, and was not sanguine about short–term nuclear development. Apart from the current 30 reactors being built, mainly in Asia, nuclear is though unlikely to grow significantly until technology has proved that the whole fuel cycle, including waste disposal, is secure, and until public acceptance returns.

Public sector financing of energy (which at 30% absorbs more finance than any other single element of infrastructure development) is in decline, and is not being matched, particularly in the developing countries, by significant increases of private sector project investment. There is, therefore, an increasingly urgent need, particularly in developing countries, to develop local capital markets from the often high savings rates, and to use this capital through appropriate institutions to finance local infrastructures, particularly energy projects. The WEC finance study on the subject is likely to come up next year with strong recommendations to mobilise local capital or pay the penalty by going without.

The Congress showed considerable interest in the process of deregulation and market liberalisation. Interestingly, it showed equal interest in some of the negative aspects of short–termism and possible limitations related to these processes. Again the many different aspects of energy market liberalisation and their advantages and disadvantages will be thoroughly analysed in another current WEC study due to be completed for the 1998 Congress in Houston.

It was generally agreed that, although some progress towards energy efficiency has been made with the development of gas turbines, etc., in general there was far too little government incentive – in an era of low prices – to achieve what is vitally required in order to institute successful energy efficiency measures. Governments and industry everywhere must realise the benefits, on a cost/benefit basis, of putting into practice real energy efficiency and conservation policies. The current global energy intensity decline of about 1% per annum is too slow and it is as much the developing countries as the industrialised countries, in their own economic interests, which should institute greater and more meaningful reductions.

In conclusion, on the whole range of major issues, the Congress felt that given the outwardly complacent current global energy situation, there is a strong temptation for governments and others to delay or avoid taking major energy decisions. The WEC recommends strongly that such organisations as its Member Committees (many of whose members are present today) should draw the attention of politicians and the leaders of industry to recognise the need to change energy systems for the longer–term and to defy the ubiquitous dilemma of the short-termism of politics versus the very long-termism involved in energy project development.

3. Congress Conclusions and Recommendations

The Congress made a number of fundamental conclusions. Let me list six of them.

(i) The Congress believed that energy subsidies should be phased out in order for suppliers and consumers alike to understand the implications of energy's real costs. It was recognised that in many areas and notably in rural areas of the lesser developed countries this process will take many years to achieve. If achieved too rapidly, consumers are likely to revert to traditional fuels with all this means in the way of additional environmental degradation. The corollary, if no such action is taken, is that governments will gradually reduce their financial ability to supply commercial energy, with the subsequent risks of much slower economic development.
Furthermore it concluded that much more should be done to raise the level of general awareness about energy and the fundamental services it provides. Many publicly held perceptions are inaccurate and there is a general need for education about the realities of energy.

(ii) Extending its reasoning further, the Congress made the interesting point that many of the facts about the energy sector which today are available to end use consumers, are wrong– and they are wrong because the signals about supply availability, costs and prices, the existing degree of environmental damage and the state of current institutions which handle such matters, are themselves inconsistent and confused. Institutional changes in a number of key areas, of which finance is one, were strongly advocated. The world will require some US$ 30 trillion in 1992 money over the next 25 years with which to develop energy projects. This equates to 50% more than the entire 1990 world GDP. If this sum is to become available, many radical changes will be needed to secure such financing, particularly by building up local capital markets.

(iii) Greater regional and transfrontier cooperation and interdependence were deemed necessary. Nowhere was this seen more clearly than in Africa where national electricity generating capacities continue to be built up despite neighbouring surpluses. Only now are the benefits of interconnection and power pooling arrangements beginning to be recognised, and this in part due to the efforts of the WEC.

(iv) Energy Research and Development budget declines in America, Europe and Japan cause widespread concern, and the Congress called for the reversal of this trend. It placed particular emphasis on its call to governments and industry to increase and refocus their R&D efforts. This was directed principally to energy efficiency measures and to the development of new renewable sources of energy.

(v) From this point was derived the fact that many energy corporations will have to reconsider their own structures and organisations, in order to become more service and customer orientated, more competitive both commercially and environmentally, and more courageous in developing projects in transitional and developing economies. But to achieve the latter will require considerably more government comprehension and change than is the case in many countries today.

(vi) There is an overwhelming need for new government, industry and consumer partnerships based on realism and real costs. As an example I quote the involvement of the WEC as a facilitator in helping to draft and implement the African Energy Commission. International agreements are required in which to incorporate such partnerships.

Above all, action is required now to avoid delays and inaction, in order to re-orientate lour energy systems for the requirements of the next century.

SUSTAINABLE ENERGY OPTIONS

5

MICHAEL JEFFERSON
Deputy Secretary General
World Energy Council

Introduction

The sustainability of energy supply may be considered from various vantage points. Fear that present abundance may be at risk because sources of supply dry up, or are subject to disruption from political or economic reasons. Concern that problems associated with supply and use – such as local, regional or essentially global pollutants – may become so severe that alternative supply options without these effects must be exercised. Realisation that nonavailability of supply means absence of energy services, which means in turn the inability of vast numbers of people to satisfy their basic needs or the preliminary requirements for economic and social development.

In what follows all three vantage points will come into play. But reflecting both fundamental human priorities and the issues as seen from South Asia, the three vantage points are taken in reverse order from that of the preceding paragraph.

I will be drawing upon earlier World Energy Council work. The WEC Commission report: "Energy for Tomorrow's World", 1993; the WEC study: "New Renewable Energy Resources: A Guide to the Future", 1994; the joint WEC/IIASA study report: "Global Energy Perspectives to 2050 and Beyond", 1995 and our ongoing work for this study; and our publications on various environmental issues, including potential climate change (upon which subject we have now issued six reports).

In particular, however, it is upon "Global Energy Perspectives to 2050 and Beyond" that I will draw.

Preliminary Assumptions

The first key assumption is that global demand for commercial energy services will rise. This will be partly due to success in reducing, over the coming decades, the number of people who cannot now access commercial energy. At present this group comprises between one-third and 40% of the world's population of 5.8 billion people – about 2 billion people therefore. No-one knows precisely how many are thus deprived.

The World Bank, in its recent report: "Rural Energy and Development", 1996 has referred to the "nearly two billion people in developing countries (whose) search for energy is a daily grind". They have neither electricity nor gas nor oil to cook their food. The World Bank notes:

"Women and children from these families often spend hours each day gathering dung and wood – hours they might otherwise be able to spend on producing work or education. Their health is damaged by the smoke given off by these fuels. Their environment may be damaged as they chop down trees for wood".

The World Bank also notes that in South Asia traditional biomass accounts for about 55% of total energy used, not much more than 25% of rural populations are connected to electricity and not much more than 50% of urban populations.

The other main reason for anticipating energy demand to increase is the expectation that world population will rise, if at a slower rate than over recent decades. Most medium projections anticipate that by 2100 world population will have nearly doubled from its current level–and that most of the increase will occur in the present developing countries.

Urbanised populations are currently increasing at a faster pace than the population at large. This is creating its own special problems, as people move from the country to the town or city. One area of uncertainty, and potential opportunity, is whether decentralised, relatively small and low cost, off-grid and locally appropriate new renewable energy schemes may slow down or avoid some of this population drift. This is one element to be included in consideration of sustainable energy supply options.

Unless commercial energy provision expands, the basic needs of growing numbers of people will remain unsatisfied. This is increasingly likely to have, first, local social and political repercussions; and then wider geopolitical consequences. The more basic needs are satisfied, through increased commercial energy availability and use, the more evident the shift from "North" to "South" in the world's energy use and geopolitical balance. This has been clearly recognised in the WEC's work, and in the Conclusions and Recommendations of its 16th Congress in Tokyo, in 1995:

"For the 40% of the world deprived of commercial energy services and supporting technology at the present time, effective measures are urgently required to bring commercial energy to them in order that they may break out of the vicious circle of poverty and deprivation"

Yet there are many obvious constraints: institutional rigidities, inappropriate policies (such as widespread and large subsidies which encourage waste and inefficiency, and discourage commercial investment), ignorance and inertia, and the present pattern of supply.

One simple point that seems too often overlooked outside the energy industry is that 76% of the world's primary energy is still supplied by the fossil fuels. Oil supplies 33%; coal 23%; and natural gas 20%. By contrast new renewable forms of energy – solar, wind, modern biomass, small hydro, geothermal – account for only 2%; and large hydropower schemes for a further 4%, of the world's primary energy. Nuclear power – from rather large and relatively expensive plants which have attracted a range of public concerns in many societies – accounts for 6%. To realise major changes in the world's regional and national energy supply options will take time. It has taken nuclear power, with substantial governmental support in many countries where it exists, 50 years to reach its 6% share.

Figure 1 provides our assumptions about world population and global primary energy use for three families of scenarios: A – High Growth; B - Middle; and C - Ecologically Driven.

Table 1 illustrates how, under these three families of scenarios, the balance of world primary energy consumption is expected to Shift from the OECD countries to the present developing countries by 2020, 2050 and 2100, respectively (Other represents centrally planned economies currently in transition).

Some Preliminary Conclusions

Some conclusions may be drawn at this stage:

— reliance on fossil fuels is likely to remain heavy for many decades (and the temptation for such reliance is likely to be particularly high in countries with large domestic reserves of one or more of these fuels);
— geological resources of fossil fuels are plentiful for the next century – and provided technological, financing and environmental challenges can be met in practice, adequately accessible;
— there is a real risk of growing import dependency and lengthening supply lines for many countries using fossil fuels. For these and other reasons there are uncertainties over the future of fossil fuel prices in the longer term. In the shorter term they are likely to pose stiff competition to alternative fuels if prices are market driven;
— local, regional and global environmental concerns and mitigation measures are likely to lead to widening efforts to incorporate environmental impacts in costs and prices;
— desires to broaden the supply base, build up indigenous capacities, reduce import dependency, and lower atmospheric emissions from fossil fuel use are likely to encourage the expansion of non-fossil fuel supply and use.

The options before us are, however, wide. Table 2 indicates what "Global Energy Perspectives to 2050 and Beyond" suggests they might be for the world in 2050. But if Scenario A2, for example, is to be realised then affordable measures will have to be in place long before 2050 to cope with sulphur, NO_x and probably several other emissions – including the economic recapture and storage of CO_2. Scenario A3 implies either great success with the development of the various forms of solar power, or capacity to cope with the environmental (e.g. loss of biodiversity) or economic (e.g. land availability) consequences of large-scale modern biomass development. And so on.

Table 3 indicates for the present developing countries how these supply options may have developed by 2050. It will be noticed how strongly the present developing countries may be expected to participate in new renewable energy developments. The benefits of local appropriateness, decentralised systems for rural and otherwise dispersed populations, smaller scale and lower costs will be evident. But this implies further technological advances, declining unit costs, and the availability of finance at the right time and place. As the WEC's 16th Congress in 1995 concluded:

> "Advances have occurred in improving the technology and reducing the costs of some renewable forms of energy – modern biomass, solar photovoltaics, and windpower particularly – but the pace is too slow".

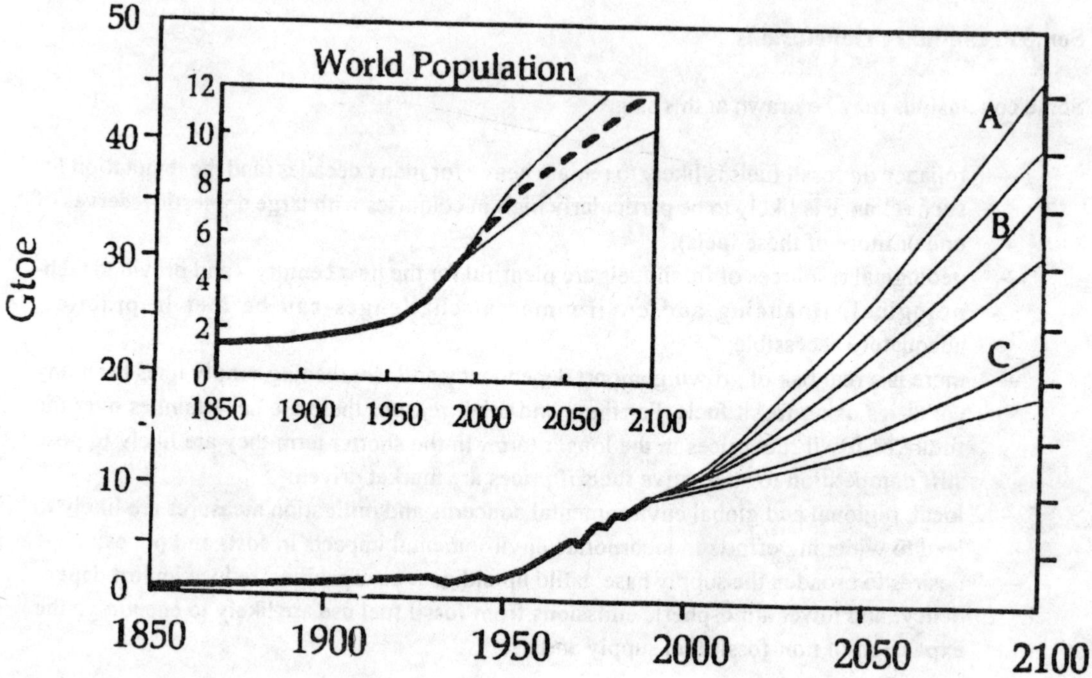

FIGURE 1 : *Global Primary Energy Use, 1850 to Present, and the Three Cases to 2100, in Gtoe*

It will also not have gone unnoticed, least of all in India, that inward investment – whether for modern power stations or large-scale solar photovoltaic projects – has not exactly been encouraged by political and bureaucratic infighting. There is a balance to be struck between successful project financing and implementation to support much-needed infrastructural investment on the one hand, and local criteria of a less strictly economic kind on the other Table 4 provides an indication of the relative contribution that the main forms of new renewable energy might make by 2050. Beyond 2050 it is believed that the various forms of solar energy will come to make the largest contribution followed by modern biomass, in Ecologically Driven scenarios. In the current phase of our work on global energy perspectives, we are hoping to refine our views on regional prospects around the world – on this and other matters.

Some Further Considerations

These are just some of the wider considerations which come into play when considering sustainable energy supply options.

TABLE 1 : **Area Shares of World Primary Energy Consumption (%) by Scenario**

	1990	*2020*			*2050*			*2100*		
PRESENT		A	B	C	A	B	C	A	B	C
OECD	47	37	38	32	27	28	21	18	16	11
DCs	34	48	49	53	58	60	67	71	74	82
Other	19	15	13	15	15	12	12	11	10	7

Source: WEC/IIASA "Global Energy Perspectives to 2050 and Beyond", 1995

TABLE 2 : World Energy Supply Mix in 2050 (Gtoe)

	A1	A2	A3	B	C1	C2
Coal	3.8	7.8	2.2	4.1	1.5	1.5
Oil	7.9	4.8	4.3	4.0	2.7	2.6
Gas	4.7	5.4	7.9	4.5	3.9	3.3
Nuclear	2.9	1.1	2.8	2.7	0.5	1.8
Renewables	4.7	4.9	6.8	3.7	4.8	4.2
Non-Comm.	0.8	0.8	0.5	0.7	0.8	0.8
Total	24.8	24.5	24.5	19.7	14.2	14.2

Note : In 1990: Coal 2.1; Oil 3.3; Gas 1.7; Nuclear 0.45; Renewables 0.5; Non-Commercial 1.1. Total 9.2

TABLE 3 : **Present Developing Countries Supply Mix in 2050 (Gtoe)**

	A1	A2	A3	B	C1	C2
Coal	2.1	4.8	2.0	2.8	1.4	1.4
Oil	4.7	2.7	2.6	2.5	1.8	1.8
Gas	2.6	2.4	3.5	2.1	1.9	1.9
Nuclear	0.9	0.5	1.0	1.0	0.1	0.7
Renewables	3.4	3.3	4.7	2.8	3.7	3.1
Non-Comm.	0.7	0.7	0.4	0.6	0.7	0.7
Total	14.4	14.4	14.2	11.8	9.6	9.6

Note : In 1990: Coal 0.8; Oil 0.97; Gas 0.27; Nuclear 0.03; Renewables 0.17; Non-Commercial 0.92. Total 3.16

TABLE 4 : **Renewable Energy Availability in 2050 (Gtoe)***

	A1	A2	A3	B	C1	C2
Modern Biomass	1.1	2.2	3.2	1.4	1.6	1.5
Solar	1.8	0.4	1.6	0.4	1.3	1.4
Hydro	1.0	1.1	1.1	0.9	1.0	1.0
Wind	0.7	0.6	0.8	0.6	0.6	0.5
Other	0.1	0.6	0.1	0.4	0.3	0.8
Total	4.7	4.9	6.8	3.7	4.8	5.2

* Excludes non-commercial energy (e.g. fuel wood, crop waste and animal dung).

For instance, simply (or, in reality, not so simply) expanding new renewable energy supplies without having paid prior and sufficient attention to raising energy efficiency is to risk putting the cart before the horse. It certainly means that expanding availability will be chasing energy demand growing faster than is necessary to supply the energy services required.

The WEC/IIASA study on "Global Energy Perspectives to 2050 and Beyond", 1995 as the WEC Commission report: "Energy for Tomorrow's World", 1993 before it did, looked rather closely at energy efficiency. Its past history and future potential were studied. At the broadest aggregate level, the energy intensity of Gross Domestic Product over time, there is evidence from many countries of long-term efficiency improvements resulting from the steady turnover of the energy-using capital stock. Considering total energy use (i.e. non-commercial and commercial energy combined) energy intensities of the USA and U.K. economies have been in decline since the middle part of the nineteenth-century. For a number of mainland West European economies this is also true since about 1920. For Japan it is the case since 1950 and also, despite a counter-movement in the 1960s, it is true of India. However, if only commercial energy is considered then in India, as in many developing countries, commercial energy intensity is increasing.

I need hardly point out to the readers that some of this increase in India's commercial energy intensity is self-inflicted, fundamentally unproductive, and partly avoidable. Low conversion efficiencies, high transmission and distribution losses, and poor maintenance procedures and performance (which negatively affect the renewable energy sector also) are a blight on performance. Experience from the industrialised countries in the period 1975-85 also suggests that if the aim is to accelerate the pace of efficiency improvement in energy conversion and end-use equipment then higher prices do help.

The recently published World Bank report on "Rural Energy and Development" has reminded us that in India the electricity tariff in rural areas bears no relation to costs. Whereas total costs are given in 1993 as U.S. ¢ 16.4 per KwH (and fuel costs alone U.S.¢ 2.1), the average agricultural tariff was U.S. c 0.5. This is not conducive to efficient use or widening availability. Thus recent efforts to raise tariffs need to be sustained despite the obvious obstacles.

As an aside, I feel moved to remark on how extraordinary it is that while we know energy subsidies encourage waste, inefficiency and low investment on economic criteria we do not know the total of energy subsidies around the world. Nothing very serious has been done at the global level since 1993. The available data, covering 28 non-OECD countries, date back to 1990/91 and probably understate the scale of subsidies even for this sample. While these partial data suggest subsidies totalled some U.S.$330 billion in 1991, annual subsidies globally may well exceed U.S.$ 600 billion thus exceeding most estimates of annual global energy supply-related investment!

Looking to the future, the WEC/IIASA study tended to mark down its energy efficiency improvement assumptions by comparison with the earlier WEC Commission report.

Three Scenarios of Energy Efficiency Improvement (primary energy per GDP mer, %/ year)[1]

	A	B	C	ETW[2]
	1990-2050	1990-2050	1990-2050	1990-2020
OECD	–1.2	–1.1	–2.0	–1.9/–2.8
REFs[3]	–2.1	–1.7	–2.2	–1.2/–2.7
DCs[4]	–1.6	–1.2	–1.9	–0.8/–2.1
World	–1.0	–0.8	–1.4	–1.3/–2.4

Note :

1. GDP mer: Gross Domestic Product at market exchange rates.
2. ETW : "Energy for Tomorrow's World", 1993.
3. REFs : Economies in transition.
4. DCs : Developing countries (at present).

It will be noted that energy efficiency is assumed to improve relatively rapidly in Ecologically Driven Scenarios (C), reflecting deliberators efforts to achieve this by policies, measures, and technology innovation and diffusion. Faster economic growth may also provide some of the conditions supportive of efficiency improvement (in the A Scenarios). A more muddled, internally inconsistent, indecisive or unfocused evolution (B) is liable to have the weakest efficiency outturn.

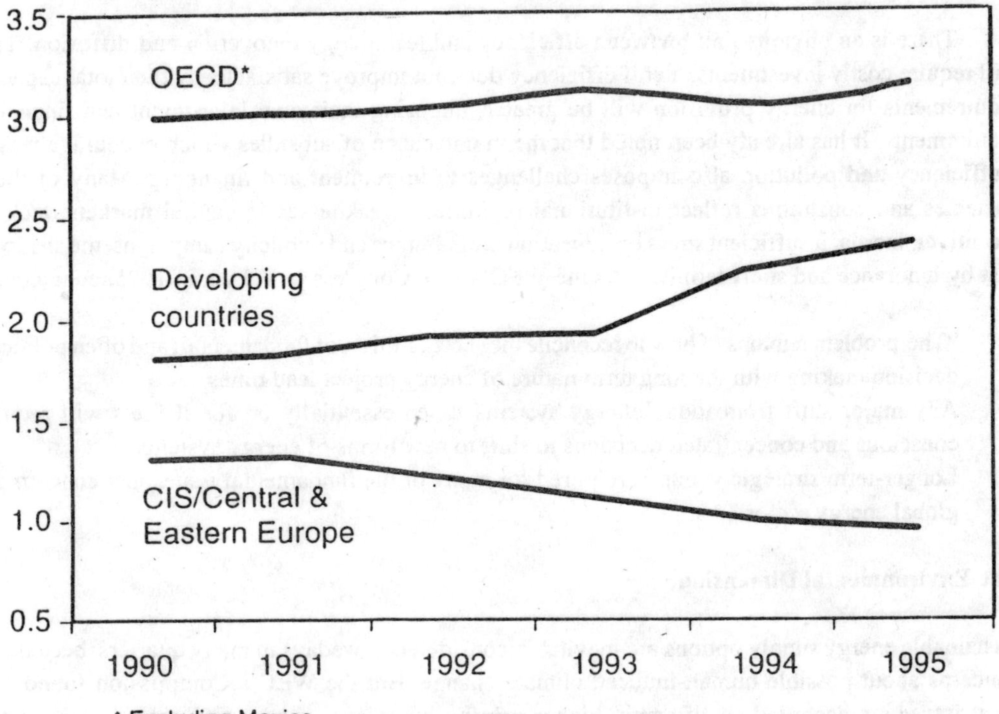

FIGURE 2 : *CO_2 Emissions from Fossil Fuel Combustion (Gigatonnes of Carbon)*

TABLE 5 : **World CO_2 Emissions from Fossil Fuel Combustion (GtC)**

Year	Scenario A1	A2	A3	B	C1	C2
1990	6.0	6.0	6.0	6.0	6.0	6.0
2020	10.2	10.8	8.9	9.0	6.8	6.8
2050	13.1	16.4	10.3	11.2	6.1	5.7
2100	17.5	24.3	8.2	15.8	2.8	2.8

Source: WEC/IIASA "Global Energy Perspectives to 2050 and Beyond", 1995.

TABLE 6 : **Present Industrialised/Developing Countries' Shares of Global CO_2 Emissions from Fossil Fuel Combustion (%)**

Year	Scenario A1 IND/DC	A2 IND/DC	A3 IND/DC	B IND/DC	C1 IND/DC	C2 IND/DC
1990				46.5/30.5		
2020	36/48	37/48	36/49	38/49	31/53	30/54
2050	25/58	27/58	21/61	24/63	15/69	13/70
2100	18/68	21/67	22/63	9/84	3/86	1/81

Source : WEC/IIASA "Global Energy Perspectives to 2050 and Beyond", 1995.

There is an obvious link bwtween efficiency and technology innovation and diffusion. This will require costly investments. Yet if efficiency does not improve satisfactorily then total capacity requirements for energy provision will be greater, imposing their own investment and financing requirements. It has already been noted that the maintenance of subsidies which encourage waste, inefficiency and pollution also imposes challenges to investment and financing. Many of these obstacles and constraints reflect institutional regidities– weaknesses in capital markets, lack of incentives, inertia, insufficient stress on education programmes and publicity campaigns, the shadows cast by ignorance and short-termism. As the WEC's 16th Congress in Tokyo, in 1995, concluded:

> "The problem remains of how to reconcile the short-termism of fundamental (and often political) decision making with the long term nature of energy project lead times.
> Any major shift from today'energy systems based essentially on fossil fuels will require conscious and concentrated decisions to shift to new forms of energy systems.
> Longer-term strategic vision is required for many of the fundamental issues that concern the global energy sector".

The Environmental Dimension

Sustainable energy supply options are inevitably considered nowadays in many quarters because of concerns about possible human-induced climate change. But the WEC's Commission found that local issues are accorded significantly higher priority by most countries and people. Among the important local issues is local (and in the case of sulphur emissions and acid deposition, regional) pollution. There is also growing emphasis on the need to prevent pollution wherever feasible, rather than merely to mitigate the effects of pollution or to control pollution.

In most industrialised countries sulphur emissions have been in sharp decline over the past two decades. Trans-boundary pollution protocols are effectively in place. Electricity generating companies have widely reduced sulphur emissions more rapidly than legislative requirements. The costs of sulphur abatement are coming down. Much has been done; more is well in the pipeline.

NO_x emissions are also beginning to be more effectively addressed by the power generation sector (if not by the road and air transportations sectors), and with new gas reburn technology NO_x emissions from coal-fired electricity generation are expected to fall by at least 50% where installed — at increasingly affordable costs.

Upgraded replacement, and new, investment in the economies in transition can also be expected to produce beneficial results for mitigating local and regional pollution.

It is in the coal-dependent developing countries, not least China and India, where the concern for the future is potentially greatest. The WEC/IIA SA study concluded that, for Scenario A2 (with its high coal utilisation), in the absence of widespread sulphur abatement measures SO_2 emissions in South and East Asia would triple by 2020. Ambient air quality would deteriorate, with sulphur deposition reaching twice the worst levels ever observed in the most polluted areas of Central and Eastern Europe. Acid deposition on economically important food crops might exceed what those crops could sustain by factors up to 10 ("Global Energy Perspectives to 2050 and Beyond", p. 84 and Figs. 2, and Tables 5 to 6).

The WEC/IIASA study in fact assumed that advanced coal technology, including scrubbers, would be incorporated in new electricity generation capacity. By comparison with an unabated scenario this means that, for Asia, emissions in 2020 are capped at 45 million tons in Scenario A2 compared to over 80 million tons if not abated.

It should be noted that this assumption is very different from that which the Intergovernmental Panel on Climate Change (IPCC) has hitherto assumed: either that sulphur emissions remain constant at their 1990 global level or change in line with CO_2 emissions (notably in their scenario IS92a)– i.e. there is a constant sulphur:carbon ratio.

Having given this prominence to atmospheric emissions from coal use and risking (some may conclude overplaying emissions from fossil fuels more generally, let me make it clear that the WEC does not hide the environmental risks which may occur with non-fossil fuels. Indeed, public concerns surrounding nuclear power and some of the environmental impacts of renewable energy were earlier touched upon.

In "New Renewable Energy Resources : A Guide to the Future"' 1994 the WEC gave due prominence to such local environmental impacts. Large structures of any kind, including large solar installations, can be intrusive. Large biomass plantations not only risk loss of biodiversity, visual intrusion and emissions (during early growth and harvesting particularly) but also compete for suitable land with the agricultural requirements of growing human populations. Wind turbines can be visually intrusive, especially in upland locations, and may cause bird fatalities. Hydropower schemes, especially larger-scale ones, can inundate large areas of land – sometimes very fertile – and cause serious populations displacement. Estuarine (tidal) barrages can destroy the local ecology, and thus the food source for significant bird numbers and species. Geothermal may also emit unattractive substances. Thus few plants in the renewables garden are entirely rosy either.

But consideration of renewable forms of energy has received an uplift with the focus on climate change in recent years. The possibility of human-induced climate change cannot, in any event, be ignored. The WEC has firmly espoused the need for appropriate precautionary measures, starting now, on the "minimum regret" principle. Industrialised countries have largely failed to do so (Figure 2). At end-1995 their CO_2 emissions from fossil fuel use were, in aggregate, 4% above their 1990 level. This was despite the agreed aim to return their greenhouse gas emissions not covered by the Montreal Protocol back to their 1990 level by the year 2000. In the period 1990-1995 developing countries increased their emissions of CO_2 from fossil use by 25.5%, reflecting their economic development; while the CO_2 emissions of the economies in transition declined some 30%.

Our scenarios indicate a wide range of possible outcomes for future CO_2 emissions from fossil fuel use. The Ecologically Driven Scenarios (C1 and 2) clearly have the lowest – reflecting lower energy consumption, greater efficiency, and a larger reliance on non-fossil fuels. Scenario A3 also has relatively low CO_2 emissions, reflecting relatively high reliance on renewables and nuclear.

What, very briefly, are the specific modelled outcomes for CO_2 emissions? First, cumulative CO_2 emissions from fossil fuel use 1990-2100 are:

Scenario	*Cumulative Emissions (GtC)*
A2	1720
A1	1350
A3	980
B	1190
C1	590
C2	580

These outcomes compare with the IPCC's scenarios on a comparable world population projection of: 1430 GtC (IS92b), 1500 GtC (IS92a); and 2190 GtC (IS92e).

Secondly, Table 5 shows the global CO_2 emissions from fossil fuel use in the selected years 1990, 2020, 2050, and 2100. In Scenarios C1 and C2 these emissions are 2.8 GtC in the year 2100, or less than half their current level. In Scenario A3 the figure is 8.2 GtC in 2100. ("Global Energy Perspectives to 2050 and Beyond", 1995. p. 87).

Thirdly, by 2100 atmospheric CO_2 concentration would be of the order of 430 ppmv (parts per million by volume) in the C1 and C2 Scenarios, and somewhat below 550 ppmv in Scenario A3 – thus still below double its pre-industrial level of 280 ppmv which some Parties to the Climate Convention refer to as a possible target. Today atmospheric CO_2 concentration is 360 ppmv.

Hitherto, most developing countries have regarded the curbing of CO_2 emissions and climate change mitigation measures as a primary responsibility of the industrialised countries. This is the underlying message of "common but differentiated responsibilities" in the UN Climate Convention. But, as Table 6 indicates, this situation will not last for very long. Historic responsibilities apart, by 2020 the present developing countries will probably account for about 50% of global CO_2 emissions from fossil fuel use (against 35% today). By 2050 the present developing countries could account for over 60% of global CO_2 emissions. Other greenhouse gas emissions, not least methane, will also be a primary developing country responsibility in future unless major changes occur in patterns of food consumption.

The global clock therefore risks ticking towards unsustainable High Noon unless effective precautionary measures are taken, and so on. The industrialised countries, whose governments agreed at Rio and with the Climate Convention to take the lead, should indeed show the way. But climate change, if it occurs and is human-induced, is a global challenge requiring a global response from which no country should stand aside. The latest scientific evidence, from the IPCC's Second Assessment Report, does "point towards" human influence (Working Group I, Chapter 8 Second Assess- ment Report, 1996). In the required global response, the pursuit of sustainable energy supply options has a key role to play. And given the lead times involved, the technology innovation and diffusion required, and the financing to be found, no further time should be lost in starting to bring the more attractive options on line.

Conclusion

I can do no better than to conclude, at this juncture, with the words from the WEC's 16th Congress in 1995:

> "The next two or three decades represent the key period of opportunity for transition to a more sustainable path of development for the long term.
>
> "Action taken, and research done, now will begin the shift of direction required of 'minimum regret' action. That direction includes greater efficiency and conservation wherever appropriate, and the bringing forward more quickly of economic non-fossil energy supplies."
>
> "Unless we begin the process now the various development paths for the long term are likely to get blocked off, and returning to them will become much more difficult and costly–even where possible."
>
> *"Conclusions and Recommendations,"*
> *World Energy Council 16th Congress,*
> *Tokyo, 1995.*

INDIAN ENERGY OVERVIEW

6

R.K. Sachdev
Former Adviser,
Ministry of Coal,
Govt. of India
New Delhi (INDIA)
J N Maggo
Joint Adviser,
Planning Commission
Govt. of India,
New Delhi (India)

INTRODUCTION

1. India is not endowed with large primary energy reserves in keeping with her large geographical area, growing population and increasing final energy needs. The current assessment in regard to the primary commercial energy resources indicates that coal is the major energy resource of the country. The gross reserves of coal are presently estimated at around 200 billion tonnes. The proven reserves of coal are placed at 70 billion tonnes which are sufficient to last the next century at an annual level of production of over 500 million tonnes. There are some lignite deposits also which are estimated to be more than 26 billion tonnes. As far as hydrocarbons are concerned, the balance of recoverable reserves are placed at 765 million tonnes of crude oil and 707 billion cubic of natural gas. The hydro-electric potential as assessed by the Central Water Commission and the Central Electricity Authority in 1987 is 600 TWh. The regional distribution of the primary commercial energy resource potential is indicated in Table 1.

Table 1
Regional Distribution of Primary Commercial Energy Reserves

Region	Coal (BMT)	Lignite (BMT)	Crude oil (MMT)	Nat. Gas (BCM)	Hydro (TWH)
Northern	1.1	1.1	—	4	225.0
Western	47.4	0.5	609	547	31.4
Southern	12.6	24.6	—	—	61.8
Eastern	138.2	—	—	—	42.5
North-Eastern	0.8	—	156	156	239.3
Total	200.1	236.2	765	707	600.0

Sources: Eighth Five Year Plan, Planning Commission.
Petroleum & Natural Gas Statistics, 1993-94, Ministry of Petroleum & Nat Gas. Geological Survey of India

2. As may be seen from Table 1 the distribution of primary commercial energy resources is quite skewed. Whereas the Eastern region accounts for nearly 70% of the total coal reserves the Western region has over 70% of the hydrocarbons reserves in the country. Similarly, more than 70% of the total hydel potential in the country is located in the Northern and the North-Eastern regions put together. The Southern region, which has only 6% of the coal reserves and 10% of the total hydel potential, has most of the lignite deposits occurring in the country.

Energy Policy

3. Energy being one of the key inputs for economic growth, due importance has been given to the development of this crucial sector of the economy over the years. A number of high level committees have been set up in the past by the government of India to examine in detail the various issues and suggest measures for ensuring the development of the energy sector for sustaining the economic growth and meeting the basic needs of the people. The recommendations made by these committees are kept in view while formulating the Five Year Plans. The energy strategy adopted for each Five Year Plan incorporates the changing needs of the economy and the society. It was felt desirable during the formulation of the Eighth Five Year Plan, which is in its last year of implementation, to adopt a long-term strategy which is consistent with the goal of sustainable development. Such a strategy was to ensure that the highest priority is accorded to meting fully the basic needs of the rural and the urban poor in the immediate future. The strategy also ensured gradual shift from non-renewable to renewable sources of energy with increasing emphasis on demand management, conservation and efficiency in both energy production and use. The short-term, medium–term and the long-term priorities as outlined in the Eighth Five Year Plan document are given as follows:

— Short-term

- — Maximise returns from the assets already created in the energy sector.
- — Initiate measures for reducing technical losses in production, transportation and end-use of all forms of energy.
- — Initiate action to reduce the energy intensity of the different energy consuming sectors of the economy and prómote conservation and demand management through appropriate organisational and fiscal measures.
- — Initiate steps for meeting fully the basic energy needs of the rural and urban households, so as to reduce the existing inequities in energy use.
- — Maximise satisfaction of demand for energy from indigenous resources.

— Medium-term

- — Initiate steps towards progressive substitution of petroleum products by coal, lignite, natural gas and electricity so as to restrict the quantum of oil imports to the 1991 level.
- — Initiate action for accelerated development of all renewable energy resources, especially the available hydro-electric potential.

— Promote porgrammes to achieve self-reliance in the energy sector.
— Promote R&D effort on decentralised energy technologies based on renewable resources.
— Initiate appropriate organisational changes in the case of different energy sub-sectors consistent with the overall energy strategy.

— Long-term

— Promote an energy supply system based largely on renewable sources of energy.
— Promote technologies of production, transportation and end-use of energy that are environmentally benign and cost-effective.

4. Electricity is generated from a number of sources including coal, lignite, natural gas, liquid fuels, hydel, nuclear and wind energy sources. The installed generation capacity in the utilities has increased from 1,743 MW in 1950, the beginning of the planning era, to 76,753 MW in 1993. The gross generation of electricity in the utilities has increased from 5.3 Billion KWh to 324 Billion KWh during the same period. The generation capacity mix presently is 72:26:2 of thermal, hydel and nuclear capacity. The share of hydel capacity has come down from an all-time high of 50.62% in 1962 to 26% in 1993 and is likely to go down further in the coming years. The progress in addition to the electricity generation capacity in the country for a few selected years is shown in Table 2.

Table 2
Progress in Addition to Installed Generation Capacity in Utilities

(MW)

Year	Hydro	Steam	Diesel	Gas	Nuclear	Wind	Total
1950	563	1028	150	–	–	–	1743
1960	1917	2436	300	–	–	–	4653
1970	6383	7508	230	168	420	–	14709
1980	11791	17122	167	274	860	–	30214
1990	18753	43004	182	2552	1565	30	66086
1993	20379	49147	302	4883	2005	37	76753

Source: General Review, Public Electricity Supply, All India Statistics, 1993-94, CEA.

5. In addition to the installed generating capacity of 76,753 MW in the utilities, the non-utilities have 10,575 MW of capacity installed with the different power intensive industries for their captive use. The annual generation from this capacity was 32.26 billion KWh in 1993. Further, the railways also have about 147 MW of installed capacity with an annual generation of 0.02 billion KWh. Almost whole of this capacity available with the non-utilities is thermal, based on either coal or diesel and gas. The changes in the gross electricity generation from different modes of installed generation capacity in the utilities since 1950 are shown in Table 3.

Table 3
Change in the Pattern of Gross Electricity Generation in Utilities

(Billion KWh)

Year	Hydro	Steam	Diesel	Gas	Nuclear	Wind	Total
1950	2.60	2.69	–	–	–	–	5.29
1960	7.84	8.73	0.37	–	–	–	16.94
1970	25.25	27.80	0.11	0.25	2.72	–	55.83
1980	46.54	60.71	0.06	0.52	3.00	–	110.84
1990	71.64	178.32	0.08	8.11	6.14	0.03	264.32
1993	70.46	233.15	0.26	14.73	5.40	0.05	324.05

Source: General Review, Public Electricity Supply, All India Statistics, 1993-94, CEA.

6. The share of hydel generation has been progressively declining over time in the total generation in the utilities. Another noticeable feature of the energy generation is the increasing level of generation based on gas which has shown an appreciable increase over the 1970 level. A small contribution also has been made in the form of wind energy generation.

Energy Production and Consumption Pattern

7. Changing pattern of energy supply in India can be realised from the fact that total energy supply, including both commercial and non-commercial forms, increased from 85.1 mtoe in 1953-54 to about 240 mtoe in 1990-91. In this the share of non-commercial fuels have declined from 76 per cent in 1953-54 to 49 per cent in 1990-91. Thus commercial energy contributed 122 mtoe and non-commercial energy 118 mtoe fuelwood accounted for 65 per cent of the total non-commercial energy consumed in the country.

8. Trends in indigenous production of commercial energy, for the period 1960-93, are given in Table 4, below:

Table 4
Indigenous Commercial Energy Production

(in mtoe)

Source	1960	1970	1980	1990	1993
Coal & Lignite	35.65	37.26	58.16	98.02	112.20
Crude Oil	0.46	7.01	10.79	33.91	27.76
Natural Gas	–	1.33	2.17	16.55	16.86
Hydro	0.67	2.17	4.00	6.16	6.07
Nuclear	–	0.67	0.78	1.60	1.40
Total	36.78	48.44	75.90	156.24	164.29

Ref: Planning Commission, Government of India – September 1995.

9. Commercial and non-commercial energy consumption proportions from 1953-54 to 1993-94 are given in Table 5 below.

Table 5
Commercial and Non-commercial Energy Utilisation

(in mtoe)

Type	1953-54 (%)	1970-74 (%)	1980-81 (%)	1990-91 (%)	1993-94 (%)
Commercial	21.0 (24.7)	45.5 (34.4)	71.9 (40.5)	121.6 (50.6)	133.4 (52.9)
Non-commercial	64.1 (75.3)	86.7 (65.6)	105.8 (59.5)	118.8 (49.4)	118.8 (47.1)
Total	85.1	132.2	177.7	240.4	252.2

Ref: Planning Commission, Government of India – September, 1995.

Commercial Sources of Energy

10. Coal, including lignite, remains the predominant source of commercial energy in India, accounting for over 62% of its energy needs. Petroleum and Natural Gas are the second highest energy contributors to the Indian economy, but there has been gradual substitution of petroleum fuels with natural gas. The share of hydel energy has diminished over the years whereas no major breakthroughs have been achieved in the nuclear energy sector. The net value of import of energy during 1993-94 was as much as 24% of the net imports of the country.

11. The share of commercial energy of different primary sources at present is:

Coal	58.4%
Lignite	3.7%
Petroleum	27.2%
Natural Gas	7.6%
Hydel	2.8%
Nuclear	0.3%

Table 6
Percentage Share in Final Energy Consumption by Different Sector (Commercial Energy)

Sector	1953-54	1960-61	1970-71	1980-81	1990-91	1993-94
Industry	39.9	41.5	49.6	49.6	44.7	42.8
Transport	43.7	40.2	28.2	23.5	19.8	20.8
Household	9.6	9.7	12.2	9.7	11.5	12.0
Agriculture	1.4	1.6	2.7	6.3	8.2	9.2
Others	2.0	2.0	1.6	1.3	2.3	2.1
Feedstock	0.5	0.3	4.2	5.6	10.4	10.4
Non–Energy Use	2.9	4.7	1.5	4.0	3.1	2.7

Ref: Planning Commission, Government of India – September, 1995.

Biomass Energy Sources

12. Biomass sources like fuelwood, agro-residue, urban/industrial waste and animal dung met 38% of India's energy demand in the year 1993-94. Table 7 gives details of contribution of different sources.

Table 7

Biomass use as Energy Fuel in India (1991-92)

(million tonnes)

S.No.	Source	Availability	Estimated Consumption	Deficit
I	Fuelwood	28.4	166.85	138/40
II	Urban/Indl. Waste	246.0	Negligible	–
III	Agro Residue	240.0	150.0	–
IV	Cattle Dung	960.0	80.0 (Dry Wt)	–

13. There are different fundamental forms of bioenergy use, but attention in India is always focused on bio-energy use in the traditional sector for household cooking and heating applications in rural areas. Fuelwood has also been used for traditional applications in rural industries and for certain industrial applications in urban areas. Traditional industrial applications include tobacco, tea, brick and pottery processing. Earlier the biomass feedstock was available at a very low cost and, therefore, there was hardly any consideration for efficiency of use. No incentive was available for its conversion in an efficient fashion. Even a 25 per cent improvement in the use of efficiency of fuelwood can mean a reduction of almost 40 million tonnes of fuelwood consumption.

14. The data on availability of biomass resources, including fuelwood, dung cake, agricultural waste and other biomass, are not regularly complied. The availability, more often than not, is estimated on the basis of sample surveys carried out from time to time by different agencies. The animal population is brought out in the Indian Livestock Census published every five years. Using the 1966 Livestock Census data and the information on feed availability from the Indian Agricultural Statistics, the estimates of dung yield were placed at around 960 million tonnes of wet dung on an annual basis in the year 1975. Since the animal population does not increase at a high rate, these estimates could even be relevant today. The yield of dry dung is roughly 20% of the wet dung. The recovery factor for dung is around 75% of the total yield by the animals.

New and Renewable Sources of Energy

15. A resource survey has indicated a large potential for exploitation of the renewable sources of energy in the country. The sourcewise potential estimates and their present status of development is given in Table 8 below:

Table 8
Renewable Energy Potential

Source/Technology	Potential Availability	Potential exploited
Biogas Plants	12 Million	2 Million
Biomass-based power	17,000 MW	Marginal
Efficient Woodstoves	120 Million	18.5 Million
Solar Energy	5 × 10^15 Whr/Year	
Small Hydro	10,000 MW	250 MW
Wind Energy	20,000 MW	250 MW
Ocean Thermal	50,000 MW	
Sea Wave Power	20,000 MW	
Tidal Power	9,000 MW	

Source: National Paper on Energy Scenario in India, 1995, India Member Committee, World Energy Council.

Investment on Energy Supply System

16. The development of commercial forms of energy is highly capital intensive and large investments are required for meeting the rapidly growing demand for these forms of energy from different consuming sectors. The relative share of expenditure on energy sector in the total plan expenditure during successive Five Year Plans as shown in the Table 9.

Table 9
Plan Expenditure and the Share of Energy Sector

Plan	Period	Plan Expend (Rs. Cr.)	Share of Energy Sector (%) Power	Oil/Gas	Coal	NRSE	Total
First	1951-56	1,960	–	–	–	–	19.7
Second	1956-61	4,670	9.7	0.8	1.9	–	12.4
Third	1961-66	8,580	14.6	2.6	1.3	–	18.5
Fourth	1969-74	15,780	18.6	1.9	0.7	–	21.2
Fifth	1974-79	39,430	18.7	3.6	2.9	–	25.2
Sixth	1980-85	109,290	19.7	7.8	3.5	0.1	28.1
Seventh	1985-90	220,220	17.4	7.3	3.2	0.3	28.2
Eighth	1992-97	434,100	18.4	5.5	2.4	0.2	26.5

17. In the Eighth Five Year Plan, an outlay of Rs. 4341 billion (US $ 124 billion) was provided for different sectors of the economy. The share of energy sector has been kept at 26.5% of the total public sector outlay. The power sector had been allocated a share of 18.4% in keeping with the requirement of investment to be made during the Plan period for implementation of the development programme envisaged for this sector.

Future Energy Demand Projection

18. The projections for future demand for commercial energy in India are given in Table 10.

Table 10
Future Demand for Commercial Energy in India

	1991-92	1996-97	2009-10
Electricity Demand (TWH)	231	336	725
Coal (Million tonnes)	229	311	650
Petroleum (Million tonnes)			
Products	57	81.2	165
Natural Gas (Billion cubic mitrs)	18.6	30.2	65

Ref: Planning Commission, Government of India – September, 1995.

An estimate of annual average future growth rate has been obtained as 8.2% for gross generation of power, 5.06% for coal and 5.54% for petroleum. In view of the high capital output ratio of the commercial energy sector, the realisation of such pressure on the financial resources of the economy.

Energy Perspective to 2020

19. A very preliminary assessment of energy requirement in the year 2020 has been attempted by taking into consideration the Business-As-Usual (BAU) approach. Such an approach assumes that the changes in the consumption patterns of energy, the efficiency improvements, etc. continue to take place in future at the rate these were taking place in the past

20. A very rough assessment of demand for electricity in the year 2020 is placed at 1300 billion units as compared to the actual consumption of about 365 billion units in 1993-94. The gross generation requirements including T&D losses and auxiliary consumption of power, plants is assessed at 1745 billion units. At the present rate of capacity utilisation, the requirement of generating capacity can be worked out to be 378,000 MW. This is roughly four times the present installed generating capacity in the utilities.

21. The demand for petroleum products is provisionally placed at 335 million tonnes in the year 2020.

22. As far as coal is concerned, the requirement will depend mainly on the demand for power generation which presently accounts for over 70% of the total coal use in the country. In a presentation made at a recent Workshop on The Indian Coal Perspective to 2010, the demand for coal in that year was projected at 690 million tonnes. If the demand for coal continues to grow at the same rate, it is projected to touch 1345 million tonnes in the year 2020. However, technological developments taking place in the coming decades in the field of renewable energy sources as well as new finds of oil and gas resources and the environmental consideration are bound to affect the coal usage in future.

23. In order to meet the rising demand for energy in the coming years it is of utmost importance that the energy sector development is supported by and adequate R&D base. The major thrust of R&D efforts in these areas should be to strengthen the indigenous base of technology and reduce the overall cost through development of appropriate materials, technologies and practices. Since in the overall context of energy planning, electricity is going to play a major role as the substitute of hydrocarbons, a high priority needs to be accorded to R&D efforts in both the coal and power sectors. While the conventional sources of electricity will continue to play a predominant role for years to come the long–term concern of energy planning should be to produce electricity more and more from renewable and cost-effective decentralised systems.

24. Since Indian coal has high ash content and is of low calorific value, its transportation is already causing severe strain on the transportation system and the problem is going to become more and more acute. From this point of view, it is necessary to locate new power plants as near to coal sources as possible. At the same time coal has to be beneficiated so that the burden on transportation may be reduced besides other benefits like improved plant performance and reduced environmental problems at the power station due to fly ash. There is also a need to evolve indigenous technologies for improving the quality of coal and also improvements in end-use efficiencies.

25. In view of the rising demand for hydrocarbons and the fast depleting indigenous resource base, the gap between demand and indigenous availability is likely to widen in the coming years. This has serious foreign exchange implications for the economy. The thrust areas in the petroleum sector are, therefore, oil exploration and development on the one hand and containment of demand for petroleum products through conservation and substitution on the other. Development of oil and gas fields, both onshore and offshore, involves heavy outgo of foreign exchange and it is important that R&D effort is directed toward indigenisation of technology in this area.

Conclusion

In the context of the steeply rising energy demand of various key sectors of economy, it is imperative that the entire energy policy is given a fresh look with a view to streamlining the procedural requirements for establishment of new capacity for production, generation and distribution of all forms of energy. Critical areas like fuel supply, transport infrastructure, transmission and distribution of electricity and tarrif/price fixation mechanism, environmental and resettlement issues also need to be addressed in great detail.

NON-CONVENTIONAL ENERGY SOURCES IN INDIA–PROGRESS AND POLICY PERSPECTIVE

7

ASHOK PARTHASARATHI
Secretary
Ministry of Non-Conventional Energy Sources
Government of India
New Delhi

INTRODUCTION

Energy is critical for economic growth, social development and human welfare. Over four billion people, constituting more than three-fourths of the world's population, live in the developing countries. Yet they use under one-third of the world's energy, with per capita levels one-sixth that of the highly industrialised countries. Electricity has yet to reach two billion people. Driven by rising populations, expanding economies, energy-intensive industries, urbanisation and a quest for modernisation and improved quality of life, energy use in the developing world has doubled in the last two decades and is expected to double again in an even shorter time span of the next 15 years.

CHALLENGES AND OPPORTUNITIES

The growth of national demand for electric power in India is likely to be around 9% per annum over the next 15 years, requiring about 10,000 MW of capacity addition every year over that period. Demand is, therefore, expected to triple by 2010 from the present installed capacity of about 82,000 MW. It has been estimated that the present energy shortage results in an annual loss of production of about 2% of the national income. By 1997, an energy shortage of 14% and peaking deficit of 28% are apprehended. Restoration of the financial health of the State Electricity Boards and improvement of their operational performance are the most critical issues in the energy sector. Improvements in the capacity utilisation of the existing power plants and in the distribution and use of electricity are required. At about 60%, the Plant Load Factor continues to be low and Tansmission and Distribution losses at 23% are very high, compared to international averages. Negligible revenue streams from the agricultural and domestic sectors have brought the SEBs to the brink of bankruptcy. The hidden subsidy for these sectors will go up to over Rs. 15,000 crore during 1996-97, with the commercial losses of SEBs going up to about Rs.10,000 crore. The main challenge facing our energy system is the bringing about of power sector reforms and restructuring of the SEBs, with emphasis on tariff reform, privatisation of distribution and establishment of independent regulatory mechanisms. This process, already initiated in a few States, needs to be expanded rapidly.

Fossil fuels will continue to be the major energy sources, and will continue to play a critical rôle in the economy. We will have to continue to rely in large measure on our vast coal reserves as a source of power generation. At the same time we will need to develop programmes and policies that will reduce dependence on fossil fuels in order to achieve sustainable economic growth and environmental stability. Three-fourths of all carbon dioxide released into the atmosphere comes from the production and use of energy. For every tonne of fossil fuel burned, at least three-quarters of a tonne of carbon is released as carbon dioxide. Other environmental concerns from conventional energy use relate to deforestation and declining soil fertility.

Our per capita consumption of electricity continues to be extremely low, only slightly above 350 KwH per annum. What is more, only one-third of the total electricity supply is consumed in the rural areas, despite three-fourths of the population living in those areas. While 86% of the villages have access to electricity, only about 31% of the rural households use electricity. 76 million rural households still do not use electricity. About 80,000 villages are yet to be electrified. A majority of these villages are located in far-flung and remote areas, with very low load densities. Extending electrification to these villages is uneconomical and would involve heavy capital investments.

ROLE AND RELEVANCE OF RENEWABLE ENERGY

These challenges of the present energy scene offer us a window of opportunity in the form of renewable energy sources to expand and diversify our energy supply and shift the development path towards greater sustainability, as well as environmental and social responsibility. Scenarios projected by reputed international organisations covering the period upto 2060 have pointed to the increasing role and growing share of renewable energy in the world's energy demand, particularly in developing countries. What is more, an increasing proportion of that future energy demand will have to be met in the form of electricity. Over the last 15 years, the technical, operational and economic viability of grid-connected and off-grid electricity production from renewables has been progressively demonstrated with increasing success. Furthermore, several renewable energy technologies are now showing signs of being able to stand on their own in the foreseeable future, without subsidies. The recent surge of interest in renewables is on account of these developments, as also their declining costs, while fossil fuels are showing rising cost trends. In addition, the steadily growing awareness of the importance of environmental protection is a major factor favouring renewables.

Renewables can create a significant impact in the generation of electricity. Several technologies for grid connected power generation such as wind power, small hydro power and biomass based power have now matured. Renewable electric technologies also offer possibilities of distributed generation, at or near actual load centres, thereby saving on costly establishment and maintenance of transmission and distribution networks. A major niche area for renewable energy technologies which should be of interest to utilities, is a range of 'demand-side' applications, particularly in our urban centres and industry, such as air and water heating, industrial process heating, solar building designs, as well as energy recovery from urban, municipal and industrial wastes. It is unlikely though that a single renewable energy technology will predominate. Variations in the availability of the energy resource, its economics and range and speed of applicability, will favour some renewables over others for different end-uses.

In our rural areas, energy for cooking, lighting, water pumping, agro and rural industry and other productive activities can be effectively provided through locally available renewable energy sources. In remote areas, where transmission of grid power has been found to be totally uneconomical, off-grid electrification can be undertaken through renewable energy systems such as solar photovoltaics. Where appropriate electricity supply requirements can also be met by hybrid systems integrating two or more sources, in conjunction with storage. It has been estimated that only another 8% of our villages can be electrified by extension of the grid. Thus, nearly 35,000 or 6% of the villages will get left out. These can be covered only through decentralised generation and distribution based on renewables.

Renewables also make economic sense, as several technologies are now competitive, and markets are growing. Besides opportunities in production of equipment, renewables offer vast employment opprtunities in the service sector, in local mangement, and in maintenance and repair. Moreover, the income and employment benefits are more evenly distributed, in accordance with the spread of resources.

The Government of India recognised this potential of renewable energy as far back as 1982, when India became the first country in the world to set up a separate Department of Non-Conventional Energy Sources. This was followed in 1992 by the establishment of a full-fledged Ministry of Non-conventional Energy Sources, under the overall charge of the Prime Minister himself.

PROGRESS IN INDIA

Considerable progress has been made over the past decade in harnessing renewable energy through a wide range of programmes. The last few years have seen particular progress in the production of elctricity from renewable energy sources. Notable among these has been the installation of a wind power capacity of about 820 MW, which has placed us among the top three countries in the world in harnessing this resource. This capacity, which has come up only over the last 3-4 years, is backed up by the world's largest wind resource assessment effort undertaken from as far back as 1985. Another priority areas is small hydro power development on canal drops, dam toe sites and in hilly areas. A capacity of about 135 MW has been installed, and another 250 MW is under implementation. An ambitious programme is being implemented to harness the 3500 MW potential available in the sugar mills in the country through advanced bagasse co-generation. 42 MW of surplus power generation capacity has been established, projects of 49 MW capacity are under construction and projects aggregating to over 250 MW are at different stages of planning and implementation.

A solar Energy Enterprise Zone is proposed for the desert region in Western Rajasthan for rapid commercialisation of various solar technologies through facilities and incentives for manufacture of equipment and installation of solar power generating capacity. A 140 MW integrated solar combined cycle power project based on parabolic trough collectors and gas and steam turbines is planned near Jodhpur. Letters of Intent have been issued for the setting up of three solar power projects of 300 MW capacity in the State based on solar photovoltaic and solar chimney technologies.

Solar photovoltaics is one of the key programmes underway, particularly for decentralised applications in rural and remote areas. Photovoltaics is a simple, reliable and environmentally benign technology which is immediately available to provide electricity to our widely dispersed households and farms. Today, India is the second largest manufacturer in the world of solar photovoltaic panels based on crystalline silicon solar cells. What is more, the technology for the manufacture of these solar cells and panels has been developed and commercialised almost entirely on the basis of domestic R & D. Industrial production has touched a level of 8 MW / year in relation to a total world production of around 85 MW/year in 1995. There are now 12 companies engaged in the production of solar cells and modules and another 70 companies designing and supplying solar PV systems. Over 300,000 solar PV systems aggregating to about 22 MW have been installed. They involve around 25 different types of systems for rural, remote areas and industrial applications. These include: stand alone village home and street lighting; water pumping for drinking and irrigation; rural telecommunication; unmanned TV transmitters in remote area; the powering of refrigerators for vaccine storage in rural health centres, to name only a few. In recent years, major market-oriented programmes have also been launched for popularisation of solar water pumps and solar lanterns.

In Karnataka, the solar Electric Light Company (SELCO), a grass-roots energy service company, has been sucessfully marketing and installing solar home systems and solar water pumps in three rural districts of the State through innovative rural credit support schemes. Tata Finance have sucessfully introduced home lighting systems in the Sunderbans areas of West Bengal and in the tea gardens of Assam through imaginative leasing packages. A group of farmers in Tamil Nadu have purchased and installed solar pumps, which have increased their yields significantly. These entrepreneurial and financing models are replicable and can lead to sustainable solutions for rural electricity from the sun. Thus, the emphasis in photovoltaics so far has been mainly on decentralised off-grid 'niche' markets which can offset grid extension. However, the programme is now being expanded to cover grid-connected voltage, support systems on weak, rural grids and roof-top 'peak shaving' systems in urban centres.

Solar thermal water heating systems are now coming to be increasingly used, particularly in commercial and industrial establishments. About 325,000 square meters of collector area has so far been installed ranging from domestic water heaters of 50 litres capacity to industrial systems of 240,000 litres of hot water per day.

A major achievement has been in the area of cooking energy, in rural areas with the establishment of about 2.2 million family size biogas plants and 20 million improved woodstoves. Nearly 400,000 solar cookers are also in use. A National Programme on energy recovery from urban, municipal and industrial wastes has been launched. Programmes are also being developed for new technologies which show long term promise, such as fuel cells, hybrid and all electric vehicles, and wave and tidal power.

Thus, our main thrust in renewable energy is to consolidate and strengthen technology development and to promote commercialisation by providing a market orientation to various programmes. The key element of the new strategy is to utilise the Government' s budgetary resources for selected demonstration projects, while creating attractive opportunities for private sector projects through a favourable policy environment and availability of institutional finance and external assistance. The development of entrepreneurship and greater competition is expected to reduce costs and improve the quality of equipment, and to lead to the establishment of a large number of market outlets, a service network, and repair and maintenance infrastructure.

PROMOTIONAL MEASURES : FISCAL INCENTIVES AND SOFT FINANCING

If renewables are to overcome the biased and distortions in the prevailing policies and practices in the energy sector which favour fossil fuels, the provision of 'incentives' to enable renewables to have a level playing field, is essential. Accordingly, Government has introduced a package of incentives for renewable energy projects which include: tax concessions such as 100% accelerated depreciation, soft loans, customs and excise duty reliefs, and liberalised foreign investment procedures. Guidelines have been sent by the Ministry of non-Conventional Energy Sources to all the States on general policies and facilities for wheeling, banking, third party sale and purchase of power by utilities from such projects. Based on those Guidelines, 11 States have announced their policy for purchase of power from renewables at a price of Rs. 2.25 per KwH. One of the States has also announced while others are considering, an annual escalation of 5% on that price. These policies are providing remunerative returns and encouraging greater commercial development of electricity from renewables.

The Indian Renewable Energy Development Agency Ltd. (IREDA), the financing arm of the Ministry is operating a World Bank line of credit amounting to US $ 195 million to promote private sector participation in renewable energy projects. Wind-power generation and canal based small hydro projects, and end-use solar photovoltaic systems are eligible for financing at low interest rates. These range from 2.5 to 5% in the case of photovoltaic systems to 18-19% in the case of small hydro and wind power projects. A line of credit from the Asian Development Bank amounting to US $ 100 million has also been recently approved for operation by IREDA for assisting bagasse-based co-generation, solar thermal systems, biomethanation and wind energy projects. Assistance from the Global Environment Facility (CEF) has been provided to the Ministry for hilly hydro and bio-methanation programmes and through USAID for advanced biomass co-generation.

POLICY FRAMEWORK

Awareness of the potential and relevance of renewable energy is steadily increasing. Yet renewables continue by and large to remain outside the ambit of national energy plans. Bringing renewables into energy planning processes, alongwith energy efficiency and rural energy, will help in the development of integrated long term energy policies and strategies.

For this purpose, Integrated Resource Planning needs to be introduced, as it considers the broadest range of supply and demand side options in selecting the optimal mix which can meet future demand at the lowest long-term environmental, social and economic cost. Opportunities which now exist, and are yet to be seriously explored, include distributed generation, 'niche' markets for off-grid rural electrification, and a range of demand-side applications.

There is urgent need for promotional and regulatory measures which can widen the role of, and demand for, renewable energy generation and use. Several initiatives and approaches are possible for this purpose. For example, a portion of all new power generation capacity can be reserved or "set aside" for renewable power projects. Or, all new fossil fuel based power capacity could be required to include suitable renewable energy power blocks. The levy of an appropriate `cess' on fossil fuel-based electricity and/or on selected hydrocarbon-based fuels could be another effective policy measure. The consequent slightly higher price of electricity or hydrocarbon fuels to industrial and commercial consumers would subsidise the development and promotion of renewable energy technologies. On the other hand, such consumers could be entitled to billing credits for renewable energy capacities created by them. Utilities could be obligated through legislation to purchase renewable energy at prices which are fair and unbiased. Wheeling and banking, which are vital for renewable energy development, need to be provided by SEBs for an appropriate charge. Rural electric enterprises, which are permitted to generate, distribute and sell electricity directly to users in rural areas, independent of State Electricity Boards need to be promoted. Mandatory use of renewables for certain applications such as water heating through solar energy in all new buildings, and in specified industries can be introduced. These are not theoretical propositions; other countries have introduced such measures and shown that they are practical and feasible.

FUTURE DIRECTIONS AND OUTLOOK

Over the last decade, in particular, Government has endeavoured to lay the foundation for a broad-based renewable energy programme. Through various initiatives, grid quality power generation from renewables has attained maturity during the Eighth Plan. By the end of this Plan, around 7% of the total addition to generation capacity achieved during the Plan is expected to come from renewables. What is more, renewables would have a share of 1.5% of the total installed power generation capacity by the end of this Plan-around 1400 MW out of a total of around 88,000 MW. There is keen private sector interest for the manufacture of renewable energy equipment, development of power projects and in the service sector. The total annual turnover in this sector has reached about Rs. 2000 crore.

In order to take a long term view, and to realise the full potential of renewables, it is necessary to formulate a comprehensive Renewable Energy Policy which will encompass the Government, industry, research institutions, non-Governmental organisations, local communities and users. This exercise is now underway. The policy will help to create awareness, foster development and demonstration and acccelerate commercialisation. It will also help to create an effective institutional framework, supporting infrastructure and delivery mechanisms for the planning, development, manufacture, installation, operation and maintenance of renewable energy systems. Government also plans to introduce special legislation to promote renewables, as the present legislation governing the electricity sector is considered inadequate for the effective and rapid development of the renewable energy sector.

Setting of broad goals in a Perspective Plan, of say, 15 years, will accelerate commercialisation and help in the development and sustained growth of a sound renewable energy sector. Government is targetting generation of an additional 3000 MW of grid power from renewables by the end of the Ninth Plan (1997-02) . If a goal of 15,000 MW of renewable electric capacity is set for the year 2010, this would amount to a share of about 6% of the total projected installed grid capacity of 240,000 MW at that time. In other words, the goal would be an addition on average of about 1000 MW or 10% of the anticipated annual conventional power increment, each year, for the next 15 years. If such an expanded programme is to be achieved, an appropriate policy, institutional, financial and technological fremework has to be put in place. It is within such a perspective that Government is promoting our renewable energy programme. The challenges are severe, but the potential is significant. Government wishes to see the active participation of all players–from Government agencies to NGOs, from manufacturers to R&D institutions, from financial institutions to developers and, hopefully a new breed of energy entrepreneurs–in this exciting endeavour.

EVOLUTION OF ENERGY EFFICIENCY INDICATORS IN ASIAN COUNTRIES—A COMPARATIVE STUDY

8

THIERRY LEFEVRE
Energy Planning Central Consultant Team
Asian Institute of Technology
Bangkok (Thailand)

1 ECONOMIC AND ENERGY OVERVIEW

1.1 Economic Development

TABLE 1: **Key Indicators of Seven Selected Asian Developing Countries**

	Unit	China	Korea	India	Indonesia	Philippines	Taiwan	Thailand
Population in 1992	Million	1161.7	43.7	876.8	186	64.26	20.65	57.8
GDP in 1992	bn US$ 1992	442	294.5	272.3	128.5	52.97	174.6	110.3
Per capita GDP	US$ 1992	381	6739	313.7	690	824	7730	1909
Per capita primary energy consumption	kgoe	655.1	2657	218.6	289	215.5	2541	625
Per capita final energy consumption	kgoe	555.4	2167	153.3	218	165.2	1723	448
Per capita electricity consumption	KwH	591.4	2266	346.8	246	304.5	3870	845
Average GDP growth rate (1980-1992)	%	9.2	8.7	5.3	6	1.4	7.6	7.9

The period 1980–1992 witnessed a considerable economic achievement in seven selected Asian developing countries, namely China, India, Indonesia, Korea, the Philippines, Taiwan and Thailand. But the rhythm of economic development and the levels that countries achieved by the end of the period were very different. China had the highest growth rate of gross domestic product, which averaged at 9.2% for the period of 12 years. Countries like, Taiwan, Thailand, Korea had sustained very high annual growth rate of GDP (7.6%, 7.9 % and 8.7% respectively) throughout the period, while other countries showed just a modest (e.g., India: 5%, Indonesia: 6%) or even very slow (e.g., the Philippines: 1.4 %) economic progress. Likewise, the GDP per capita of countries was within a wide range from US$ 304 (India), to US$ 7,730 (Taiwan) (Figures are in US$ 1992 current price). Countries, enjoying high economic growth, have a higher per capita energy consumption; on the contrary, the lower level of per capita energy consumption was observed for countries with low GDP growth. For example, per capita energy consumption of Taiwan and Korea was almost ten times that of countries like India, Indonesia, and the Philippines (values in 1992). The remarkable disparity in level of development from one country to another can be clearly seen in Table 1, which summarises key indicators of seven countries under this study.

1.2 Energy Development

1.2.1 Primary Energy Supply

Countries are quite different in terms of energy resource endowment. China, and Indonesia have abundant energy resources such as coal, natural gas and crude oil. They are net energy exporters. China is one of the few countries in the world that uses coal as principal energy source. Coal in China accounted for 76% of primary energy supply and 68% of final energy consumption. China is also an important oil producer, which ranked sixth in the world. In addition to its crude oil resources, Indonesia enjoys large reserves of coal and natural gas. Thailand and India possess considerable indigenous resources like coal, crude oil in India, lignite and natural gas in Thailand, but they had to import substantial amount of energy. Meanwhile Korea, the Philippines and Taiwan are very poor in energy resources. The energy import made up almost 99% of total primary energy supply for Korea, 97% for Taiwan and more than 70% for the Philippines in 1992.

The availability of domestic energy resources or lack of them affects, in great extent, energy policy of a given State. Countries without considerable energy resources tended to manage their energy development in such a way that reduces energy import. On the other hand, countries with abundant domestic energy resources tended to use extensively their own energy, for example coal in China, India and oil in Indonesia.

A tendency of reducing the share of oil in primary energy supply was clearly observed in all countries even in oil exporting country like Indonesia. Oil had been substituted by natural gas and lignite in Thailand, by coal and natural gas in Indonesia, by hydro and geothermal in the Philippines and by nuclear power in Taiwan and Korea.

1.2.2 Final Energy Consumption

The same trend of oil substitution was found in final energy consumption of Asian countries. Thailand, the Philippines and Indonesia had substantially reduced their consumption of petroleum products by about 10 %. Still, oil products accounted for biggest part of final energy consumption of majority of countries except China. This share was 74 % for Indonesia in 1990; 67% for Korea in 1992; 77% for the Philippines and Thailand in 1992; even for India, country with big share of coal consumption, the share of petroleum products was as high as 43 % in 1991. Share of electricity in final consumption was just around 10 % to 20 % but growing rapidly, reflecting an active electrification process in these countries. The highest share of electricity was that of Taiwan (20 %) and Thailand (16%). Coal, lignite, natural gas have also been actively promoted for final uses in countries under study.

Among economic sectors, industry was the biggest energy consumer in China, Taiwan, Korea and India, where its energy consumption exceeded 50 % of total final consumption.

The situation was different in Thailand and the Philippines where the transport sector outweighed the industry in terms of energy consumption. On the opposite extreme, China's transport sector was the smallest, which consumed only 5% of final energy consumption.

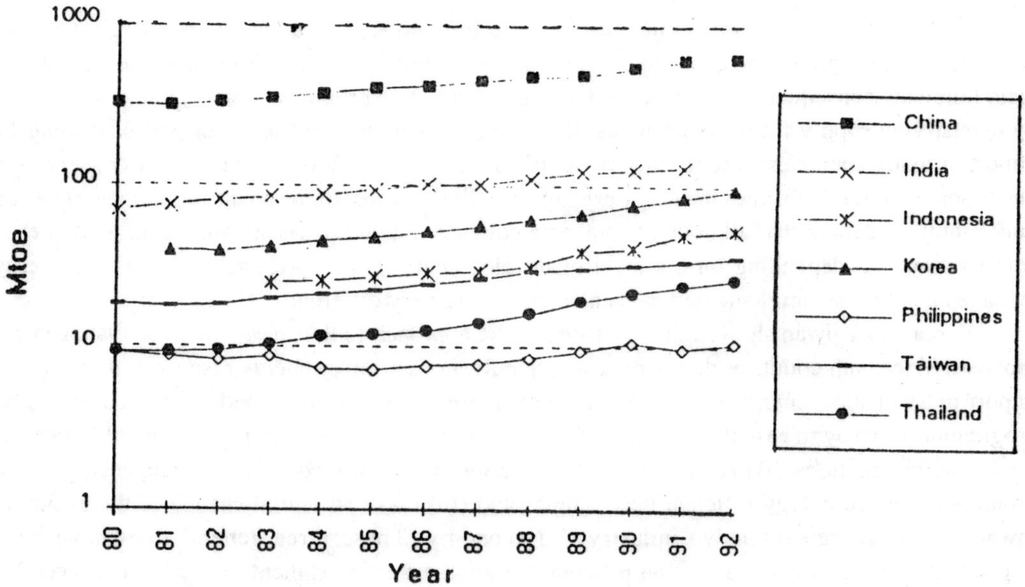

FIGURE 1: *Final Energy Consumption*

China had the biggest final energy consumption, followed by India and then Korea, Indonesia, Taiwan, Thailand and the Philippines by that sequence (Figure 1). Thailand, Korea and Indonesia showed high (above 8% per annum) and increasing rates of energy consumption growth, while China, India and Taiwan maintained constant growth rates of their final energy consumption (about 5% – 6% per annum). The Philippines' final energy consumption increased insignificantly over the study period.

The ranking of the per capita final energy consumption took another order due to difference in population. The highest per capita final energy consumption was that of Korea (about 2.2 toe/cap), followed by Taiwan (1.7 toe/cap). Thailand and China have comparable final energy consumption per capita which was close to 0.5 toe/cap. The values for India, Indonesia and the Philippines were below 0.2 toe/cap.

Regarding the growth rate of the final energy consumption per capita, one can categorize as follows: the high growth rate was seen in countries like Korea, Thailand, and Taiwan, which more or less doubled their consumption per capita over the study period. China, India and Indonesia showed smaller growth rate, while the Philippines merely kept its energy consumption per capita at the level of the beginning of the period. The gap in absolute value of energy consumption per capita between countries was quite significant, as mentioned before: 0.15 toe/cap (India) compared to 2.2 toe/cap (Korea) (Table 1).

1.3 Energy Demand Management Policies

Energy management policies of countries under study were strongly influenced by their resource endowment, socio-economic conditions and goals defined by the respective government.

In all seven countries, the prime objective of energy policy was a security of energy supply to meet fast increasing demand so as to sustain high economic growth. This situation let to the two main following consequences: first, reduction of the share of oil products in supply and consumption by diversifying supply forms and sources. Because oil accounted for the major part of the energy import of many countries; second, lack of stimulus for energy efficiency and energy conservation due to some kinds of subsidies in energy pricing prevailing in most of the countries. In the last period under study it seemed that all countries become convinced about necessity and benefits of energy conservation, but depending on the level of development, the comprehensiveness of the energy management policies implemented in each country were very different.

Korea and Taiwan showed the most systematic approach to the energy demand management problem. Since both countries do not have much indigenous primary energy resources, they have to import most of their energy requirements. Thus, these countries developed energy conservation programme as a way to ease this supply difficulty. Current energy management policy of Korea has three general directions: (i) to continue public education about the needs of conserving energy, (ii) to promote advanced energy efficient technologies and (iii) make structural change of the economy toward a less energy intensive industry and economy. These three general lines have been implemented through concrete action programmes such as: establishment of 5 years conservation programme in energy intensive industries; special plan to improve energy efficiency of boilers and electric motors; improvement of building efficiency; demand side management programme, etc.

Similarly, Taiwan designed its energy conservation strategy under the form of Measures, which include five major issues: education and dissemination; technical services; technological transfer; efficient management and financial support for penetration of energy conservation technologies into the market. Each of these measures has a specific action plan and goals.

Thailand embarked on a very ambitious strategy for energy conservation, which includes a set of measures such as energy audit, energy conservation training programme; R&D programme; low loan interest for investment in energy saving projects, and import duties reduction for energy saving equipment. The Energy Conservation Law was promulgated in 1992 and Demand Side Management Programme was set forth in 1993.

Recognizing the importance of the energy conservation potential, Chinese Government has strengthened energy efficiency policy by implementing administrative, legislative, technological and economic measures. It also gradually shifted from a chiefly administrative management to more economic and legislative management system.

In Indonesia and the Philippines the main energy efficiency management activities included public education campaign, furthered by energy auditing and technical services. More rigorous programmes are on the process of formulation, which will include the setting up of standards, regulations, energy conservation law, etc. Apparently these two countries are far behind as compared to Korea and especially Taiwan, in the formulation and implementation of concrete action plans for energy conservation.

In summary, the energy conservation management policies in the countries under study include part or all of the following components:

- Public education campaign.
- Energy audit/monitoring to identify saving potential for given country.
- Technical consultancy service to help interested parties to achieve energy saving.
- Promotion of efficient technologies through R&D, demonstration, and application of financial privileges in new technologies.
- Restructuring of energy pricing system.
- Establishment of a legal environment for the implementation of energy conservation policies by setting up law, regulation and standards.
- Formulation and implementation of concrete action programmes.

Depending on the countries; the energy conservation strategy could embody several or all of these components. Table 2 gives some indications about the extent to which policy measures are implemented in each country.

TABLE 2: **Summary of Country-wide Energy Efficiency Policy Measures**

		China	India	Indonesia	Korea	Philip-pines	Taiwan	Thailand
1	Public education campaign	S	n.a	M	M	M	S	n.a
2	Energy auditing/ monitoring	n.a	n.a	S	n.a	S	S	S
3	Technical consulting service	n.a	n.a	S	S	S	S	S
4	Support for R&D	n.a	n.a	M	S	n.a	S	M
5	Technology promotion	S	n.a	W	S	M	S	S
6	Economic incentives	S	W	W	S	W	S	S
7	Standards, law, regulation	S	n.a	M	S	M	S	S
8	Specific action programmes	Yes	No	No	Yes	No	Yes	Yes

Note: S– strong

M – medium

W – weak

n.a – information is not available

2. EVOLUTION OF ENERGY EFFICIENCY

2.1 Energy Intensity of GDP

China[1], Korea, India and Indonesia form the first group with higher intensity. Highest final energy intensity belonged to China, which was still as high as 1.1 toe/'000 US$[2] by the end of the study period even if a substantial reduction of 50% was achieved. In the second place was Korea, which had final energy intensity equal to about 0.55 toe/'000 US$ in 1992. India was in third position with an energy intensity of GDP within the range of 0.4 to 0.45 toe/'000 US$, Indonesia's final intensity exceeded 0.35 toe/'000US$ in all years. Taiwan, Thailand, the Philippines form the second group with final intensities below 0.35 toe/'000 US$.

Among countries, China was singled out as the one which showed the most rapid reduction of final energy intensity. The 50% decrease over the period reflected the outcome of conservation policy, that was implemented together with China's economic reform. However it is necessary to note that even if the pace of intensity reduction was fast, the level of the final intensity was still very high. Taiwan is another country that showed a continuous decline of its final energy intensity. Over twelve years the island had successfully cut down the final intensity by almost 20% (Figure 2), and its real final intensity declined further than that of constant structure (Figures 2 and 3), indicating that there have been structural changes in the economy in favour of less energy intensive sectors.

Korea, on the contrary, displayed different evolution of its final energy intensity with significant decrease till 1989, followed by a swift increase as a result of a decline in its GDP growth, while energy consumption growth remained high.

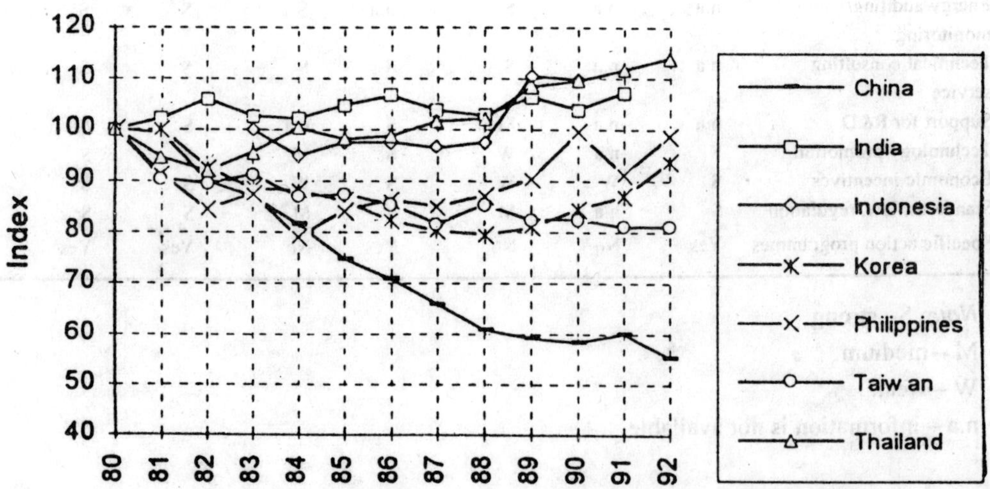

FIGURE 2: *Evolution of Final Energy Intensity*

For other countries, the final intensities increased continuously (India, Indonesia) or fluctuated with upswing in the end of study period (The Philippines, Thailand). The most probable reason being that these countries have been on an intensive process of industrialization during the period under study, expansion of industry, increase of transportation system, improvement of living standards, while infrastructure and technologies were still not yet developed, all factors leading probably to an increase of energy intensities. India has actively promoted the energy intensive industries such as fertilizer, steel, cement, aluminium, heavy chemicals. Structure of Thai and Indonesian economies have also changed in favour of industry; by the end of the period the share of industry in total GDP was 41% and 44%, respectively.

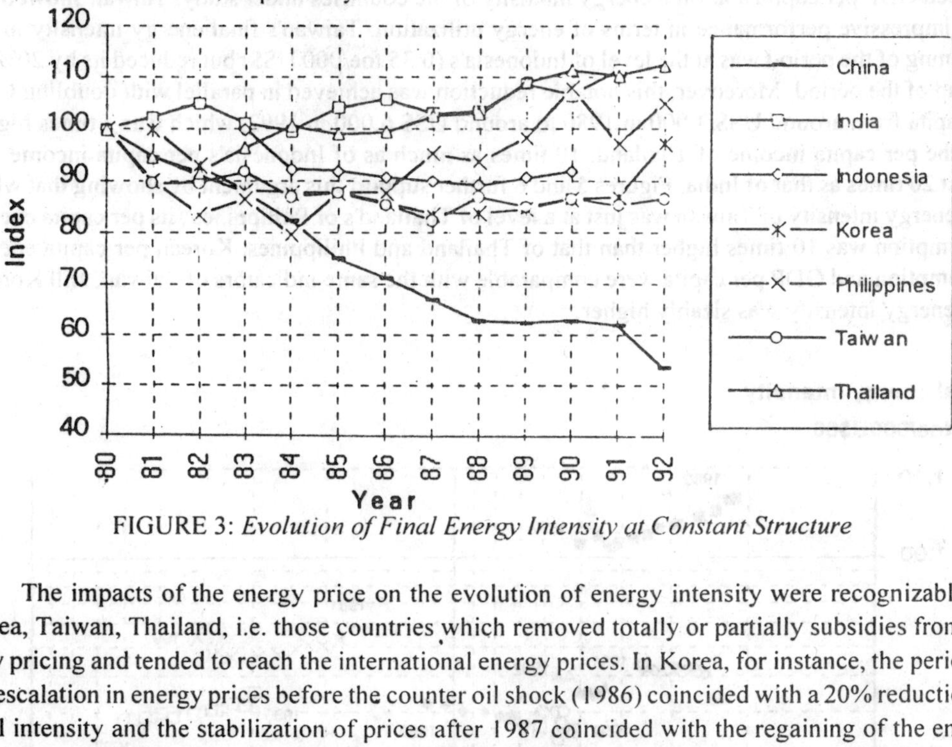

FIGURE 3: *Evolution of Final Energy Intensity at Constant Structure*

The impacts of the energy price on the evolution of energy intensity were recognizable for Korea, Taiwan, Thailand, i.e. those countries which removed totally or partially subsidies from energy pricing and tended to reach the international energy prices. In Korea, for instance, the period of the escalation in energy prices before the counter oil shock (1986) coincided with a 20% reduction of final intensity and the stabilization of prices after 1987 coincided with the regaining of the energy intensity. The same can be seen in Thailand, where the sharp increase of energy prices in the early 80's (by an average factor of 1.38) led to a 10% contraction of final energy intensity in that period. It is interesting to note that, the impact of energy prices on energy intensity vanished as energy prices become stable. In Taiwan, one can see that the liberalization of energy pricing coupled with a strong management policy resulted in constant decrease of energy intensity (20% reduction by the end of study period). The effect of high prices in first–half of 80's lasted beyond that period because the adjustment in direction of more efficient use of energy had been realized in the Taiwanese economy. (Figures 2 and 3 and Annexure B).

For other countries price effect was not as clear as compared to three countries mentioned above; Energy price indices of India, Indonesia. and the Philippines displayed upward trend, in the meantime, their final energy intensities showed the increasing trend as well. The average overall energy commodities price of Philippines, India and Indonesia had increased by factors of 3.0 2.6, and 1.6 respectively during study period (See Annexure B), even though energy prices in these countries were still considered to be low compared to other developing countries. In India for example; subsidies were applied to electricity and middle distillates. In such a case the dynamic impact of energy prices on energy intensity is difficult to monitor. China has exercised the so called "dual pricing", the effect of which on the China's energy intensity is yet to be identified. On the one hand, the dual pricing system corrects the deficiency of the centrally regulated prices and fosters the energy conservation, but on the other hand, some scholars argued that the dual pricing could undermine the implementation of the new and energy efficient technologies.

Figures 4, 5, and 6 symbolize the discussion presented so far. Figure 4 depicts a relationship between GDP per capita and final energy intensity of the countries under study. Taiwan showed the most impressive performance in terms of energy utilization. Taiwan's final energy intensity in the beginning of the period was at the level of Indonesia's (0.35 toe/000 US$) but reduced its by 20% by the end of the period. Moreover, this notable reduction was achieved in parallel with doubling GDP per capita from around US$ 3,000 in 1980 to around US$ 6,000 in 1992, which was 5 times higher than the per capita income of Thailand, 10 times as much as of Indonesia's per capita income and almost 20 times as that of India. Figures 5 and 6 further support this argument by showing that while final energy intensity of Taiwan was just at a level of Thailand's or Philippines', its per capita energy consumption was 10 times higher than that of Thailand and Philippines. Korean per capita energy consumption and GDP per capita were comparable with the same indicators of Taiwan, still Korea's final energy intensity was sizably higher.

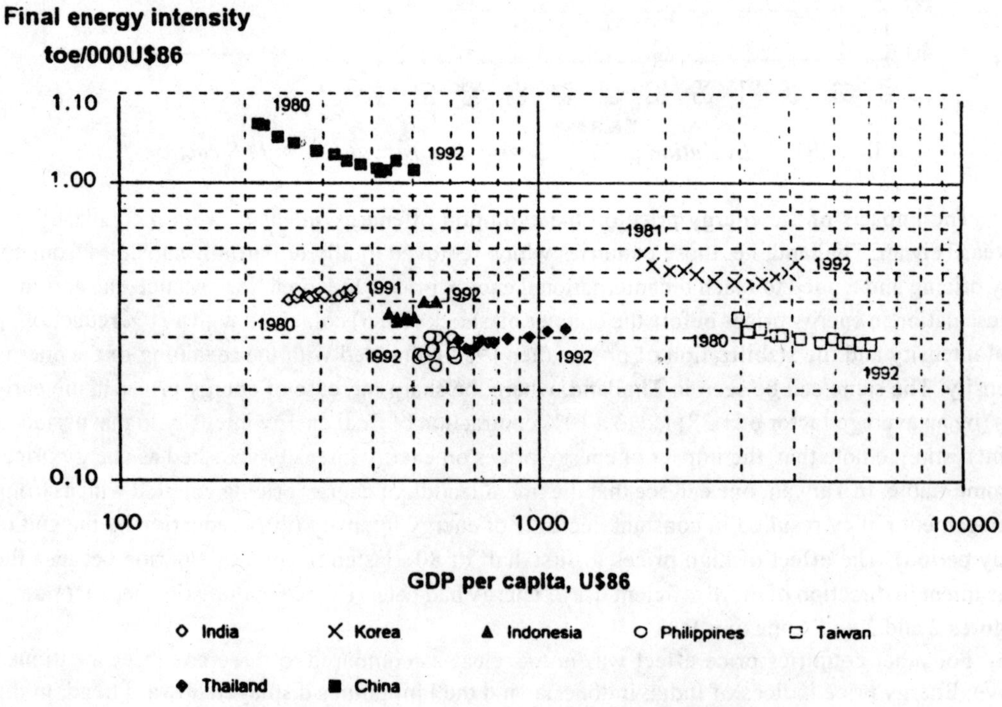

FIGURE 4: *Energy - Economy Trajectories of Seven Countries*

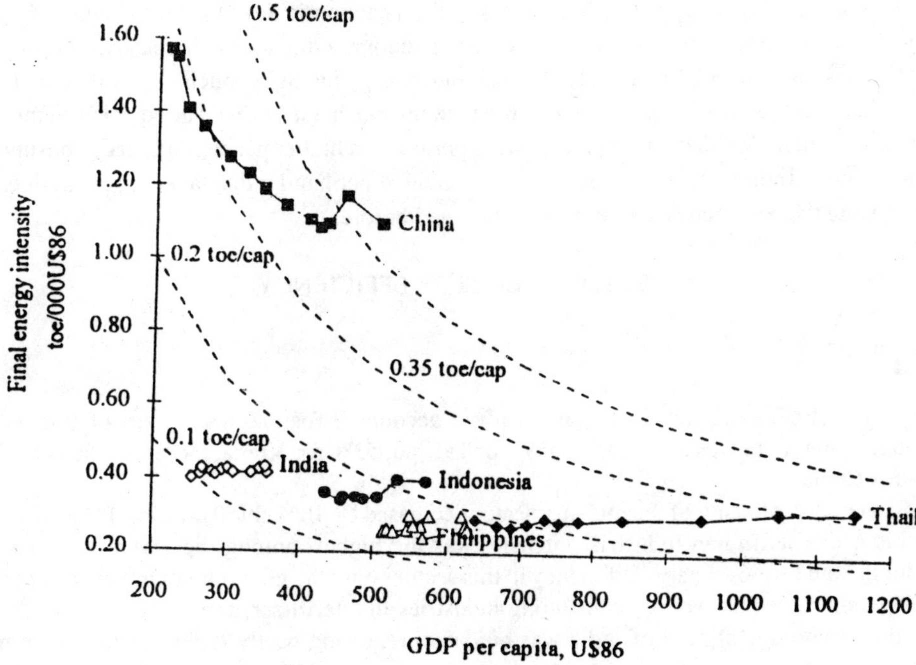

FIGURE 5: *Detailed Presentation of Energy – Economy Trajectories of Five Countries: China, India, Indonesia, Philippines, Thailand*

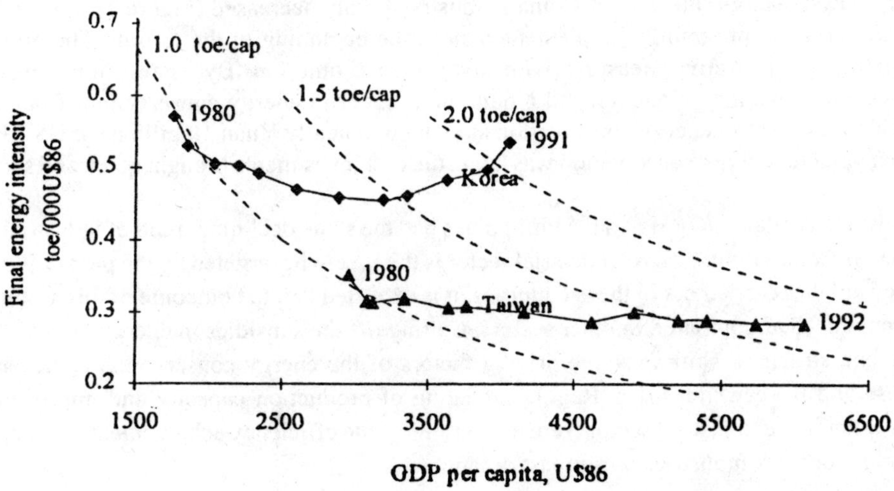

FIGURE 6: *Detailed Presentation of Energy – Economy Trajectories of Two Countries: Korea and Taiwan*

Although there had been a visible progress in their economic development (Figures 4 and 5), India, Indonesia and Philippines were in the similar situation, which was characterized by low per capita indicators. In addition, India experienced a final energy intensity much higher than that of the other two. The energy economy pathway of China, on the one hand, showed good performance with GDP per capita which doubled on the twelve years period, and higher per capita energy consumption compared to India, Indonesia and Philippines. But on the other hand it displayed a very high energy intensity despite the spectacular decrease as mentioned before.

3. SECTORAL ENERGY EFFICIENCY

3.1 Industry

Industry sector of China, India, Korea and Taiwan accounted for the largest part of final energy consumption. This share was averaged at 55% for Taiwan, 53% for Korea, more than 50% for India and 59% for China.

Indian energy intensity of the industry sector decreased by 10% chiefly during 1986-88 period. Despite this decrease, Indian industrial intensity was the highest among the seven countries. The main factor behind this low energy efficiency in this sector being the government policy to encourage the development of heavily energy consuming industries like fertilizer, cement, aluminium, etc. In addition, the technological pool of India was outdated, requiring costly replacement, which many enterprises could not afford.

Traditionally, industrial sector in China had high energy intensity due to the following factors: (i) technologies used were old; (ii) industry sector consisted of a great number of local small scale enterprises, that hindered the economy of scale, leading to lower energy efficiency. However during the study period, energy intensity of China's industry sharply decreased (Figure 7) as a result of an energy conservation programme implemented since the beginning of the reform. The programme incorporated administrative measures with investment promotion. By establishing special loan incentives, the Government encouraged a huge investment in energy conservation: from 1981 to 1985 total investment in energy conservation amounted to nearly Yuan 10 billion (or U$ 3 billion). As the potential for energy conservation was large, the first investments brought quick and significant results.

Taiwan, Thailand, Korea and the Philippines had the same declining trend of industrial energy intensities as China (Figure 7). As industrial sector is the first to be targeted by the pioneering energy efficiency policies carried out in these countries, it is expected that the outcome of this sector will re more pronounced as compared to other sectors. Cutting off the subsidies in energy prices; standard enforcement, structural shift were the driving factors of the energy conservation achievement in Taiwanese and Korean industries. Regular updating of production capacity and improvement of house keeping in the industrial sector were the reasons of the efficiency achievement in Thai, Philippines and Indonesian industries during last decade.

Energy intensive subsectors were the focal point of the conservation efforts. Unit energy consumption of steel and cement production was found declining in most of cases, except India, whose unit energy consumption experienced little or no improvement. (Annexure C).

It is remarkable that in six out of seven countries under this study, industrial energy intensity exhibited a declining tendency. The rate of reduction was high in case of China. Taiwan, moderate in

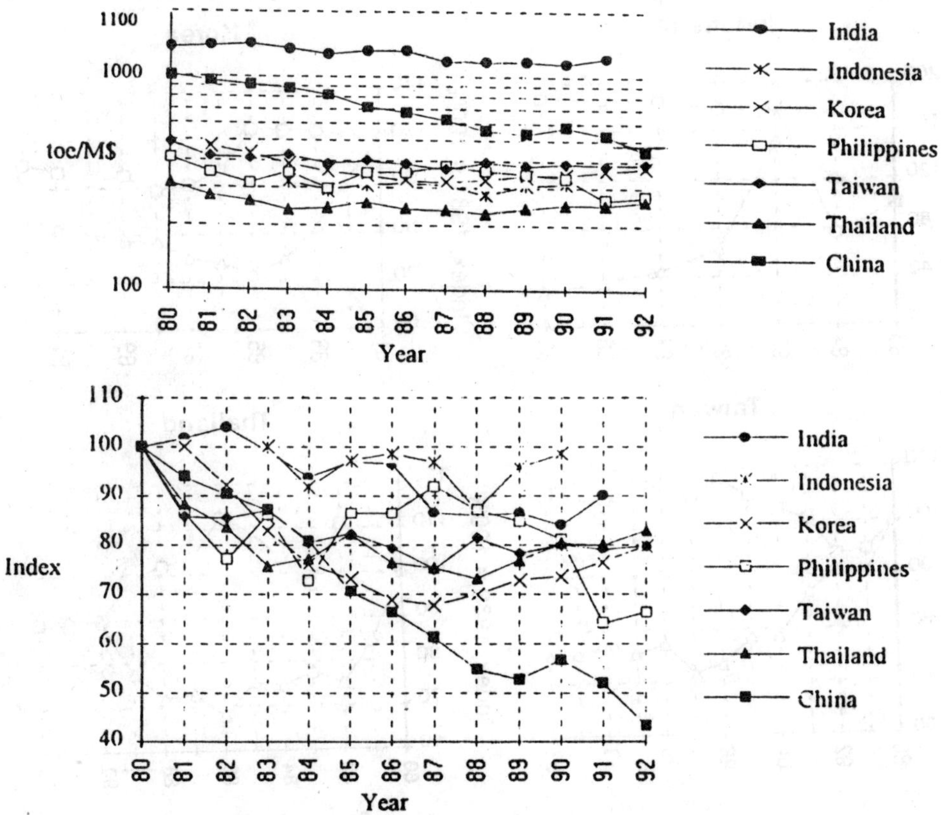

FIGURE 7: *Industrial Intensity of Value Added (Absolute Value and Index)*

case of the others. Exception is Indonesia, whose industrial intensity was constant over the period. (See Figure 7). This fact suggests that even if the energy efficiency management policies were not completed in the majority of the countries, they did bring some results in the industrial sector, which was the prime target of energy conservation policy in all countries. This conclusion is backed by Figure 8, which shows decomposition of the industrial intensity variation with Divisia index of four countries. The figure shows that all efficiency components of Divisia index were decreasing, reflecting a notable achievement in terms of energy conservation in the industrial sector of these countries. In the case of Taiwan, the technological change effect was so strong that it compensated for the effects of structural changes, which tended to push up industrial intensity (see Table 3 and also Figure 3).

The achievement was attained through implementation of the following measures:

- Adequate energy pricing
- Energy auditing and providing technical services for industrial enterprises
- Technological refurbishment, stock renewal
- Enforcement of conservation standards
- Investment in new energy efficient technologies.

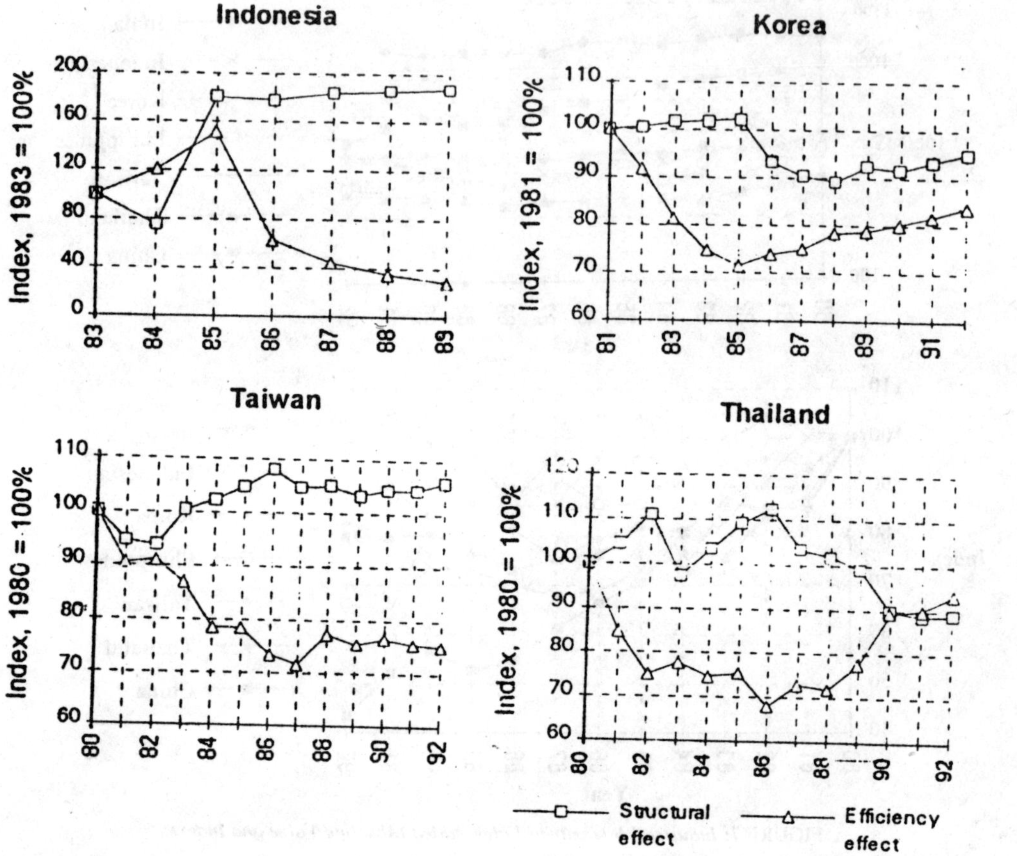

FIGURE 8: *Decomposition of Variation of Industrial Energy Intensity with Divisia Index (4 countries)*

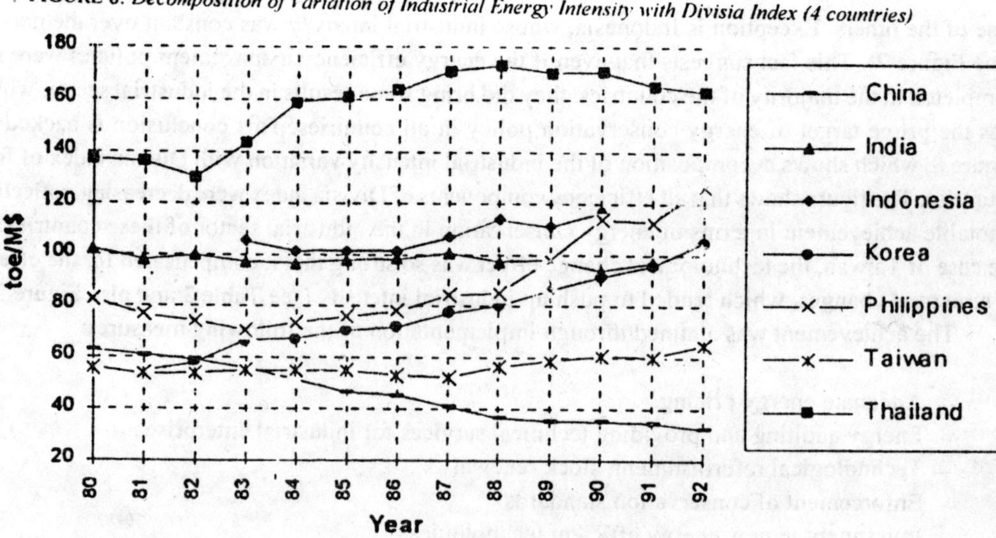

FIGURE 9: *Transport GDP Intensity*

TABLE 3: **Variation of Structural and Technological Effects of Industry Sector in Three Selected Countries Over the Study Period**

	Korea	Taiwan	Thailand
Structural effects	— 3%	+ 6%	— 16 %
Technological change effects	— 18%	— 25%	— 1%
Variation of industrial intensity	— 20%	— 20%	— 17%

3.2 Transport

Transport energy intensity of GDP had an ascending tendency over the study period for most of countries under study (Figure 9). Again there is an exception, notably in the case of China, which had a decreasing intensity. The highest increasing rate (200%) belong to Korea, followed by the Philippines (about 40%) and then by Taiwan and Thailand (around 20%). Indonesia, India had rather constant transport energy intensity.

For all countries the energy consumption of road transport was predominant. For example, at the end of the study period, the share of energy consumption in road transport in total energy consumption of this sector was 86% for Philippines and Indonesia, 75% for Thailand and India. Furthermore, number of vehicles in road transport had dramatically expanded in all countries especially motorcycles and cars (Annexure D). So the combination of these two factors virtually can explain the continuous increase in transport energy intensity of majority of countries.

The energy efficiency of road transport is related to many factors such as: (i) infrastructure of the road system; (ii) vehicle model; (iii) frequency of usage; (iv) traffic condition; (v) driving behaviour, etc. In Asian developing countries, the road infrastructure is quite different from country to country. However, the vehicle fleets are of comparable to developed countries, models. So energy conservation measures are chiefly focused on improvement of road infrastructure, traffic condition and driving behaviour.

The effect of traffic condition on per vehicle energy consumption per annum is difficult to allocate because, on the one hand traffic congestion leads to higher fuel consumption once the vehicle is on the road; but it also leads to infrequent use of the vehicle which in turn prompts a lower consumption per annum. In general, total road fuel consumption is proportional to the stock of vehicles, but one can expect that the more vehicle owned by a family the rarer it being used. Annexure E shows the per vehicle fuel consumption for some countries.

3.3 Residential and Commercial Sector

Data on energy consumption of these two sectors were recorded together in most cases, therefore it is necessary to look at both sectors simultaneously.

The first thing to note in this sector is that the important part of its energy consumption was traditional fuel in all countries except Korea and Taiwan. The second common feature of this sector is the noticeable increase of the average electricity consumption per household. In some cases the average per household's electricity consumption grew even faster than the one of urban household's (Philippines, Indonesia). Over the twelve years period, per household

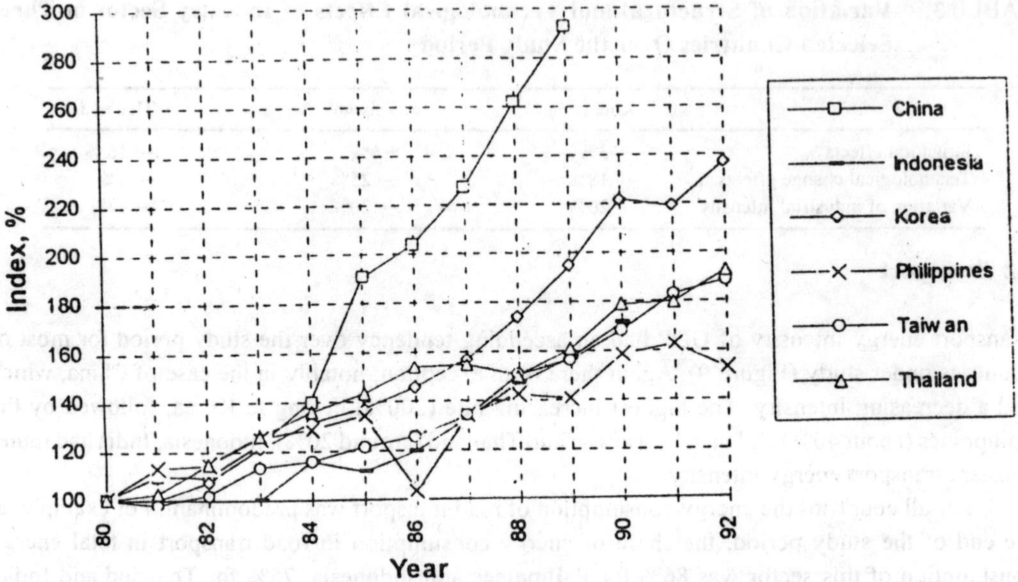

FIGURE 10 : *Evolution of Average Per Household Electricity Consumption*

electricity consumption grew by a factor of 2.0 for Thailand, Indonesia, Korea; by a factor of 1.5 for Philippines and Taiwan. (Figure 10)

Among countries, Korea was the one that has achieved tangible results in energy conservation in this sector. The Korean government had implemented several measures to achieve energy efficiency in the residential and commercial sector, including: investment in constructing district heating system; improvement of energy standards; shifting from low quality energy form (coal, wood) to high quality energy form (gas, petrol).

Energy conservation in this combined sector focused mainly on the commercial side, e.g. commercial building rather than residential side. Many countries have formulated standards for energy efficiency and conservation in building from design to construction and operation stages. (Thailand, Korea, Taiwan) Other countries are preparing to implement these standards. Because cooking using biomass was the main end use omnipresent in rural households of all countries, the improvement of cooking stoves have been a prominent measure in many countries. Introduction and dissemination of other efficient equipment and appliances virtually has not been done in most of countries.

3.4 Agriculture

Agricultural sector plays a considerable role in Asian developing economies in terms of value added, but in terms of energy consumption this sector does not appear to be an important one, especially if only commercial energies are taken into account. The share of agricultural energy consumption in total final commercial energy consumption was just around 5% to 6% for all the countries under study. Mor}over data on energy consumption of this sector was poorly recorded in all countries. Energy conservation measures implemented in this sector were not reported in available documents.

Thus, unavailability of primary information precludes us from making even a rough analysis about efficiency indicators of agricultural sector.

4. CONCLUSION

From the analysis presented so far one can see that socio-economic development of seven selected countries formed a wide spectrum so does the state of the art of their energy efficiency and conservation.

Energy management policy is closely linked to the energy resource endowment situation of a country, which defines its energy management strategy. It is observed that countries, which faced difficulties in meeting energy demand by their own resources tended to adapt energy conservation policy earlier and more effective (Taiwan, Korea). In contrary, countries with sizable energy resources are still in the stage of formulating and implementing energy conservation strategy (Indonesia). It should be stressed on the necessity of persistent pursuit of a rational use of energy in all sectors of the economy.

Energy demand management policy also differs from one country to another. For some countries, the policy is a coherent set of administrative, economic, legislative measures enabling to achieve step by step efficient energy utilization. The other countries have just started with inaugural policy measures such as: propaganda, economic incentives or the like.

In all countries, the enrgy consumption increased rapidly in accordance with the economic expansion but energy utilization in terms of the energy intensities of GDP differs from one country to another. For example, China and Taiwan exhibited declining final energy intensities, while other countries displayed increasing (or sporadic with an increasing end) trend of the final energy intensities. Even in case of China and Taiwan, whose final energy intensities evolved analogously, the absolute values of their intensities were quite different, i.e. the final intensity of Taiwan was just one-twentieth of that of China. Meanwhile, the GDP per capita of China was much lower than the GDP per capita of Taiwan. Also the extensive industrialization process in many countries has contributed to ascending trend of their energy intensities, as observed in this study.

A visible impact of the energy price evolution on the intensity evolution has been monitored in the case of Korea, Taiwan and Thailand in some extent. For other countries the price impact was not evident, because existing pricing mechanism seemed to distort prices of energy commodities.

The energy conservation achievements were more noticeable in Taiwan, Korea and China than in the others, because these cquntries have strong energy efficiency management policies. Other countries have not achieved much result in energy conservation due to their immature management policies.

Among economic sectors, industry had demonstrated considerable achievement of energy efficiency, especially through technological changes within the sector. The efficiency effects of all four countries shown in Figure 8 led to reduction of industrial energy intensity. Commercial sector was being targeted in second priority. Transport and Residential sector are still free from vigorous energy conservation measures. It seems that aside from Taiwan and Korea, promotion, dissemination of energy efficient technologies proceed slowly due to financial constraints. Moreover, the task of industrialization sometimes downplays the priority of energy conservation and the expansion of vital to the economy industries sometimes contradicts the need to conserve energy though in long term they should go hand in hand in an optimal development strategy.

NOTES

1 Chinese statistics records Gross Output value instead of GDP (See more detail in "Report on China").

2 From this point onward monetary unit is in US$ 1986 Constant price.

REFERENCES

ADB, 1992 "Energy indicators of developing members countries of ADB".

ADB, 1994 "Key indicators of developing Asian and Pacific countries".

Akiyama, Ishiguro, 1994 "Structure of and prospects for energy demand in five major Asian developing countries", The World Bank.

DEDP, "Thailand Energy Situation" (various issues).

Department of Energy, 1992 "Philippines Energy Plan 1993 — 2000".

Energy Commission, Ministry of Economic Affairs, 1994 "The Energy Situation in Taiwan Republic of China".

Korea Energy Economic Institute 1993 (a) "Energy conservation study for Korea" Research Report, KEERI.

Korea Energy Economic Research Institute 1993 (b) "Year Book of Energy Statistics".

Lefevre T., Chen X., N.C. Thanh, Bui D.T. 1995 "Cross Countries Comparison of Energy Efficiency Indicators — Report on China", *ADEME/AIT Project*, Bangkok Thailand.

Lefevre T., Chen X., N.C. Thanh, Bui D.T., 1995 "Cross Countries Comparison of Energy Efficiency Indicators — Report on India", *ADEME/AIT Project*, Bangkok Thailand.

Lefevre T., Chen X., Bui D.T., N.C. Thanh, 1995 "Cross Countries Comparison of Energy Efficiency Indicators — Report on Inelonesia", *ADEME/AIT Project*, Bangkok Thailand.

Lefevre T., Chen X., N.C. Thanh, Bui D.T., 1995 "Cross Countries Comparison of Energy Efficiency Indicators — Report on Korea", *ADEME/AIT Project*, Bangkok Thailand.

Lefevre T., Chen X., Bui D.T., N.C. Thanh, 1995 "Cross Countries Comparison of Energy Efficiency Indicators — Report on The Philippines", *ADEME/AIT Project*, Bangkok Thailand.

Lefevre T., Chen X., N.C. Thanh, Bui D.T., 1995 "Cross Countries Comparison of Energy Efficiency Indicators — Report on Taiwan", *ADEME/AIT Project*, Bangkok, Thailand.

Lefevre, T., Chen X., N.C. Thanh, Bui D.T., 1995 "Cross Countries Comparison of Energy Efficiency Indicators — Report on Thailand", *ADEME/AIT Project*, Bangkok, Thailand.

Lefevre T., Chen X., Bui D.T., N.C. Thanh. 1995 "Synthesis Report on Cross Countries Comparison of Energy Efficiency Indicators in Asian Countries" Revue de L'energie, No. 470, 1995.

MOE, 1992 "Energy in China", Ministry of Energy.

TERI, 1994 "TERI Energy Data Directory and Yearbook 1994/1995", Tata Energy Research Institute.

UN 1993, "Statistical Yearbook for Asia and the Pacific".

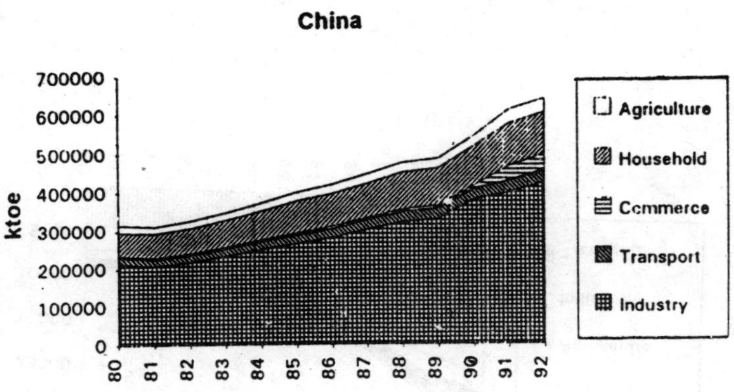

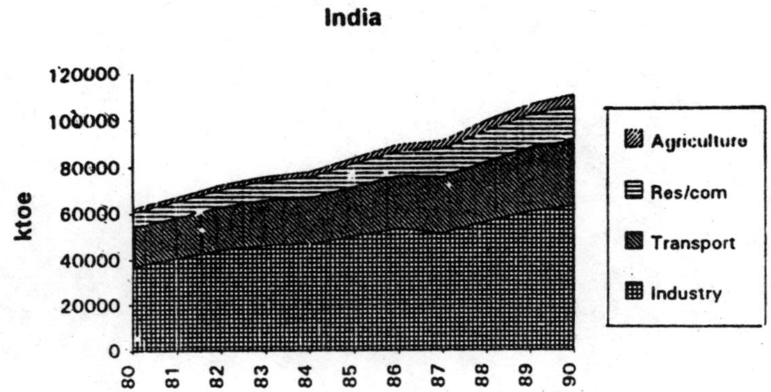

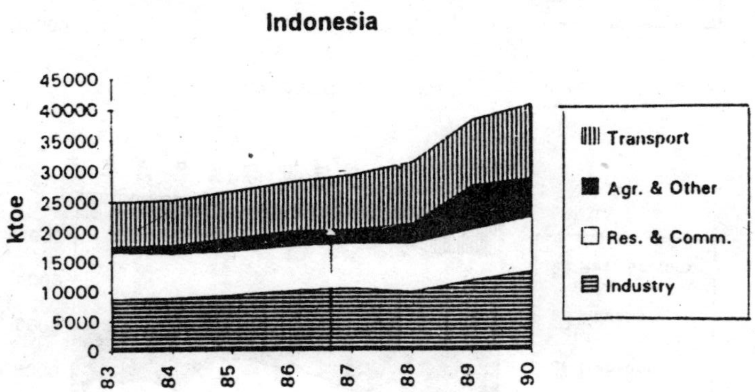

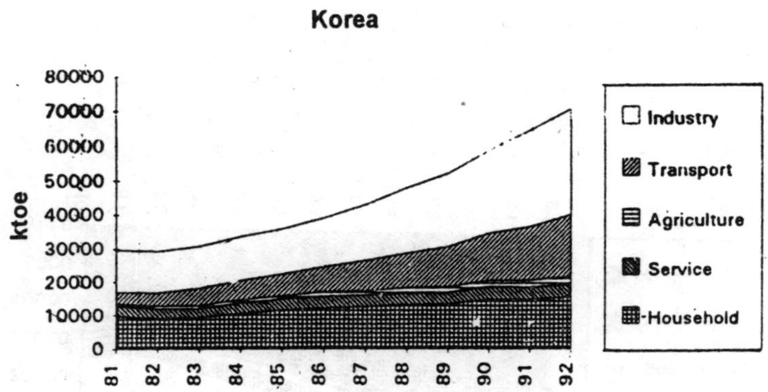

ANNEXURE 'A'

FINAL ENERGY CONSUMPTION BY SECTORS

ANNEXURE 'A' — *Contd.*

Taiwan

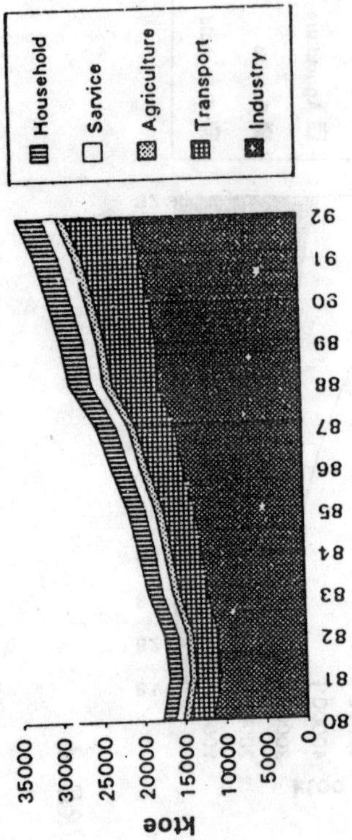

Philippines

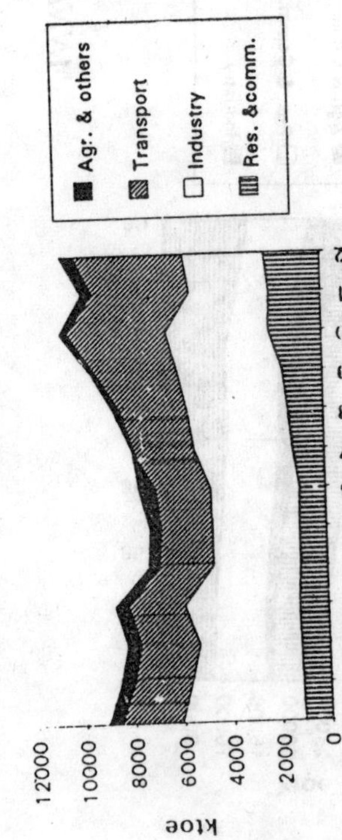

Thailand

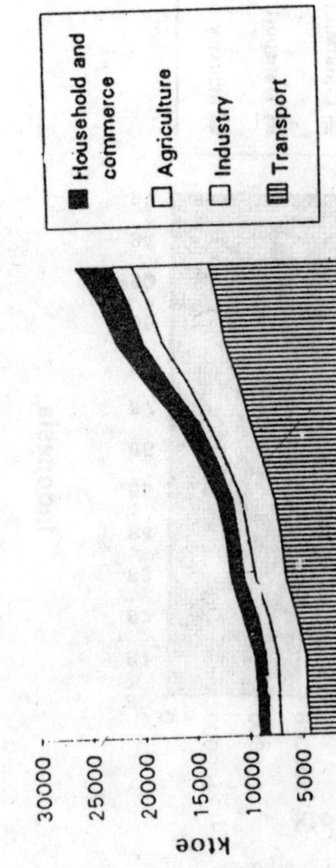

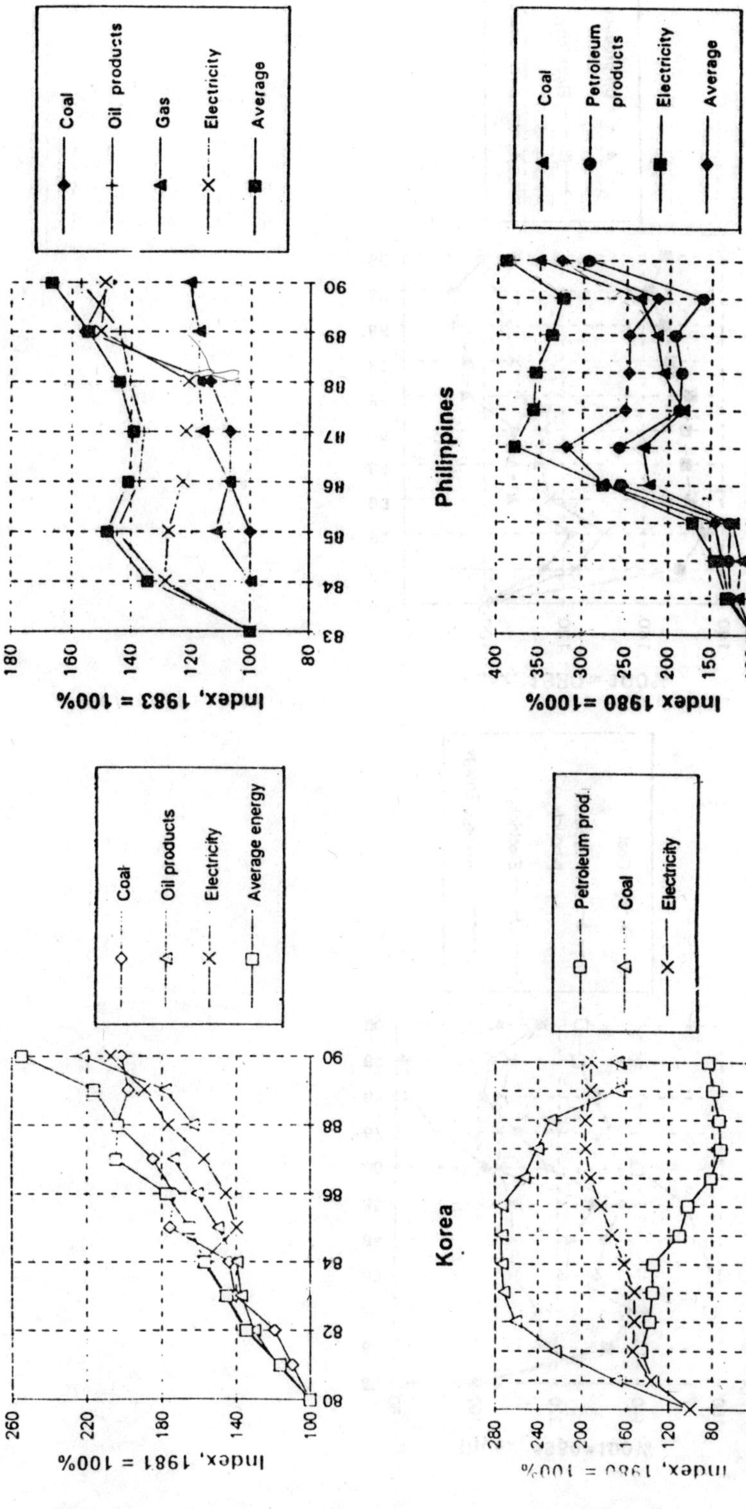

ANNEXURE 'B'

ENERGY PRICE INDEX

ANNEXURE 'B' — *Contd.*

Taiwan

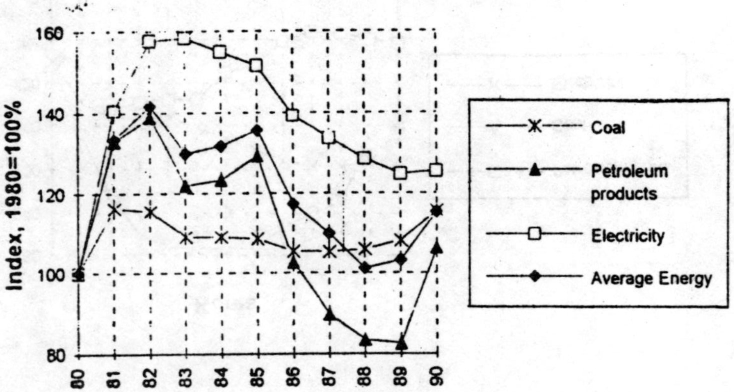

Thailand

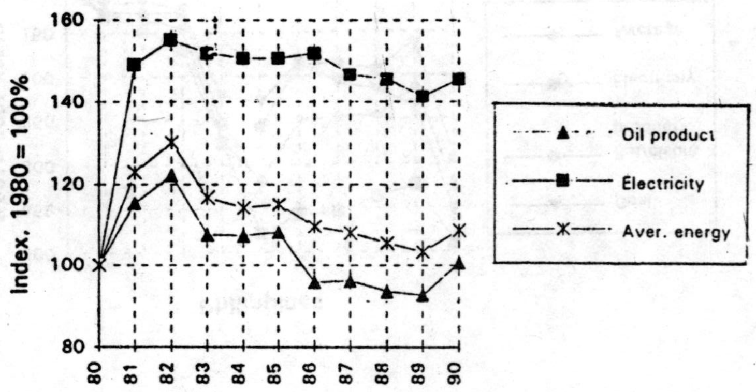

ANNEXURE 'C'

UNIT ENERGY CONSUMPTION IN STEEL AND CEMENT INDUSTRIES

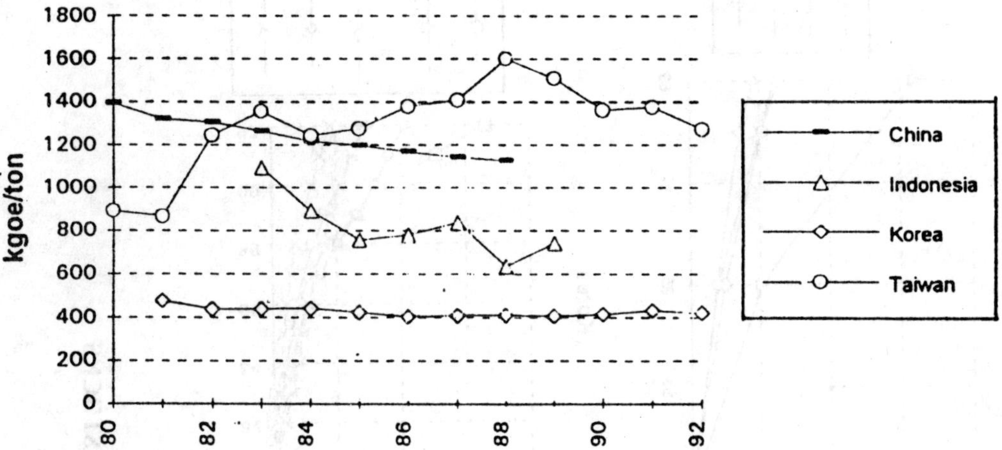

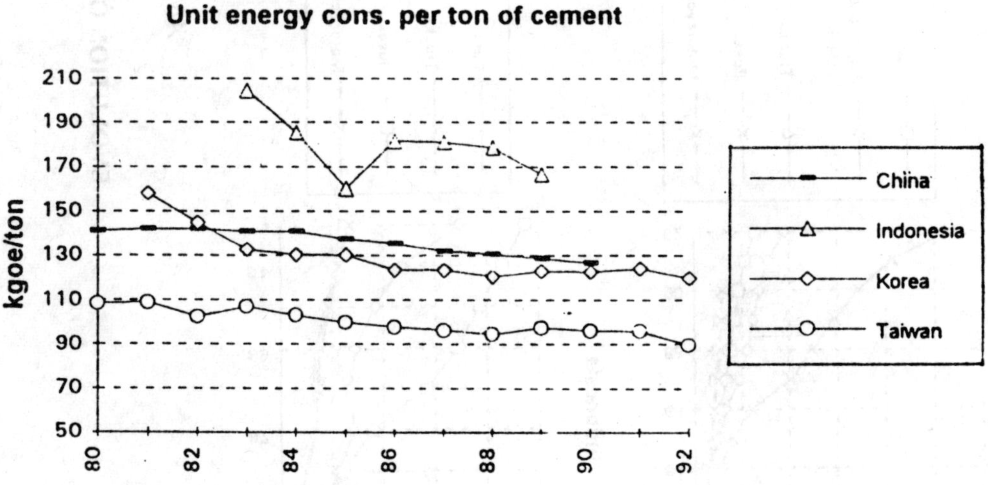

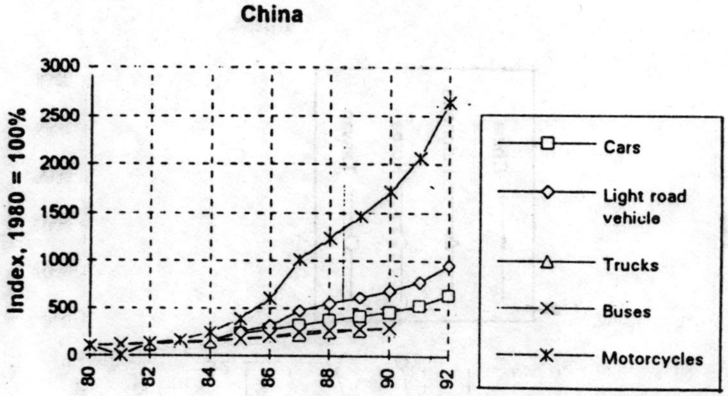

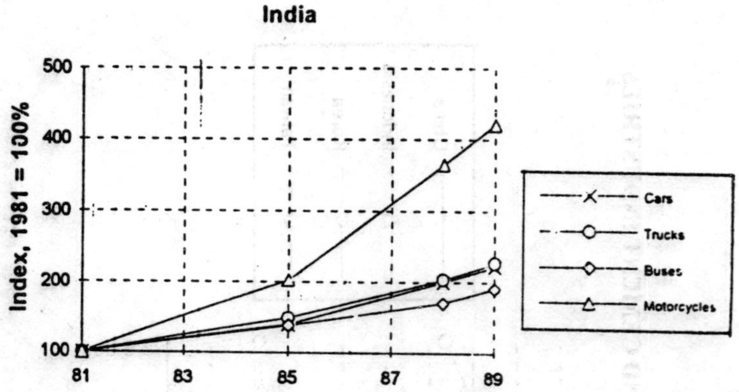

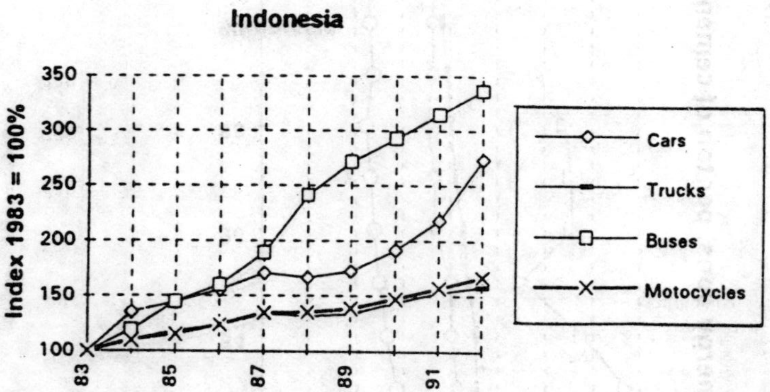

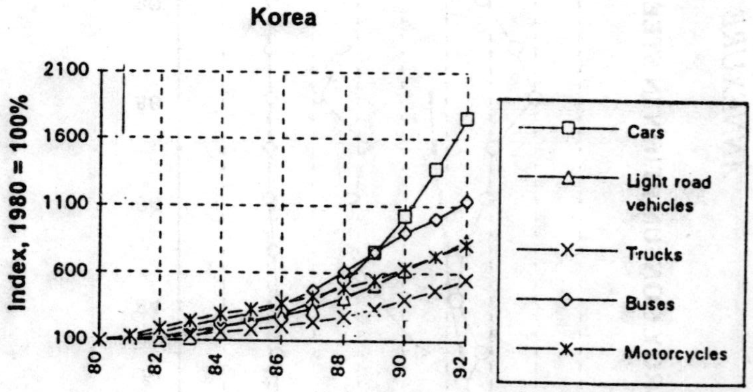

ANNEXURE 'D'

EVOLUTION OF THE VEHICLE STOCKS

ANNEXURE 'D' — *Contd.*

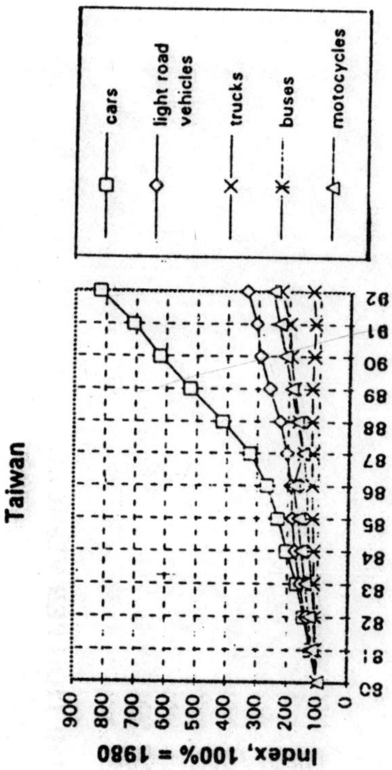

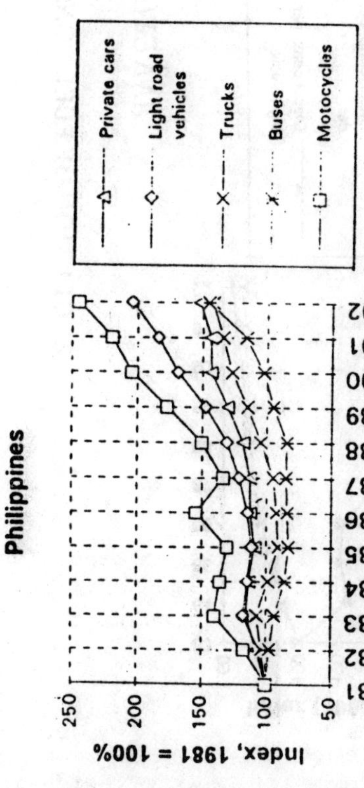

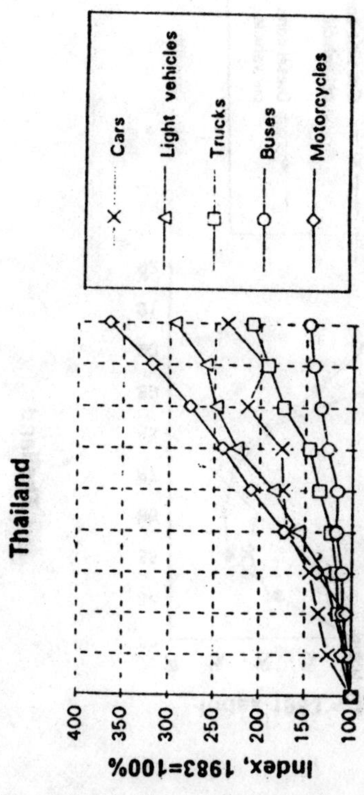

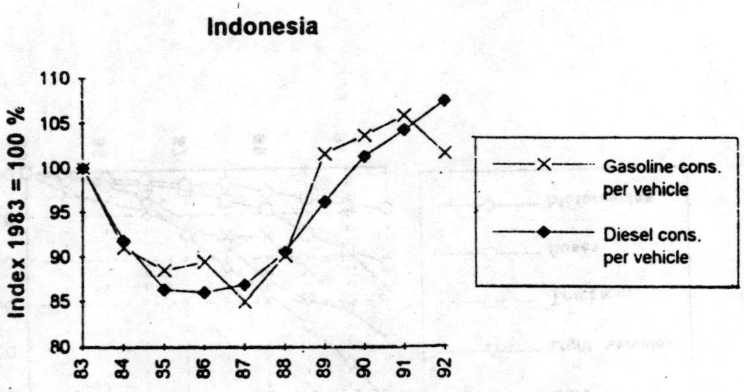

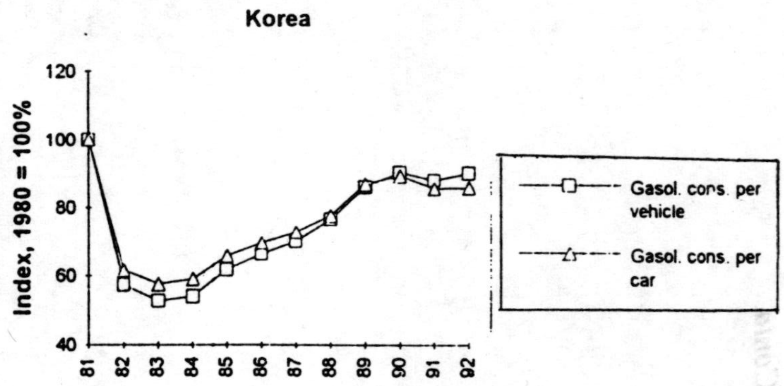

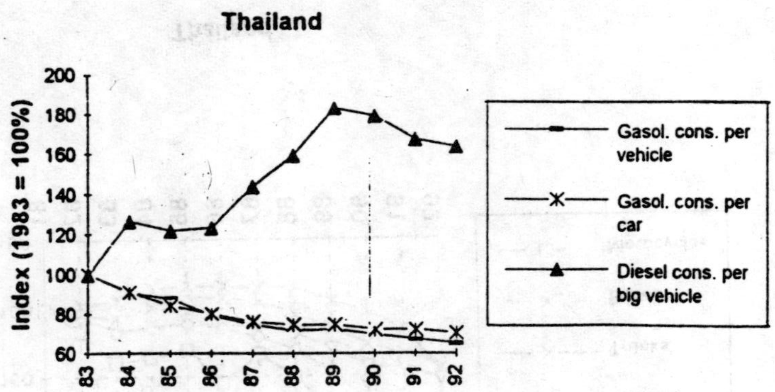

ANNEXURE 'E'

EVOLUTION OF FUEL CONSUMPTION PER VEHICLE

PETROLEUM CONSERVATION AND PCRA'S ROLE— INDIAN PERSPECTIVE

9

K.K. DHINGRA
Executive Director

INTRODUCTION

In India, the energy sector has been the centre of much debate in recent years. The debate reflects the importance of the sector, the amount of the investments envisaged and the size of the market for energy in India. Nearly 30% of the total plan outlay of the government are earmarked for the energy sector every year and still traditional and inefficient sources of energy account for nearly 40% of the total energy consumed in the country.

ENERGY SCENARIO IN INDIA

The requirements of energy in India are met through a combination of commercial energy (electricity, coal, oil and gas) and non-commercial energy. In India, commercial energy consumption is concentrated in the urban and sub-urban areas for industrial, commercial and even domestic use. The non-commercial energy sources are widely distributed in rural areas in the form of firewood, agricultural residues, animal manure, and human and animal power.

While during 1994-95 total energy requirement was to the tune of 219 Million Metric Tonne Oil Equivalent (MMTOE), as per the projections of future requirements, this will exponentially increase to 537 MMTOE in year 2010 in a business as usual scenario. However, with serious conservation efforts, the same can be reduced to 442 MMTOE, giving a saving of around 95 MMTOE.

In India, future economic progress and quality of life of the masses will depend on how we can successfully implement an effective petroleum conservation programme. To understand the subject in national perspective, vital issues such as importance of energy for growth and development, demand-supply position of petroleum products and importance of oil conservation and substitution etc. are important.

PETROLEUM CONSUMPTION IN INDIA

The high rate of growth in the consumption of petroleum products in India has outstripped the indigenous production of crude oil, causing sharp decline in the level of self-sufficiency. With the possible hardening of the oil prices, depreciation of Rupee, the enlarging proportions of the oil import bill will have a staggering impact on the balance of payments position. The crude oil production in the country and the oil import bill to bridge the supply-demand gap for 1995-96 and projected figure for 1996-97 show the following alarming situation on the position of balance of payment.

Year	Crude oil Production (MMT)	Petroleum Products Consumption (MMT)	Percentage Self Sufficiency	Import Bill (Rs. Crores)
1995-96	34.6	72.5	48	24,095
1996-97 (Projected)	33.0	80.9	41	34,550

The demand for petroleum products is expected to further increase sharply to around 113 MMT by the terminal year of Ninth Plan (2001-02) and 155 MMT by the terminal year of Tenth Plan (2006-07).

Moreover, the position of recoverable crude oil reserves have shown a declining trend and currently the country has reserves of around 739 MMT, which would sustain production flow for only 21 years at the current rate of depletion. This underscores the imperative need for maximising petroleum conservation in addition to resorting to major technological changes and use of alternative forms of energy.

PETROLEUM CONSERVATION AND PCRA'S ROLE

The spiralling growth in the consumption of petroleum products with its consequential impact on the foreign exchange outflow has focused the attention of the entire country to take immediate steps to reduce our dependence on liquid hydrocarbons. Conservation and efficient use of petroleum products can be very effective in curbing this ever increasing demand of petroleum products and has emerged as a major supply option.

In India, the focus on conservation of petroleum products was initiated with the setting up of the Petroleum Conservation Action Group (PCAG) in 1976 which was later registered as a full-fledged organisation in 1978 called "Petroleum Conservation Research Association (PCRA)" to undertake studies for identifying the potential and to make recommendations for achieving conservation of petroleum products in various sectors of the economy. PCRA was also entrusted with the task of sponsoring R&D activities for the development of fuel efficient equipment/devices and running of multi-media campaigns for creating mass awareness for the conservation of petroleum products.

PCRA IN INDUSTRIAL SECTOR

Industrial sector consumes more than 50% of the commercial energy used in the country. Indian industries have higher energy intensity compared to their counterparts in developed countries. In India, the average energy consumption for steel production is 9.5 million Kcal/Tonne whereas in the developed countries, the figure is 4-6 Million Kcal/Tonne. Likewise, in India, the specific energy consumption for cement production is 2.0 Million Kcal/Tonne against 0.82—1.4 Million Kcal/Tonne in developed countries.

To help the industries in saving the precious fuel, PCRA offer free consultancy services. The services are open to all industrial units consuming petroleum fuels. Over 6000 such units have already been surveyed. The result of these studies is that every year, fuel worth about Rs. 80 crores is saved.

The activities undertaken by PCRA for conservation of petroleum products in the industrial sector are as follows:

- Fuel Oil Diagnostic Studies
- Energy Audits
- Helping small scale industries
- Subsidy towards conduct of energy audit in the industries
- Empanelment of energy auditors
- Adoption of selected industries for continuous monitoring of their operations with the objective of achieving conservation
- Promotion of fuel-efficient equipment such as boilers, low excess air film burners etc. to achieve fuel economy.
- PCRA Boiler Modernisation Scheme under which soft loans are extended to industrial units by oil companies for replacement of oil fired inefficient boilers with efficient boilers.
- Development of specific oil consumption norms in industries such as fertilizers, steel rerolling mills, glass, ceramics etc.
- Human resource management—focus on training and retraining of energy managers.
- Substitution of liquid hydrocarbon by coal, natural gas and other products.
- Promotion of fuel-efficient systems, equipment/appliances.

PCRA IN TRANSPORT SECTOR

This is the largest consuming sector of petroleum products and cosumes about the entire quantity of petrol and about 70% of diesel. In the transport sector, the main emphasis is on:

- Development of fuel-efficient engines
- Setting up of fuel efficiency norms

— Training of drivers and trainers in acquiring and applying improved driving skills
— Establishing model depots
— Establishing model garages
— Re-refining and recycling of used lubricating oils, promoting high performance oils etc.
— Instituting diesel conservation awards for State Road Transport Undertakings
— Setting up of Action Groups and making them result-oriented
— Development and promotion of fuel-efficient technologies
— Inter-fuel substitution etc.

PCRA IN AGRICULTURAL SECTOR

More than 6 million tonnes of diesel is consumed annually by the agricultural sector, mainly through tractors and lift irrigation pumpsets. In the agricultural sector, the areas of stress are:

— Setting up of demonstration centres
— Partial or complete rectification of diesel-powered fuel-inefficient irrigation pumpsets
— Replacement of fuel-inefficient footvalves by those bearing ISI mark
— Upgradation of footvalve testing facilities of various manufacturers to meet BIS requirements
— Imparting education to farmers on oil conservation through van publicity campaigns, exhibitions, kisan melas, workshops, clinics etc.
— Better maintenance and operation of tractors, agricultural implements etc.

PCRA IN DOMESTIC SECTOR

The domestic sector where nearly 20% of our total petroleum products are consumed for cooking and lighting of rooms is another area where tremendous savings are possible.

PCRA has conducted a series of experiments on how to save cooking gas (LPG) and kerosene in Indian homes. Certain cooking habits cause substantial waste of fuel. PCRA studies have revealed that it is possible to save about 30% of cooking gas or kerosene by following good cooking practices.

In the demestic sector, the focus is on :

— Better cooking habits aimed at the conservation of LPG and kerosene
— Development of fuel-efficient kerosene wick stoves, LPG stoves and lighting appliances as also promoting their use
— Encouraging the use of alternative sources of energy such as solar, bio-gas etc.

PCRA IN MASS AWARENESS

PCRA has identified various measures to propagate the cause of petroleum conservation. One of them being public awareness campaigns through mass media and printed literature.

Millions of housewives, farmers, motorists, truck/bus drivers, fleet operators and industrial units account for consumption of petroleum products. Education of these widely dispersed millions is a challenge which PCRA is meeting through the publicity campaigns in the mass media and distribution of booklets in various languages in transport, industrial, domestic and agricultural sectors.

CONCLUSION

The developed countries have reviewed the energy scenario on a systematic and all-round basis introducing new, more energy-efficient technologies, improving the energy efficiency of equipment and systems and also introducing efective petroleum conservation measures in the existing and new plants and equipment.

TRAFFIC, ROADS AND THE ENERGY INTER-RELATIONSHIP

10

PROF. (DR.) A.K. GUPTA
Director,
Central Road Research Institute,
New Delhi 110 020

INTRODUCTION

Transport sector bears a close and complex relationship with all other sectors of the economy. It is a vital infrastructure for rapid economic growth of the country. Development of agriculture, industry, mining, tourism, forestry, and national as well as international trade depends upon an efficient and speedy transportation network. Socio-economic planning begets structural shifts in the economy, greatly influencing the pattern of movement of people and goods; and, therefore, the nature and quantum of demand for transportation, social activities such as education and health, strategic and defence requirements also ask for good transportation infrastructure.

While the transport sector tends to act as the prime mover of the development process, it must also respond to the development process, if the latter has to be smooth and sustainable. It is necessary to anticipate the demand on the transport system ahead of socio-economic changes. However, the transport infrastructure has expanded considerably in terms of network and services. With economic growth as projected from the 75% increase in electricity generation by 2010 AD, the transport activity is further expected to rise significantly. Thus, the transport planning has to be integrated with overall development plans.

The strategy for energy development also forms an integral part of overall economic development strategy. Efficient use of resources and long-term sustainability are the two important objectives of economic planning. The transport sector accounts for a substantial share of total commercial energy consumption. In India, Transport sector had its share of 46.2% in 1953-54 which reduces to 24.5% in 1990-91 from final commercial energy consumption. The decrease in percentage is attributed mainly to the replacement of coal in railways, by diesel and electricity which are more energy efficient than coal for transportation. Among various modes, roads and road transport occupy significant place in transportation system due to distinctive features. It consumes 25% of total oil consumed in India which was 89% (in 1990-91) of energy in transport sector.

TRENDS AND PROJECTIONS

The total energy supplies increased from 82.7 MTOE (Million tonnes of oil equivalent) in 1950-51 to about 291 MTOE in 1990-91. Final commercial energy consumption increased from 19.76 MTOE in 1953-54 to 101.5 MTOE in 1990-91 at an implicit average annual growth rate of 4.52%.

The Oil Application Ratio (OAR) i.e. ratio of oil consumption in different sectors to the total consumption of oil in the country has increased from 0.30 in 1970-71 to 0.35 in 1990-91 for the transport sector. The Oil Use Ratio (OUR) i.e. ratio of oil consumption in that sector total energy consumption in that sector has increased from 0.38 in 1970-71 to 0.89 in 1990-91 for the sector.

The significance of road network can be inferred from the increase in freight traffic from 34.5% in 1970-71 to 54.0% in 1988-89 and in passenger traffic from 67.1% in 1970-71 to 77.0% in 1988-89.

TABLE-1: Shares of Rail and Road Traffic: 1970-71 to 1991-92

Mode	1970/71	1973/74	1976/77	1979/80	1982/83	1985/86	1988/89
Freight traffic							
Total (billion tonne-km)	194.4	199.1	251.5	285.8	365.9	422.7	505.0
Rail (%)	65.5	61.5	62.3	54.6	48.6	48.7	46.0
Road (%)	34.5	38.5	37.7	45.4	51.4	51.3	54.0
Passenger traffic							
Total (billion passenger-km)	359.1	427.6	549.3	710.4	955.9	1079.2	1169.0
Rail (%)	32.9	31.6	29.7	27.9	23.7	22.2	23.0
Road (%)	67.1	68.4	70.3	72.1	76.3	77.8	77.0

Source:

1. *Annual Statistical Statements* (various issues) Railway Board, Indian Railways.
2. *Basic Statistics Relating to the Indian Economy*: India, Aug. 1993. p. 5.8. Bombay: Centre for Monitoring Indian Economy.

Considering growth rate of 5.5% during Eighth Plan, road transport is expected to be increased by three times.

Table 2: Growth of Road Traffic

	1950-51	1988-89	1990-91	Projected for 2002-03 Low	Medium	High Growth
Passenger (BPK)	31	1000	1200	2309	3001	3844
Freight (BTK)	5.5	250	295	791	904	1030

Value added in the transport sector also symbolises the importance of road over other transportation modes. Road sector has increased from 49% in 1980-81 to 58% in 1989-90 (considering

1980-81 prices). The public sector investments in the roads has decreased from 6.9% during First Five Year Plan to 3.0% during Eighth Five Year Plan. But the annual revenue from road users showed buoyant growth from 47 crore in 1950-51 to Rs. 8,500 crore recently. A total investment of Rs. 20,000 crore has gone in so far on the road network. The present revenue earnings imply a 42% return per annum. Vehicle operating costs (VOC) on roads in India are of the order of Rs. 100,000 crore annually and according to some estimates, 15% of such VOC (i.e. Rs. 15,000 crore) can be saved provided the roads are improved and maintained to required level. The saving on account of fuel alone can be of the order of Rs. 2,250 crore.

The length of our road network has increased five-fold from 4 lakh km in 1951 to 20 lakh km in 1991. While the vehicle population has grown from 3 lakh in 1951 to about 213 lakh during the same period, registering a 70 fold increase.

TABLE 3: **Growth of Road Network and Vehicle Population**

Year	Road Network lakh km	Of which National Highways plus State Highways lakh km	No. of Vehicles (lakh)	Vehicle per km
1950-51	4.00	0.65	3.06	0.76
1970-71	9.17 (129%)	1.04 (60%)	18.65 (509%)	2.03
1990-91	20.37 (122%)	1.61 (55%)	213.10 (1042%)	10.47

Note: Figures within brackets indicate increase over the previous 20-year period.

According to the Association of Indian Manufacturers, the estimates of the likely vehicle population in India by 2001 AD are as given in the Table 4.

TABLE 4 : **Projected Vehicle Population in India in 2001 AD**

In Numbers

Year	Two Wheelers	Three Wheelers	Cars	Jeeps	Light M & H C. Vs	Total
1992-93	17,025,582	861,743	2,828,992	501,388	1,980,385	23,198,090
1993-94	18,711,683	941,362	3,023,597	548,013	2,118,771	25,343,426
1994-95 $	20,689,449	1,010,735	3,198,325	595,853	2,267,496	27,761,857
1995-96	23,305,449	1,108,835	3,459,925	661,253	2,480,046	31,015,507
1996-97	26,183,049	1,216,745	3,759,925	733,193	2,713,851	34,606,762
1997-98	29,348,409	1,335,446	4,159,925	812,327	2,971,036	38,627,143
1998-99	32,830,305	1,466,017	4,709,925	899,374	3,253,940	43,159,562
1999-2000	36,660,391	1,609,645	5,409,925	995,126	3,565,135	48,240,222
2000-2001 #	40,260,383	1,741,122	6,166,026	1,083,947	3,848,837	53,100,314

$ Population arrived at after reducing scrappage @ 2.0% Data published by the Ministry # Population arrived after reducing scrappage @ 1.5% of Surface Transport in Motor Transport Yearly sales for 94-95 to 2000-2001 are estimated Statistics of India, they Include data on by AIAM Secretariat based on statistical trends, for Trailers & Other Vehicles which are not arriving at population of vihicles Included above.

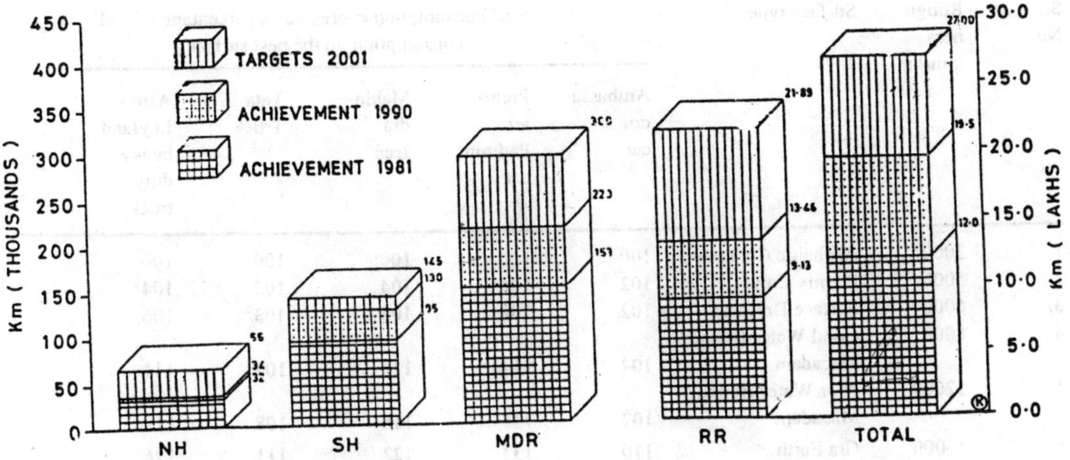

TABLE 5 : **Road Targets (2001) and Achievements (1990)**

TABLE 6 : **Fuel Consumption Savings (in per cent) Possible by Pavement Widening**

Terrain	Widening single lane to intermediate lane			Widening single lane to two-lanes			Widening intermediate lane to two-lanes		
	Car	Truck	Bus	Car	Truck	Bus	Car	Truck	Bus
Plain	1	1	1	2	2	2	1	1	1
Rolling	5	3	2	8	3	2	2	2	2
Hilly	15	8	10	20	10	14	10	3	3

TABLE 7: **Fuel Savings by Various Road Improvement Schemes**

S. No.	Nature of improvement	Fuel saving (%)
1.	Widening single lane pavement to intermediate width and improving its riding quality	5
2.	Widening single lane pavement to two-lanes with minor improvement in riding quality	6
3.	Widening single lane pavement to two-lanes. strengthening the pavement and improving riding quality.	10
4.	Providing thin bituminous surface on water-bound macadam roads	10
5.	Paving earthen and gravel roads with a water-bound macadam, surfaced with thin bituminous course	15
6.	Strengthening weak bituminous pavements and providing superior bituminous surfacing	5

TABLE 8

S. No.	Roughness mm/km	Surface type	Fuel consumption expressed as percentage of fuel consumption on the best surface				
			Ambassador car	Premier Padmini car	Mahindra Jeep	Tata Truck	Ashok Leyland heavy duty truck
1.	3000	Asphaltic Concrete	100	100	100	100	100
2.	5000	Premix Carpet	102	106	104	102	104
3.	6000	Surface Dressing	102	108	105	103	106
4.	8000	Good Water-Bound-Macadam	104	114	110	105	111
5.	12000	Poor Water-Bound-Macadam	107	125	116	108	119
6.	15000	Gra Earth	110	133	122	111	126

TABLE 9: Effect of Upward Gradient on Fuel Consumption

Gradient (Per cent)	Fuel consumption expressed as a percentage of fuel consumption on level road				
	Ambassador car	Premier Padmini car	Mahindra Jeep	Tata Truck	Ashok Leyland heavy duty truck
0	100	100	100	100	100
1	118	113	118	125	114
2	137	126	137	150	128
3	155	140	155	175	142
4	173	153	174	200	156
5	192	166	192	227	170
6	210	179	210	250	184
7	228	193	229	275	198

TABLE 10: Effect of Speed on Fuel Consumption of Vehicles

Speed km/hr.	Ambassador car	Premier Padmini car	Mahindra Jeep	Tata Truck	Ashok Leyland heavy duty truck
10	239	126	187	298	145
15				203	120
20	134	105	139	156	108
25				131	102
30	106	100	104	115	100
35			100	107	100
40	100	100	102	102	102
45				100	104
50	105	102	119	102	108
60	117	106	148	108	118
70	134	111	188	122	132
80	156	118		140	148
90	183	126		162	
100	214	135			

TABLE 11: Idle Fuel Consumption

S. No.	Vehicle	Idle fuel consumption (cc/min)
1.	Ambassador Car	13.0
2.	Premier Padmini Car	10.5
3.	Mahindra Jeep	12.3
4.	Tata Truck	15.3
5.	Ashok Leyland heavy duty truck	35.4

TABLE 12: Deficiencies in the National Highways, Cost of their Removal and Fuel Economy Possible

S. No.	Deficiency	Length (km)	Cost of removal of deficiency (Rs crores)	Fuel economy possible for year excluding of taxes (Rs crores)
1.	Low grade sections to be upgraded and widened	671	100	4
2.	Single lane road to be widened and strengthened	7371	737	52
3.	Strengthening two-lane pavements	19291	1450	182
			2287	238
		Say	2300	240

TABLE 13: **Investments on Roads other than National Highways and the likely Fuel Savings therefrom**

S. No.	Road category	Total Length (km)	Min. Investment needed for fuel savings (Rs crores)	Likely fuel savings (excluding of taxes) (Rs crores)
1.	State Highways	96,000	960	100
2.	District Roads	6,00.000	1000	55
3.	Village Roads	4,00,000	500	30
4.	Urban Roads and others	3,00,000	500	70
			2960	255
		Say	3000	260

TABLE 14: **Energy Efficiency of Different Modes of Passenger Transport in Urban Areas**

Mode of transport		Capacity in persons per metre	Operating speed in km/hr	Energy consumption in KwH per seat kilometre
Car	Mixed traffic	120-220	15-25	0.29
	Motorway	750	60-70	0.23
Taxi		400-720	15-25	0.30
Bus :	Mixed traffic	2700	10-15	0.12
	separated busway	5200	35-45	0.09
Mini bus		1100	12-20	0.17
Tram		3000	12-15	0.13
Rapid rail transit		9000	35	0.15

The road targets (2001) and achievements (1990) can be visualised from the Table-5.

ENERGY, ROADS AND TRAFFIC

Various road factors attributable to the road and traffic govern the fuel consumption rate of motor vehicles. The following are the important ones:

1. Pavement width
2. Pavement specification
3. Pavement roughness
4. Vertical profile of the road
5. Shoulder width, type and condition
6. Vehicle speed
7. Traffic conditions
8. Vehicle type

In a study, four energy demand scenarios to the year 2005-06 are made:

	Road	Transport trend Rail	Energy conservation measures
Scenario I	Unchanged	Unchanged	No specific
Scenario II	Unchanged	Unchanged	Specific
Scenario III	Unchanged	Increased share	No specific
Scenario IV	Unchanged	Increased share	Specific

TABLE 15: Transport - Energy Demand Projections (MTOE)

			2005/06			
		1985/86	Scenario I	Scenario II	Scenario III	Scenario IV
a.	**Total**	19.98764	47.35720	42.08189	40.96492	36.97515
b.	**Petroleum Products**	13.34130	43.26592	37.99061	35.17030	31.18053
b.1	**Petrol**	1.96493	6.02749	4.82021	3,72157	2.97616
b.2	**ATF**	1.46429	3.20775	3.20775	3,20775	3.20775
b.3	**Diesel**	9.55944	33.70676	29.63873	27.91706	24.67270
b.4	**Furnace Oil**	0.35264	0.32372	0.32372	0.32372	0.32372
c.	**Electricity**	0.94598	4.09148	4.09148	5.79482	5.79482
d.	**Coal**	5.70036	—	—	—	—

R.K. Bose has given the energy demand for passenger transportation in Delhi using LEAP-EDB model. Another report of the study group on Alternative System of Urban Transport has given the projections of urban energy demand for passenger transport in 2001 AD.

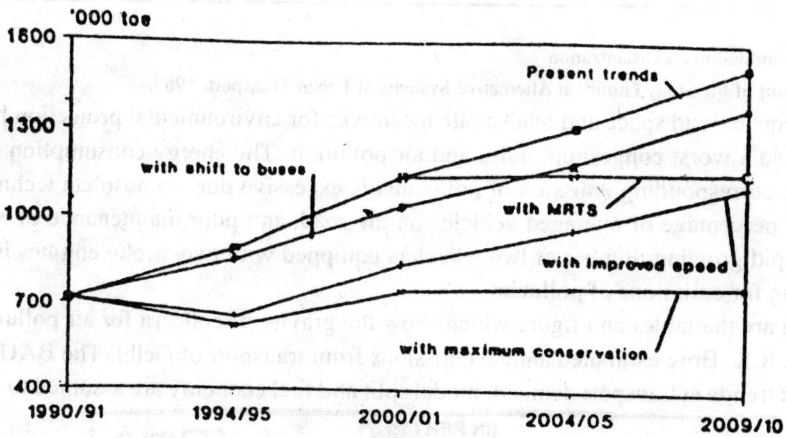

FIG. 1: *Energy Demand for Passenger Transportation in Delhi*

ENERGY, ENVIRONMENT AND TRANSPORTATION

The interface of transportation with energy and environment is primarily in the cities where the use of energy in transportation develops concerns for the environment. Although the overall levels of personal mobility in the developing world are lower, the concentrations of populations in very large

TABLE 16: **Projections of Urban Energy Demand for Passenger Transport in 2001 AD**

Parameter	Scenario 1 as per present trends					Scenario 2 as per NCU* recommendations				
	Buses	Cars	Two-wheelers	Three-wheelers	Total	Buses	Cars	Two-wheelers	Three-wheelers	Total
Model split (%)	57	9	22	12	100	80	4	10	6	100
Annual Passenger-km (billions)	20.5	3.3	8.0	4.0	35.8	28.8	1.4	3.6	2.0	35.8
Number of vehicles (thousands)	62	1173	7809	532	9576	87	488	3592	260	4377
Passenger car units (thousands)	186	1173	3904	532	5795	261	488	1771	260	2780
Fuel demand (thousands tonnes of oil equivalent)	1270	1012	1277	733	4292	1786	421	579	359	3145

*National Commission on Urbanization

Source : Report of the Study Group on Alternative Systems of Urban Transport, 1987.

cities with inadequate road space and inadequate incentives for environmental protection has led to some of the world's worst congestion, noise and air pollution. The energy consumption of motor vehicles and the corresponding emission of pollutants is excessive due to obsolete technology of vehicles, higher percentage of overaged vehicles on the road, and poor maintenance of roads and vehicles. The rapid growing number of two wheelers equipped with two-stroke engines is a major factor accounting for emissions of pollutant.

Following are the tables and figure which show the gravity of concern for air pollution from road and traffic. R.K. Bose estimated annual emissions from transport of Delhi. The BAU scenario in which present trends in transport demand, modal split and fuel economy are assumed to continue,

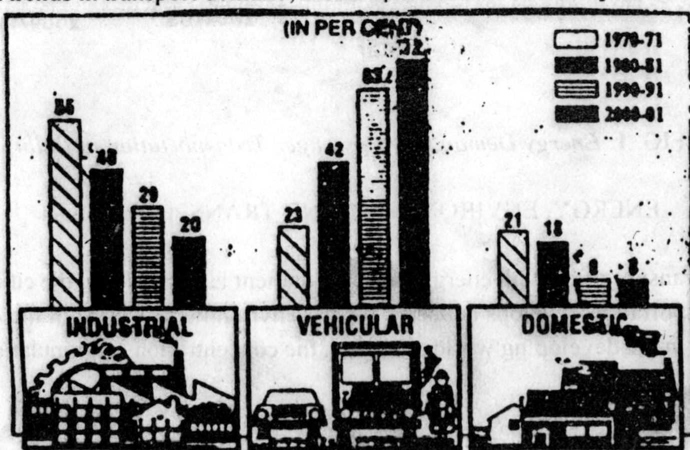

FIG. 2: *Air Pollution Trend*

TABLE 17: Importance of Energy Activities in the Generation of Air Pollutants

Pollutant	Man-made as % of total	Energy as % of total	Activities as % of man-made	Contributions as % of energy related releases
NO_x	75	64	85	Transport 51 Stationary source 49
CO	50	15-25	30-50	Transport 75 Stationary source 25
Lead	100	90	90	Transport 80 Combustion in stationary sources 20
PM	11.4	4.5	40	Transport 17 Electric utilities 5 Wood combustion 12

TABLE 18: Estimation of Annual Emissions

(in '000 tonnes)

Alternative Scenarios	1990/91	1994/95	2000/01	2004/05	2009/10
Scenario 1: As per present trends (BAU)					
CO	179.59	222.28	286.33	329.05	382.42
HC	73.00	90.35	116.39	133.75	155.44
Pb	0.10	0.13	0.17	0.19	0.22
NO_x	16.12	19.95	25.70	29.53	34.32
SO_2	2.27	2.81	3.63	4.17	4.84
TSP	0.36	0.45	0.58	0.67	0.77
Scenario 2: Improvement in vehicular speed					
CO	179.59	184.21	237.29	272.69	316.92
HC	73.00	76.58	98.64	113.35	131.74
Pb	0.10	0.11	0.14	0.16	0.18
NO_x	16.12	12.88	16.59	19.06	22.16
SO_2	2.27	1.73	2.22	2.55	2.97
TSP	0.36	0.26	0.33	0.38	0.44
Scenario 3: Increase the share of bus					
CO	179.59	203.22	268.22	268.01	311.48
HC	73.00	82.88	94.72	108.85	126.51
Pb	0.10	0.12	0.13	0.15	0.18
NO_x	16.12	19.86	25.42	29.22	33.95
SO_2	2.27	2.86	3.74	4.30	5.00
TSP	0.36	0.46	0.62	0.72	0.83
Scenario 4: Introduce mass rapid transit system (MRTS) using electricity					
CO	179.59	222.28	286.33	288.24	285.13
HC	73.00	90.35	116.39	117.16	115.90

Alternative Scenarios	1990/91	1994/95	2000/01	2004/05	2009/10
Pb	0.10	0.13	0.17	0.17	0.17
NO_x	16.12	19.95	25.70	25.87	25.59
SO_2	2.27	2.81	3.63	3.65	3.61
TSP	0.36	0.45	0.58	0.58	0.58
Scenario 5: Maximum conservation scenario					
CO	179.59	168.73	192.43	193.71	191.62
HC	73.00	70.10	79.89	80.42	79.55
Pb	0.10	0.10	0.11	0.11	0.11
NO_x	16.12	12.70	16.06	16.17	15.99
SO_2	2.27	1.74	2.26	2.28	2.25
TSP	0.36	0.27	0.36	0.36	0.35

has shown a dramatic increase in energy demand and the resultant emissions. The alternative scenarios have shown how the increase in energy consumption and pollution levels can be limited through measures such as increasing the average speed of traffic, encouraging shift to buses, the introduction of new form of mass transit and through conservation efforts.

SUSTAINABLE TRANSPORTATION SYSTEM

Transportation is inextricably linked not only to our private well-being but to a host of societal problems. Some of the social costs of transportation include air pollution, lost productivity due to traffic congestion, death and disability resulting from traffic accidents, water pollution caused by spilled petroleun, and global warming. It represents that our transportation system is not sustainable. Thus to define and implement sustainable transportation system, we need to change people, technology and prices.

Changing People

The main societal goal must be to reduce the need for transportation. We must permanently change our transportation habits if we are going to maintain our mobility and preserve the environment that sustains us. All needs work, shopping, personal services and other social and recreational activities– would be met close to where we live, within a comfortable walking distance. It requires different land-use pattern characterised by compact, multifunctional, pedestrian-and bicycle-oriented urban development.

Changing Technology

Attaining sustainability through changing the technology to reduce the impact of transportation on society is more promising. It relieves the transportation's burden on society while maintaining individual mobility and preserving our current lifestyles. It includes use of cleaner, alternative fueled vehicles, enhanced vehicle fuel economy, ultralight composite, non-fossil-fueled cars, tele-communications and other advanced technologies.

Changing Prices

To get the prices right and let the market solve our transportation problems is another approach. The goal of this vision is to modify transportation demand through the use of market forces to enhance overall system-wide transportation efficiency. It can be achieved through developing a least-social-cost planning approach to transportation decision-making.

The ultimate soulution to sustainable transportation system is to use combination of these approaches in decision-making and policy formulation. The effective policies must effect all the three approaches for optimum results.

ALTERNATIVE VEHICLES

Conventional cars, like other technologies, have entered their era of greatest refinement just as they may become obsolete. Advanced materials, software, motors, microelectronics, power electronics, electric storage devices, and computer-aided design and manufacturing will bring a revolution in both vehicles and auto-manufacturing. Well before 2003, competition, not government mandates, can bring to market cars efficient enough to carry a family coast–to–coast on one tank of fuel, more safely and comfortably than present vehicles, and more cleanly than electric cars plus the power plants needed to recharge them. This revolution in automaking will require leaping forward to a completely new car design, the ultralight hybrid-electric hyper car.

The keys to a superefficient car are to:

— cut weight (hence the force required for acceleration) by three to four-fold by using advanced materials, cheefly synthetic composites, while improving safety through greater strength and sophisticated design.

— cut aerodynamic drag by two-and-a-half to six fold through sleeker streamlining and more efficient packaging.

— cut tire and road losses by three to five fold through the combination of better tires and lighter weight.

However in existing condition, introducing or changing the vehicle mix from 2-axle to 3-axle rigid vehicles would result in economy of over 26% in total transportation cost. Similarly, road costs which is 3.5% of total transportation cost could bring very high economic returns, since a little investment on roads bring down the vehicle operating cost which is 95-97% of total cost.

ALTERNATIVE FUELS

The International Energy Agency has recently carried out a study of the potential of alternative fuels to reduce greenhouse gas emissions from the use of cars. The use of alternative fuels and electricity from renewables has the technical potential to reduce the life-cycle greenhouse gas emissions from car-use by 80% but these energy sources are more expensive than gasoline. The economic potential for greenhouse gas emission reduction is smaller. Alternative fuel cars using diesel, liquified petroleum gas (LPG) and compressed natural gas (CNG) can have life-cycle greenhouse gas emissions 10 to 30% lower than those from gasoline cars.

The use of alternatives available now or expected to become available by 2005, included diesel, LPG, CNG in optimised engines, and electric vehicles can reduce life-cycle greenhouse gas emissions by 10 to 25%. Synthetic fuels from wood or other low-input biomass feedstocks are not yet technically demonstrated but the fuels could offer 60 to 80% life-cycle greenhouse gas emission reduction. Recent studies identify the CO_2 equivalent green house gas emission of alternative fuels relative to comparable gasoline powered vehicles.

TABLE 19: **Greenhouse Gas Emissions**

	Greenhouse Gas Emissions relative to gasoline
Natural gas vehicle	–25 to –5%
Electric vehicle (natural gas)	–50 to –25%
Methanol from matural gas	–10 to +10%
Electric vehicle (current energy mix)	–5 to 0 %
Electric vehicle (coal)	0 to + 10 %
Methanol from coal	+30 to +80%

Electric and natural gas vehicles are the most attractive near-term options for reducing greenhouse gas emissions.

Policies and Strategies

Obtaining overall sustainable transportation system is going to involve the coordination of a variety of programmes.. Any successful policy for a more sustainable transportation system must balance energy security, energy efficiency and environmental soundness as sometimes completing yet ultimately interconnected goals. Following are the policies and strategies which can help in designing and implementing and effective sustainable transportation system.

Pricing Policies— Include taxes, subsidies, user fees, allowances, variable interest rates, buyback programme and least social cost planning framework.

Technology Policies— Include investing in low-tech transportation alternatives, such as bicycle and pedestrian services, developing ITS technologies for public transit and ride sharing vehicles, developing resource-efficient transportation technologies that minimise waste by-products.

Land–use Policies— Include designing urban villages, employing mixed use development around transit, instituting traffic calming, reconfiguring zoning ordinances, adopting land-use investment strategies, instituting regulatory strategies and revisiting parking requirements.

Behavioural and Educational Policies— Include promoting voluntary non-drive days, increasing the distribution of the traffic guide, launching statewise promotional compaigns, using the media to encourage shifts to travel behaviour, providing training for the general public, and educating children.

Restructuring Policies— Include conversion or construction of high, medium and low-traffic corridors for more than 15000 PCUs/day between 5000 to 15000 PCU/day and less than 5000 PCUs/day respectively; designing, developing and maintaining the pavement management systems, creation of suburban centres with integrated housing, employment and recreational facilities, and shifting from personalised transport to public transport.

EMERGING MANPOWER NEED OF THE POWER SECTOR IN INDIA

PRAN NATH
Director (Personnel)
NTPC

11

"It is never too late to
what you might have been"

–T.S. Eliot

1. INTRODUCTION

1.1 "May you live in changing times"– is an ancient Chinese saying. We are lucky to be witnessing the changes around us. The global economic map is being redrawn and Indian economy has also been undergoing a quiet economic revolution of sorts. This major economic transition is arousing wide-ranging thrills and anxieties across widely differentiated socio-economic groups. The changing scenario of Indian Power Sector necessitates a re-look into existing systems and practices of human resources planning to make in more objective, need-based and result oriented. In this chapter, an effort has been made to project the existing manning practices for self–retrospection. This should help Indian power sector to prepare for the future.

1.2 An estimated 10.5 lakh of total manpower is deployed in power sector including tansmission and distribution as per the Annual Report on the working of SEBs and electricity departments, October 1995 prepared by the Planning Commission. The break-up of manpower deployment in SEBs and PSUs under sectors of Thermal, Hydro and Transmission/Distribution and the installed capacity is given in Table-1. This table also gives the Man/MW ratio for these sectors i.e. man deployed/MW. The less is this ratio, the more is productivity of manpower.

TABLE-1: **Power Sector: Manpower and Installed Capacity**

Description	Manpower		Capacity		Man/MW
	Nos.	% age	MW	%age	
Thermal	133207	12.7%	60171	72.3%	2.21
Hydro	55699	5.3%	20829	25%	2.67
Nuclear	5000	0.4%	2225	2.7%	2.25
Sub-Total	193906	18.4%	83225	100%	2.33
TL & Dist.	858213	81.6%	--	--	10.31
Total	1051119	100%	83225	100%	12.63
ESKOM*	40000		37840		1.05

* ESKOM produces nearly 60% of all electrical power generated in Africa meets requirement of more than 95% of South Africa's electrical energy. ESKOM's 19 power stations generate into a national grid comprising some 241802 km of power lines.

1.3 The man/mw ratio on generation side varies marginally in these three sectors (Thermal, Hydro and nuclear) from 2.21 to 2.67. Hydro sector deploys more manpower per MW compared to other two sectors.

1.4 It may be seen that only 18.4% of total manpower in power sector is engaged in generation and the balance 81.6% is tied-up with power system i.e. transmission and distribution. Seventy–five per cent of total power generation is accounted for thermal and Nuclear and remaining 25% in the Hydro sector. The average Man-MW ratio in power generation is 2.33 against 10.31 in Transmission and distribution. The overall Man-MW ratio for power generation, transmission and distribution is around 12.63. The ESKOM of South Africa having installed capacity of 37840 MW and 241802 kms of power lines in deploying just 40,000 manpower. This works out to a Man:MW ratio of 1.05 against man:MW ratio of 12.63 for the power sector as a whole in India.

2. MANPOWER DEPLOYMENT

2.1 Manpower in Central PSU and SEBs

2.1.1 The total manpower including tansmission and distribution system corresponding to the installed capacity of NHPC, NTPC and DVC in Central PSUs and some of the SEBs has been analysed as given in Table 2.

TABLE 2: **Organisations: Total Manpower Deployment**

Organisation	Man MW/Ratio		
	Generation	TL& Distribution	Overall
1. PSUs			
NHPC	7.37	—	7.37
NTPC	1.43	—	1.43
DVC	6.08	1.23	8.27
2. SEBs			
WBSEB	3.82	28.15	34.37
MSEB	1.99	10.50	14.46
MPSEB	3.02	20.53	23.55
KSEB	—	14.25	16.87
KPC	2.33	—	—
PSEB	2.63	13.67	22.50

2.1.2 The Man:MW Ratio of NHPC works out to 3.37 if capacity under execution is also taken into account. Similarly the Man/MW ratio i.e. 1.23 in case of DVC pertains to the transmission line system including capacity under execution.

2.1.3 Table 2. indicates organisation-wise overall manning level in terms of Man:MW ratio in PSUs and SEBs giving break-up in generation and TL, and distribution areas.

2.1.4 It can be seen that in case of generation area, the manpower per MW deployed by NTPC in terms of Man:MW ratio is lowest at 1.27. In other words manpower productivity is the highest. Next comes MSEB which is 1.99. The man:MW ratio is NHPC is 7.37 closely followed by DVC at 6.08.

3. ORGANISATIONS: BREAK-UP OF GENERATION MANPOWER

3.1 Table 3 below gives further break-up of manpower per MW deployed in Hydro, Coal and Gas based stations. The manpower is also deployed in transmission and distribution set-up.

TABLE 3: **Organisation Level Manpower in Hydro and Thermal Generation**

Name of the organisation	Man-MW Ratio in Generation area				
	Hydro	Coal	Gas based	Overall (Gen)	Base (100)
Central PSUs					
DVC	19.69	3.18	5.18	6.08	425
NHPC	7.37			7.37	515
NTPC		1.50	0.38	1.43	100
NLC		4.65		4.65	325
SEBs					
WBSEB	7.26	3.49		3.82	267
MSEB	1.37	2.39	0.45	1.99	139
MPSEB	0.57	3.89		3.02	211
KPC & KSEB	2.30	2.41		2.33	163
PSEB	2.43	2.54		2.63	184
TSEB	2.43	2.94		2.78	194

TABLE 4: **Manpower in TL and Distribution in SEBs (ckt-km/man)**

Name of the Organisation	TL Network (ckt-kms)	Manpower in Transmn & Distbn.	Installed Capacity (MW)	Ckt-Km/ Per man	(TL/Dist) man/MW	Man/ Ckt-KM (Base 100)
Maharashtra SEB	571653	80953	7713	7.06	10.5	0.142 (100)
Tamilnadu SEB	482870	76759	6158	6.29	12.46	0.159 (112)
Madhya Pradesh SEB	556420	85893	4184	6.48	20.53	0.154 (108)
Punjab SEB	228317	48000	3510	4.76	13.67	0.210 (148)
West Bengal SEB	142444	31424	1116	4.53	28.16	0.220 (155)
Karnataka SEB/ KPC*		44542	3126*	N.A.	14.25	N.A.
Power Grid*	27250	6751	—	4.04	—	0.248 (175)
DVC*	5256	2621	2137	2.0	0.82	0.499 (351)

* Manpower in Transmission Line network only.

4. MANPOWER DEPLOYMENT AT STATION/UNIT LEVEL

4.1 **The deployment of manpower and actual manning pattern in the concerned areas can be better understood and differentiated if the manpower deployment is compared at the gross-root level i.e. stationwise/unit-wise in the power stations of various organisations. Station-wise comparison of coal based stations having similar capacity and set-size, closer range of set-size with varying number of units can be seen in Table-5. The Man/MW ratio for the station having same number of units with varying capacity is an indication of economy in manpower for the station of higher capacity and manpower per unit is indication of extent of manning level, manning pattern and practices followed in different organisations.**

TABLE 5: **Station/Unit Level Manpower in Power Generation**

Sets/Units Size	No. of Units	Station cap. (MW)	Total Manpower	Station Man/MW Ratio	Man/ Unit	Remarks	Type of Organ.
Below 100/110MW	4	270	489	1.81	122	BCPP, NTPC	PSU
sets and combination	9	600	1886	3.14	210	NLC	PSU
with 200/210 MW	5	540	1956	3.62	391	BTPS, WBSEB	SEB
	5	460	2124	4.62	425	TTPS, OSEB	SEB
200/210MW	2	420	320	0.76	160	CEA, Jt. Ven.	IPP
Sets	6	1260	1658	1.31	276	VTPS, APSEB	SEB
	4	840	1217	1.45	304	Kh, NTPC	PSU
	3	630	1373	2.18	458	Mezia, DVC	PSU
250/500MW	2	1000	416	0.42	208	Iron Bridge UK	UK
Sets	2	500	674	1.35	337	Dahanu, BSES	PVT.
	2	1000	800	0.80	400	Trom. TEC	PVT
	2	1000	1021	1.02	510	Rh, NTPC	PSU
Combination of	7	2000	2227	1.11	318	SSTPS, NTPC	PSU
200/210MW & 500 MW	6	2100	2226	1.06	371	KSTPS, NTPC	PSU
COMBINED CYCLE POWER STATION							
Gas/Oil fired	6	700	33	0.05	6	Power Gen, UK	IPP/UK
combined cycle	6	180	112	0.62	19	DESU	PSU
station	10	912	277	0.30	28	Uran, MSEB	SEB
	6	652	295	0.45	49	AuGPP, NTPC	PSU

4.2 **Analysis of manpower deployed in various power stations of different configuration grouped on similarity of unit size is given in Table 5 which further reveals as under:**

4.2.1 **The manpower per unit of even smaller capacities sets deployed in a station run by old PSUs and SEBs in more than the manpower per unit of higher capacity sets in stations run by private sector, IPPs and sets commissioned by NTPC stations.**

4.2.2 **The per unit manpower deployed by private sector power station is less than the manpower deployed in comparable stations having similar sets run by PSUs and SEBs.**

4.2.3 The per unit manpower deployed by power stations abroad is much less than the comparable station capacity with similar sets run by PSUs and SEBs in India.

5. ANALYSIS OF DATA

5.1 Analysis of Tables 1, 2, 3, 4 and 5 above clearly indicates too excessive and very liberal manpower deployment in power stations in the country. Therefore, there exists a vast scope to improve our manning practices, systems and procedures to achieve economy in running the power station and improve productivity of manpower. If the manpower productivity as achieved in NTPC at organisation level in the area of generation with overall Man:MW ratio of 1.43, is extended for projecting the total manpower need of the country for power generation alone (with respect to the existing installed capacity of 83,225MW and manpower to the tune of 1,93,906 numbers estimated in generation), about 40% surplus manpower may be generated. Similar surplus may also emerge in Transmission and Distribution sector where around 82% of the total manpower in power sector is tied up.

5.2 In a recent study conducted by IBM, the staffing analysis in power sector has been done by comparing power projects from seven companies in Asia. A comparison indicates staffing levels evaluated on the basis of the Full Time Equivalent (FTE) which represents 2080 hours i.e. 260 days of work performed by an employee. The staffing analysis gives manning level in O&M functions as well as on overall basis of various power stations including that of NTPC power station at Rihand. Accordingly, staffing level in NTPC power station is found to be 10 to 20 times more than the US average and 13 to 32 times more than the US Best when compared with respect to plant size as well as with respect to number of units on site on FTE basis.

6. FACTORS CONTRIBUTING TO MORE MANPOWER IN POWER STATION

6.1 Most of managers in power stations came from SEBs who had the experience and habit of working with more manpower. They were not exposed to the practices followed in good power utilities for achieving higher productivity level. The other reasons for excessive deployment of manpower in power stations are as under:

6.1.1 The mindset prevail in the managers of India that the manpower is very cheap in this country. So, there is no harm in deploying more manpower.

6.1.2 The objective of PSUs and SEBs in the past were not very clear whether they are required to function as a social organisation or commercial/profitable venture.

6.1.3 The Indian power sector was living in the world of their own and felt very happy while comparing their achievements with bad performing SEBs. The benchmarking was done with reference to SEBs. So, improvement with respect to SEBs gave them a sense of complacency.

6.1.4 More numbers of small size sets for the same capacity of station rather than fewer large size sets resulted in more manpower.

6.1.5 Number of small capacity units are installed which resulted in lower productivity.

6.1.6 Plant lay-out of the station is not made as integrated system rather it duplicates the common facilities as mirror image in subsequent stages of expansion programme.

6.1.7 Non-adoption of latest technology which is not cost effective.

6.1.8 Excessive manning points in operation requiring more manpower.

6.1.9 The perception of over insurance for manning a location/equipment–"fear that something unforeseen may happen".

6.1.10 Non-functioning of control and instrumentation facilities compelling to resort to manual operation.

6.1.11 Creation of more and more specialists trade/craftsman irrespective of workload.

6.1.12 Working in a water tight compartment which does not allow pooling of resources leading to gross under utilisation.

6.1.13 Varieties of functions got introduced as a statutory requirement operating as a full fledged group/department irrespective of workload.

6.1.14 Non-adoption of simplified systems and procedure.

6.1.15 Union and political interference due to vested interests and unemployment problem.

6.1.16 Absence of scientific approach/norm to standardise resource requirement and tendency to build-up empire around one self.

6.1.17 Absence of reward and punishment system as well as security of jobs in a State run organisation leading to gross under utilisation of resources and unaccountability of the person and thus, more manpower requirement.

6.1.18 Resorting to deployment of contract labour directly or indirectly resulting in under-utilisation of departmental workforce.

7. STEPS TO ACHIEVE ECONOMY IN MANPOWER REQUIREMENT

7.1 Since IPPs are coming, it is imperative for PSUs and SEBs to face competition. Therefore adoption of the best practices would be absolutely necessary. Following steps are required.

7.1.1 Development of systems and procedures and continuous review of the manpower deployment to improve the utilisation.

7.1.2 Development of standard job specification for each position.

7.1.3 Standardisation of designations and reduction in varieties to facilitate multi-skilling.

7.1.4 Clubbing the responsibility of more equipments under each manning location and elimination of undesirable manning points.

7.1.5 Automation, computerisation and simplification of systems and procedure to reduce manpower in offices and staff function.

7.1.6 Proper maintenance of control and instrumentation system and emphasis on house-keeping and proper up-keeping of plant.

7.1.7 Interchangeability of operative staff from one unit to another unit and also pooling operation staff for maintenance jobs during shut-down etc.

7.1.8 Adoption of standard oporeating practices and maintenance practices. Units commissioned in more than one stages to be managed as integrated constituents of a single station with marginal increase in manpower and no duplication of manpower for common facilities to be allowed.

7.1.9 Merger of C&I and Electrical maintenance groups for greater utilisation of common expertise and avoiding creation/multiplication of too many specialised groups.

7.1.10 Introduction of incentive and disincentive systems linked to productivity.

7.1.11 Development of good recruitment systems for Executive, Supervisor and Artisan Trainees.

7.1.12 Good training and management development supports for all categories of manpower.

7.1.13 Launching of elaborate Employee Development Scheme for acquiring qualifications for making the needs of organisation and individual's growth compatible.

7.1.14 Directing efforts in simplifying personnel systems/policies in order to make them more transparent and understandable.

7.1.15 Development and improvement of good manning system and practices from time to time and rationalisation of manpower resources for organisational efficiency.

8. CONCLUSION

8.1 The Indian position in terms of the manpower productivity in the power sector is very dismal. We in India deploy 32 times more manpower for the same facilities as compared to USA's best power utility. As compared to ESKOM which meets the power requirement of more than 95% of South Africa, Indian power sector engages 12 times more manpower. In today's competitive economy, we cannot close our eyes. The opening of economy, IPPs already making their appearance in India and many changes happening all around us, make our environment very uncertain and unpredictable. Prior to this, we had practically stable era and during these times the Corporate direction was continuous and incremental change was acceptable. It was good enough to make token reduction of manpower here and there. Now it is the uncertainty and volatility of business environment that make far discontinuity in corporate direction and demand rapid and not incremental change in action. With this intention this chapter brings out the gaps between productivity figures of Indian power sector and power sector abroad. These need to be bridged. And the first condition for bridging the gap is to create a sense of discomfort and frustration with the current way of work. Comparison of Man:Megawatt ratio in this chapter is an attempt in that direction which brings out that Indian power sector have to travel a very long distance and travel it very fast.

8.2 Organisations now need to set ambitious and new targets in all the areas including labour productivity and have to try harder to attain them. I would like to conclude this chapter with a quotation from the *Bhagwat Gita* where Mother Kunti exhorts the Pandavas and Lord Krishna "Only a mouse finds its hands full with a couple of foodgrains. Small desires do not behove you. Commit yourself to a lofty target worth of you. Be conceived and you will succeed."

ACCESS TO A LARGE DATABASE ON ENERGY SECTOR— A MENU BASED APPROACH

12

DR. R.P. SAXENA
Senior Technical Director & Head,
Energy Information Systems Division,
National Informatics Centre, Planning Commission,
CGO Complex, A-Block, Lodi Road,
New Delhi - 110003.

INTRODUCTION TO ENERGY SECTOR

State Electricity Boards (SEBs), Power Departments, Power Corporations and the State Nodal Agencies are responsible for exploiting the sources of energy through latest state of the art technology. SEBs cater the needs of electrical energy of lacs of consumers in the State. The transmission and distribution network spread over all the districts. The infrastructure set up by SEBs can be divided into generation, transmission and distribution which can be strengthened if all the projects of SEBs/ corporation may be linked to their HQ and all the organisations relating to power sector are interconnected.

IMPLEMENTATION OF POWERNET

National Informatics Centre has constituted a Steering Group V on Energy Sector on 25th June, 1987 to draw a long term computerisation plan and implementing the computerised information system. The group felt that there is an immediate need for developing a common database on power sector which could be accessible to various government agencies for meeting their requirement. Consequently, a subgroup of NIC Steering Group V was formed under the chairmanship of Advisor (Energy) for the development of computerised power information system under POWERNET. The sub-group had recommended a common power database (CPDB) which could meet the requirement of various agencies in central government dealing with power sector. The report No. 1 of the Sub-Group on Construction data has already implemented.

NEED FOR A VERY LARGE DATABASE

In the management, there is a recognised need of objective approach to decision making in respect of planning, controlling, execution and problem solving. Many agencies are involved in monitoring of power projects but data inconsistency exists in the data items maintained by them. To meet the requirement of top managers of various agencies, there is a need to develop (1) mechanism to store data in structured way, (2) Intelligent retrieval routines to allow more efficient retrieval and (3) Intelligent Front End (IFE) to improve interactions. Storing of data in a structured way has been made possible by designing a VLDB, called ENEDB, using NETWORK approach. ENEDB covers many areas concerning with operation monitoring of power projects.

ENERGY DATABASE (ENEDB)

The ENEDB design provide a comprehensive and up-to-date information system for the experts/ advisors of Energy sector. Menu based query system interact with ENEDB through Information Retrieval System (IRS). The information system containing MBQS, ENEDB and IRS is named as Global Energy Information System (GEIS). The model permits sharing of information amongst managers, and provides multiple computing facility across multiple sites, while deciding on long range strategies for planning allocations, projection and decision making. The software is engineered in such a way that user views the database as a single database and he does not have to be concerned with the geographical location of database.

ENEDB database covers information on the areas of Installed Generating capacity, Gross Electricity Generation, oil and auxiliary consumption, coal field, coal utilizing power stations, company involved in supply of coal, outages inpower plant and inter system exchange of energy (Ref. Figure 1).

MENU BASED QUERY SYSTEM (MBQS)

Menu Based Query System (MBQS) is an approach through which user of computer system becomes free from learning tedious query language supplied with the system. Also, he does not bother how to form a query or what area to select for getting information. MBQS permits information handling between user and database with the help of 'MENUS'. These menus are controlled by a program called *Intelligent Assistant* (IASSTT).

MBQS retains the main goal of 'Natural' language as query in natural language is inbuilt into the system and is presented to user along with text and code. It is built as an upper layer of the query language provided with the system. It uses tree structure like menu system in which each choice of a word or phrase from a menu causes the display of a new menu dependent on that choice. When we select final menu, query is formulated automatically and the system displays the report on the monitor. User has been given freedom to store the output in a file for use at a later stage or take a hard copy on printer. Under each menu, there exists a menu to quit or to view previous menu. It serves as an improved interface for storing, organising, analyzing and retrieving information with flexibility, learnability, mnemonicity. It provides format variation, responsiveness, and customization. It solves most of the problem inherent to conventional database system and user is not bothered of learning the tedious query commands supplied with the system. It is very helpful to users who are new to sector or who simply wants to get information based on pre-defined queries. The problem posed by the user due to unambiguous questions is automatically solved as user is able to navigate through queries which are in-built. Also, after processing the query user gets output in a tabular format.

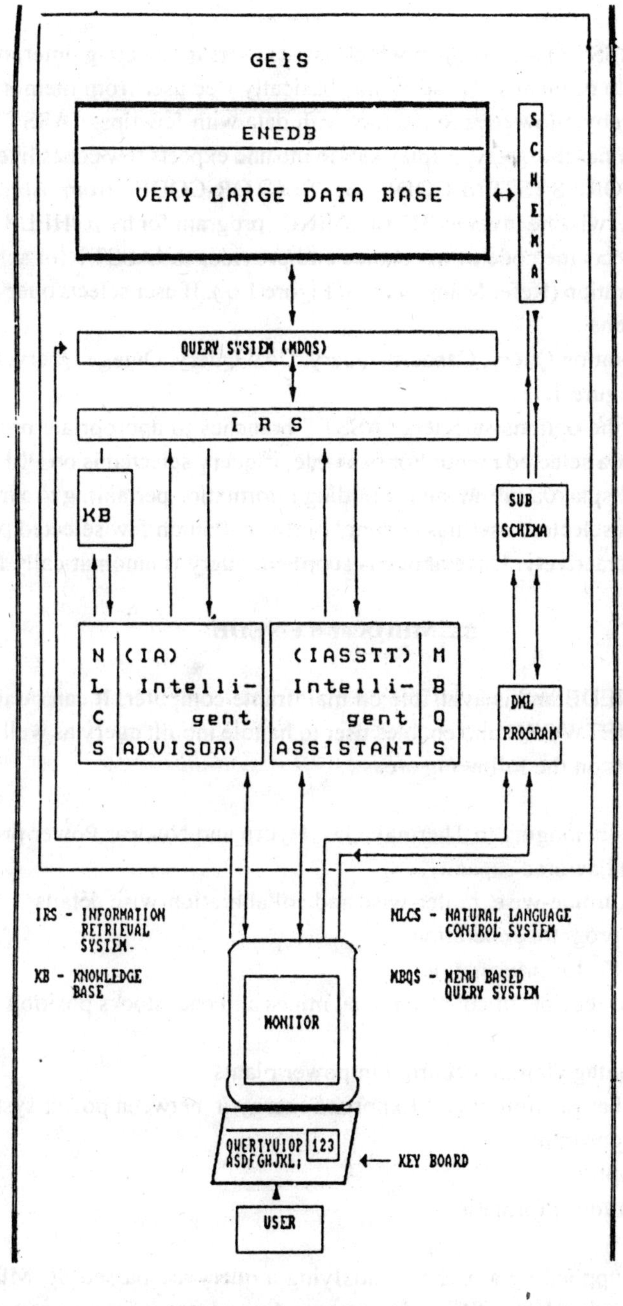

FIGURE : 1 : GLOBAL ENERGY INFORMATION SYSTEM AND USER

5.1 Intelligent Assistant (IASSTT)

Intelligent Assistant (IASSTT) is a program which assists users in selecting, interpreting and executing a query with the help of menus. These menus basically free user from memorising commands, rules of syntax and thereby allow user to interact with data with few tips.. IASSTT reads the codes supplied by user, determines the query, display sub-menu and expects few codes like 'STATE-CODE' or 'STATION-CODE ' OR ' SYSTEM-CODE ' or 'SECTOR-CODE ' from user. If user knew the code, he supplies it, otherwise he invokes 'HELP MENU ' program for help. 'HELP MENU ' displays a list of codes. User selects the code of his choice and provides to IASSTT for selecting the option for query or report generation (Refer Main menu of Figure 1.0). If user selects query option, IASSTT displays following options :

PLF Query, Generation Query, Capacity Query, Coal Query, Outage Query, Quit and move to previous menu. (Ref. Figure 1.1).

Depending upon the options selected, IASSTT branches to appropriate program to perform the task displayed under a selected menu. For example, if query selected is on PLF (menu option-1, Fig 1.1), IASSTT will display0a sub-menu for finding information pertaining to a month or duration. if option 1 of Fig. 1.1 is selected, user has to supply information on few selected parameters (Refer Section 5.3.1.1.1.1) interactively. Once above is supplied, query is automatically formulated.

5.2 MBQS and ENEDB

MBQS operates on ENEDB and is available on mainframe computer. It can be invoked from any terminal connected to NETWORK and enables user to handle inbuilt query as well as view/generate different types of reports on the following areas :

- Power-units belonging to Thermal, Gas, Hydro and Nuclear Power projects.
- Installed and derated capacity
- Unit-wise, turbine-wise, boiler-wise and collaboration-wise details
- Actual and program generation
- Plant Load Factor information
- Linkage and receipts of coal from coal mines and coal stocks position in thermal power plants
- Details of outage/failure occurred in power plants
- Energy exchange (Import and Export of energy in between power systems)
- Auxiliary consumption
- Oil consumption
- Hydro reservoir information

The parameters supplied by a user for satisfying a query are passed by 'MBQS' to 'IRS' and MDQS which invokes sub-schema file and corresponding report writer procedure for generating results. It uses access path (Path of accessing records) defined in schema for storing and retrieving of data from database. IRS helps in retrieving and computing information contained within data files.

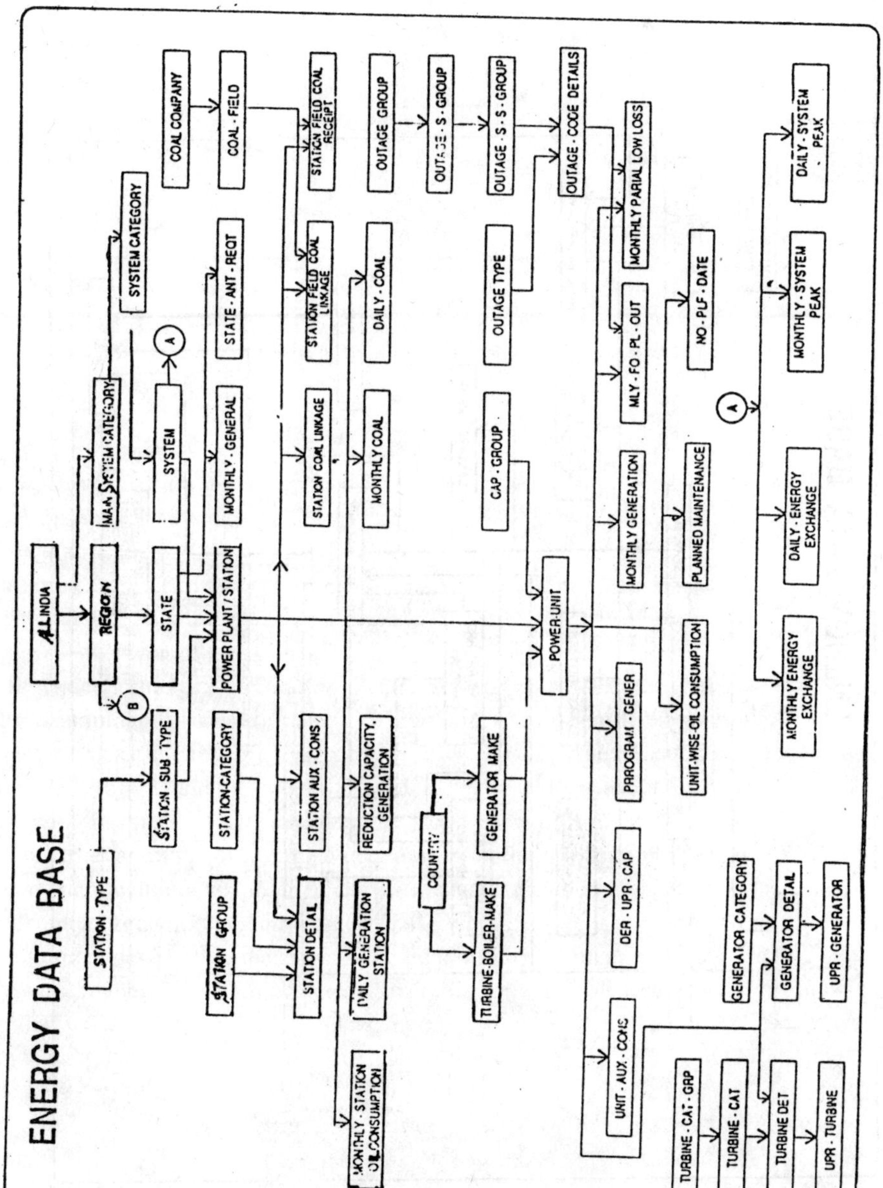
ENERGY DATA BASE
ALL INDIA
REGION
STATE
POWER PLANT / STATION
SYSTEM CATEGORY
COAL COMPANY
COAL - FIELD
STATION FIELD COAL RECEIPT
OUTAGE GROUP
OUTAGE - S - S - GROUP
OUTAGE - CODE DETAILS
MONTHLY PARIAL LOW LOSS
DAILY - SYSTEM PEAK
STATE - ANT - REQT
STATION FIELD COAL LINKAGE
DAILY - COAL
OUTAGE TYPE
MLY - FO - PL - OUT
NO - PLF - DATE
MONTHLY - SYSTEM PEAK
SYSTEM
MONTHLY - GENERAL
STATION COAL LINKAGE
MONTHLY COAL
CAP - GROUP
MONTHLY GENERATION
PLANNED MAINTENANCE
DAILY - ENERGY EXCHANGE
POWER-UNIT
UNIT-WISE-OIL CONSUMPTION
MONTHLY ENERGY EXCHANGE
STATION - SUB - TYPE
STATION-CATEGORY
STATION AUX - CONS.
REDUCTION CAPACITY, GENERATION
GENERATOR MAKE
PROGRAM - GENER
STATION - TYPE
STATION GROUP
STATION DETAIL
DAILY GENERATION STATION
COUNTRY
TURBINE-BOILER-MAKE
DER - UPR - CAP
GENERATOR CATEGORY
GENERATOR DETAIL
UPR - GENERATOR
MONTHLY - STATION OIL CONSUMPTION
UNIT - AUX - CONS
TURBINE - CAT - GRP
TURBINE - CAT
TURBINE DET
UPR - TURBINE
A
B

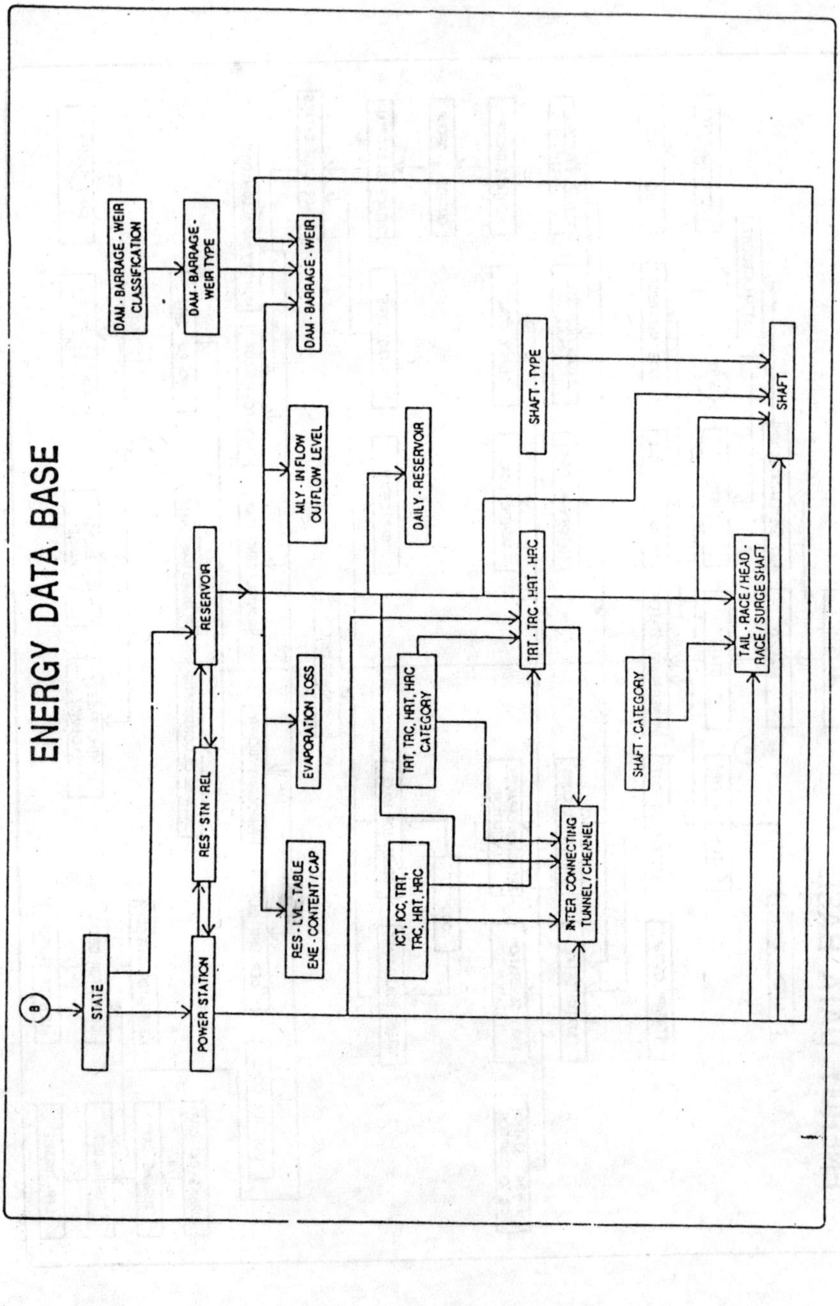
ENERGY DATA BASE
B
STATE
POWER STATION
RESERVOIR
RES - STN - REL
RES - LVL - TABLE ENE - CONTENT / CAP
EVAPORATION LOSS
MLY - IN FLOW OUTFLOW LEVEL
DAM - BARRAGE - WEIR CLASSIFICATION
DAM - BARRAGE - WEIR TYPE
DAM - BARRAGE - WEIR
DAILY - RESERVOIR
ICT, ICC, TRT, TRC, HRT, HRC
TRT, TRC, HRT, HRC CATEGORY
TRT - TRC - HRT - HRC
INTER CONNECTING TUNNEL / CHENNEL
SHAFT - TYPE
SHAFT - CATEGORY
TAIL - RACE / HEAD - RACE / SURGE SHAFT
SHAFT

6. PROBLEMS ENCOUNTERED DURING IMPLEMENTATION

The problem in implementing a database is firstly, the collection and keying in of past data at the start of the system. The second problem is the full understanding of the system by all concerned users and their effectively making use of the system in their day-to-day work. Thereafter, regular updation of the system as well as its use can be achieved with relative ease. The requirements of user have been met by developing menu based interface and also by providing necessary training.

REFERENCES

Anderson, J.R. *Language, Memory and Thought,* Lawrence Erlbaum Hillsdale, NJ - 1976

Babrow, D.G. *Natural Language Input for a Computer Problem Solving System,* MIT Press, Cambridge, Mass - 1968

Codd, E.F. *Relational Completeness of Database Sub-language in Database Systems,* Prentice - Hall, Englewood Cliffs, N.J., 1972

Codd, E.F. *A Relational Model for Large Shared Databases, Communications of ACM*, 13, 377-387 - 1970

Date, C.J. *An Introduction to Database System,* Vol II, Addison Weslay reading, MA., U.S.A. - 1983

Fry, J.P., Scbhey, E.H. *Evolutionary DBMS,* ACM, Computing Survey, 8 (1), March - 1976

Hafner, Curok D., S. Kurf, S. Gpdden. *Design of Natural Language Interfaces: a case study,* Research Publication GMR-4567, General Motors Research - 1984

Heraig Howard, Daniel R., Rehak. *Interfacing Expert Systems with Database,* IEEE Expert, 65-76 Fall - 1989

Hobbs, J. Building a large knowledge base for a Natural Language System, Proceedings of the Tenth Int. Conf. on Computat. Ling., CA - 1984

Kaplam, S.J. *Designing a Portable Natural Language Data -Base Query System ACM Trans Database System*. vol. 9, No. 1, pp (1-19) - 1984.

Kent, Allen, J.G. Williams Albert G. Holzman. *Data Base Design: Encyclopedia of Computer Science and Technology,* Supplement 4, University of Pittsburgh, Perusylvanca Page 84-113, Sept. - 1988

Martin, J. *Principles of Database Management.* Prentice Hall, Engelwood Cliffs, NJ - 1976

Phillip EIN-DOR Israel Spiegler. *The Answerability of Database Queries,* Information systems, USA, vol 10, July - 1983

Saxena R.P.; Saxena P.C. PROLOG - A Comprehensive Study, *Indian Journal of Management and Systems*, Vol. 3, ND 3, Sept - Dec., - 1987

Saxena, R.P.; T.A.V. Murthy; D.K. Agrawal. *Data Base for Small Industries– A Network Approach, SEDME Journal*- 1979

Saxena, R.P.;T.A.V. Murthy. Data Base for Enterprises: A Relational Approach, *SEDME Journal*, - 1978

Sexena R.P. Computerisation in Energy Sector, 9th conference of central and state statistical organs. 15-19th March – 1990

Sexena R.P. Revenue Realisation Monitoring system for SEBs & POWERNET, National Conference on cost deduction in power generation and distribution, New Delhi. April 26-28, 1994

Saxena R.P. Natural Language Access to a Large Data, — An Engineering Approach, Ph. D. Thesis - 1992 J.N.U., New Delhi

Sexena R.P. IT in Government; CSI Conference held at Bangalore from 30th October to 3rd November, 1996.

Wolfgang Effelsberg, Mary E.S. Loomis. Logical, Internal and Physical Reference, Behaviour in CODASYL Database System, *ACM Transaction on Data Base Systems,* vol 19, No. 2, June - 1984

Woods, W. An Experimental Parsing System for Transition Network Programs, in *NLP Algorithmics,* Press, New York - 1973

INDIGENOUS DEVELOPMENT PROGRAMME AND TECHNOLOGICAL ADVANCES IN NUCLEAR ENERGY IN INDIA

13

ANIL KAKODKAR
Director, Bhabha Atomic Research Centre,
Mumbai 400 085 (India)

1. INTRODUCTION

Considering the limited resource of fossil fuels and hydro potential available in the country, a long range energy plan for India has to be based on exploiting the nuclear and renewable energy resource potential available to us. The renewable energy options such as solar and wind need to be exploited to the fullest extent, but it can be easily shown that these options will be unable to cater to Indian industrial requirements in a major way in the forthcoming fifty years.

The only option available for meeting long term energy requirement of India lies in utilisation of indigenous resources using nuclear fuel. India has rather meagre resources of uranium (energy content of 500 GWe-years in thermal reactors, and 16,000 GWe-years in fast reactors) and very large reserves of thorium (energy content of 200,000 GWe-years). Assuming that thorium has to totally replace coal by the year 2044 AD (economically mineable reserves of coal will be exhausted by that time if we have an installed capacity of 326 GWe of coal based thermal power plants in the year 2020 AD), the currently envisaged indigenous reserves of thorium would last well into the 22nd century and much beyond. Therefore our focus lies in development and deployment of technologies towards utilisation of thorium at an early stage. Most of the other nations of the world, which are advanced in nuclear technology, have access to fairly large reserves of uranium. Therefore, the technologies being pursued by these advanced nations today mostly revolves around uranium based reactor system, with little emphasis on reprocessing to extract plutonium and a negligible programme of thorium utilisation. At present India is subject to a widespread technology control regime as far as importing nuclear technologies and nuclear fuel iş concerned. Even if it is postulated that at some time in the future technology control regimes will relax, any hypothetical external assistance, will not be able to address and long term Indian need to be able to exploit the Indian thorium resources, since as already mentioned, other advanced countries do not have an immediate interest in this technology. On the other hand, however, if India continues to pursue the path of development of indigenous nuclear technology, it will be in a position to play a leading role to provide thorium utilisation based technologies not only to meet its own needs, but also similar needs of other nations of the world, when the world uranium resources eventually become uneconomical for being exploited in the existing manner.

To sum up, India does not have any option other than having self-reliance by developing nuclear technologies tailored to exploit indigenous nuclear fuel resources to ensure its long term energy security and well-being of its population.

2. THREE PHASES OF INDIAN NUCLEAR PROGRAMME

To meet the goal of self reliance in exploiting indigenous nuclear fuel resources, a well thought of strategy in the form of three phases of the Indian nuclear programme has been evolved, and is being pursued vigorously. The three phases of the Indian nuclear energy programme are pictorially depicted in Figure 1.

In the first phase of this programme, Pressurised Heavy Water Reactors (PHWRs) using natural uranium as fuel and heavy water as coolant as well as moderator are planned to be constructed. This phase can support an installed capacity of 10 GWe for about 50 years. The spent fuel coming out of the PHWRs will be processed to extract Pu^{239} which is a very efficient man-made nuclear fuel material. Using this plutonium in Fast Breeder Reactors along with fertile materials, viz., depleted uranium and thorium, leads to not only generation of additional energy but also breeding additional Pu^{239} and U^{233}. Therefore, the second phase of the Indian nuclear power programme is based on Fast Breeder Reactors, utilising plutonium as the major constituent of nuclear fuel.

In the third phase, reactor systems based on utilisation of U^{233} and thorium will be developed and constructed. It is to be noted that the growth of Fast Breeder Reactors has been considerably slowed down all over the world on account of some adverse perceptions about their economics and safety, and a greater degree of familiarity with the thermal reactor systems using uranium. In India, currently we have a growing programme for the construction of 220 MWe and 500 MWe Pressurised Heavy Water Reactors. This programme has already been generating spent fuel from which the plutonium required for fuelling the second phase can be, and is being recovered. A 40 MWTh Fast Breeder Test Reactor is in operation at Kalpakkam. Aprototype Fast Breeder Reactor of 500 MWe capacity has been designed and is awaiting sanction for funding its construction. In the Indian context, with this backdrop, our activities towards utilisation of plutonium in Fast Breeder Reactors may reach a level of maturity only by the end of first decade of the twenty-first century. On the other hand, technologies for the utilisation of thorium in the third phase must be fully mastered by the third decade of the next century. Keeping this in view multiple approaches need to be pursued towards development of indigenous capability for utisation of thorium in our existing heavy water reactors, in our existing and forthcoming FBRs and also in Advanced Heavy Water Reactor, which is being developed actively in BARC. The Advanced Heavy Water Reactor (AHWR) derives its base from the technologies already mastered towards the deployment of PHWRs. It incorporates a large number of inherent and passive safety features either fully meeting or exceeding the requirements which are now internationally being stipulated for the next generation advanced nuclear power plants This reactor will require a very small feed of plutonium as fuel while most of the power will be generated from thorium in the core. This reactor can be perceived,

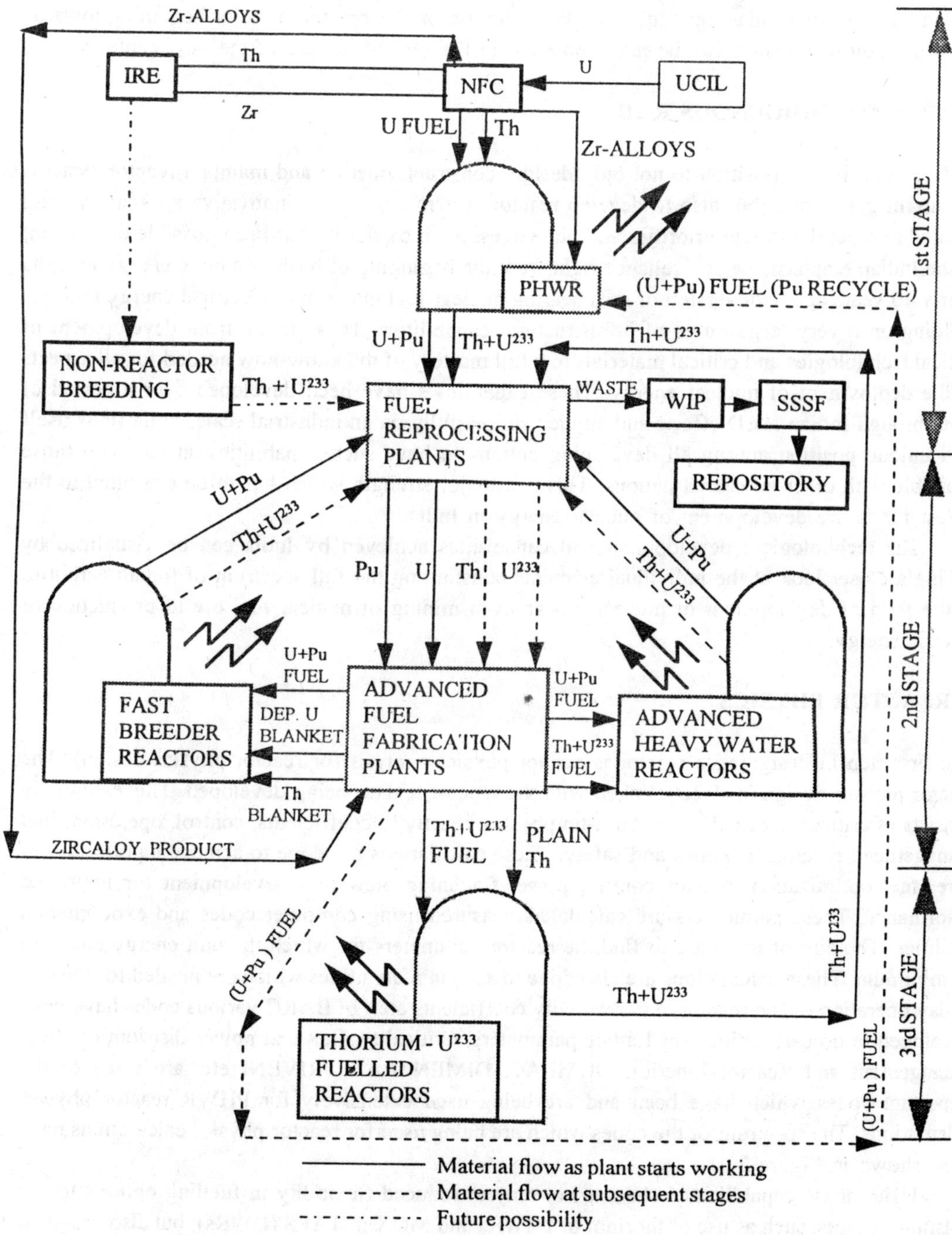

FIG. 1 : *Phases of Indian Nuclear Programme*

to begin with, as an important means of bridging the gap between phase-1 and phase-3 of the Indian nuclear power plan expeditiously. With the use of this reactor, a major beginning towards the utilisation of thorium can be made possible in the second decade of the next century.

3. ROLE OF INDIGENOUS R&D

Today, India is in a position to not only design, construct, operate and maintain reactor systems of current generation, but also to develop reactor systems, based on innovative approaches, and tailored to meet the Indian priorities and objectives. Such capability has been possible on account of the Indian emphasis on self reliance right from the beginning of birth of atomic energy in India nearly 40 years ago. Conversion of ores bearing nuclear fuel material to electrical energy requires building up a very large range of infrastructural capabilities. These range from development of critical technologies and critical materials to a full mastery of the know-how needed for all aspects of the deployment of nuclear energy. These capabilities have been developed over a period of time through Indian R&D efforts and applied successfully on an industrial scale. India finds itself in a unique position among all developing nations to have such capabilities, at par with those available with other advanced nations. This is a major strength which has to be exploited to the fullest for future development of nuclear energy in India.

The technological development and capabilities achieved by India can be visualized by having a closer look at the individual elements constituting the full spectrum of Indian activities in the field of development of nuclear power from mining of nuclear fuel ore to production of electric energy.

4. REACTOR PHYSICS

The first step in designing a reactor is reactor physics analysis (or reactor physics design). The reactor physics design is closely linked with the type of reactor being developed. This essentially consists of criticality calculations, burn-up physics, reactivity coefficients, control, operation, fuel management, reactor dynamics and safety. These calculations are done to achieve equilibrium of core, fuel optimization, reactor control, power flattening, new fuel development for improved peformance. These parameters are calculated/measured using computer codes and experimental facilities. The aim of these are to find the reactor parameters for which the unit energy cost will be minimum. These calculations are also done to ascertain quantities which are needed for the day to day operation of the reactor, e.g. reactivity coefficients etc.. In BARC various codes have been developed to do calculations for Lattice parameters, Criticality, flux and power distribution, Fuel Management and Reactor kinetics. DUMLAC, DIMENTRAI, TRIVENI etc. are some of the important codes which have been and are being used extensively for PHWR reactor physics calculations. The spectrum of the codes which are being used for reactor physics calculations have been shown in Figure 2.

Using these capabilities India has not only introduced flexibility in fuelling options for its existing reactors, such as use of thorium in PHWRs and MOX in TAPS (BWRs), but also designed new reactors DHRUVA, PURNIMA and KAMINI.

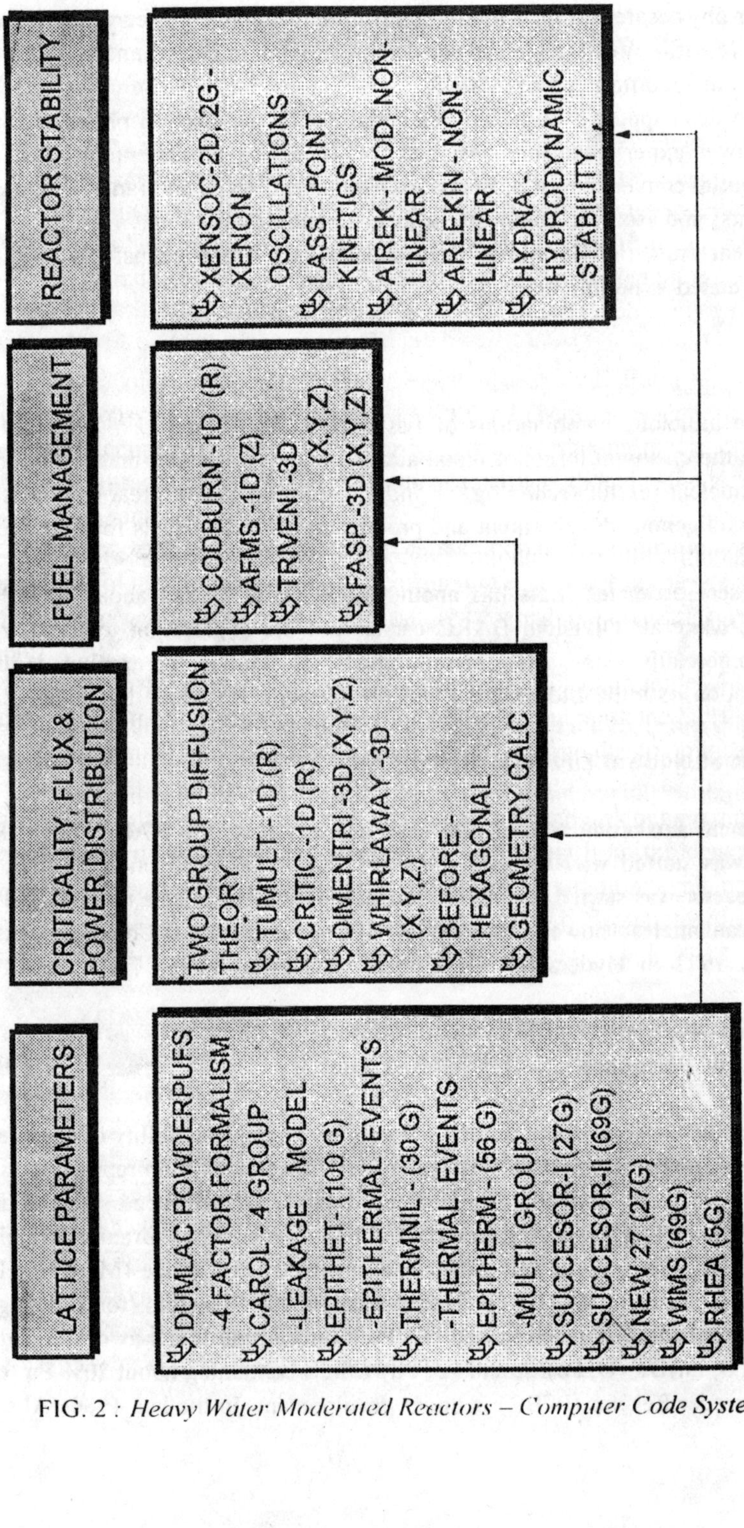

FIG. 2 : *Heavy Water Moderated Reactors – Computer Code System*

The reactor physics design for a totally innovative fuel concept is currently being developed for AHWR. The versatile WIMS code has been commissioned in BARC and certain modules were modified to suit the specific need of AHWR.

Reactor physics support is absolutely essential during the start-up phase of any reactor. In PHWRs, low power experiments are designed for buckling measurements, fast fission ratio measurements, Initial conversion ratio, Spectrum parameters, Axial and radial flux measurements, Power coefficients, and Isotopic composition as a function of burn-up.

Research reactors in BARC like CIRUS and DHRUVA are extensively used to carry out reactor physics related experiments.

5. FUEL

Nuclear fuels are judicious combinations of fissile U^{235}, Pu^{239}, and U^{233} and fertile U^{238} and Th^{232} isotopes in the chemical forms of metal alloys, ceramics or metal-matrix dispersions. From the inception of nuclear reactor technology in India in the mid 1950s, great emphasis was laid on self reliance and indigenous development and production of nuclear fuels for both "research" and "power" reactors.

As per present estimates India has about 76,000 Te of U and about 400,000 Te of Th reserves. Atomic Minerals Division, BARC carries out the exploration to identify technically feasible and commercially viable sites for mining of ores. Mining for uranium is then done by Uranium Corporation of India Ltd. (UCIL) and for Thorium it is done by Indian Rare Earths.

5.1 Development of Natural Uranium Based Fuel

Design, development and production of fuel from ore was started in BARC. For CIRUS reactor fuel production was started way back in 1959. In late sixties production of fuel bundles for RAPS-1 power reactor was started in BARC. Considering the large scale need of fuel bundles for the growing Indian nuclear power programme, NFC (Nuclear Fuel Complex) was setup and commissioned in 1971 in Hyderabad. Since then NFC is producing fuels for all our power reactors.

5.2 Development of Plutonium and U^{233} Based Fuel

Considering the small reserves of Uranium in our country and unavailability of enriched Uranium, parallel work was started to develop fuels based on Pu^{239} (produced from reprocessing of uranium Based spent fuel) and vastly available Th^{232}. These fuels are in advanced stage of development and trials. In India only TAPS reactor needed a fuel based on enriched Uranium which was being imported from France. To achieve self sufficiency a mixed U/Pu oxide (MOX) fuel containing 2-5% PuO^2 was developed in BARC. To cater to Fast Breeder Reactors (for 2nd stage of Indian Nuclear Energy Programme) Pu rich mixed U - Pu monocarbide fuel was developed for FBTR. Development of (U, Pu)O_2, (U, Pu) C, and (U, Pu)N fuels containing about 20% Pu for 500 Mwe Prototype Fast Breeder Reactor (PFBR) is under development. In the late 1980s Al clad Al-U^{233}

and A1-23% Pu plate fuel elements was developed for the 30kWt KAMINI reactor. BARC has a plutonium fuel production facility which could produce about 200 Kg annually of monocarbide fuel>for FBTR.

5.3 Development of Fuel Engineering Capabilities

Fuel engineering involves the design of fuel assembly, in-pile and out-of-pile testing. The range of fuel assembly designs handled include 19 and 22 element fuel bundles for 220 MWe PHWRs, 37 element fuel bundles for 500 MWe PHWRs, fuel assembly designs for TAPS with (6×6) and (7×7) array, MOX fuel assembly for TAPS and AHWR fuel assembly. A number of computer codes for the analysis of fuel pin and fuel assembly have been developed and are being regularly used for fuel design verification. The out-of-pile testing of fuel bundles is carried out in an Integral Thermal Facility for Pressure drop measurements over a wide range of Reynolds numbers, single pin and bundle strength test, wear test and endurance test.

5.4 Irradiation and PIE Tests of New Fuel Designs

Irradiation testing and.Post Irradiation Examination (PIE) are important activities pertaining to fuels development. Accordingly a pressurised water loop (PWL) was commissioned in CIRUS and two more are in an advanced stage of commissioning in DHRUVA. Some of the experimental fuels which have undergone irradiation in CIRUS-PWL include; Nat. UO_2 , UO_2-4% PuO_2 , and ThO_2-4%PuO_2 based fuel for BWRs and UO_2-3%, PuO_2 , UO_2 , ThO_2 , and ThO_2 -6.75%, $P\mu O_2$ for PHWRs.

Based on the confidence generated through these in-pile and out-of-pile tests MOX assemblies for TAPS could be designed. The PIE results of failed fuels from TAPS, DHRUVA and MAPS have provided useful feedbacks for making necessary modifications and alterations in fuel design, in quality control steps in fuel fabrication plants and in fuel management during in-pile operation.

5.5 Development of New Fuel Manufacturing Processes

The conventional and advanced ceramic nuclear fuel pellets for PHWR, LWR and LMFBR are fabricated by the "powder pellet" route starting from UO_2 , PuO_2 , and ThO^2 powders. The major drawback of this process is "radioactive dust hazard" arising from the generation and handling of fine dust particles. In the case of carbide and nitride fuels, additional problem of "pyrophoricity hazard" also exists. Further, fine powders are not free flowing and are not suitable for remote fabrication. Sol-gel microsphere pelletisation (SGMP) process developed in BARC is an advanced method for the fabrication of ceramic nuclear fuel pellets [namely, UO_2 , $(U,Pu)O_2$, $(Th,U)O_2$, (U,Pu)C and (U,Pu)N] of controlled density and microstructure by using "dust free" and "free flowing" sol-gel derived coarse (100-100 microns) microspheres as the starting material for pellet making and sintering. This process removes above - mentioned problems and facilitates remote and automated pellet fabrication (attractive for highly radiotoxic Pu and U^{233} bearing fuels). The production route becomes simple and it has other features which make fuel excellent microhomogeneous. By this process it is possible to tailor make both high density fuel pellets for PHWR and LWR and relatively low density fuel pellets with "open" pore structure recommended for LMFBR. This process has been demonstrated on laboratory scale and plant scale demonstration facility is being developed.

A combined sol-gel microsphere pelletisation (SGMP) low temperature oxidative sintering (LTS) process has been developed for the fabrication of high density UO_2 and $(U,Pu)O_2$ containing upto 3% Pu. This novel process, apart from avoiding "radiotoxic dust hazard" and facilitating automation and remotisation, leads to considerable energy saving during the sintering process. Several kilograms of high density UO_2 and $(U,Pu)O_2$ fuel pellets have been fabricated by the SGMP-LTS process. Fuel bundles containing high density UO_2 pellets prepared by this process have undergone irradiation in MAPS-1.

During the last three decades BARC has been successful in achieving indigenisation in all aspects of nuclear fuels for our research and power reactors. Various types of fuel developed, and under development, for our research and power reactors are shown in Figure 3.

6. SPECIAL MATERIALS

Apart from fuel, alloys of zirconium (used for fabrication of reactor core components) and Heavy water (used as moderator and primary coolant in the reactor) are other important materials used in the reactor.

6.1 Zirconium

The ore for production of Zirconium is Zircon. In India it is estimated that about 12.4 million Te of Zircon reserves are available in beach sands of Kerala, Tamilnadu and Orissa. Mining and processing for Zircon sand is done by Indian Rare Earths (IRE). Conversion from Zircon sand to finished product such as Pressure tubes, Calandria tubes, fuel cladding, sheets, etc. is done at Nuclear Fuel Complex (NFC), Hyderabad.

From the very beginning BARC is engaged in development of Zirconium based nuclear reactor components. BARC was instrumental in development of parameters for the manufacturing process of Zr-2.5%Nb pressure tubes. The cold working and heat treatment parameters were optimised in BARC.

Indigenously developed high technologies process equipments such as Electron beam welding unit for making Zirconium alloys electrodes for vacuum arc melting, microprocessor operated horizontal vacuum annealing furnace etc. are used to meet the end-requirement of the product. The high level of performance required of all these components during reactor service calls for comprehensive and stringent quality control at all stages of production. NFCs quality assurance programme is organised to achieve the desired objectives and incorporates a large variety of analytical testing and inspection procedures.

Presently plans are there for increasing the production capacity so as to meet requirement of installed capacity (nuclear) of 6000 MWe. This will have to be further expanded to meet requirement of 20,000 MWe in the year 2020 AD.

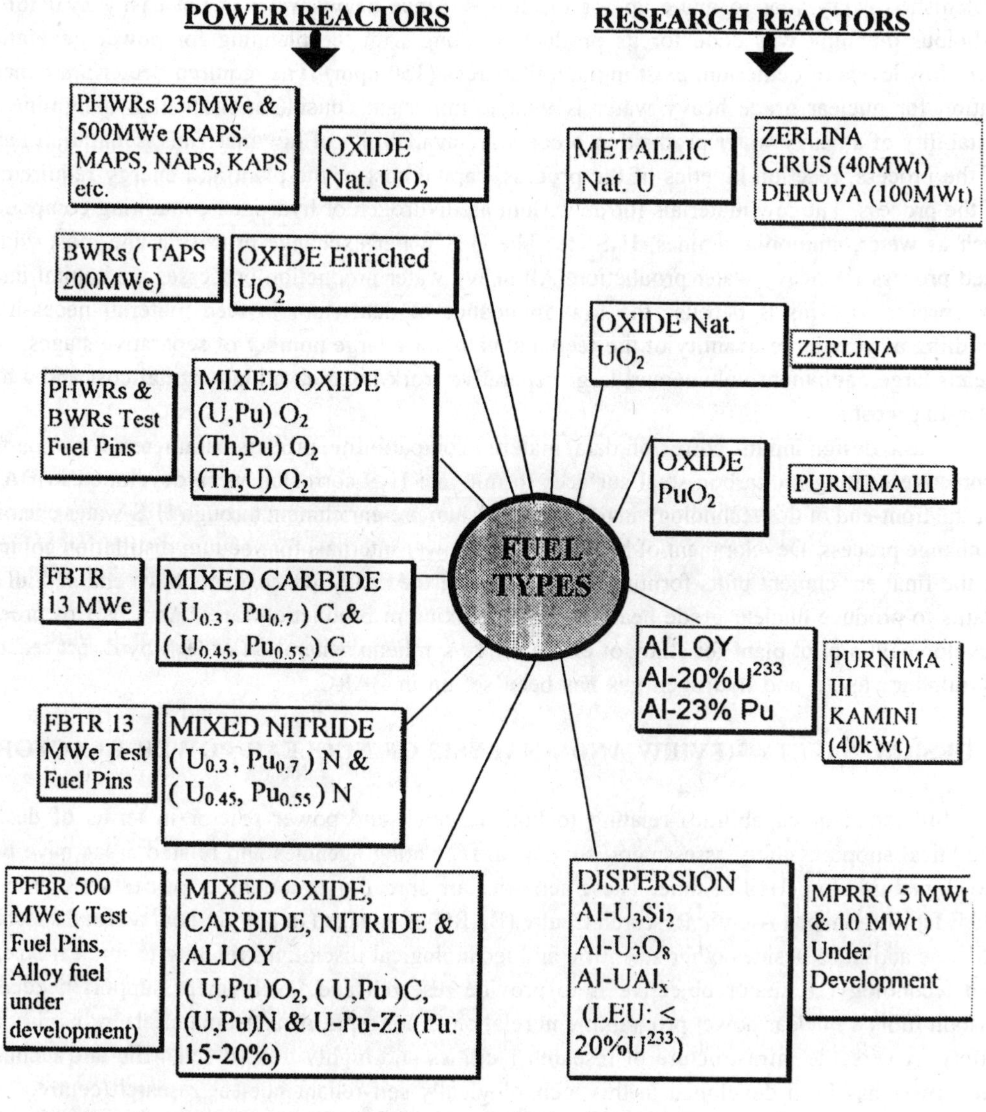

FIG. 3 : *Nuclear Fuels Developed and Fabricated Indigenously in India*

6.2 Heavy Water (D_2O)

Heavy water is an essential input for our nuclear reactors. A PHWR of 235 MWe needs about 220 Te of heavy water as initial inventory. About 10% of the initial inventory is required to be replenished every year to make up for the losses. Since heavy water is not easily available, a judicious planning was done for its production along with the planning for power generation. Very low levels of deuterium exist in natural sources (150 ppm). The required deuterium concentration for nuclear grade heavy water is 99.8%. Important considerations which determine the suitability of a heavy water production process are availability of raw materials, separation factor of the process, reaction kinetics of the process, capital cost of the plant and energy requirement of the process. The raw materials for deuterium are hydrogen or hydrogen containing compounds such as water, ammonia, amines, H_2S etc. The H_2S/Water exchange process is the most widely used process for heavy water production. All heavy water production processes are capital intensive processes. This is because the low abundance of deuterium in feed material necessitates handling of very large quantity of the feed and calls for a large number of separative stages. This means large equipment volume and large separative work i.e. both a high capital cost and a high operating cost.

Basic design inputs, corrosion data, material compatibility, process parameters for imparting protective coating on carbon-steel surfaces to mitigate H_2S corrosion were developed in BARC for the front-end of this technology involving deuterium pre-enrichment through H_2S-water chemical exchange process. Development of highly efficient tower internals for vacuum distillation columns of the final enrichment units forming the rear end of the heavy water plants have enabled all the plants to produce nuclear grade heavy water at maximum production rates. As a part of process development a pilot plant for study of effluent free synthetic generation of H_2S by direct reaction of sulphur vapour and hydrogen gas has been set up in BARC.

7. DESIGN, SAFETY REVIEW AND ANALYSIS OF NUCLEAR POWER REACTORS

The full range of capabilities relating to both research and power reactor in terms of design, analytical support, safety assessment, support to regulating agencies and related areas have been developed over a period of time. These activities are spread over a number of centres within the DAE family. Bhabha Atomic Research Centre (BARC) is engaged in carrying out a broad spectrum of these activities besides other scientific and technological disciplines relating to nuclear science and technology. A major objective is to provide research and.development support needed to sustain India's nuclear power programme in relation to concepts, designs, materials, reliability and safety. A versatile infrastructure of research facilities and highly trained scientific and technical manpower has been developed in this technologically self-reliant nuclear research centre.

In order to provide R&D support to the nuclear power programme, particularly the thermal reactors based programme currently under way, a number of major reactor engineering facilities have been set up in BARC. In addition to the experimental investigations and R&D work for the nuclear reactor technology and reactor safety, investigations using specially developed computer codes are also carried out to analyse complicated design problems and to model safety related phenomena.

The current activities include the design and development of an Advanced Heavy Water Reactor (AHWR), which is being designed to utilise thorium based fuel. The AHWR has a number of passive safety features including the use of natural circulation in primary heat transport system during normal operation.

The following are some of the typical fields in addition to the standard practice relating to the aforementioned fields where sufficient expertise has been developed in terms of trained manpower, tools, techniques, test facilities and computer codes :

(1) Thermalhydraulic studies and simulation of nuclear reactor heat transport system relating to :

(a) Normal operation and transient conditions.
Codes developed : HCHF. H_2O-PROP, D_2O-Prop, CODA-220, CODA-500 etc.
Codes adapted: APCIGS, COBRA3C, COBRA-IVM

(b) Loss of Coolant Accident (LOCA) conditions, Thermosyphon cooling of reactor core and scenario following station blackout.
Codes developed: COHRA, RELAP-PHWR, NODE-1, TINFLO-III.
Codes adapted: ATHLET-MOD1-Cycle-D & Cycle -E, RELAP4 / MOD6.

(2) Study of Fuel and coolant channel behaviour following LOCA and Emergency Core Cooling System (ECCS) failure (severe accident condition)—thermalhydraulic and thermomechanical aspects.
Codes developed: HT/MOD4, TABS, WELTEM

(3) Studies and experimental techniques developed for study of thermal hydraulics of PHT system under two phase flow conditions.
Codes developed: HFLOWR, VFLOWR
Instrumentation developed: Drag disc flow meter.

(4) Fluid-structure interaction modelling studies to simulate sloshing and dynamics of submerged structures.
Codes developed : FLUSHEL, FLUSOL, SLOSH3D

(5) Study of Pressure and temperature transients within the nuclear reactor containment.
Codes developed : CONTRAN

(6) Studies relating to Hydrogen generation following metal water reaction in PHT, its transport, combustion and mitigation.
Codes developed : HYCOMB, HYRECAT

(7) Studies relating to Aerosol transport in PHT system and containment.
Codes available : NAVA, TRAPMELT

(8) Containment isolation studies.

(9) Modelling and simulation of Static and Dynamic response of RC/PC reactor containment.
Codes developed : ULCA, ARCOS-3D

(10) Studies relating to Earthquake engineering-ground motion and response of structures.

(11) Studies relating to Fuel performance under normal and transient conditions
Codes developed : FAIR, HANU

(12) Studies and experimental set-ups for Probabilistic Safety Assessment and ageing research:
(a) Studies relating to design evaluation of standardised PHWR.
(b) PSA level-1 studies of operating reactor systems.
(c) Development of operator support systems to enhance safety and reliability during plant operation and maintenance.
Codes developed : LIVING-PSA, Symptomatic Diagnostic System.
(d) Development of facilities for accelerated ageing studies including qualification of components under normal and accident conditions like LOCA for process monitoring devices, I&C cables, elastomeric materials, containment surface coatings and various electrical and electronics items.

(13) Studies on degradation mechanisms affecting coolant channels.
Codes developed: SCAPCA, HYCON, BLIST-1D, BLIST-2D, SPARSH, TABS

(14) Development of Coolant channel inspection, life extension and replacement technologies.
Tooling developed for inspection : BARCIS, Scrape sampling tool, DRYVIS
Tooling developed for life extension : Mechanical Flexing Tool, Pulsed EMD, INGRES
Tooling developed for large scale reactor retubing : CCRM

(15) Fracture mechanics and Leak-before-break studies :
Codes developed : FABS, JINT, BARC-R6, LEAK, JINT, THESIS, CEAL

(16) Development of systems for on line fatigue life monitoring of structures.
Codes and instrumentation developed : FMS.

(17) Studies related to Life extension of power plant components.

(18) Development of Nuclear Power Plant Software Simulator.
Codes developed : Simulator For RAPS Type of 220 MW(e) PHWR.

(19) Thermal Shock and Fatigue Studies.
Codes Developed : Heat-2D

(20) Experimental and Analytical evaluation of residual stresses in Roller Joints of PHWR coolant channels and their design.
Codes developed : ROLLER, SPARG.

(21) Development of Super Computer and utility software to model, simulate and solve amongst others, reactor problems requiring fast data processing.
Hardware developed : ANUPAM-64 node parallel processing Super computer having peak speed of 5.12 GFLOPS based on i860/XP RISC microprocessors on industry standard.
Multibus-II backplane having UNIX as the operating system.
Software developed : modelling of impact of aircrafts on outer containment of reactors, FAIR, Equation solver for FEM codes.

7.1 Facilities for Reactor Engineering and Reactor Safety Research

BARC has the following major experimental facilities available for :

Thermal Hydraulic Studies,
Heat Transfer Studies,

— Containment Studies, and
— Component Qualification and Testing.

7.1.1 3 MW Boiling Water Loop

The 3 MW Boiling Water Loop is a high pressure, high temperature test facility for thermal hydraulic experiments. In this facility a centrifugal pump delivers water to the test section inlet through a preheater. The loop pressure is maintained by a surge tank in which steam is produced by immersion heaters. Water or steam-water mixture exiting from the test section goes to the separator. Steam from separator is condensed in jet condenser by water which comes from a pool boiling cooler. Loop heat is dissipated to the atmosphere through the pool boiling cooler. A feed and bleed system and a purification system are also provided in the loop. This loop can be used for heat transfer studies in horizontal as well as vertical fuel channels, and even with the coolant boiling. The data generated are needed not only for the analysis of the reactor under normal operating conditions; they are also essential for the analysis of the safety of the nuclear reactor under various postulated accident conditions.

The loop parameters are as follows :

— Design pressure : 124 kg/cm^2
— Maximum operating pressure : 100 kg/cm^2
— Design temperature : 315°C
— Pump flow : 136.5 m^3/hr
— Pump head : 300 m
— Maximum test section power : 3 MW

In this loop thermal hydraulic experiments can be conducted on nuclear fuel simulators. Typical experiments which can be performed are dryout and post dryout experiments, subchannel mixing experiments, experiments to study parallel channel instability, blowdown, rewetting experiments etc.

7.1.2 High Flow, Low Pressure Test Facility

In the High Flow Low Pressure Test Facility, water from the basin of a cooling tower is led to the test model using centrifugal pumps. A valve station is incorporated in between pumps and the test model to control the pressure in the test model. From the test model water returns to the top of the cooling tower by gravity. This facility is meant for fluid flow experiments at low pressure and temperature, but requiring large flows.

Further details of the facility are :

— Maximum flow rate : 5000 litres/min
— Maximum pressure : 140 m
— Minimum pressure required at test model : 2 m

In this facility experiments such as those to study moderator flow field in a PHWR calandria, tests on moderator diffuser nozzle and experiments requiring free water surface in the test model can be conducted.

7.1.3 Facility for Integral System Behaviour Experiments (FISBE)

FISBE is an Integral Test Facility being setup to study PHWR system behaviour under LOCA and non-LOCA transient conditions. Experiments are proposed to be conducted on the facility to generate data for the validation of computer codes for safety analyses. The facility will have 1:1 elevation scaling. The volumetric scaling is about 1:75. Apart from generating data for code validation, the facility can also be used for testing operational and emergency procedures. The facility is being set up in two phases. Construction work for Phase-I, in which experiments on natural circulation, limited blowdown and station blackout are to be conducted, is nearing completion.

The major parameters of the test facility are as follows :

- Design pressure : 125 Kg/cm^2
- Maximum operating pressure : 87 Kg/cm^2
- Design temperature : 315° C
- Maximum power 3 MW

7.1.4 Heat Transfer and Fluid Flow Laboratory

The Heat Transfer and Fluid Flow Laboratory contains equipments and instruments required for flow visualization studies and for the measurement of important thermal hydraulic parameters.

The laboratory at present has the following main facilities :

- Optical system for flow visualization
- Equipment required for visualization of flow using tracer particles, dyes, surface paints, electrodes etc.;
- Two-phase flow parameter measuring devices, e.g. gamma ray densitometer, drag disc flowmeter etc.; and
- Transparent set-up for closed loop thermosyphon studies.

The facilities in the heat transfer and fluid flow laboratory are useful for basic studies to understand the thermal hydraulic phenomena in complicated geometries.

7.1.5 Facilities for Component Qualification and Testing

These include :

- Thermal ageing test facilities for tests ranging from ambient to 400° C;

- LOCA Environment Simulator-Design pressure of this is 50 psig and steam temperature can be upto 150° C; and
- Synergism simulator will be used for studying the combined effects of temperature, humidity and radiation on the ageing of components used in nuclear power plants. Design of the facility is in progress.

7.1.6 Integral Thermal Facility (ITF)

A testing facility called Integral Thermal Facility (ITF) provides water flow at a temperature of 270° C and pressure of 110 bars (simulates PHT system conditions) through coolant channels identical to those in a power reactor. The pressure is built up by an electrically heated surge tank, and temperature is raised by electrical heating of the circulating water. One noteworthy feature of ITF is that, unlike in many engineering systems, there is no special heating element in ITF. Electric current is passed through the appropriate portion of the piping itself so that the ohmic resistance of the pipe leads to the generation of the required quantity of heat.

ITF is Utilised for Following Purposes

- ITF is a versatile facility which can be used for testing and establishment of thermal hydraulic parameters relating to any component or sub-system of PHT system. ITF has been extensively used for tests on fuelling machines. Before the fuelling machine is installed in a power reactor, its ability to perform all the operations in a smooth fashion, and in the desired sequence, has to be ascertained. ITF is mainly used for simultaneously testing the performance of the two fuelling machines, operating exactly in the same way as they are expected to function in an operating power reactor, before they are despatched for installation at operating power reactors.
- ITF has been very useful in trying out and perfecting effective trouble-shooting techniques whenever the fuel handling system in an operating power reactor experiences any serious problem.
- ITF has been used for measuring pressure drop across fuel bundles in reactor simulated conditions.
- ITF has been used for carrying out endurance test of fuel bundles.

7.1.7 Burst Test Facility

This facility which is under advanced stage of erection will be used for following purposes :

- Biaxial burst testing of pressure tube along with rolled joint and other pressure retaining component of coolant channels of PHWR at room and elevated temperatures. The objective being determination of ultimate pressure retaining capability of these components and available margin of safety. This is also needed for characterisation of new pressure tube material.

- For slit burst test of PHWR components like pressure tubes, end fittings and other components of cylindrical shape. The aim being to determine the critical crack length and crack propagation rate of these components under various metallurgical conditions at elevated temperature (300°C) and at room temperature.
- For programmed cyclic pressure loading of the pressure retaining components. The aim being to find the effects of such loading on flaw propagation, long term integrity and leakage behaviour of rolled joints.
- For development of water jet cutting system for remote cutting of non-metallic and metallic components. This will be used for reactor maintenance and repair tasks.
- For use as high pressure water jet for descaling and decontamination.
- Application of high pressure for hydraulic expansion of large dimension tube to tube sheet joints.
- Hydro forming of thin walled components
- For Autofrettage of cylinders.

The key element for all these tests is an High pressure intensifier pump which has a working pressure of 4000 bars had a operating flow rate of 2 lpm. This high pressure pump has been commissioned. Other related components are being erected.

7.1.8 Bi-Axial Creep Test Facility

This facility which is under advanced stage of erection will be used to study out of pile bi-axial creep properties of Zr-2.5% Nb pressure tubes at various stresses and temperatures. Following are the objectives of this test facility;

- Biaxial creep rupture test: The measurement of the time for fracture to occur under biaxial stresses and temperatures.
- Measurement of biaxial creep rate room temperature, and higher temperature under biaxial stresses.
- Study of propagation of partial depth cracks
- Verification of design stress intensity
- Data generation for mechanical properties in biaxial loading.

7.1.9. Pressure and Thermal Cycling Facility

This facility is under development for carrying out certain tests on reactor components under specified cyclic conditions of temperature and pressure. The range of test pressure being 50 to 150 Kg/cm^2. The range of test temperature being 50°C to 330°C. Water will be used as working fluid. The test facility consists of a reciprocating pump, control valves for hot and cold legs, heat exchanger, heater, mixing chamber, test section etc. The reciprocating pump delivers water at 30°C and 150 Kg/cm^2 to hot and cold leg. The water in hot leg is passed through shell side of the heat exchanger and heated to 320°C. Temperature is further raised to 330°C by a heater. After passing through mixing chamber and test section, this hot water is circulated through tube side of heat exchanger which heats the shell side cold water to 320°C maxm. The tube side water gets cooled to 50°C and is further passed through another set of valves, low temperature heat exchanger before it is discharged to supply vessel at 30°C.

8. SPECIAL CONSTRUCTION/FABRICATION TECHNIQUE DEVELOPMENT

8.1 Development of Rolled Joints

In PHWRs the Zr-2 pressure tube is joined to AISI403 stainless steel through "roll expanded" joints (because of dissimilarity of the metals other joints such as welded, friction bonded etc. will not be effective). These joints, during reactor service, are subjected to high pressure, high temperature and high neutron flux and as such should be a highest order with no leakage and a very high pullout strength. During the rolling process the tube expander (also called rolling tool) exerts a very high force on the tube through each of its rollers so that tube metal is plastically deformed under the zone of contact with the rollers. In order that the joints maintain their integrity and do not deform permanently under these three or five line loads it is the outer end-fitting is made strong and rigid.

In BARC 12 different types of rolled joints were developed for DHRUVA research reactor. Because of this developmental work a few novel design concepts were evolved Subsequently joints were developed for pressure tubes and calandria tubes of 235 MWe and 500 MWe PHWRs. Fig. 4 depicts the range of reactor grade rolled joints developed with Zr-2 pressure tubes at BARC. Each point in this Fig. represents a rolled joint design having a particular diameter and thickness combination which has been developed.

Thin sleeve type rolled joint was initially developed in BARC for DHRUVA coolant channels. Subsequently this concept was used in the design of rolled joints for the replaceable liquid poison tubes for NAPP and subsequent Indian 235 MWe PHWRs.

Since Zr-2.5% Nb pressure tubes have been adopted in current generation of PHWRs (KAPP-2 onwards) it was needed to develop the rolled joint for this material. This material has higher room temperature yield stress and lower wall thickness than Zr-2 tubes a new design of the rolled joint had to be carried out. This involved optimisation of groove geometry and percentage tube wall reduction so that the joint could meet the stipulated specifications. Moreover since the clearance between the end fitting and pressure tube has been reduced to nominal clearance/ interference it was necessary to heat the end fitting sleeve before assembling the joint. This type of joint called "Zero clearance type rolled joint" was developed in BARC. 58 such joints were made and successfully tested during the developmental programme for 235 MWe coolant channels. The optimised design met the stipulated specifications.

8.2 Electron Beam Welding

The welding of zirconium alloys for reactor components is exceptionally critical work. To avoid contamination welding in a controlled environment i.e. electron beam welding in vacuum is preferred. Secondly electron beam welding has the characteristics of low heat input, narrow heat affected zone and deep penetration. Extensive trials were conducted in BARC not only to finalise the weld parameters but also to facilitate machining to obtain the required critical groove patterns for the zircaloy tubes. By using electron beam welding in medium level vacuum oxygen H_2 and N_2 pickup can be avoided. Further annealing of the weld and heat affected zone which would otherwise would have been adopted for other types of welding process is avoided since EB welding has the characteristic of deep penetration and very narrow heat affected zone. Tests were carried out in BARC to perform EB welding of Zr-2 in a vacuum chamber producing a vacuum level of 10^{-4} to 10^{-5} torr. For this all the developmental work like design and fabrication of vacuum chamber, driving system etc. were done.

9. INSTRUMENTATION AND CONTROL SYSTEM DEVELOPMENT

9.1 Development of Control Systems for Nuclear Reactors

Control systems for reactors demands high reliability, operational flexibility and fail safe features. Such systems have to cater to operational needs and safety in all regimes of plant operation. Several comprehensive control systems have been designed for a variety of research reactors like APSARA, ZERLINA, PURMNIMA-I, II, & III, DHRUVA and KAMINI as well as for the various nuclear power plants. These comprehensive control systems generally comprise the following :

(a) Control Systems:
- (i) Process instrumentation and selection of sensors.
- (ii) Interlocks and logics for valves, pumps, compressors, fans, motors etc.
- (iii) Close loop system-reactor regulation system, pressure, flow, level and temperature controls for various processes.

(b) Safety Systems:
- (i) Reactor protection systems
- (ii) Emergency core cooling system
- (iii) Containment and isolation systems
- (iv) Decay heat$removal system

(c) Control Centres:
- (i) Main control room
- (ii) Auxiliary control room
- (iii) Local control rooms
- (iv) Equipment rooms:

(d) Operator support systems:
- (i) alarm annunciation systems
- (ii) Data acquisition and process information system.

(e) Reactivity Control Mechanisms:
- (i) Shut off mechanisms
- (ii) Control
- (iii) Core drop mechanisms

(iv) Reflector/burnup plock drivers
(v) Liquid poison injection system.

10. REPAIR AND MAINTENANCE SUPPORT TO OPERATING REACTORS

The rapair and maintenance problems of a nuclear reactor are specially challenging ones and quite complex considering radiation environment and access problems. Lot of R & D efforts have been put in to solve these problems. A number of these problems were anticipated and accordingly the development of a remedial system for the same was done before the problem actually occurred in a reactor. Some other systems were developed and successfully used after facing unanticipated problems in our reactors. Coolant channels are central to the design of PHWRs. It would not be wrong to say that the success of such reactors depends upon the success of coolant channel design and associated technologies. Coolant channels represent individual pressure boundaries influenced by the reactor environment in terms of neutron fluence and hydrogen effects. As a result they undergo degrading phenomenon such as creep, growth, reduction in fracture toughness etc. The problem of significant displacement of garter spring spacers from design locations and resulting phenomenon of pressure tube deflecting and contacting outside calandria tube is especially adverse to life of coolant channels. To tackle to all these problems in addition to good design, selection of right material, and proper operational control, it is important to have good inspection and surveillance as well as corrective capability available. Several techniques have been developed for this purpose. Following paragraphs describe some of the major systems developed to cater to repair and maintenance needs in our power reactors, and a number of the items described are related to coolant channels.

10.1 Development of Inspection Techniques

10.1.1 Post Irradiation Examination (PIE)

Post Irradiation Examination and failure analysis of fuels and critical core components and out-of-core components of zirconium alloys, aluminium alloys and alloy steels from water-cooled research and power reactors are being done in BARC since last two decades. The PIE facility in BARC consists of the following:

— Six interconnected concrete cells housing: (i) a wall periscope, (ii) NDT (UT, EC, profilometer, gamma scanner) and fission gas analysis instruments, (iii) sample cutting tools, (iv) a universal testing machine for ambient and elevated temperature mechanical properties of structural materials, and (v) metallography and autoradiography set-ups.
— A large lead cell for visual examination and non-destructive evaluation of full length pressure tubes from Indian PHWRs. This cell is also used for trepanning and cutting of blanks and specimens for metallography, hydrogen estimation and mechanical testing.
— A number of experimental set-ups housing; (i) low active metallography, (ii) a hydrogen analyser based on inert gas fusion, (iii) an instrumented Charpy impact testing machine, and (iv) an electron probe microanalyser.

PIE of Fuel

Reliable performance of the nuclear fuels has a large bearing on economics of nuclear power and radiation safety of plant operating personnel. Fuel performance in the reactor is, therefore, routinely monitored. On selected fuel bundles discharged from the reactor, PIE is periodically carried out to generate feedback information which is used by the fuel designer, fuel fabricator and the reactor operating personnel to bring about changes for better performance of the fuel. Since last 20 years PIE of fuels from research and power reactors are being carried out in Hot cells in BARC. PIE has played an important role in development of fuel for DHRUVA reactor and more recently MOX fuel. As a result of feedback from PIE at various stages, the various problems occurring were tackled by making necessary modifications in the fuel fabrication to arrive at the final design. Various problems faced by TAPS fuel such as internal hydriding at the end plug weld, fretting at the spacer grid site and pellet clad interaction were identified and necessary action was taken to avoid this problem. MAPS fuel was facing low burnup fuel failures and the resulting rise in coolant activity. To understand this PIE for MAPS fuel was done and fuel failure modes were identified as fretting caused by foreign debris and end plug separation due to hydriding near weld line. The tests which are routinely done for PIE of fuels include; visual examination, dia measurement, eddy current test, gamma scanning, gamma spectrometry, fission gas analysis, leak testing, macro examination, auto radiography, metallography, ring tensile test, micro probe analysis etc.

PIE of Coolant Channels

The PIE activities for coolant channels was started in mid 1980s when specimens of Zr-2 pressure tubes from RAPS-1 were evaluated in terms of hydrogen content, thickness of oxide layer, yield strength, ultimate tensile and ductility. The garter springs from these pressure tubes were also inspected. Then cut pieces of some of the MAPS coolant channels were examined. Two full length channels from RAPS-2 reactor have recently been examined. Examinations which are done on coolant channels include: visual examination, measurement of diameter and sag profiles, hydrogen content and oxide layer thickness, tensile properties and fracture toughness by making tensile specimens and compact tension respectively.

Keeping in mind the need and urgency of a data on PIE of irradiated PTs from Indian PHWRs to verify the results from in-service inspection and to validate different performance codes developed in BARC, the PIE facilities are being augmented to include advanced NDT instruments for monitoring oxide layer and hydride blisters, a differential scanning calorimeter (DSC) for hydrogen estimation of scrape or trepanned samples from PTs and a computerised remote cutting, milling and drilling machine for preparation of specimens for fracture toughness evaluation.

10.1.2 Non-Intrusive Diagnostic Technique

Coolant channel inservice inspection provide the capability to carry out detailed inspection of a coolant channel. But the process of inspection is slow and it is not possible to cover the entire core of the reactor in shut down of the plant. It is therefore necessary to have a supplementary scheme which can quickly identify a coolant channel in contact with its calandria tube, or a coolant channel with significant displacement of its garter springs. A technique based on measurement of vibration at either end of the coolant channel, caused by the disturbance created by shutdown cooling flow, has been devised to identify the coolant channels in contact with the calandria tubes. This procedure enables screening the entire reactor core in 3 to 4 days time. This technique has been used as a screening method to identify channels which are likely to be in contact so that further inservice inspection efforts and subsequent garter spring relocation efforts, if necessary, could be applied more effectively.

While the above technique has already been applied in the operating PHWRs, a technique to identify channels with significantly displaced garter spring is being developed. This technique utilises excitation by a shaker mounted at one end of the channel and measurement of the response at both the ends.

10.1.3 In-Service Inspection

Several complementary techniques have been developed in BARC to carry out in-service inspection of the coolant channels. They include:

—*DRYVIS:* This is Dry channel Visual Inspection System for coolant channels. In case it becomes necessary to examine the bore of a channel visually this system carries a readiation resistant micro video camera and propels itself along the entire 5 m length of coolant channel, enabling the internal surface to be seen on a video monitor. With this system it is possible to visually inspect pressure tube and calandria tube. This system has been successfully used at reactor site.

—*BARCIS:* A coolant channel in-service inspection system (BARCIS) has been developed and deployed at power stations for carrying out non-destructive inspection activities such as detection of flaws using eddy current and ultrasonic testing, measurement of wall thickness, measurement of position of garter springs, measurement of gap between coolant tube and calandria tube etc. under wet condition. In this scheme, the fuel is temporarily taken into the fuelling machine at one end and brought back into the channel after completion of the inspection. More advanced versions of this device are being developed.

—*Garter Spring Detection Technique* Eddy current based detection technique was developed in NARC and deployed at reactor sites to carry out Pre-service inspection of coolant cannels to identify location of garter springs before the reactor is commissioned into service. Necessary garter spring relocation devices then move the garter springs towards their design location if they are appreciably shifted.

10.1.4 Remote Creep Measurement of PHWR Coolant Channels

In order to detect ay anomaly in the creep deformation of coolant channels, and also to enable periodic adjustment of energy absorber devices located at the ends of each channel, monitoring of axial creep at regular intervals is necessary. Previously this measurement used to consume significant man-rem expenditure. A scheme has been implemented wherein this measurement is now done using measurement of the 'Z' movement of the fuelling machine With this scheme it is now possible to carry out measurement of creep across the entire reactor face in a very short time, with practically no man-rem expenditure. This scheme also enables identification of anomalies caused by problems like bearing seizure. An automated device for adjustment of energy absorber device has also been developed.

101.5. Hydride Scrape Sampling

Estimation of amount of H2 present in the coolant tube and to establish the trend of its pickup in-service is an important parameter related to its service life. Analytical methods have been developed to estimate the hydrogen level in the pressure tube have a considerable scatter band associated with it. A scrape sampling tool has been developed for getting actual values of the H2 level at any axial location of the pressure tube. Using this tool a small amount (about 50 mg) of material is removed for deuterium analysis from the top inside surface of the pressure tube. These slivers are then analysed later for hydrogen content. The contour of cut is such that the tube integrity is preserved and no crack initiating notches of sharp corners are created. This tool has been successfully tried out in PIE facilities in BARC for removal of scrap samples from removed pressure tube of reactors. All modules for its deployment at operating stations have been completed.

10.1.6 Inspection of MAPS Calandria Internals

After detection of moderator heavy water leak in MAPS reactors and in order to carry out video (visual) inspection inside calandria vessel to ascertain health of the calandria internals a scheme was developed and in BARC and successfully used in both units of MAPS. This scheme consisting of radiation resistant miniature video camera, a camera positioning manipulator, lighting and other fixtures were used to carry out inspection of moderator inlet manifold, calandria tubes, rolled joint grooves of calandria tube sheets, dump ports, moderator outlet manifold and adjuster rods.

10.1.7 Visual Inspection of RAPS-1 OPRD and Top of Calandria

For detection of source and location of heavy water leakage into calandria vault of RAPS-1, remote video inspection of calandria vault was necessary. Considering the availability of limited access through inspection openings in hatch blocks from top side of calandria vault, a remotely operable camera manipulator was designed to carry radiation resistant camera, high power lighting into the vault to scan various areas of the vault. Arrangements were made to tilt the camera, lighting head so as to get clear views of different areas. Actual inspection was carried out from a central opening and two side openings to cover largest possible area in the vault . Calandria over pressure relief device (OPRD) including its top cover and other constructional members like holding chains, shear pins, lugs could be closely scanned. Views of spray header and branches, helium line and vault wall liner plates were also obtained. The source and location of heavy water leak was detected. A volume of debris powder was also detected on OPRD cover.

10.2. Development of Tools and Techniques for Judging Fitness for Service of Coolant Channels

A number of computer codes have been developed in BARC to study various phenomenon which degrade the service life of coolant channels of Indian PHWRs. The following paragraphs describe some of the computer codes in brief :

10.2.1 SCAPCA : A computer code 'SCAPCA' (Static and Creep Analysis of Pressure tube and Calandria tube Assembly) has been developed in BARC to simulate the creep-growth behaviour of the coolant channel assembly. This code estimates all the creep-growth limited life limits and other relevant parameters based on channel specific design, operation, environment and material inputs.

This code has been validated by comparing the gap profile obtained from in-service-inspection of PHWR coolant channels with those obtained from theoretical estimates obtained from SCAPCA. With the help of such a code it is possible to predetermine the creep-growth response of a coolant channel, estimate its creep-growth limited life and use this information to design a channel with a suitable life provided the relevant channel specific information like channel dimensions, spacer locations, fast neutron flux (> 1 Mev) profile and other environmental and operation variables are known.

10.2.2. HYCON: To assess the integrity of the prssure tube it is necessary to estimate the bulk hydrogen level either through analytical means or through actual measurements. Using the correlations given in published literature a computer code 'HYCON' (for Hydrogen Concentration estimation) has been developed which gives the variation of hydrogen concentration along the length of the pressure tube after any period of its in-reactor operation at varying power levels.

10.2.3. CEAL: Using criterion for LBB, correlations for the estimation of various terms in the criterion, fracture toughness data taken from the published literature for NPD pressure tube, and other inputs relevant for the cold worked zircaloy-2 pressure tubes of Indian 235 MWe PHWRs, a computer code 'CEAL' (code for the Estimation of Assurance of Leak before break) has been prepared for the following illustrative studies have been carried out using this code:

- Assessment of required operator response time for different minimum leak detection levels and fluence seen by pressure tube.
- Determination of the operating pressure and temperature domain required for LBB at different fluence levels.

10.2.4. SPARSH: In zirconium alloys, as well as in several other metals, hydrogen is known to migrate down concentration and temperature gradients. A 3-D Finite Element code SPARSH has been developed in BARC for modelling migration of hydrogen due to concentration and thermal gradients in materials such as Zircalloy-2, Zircalloy-4, Zr-2.5% Nb etc. The code also can calculate 3-D temperature calculations. The code has been tested against certain benchmark cases quoted in literature.

10.2.5 BLIST2D: After migration of hydrogen in zirconium alloys and several other metals, under suitable conditions this mechanism leads to formation and growth of hydride blisters at cold spot of a pressure tube in contact with the calandria tube. A 2-D Finite Difference code BLIST2D has been developed to model the diffusion of hydrogen under the given temperature distribution, given initial concentration and the given ingress rate etc. The code has been validated against benchmark cases quoted in literature. This code is being used for analysing the channels of our operating reactors.

10.3 Development of Repair and Maintenance Toolings and Techniques

10.3.1 Decontamination of PHT System

Within a few years of operation of a nuclear reactor, the pipe work exposed to the reactor coolant can become substantially contaminated giving rise to the problem of high man-rem consumption in maintenance, operation and inspection activities. The primary cause of the radioactive fields in the corrosion of structural materials in the primary heat transport (PHT) system. Corrosion products get transported to the reactor core and get activated. Another source of contamination is failed fuel bundles which release fission products. Full scale decontamination of PHT system in a PHWR with the presence of the fuel bundles a particularly tricky problem. Indigenous decontamination technique was developed using dilute chemical decontaminants. The aim was to achieve a high decontamination factor but the decontaminant should be compatible with system materials. In addition the process should not cause draining or downgrading of the heavy water coolant and should generate minimum solid waste.

Extensive laboratory tests were conducted for optimising the chemical composition of the decontaminant formulation and other process parameters. A mixture of organic acids containing acid, reductant and complexant whose total concentration was 0.1% by weight was used in heavy water medium. The acids provided by the organic formulation attacked the oxide on CS surface effectively and dissolved it. The resulting solution was circulated through ion exchange resin, which removes the complexed metals and the reagents are regenerated. After decontamination is over, the reagents and the remaining corrosion products are removed from the solution by passing them in mixed bed ion exchange columns containing strong anion and cation resin. This technique was successfully used in MAPS-1 and MAPS-2. The decontamination factor ranging from 1 to 30 was achieved.

10.3.2 Design Improvements to Improve Maintainability

Modular Insulation Panels

Insulation cabinet are provided in the fuelling machine vaults to prevent excessive heat loss from bare end-fittings and feeders. Thereby the insulation cabinets also reduce the fuelling machine vault system heat load and protect the fuelling machine vault-liner and concrete from excessive heating. In the region of end-fitting the insulation cabinet has to be removable type to facilitate work on reactor face. It is also necessary to provide seals for end-fittings penetrations in order to prevent leakage of hot air into the fuelling machine vault. The construction of insulation cabinet in this area thus consists of an outer sheet steel ring section (angle ring), which is fastened to the endfitting and a number of sheet steel plates called Jig-saw panel which will fill in the area between end-fittings.

The present Jig-saw panel requires considerable Man-Rem consumption for assembly/disassembly at the time of creep-adjustment or any other repair/maintenance works to be done on coolant channels. In view of this, development of a new design of Jig-saw panel was done in BARC. The new design, named as MODULAR INSULATION PANEL, is of modular construction in place of interlock construction. This consists of a tension spring and modular panels. These panels can be removed and installed using a pneumatically operated tools. Such panels have replaced the existing Jig-saw panels from MAPS reactor.

Compact Mechanical Snubber

Our reactors employ energy absorber devices at either end of the coolant channel. These devices require periodic adjustment. Consideration of a double ended guillotine break in the coolant tube or the rolled joint dictate availability of such a device which will arrest the flying missile that would arise in such an event. We have also noticed that under postulated conditions of severe channel heatup, such energy absorbing devices can become an axial restraint against free thermal expansion of the coolant tube. As a result, the coolant tube tends to sag and make contact with the calandria tube, in preference to the more desirable ballooning mode which has the potential of making a larger heat transfer area available for dissipating the decay heat. It has also been noticed that for coolant tubes with no axial restraint, the sagging threshold increases to above the threshold for ballooning. Thus, from the point of view of safety under severe channel heat up, it is desirable to allow free axial expansion of the coolant tube. A compact mechanical snubber which will allow such free thermal movement but would restrain a flying resulting from a double-ended guillotine break in a coolant channel offers an ideal solution to meet such conflicting requirements. A compact mechanical snubber which can fit in the space of the existing energy absorber device has been developed.

Tight-fit Garter Spring

Garter Spring spacers are provided as supports in the horizontal coolant channels of Indian PHWRs to maintain the annular gap between pressure tube and the calandria tube and prevent premature contact of the two tubes. The structural integrity of these garter spring spacers and their axial locations within the coolant channel are important factors deciding the creep limited life of coolant channels. To avoid possible shifting of loose-fit garter spring spacers from the design locations, it has been contemplated to adopt a tight-fit design of garter spring spacers. In order to carry out design evaluation of tight-fit garter spring spacers a series of computer codes were developed, which are as follows:

(a) Computer code 'GSLN': This code evaluates the load acting on the garter spring of two different designs.

(b) Computer code 'GSSA': Stresses at the critical sections of the coil caused by different loads acting on the critically loaded coil of the garter spring are calculated by this computer code using strength of material approach.

(c) Computer code 'GSOD': This code evaluates all feasible candidate designs of garter spring by varying different values of spring dimensions viz. breadth, thickness, pitch and coil outside diameter within a range such that the constraints on the maximum and minimum axial stiffness are satisfied.

10.3.3 Development of Tolings for Life Extension of the Reactor

Life Extension of Coolant Channels by Garter Spring Repositioning

Early generation PHWRs use loose fit garter spring (GS) spacers between pressure tube (PT) and calandria tube (CT), which are prone to displacement from their design locations resulting in significant lowering of creep contact limited service life of coolant channels. Tools and tehniques have been developed for extending service life of coolant channels by carrying out the task of repositioning of these garter springs in new as well as operating reactors.

Gai er Spring Repositioning in New Reactors using Mechanical Flexing Technique and Pulsed Electromagnetic Technique : Garter Spring repositioning was carried out in NAPP-1, NAPP-2 and KAPP-1 reactors after their hot conditioning using Mechanical Flexing Technique (MFT) and pulsed electromagnetic technique developed in BARC. MFT employs a specially developed hydrauvelically actuated Pressure Tube Flexing Tool (PTFT) which, by causing gentle cyclic flexing of the pressure tube, enables the displacement of garter spring by walking mechanism. The gentle cyclic flexing is produced by three load return type hydraulic jacks. A personal computer based computer program RESET was also developed to simulate operation of PTFT. In the pulsed electromagnetic technique a time varying magentic field produced by a field coil inserted into the coolant tube near the garter spring to be moved induces an eddy current in the garter spring. The eddy current interacts with the radial component of the magnetic flux producing a Lorentz force on the garter spring which makes it to move or causes a tilt reversal. By using these methods in the above three reactors life of about 50 channels were increased beyond 7 years.

Ingres: Integrated Garter Spring repositioning System

In order to reposition the loose-fit garter spacers in the affected coolant channels of early generation Indian PHWRs which are under operation, an Integrated Garter Spring Repositioning System has been developed. The INGRES consists of the following four devices working inside the pressure tube.

(i) Garter spring detection probe (GSDP),
(ii) Concentricity detection probe (CDP),
(iii) Pressure tube flexing tool (PTFT),
(iv) Garter spring repositioning device (GSRD).

The channels for repositioning operation are identified on the basis of in-service inspection. During an INGRES campaign the axial position of the garter spring to be repositioned is first detected. Concentricity between pressure tube and calandria tube is assessed and accordingly the pressure tube is flexed to make both the tubes concentric. Then repositioning device is activated to move the garter spring either in forward or backward direction as needed. A feeding mechanism is used for movement of the tool inside coolant channel. All the operations are controlled remotely from non-active areas outside the reactor containment. A new position of the relocated garter spring is detected again using the detection probe. INGRES has been used for garter spring repositioning in drained and dried coolant channels in a MAPS reactors. Now the system has been upgraded to make it operable in wet channels.

Coolant Channel Replacement Machine

The in-reactor service life of the Zircaloy-2 pressure tube is expected to be around 8 to 10 full power years. In order to enhance the life of the reactors using these pressure tubes, it is necessary to replace all the tubes when their service life is over. To carry out large scale replacement of these channels a semi-automatic, remotely operable Coolant Channel Replacement Machine (CCRM) consisting of large number of subsystems, tools and components has been developed.

The most important element of the design of CCRM is the remotisation of tool alignment and positioning, and channel removal operations. This helps in severely reducing the man-rem expenditure otherwise required for this exercise.

The CCRM mainly consists of a self elevating work table, tool positioning and aligning system, servo manipulator, remotely operated EOT crane and large number of tooling to perform various tasks for component removal and refurbishing. This system has been fully commissioned and is undergoing trials on a mock-up facility.

10.3.4. Development of Repair Toolings

Manipulators for MAPS Rehabilitation (Debris relocating inside calandria)

Loose debris pieces were detected inside calandria during remote video inspection of calandria internals. It was necessary that before the reactor is started again, these pieces be moved to corner pockets inside calandria such that they do not come in the flow path of moderator. With a view to reduce down time of the reactors to minimum possible, a scheme based on telescopic manipulator was quickly finalised for remote handling of the loose pieces. The limited access through 118 mm diameter lattice bore, ease of insertion and withdrawal during regular operations, safe removal in case of snapped wire ropes of the manipulator and maximum weight of the debris pieces to be lifted were the considerations taken into account while designing. A special pneumatically operated gripper was devised to lift the debris pieces. Provision of telescopic arm enabled the manipulator to reach all the debris pieces of different dimensions lying at different locations. This was successfully used in relocating the debris pieces in both the units of MAPS.

Manipulator for RAPS-1 OPRD Debris Cleaning

During inspection of OPRD it was seen that it is covered by about 50 mm thick layer of debris. This debris mainly consisted of ferrous/ferric salts with traces of cobalt and various other elements. It was also observed that the debris was radioactive. It was necessary to clean this debris before a repair work is carried out on OPRD. A system was developed in BARC for this purpose and was successfully implemented at site. The system developed consists of a remotely operable manipulator carrying the vacuum hose fitted with suitable attachments to suck the debris. The air such sucked is then passed through cyclone separator, pre-filter and HEPA filter. These filtration units prevents contamination of the vacuum system.

Development of Remote Soldering Equipment for RAPS-1 OPRD Leakage Rectification

The leakage of OPRD spray cooling heavy water was detected to be through gasket between OPRD nozzle and its cover. For sealing the leakage a scheme has been devised to deposit low melting eutectic alloy all along the joint between OPRD nozzle and its cover. The advantage of this scheme is that no removal of components of OPRD is needed hence saving time and thereby reducing man-rem consumption. Also as the sealing material is deposited from outside, the calandria internals, especially thin calandria tubes are not exposed to any hazard. The system designed for performing the task remotely is capable of prforming the following functions under CCTV based visual feedback.

(i) Cleaning and grinding of the 1 metre dia. joint to be sealed,

(ii) Preheating the sealing material and the joint locations to be sealed,

(iii) Depositing the sealing material all around the joint,

(iv) Rinsing the soldered region to remove flux, and retrieving the rinse water.

The operating parameters and procedure were finalised on full scale mock-up.

11. BACK END OF NUCLEAR FUEL CYCLE

While mining of uranium, its fabrication into a fuel form and its subsequent utilisation in a nuclear reactor for the extraction of electrical power constitute the front-end of the nuclear fuel cycie, fuel reprocessing and high level waste management are the two major back-end activities. Fuel reprocessing aims at recovering these reusable fissile components of the spent fuel while high level was to management consists of treatment and safe disposal of those components of spent fuel that have no immediate and ready use and which are also quite hazardous in nature.

11.1 Fuel Reprocessing

India's Nuclear Power Programme envisages the use of PU^{239} and U^{233} as fuel materials for the breeder power reactors to be set up in the second and third phase respectively. This would require the separation and recycle of the Plutonium produced in the thermal power reactors of the first phase and the separation and recycle of the U^{233} produced in the Thorium blankets of the plutonium fuelled fast reactors of the second phase. Separation of the Pu or U^{233} from the spent fuel would inevitably result in the generation of a concentrated stream of fission products that would require treatment and safe disposal. Thus spent fuel reprocessing and high level waste management hold the key to the success of our long term nuclear power strategy—a factor recognised by our policy planners long ago. This recognition led to our decision to embark on fuel reprocessing as early as the sixties.

Fuel reprocessing started in India at a time when it was regarded as an exclusively neclear weapon related activity. Understandably, no collaboration would be forth coming in this field and so we had to develop this technology entriely by indigenous efforts.

In India today we have three reprocessing plants to extract Plutonium from spent fuel, one at Trombay, second at Tarapur and recently the third one was cold commissioned at Kalpakkam. The third plant at Kalpakkam incorporates a number of innovative features such as hybrid maintenance concept in hot cells using servo manipulators and engineered provision for extending the life of the plant. This plant will cater to needs of reprocessing fuel form MAPS as well as FBTR.

Simultaneously, the technology for processing irradiated Th fuel elements to recover the U^{233} formed in them was also developed in small experimental rigs. The fuel for research reactors like PURNIMA II & III and KAMINI was made out of the U^{233} separated in these facilities located at Trombay and Kalpakkam. Based on these successes an engineering scale facility is being built at Trombay to process the Th fuel elements undergoing irradiation in various reactors like CIRUS, DHRUVA, and KAPP-1.

11.2 Radioactive Waste Management

With total protection of the environment as an overriding consideration, management of the radioactive waste generated in the fuel cycle has received high priority in our nuclear programme right from its inception. An entirely self-sustained fuel cycle based on indigenous resources necessitated evaluation of hazard potential vis-a-vis radioactive wastes at different stages of the cycle, starting from mining and milling, fuel fabrication through the stages of reactor operation and finally the reprocessing of the spent fuel. More than a decade before commissioning of the power reactors emphasis was laid on studies relating to impact of radioactivity to the environment and on developing technologies to effectively isolate and contain them.

In principle the Indian programme envisages two distinct modes of final disposition in respect to radioactive wastes;

- Near surface engineered extended storage for low and intermediate level wastes
- Deep geological disposal of high level wastes.

The broad outline of the management policy are :

- Any discharge of radioactive liquid or gaseous wastes to the environment should be as low as reasonably achievable, economic and social factors being taken into account.
- Conditioned primary solid wastes and waste products, resulting from conditioning of liquid wastes generated due to operation of reactors and research laboratories are to be stored in near surface facilities specially engineered for this purpose.
- High level liquid wastes from reprocessing plants are initially stored for an interim period underground in stainless steel tanks prior to their conditioning. These wastes will be vitrified and these solidified waste products will be provided interim storage in near surface engineered facilities. The interim storage will allow reduction in heat level and

also facilitate quality assurance of the waste products prior to their transportation and emplacement in a centralised repository.

— Conditioned high level and alpha wastes are to be deposited in a suitable deep geological formation forming the centralised repository.

Based on indigenous materials and capabilities, technology has been developed and is in routine application for the management of low and intermediate level wastes meeting the stringent regulatory requirements and standards. Plants for disposing of all types of wastes have been set up and are operating successfully along with evey nuclear facility in the country.

Treatment of long reprocessed wastes has received considerable attention because they contain nearly 99% of the activity generated in the entire nuclear fuel cycle. Based on years of developmental studies a long term action plan has been formulated for the management of these wastes. It consists of;

- Solidification of the medium level wastes along with suitable matrix in reliable containers and burying them in totally water-proof concrete tile holes with protective barriers.
- Immobilisation of the high level wastes by vitrification in glass matrix and encapsulating the solidified mass in corrosion resistant containers called canisters.
- Interim storage of the sealed waste canisters under continuous cooling and surveillance in an engineered storage facility for 20 to 30 years.
- Ultimate disposal of the cooled solidified waste in deep underground geological formations with added protection barriers.

A waste immobilisation plant (WIP) has been setup at Tarapur incorporating high technology features like complete remote operation and maintenance. A facility for interim storage of the vitrified waste from this plant has also been built. Based on the initial successes achieved during the early runs, one more waste immobilisation plant is under construction at Trombay and another one at Kalpakkam.

12. RADIOLOGICAL, HEALTH, SAFETY AND ENVIRONMENT

The responsibility for ensuring safe working conditions for all radiation workers in India rests with the Indian Department of Atomic Energy, and this responsibility is discharged through Health, Safety and Environment (HS & E) Group of BARC. The activities of this HS & E Group, BARC extend to the following major specific areas:

- Providing radiation and radiological protection services in the use of ionising radiations in India
- Development of instruments and systems for the above-mentioned assignments.
- Personnel internal and external dosimetry
- Radiological surveillance in and around nuclear installations

- — Design, authorisation and inspection of installations handling radiation and radioisotopes in industry, medicine, agriculture and research.
- — Countrywide environmental monitoring
- — Technial assistance and support to atomic energy regulatory board in safety reviews, development of codes and guides and safety standards
- — Review/audit of safety regulation and implementation
- — Training in radiological protection
- — Monitoring of non-ionising radiation sources like microwaves, lasers, radiofrequency and ultrasound.

Health physics instrumentation, medical instrumentation and environment monitoring systems developed by BARC include the following.

- — BARC personnel monitoring badge
- — Auto TLD badge reader
- — Health Physics Instrumentation for Internal Dosimetry Programme:
  - ⇨ Shadow Shield whole body counters foe gamma emitting radionuclides
  - ⇨ In-vivo measurement systems for actinides
  - ⇨ Radon dosimetry systems
  - ⇨ Bio-assay techniques and facilities
- — Environmental Radiation Measurements instrumentation
  - ⇨ External radiation exposure levels using scintillometers, high pressure ionisation chambers, thermoluminescence dosimeters.
  - ⇨ Radioactivity levels in a variety of samples such as air, water, milk, meat, vegetables, sediment, salt and food-stuffs.
- — Aerial gamma spectrometric survey system and methodology
- — Radiation emergency early warning system.
- — BARC dry aerosol generation/inhalation system.

BARC also conducts training programmes to qualify personnel for radiation protection jobs. It conducts workshops on planning, preparedness and response to radiological emergencies for DAE officials, medical offcers and public authorities. Under RAC of IAEA several training courses have been organised.

13. DESIGN, CONSTRUCTION AND OPERATION OF RESEARCH REACTORS:

India has designed and built through indigenous effort a host of research reactors primarily to support its nuclear power programme apart from supporting the requirement of nuclear science and technology together with applied research. Sufficiently large manhours of reactor operation experience has been accumulated and a large number of personnel have been trained in the various aspects of reactor operation. APSARA, ZERLINA, PURNIMA-I, PURNIMA-II, PURNIMA-III, DHRUVA, FBTR and KAMINI are the research reactors designed, built and operated through indigenous effort while CIRUS reactor has been built in collaboration with Canada.

14. ROLE OF R&D FOR FUTURE DEVELOPMENT OF NUCLEAR POWER IN INDIA

14.1 PHWRs

Design of our PHWRs. which are important to first phase of our nuclear energy programme, though standardised, will need continuous support from R&D in order to improve their economics and safety. These aims can be broadly classified as shown below. Some examples which cater to above needs where work has been completed/initiated/planned and are futuristic goals have also been shown.

- *Decrease in capital cost per Mwe installed by*
  - Reduction in PHT components to reduce equipment cost
  - Use of integral calandria-ball filled end shield to reduce construction time
  - Advanced heavy water production processes (H_2/H_2O exchange)
  - Nearly shop assembled coolant channels to reduce construction time
  - Modular Instrumentation and control system to reduce construction time
  - Increase thermal power for same investment by
    - Producting high power through increased flux form factor
    - Introduction of partial boiling in coolant channels
    - Having higher termperature PHWRs
- *Reduction in operating and maintenance costs per KWh*
  - Reduction in fuel cost by proper fuel management
  - Development of graphite coated fuel clad
  - development of high burn-up fuels
  - Use of denatured zirconium
  - Enhanced operating margin for fuel through improved fuel fabrication processes
  - Reduction in potential D_2O leak points
  - Reduction in maintenance cost by
    - Decontamination
    - Use of modular insulation panels
    - Use of compact mechanical snubber
    - Moderator introduction in calandria through sparger channels
    - Having easily replaceable coolant channels
- *Improvement of safety of txe reactor systems.*
  - By enhanced engineered safety features and defence in depth
    - Substitution of dumping by two independent fast acting shut-down systems
  - By Providing Annuals goes monitoring system.
  - Providing high pressure ECCS
  - ECCS simplification through rupture disc
  - Higher heat transfer from pressure tube to moderator during severe accidents
  - Tailored negative void coefficient fuels
  - By improved man-machine interface
  - By layout improvements in electrical system
  - Improved containment such as double containment

- Containment designed for severe accident
- Passive containment isolation and cooling
- Improved waste management technologies
  - Development of solvent extraction processes for partitioning of long-lived actinides
  - Development of borosilicates glass formulations as vitreous matrices
  - Final disposal of HLW in deep geological repositories
  - Detriation of heavy water
  - Actinide partitioning by photochemical processes
  - Actinide burners
- *Inspection and Maintenance Technologies to reduce downtime of the reactor*
  - Development of channel inspection capability—BARCIS
  - Development of visual inspection capability—DRYVIS
  - Garter spring repositioning in dry and wet channels
  - Hydride scraping tool
  - Coolant channel replacement machine
  - Advanced servomanipulators
  - On-line assessment of fatigue usage factor for critical components
  - Robotic systems for inspection and maintenance in inaccessible areas
- *Technologies to increase service life of critical components*
  - Change of pressure tube material to Zr-2.5% Nb
  - Tight fit garter spring
  - Zero clearance type rolled joints
  - Zr-2.5%Nb pressure tubes with low initial hydrogen, chlorine, other impurities to improve its post-irradiation fracture behaviour
  - Use of hydrogen getters/hydrogen barriers in coolant channels
  - Modified pressure tube fabrication route to minimise dimensional changes.

14.2 Fast Breeder Test reactor (FBTR) and Prototype Fast Breeder Reactor (PFBR)

Fast Breeder Reactors are for the effective utilisation of Pu from the second stage of Indian Nuclear Energy Programme. They are key to the full utilisation of the country's uranium resources and prepare the way for the long term utilisation of the more abundant thorium resources. Because of a better neutron balance in FBRs more fresh fissile material is produced than consumed for power production. While a PHWR produces only about 0.6 Kg of Pu for every Kg of U^{235} cosumed, a large FBR could typically produce 1.2 Kg of fissile Pu for every Kg of such Pu consumed.

At IGCAR in Kalpakkam, near Madras Fast Breeder Test Reactor (FBTR) with a capacity of 40 MWt is in operation. The purpose of this reactor is to use it as an irradiation facility for the development of the FBR fuel cycle, and to gain experience in the design, construction and operation of LMFBRs.

14.3 Advanced Heavy Water Reactor (AHWR)

Thorium utilisation has been an important aspect of Indian Nuclear Energy Programme. The strategy of realising large scale power generation from thorium has been clearly laid out right from the inception of Indian Nuclear Power Programme. As a part of this strategy an Advanced Heavy Water Reactor (AHWR) is being developed. This reactor would generate most of its energy form U^{235}, bred *in-situ* from thorium. The reactor is based on well proven water reactor technologies and would enable a number of passive systems to be incorporated in the design. Some important design features of this reactor are brought out in the following description:

Emphasis on Utilisation of Thorium

Thorium works almost as well in thermal neutron spectrum as in fast neutron spectrum. This is due to the eta value (number, of neutrons liberated for every neutron absorbed in fuel) of U^{233}, which is nearly constant over thermal to fast neutron energy range. In thermal neutron environment the eta value of U^{233} is highest and is also marginally higher than the eta value of U^{235} in fast neutron spectrum. Heavy water reactors are therefore good candidates for thorium utilisation. As regards safety the trend in the world today is to incorporate greater degree of passive safety systems for carrying out different safety functions. Thorium is capable of producing large amounts of energy through burning of *in-situ* bred U^{233}. Thorium-U^{233} fuelled lattices can achieve this characteristic without a very large change in reactivity. Thus while the burn-up realisable in Self Sustained Equilibrium Thorium (SSET) mode is moderate, with a small reactivity boost, the burn-up can be augmented considerably. This requires a net fissile input into the system on a sustained basis. Pu is logical choice for this. But Pu has a relatively high capture-to-fission ratio in soft neutron spectrum. Therefore it is necessary to locate Pu in regions where spectrum is somewhat harder. This would keep the consumption of Pu as well as power generated from Pu to a low level whereas the power generated from thorium is maximised. This is the genesis of AHWR.

- ***Core design specifications***
  - Most (75%) of the power generation from thorium:
  - Negative void coefficient under all operating conditions. This will bring down the reactor power safely whenever the coolant supply becomes inadequate irrespective of the availability of the power supply, instrumentation and control systems.
- ***Reactor configuration***
  - Vertical tube type construction
  - Heavy water as moderator
  - Boiling light water as coolant: This enables heat removal from core through natural circulation.
  - All safety systems in four identical loops each rated at 50% of overall plant requirement
  - Reactor vessel housed in a water filled reactor vault acting as effective radiation shield
  - Double containment for the reactor
  - Gravity Driven Water Pool to cool the reactor in case of an accident
  - Isolation condenser (IC) pool for decay heat removal
  - Another set of IC pool to cool the primary containment during LOCA.
  - ***Simple design :*** Minimum possible components constituting the reactor. Reactor oulet headers, secondary steam generators, primary coolant circulation pump, moderator heat

exchangers (all are present in PHWRs) have been eliminated. Minimum possible large diameter pipes to minimise probability of LOCA.

— ***Absence of high pressure heavy water :*** This is one of the major difference from PHWRs. Light water has been substituted in place of heavy water as coolant. Heavy water inventory has come down and is confined to low pressure moderator system only. This realties in considerable reduction of heavy water loss due to system leakages, thus reducing the plant operating cost.

— ***Conventional equipment :*** An important advantage of assigning the safety function to passive systems and adopting the 4 × 50% policy is relaxation of performance demand on systems. These equipments will be moved away from high radiation field thus will be easily accessible for maintenance without affecting reactor operation. Thus conventional equipment can be used causing reduction in its cost as well as procurement time.

— ***Maximum shopwork :*** AHWR systems will be subdivided into suitable units/components. This will result in speedier construction schedules and better quality control. e.g. shop assembled coolant channels.

— ***Prudent Energy Management*** : AHWR will utilise heat generated in reactor shielding and moderator (totalling 50 MW) to preheat feed water which otherwise is wasted in most nuclear power plants.

— ***Passive safety features :*** All important safety functions have been assigned to passive systems. These systems do not depend on external power for their opearation and have none or minimum number of moving parts. They are directly actuated by an event or a process without depending on instrumentation or control interface. These safety features are as mentioned below :

— ***Pumpless primary circulation :*** The full power primary circulation is obtained by natural circulation caused by thermosiphoning phenomenon. As the core cooling is not dependent on external pumping power or instrumentation, the system safety and reliability is increased considerably.

— ***Thermosiphon based decay heat removal system :*** In the event of reactor shut down or scram, the reactor will be cooled by thermosiphoning. The steam flow from steam drum will be diverted to a bank of isolation condensers which subsequently will dissipate the reactor heat to IC pool water. To cool the coolant system, a pumped circulation through reactor core via a set of heat exchanger will be utilised.

— ***Isolation condenser based containment cooling system :*** Passive containment coolers loated in IC pool will cool the containment and mitigate pressure build up in case of LOCA. Then the condensed steam will flow down to the Gravity Driven water Pool (GDWP). In addition regular containment coolers are also provided.

— ***Advanced accumulator based ECCS:*** Under postulated break in the PHT system, accumulators storing water under high pressure nitrogen will cool reactor core. A series of rupture discs which isolate the accumulators and reactor core rupture when post LOCA depresssurisation of PHT system reaches a pre-set level. This water gets sprayed of the individual fuel pins as well as the coolant tube through eight perforated tubes running along the periphery of fuel cluster coolant tube to provide efficient emergency cooling.

— ***Gravity driven water pool (GDWP) :*** The GDWP is used for core injection by gravity when PHT system pressure is reduced close to the containment pressure in the aftermath of a LOCA. This is connected to the core through a series of rupture discs. This can cool the core for three days after LOCA.

— ***Core submergence :*** The water spilling out from the PHT system, accumulators and GDWP, as a sequel to LOCA, will be guided to collect in the space called reactor cavity to submerge the core. After submerging the core, any excess of this spillage will be directed to the suppression pool to prevent unwanted flooding of other areas of the reactor building.

— ***Suppression pool :*** The suppression pool is located on the lowest floor of the primary containment building. In the event of a LOCA, the emanating stream from the PHT system is condensed in the suppression pool and the pressure rise in the reactor building is prevented.

— ***Passive containment isolation :*** In a post LOCA scenario, considerable radioactivity might be expected to be released in the containment building. In order to prevent the release of this radioactivity the containment must be isolated within specified time of initiation of a LOCA. This function is achieved in a passive manner. The reactor building air supply and exhaust ducts are in the form of U-bends. In the event of a LOCA, the steam pressure acts on the GDWP inventory and pours water by Swift establishment of a siphon in to the U-bend ventilation ducts and this water seals between the containment and the environment.

— ***Cooplant channels*** : It has following features;
— Uniform coolant channel design for all lattice location.
— They will be shop fabricated
— Easily replaceable coolant channels with its calandria tube.

— ***Fuel Elements :*** It has following features;
— 52 fuel pins arranged in a square array
— cluster has MOX surrounded by ThO_2 fuel pins
— All clads are of zircaloy-2
— The fuel pins will be supported at regular intervals along their length by spa{ers
— Burn-up: 20,000 MWD/Te
— Provision to direct emergency core cooling flow to the individual pins and to the coolant tube. Eight perforated tubes border the fuel pins and run along the entire length of the cluster so as to direct the water jets at different elevations and from different directions.

14.4 Development of New Technologies

New Reactor Concepts

These are reactor technologies to meet future national needs when it is postulated that by the end of the second decade of the 21st century the society at large shall have to restructure its away of life following the depletion of world reserves of petroleum. The sectors needing alternative energy sources will be electrical power generation, process heating, and transportation. In the Indian context, our strategies to meet this challenge must take into account our nautral resources, our perceived future needs, and possibility of continued embargo on imports of strategic materials and technologies. Keeping these in view, thrust in R&D will be required in the following frontier areas of reactor technology:

(i) Hydrogen Generation

With its meagre petroleum reserves India may face serious difficulties in mobilizing resources needed for adequate imports of crude oil in the very near future. During the early 21st century the single most important item, needed to secure a self-reliant national economy, is expected to be an alternative to petroleum based fuel required in the surface transport and aviation sectors, in civilian as well as military domains.

Out of the three probable liquid fuel alternatives, viz., Ethanol, Methanol and Hydrogen, the first–two are uneconomical to harvest in large quantities, compete with agricultural food production in terms of utilisation of cultivable land and leave behind pollutants in their production and consumption cycle. These drawbacks may make them unsustainable in the long run.

Hydrogen as a fuel has a greater potential. The only clean and reliable means of providing energy, required for generation of large quantities of hydrogen, in the long run is nuclear energy. We must therefore contemplate reactor systems in which high temperature, radiation and electricity could be used to generate hydrogen in a cost-effective manner.

(ii) Small Portable Power Packs

Another important area in which non-availability of fossil fuel could cause severe constraints is the on-the-spot generation of elctricity through small power plants at remote locations which may be strategically important but not serviceable through electricity grids. This requirement can be fulfilled by small nuclear power packs which may be transported by land or waterways and made to run in an unattended manner.

(iii) Space Power Packs

Space will continue to contribute substantially to the strategic and economic interests of the nation. In times to come our presence in space will have to be augmented in order to effectively harvest the benefits. This would require a large number of artificial satellites and eventually space-stations in orbit around the earth. Compact nuclear power units will be required to feed the power requirements in such strategic space applications.

(iv) Desalination

Potable water is yet another essential resource necessary for human survival. It is likely to become scarce in course of time due to the ever increasing population pressures and gradual depletion of naturally occurring sources of fresh water and hence likely to become precious. In such a situation conversion of sea water into potable water would be essential to meet the requirement in deficient regions. Specially designed reactor systems would be ideally suited to provide the process heat and/or electricity for these desalination plants and for pumping the potable water to mainland consumers.

CONCLUSION

The conversion of nuclear fuel in the mineral form to nuclear electricity demands the availability of capabilities in a large number of critical technology areas. India is one of the few countries in the world to have mastered the entire range of these technologies through indigenous R&D support. With this major strength India can take the challenge of deploying nuclear power as a major contributor to the national energy mix, in the next century, when the conventional fuel resources are depleted.

ENHANCED COAL PRODUCTION FROM UNDERGROUND MINES IN INDIA

14

DR. MAHENDRA K. PANDEY
Joint Adviser (Coal)
Planning Commission, Government of India
and
VIJAY K. TIWAR
Superintending Engineer
Coal India Limited

ABSTRACT

The percentage share of UG coal production has been declining over the years. During the early years following the nationalisation of coal mines, the share of UG and open cast (OC) coal production was 75 per cent and 25 per cent respectively. Today, however, the situation has just reversed. In Coal India Limited (CIL), while the production from OC mines increased from 20.77 MT in 1974-75 to 167.46 MT. in 1994-95 and is expected to increase further to 194 MT in 1996-97, the coal production from UG mines, which usually produce superior quality premium coal, has declined from 65.21 MT in 1975-76 to 55.61 MT in 1994-95 and is expected to be around 58 MT in 1996-97. This is against a backdrop of more than 55 per cent of country's proven coal reserves being amenable to mining>solely by UG mining methods. The UG coal mines in India employ 80 per cent of manpower, but their share in total output is less than 30 per cent. The UG mines are characterized by low productivity which has been stagnating at an output per man shift (OMS) of 0.55 tonne for the last two decades. This has caused one of the major constraint on the profitability of the coal sector. Thus, therefore, there is an urgent need to increase good quality coal production through high productivity in UG mines.

Bulk of India's underground coal is produced by Manual Board & Pillar (B&P) method and it accounts for nearly 75% of the total UG production. Whereas, in the U.S.A. 60% of total UG coal production is obtained by means of B&P mining. In South Africa, of the total UG coal production of 120 million tonnes annually, 111 million tonnes (92.5%) is obtained by B&P Mining methods, of which 53 million tonnes is produced by means of continuous miner and 58 million tonnes is mined by mechanised undercutting and drill and blast methods. In Australia, of the 56 million tonnes of underground coal production annually, 45% is obtained by means of longwall and 55% by B&P methods. Whilst the longwall, when working properly, is the most productive method, the B&P methods with continuous miners, shuttle cars and feeder breakers can produce coal at similar low costs per tonne, with comparatively much lower capital investment in equipment.

A two-pronged strategy has been suggested for improving quality coal production in India. The first part deals with the running mines. The second part of the strategy would take care of the new coal projects work on which is yet to be initiated.

INTRODUCTION

Coal is the main source of India's commercial energy, accounting for about 67 per cent of total energy consumption in the country. Although the present proven reserves of coal are estimated at 70 billion tonnes, keeping in view the size of the extractable reserves and the rapidly increasing demand for coal in the country, there is greater need than ever before to enhance coal production from the underground (UG) mines by means of efficient mining technologies. According to an estimate by Ministry of Coal (MOC), coal consumption in the country is projected to increase from a level of 318 million tonnes (MT) in 1996-97 to 400 MT in 2001-02 and further to 560 MT in 2009-10. The indigenous production of coal, according to MOC's projections, is expected to reach the level of 331 MT in 1996-97, 348-385 MT in 1999-2000 and 545-658 MT in 2009-2010.

However, as against a targetted growth rate of 6.3 per cent in coal demand during the Eighth Five Year Plan (1992-97), assuming an average annual compounded growth rate of 6.5 per cent in future, based on a demand level of 332 MT in 1996-97, the overall demand for coal will reach a level of 445 MT in 2001-02, 609 MT in 2006-07 and 834 MT in 2011-12. It would, therefore, call for herculean efforts to meet the projected levels of coal production in the future years. The percentage share of UG coal production has been declining over the years. During the early years following the nationalisation of coal mines, the share of UG and open cast (OC) coal production was 75 per cent and 25 per cent respectively. Today, however, the situation has just reversed. In Coal India Limited (CIL), while the production from OC mines increased from 20.77 MT. in 1974-75 to 167.46 MT in 1994-95 and is expected to increase further to 194 MT in 1996-97. The total coal production in India in 1995-96 was 270.13 million tonnes. The production from UG mines, which usually produce superior quality premium coal, has declined from 65.21 MT in 1975-76 to 55.61 MT in 1994-95 and is expected to be around 58 MT in 1996-97. This is against a backdrop of more than 55 per cent of country's proven coal reserves being amenable to mining solely by UG mining methods. The UG coal mines in India employ about 70 per cent of manpower, but their share in total output is less than 30 per cent. The UG mines are characterized by low productivity which has been stagnating at an output per man shift (OMS) of 0.55 tonne for the last two decades. This has caused one of the major constraints on the profitability of the coal sector. There is an urgent need to increase good quality coal production through appropriate mechanisation and high productivity in UG mines

Bulk of India's underground coal is produced by manual Board & Pillar (B&P) method and it accounts for nearly 75% of the total UG production. As such even in U.S.A. 60% of total UG coal production is obtained by means of B&P mining. In South Africa, of the total UG coal production on 120 million tonnes annually, 111 million tonnes (92.5%) is obtained by B&P mining methods, of which 53 million tonnes is produced by means of continuous miner and 58 million tonnes is mined by mechanised undercutting and drilling and blasting methods. In Australia, of the 56 million tonnes of underground coal production annually, 45% is obtained by means of longwall and 55% by B&P methods. Whilst the longwall, when working properly, is the most productive method, the Board and Pillar methods with continuous mines, shuttle cars and feeder breakers can produce coal at similar low costs per tonne, with comparatively much lower capital investment in equipment.

UNDERGROUND MINING METHODS

The majority of coal mined underground in India is by manual methods. Coal India Limited (CIL) a Govt. owned company, which accounts for approximately 90% of national coal production, is the second biggest corporate employer in the world employing approximately 660 thousand employees.

The underground productivity results are in keeping with low level of mechanisation and have been almost stagnant at 0.55 tonne per manshift. Each manshift is for approx. 7 wokring hours. This does not compare well with other coal producing countries who normally exceed 2.5 metric tonnes for each hour worked.

The other methods of work used in India are the "semi-mechanised" workings and the fully mechanised "longwall face". Though the contribution from these methods is growing, it still represents less than 25% of the total underground coal production.

WHY MINING COAL BY UNDERGROUND METHODS

More than half of proven coal reserves – in India – are available in Jharia, Raniganj, North and South Karanpura coalfields lying by and large, in State of Bihar, and adjacent areas. At the same time, almost entire steel grade coal and superior grade non-coking coals lie in the above region. Good quality coals are generally extractable only through underground coal mining methods. These coals find applicability in production of steel, heat intensive industries and coke production. In the last 20 years since coal nationalisation, major thrust has been for quantitative production of coal required for generation of electricity. Normally these coals are of inferior grades, usually available at shallow depths in Singrauli, Korba, Talchar and IB-valley coalfields. Due to falling standards of quality in coal produced, there has been general feeling amongst the consumers to organise better quality coal, required to suit needs in steel, chemicals, fertilizer, textile, cement, paints and many other industries. In absence of quality coal, consumers are looking towards coal imports to make up the needs. Already total coal imports for steel sector is expected to cross 10 million tonnes during 1996-97 in addition to non-coking coal imports crossing over 2 million tonne mark. In due course, these figures are quite likely to rise. This trend can be checked by appropriate development of new UG coal mines indigenously for enhancing production of good quality coal. Large scale unnecessary coal imports may also put strains on country's foreign exchange reserves.

Opencast (OC) mining methods may not be suitable to mine coal from highly populous Jharia and Raniganj coalfields and other coalfields in the vicinity. OC mining is certainly more environmental unfriendly, leads to large land degradation and gross ecological imbalances. Land requirement for operating OC mines usually lead to human problems on account of land acquisition and mass resettlements. Most of the big OC projects in the above coalfields are delayed beyond economical limits, largely on account of such constraints.

It is also well known that in OC mining very large volume of over burden has to be removed to expose coal any further.

With all the above in background, it is sensible to develop and concentrate more focus on operation of underground coal mines in an economically rewarding manner.

FULLY MECHANISED LONGWALL MINING

As is known, among all the UG mining methods, fully mechanised longwall instalments are the most productive bulk tonnage producers of coal as proven in several coal producing countries. The choice of longwall as the system of working is made however, on several criteria including the high capital investment, the long set/up time, the lack of flexibility, adequate geological and geomining information and other factors. The results from Longwall can be excellent, provided all the above elements are appropriately covered.

In the longwall system of mining the equipment is laid out along the coal face, possible 250 metres in length and the shearer cuts in both the backward and forward directions loading the coal on to a steel conveyor. The full face slowly advances as the coal is cut along the strata to collapse behind.

Fully mechanised Longwall faces have been tried and are still being tried in India. A detail examination of the various longwall faces installed in different mines and their present status reveals that with one or two exceptions, the results so far have been rather disappointing. Fresh initiatives have, however, been undertaken to involve Chinese expertise with the presumption that their roof and strata conditions alongwith other geo-mining factors are similar to Indian conditions.

Reasons for the disappointing performance are thought to include the inadequate specifications of machinery, lack of mine infrastructure, lack of equipment/spare parts back-up support from suppliers, lack of maintenance and the inappropriate deployment of equipment geologically, etc.

With the present performance the very high purchase and running costs of longwall equipment make the coal produced highly expensive to be economically viable. This, however, may change and coalfields suited to longwall mining may eventually perform economically provide all factors mentioned above are efficiently covered. Even then, many of the coal deposits in India are not suitable to this method of mining and could be more successfully mined by mechanised room and pillar method.

ROOM AND PILLAR METHODS

The most popular method used in India to produce coal by manual or semi-mechanised machinery is known as "Room" (or "Board") and Pillar "mining. The B&P method is the basic method of mining which is used throughout the world. In the U.S.A., which is the most competitive mining market in the world, the B&P method accounts for over 65% of the underground production.

The Development Phase

The basic concept of room and pillar methods is that the coal seam is divided into a regular block by driving primary headings or tunnels parallel to each other. These are intersected at regular intervals by connecting tunnels, forming a regular grid pattern. The headings are known as the "rooms" "boards" and the blocks of coal bounded by them are the "pillars". The coal is extracted as the roadways are developed and the pillars support the overlying strata. The relationship between the width of the roadways and the pillar size being a function of the depth of the seam and the stress within the strata. Typical sizes in India would be 20-45 sqr. metres, separated by 4.8 metres wide roadways. As the roadways are driven, systematic support, of the roof is installed. This is accomplished normally by means of a pattern of steel bolts, 1.5 to 2.5. metres long and emented into the roof of the tunnel in order to "bolt" the strata, forming the roof, together into a competent beam.

This forming of the grid of rooms and pillars is the development of the district and will produce upto 40% of the extractable coal within the district. The effect on the surface environment is at this stage minimal.

The Retreating Phase

When the district has been driven as far as intended, to the boundary, the work direction is reversed. The coal is now extracted from the pillars, starting at the boundary and working back along the already driven roadways, along the roof to collapse behind the working place. The systematic collapse of the roof is a necessity, the relieve of stress building up in the pillars must be done in order to enable the pillar extraction to continue.

This second stage where practiced is considered by the leading coal producers to be the most lucrative part of the operation as minimum roof support needs to be established and upto 95% of the coal within the district may be extracted.

MANUAL OR BASKET MINING

The predominant means of extracting coal underground in India, uses the Room and Pillar method. The breaking of the coal and initial transport is done by hand and is often referred to as "basket mining". The grid of rooms and pillars is developed by systematically driving forward the roadways. For each roadway to advance, a cycle of operations must take place as follows:

Drilling

A team of men, using a hand-held electric boring machine, drill a pattern of holes in the face of the tunel to a depth of frequently 1.8 metres. Twenty to thirty holes could be necessary, after which the drilling team moves on to the next heading to drill that.

Blasting

After testing for gas, the holes are charged with explosives and fired. The coal will be broken and the tunnel advanced by approximately 1.5 metres.

Loading

This operation is where men with shovels load the broken coal into baskets. The baskets are then lifted onto the shoulder of the basket carriers and carried to a central point some distance away and tipped into mine tubs. The mine tubs are fastened to a haulage rope and winched either to a shaft or drift and thence to the surface.

Supporting

Systematic support of the roof is accomplished by drilling holes in the roof with an electric drill upto a length appropriate to the strata, frequently 1.2 metres plus. Cement is inserted into the hole by hand and a steel bolt forced into the hole. When the cement has hardened the bolt helps hold the strata in the immediate roof together into a beam and prevents its falling away.

The four steps above are physically demanding and offer little opportunities for productivity improvements. It is hardly surprising that the average absenteeism is 30%, rising to almost 40% in some areas.

A typical district manpower consists of 100 men per shift, 300 men per day producing on an average 300 tonnes of coal each day. Giving a `district output per manshift' (OMS) of 1 tonne.

Outside the district, further personnel is required for transport, ventilation, repairs, track laying and many other function associated with the method of work. The overall output per manshift is nearer to 0.5 tonne.

The manpower, therefore, necessary for a mine producing 250,000 tonnes per year would be:

$$\frac{250{,}000 \text{ tonnes} \times 30\% \text{ absenteeism}}{300 \text{ working days} \times 0.5 \text{ tonnes per man}} = 2{,}166 \text{ men}$$

SEMI-MECHANISED MINING

This is essentially the four stage cycle similar to manual mining. The difference being that machines are introduced to do some of the manual work.

An electrically powered vehicle [Side Discharge Loader (SDL) or Load Haul Dump (LHD)] loads and carries the coal, once it has been blasted, from the heading to a conveyor which transports it out of the mine.

This removes the most physically demanding part of the mining cycle but does not on an average produce more coal per day from the district.

A typical semi-mechanised district has three SDLs or LHDs, each of which will move up to 100 tonnes per day, giving a daily district production of 300 tonnes. Though the production is similar to that to `basket mining' the manpower required is significantly less. A typical semi-mechanised district will employ 40 men per shift, or 120 men per day.

A mine producing 250,000 tonnes per year using semi-mechanised equipment would employ approximately 1,000 men. This is significantly less than with manual methods but leaves the only means by which production can be increased at a mine to be the increasing of districts which is not always possible due to limitations in the mine infrastructure.

The continuing introduction of semi-mechanised equipment certainly reduces the manpower at a mine that is exposed to hostile and physically demanding conditions. It will though certainly improve the productivity of a district but will not increase the production.

FULLY MECHANISED ROOM AND PILLAR MINING

When we look at the leading underground coal producers, three of the top four are market-led economies: USA, Australia and South Africa. The fourth, China, being demand-led.

It is in the market-led economies whe . underground maining has been forced to undergo rapid development in order to achieve high productivity and low ownership costs and enable them to compete with cheap opencast coal on the world market.

The production methods used in these competitive markets are highly efficient longwall mining and significantly and mainly room and pillar mining. In fact room and pillar mining still accounts for 68% of underground coal production in the major, coal exporting market economies.

The equipment that has made this possible is as much a product of this competitive environment as it is the means of success, namely, the Continuous Miner and associated systems.

CONTINUOUS MINER SYSTEMS

A continuous miner is an electrically driven machine designed to cut and load coal simultaneously. As may be gathered from its name, it is designed to cut coal continuously and is independently driven on tracks. The only limitations on it being able to cut and load continuously are the efficiency by which the cut coal can be transported away and the need for the newly exposed roof to be systematically supported.

Cutting the Coal

The cutting is done by a rotating drum, typically 1 metre wide and 3 metres long which is driven by two electric motors. The drum is mounted on a boom which allows it to cut varying seam sections by raising the drum to the roof and then lowering it to the floor cutting coal all the while.

Typical cutting rates may be 5 to 10 tonnes per minute in a moderately thick seam. With reasonable coal clearance and organisation, the continuous miner should cut for 60 to 120 minutes per shift.

Allowing for the worst situation where the machine is cutting at only 5 tonnes per minute and works for only 60 minutes out of the seven hour shift, then the coal cut and loaded should be:

= 5 tonnes × 60 minutes = 300 tonnes pr shift.

Thus, just in one shift the continuous miner has easily produced the same tonnage as 300 men do in a day by the manual methods or 120 men do in a day using semi-mechanised equipment.

A continuous miner is driven by one man though he would normally have a helper to assist with the cable management and ventilation system.

9.3 Transporting the Coal

After the coal is cut by the continuous miner, it is loaded on to a central conveyor by gathering arms on the front of the miner. The conveyor then loads the coal in a shuttle car.

The shuttle car is an electrically powered, rubber tyred vehicle that is designed to carry the coal from the back of the continuous miner to the district conveyor belt and thence out of the mine. The payload of typical shuttle car is approximately 10 tonnes of broken coal in a mdium seam of 2 metres thickness.

Clearly, when a shuttle car is loaded, it must carry the coal away to the belt. During that time the continuous miner is unable to load out. Therefore, a second shuttle car is normally used behind each continuous miner and in certain circumstances a third may be used to improve efficiency. Each shuttle car waits for its turn to enter the heading, fill with coal and then transport the coal away to the belt. Each shuttle car deployed has one driver

Feeder Breaker

9.4.1 When the shuttle car arrives at the belt, by means of a centre conveyor, it is able to empty its load of 10 tonnes in less than one minute. As this sudden surge of coal is greater than most conveyor belts are capable of, a feeder breaker is placed at the end of the belt conveyor to size the coal to a uniform size and arrange the coal on to the conveyor at the speed that the belt can cope with it. The feeder breaker is unmanned.

Roof Bolter

After the coal has been cut and loaded by the continuous miner, the roof exposed must be systematically supported by the roof bolts. So that this task is done at the same rate as the cutting and loading, a machine is deployed to the task.

The roof bolter is a free steered, electrically powered vehicle that can drill a 2 metre roof hole in less than a minute, insert and spin the bolt in a further minute and provide a faster, better supported roof. Normally two men operate the roof bolting machine.

The Mining Sequence for the Development Phase

A typical room and pillar design for continuous miners comprises of five headings and, therefore, offers a minimum of five places at which the continuous miner may cut coal and advance. At cut throughs this number increases upto thirteen places that the miner can, if supported, cut coal.

The sequence of operations for the continuous miner to cut forward in one heading for a distance is dictated by the roof conditions. For efficiency this distance should be as far as possible. In the USA, a distance of 12 metres is considered normal using remote control machines. After cutting out the first heading the miner tracks to the second heading and cuts out there. Meanwhile, the roof bolting machine enters the first heading and supports the roof. When the continuous miner has finished in the second heading it tracks to the third and the roof bolter enters the second heading. This sequence is repeated until the continuous miner is back in the first heading and the sequence starts again.

The Mining Sequence for the Retreating Phase

Having driven the district out to the boundary, natural or otherwise, and extracted possibly 40% of the available coal, the continuous miner can very efficiently extract the pillars, starting at the boundary and retreating back to the main mine roadways.

The exact method by which the pillar is extracted varies depending on the situation and have names like Wongawalli, Rib Pillar, Old Ben. All the methods attempt to maximise the amount of coal that can be safely extracted. A maximum extraction up to 95% can be achieved

The big advantage of the Retreating Phase which makes it particularly economically viable is that the coal transport, ventilation, power and other services are already established and in particular that little further support of the roof strata is necessary. The collapse of the roof, in a controlled manner behind the pillar extraction, is a necessity of the method and relieves the stress from the remaining pillars.

Pillar extraction by hand or semi-mechanised method exposes the employees to risks of sudden and sometimes fatal collapses of roof. The use of continuous miners, fitted with remote control, keeps the employees well back and away from the cutting face and constantly under supported ground.

Manpower Requirements for a Continuous Miner Section

In a fully mechanised district there must be sufficient manpower to operate the equipment, ventilation, maintenance and services advancement. Typically, this would require 8-13 men, as follows:

Continuous Miner Section, Manning Levels

Continuous Miner Operator	x	1
Continuous Miner Helper	x	1
Shuttle car Operator	x	2
Roof Bolter Operator	x	2
Mechanic	x	1
Electrician	x	1
Sub-total		8
Possible additional:		
Utility Man	x	1
2nd Roof Bolter	x	2
3rd Shuttle Car	x	1
Foreman	x	1
Total per Shift		13

9.8.2 In India, reducing the manpower from 300 to 39 per day per district certainly increases the productivity but more importantly dramatically increases the district production very sgnificantly.

THE ADVANTAGE OF USING CONTINUOUS MINER SYSTEM SAFETY

The introduction of the Continuous Miner equipment into the underground environment represents a major opportunity to improve the safety of the personnel in the mine, as shown below:

(i) The use of continuous Miners, especially by remote control, takes men away from the coal face where most falls of ground occur.

(ii) As there is no longer a need to drill and blast the coal, men are no longer handling and using explosives or breathing in the fumes.

(iii) The roof and sides of the headings are no longer weakened by blasting operations.

(iv) The working environment with respect to dust is improved by the integral dust scrubbers which draw the dust laden air through the machine where it is removed by filter and water spray.

(v) Though the men must work hard, continually, they are operating machinery, not doing hard physical work as before.

(vi) The standard and quality of the roof control is greatly improved by using a roof boring and bolting machine.

(vii) The fewer men underground directly reduces the scale of the catastrophe, should one occur.

Production and Productivity

The world's cheapest coal producers are the USA, South Africa and Asutralia. It is interesting to note that in these countries room and pillar mining still accounts for on an average 68.5% of the underground coal production with longwall producing the remainder.

In the USA, of the 387 million tonnes of underground coal produced annually, over 60% is from room and pillar methods. This is virtually all won using continuous miner and shuttle car systems.

In South Africa, of the 120 million tonnes of underground coal produced annually, 111 million tonnes (92.5%) is from room and pillar methods of which 53 million tonnes is produced with continuous miner and 58 million tonnes from mechanised undercutting and drill and blast methods.

In Australia, of the 56 million tonnes of underground coal produced annually, 45% is from longwall and 55% from room and pillar (partial extraction) methods.

While longwall, when working properly, is the most productive method, room and pillar methods with continuous miners, shuttle cars and feeder breakers can produce coal at similar low costs per tonne and the capital investment in equipment is much less.

CONTINUOUS MINER PRODUCTION WORLDWIDE

U.S.A.

Continuous miners account for over 60% of underground production in the USA which is the world's cheapest and largest underground coal producer.

The competitiveness of this market has led to continuous improvements in the machines which is reflected in the output.

The table below shows the productivity in USA underground mines utilising continuous miners which compares well with the present underground OMS in India of 0.55 tonne.

Year	No. of Mines	No. of Employees	Production Tons (000)	Tons per Manshift
1988	1,690	43,500	244,560	18.82
1989	1,602	47,600	254,491	19.82
1990	1,589	54,300	266,337	20.94
1991	1,401	55,700	245,580	21.95
1992	1,268	58,000	234,369	23.33

The performance of some continuous miners is very impressive, like the super sections at Marrowbone colliery in the Coalberg seam. Each section has two continuous miners (either 12CM12 and 14CM15) and three shuttle cars (10SC22's). Extraction heights vary from 1.8 m to 3.0 m and support is by roof bolts. A seven entry place changing section with 6 m wide roadways and 12m×12m pillars averages 63,000 to 72,000 tonnes per month. The best production has been 115,000 tonnes per month.

At Loveridge mine in the 2.1m thick Pittsburgh No. 8 seam, a 4 entry system and 4.7 m wide roadways and 28 m between pillars with 61 m between crosscuts is driven by one 14×12 satellite continuous miner and two 10SC22 shuttle cars. The continuous miner has roof bolters mounted over the machine providing support and places two bolts while cutting out. Average advance achieved is 44 m per shift (550 tonnes) and best is 95 m per shift (1,300 tonnes per shift).

At Martin County, Kentucky in the Stockton seam (1.5m – 1.8m thick), a supersection with two 14CM9's working a entry place changing system with 15 m centres and 15 m crosscuts and 6 m roadway width average 2,300-2,500 tonnes per shift with a best of 3,100 tonnes per shift (availability 97-98%).

In Monterey No. 2 Mine Illinois in the Illinois No. 6 seam (1.8m thick), two 12CM12's are operated alternately in a 10 entry place changing supersection (22 m centres × 22 m crosscuts) 7 m roadway width to give an average of just under 1,000 tonnes per shift and a best production of 2,500 tonnes in 8 hours on 17th October 1990 (availability 97.6%).

South Africa

92.5% of underground output in South Africa is from room and pillar methods and continuous mines cut approximately half of this.

At Khutala Mining Services which supplies coal directly to the Kendal power station in South Africa output is 370,000 tonnes/month from 3 double header sections, two of which are producing approx. 125,000 tonnes per month (tmp) and one approx. 120,000 tpm.

Australia

In Australia longwall only makes up 45% of the underground production. 55% is still from partial extraction system mainly using continuous miners. The number of continuous miners in operation in Australia has remained at the 1985 level of about 250 units, most employed in high production room and pillar sections.

The examples above illustrate that relatively low capital investment in room and pillar methods can achieve excellent productivity and very high production, matching longwall systems on cost per tonne. These systems are also extremely flexible, able to mine small coal pillars between longwall sections as well as full room and pillar sections.

In India over 80% of the underground output is from room and pillar mining, using manual or semi-mechanised methods. By building on the expertise and experience in room and pillar and adding the massive potential of the Continuous Miner technology, huge increase in output are possible together with a reduction in heavy manual labour and accidents.

A comparison of manpower requirement and daily production achieved by various UG mining methods is as given below :

Method of work	*Manpower*	*Tonnes/Day*
Manual Mined District	300	300
Semi-mechanised District	120	300
Continuous Miner District	39	>900

SIMPLICITY

The Room and Pillar system is relatively simple and relies on establishing the correct work patterns and routines for the individual situation. As room and pillar is the normal method of work in India, the introduction of Continuous Miner equipment is not a major departure from the systems presently in operation. Under the Continuous Miner system, the manual steps of Drilling, Blasting, Loading, Supporting are replaced by Cut and Load, Transport and Roof Bolting.

REASONS FOR SLOW DEVELOPMENT IN UNDERGROUND COAL MINING IN INDIA

Although approximately Rs. 4000 crore has been invested in various modernisation programmes in the last two decades or so the net production from underground coal mining in India have actually declined. However, it is a fact that an underground coal mining can only maintain its production on the basis of continuous developments. It is an understood feature of an underground mine quite different from a factory, that constant feeding of capital is required in areas moving further and further away or going deeper down thereby necessitation extra cost on haulage, pumping, ventilation etc. It may also be noted that presently about 50% of underground coal production comes from development only. However, most of the Indian underground mines face following problems:

(i) A large number of taken over underground mines were very old. Many of these mines had extended beyond optimum dimensions. As a result, production from such mines declined gradually over the years.

(ii) Due to exhausting coal reserves and because of closure due to uneconomic operation, there has been loss in production from such underground taken over mines to the extent of about 20 million tonnes during the last two decades.

(iii) Many of the shallow underground mines were converted to opencast for improved coal recovery.

(iv) A large number of old mines were integrated in the re-organisation projects for achieving higher production with improved technology. However, presence of multi-section development in thick seams, old waterlogged workings, fire, presence of built up area with large scale habitation are some of the factors which slowed down implementation of these re-organisation projects. While there has been steady decline in production from the old takenover underground mines, increased contribution from the underground projects could not match with this declining trend.

(v) Infact, there is a need for regular investment in the existing mines for maintaining coal machinery required for extension of haulages, electricals, pumping arrangements, stowing etc.

(vi) Most of the underground coal mines are located in Jharia, and Raniganj coalfields in Bihar/West Bengal region. The two companies, namely, Bharat Coking Coal Ltd. (BCCL) and Eastern Coalfields Limited (ECL), have been incurring perpetual losses since their inception. As of now, their accumulated losses surpass their total paid-up capital and as a result thereof, these are declared sick under "Sick Industries Act". Infact, they have already been referred to Board of Industrial and Financial Reconstruction (BIFR) by Government of India. It has been difficult for such companies to arrange capital needed for investments required for initiating new underground coal mining projects. They could barely manage to somehow fend for reorganisation projects to atleast maintain coal production level. Of course, that also could not be done and actual losses increased year after year.

(vii) There have been chronic power shortages in the eastern region affecting underground coal production of ECL and BCCL. Such power shortages have also affected implementation of underground projects in these regions.

THE WAYOUT AND CONCLUSION

The wayout from this situation is not very so simple. There has to be two pronged strategy for improving quality coal production in India. While the Part One of the strategy shall take care of already running mines, under the Part Two new coal underground projects are to be initiated. The first strategy has to deal with problems of running coal mines like tough geo-mining conditions, mine fires, water loggings, human re-settlements and other environmental implications. The second one has to be totally programmed towards production of quality coal with latest technology and modern concerns and optimum investments to yield globally competitive productivity. Powered Support Longwall Technology is obviously the best choice for all new underground coal projects. the potential of this technology has already been established in the world with enough examples of high production longwall faces hving surpassed the productivity of opencast operations as in Australia and USA. New systems will focus on integrated face concepts with longwalls producing 33,000 tonnes of coal per day from the new generation of heavy duty faces by the year 2005. Due consideration may be needed for increasing applications of single-pass longwall systems in seams upto 5 metres.

The first strategy which is exclusively for the existing mines, new types of development equipment may be required to significantly increase the development trends. Continuous miners may find wide usages in multi-entry systems with average drivage rates of approximately 400 metres per week in conjunction with roof bolters. The continuous miners may be backed by either shuttle cars or coal haulers or continuous flexible conveyors. Existing reserves of developed pillars may also be extracted by using continuous miner system and breaker line supports.

One common essential input for both the strategies is complete renovation of information technology in underground coal mines. Application of wireless telephony with radio trunking facilities from coal to surface office, workable even under worst circumstances, has to be made available in Indian coal mines. Computer controls for improving the underground operational systems would be obviously required to cope up with the new situations. This technology has already been established in World with enough examples of high production longwall faces having surpassed the productivity of opencast operations as in Australia and USA.

REFERENCES

1. CMPDIL, 1984, Coal Mining in India, Proceedings of the World Mining Congress, Ranchi.
2. Coal India Ltd. (CIL), Annual Accounts Reports, 1991–1994.
3. *Coaltrans International*, Monthly Journal of Coal, London, April, 1994, July/Aug., 1995.
4. Ghosh, A.K., 1987, *Coal Industry in India.*
5. Ministry of Coal, Annual Reports, 1991–1994.
6. National Paper on Energy Scenario in India, 1995, Indian Member Committee of the World Energy Council and Institution of Engineers (India), New Delhi.
7. Proceedings of Pacific Rim Coal Conference, New Delhi, 1994.
8. Report of the Parliamentary Committee on Energy Development, 1994, New Delhi.

FINANCE FOR INDIAN POWER PROJECTS — AN OVERVIEW

15

A. MUKHERJEE
Vice President
Corporate Planning
CESC Limited
D. GHOSH
Dy Manager
CESC Limited

ABSTRACT

Development of the power sector holds the key to the integrated development of any economy. This is particularly true for nations poised at 'take of' stage as India is. However, given the socio-economic ground realities of the day, the need for restricting the ultimate cost of electricity to the bare minimum cannot be overemphasised.

The chapter aims to take a close look at the various issues involved in optimising the financing structure, especially of large power projects. It also attempts to focus on some of the major issues associated with raising and servicing cross border funds, which today is not just an alternative but a necessity.

Amongst the issues covered are the following :

(a) Eligibility norms of the power agency,

(b) Rating mechanism of the power agency to arrive at the conditionalities and cost of finance,

(c) Guarantees and assurances required from the Central and State Governments,

(d) Concessional credit from suppliers of major plant and equipment and a close look at some of the present hurdles which increase the ultimate cost of import,

(e) Impact of foreign currency exposure and a suggested hedging mechanism, and

(f) Analysis of an increasingly relevant low cost options e.g. utilisation of second-hand plants, mothballed generating sets and renovation and refurbishment of old power stations.

The benefits that will accrue to the industrially developed nations from a healthy and vibrant Indian economy in general, and the power industry in particular, have been outlined. To achieve this, an integrated development programme of the Indian power sector is of critical importance. This issue has significantly gained importance due to the depressed capital goods market in the West. With globalisation no longer being just a buzz word, it must be appreciated that 'win win' approach is the need of the hour. Only if the international economic community as a whole adopts this credo, can the growth of the global economy keep pace with its true potential.

INTRODUCTION

The accelerated growth in the demand for power is a direct fallout of the expansion of the economy where growth in industrial production, commercial activities and domestic consumption need more and more of electricity to sustain the progress. The unique properties of portability, efficiency, cleanliness and ease of transformation into other forms of energy together make electricity the most convenient form of usable energy known to mankind. The social and economic development of a country is largely dependent upon the use of energy and per capita consumption of electricity is a widely accepted index of the economic prosperity of a nation.

Following the initiation of planned development of the economy in 1951 to improve the socio-economic condition of the masses, power development was always given its due importance in the plan programmes. Capacity additions over successive Five Year Plan periods, to the extent that was possible with due consideration to resource availability, reflect the perceived need for accelerated development of this core sector by the nation's apex body of planners. The installed generation capacity which was a meagre 1,362 MW at the time of Independence in 1947 has increased to about 82,000 MW at present.

The allocation for the power sector increased from about 13.3% (Rs. 2,600 million) of the total Plan allocation in the First Five Year Plan to about over 18% (Rs. 795,890 million targeted) of that in the Eighth Five Year Plan. The following table shows the chronological growth in the countrywide installed capacity (MW), energy generation (Million units) along with their average growth rates during the first-four and a half decades of planned development.

Growth of Installed Capacity and Energy Generation

Year	Installed Capacity (MW)	Average annual rate of growth	Energy Generation (Million units)	Average annual rate of growth
1950-51	1,835		5.858	
1955-56	2,886	9.5%	9,662	10.5%
1960-61	4,653	10.0%	16,937	11.9%
1965-66	9,027	14.2%	32,990	14.3%
1970-71	14,709	10.3%	55,827	11.1%
1975-76	20,117	6.5%	79,231	7.3%
1980-81	30,214	8.5%	110,844	6.9%
1985-86	46,769	9.1%	170,037	8.9%
1990-91	66,055	7.1%	264,231	9.2%
1994-95	81,164	5.3%	351,020	7.4%

While the progress made so far is commendable, there has been a sharp deceleration in this sector over the last five years and the addition to generation capacity, in absolute terms, has been lower in the Eighth Plan period as compared to the Seventh Plan period. In a situation where the demand for power has been growing exponentially, this has necessitated a mid-course review of our power sector, especially the participation of the private sector.

Inspite of the Government's best intentions and the interest shown by prospective independent

power producers (IPPs), little headway has been made in this respect. Almost all the power projects in the ambit of MoU, PPA and compliance with other formalities and none of them have crossed the limit of the drawing board.

SLOW PROGRESS OF PRIVATE SECTOR POWER PROJECTS

The single most important factor behind the tardy progress in this area has been the tariff which the new power generating companies are being compelled to demand from the concerned State electricity Boards (SEBs). The price being asked by the IPPs appears to be extremely high to the SEBs, especially because this is often compared with the average cost of generation from the SEBs' existing power plants or at times, with their current average selling price.

It would, however, not be fair to say that the slow pace of implementation of green field power projects is soley due to the inflexible stand adopted by the SEBs. In many a case, the funding package and commercial terms of sale formulated by the power generating companies do leave some scope for improvement from the point of view of the ultimate consumer. Electricity is today an expensive item and, therefore, all care needs to be taken to arrive at the least cost solution, tailored to individual needs.

FINANCIAL STRUCTURE OF POWER GENERATION PROJECTS

In view of the urgent necessity to minimise the overal cost of generation, the issue of financial leverage has gained importance. Although the Government guidelines permit a Debt-equity ratio of upto 4:1, a few issues should be considered while drawing the financial structure for specific power projects.

The higher the debt component in the total project cost, the greater will be portion that will not have to be serviced once the borrowed funds are repaid. considering that the life of a new power station is expected to be at least 25 years after commissioning and a loan repayment period of 10 years, a high debt-equity ratio will substantially reduce the cost of generation in the latter half of its life. However, if the debt component of the project cost is very high, financial institutions are likely to feel uncomfortable about the commitment of the promoters to the project.

As per our income tax laws, payment of interest and other associated financial charges are charged to the company's Profit and Loss account and hence have to be paid out of the **pre-tax profit.** On the other hand, return on equity is a **post-tax** item. On the basis of present income tax rates, the generating company will have to earn almost twice as much additional operating surplus to be in a position to earn one extra Rupee as return on equity compared to that required to pay one extra Rupee as interest i.e. to earn Rupee one by way of return on equity, the company will have to earn Rupees two as operating surplus against Rupee one to be able to pay Rupee one as interest.

Piecemeal funding of the foreign exchange requirements of mega power projects, be it through equity or debt, is not a practicable option. Rather, a major player—be it the major supplier of plant and equipment, the Engineering, Procurement and Construction contractor or a financial institution should agree to syndicate the mobilisation of funds. This will greatly cut down the lead time and effort involved in repeated project appraisals alongwith other allied activities.

OFFSHORE EQUITY FUNDS

The directive of Foreign Investment Promotion Board (FIPB) that foreign debt should not be more than twice the foreign equity is restrictive. While raising foreign loans is relatively easy on account of the large number of players in the international loan capital market, there are relatively fewer players in the equity market. Segmenting debt and equity into local and foreign is not desirable and stands in the way of evolving an optimal cost-effective project financing plan.

Apart from meeting FIPB conditionalities, part of the cost of imported plant and equipment is required as foreign equity capital due to paucity of resources within our country. This will also ensure commitment towards the project by the equipment suppliers, O & M contractors and EPC (Engineering, Procurement and Construction) contractors.

The large foreign players in the equity Market would like to have a controlling stake in greenfield projects and often this is not acceptable to the Indian promoter. In case of small players, the extent of their equity participation adds little value in the funding of mega projects.

The relatively cheaper Euro issue option is not available for greenfield projects as under the existing guidelines, only existing profit making companies are allowed to tap the Euro issue market. In an era of liberalisation, it is perhaps time that this restriction be removed. Investors, be they Indian or overseas, should be free to participate in power projects.

RETURN ON FOREIGN EQUITY CAPITAL

As in case of foreign currency loans, the ultimate cost of foreign equity is largely dependent on the actual escalation in foreign exchange rates. In fact, the impact of exchange rate variations is far more pronounced in case of foreign currency equity than for foreign currency borrowings as the base amount (in foreign currency) remains constant over the life of the project which is not the case for offshore borrowings. Moreover, as for all other economic resources, the permitted return on equity should be directly related to the opportunity cost of the same suitably adjusted for inflation.

As the long term inflation rate in international currencies is significantly lower than that in India, it follows that the opportunity cost for US dollars is lower than that of the Indian Rupee. This is further corroborated by the differential between the LIBOR for US Dollars and the Prime Lending rate for Indian Rupees which currently is in the region of 10% per annum. In view of the above, it is difficult to justify a 16% return on equity, denominated in foreign currencies. It is felt that either the 16% return on equity should be uniformly denominated in Rupees or if part of it is to be denominated in foreign currrencies, the permitted rate of return should be substantially lower on that portion of the equity whose return is proposed to be denominated in foreign currencies.

OFFSHORE LOAN FUNDS

Issuing bonds in line with Section 144A bonds in USA to qualified buyers is another option. Moreover, compulsory credit rating, required from internationally reputed credit rating agencies, is a time consuming and expensive exercise. This route is not yet widely prevalent in India. The rates of interest and other finance charges, which depend on the company rating as well as the country rating, are expected to be quite high for Indian power companies.

Syndicated loans are being proposed by offshore funding agencies. The major problem is the relatively short maturity period associated with these loans since these are mostly raised from commercial banks. To make lending by offshore commercial banks more lucrative, it may be worthwhile to explore the possibility of the participation of these funding agencies in the Export Credit Authority loans on a *quid pro quo* basis.

It is felt that the unhappy experience of the World Bank and their affiliated organisation with various State Electricity Boards has queered the pitch for private sector projects. Considering the excellent track record of the existing power supply agencies in the private sector, it is felt that the appraisal of their projects should be influenced by their own credibility and performance rather thank those of the State Electricity Boards and other public sector undertakings in power generation, transmission and distribution.

OFFSHORE SUPPLIERS' CREDIT

While evaluating the true cost of foreign currency loans, a realistic assessment of the impact of the likely devaluation of the Rupee is of critical importance. What appears to be, *a priori*, very attractive rates of interest—especially when one compares them with the present rates of Indian financial institutions—often turns out to be far more expensive in the long run.

Overseas supplier's credit or buyer's credit is associated with a premium charged by Credit Authorities and counter guarantees from local financial institutions. This raises the effective cost of credit quite substantially. To reduce the effective cost of these funds, the Export Credit Guarantee Department of UK is currently looking into the possibility of waiving the requirement of counter-guarantees from Indian financial institutions. This, if implemented, will be step in the right direction.

RATING MECHANISM OF OFFSHORE FUNDING AGENCIES

As the business of Independent Power Producers (IPPs) is characterised by long lead times, relatively high uncertainty of success and large front-end costs, the project financiers carry out a rigorous project appraisal. A further complication of IPPs is their increasing international focus. The huge investments required by the power sector in India cannot be met from within the country and inevitably calls for large foreign investments. The foreign investors including all the multilateral agencies and financial institutions thus have to carry out the onerous task of determining whether the IPP has the potential to yield an acceptable return to meet their specific goals and objectives.

A long term debt financing agreement with foreign and local financial institutions, multilateral agencies and banks is necessary. The largest call on the cash flow of a power plant are usually fuel costs and debt service. The long term lenders often have more money at risk in a power project than the owners, creating a need for high degree of security for the lenders. Usually, common loan term sheets are drafted in respect of IPPs' projects for all the senior lenders in such

a manner and with such covenants which makes the IPP legally bound to repay from their future cash flow streams.

The key elements that the offshore agencies are interested in closely evaluating, apart from the return on borrowed funds, are diverse ranging from the offshore investors' interest in equity participation to various social, political, economic, legal and environmental considerations. The lenders, as well as the equity investors, usually carry out a 'due diligence' exercise on the basis of which they try to ascertain what is termed as 'Departmental Maximum Liability (DML). While ascertaining the DML, due weightage is given to the project risks as well as the country risks.

The following paragraphs summarises the major areas of interest of offshore agencies while determining the project viability.

The basic agreement is usually the partnership or joint venture agreement with the local entity. Quite some time is spent in getting to know the potential partners, the government and the local environment. The local project sponsor's ability to push the project through, their past experience, the credibility of the fuel supplier coupled with the firmness of the fuel supply agreement are vital elements which the offshore agency takes care to evaluate.

Another critical contract is the Power Purchase Agreement (PPA) entered into by the IPP with the respective authorities—normally the State Government and/or SEB. The PPAs are usually of a long tenure with the selling price often being higher than the present average rates. In most of the cases, the higher price is necessary since historically, most of the SEBs have been selling electricity at less than the required tariff. Consequently, in addition to having to negotiate a PPA that will stand up to the scrutiny of international lenders, the IPPs are normally burdened with the task of having to satisfy the local authorities the justifiability of relatively high selling prices. For the international lender, the creditworthiness of the power purchaser, treatment of deemed generation and liquidated damages clauses to be borne by the power purchaser are issues of great importance.

In view of the poor track record of most of the SEBs, what often makes or breaks the project is the payment guarantee mechanism envisaged. As it has been clarified that no further counter-guarantees will be available from the Government of India, the long term reliability of the payment mechanism—be it through the escrow account route, the irrevocable revolving Letter of Credit route backed by State Government guarantees or any other—is a matter of great concern.

Another key area is the package of design and construction contracts to ensure timely completion of the power project. Usually, the preferred route is that the IPP should enter into EPC contracts with reputed foreign or local parties who have already commissioned similar projects in other countries. What are considered to be very important are the credibility of the EPC contractor, the nature of the EPC contract draft, the stiffness of penalty clauses and the court of law in which disputes between the parties will be settled.

A firm long-term fuel supply contract for the proposed power plant which contracts for the delivery of the necessary fuel(s) covered by adequate penalty clauses for shortfalls in performance is also a prerequisite for overseas lenders.

Apart from the ones mentioned earlier, the project management and operation and maintenance agreements are also subjected to strict scrutiny. It is insisted that each of these agreements need to be drafted as per international norms alongwith enforceable penalty clauses. Moreover, implementation agreements with the SEBs are often required. If the project is sought to be

financed by those offshore agencies who have little or no prior experience in India or if the funding consortium includes multilateral agencies, letters of comfort from the State governments are absolutely necessary.

INDEPENDENT APPRAISAL BY FUNDING AGENCIES

Most financial institutions insist on having even the technical particulars of the project appraised by independent engineers from scratch which is a time consuming affair. In view of the multitude of funding agencies involved with large projects, this exercise substantially adds to the gestation period and consequently, the project cost.

As far as practicable, the detailed assessment of CEA should be made acceptable to funding agencies.

Let us now take a close look at an issue which holds the key to formulation of any cost-effective strategy for financing the highly capital intensive power projects of the day i.e. the projected rate of devaluation of the Indian Rupee vis-a-vis major international currencies to be used for determining the Net Present Value of future revenue and expenditure streams.

PROJECTED RATE OF DEVALUATION OF THE INDIAN RUPEE

Since substantial forex exposure of the power project is now the rule rather than an exception, a rational estimate of projected forex rate is of crucial importance for optimising the financial structure of the package.

The projected rate of devaluation of the Indian Rupee in the international market, economic theorists are near unanimous that, in the long run, the Law of Purchasing Power Parity holds. This law states that other factors remaining constant, the rate of devaluation of any currency vis-a-vis another will be equal to the difference in their domestic rates of inflation. With the speedy opening up of our economy, our country is not likely to be an exception to this rule. Based on our traditional inflation rate of about 9% per annum compared to about 4% per annum internationally, it would be reasonable to assume that the Indian Rupee is likely to depreciate at about 5% per annum vis-a-vis major international currencies in the foreseeable future.

Graph below shows the rate of interest on Rupee loans that would be equivalent to different rate of interest on foreign currency loans based on the following assumptions for different rates (5.0% p.a., 6.5% p.a. and 8.0% p.a.) of escalation in foreign exchange rates :

(a) Export Credit Authority (ECA) premium, which is an up-front payment, will be the same (in certain cases it is higher), in percentage terms, as the annual rate of interest charged by the offshore supplier of plant and equipment. This premium is charged by the ECA of the country from where the major plant and equipment is imported as an insurance premia guaranteeing payment by the buyer to the supplier.

(b) Loan Repayment period is 10 years commencing after 3 years construction period.

(c) Annual discount factor of 12% on all Rupee/Rupee equivalent cash outflow figures.

Evolution of a mechanism whereby Independent Power Producers could have access to World

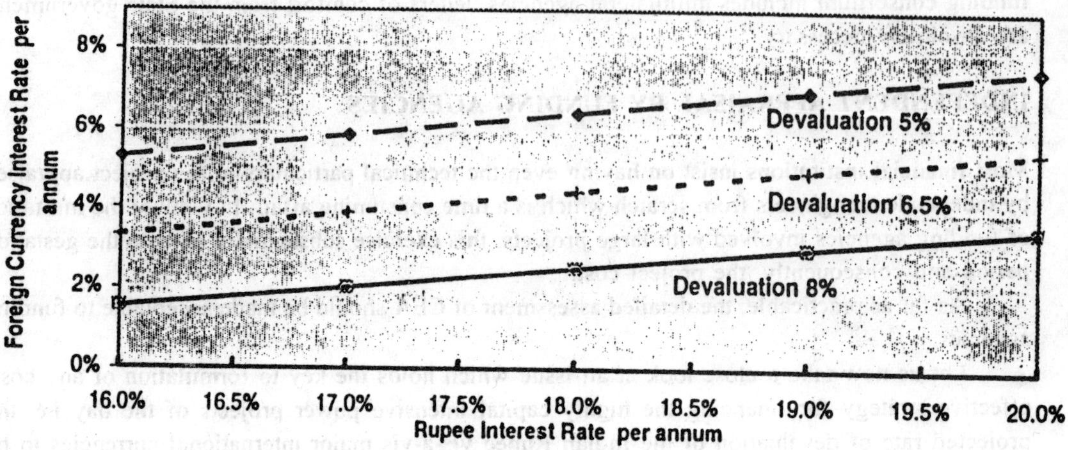

Equivalent Interest Rates : Rupee vis-a-vis Foreign Currency

Bank funds, converted into Indian Rupees, through a government agency at a reasonable premium over the interest rates actually charged to cover forex risks, would provide much needed relief to the power sector.

USE OF SECOND HAND PLANT AND/OR 'MOTH BALLED' POWER PLANT AND EQUIPMENT

There has been sharp escalation in the cost of major plant and equipment coupled with the 100% increase in the cost of foreign exchange that has occurred over the last five years. An option which till recently was not considered seriously by most generating companies, is now being explored — that of importing second hand plant or 'moth balled' generating sets. If they can be procured at considerably lower capital costs, this will reduce the cost of generation, provided that they can be re-engineered and re-designed to suit the local ambient conditions and fuel quality.

A number of these plants have had to be shut down within a short period from their commissioning due to the strict environmental norms that have been laid down by the densely populated countries of Western Europe. In this connection, it may be mentioned that international funding agencies have taken a tough stance in respect of the environmental damage associated with thermal power stations. While the benefits of pollution control are well known, it should be acknowledged that, especially for a LDC (Less Developed Country) such as India, the marginal benefits derived from increasingly strict norms should be weighed against the economic costs to ensure a balanced perspective in this issue.

There is however, a flip side of the coin. It must be ensured that the remnant life of second hand plant and equipment fully justifies its price and in case of moth balled sets, there is no serious design, engineering or metallurgical flaw which may jeopardise the continued operations of the power station. A comprehensive assessment of the plant and equipment by internationally reputed chartered engineers will be utmost importance in this respect. The plant should also be backed by a firm guarantee from the original plant manufacturers regarding both its performance and availability of spares including those which may be damaged during the process of dismantling, transportation and erection. A fair amount of re-designing and detail engineering may also be necessitated by the difference between the Indian conditions and the conditions in which the equipment was originally intended to operate, the cost for which will have to be built into the overall project cost.

RENOVATION OF OLD POWER PLANTS

In the context of the high cost of fresh generation capacity, the importance of renovation and modernisation schemes needs to be underlined. As per the Ministry of Power's Annual Reort for the year 1994-95, an additional generation of 7000 million units per annum was anticipated at a cost of Rs. 1174.01 crores under Phase I and 8750 million units at a cost of Rs. 2105.23 crores under Phase II. Considering an annual capital servicing cost of 25%, the average cost of the incremental generation works out to little more than 50 Paise per unit (excluding fuel cost) as compared to the present day figure of around 160 paise per unit. Of course, proper evaluation of the benefits of such renovation programmes is of vital importance and should be estimated on a conservative basis.

MAJOR STUMBLING BLOCKS

It is now clear that the private sector will have to contribute substantially if we are to reach anywhere near the ambitious targets set for the Ninth Plan period. However, if the efforts in this direction are to bear fruit, a number of issues have to be resolved immediately, of which.the major ones have been outlined in the following paragraphs.

COMPLEXITY OF LEGAL DOCUMENTS

The complexity and scope of legal documents required to be executed before a power project can be implemented is threatening to go out of control. In addition to the clearances and approvals, the volume of legal documents has grown sharply, especially with the participation of offshore funding agencies and export Credit Authorities. Moreover, these agreements have to be drafted by legal experts at a high cost. Amongst the documents which have to be executed by the IPPs are :

(a) Power Purchase Agreement
(b) Engineering, Procurement and Construction (EPC) Contract,
(c) Operation and Maintenance (O & M) Contract,
(d) Fuel Supply Contract
(e) Fuel Transport Contract

To make things worse, the different parties involved often have conflicting requirements. As a result, more than two years go by before financial closure can be achieved, if at all.

It has, therefore, become necessary to develop a standard set of contracts encompassing the above, which have to be accepted by all concerned parties with modification required due to site specific conditions. These model contracts should be drawn up with the active involvement of Independent Power Producers (IPPs), Indian Financial Institutions (IFIs) as well as offshore funding agencies and suppliers of major plant and equipment.

ABSENCE OF STANDARD BIDDING PROCEDURES

The absence of standard bidding and evaluation procedures makes it difficult for IPPs to respond quickly with the optimum package as considerable time and effort is required to tailor their offers as per the requirements of the individual SEBs. Moreover, it is observed that some SEBs are forced to devote a lot of time and manpower to explain their requirements to the bidders and then to respond to innumerable points raised by them. The entire bidding and evaluation procedure needs to be standardised and made as transparent as possible.

RISK AVERSION

Total risk aversion on part of both the SEBs as well as the IPPs has often led to a stalemate at the project formulation stage itself. For example, some SEBs insist that they will not accept the risk arising out of forex exposure of the IPP in respect of borowings, without which no project can ever be viable. On the other hand, some IPPs insist that if extensive repairs have to be carried out to a power station for reasons beyond the control of the recipient SEB, the SEB will have to bear the cost of such repairs. Both the above appear to be quite unreasonable.

Both SEBs and IPPs have to approach the issue with a spirit of risk sharing on a equitable basis. It should be accepted unless capital intensive power projects are approached with a 'win-win' mind-set, little progress can be made.

TARIFF REVISION

Most IPPs are sceptical of getting the necessary tariff revisions as may become necessary during the tenure of the project and therefore seek not only high rates of return but in some cases, multi-layered guarantees and counter guarantees from both the State and Central Governments. For reasons well known, this is not a sustainable path.

As a concrete step towards formulation of a uniform national tariff, a National Tariff Regulatory Commission should be formed which will dictate the tariff payable by the recipient SEBs to the IPPs. Acceptance of whose directives will be mandatory on all parties concerned. We understand that the process is already under way, but perhaps, the urgency needs to be underlined.

ADOPTION OF 'TARIFF-BASED' RATHER THAN 'COST-PLUS' APPROACH

The present 'cost plus' approach often leads to suspicions of padding of costs. This which has politicised awarding of projects with the consequential fall-outs of major time and cost overruns.

To do away with problems associated with the traditional approach, some of the SEBs adopt an alternative 'tariff based' approach which will focus on project performance rather than project cost. The tariff may, however, need to be denominated partly in Indian Rupees and partly in forex. In the light of its increasing popularity, this approach needs to be formalised through clear and specific guidelines from the Ministry of Power.

HIGH REAL RATES OF INTEREST

IPPs are required to procure funds at very high rates of interest, be it from local or from foreign sources. While Rupee funds are not available at a real interest rate (nominal rate of interest less rate of inflation) of less than 14-15% p.a., the effective real rate of interest on foreign funds is even higher if one accounts for the multi-layered guarantees and counter-guarantees and the projected devaluation of the Indian Rupee.

In view of the critical importance of the power sector, IPPs should be allowed access to IFI and international funds on relatively soft terms with long maturity periods.

CONCLUSION

Resource constraints, caused by decades of neglect of the commercial viability of SEBs, was the major reason for opening up this core sector. Offshore funds are a necessity. Yet, foreign manufacturers and lending agencies must realise that the Indian market is perhaps, their best long-term option in view of the near saturation of this sector in their own countries. The stable political environment coupled with a sound judicial framework endows India with a great potential for investment.

The road towards rejuvenation of the power sector is now open. The path is strewn with hurdles as outlined above which must be overcome to reach at the journey's end. The stakes are enormous and the pitfalls many. It is only through a clear focus of the objectives and capability to take considered decisions that the goal can be attained.

NOTE : *The views expressed in this article are the authors own and do not necessarily represent those of the organisation to which they belong.*

COAL AS A SMOKELESS FUEL

16

Col. S.N. Malhotra (Retd.),
Chairman & Managing Director
M/s SAM Smokeless Fuel (Pvt.) Ltd.,
Gurgaon (Haryana)

ABSTRACT

The paper is devoted to making coal as eco-friendly source of energy in domestic sector. Derived from coal SSF is free from volatile matter. The volatile matter has been entrapped, and cooled down to 'Coal Tar' as a by-product in SSF industry. This paper is of interest to third world countries whose coal deposits are large but calorific value poor. Compared to oil whose import bill for India is mind-boggling but coal deposits are natural assets and cheap. Popularising SSF is the essence. The traditional drawbacks of coal emitting large smoke and low thermal efficiency (compared to oil or gas) have been suggested to be converted to its advantages by using modern technology. The paper is thought provoking. To improve the shape of SSF coal ball by giving 19° slope to its faces (like pitting on Golf ball) so as to provide best angle of attack for oxygen, will give maximum heat-efficiency. Novel ideas for its marketing has been suggested in user friendly contents of packaging.

The process of producing SSF being continuous, the escaping steam can be used as useful and cheap power source for 'Light Rail Systems (LRS)' which are crying needs of metropolis cities of India today. Thus making these projects viable. Paper suggests that LRS follow ground contours. The principle of overcoming inertia and gaining momentum by following ground contours has been suggested for LRS, working with steam as power source from such projects.

The paper also suggest a concept of 'Urja Gram' (Energy Farms) for rural and semi-urban areas of India around developing metropolis, with open ended coal-linkages.

A conglomerate of conventional and non-conventional energy, projects can be raised and then coordination by Ministry of Energy is advocated to overcome pitfalls in financing these projects by present system of water-tight compartments. **An interesting paper for 21st century needs of India and South Asia.**

INTRODUCTION

It had been brought out at the 1994 conference of World Energy Council held at New Delhi that Low Grade Coal is available as 2/3rd of Coal Deposits in India. All of it is located in the East, but its potential users are in the West and North-West. As a part of Reforms, this Low Grade Coal is being offered to Private Sector, to produce Special Smokeless Fuel. The Technical Know How is available with CMPDIL, for converting Low-grade coal (Washery Grade IV), to High Grade SSF.

As per Central Mine Planning & Design Institute Limited (CMPDIL), a subsidiary of Coal India Limited, Technical know-how has been perfected and patented (ISI No. 156855) to manufacture SSF. The State Governments have banned the Financing of SSF projects. The normal feeling is that coal creates pollution. It is not environment friendly. Its supply by Coal India is uncertain. Jharkhand movement is a damper, in movement of coal from East to West. Non-availability of Railway Wagons by Indian Railway also dissuades, State financial corporations to invest in coal based projects, due to uncertain performance prospects. Added to all this, is the hunch of poor marketability of finished product SSF, whose track record is not known.

MARKETING STRATEGY

In the current scenario of popularising the use of coal, as its reserves are of greater longevity than that of oil or gas; a crusade for popularising coal has to be undertaken. It cannot be done by private entrepreneurs, in this nascent stage. Hence Coal India should take up the task of Advertising in media and TV, as part of National Policy. During oil crises a Gazette of India had also been issued towards this end by Ministry of Energy, in 1991.

The following theme can be highlighted in favour of coal as a Domestic fuel, "the aim of SSF is to minimise dependency on LPG and ultimately replace it, if we do'nt find oil reserves".

Special Smokeless Fuel is the Fuel for 21st Century.
Say 'No' to Cooking Gas Use 'SSF'.
SSF is the most environment friendly fuel.

— It is cheaper	While cost of LPG is Rs. 8.00 per Kg. (Rs. 133 for cylinder of 16.5 Kg.) SSF is a mere 0.45 Paise per Kg. at 1996 prices.
— It is safer	As compared to LPG, cylinder filled at high pressure, coal ball is safer.
— It is healthier Gas causes gas.	While LPG has hydrofloro carbons, SSF is smokeless.
— It is Re-useable	The simmering coal balls can be covered by ash, for use a little later.
— It causes no waste	Even the ash can be used to wash utensils or even use it as tooth powder.
— It gives charcoal tablets	Under sized coal can be used to make charcoal tablets for use as anti-flatuant to kill acidity in stomach as done in World War II.

Packaging

SSF can be marketed as :

— The Tea for Two Fuel	A pack of 2 SSF coal balls.
— The One Meal Fuel	A pack of 4 SSF coal balls.
— The One Day Fuel	A pack of 12 SSF coal balls.
— The 'Use as you please' Fuel	Fuel cakes with a portable collapsible burner (*unnat chulla*) & SSF balls.

The spherical balls of SSF, with black Diamond shapes of 19^0 slope, can then be packaged in attractive packing as suggested above.

Thus a great flexibility can be achieved in meeting the market demands of rural India, road-side '*Dhabas*' poor people as well as the 5 star hotels, who may demand more quantity for a better quality cooking of chicken and chapati, where LPG cannot give same quality of cooking.

Improving Thermal Efficiency of SSF

The special smokeless fuel (SSF) can be produced in the shape of 'Golf Balls'. The ball instead of pitting can have multiple angular surface. Each face can be 19°. By Army's R&D it has been found that 19° is the best angle for recochet to defeat attack. Hence armour plating on tanks takes 19° slopes at all points. In SSF spherical balls, the 19° sloping will make it most fuel efficient when the oxygen attacks the burning surface it will increase fuel longevity and cooking efficiency.

Indigenous Opportunities

Considerable savings in cost can be effected by using new material technology substitutes, like instead of Refractory Brick lining of Retorts, use Micro Light Ceramic Compound, (Exhibited in 1994 World Energy Council) whose rate is Rs. 60 per kg. as compared to Rs. 80 per refractory brick of one kg. at 1996 prices.

Management of Ash

The ash coming out of controlled burning of coal in retorts and domestic burning of SSF can be managed as brought out by Advisor to Coal India, some methods can be as under.

In Domestic use of SSF the SSF ash can be used to wash utensils as in yester years. The ash can also be used to cover and retain heat of partly burnt SSF. So as to subject it to re-use. This used to be practice in olden days. SSF, compared with LPG needs no high pressure container, or special burner, or safety devices like valve. In these fields it is a clear winner over LPG, as a domestic fuel.

The industrial use of Ash coming out in the plant can be to make 'Ash Candles' or 'Ash Screens' which can be used in Delhi (or Metropolitan cities) where pollution level is high. The exhaust of cars, buses and automobiles emanates burnt hydrocarbons which are health hazards. The porous quality of ash, if used with a suitable binding agent can be made into candles and erected along electric poles or visual display screens on roads, which can absorb these emanating polluting gases, then environment will be cleaner. This idea needs development.

In the field of construction activity bricks made of ash are stronger, cheaper and offer smooth surface for house building. Technology has been developed for mass production. A mini plant to manufacture flyash bricks alongside SSF plant can open profitable business avenues.

Put Byproduct to Use

The by-products of this non-conventional source of energy viz. SSF are even more lucrative than main product going by present day popular usage. Hence coaltar for road metalling, Fly Ash to fill side berms and shoulders of roads, the platerscine like waste (brown in colur) for Art and Sculpture work, and refined impurities for Napthelene Balls, are lucrative uses. Praffin wax derived from this industry as byproduct can be mixed with camphor can even be used for fuel cakes, used in survival rations of Army for High Altitude Areas, where no other fuel is available. We need to undertake an awareness campaign, while we undertake production. We may leave speed to oil and bring safety by coal in domestic fuel usage.

Scope of Foreign Collaboration

Since, the process to manufacture SSF is continuous carbonisation, the plant once started cannot be (or at least should not be) stopped. To utilize this continuous source of energy supply, a suitable field is Local Rail System (LRS). A blue print for same has been developed by Andhra Pradesh Govt. for the Twin City of Hyderabad–Secundrabad. Similar demand is for Delhi. It was in press news that CM visited Singapore and Hong-Kong to introduce a Tram-Car System for Delhi based on electricity. The slow stream of energy so generated can be sufficient to overcome inertia and tudge along a tram car to run a transport system following contour levels of ground, so as to keep following lower gradient. In this manner the momentum generated will provide it speed to overcome minor crest and fall of ground.

Japan's Experience in using gas for Refrigeration and Air Conditioning was highlighted at the Congress World Energy Council held at Tokyo from Oct. 8-13-1995. We can gain from their experience.

Power Source for Light Rail Systems

A suggestion is that between Gurgaon and Delhi this light rail system can be developed. The energy base will be continuous process of SSF manufacture. Since Gurgaon is on a Higher Contour level than Delhi, the geographic layout of contours can be used as a Rollar - Coaster, to allow the LRS to follow the natural slope of the ground. Whenever braking brings in intervals of stoppage, spurts of energy can be used to overcome inertia. Since in past 100 years, even in worst floods, never has Delhi been cut-off from Gurgaon. Thus Gurgaon would be ideal for LRS to be developed, first within Gurgaon then graduate to Mehrauli and Finally to Delhi–New Delhi.

Open Commercial Avenues

The Gurgaon region being on higher contour level than Delhi and a Japanese City coming up on NH No. 8, Near Manesar use of computers in the Light Rail System can open new vistas in punctuality and safety. Delhi Chief Minister has expressed his willingness to use the Centre of the Road Dividers in Delhi for Tramways or Mono-rail, light rail system. This is an exciting area for non-polluting new source of energy. Efforts should be made to promote special SSF smokeless fuel and its by-products.

Cut Cost of Transportation

Since American Coal Reserves are distributed, their haulage is short whereas, our known coal deposits are concentrated in East, their haulage is long. We should undertake geological survey by our indigenous satellites to locate coal reserves if any in west or north-west to overcome socio-political pressures of present coal producing state. To make transportation of raw material cheaper and minimise pilferage which is inherent in long haulage, Thermal maps by satellites should be obtained.

To overcome Jharkhand (Bihar) political problem, the coal linkage to Northern States can now be allotted from Madhya Pradesh, from new coal fields identified by Geological Survey of India. This would be economically more viable and cheaper, due to gauge conversion by Railways from MG to BG in the Rajasthan–Haryana Sector, like Gurgaon–Rewari belt. Here coal linkages be open-ended for greater duration then mere six months to give gestation time for new projects.

Establish 'Urja Grams'

Working in water tight compartments for conventional Energy sources like coal and Non-conventional energy sources will not allow proper viable solution for either. At grass-root level a mis-match is already occurring. While Ministry of Non-conventional Energy sources refuses to acknowledge SSF as a non-conventional source and hence not release fund for it. Unless Energy is seen as a whole, and a conglomerate of source develop a concept of Energy Farm or 'Urja Gram', rural and semi-urban India cannot benefit. A nodal agency needs to be developed for co-ordination

Tap Export Potential

The Broad Guage Railway line now operational and extended from Delhi to Rewari has eventually linked Mumbai to Delhi. This BG route will be three hundred Km. shorter than via Kota, and hence will be made main route to connect Delhi-Mumbai. It will soon be electrified to gain speed. Then we can think of even exporting SSF to South Africa or middle East via Okha Port or Bombay Port.

CONCLUSION

India is a poor country. Hallmark of poverty is that it makes one possessive of whatever one has Bihar's desire to control coal or Assam's desire to hold oil are understandable. So is West possessive of its Technology. In joint venture, we need a collaborator who can give life to our dreams of better utilisation and greater remuneration for our labour. Our coal reserves are just 1% of world, but put together with South Africa and China, our immediate neighbours, we are a formidable lot for coal technology to develop in private field. The ideas of utilising flyash to control pollution, continuous energy for Light Rail System and Special Design for greater efficiency of coal based SSF, are all areas where advanced technologies are needed. For this purpose indigenous private entrepreneurs as well as foreign collaboration is urgently needed.

FUEL SUPPLY STRATEGY FOR POWER PROJECTS IN ASIA

17

DAVID BOYD
General Manager,
Energy Development
RTZ – CRA,
15-17 November 1996
New Delhi

Fuel Supply: The Importance of Long-Term Planning

Electricity generating capacity is forecast to grow in Asia over the next twenty years faster than any other region in the world. This is particularly so in China, South Asia and South East Asia where there are many developing countries with forecast high increases in population and standard of living.

This will bring with its significant growth in demand for energy resources. Long-term planning is, therefore, critical not only for each power project, but also to electricity supply and growth prospects of the countries concerned.

Significant increases in demand will continue from the developed countries of Japan, Korea and Taiwan. This means that it is very important for developing countries such as China, India, Indonesia, The Philippines, Pakistan and Thailand to prepare fuel strategies to ensure that their fuel needs are met. Early planning in conjunction with fuel suppliers will be necessary to endure that these countries aren't caught short of supply by the continuing increasing demand from the developed nations, which have long-term contracts and supply relationships in place.

The rapid growth in power demand in Asia has facilitated dramatic change in the nature of the power generation industry in Asia in recent years. The growth of Independent Power Producer (IPP) projects in the region has been significant to date, and is expected to further acclerate. This is largely driven by the enormous capital required to meet increasing power demand, with forces enormous capital required to meet increasing power demand, which forces Governments and traditional public utilities to rely more and more on private investment.

This trend also drives a change in the approach required to fuel supply arrangements. The long-terms of the Power Purchase Agreement and the loans which support IPP projects require equally long-term, secure fuel supply arrangements.

Fuel Supply is currently one of the most difficult issues faced by Independent Power Producers. Regardless of the choice of fuel, long–term, secure arrangements are critical to ensure that IPP projects are financeable, and able to provide competitively priced power over the long-term. Financing of IPPs has frequently been delayed for extended periods by the need to satisfy lenders of secure fuel supply arrangements.

The time required to develop new fuel sources and being them into production makes long–term planning, and long–term contracts essential. Fuel suppliers will not develop resources unless there is a reasonable certainty of finding a market and achieving a price which will cover costs and provide a return on investment. This means that it is in the interests of consumers as well as suppliers to work together in planning fuel strategies.

Choice of Fuel–the outlook for Asia

The choice of fuels for power projects also requires planning. Power project sponsors requires certain characteristics in their fuel supply arrangements. Fuels will be preferred which offer;

— abundant supply;
— long–term commitment from reputable, financially strong suppliers;
— multiple sources to cover contingencies;
— agreed pricing mechanisms which ensure competitively priced fuel over the long-term.
— consistent quality;
— acceptable environmental performance;
— efficient, economical, reliable transport options.

Coal is currently the largest source of energy fuel in Asia, accounting for about 70% of current fuel supplies. Choice of fuels in the future will continue to depend on the ability of each fuel to meet the above requirements. The growth of IPPs will place further emphasis on the security of supply, through the complete fuel supply chain.

Oil, gas, hydro, nuclear and coal have all enjoyed the limelight as the "new age" energy source at some stage. All will continue to play major roles in Asia's power generations for many years to come.

Which fuel, or fuels, will best match the expectations of tomorrow's power developers and increasing public scrutiny.

Oil

Oil's economic and political volatility will continue to cloud any charisma that was attached to this commodity back in the late 1960s/early 1970s. This fuel is still perceived as being scarce, and therefore, unreliable. It remains subject, albeit to a lesser extent, to cartelised pricing. The oil industry structure highlights the limited supply options available to buyers. As we all know, a significant proportion of Asia's oil supplies come from politically sensitive countries within the Middle East, many of which have experienced military aggression and hostility in recent years. Oil has also had some bad publicity of late regarding its suitability as an environmentally sensitive energy source. Regular spillage's continue and power developers may well find it increasingly difficult to disassociate themselves form accidents that may occur within a host country's waterways and wildlife habitats.

Gas

Gas has without doubt been portrayed as the sexiest fuel of the day. Impressive promotional campaigns by large multinational concerns have convinced many governments and project administrators that gas in the "clean" fuel of the future. There is no doubt that, when in use, gas look good. There are not stack emissions to speak of, and operating environments at utilising power plants are usually sparklingly clean. This all adds up to something which is very easy to "sell" politically.

When burned efficiently, gas is a clean fuel indeed. The exploitation of gas reserves, its transportation to market and storage can be, however, far from efficient. A large percentage of all liberated gases are lost to the atmosphere prior to ignition. Such unburnt gases can be many times more detrimental to the environment in terms of contribution to the greenhouse effect than export quality black coal.

CO_2 emissions at the well-head is one example. The CO_2 released upon combustion at a gas-fired plant is only a small part of the problem. One should note that the CO_2 content of gas at the well-head can be anything up to 20%. This percentage, however, drops to the order of 2% once placed in the pipeline; the rest in disposed of–released into the atmosphere. Of course, these losses are invisible to the eye and occur chiefly in the pre-delivery supply chain.

Gas is also expensive to transport and is dependent on specialised vessel or dedicated pipelines. Given the enormous capital costs associated with exploiting, transporting and utilising this fuel, it is understandable that suppliers tend to the oligopolistic, if not monopolistic in nature. The cost constraints infrastructure from a degree of natural protection over established markets restricting buyers' supply options.

Exacerbating this natural monopoly characteristic of gas supplies and its associated physical infrastructure are the pricing structures which major oil companies have historically imposed over gas supplies. Gas prices are usually pegged to the price of oil in some way, subjecting it to the same volatility. This factor alone has reduced the attractiveness of gas for many power projects, especially private ventures which depend greatly on tariff stability over long time frames for their success.

Water (Hydro)

Most, hydro potential in Asia, with a few notable exceptions, has already been identified and incorporated into power generation strategic plans. In short, suitable locations are extremely scarce and require detailed economic, environmental and political impact analyses to determine net benefit.

Pollution is not a problem normally associated with hydro schemes, but many other factors can qualify the appropriateness of this energy source for power projects. Distances to major power load centres, damage to existing ecology and devastation of important agricultural industries caused by waterway diversion are always issues.

Uranium (Nuclear)

While still going through an image crisis, nuclear power will continue to struggle to win market share in the short to medium term. Despite the general cleanliness of this fuel in use, the radioactive by-products and the very public accidents and near-misses continue to make the chose of nuclear a "courageous" political decision. If the trend towards political democracy and environmental awareness continues in Asia, it is difficult to see uranium winning the "most preferred fuel" award in, at least, the medium term.

Coal

Of all fuels available in Asia, coal has the greatest potential to meet the requirements of the bulk of future power projects.

Coal is a commodity that has significant market acceptance, is internationally traded and subject to open, competitive market forces whereby price is determined by a true reflection of demand and supply conditions. There is not cartelisation of supply and there are many independent suppliers exporting hundreds of coal brands to Asia from as many as ten diverse countries. Coal is abundant, economic and will be available in large quantities for hundreds of years to come.

This gives power project developers and financiers and great deal of comfort. Not only is the market-driven price self-regulating and less volatile, the broad supply options give reliability and security to the fuel supply and underpin power project viability.

Coal is available in a variety of qualities and can be beneficiated, if necessary, by a variety of methods to meet environmental targets. Clean coal technologies will ensure that coal remains an environmentally acceptable fuel. While CO_2 remains a concern, there is little prospect of fuels which produce lower CO_2 emissions being able to meet the needs of the Asian power industry. As was discussed previously, the perceived advantage of gas in this area is debatable and limited.

Manly countries in Asia have their own coal reserves, but, with the exceptions of Australia, Indonesia, China and India, most countries have either exhausted their indigenous supplies or are only endowed with high sulphur, lignite quality coals. Export quality coals are relatively high in energy content, low in sulphur and low in ash; all in all, economic to transport long distances and environmentally acceptable.

Current Asian demand projections support the assessment that coal will continue to be the major fuel source in Asia for at least the next twenty years.

What is likely to change?

Aside from the huge commitments in terms of quantities of fuel required, how is the market changing? The transfer of power supply risk to private investors and the global movement towards more stringent environmental controls will see buyers looking for much more from all their input suppliers. Developers are increasingly seeking to push as much of the risk as possible back onto the suppliers with the objective of increasing the competitiveness of their power tariffs. In the case of fuel, suppliers are being asked to absorb by offering :

— complete product packaging, including multiple sourcing, transportation, handling and overall fuel management.
— back-up supply arrangement.
— technical input into plant design, handling facilities and operation logistics;
— longer term, large tonnage, contractual arrangements, that more closely support PPA terms with increased certainty and assure financial backers of supply security;
— the formation of closer relationships with projects developers (occasionally to the point of Joint Venture partnerships).

Not all fuel suppliers and positioned to offer such services and certainly even fewer could respond to all of these concerns. Large, multinational companies with diversified deposits and supporting infrastructure are particularly well suited to these changes in demand. Smaller, well organised suppliers could also capture strategic market shares by working in a co-operative manner to compile appropriate packages. In the case of indigenous fuel, new mines will need to be developed to match the large new power projects.

How much Coal will be Required?

World annual demand for seaborne traded steam coal is expected to increase by more than 100% by 2010. Asia's consumption will be even more spectacular at 145%. This translates into annual volumes approximating 480 and 280 million tonnes respectively, per years by the year 2010.

Where will the Coal Come From?

Of all potential supply sources, Australia appears to be best positioned to meet the demand growth anticipated in Asia. Detailed analyses of the world and regional steam coal markets consistently advocate Australia as the most logical and likely supply source due to its surplus-to-domestic demand production capacities (*i.e.*) its natural export orientation), its strategic positioning in the region, and its excess capacity in available (and planned) supporting infrastructure, such as rail and ports.

Only Indonesia and China could match the locational advantage of Australia, but both will be constrained by their own demand requirements within the foreseeable future. The vast majority of other world coal supplies form USA, South Africa, South America, Poland and Russia will be directed to their own consumption and their natural markets within Europe.

India will increase production considerably but will consume all this and possibly need to import some coal for coastal based power stations. Indian demand is forecast to increase by 150-200 Mtpa in the next 10 years. Production would almost certainly need to be boosted by private participation in coal mining to meet this.

What will Encourage Coal to Become Available in such Quantities?

There is absolutely no shortage of coal as a commodity. The main challenge is now to provide an environment which will stimulate the investment in new mines, to expand existing capacity and to extend and improve the infrastructure necessary to meet the substantial growth in demand into the 21st Century.

Over-capacity during the past decade resulted in depressed market prices and a consequent lack of new investment. It would be expected that recent increases in the Japanese Benchmark price will stimulate some mine expansion or new mines, although the majority of coal miners will remain cautious given recent experiences.

Three major factors will influence the coal supply responses:

— stable, long-term pricing,
— secure markets through long-term contracts,
— an increasing emphasis on all quality parameters in pricing.

Major Asian utility coal procurement strategy to date has aimed at securing a balance between two conflicting needs-security of supply and low cost of supply. Despite the oversupply of coal in the eighties, the two countries experiencing significant growth in demand, Japan and Korea, planned their fuel strategies to ensure security of supply. The supply shortage over the past two years has justified their forward thinking in keeping long term coal contracts in place.

Changes are already happening. Recent developments are seeing a move towards the positive attributes of low ash/low sulphur coals appropriately focusing suppliers thoughts on the need in the future to develop and concentrate on environmentally attractive coals to achieve greater returns. Indeed, generally there is a trend in North Asia to demand coal with lower sulphur, and lower ash. The aim now should be to persuade consumers to amend the narrow benchmark view in favour of one that considers the value in use of a particular coal to the customer.

The method of pricing thermal coals is also likely to be influenced by the liberalisation of the electricity market itself. The introduction of competition amongst generators and the appearance and growth of the Independent Power Producer could well see further pressure on the current simplistic Benchmark system. Those financing such projects will seek to reduce risk by every means possible and, in particular, will want to be confident of the long–term security of coal supplies.

Independent Private Power Producers

The privatisation of power generation in developing nations in Asia is a relatively recent development. The problems facing potential investors are formidable in terms of the complexities involved in arriving at long-term agreements. The impact of these long-term agreements on fuel supply strategy is to reinforce the need for close, long-term relationships with fuel suppliers. This may be achieved through long-term contracts, or, in some cases, a coal producing company will itself become a participant in the power project through joint venture arrangements or vice versa.

The advent of IPPs also means that any one utility is likely to focus on a smaller number of suppliers. This is partly the result of coal supplies being identified and guaranteed prior to construction. But it also happens when the power producer and the fuel supplier collaborate in the design of the power station. This can allow power stations to be designed to run on a limited range of fuel specifications with consequent saving in construction costs. The corollary is fuel sources become limited to those who can meet the specifications of the power generator.

The closer relationship between supplier and customer has significant benefits in that it allows fuel sources to be specifically developed for a particular power stations whilst it also gives greater security and opportunities for financing.

In the case of India, which is well endowed with coal resources the captive coal mine concept has been in place for a couple of years. In most instances however, the power projects have not moved ahead for a variety of reasons so it is still too early to gauge the effectiveness of the concept.

What do Suppliers Require of IPP Developers?

Just like power developers, fuel suppliers need to minimise risks associated with their own multi-million dollar investments. Most new power projects (of reasonable size) will require quantities of fuel beyond most suppliers' current production capacities. New investment will be required, and new resource developments planned and constructed. Long-term contracts will be required to give financiers security and comfort and enable producers to plan efficiently. Minimum take arrangements will remain a critical issue.

Sufficient lead time is also required in order to identify and develop the most suitable fuel supply project for a specific power plant. Time to prepare feasibility studies, acquire the necessary approvals and allow pre-production development of mines or other production facilities is crucial. This can all take several years, but is vital to ensure the viability of the supply source and reliability. Power project time tabling and consultation with suppliers is imperative to successful working relationships. Demand for coal in particular will be such that, it is essential for large power projects to make "advance booking".

It will also be important for fuel suppliers to take a more active role in power projects and *vice versa* whereby strategic alliances are formed. It will be increasingly important for suppliers to be constantly informed of all aspects of the power purchase agreements, and the impact of changes in planning and plant performance, to ensure a sympathetic supply response wherever possible.

This seems to be recognised by the Ministry of Power in India recently with the calling for birds for a combined power/coal mine project in Bihar. There will undoubtedly be many issues that this will bring out, but it is certainly moving in the right direction.

Conclusion

The social and economic goals of Asia make the timely and rational development of new fuel sources imperative. Both power producers and fuel suppliers have a common interest in ensuring that the next generation has access to adequate electrical power supplies as efficiently and as clean as possible. Advance planning of fuel supply strategies is crucial in achieving these goals.

Longer term and more comprehensive relationships between buyers and sellers, and a greater understanding of their mutual technical and financial needs, will be the key to maintaining an expanding profitable power and fuel supply industries into the early part of the 21st century.

RURAL ENERGY SUPPLY IN INDIA

18

PRADEEP CHATURVEDI
Secretary General Indian Association
for the Advancement of Science
DII/55, Kaka Nagar,
New Delhi-110 003

ABSTRACT

Rural energy basically refers to fulfilling the needs of millions of people in rural India who have a subsistence living. Large proportions of this population do not have an access to commercial energy. Efforts at the national level have improved the situation, but still a lot more needs to be done. The Paper deals with the constraints and issues essential to be dealt for ensuring energy supply to rural areas.

1.0 Introduction

The economic development in India under recent thrust on market oriented approach has opened avenues for larger involvement of private sector in creating necessary infrastructure. The economic reforms of the past five years have got around a recovery in the growth of production sector and employment and ushered in far-reaching changes in agriculture, industry, finance, capital markets and the tax structure. The Economic Survey 1995-96 published by the Government of India in July 1996 has focused on the major challenges and policy priorities confronting the Indian economy. As the government's approach is to reduce subsidies there is an urgent need to create more jobs in the rural areas and improve rural income generation by creating necessary infrastructure.

Two-thirds of India's labour is employed in agriculture, and it is crucially important to broad-base agricultural growth for raising rural living standards, ensuring basic food security for the nation and strengthening the basic markets for industrial and service sectors. To succeed in this endeavour, public policies for agriculture are being pursued by the government with intense vigour.

Poverty alleviation programmes have received an overall support at various national and international conferences that took place during the year. These include the International Conference on Habitat, the World Solar Summit and the Preparatory Meetings of the World Food Summit. There is a general agreement that there is a need for an overall improvement of the rural sector.

The strategy for energy development forms an integral part of the overall economic development strategy in India. Efficient use of resources and long-term sustainability are two important objectives of the Economic Planning. The concept of sustainability takes account not only of natural resource and ecological balance, but also economic equity and self-reliance. Any strategy for energy planning, therefore, is dovetailed with these broad objectives.

Indian economy continues to be primarily based on agriculture, even though its share in the GDP at constant prices has been declining over time. The share of agriculture sector in the GDP has declined from 39.7 per cent in 1970 to 29.5 per cent in 1993; whereas that of the industries sector has gone up from 17.4 per cent to 22 per cent during the same period. The average rate of growth in GDP, which was about 3.5 per cent up to the 70s, has GDP growth is to improve in the years to come. Overall economic growth of GDP at factor cost, accelerated to 7% in 1995-96. The growth rate was not due to exceptional agricultural growth, the high growth rate was mainly due to the 11.7% growth of industry and construction sector and the 7% growth of the services sector. The overall economic growth during the first four years of the 8th Five Year Plan (1992-1997) averaged 5.9% per annum, a rate significantly higher than the Plan target of 5.6%.

The situation is building up to give prominence to rural sectors in India. Therefore, rural energy is becoming important in India's economy.

2.0 Primary Energy Endowments

Following primary energy recoverable reserves of different sources of commercial energy in India were identified in 1993(3).

(i)	Coal	68,047 mt
(ii)	Crude Oil	757 mt
(iii)	Natural Gas	686 bcum

3.0 Long Term Energy Planning

It has been recognised that the long–term energy strategy should be consistent with provision of high priority to meeting fully the basic energy needs of the rural and the urban poor in the immediate future. It should also ensure gradual shift from non-renewable resources to renewable ones with increasing emphasis on demand management, conservation and efficiency.

4.0 Energy Production

Changing pattern of energy supply in India can be realised from the fact that total energy supplies, including both commercial and non-commercial forms, increased from 85.1 mtoe in 1953-54 to about 240 mtoe in 1990-91. The commercial energy contributed 122 mtoe and non-commercial energy 118 mtoe. Fuelwood accounted for 65 per cent of the total non-commercial energy consumed in the country.

Amongst the indigenously produced primary commercial fuels, the relative share of oil and gas has increased from 1.2 per cent in 1950-51 to 32 per cent in 1990-91; whereas the share of coal has declined from 98 per cent in 1950-51 to 63 per cent in 1990-91.

5.0 Future Energy Demand Projection

The projections for future demand for commercial energy in India are as given below:

Fuel	1991-92	1996-97	2009-10	2020
Electricity Demand (TwH)	231	336	725	1 300 (1745)
Coal (Million Tonnes)	229	311	650	1 345
Petroleum Products (Million Tonnes)	57	81.2	165	335
Natural Gas (Billion cubic metres)	18.6	30.2	65	135

Ref: J.N. Maggo – Planning Commission (Sep. 1996).

6.0 Investment on Energy Supply Sector

The development of commercial forms of energy is highly capital intensive and large investments are required for meeting the rapidly growing demands for these forms of energy from different consuming sectors. The relative share of expenditure on energy sector in the total Plan expenditure during successive Five Year Plans is as shown in the Table below.

TABLE 1: **Plan Expenditure and the Share of Energy Sector**

Plan	Period	Plan Expend (Rs.Cr.)	Share of Energy Sector (%) Power	Oil/Gas	Coal	NRSE	Total
First	1951-56	1 960	–	–	–	–	19.7
Second	1956-61	4 670	9.7	0.8	1.9	–	12.4
Third	1961-66	8 580	14.6	2.6	1.3	–	18.5
Fourth	1969-74	15 780	18.6	1.9	0.7	–	21.2
Fifth	1974-79	39 430	18.7	3.6	2.9	–	25.2
Sixth	1980-85	109 290	19.7	7.8	3.5	0.1	28.1
Seventh	1985-90	220 220	17.4	7.3	3.2	0.3	28.2
Eighth	1992-97	434 100	18.4	5.5	2.4	0.2	26.5

Almost 26 – 28 % of the plan allocations (in different Plans) have gone to energy sector, and two–thirds of this allocation has been for the power sector.

7.0 Rural Energy Planning

Rural energy planning is basically the concern for the minimum domestic cooking energy, which comes mostly from biomass fuels. The data base for energy consumption in the household sector especially for the non-commercial fuels is not comprehensive. This is because non-commercial fuels are mostly gathered outside the market place and the transactions are rarely recorded. Primary data are available from field surveys conducted by the National Sample Survey Organisation (NSSO), National Council of Applied Economic Research (NCAER) and other research institutes and individuals.

The methodology of estimation of household energy demand was first given by the Energy Survey of India Committee (ESIC) in 1965. Based on the results of the sample survey carried out by the agencies mentioned above, the ESIC adopted a norm of 0.39 tonnes of coal replacement (tcr) of energy consumption reuirement per person per year in the urban households and 0.38 tonne of coal replacement of per capita energy consumption per annum in the rural areas. After so determining the total energy requirement in the household sector, the projected contribution of commercial sources of energy was subtracted from it to arrive at the share of non-commercial fuels. The same was then apportioned among fuelwood, agricultural waste and cow dung in the proportion of 65:20:15. This is only to indicate that the available estimates of non-commercial energy consumption in the household sector yield only rough estimates rather than precise estimates. This methodology continued to be adopted till the mid-80s. The Advisory Board on Energy adopted the concept of useful energy requirement per person in the household sector. In the report on 'A Perspective on Energy Demand and Supply in India in 2004-05' published in 19985, the ABE adopted a norm of 650 Kilocalories of useful energy requirement per person per day for cooking and space heating in the year 2004-05. The fuel mix for the year under consideration was then worked out on the basis of population, the rural-urban distribution of population etc. The fuel mix adopted for the exercise was as per the Domestic Fuel Survey of the NCAER for the year 1978-79. Later in 1986, the Energy Demand Screening Group (EDSG) set up by the Planning Commission modified the consumption requirement of useful energy to 520 kilocalories per person per day in the rural areas. Other studies carried out later like the Sectoral Energy Demand in India also estimated the useful energy requirement in 2009-10 as 641 kilocalories per person per day. In a number of studies conducted by IAAS, it was observed that the rural domestic energy consumption in the final form varied from 420 kilocalories to 620 kilocalories per capita per day, depending on the affluence of the family.

The Sixth Five Year Plan (1980-85) recognised fuelwood as the most important non-commercial fuel which was becoming increasingly scarce and also getting commercialised. Energy forestry was also recognised as an important component of the energy strategy as it was felt that the rural communities will continue to depend on firewood for several decades to come. A large programme of fuel and farm forestry was taken up in the Plan. Also, an Integrated Rural Energy Programme (IREP) was initiated in the Sixth Plan on a pilot scale with an objective of providing a mix of energy options for meeting the energy needs of the rural areas on an area-wise basis in the most cost-effective manner. To implement this programme, the Planning Commission set up a separate Rural Energy Division. A Rural Energy Planning Exercise was commissioned by the Planning Commission to develop the design and approach of planning and implementing area-based block-level integrated rural energy projects.

The Advisory Board on Energy brought out a report 'Towards a Perspective on Energy Demand and Supply in India in 2004-05' in 1985. The main conclusions of the report pertaining to supply options, analysed Statewise, in the case of firewood were that at-least 60 million hectares of land (comprising open/degraded forest land, barren and unculturable land, culturable wasteland, land other than current fallows and farm perimeter land) might be available for raising fuelwood/ fodder plantations. Given the acute pressure on land and the other potential uses it could be put to, not more than half of the land area thus identified could actually be used for plantation purposes. The study brought out that the productivity of the Indian forest land is quite low and need to be increased.

The Government of India also set up a Wastelands Development Board in 1981 to explore the possibility of greening wastelands for energy plantations. Though 92 mha are reported available, only about 20 mha can be economically exploited. Expert Consultations, including those under

FAO and IAAS, have indicated that energy plantations with long-term yields of 4 to 5 tonnes per ha per year can be certainly grown on various wastelands and village lands.

Yields of upto 40 tonnes per ha have been recorded in test fields. Subsequent policy decisions to separate community based/farm forestry energy plantations from large forest lands has been found a good option. Various studies in places like Haryana, Kerala and Western Uttar Pradesh have indicated that the local availability of fuelwood from village woodlots has resulted in stabilising the local market price of fuelwood.

7.1 Rural Electrification Programme

The First Five Year Plan considered agricultural pump energisation as an important thrust area. It was initially envisaged to provide electricity as a social amenity to rural areas and was confined to a few States. Subsequently, it was extended to cover all the States. The importance of this programme was especially recognised during the drought in mid-60s when lift irrigation had to be resorted on a large scale to several subsistence crops. This programme was subsequently integrated with Minimum Needs Programme for the rural poor.

The Programme was further strengthened by the formation of Rural Electrification Corporation in 1969, which now provides over 90% of the funds to the State Electricity Boards for rural electrification, as concessional loan. In addition to energisation of pump sets, emphasis was also laid on promotion of industrial load in villages by extending electricity connections for setting up low tension cottage/home based industries.

The present thrust under the Programme is as follows:

(i) Improvement in the system of power distribution in the electrified villages to ensure quality and reliability of supply.

(ii) Load development activities in electrified villages to promote the use of electricity for productive purposes other than irrigation, so as to provide wider opportunities for the rural communities for gainful employment and to improve their incomes. In this context rural and agro-based industries will be specifically promoted in coordination with the rural electrification programme. Presently growth of rural industries average around one small industry per village in backward areas and about two per village in advanced areas with power capacity ranging from 5 H.P. to 25 H.P. Each industry employs around 1.45 persons on an average.

(iii) Development of co-operatives, preferably operated by the panchayats and other local bodies for distribution of electricity in the rural areas so that there is an improvement in collection of dues, reduction in thefts of electricity and T&D losses.

7.2 Integrated Rural Energy Programme

Besides the rural energisation programme, a new approach of integrated rural energy systems was also promoted. The Planning Commission initially started an Integrated Rural Energy Planning Programme (IREPP) which was taken up as a plan scheme in the 7th Five Year Plan.

Most of the energy consumed in rural areas does not enter the organised market place, and therefore, there is no accurate data on patterns of supply and consumption of energy in the rural areas. These patterns also often vary from one agro-climatic region to another. Therefore, a need was felt to understand patterns of energy consumption at the micro-level, through decentralised energy planning exercises, in order to provide sustainable and affordable supply of energy sources for meeting the growing energy needs of the rural population. While locally available renewable energy sources based on non-conventional technologies have been projected to play an increasingly important role in the coming years in meeting rural energy needs, contribution of commercial energy for productive and income generating activities was also realised.

The Central Plan component was utilised for setting up the institutional mechanisms in the State/UT through funds for professional and support staff in the IREP cells at the State level and in the selected districts/blocks, as well as for the training. The State component of IREP funds was utilised for project preparation and implementation and financial incentives for the promotion of rural energy technologies, as part of the block level IREP projects.

The training component involved national, regional, and State level academic professional institutions, besides technical colleges/universities which were provided technical and financial support for conducting training courses on various aspects of the IREP programme. The setting up of one national and four regional training-cum-R&D centres was also taken up under the programme in the Seventh Plan. The first national level training-cum-R&D centre (Centre for Integrated Rural Energy Planning, CIREP) was set up in Bakoli village in Delhi in cooperation with Delhi Administration and with the technical and financial support of the Planning Commission under the centrally sponsored scheme for IREP. The other four regional centres are located in Lucknow (UP), Bangalore (Karnataka), Kheda (Gujarat), and Shillong institutions, State level technical back-up units have been set up in selected State level institutions and for implementation of the IREP programme. Also, district level IREP technical back-up units have been set up through ITI/ Polytechnics in selected IREP blocks to provide technical assistance especially in the selection, operation and maintenance and follow-up on applied R&D on rural energy technologies in the IREP projects.

8.0 Consumption of Cooking Energy and Expenditure

A comparison on the basis of dietary energy intake alone may disorient the projection. Comparison has also been made on the basis of total expenditure on the coked foodstuff, i.e., on the dietary energy intake and on the cooking energy source. It has been observed that in the tribal, desert and the hilly region, of coastal area and plains it is dung cakes and LPG. Reasons are simple that in cases where the people are poor or remote from the urban centre LPG supply is not feasible. On the other hand in case of Singhu it is a part of a large metropolitan city of Delhi and there LPG is abundantly available and the income levels are also such that LPG can be afforded. Another observation is that the firewood in the metropolitan city cost Rs 1.50 per kg and, therefore, use of firewood is non-existent and only dung cakes are being used.

At present there is no commercial value attached to firewood taken out of forests by unauthorised tree felling; and also no commercial value of dung cakes as the cow dung is freely available and the labour for making cakes out of it is from the family itself. Therefore, there is no effect of cooking energy cost in the tribal, desert and hilly villages but there is a small component of 14 paise per capita per day alone in case of Singhu village. If actual price of the cooking energy source was taken then it will be observed that in the tribal village of Peepalwas cooking energy costs Rs 1.22 per capita per day; in desert village of Madni the cooking energy costs Rs 0.79 per capita per day; in case of hilly village of Niglat the cooking energy costs Rs 1.10; and in case of plains village of

Singhu the cooking energy cost is Rs 1.24 per capita per day. Refer Table 5.

TABLE 2: Consumption of Cooking Energy and Expenditure

	Site	Cooking		Lighting		Total Expenditure ture (Rs)
		Energy (Kcal)	Expenditure (Rs)	Energy (Kcal)	Expenditure (Rs)	
1.	Singhu, Delhi	598	1.24 (0.14)	391	0.45	1.69 (0.59)
2.	Niglat, Nainital	523	1.10 (-)	189	0.13	1.23 (0.13)
3.	Madni, Sikar	373	0.79 (-)	107	0.09	0.88 (0.79)
4.	Peepalwas, Udaipur	581	1.22 (-)	266	0.09	1.31 (0.09)
5.	Rajpuriya, Dhar	418	0.58 (-)	292	0.29	0.87 (0.29)
6.	Pallippuram, Ernakulam	470	0.94 (0.19)	293	0.33	1.27 (0.52)

Cooking Energy Sources used are firewood and dung cake only in case of 2, 3 & 4; and dung cake and LPG is used in case of 1. Users in 1 pay only for LPG. Figures in brackets indicate actual paid and others represent actual market price of source use.

9.0 The Future Course

The future course of rural energy consumption will depend on government's policy options. Under present conditions, the rural India will see emergence of rural based industrial units and increasing contribution to nation's economy.

9.1 Rural Industry

Benefit of industrialisation can be distributed to the rural areas by setting up or promoting small scale but relevant industries. Apart from basic infrastructure elements such as energy and transport, the necessary technological infrastructure must be created in terms of entrepreneurial capability, growth of skills and the like. Despite its complexity greater industrial dispersal may have to be undertaken leading to greater income and employment in the rural communities.

Nature and pattern of rural industrial activity could cover a wide range, including the following:

(i) Repairs and servicing of transport and agricultural equipment;
(ii) Establishment of low capital and labour intensive ancillary industries;
(iii) Setting up processing units for both perishable and non-perishable local produce;
(iv) Production of simple agricultural equipment, such as farm implements and pumps;
(v) Manufacture of products to meet local needs, such as building materials, cloth and shoes.

9.2 Rural Energy Supply Options

Realising that the conventional energy supply systems that have evolved in conjunction with large scale, highly concentrated production and distribution units have often proved to be inadequate for the needs of the rural population in India. They have often failed to yield adequate electrical power in relation to the micro-economic needs; and have usually not met the total specialised demands of agriculture or of rural and small scale industries.

When properly used, small and perhaps intermittent amounts of energy can be of critical value to the rural economy. The energy planning for rural areas development of self sufficient and self-supporting local systems. In some cases, such a programme might even involve suitable phasing and postponement of centralised rural electrification schemes in order to release more resources to implement more quickly maturing decentralised schemes.

The Planning Commission of India has, for the first time, considered the possibility of introducing contribution of fuel wood in the policy statement. The Working Group for the IX Five Year Plan 1997-2002) includes the same.

9.2.1 Objective

IAAS and other Expert Groups have realised that the objective of Rural Energy Planning in India, under present trend of economic growth, should be:

(i) To increase the energy availability to the rural areas so as to reduce human drudgery, especially of women, such as the long hours and physical effort required for gathering fuel, fetching water and carrying heavy loads;

(ii) To give spur to the economically productive activities necessary for rural development; and

(iii) To create an organisational framework to initiate, introduce, implement and maintain conventional and new energy technologies.

The Strategy to meet rural energy needs for the rural and remote area communities should have the following components:

(i) Harmony between planning for rural energy needs and ongoing activities of rural development agencies.

(ii) Planning for energy needs in an integrated fashion so that the community needs and the income generation needs are met in a balanced fashion; and implementing programme and projects accordingly.

(iii) Planning to meet energy needs of a cluster of villages, to achieve cost effectiveness; and implementing them through community based infrastructure.

9.2.3 Benefits

Benefits from the outcome of such a strategy are expected to be as follows:

(i) Solution to employment generation, as more self-employment programmes will be supported in rural areas.

(ii) Rural urban migration to be restricted as employment opportunities in the rural areas will be available for larger numbers.

(iii) Sustainable development of the area possible as most of the locally available resources will be utilised with total participation of local communities.

(iv) Higher incomes due to value addition at source and thereby better prosperity for the people of the area.

10.0 Conclusion

The situation in India is no longer of complacency. A decision has been taken to allow smooth transition to urban-rural continuum, as projected at Habitat II. Whereas rural urban migration will not stop, wealth from urban areas will be transferred to the rural areas through government and private agencies and individuals. But their sustenance will depend on affordable core infrastructure, including energy. Energy supply in future in the rural areas in India will be dictated by market forces, that are evolving now. The booming small scale industry sector and the service sector will dictate rural energy mixes and rural energy supply options. Government will not be the only actor to fulfil these needs, and be overburdened by social needs of rural areas. The economic growth will by itself set a pattern for fulfilling rural domestic energy needs. It would not be a surprise if the domestic energy needs in the rural areas match or surpass the urban area needs. That may take time, but the process has already set in motion.

A shift towards liquid fuels and electricity is unavoidable. That does not necessarily mean supply though commercial, centralised sources of energy. That will, in all possibility, come through environment friendly small scale bioenergy based conversion devices, operated on a large scale by young entrepreneurs in rural areas.

The future in India will see a paradigm shift in energy consumption patterns. Commercial sources will have a larger role to play in rural areas because of transfer of wealth and prosperity and the urban areas will see higher consumption of fuelwood based conversion devices, obviously more efficient than now.

ENERGY SECTOR DEVELOPMENTS IN SRI LANKA

19

G.B. A. FERNANDO
Secretary, Sri Lanka National Committee of WEC

1. Introduction

Sri Lanka has a population of 18 million in an area of 65,610 sq. kms. with a per capita GNP (Current) (1994) of US $652. About 76% of the population live in the rural areas where their main activity is agriculture. There are two monsoons. The South West monsoon from May to September and the North-East Monsoon from December to February. The annual rainfall varies from below 100 cms. in the low lands to 500 cms. at certain places the north-western slopes in the hills. The relative humidity varies generally from about 70% during day to about 90-95% at night.

The present composition of energy sources available to meet the energy needs of Sri Lanka are mainly:

Fuelwood and biomass	(70%)
Petroleum Products	(20%) and
Hydro electric resources	(10%)

Annual consumption of fuelwood and agricultural residues amounts to around ten million metric tonnes. The bulk of fuelwood (about 85%) is consumed in the domestic sector at a low efficiency, whilst the remainder is used in the industrial sector. Energy efficient wood stoves were developed and promoted in order to popularise fuelwood conservation. A Forestry Sector Development Programme is presently underway in order to arrest the depletion of the forest cover and meet the industrial timber and domestic energy requirements.

Energy conservation programmes successfully initiated in the oil refinery in 1974, were persued in several other large energy comsuming industries. These efforts reduced industrial consumption of petroleum products from 20% in 1972 to 14% in 1990. With the shift of Kerosene to fuelwood and LPG for cooking, domestic consumption of petroleum products reduced from 30% in 1972 to 12% in 1990. Almost two-thirds of petroleum product consumption is in the transport sector. A Road Development Authority and a Transport Studies and Planning Centre have been established in order to carry out studies and implement programmes to increase the efficiency of use of petroleum products in the transport sector by easing traffic conjestion, improving road maintenence and traffic flow within cities and plan intercity highways. Proper energy utilization in the transport sector will considerably increase efficiency of imported energy use and reduce adverse impact on the environment. Domestic sales of petroleum products in 1995 amounted to around 2 million metric tonnes.

Hydro electric sources have been providing the main source of power generation in the past. During the period 1972-1990, almost 90% of power generation has been from hydro electricity. In 1995, the share of hydro electricity in power generation has been around 94%. During periods of drought, recently experienced in 1992 and again in early 1996, the reduced contribution from hydro electricity have lead to power shortages due to an insufficient thermal back-up. The Ceylon Electricity Board Generating system has an installed capacity of 1409 MW of which hydro is 1137 MW and Thermal 272 MW connected to the national grid which operates at 220 KV, 132 KV and 66 KV. The Maximum demand recorded was 679 MW. It is expected that diesel, gas turbine and combined cycle systems amounting to a total capacity of 530 MW would be installed during 1998-2000 in order to avoid the expected power shortages in the near term. Per capita electricity consumption in 1995 was 216 KWh. As a result of a major rural electrification programme, a large number of villages received electricity resulting in around 45% of households in Sri Lanka having access to electricity at present.

Around 80% of National Energy supply is contributed from New and Renewable sources of Energy. The main contribution is from biomass and major hydro electric resources. Attempts have been made to promote photovoltaics for rural lighting, wind energy for lift irrigation, agricultural residues for industrial heating, biogas generators for domestic use; and solar water heaters for industry, hotels and domestic use; but their contributions have still remained small. The technology of solar water heaters and biogas generators have reached a stage where its application can be used with advantage. A large number of small hydro power plants are now being developed by the private sector although their total contribution still remains small when compared to conventional hydropower development.

2. Prospects in Energy Development

2.1 Biomass Sector

A National fuelwood shortage is expected by the year 2001. The presently available supply of fuelwood could hardly be expected to increase. The forestry master plan has been developed to ensure environmentally sound practices to meet the increasing demand for biomass for timber and wood fuel needs. The increase in fuelwood demand should be met by improving fuelwood efficient stoves.The non-forest sector supplies 80% of the fuelwood requirements and 50% of the timber requirements of the country and 67% of the fuelwood requirements in the rural areas of the country. Special attention should be given to increasing the non-forest vegetation in the country. This would help the country safeguard its natural forests. A rapid increase in use of LPG for domestic cooking has been observed among the more affluent households, in replacement of fuelwood. Nevertheless, tree planting to provide the fuelwood requirements of the development projects should be made mandatory. Reforestation programmes country wide to increase biomass supply should be stepped up. Agroforestry and homestead forestry should be increased to make people self sufficient in their cooking needs. Lands should be parcelled on a larger scale for management of energy plantations. Marginal lands should be put under tree cover. Scrubland may be used to grow trees for dendropower.

There is scope for recycling Unban. Municipal and Agricultural wastes, through recycling plants, and biogas generators. Although many attempts have been made to recycle wastes, no satisfactory programme has yet been evolved. Collection and disposal of garbage has to be planned for a complete cycle with the cooperation of all concerned for maintaining a clean environment.

2.2 Petroleum Sector

The present demand for petroleum products is met by retining imported crude oil at the Ceylon Petroleum Corporation Refinery, whilst the balance requirements are imported. The refining capacity has recently been increased. Domestic sales of petroleum products in 1995 amounted to around 2 million metric tonnes. Programmes to improve the mass transit system and improve the road and rail network, in order to reduce traffic conjestion and meet increasing demand for transport are being planned. The demand for petroleum products would rapidly increase to meet the increasing demand for industry, transport, consequent expansion of the road netword and future rail network, power generation and energy for domestic cooking in a clean environment.

Petroleum products are expected to be available at current prices even by the year 2020. The availability of oil reserves are expected to deplete beyond the year 2040.

2.3 Power Sector

The increase in demand for electricity for the predominantly hydro electricity system has to be met by the addition of thermal power plants run on coal and oil. The development of available hydropower resources to meet the increasing demand for power generation is limited. The private sector is expected to make a significant contribution to power generation projects on BOO/BOT basis. A large number of small hydro power plants are being developed to supply electricity to the national grid and their own use. The transmission and distribution network is being improved to meet the increasing demand for electricity.

The Generation Expansion Planning Studies of the Ceylon Electricity Board indicate that, for the base case:

1997-2003 Oil based power plants with capacity totalling 702 MW, coal based power plants with capacity totalling 300MW, and hydro based power plants with capacity totalling 70 MW, need be commissioned. The capacity of a single plant will not exceed 150 MW. The addition installed capacity (including refurbishing) in this period amounts to 1072 MW.

2004-2010 Oil based power plants with capacity totalling 360 MW and coal based power plants with a capacity totalling 1200 MW need be commissioned. The capacity of a single plant will not exceed 300 MW. The additional installed capacity (including refurbishing) in this period amounts to 1560 MW.

The use of LPG for power generation dujring this period has not been discussed. LPG and LNG is increasingly being used for power generation in many countries. LNG could be used with the availability of gas pipelines. In the absence of gas pipelines, LPG could be shipped for power generation, if it is found to be economically feasible. Nuclear power plants can be introduced only when a single plant of capacity 600 MW can be accomodated in the grid. This may be possible around the year 2015, if the problems of storage of radio active wastes and safety are resolved.

2.4 New and Renewable sources of Energy

The Government has supported measures for the development and utilization of new and renewable energy over many decades. These include solar (photovoltaic and thermal), wind and biomass energy sources; biogas and waste (urban, industrial and agricultural) recycling; wave and ocean energy. There is much scope in the promotion of solar water heaters and biogas generators, if the proper technical backing is established.

2.5 Impacts on the Environment

With increase in use of energy for development, a certain amount of environmental degradation takes place. Environmental Impact Assessments are now being requested when proposals are being appraised in the energy sector. A National Action Plan on Climate Change is under preparation in Sri Lanka.

3. Sri Lanka Energy Vision 2020

3.1 Biomass Sector

The Biomass sector would have settled to a dynamic equilibrium by year 2010 with the correct steps being taken as described in Section 2.1. It will be difficult to regenerate additional biomass resources more than what could be expected by year 2010. LPG will be widely used to meet domestic cooking and industrial heating requirements that cannot be met by fuelwood.

3.2 Petroleum Sector

Although Petroleum prices are expected to be stable, the horizon for depletion of oil resources could be seen. Unless development of alternative sources of energy to meet basic energy needs have been harnessed in an environmentally sustainable manner, which is also economically acceptable to the average citizen, there will be a run on the available petroleum resources. The technology of coal liquifaction as an energy source should find wide application. Efficient mass transit systems would be essential to solve the increasing needs of transport. There will be a wide use of telecommunication systems in order conduct those activities that do not neccessiate travel. The use of private motor vehicles would decrease.

3.3 Power Sector

The demand for electricity would continue to rise at 9% per annum or more from 2010 onwards to meet the ever increasing needs of society. There would be a wider use of electricity for telecommunication purposes because telecommunication would be seen as the cheapest energy cost of communication.

The projected demand for electricity would be as follows:

Year	Generation (GWh)	Peak Demand (MW)
2010	17096	3365
2015	26304	5177
2020	40613	7965

In order to meet an aditional generation of 23517 GWh of electricity over the 2010 demand, the equivalent of 12 units of 300 MW Coal Power stations need be installed between years 2010 and 2020.

Nuclear Power plants, if feasible, can be installed between 2015 and 2020. In this case, University undergraduate courses in nuclear power technology should be available as an optional course for the better students by the year 2000.

If NRSE is sufficiently developed, a part of this demand could be met by NRSE, ESCAP ENERGY NEWS (Vol xii. No 1, July 1995) reports that remarkable technological advancements made in certain renewable technologies during the past decade indicate that an energy future making intensive use of NRSE is technically feasible. But the transition to NRSE has been progressing slower that envisaged especially in the context of developing countries. Inefficient outreach activities due to inadequate manpower capabilities have been widely considered as one of the major reasons for this slow progress in NRSE diffusion. A longer term sustainable development effort through training, capacity building and education at the grass-roots level needs to be encouraged to overcome these difficulties. The use of Solar water heaters in almost every household would be a reality.

The use of LPG and LNG for power generation and domestic heating as a clean source of energy could become a reality if gas pipelines are established. These sources of energy would still be available in the year 2020.

If population stabilises around 25 million by year 2020, the per capita electricity consumption in year 2020 would be 1625 KWh which is less than the per capita annual electricity consumption in most of the developed countries of today.

3.4 Impacts on the Environment

It would be useful to recollect the statement made by Dr. Henrik Ager-Hanssen, Chairman, WEC

The future global energy demand in Giga tonnes of Oil Epuivalent (Gtoe) in year 2020 would be

13.3 for the Reference Case (updated WEC assessment for the 14th Congress at Montreal).

11.2 for the Ecologically Droven Case which assumes the same economic growth as the Reference case but with a sharper improvement in energy efficiency.

17.2 for the Enhanced Economic Development Case which assumes a somewhat hegher rate of economic growth in the developing countries than the Reference Case but with somewhat reduced energy efficiency.

as compared to

8.7 for the global energy demand in 1990.'

In order to reduce Global Energy Demand, Developed countries should contribute to the Developing countries the transfer of technology on energy efficiency.

This conference is to discuss the Asian Energy Vision in year 2020. In this context, I wish to make a few recollections in the past. I recall the Late Dr. Homi J. Bhabha, the Asian Nuclear Visionary, address the then Ceylon Association for the Advancement of Science as the Chief Guest in 1957 in Colombo. I was then an undergraduate at the university of Ceylon. He was convinced that Energy from Nuclear Fusion will have commercial application by the year 1966. We are now 30 years past that date and still there are no signs yet of stable magnetic bottle to sustain the nuclear reaction. I remember the Late Mr. Rajiv Gandhi as an undergraduate at cambridge university attending lectures at the Cavendish Laboratory when I was an affiliated student at that University in the early 1960's. He was known as Nehru's grandson because the Late Pundit Nehru was then the Prime Minister of India. Then, in 1966, Prof. Abdus Salam, the Nobel Laureate, on return from a visit to New Delhi scribbled on the Notice board at the International Center for Theoretical Physics in Trieste, Italy (I was then a student of Prof. Salam on the Physics of the Elementary Particles) :

'Heard in the streets of New Delhi:
The first Prime Minister was a VISIONARY
The Second, a RE-VISIONARY
The third, a TELE-VISIONARY'

I recall the Late Mrs. Indira Gandhi inaugurating The First Meeting of the Commonwealth Regional Consultative Group on Energy at the Parliament House Annexe in October 1978 in New Delhi; and her Address at the United Nations Conference on New and Renewable Sources at Nairobi in August, 1981. There has been much hope for mankind at these conferences to resolve future energy needs in the Globe in a sustainable manner and many resolutions and actions have been traken at these conferences. Much has been achieved and still much remains to be done. I should recall the efforts made by the Late Mr. Maheswar Dayal, the able advisor to the Late Mrs. Indira Gandhi, contributing much to these Programmes. Incidentally, his brother, Nareshwar Dayal is presently the High Commissioner for India in Sri Lanka.

In conclusion. I wish to repeat a statement made by Ms. Menka Gandhi, at the 15th WEC Congress in Madrid in September 1992 during her address on 'Energy Horizons in a world of 9 billion Inhabitants'– the statement issued by the conference on 'Global Warming and Climate Change: Perspectives for the Developing Countries' held here in New Delhi in February 1989:

> The developing countries' contribution in response to the greenhouse challenge should be carried out in a way that enhances, rather than diminishes, development prospects. Where these are in conflict, priority should be given to development, which brings so many clear and needed benefits. particularly for the poorest 60% of the population in developing countries. Only in this way can these populations be brought to the minimum level of health and resilience needed to cope with environmental stress and stabilise population sezes.
> When resources are inadequate for mounting programmes needed both for development of the poor and for achieving globally a desiarable reduction of greenhouse gas emissions, developed countries should be asked to contribute the difference. Climate protection should be seen as a challenge to meet in partnership with the development assistance community and the industrial countries and not simply as another problem for the developing world. Having caused the major share of the problem and possessing the resources to do something about it, the industrial countries have a special responsibility to assist the developing countries in finding and financing appropriate responses.'
> Thank you.

4. References

1. Ceylon Electricity Board. Statistical Digest 1995.
2. Ceylon Electricity Board. Generation Expansion Planning Studies 1995.
3. Mr. S. Sivasundaram, Ceylon Petroleum Corporation. Personal Communication.
4. Sectoral Energy Demand in Sri Lanka. United Nations 1992. ST/ESCAP/1193.
5. Forestry Master Plan for Sri Lanka Sept. 1986. (World Bank, FINNIDA & GOSL)
6. Towards an Energy Policy in Sri Lanka. T.L. Sankar & G.B.A. Fernando (UNDP, GOSL, 1978) DP/SRL/74/065.
7. The feasibility of a Rural Energy Centre for Sri Lanka. Jack Allison. Energy Engineering Laboratory. Oklahoma State University 1976 (UNEP/GOSL).
8. Kyi Lwin and S.J. Torok. Hope for longer term sustainable development and utilization of NRSE in Asia and the Pacific via training, capacity building and educational efforts. ESCAP ENERGY NEWS Vol. XII. No. 1, July 1995.
9. Energy for Tomorrow's World. World Energy Council 1993.
10. Dr. Henrik Ager-Hanssen, Chairman WEC Commission. 15th WEC Congress, 1992, Madrid.
11. Ms. Menaka Gandhi. Energy Horizons in a World of 9 billion Inhabitants. WEC 15th Congress, 1992 Madrid.

CONCLUSIONS AND RECOMMENDATIONS NEW DELHI DECLARATION — 1996 20

SUSTAINABLE ENERGY SUPPLY AND ASIAN DEVELOPMENT

SUMMARY CONCLUSIONS AND RECOMMENDATIONS

Recognising the plight of 2 billion people without adequate energy supplies, and unable to meet their basic needs or begin economic development – 800 million of them in the South Asia Region – this Declaration calls for urgent action focussed on:

1. **Tackling rural poverty and energy deprivation through the installation of decentralised, locally appropriate, and affordable renewable energy systems.**
2. **Raising the efficiency of energy provision and use at every point – from modern, centralised power generation systems to end uses – and to provide the necessary support for more environmentally attractive options.**
3. **Addressing, much more vigorously than at present, the problem of local and regional pollution – from the use of fuelwood through the cleaner fossil fuel provision and reduced emissions from road transport.**
4. **Building up the necessary infrastructure and attract the required investment and its financing – which requires in turn realism about economic decision-making and how investible funds are successfully mobilised – domestically and internationally.**
5. **While recognising the inevitable realities for the poor of inability to pay, which requires innovating financing schemes, phasing out subsidies and eventually to price energy at a level which covers all costs.**
6. **Liberalising energy markets and privatise energy institutions in order that efficiency, availability and investment can be increased and the benefits of competition and market signals realised.**
7. **Raising the level and effectiveness of Research, Development and Demonstration expenditures.**
8. **Reinforcing the effectiveness of international co-operation, particularly at the Regional level.**
9. **Requiring the industrialised countries to fulfil their agreed commitments to contribute towards technology and financial transfers, to meet their agreed environmental aims, and to demonstrate their wider obligations to humanity.**

NEW DELHI DECLARATION - 1996
SUSTAINABLE ENERGY SUPPLY AND ASIAN DEVELOPMENT

This New Delhi Declaration forms the Conclusions and Recommendations of the International Conference: Asia Energy Vision – 2020, held in New Delhi 15-17 November, 1996. The Conference fully confirmed the World Energy Council's focus in its "Message for 1997" on the plight of 2 billion people in the world now without adequate energy – 800 million of them in the South Asia Region – and the challenge of meeting future requirements in a world where population is likely to double over the next century.

The challenges of inadequate water and food supplies are closely linked to those of reliance on fuelwood and agro wastes. Failure to meet them results in comprehensive deprivation. More than half these deprived people live in the South Asia Region.

Policymakers have, as their prime duty, to create the necessary conditions for social development and economic output while conserving essential environmental features. This is the essence of sustainable development. For sustainable development to occur, a clear strategy is required. The widest possible range of economic and publicly acceptable options needs to be kept open – including accelerating renewable energy development and maintaining the nuclear option, as well as recognising fossil fuel dependency. That strategy needs to be implemented without further delay. The necessary actions must start now. The lead times to achieve major change are long. Policies need to be expressed with clear timetables and action programmes. The appropriate actions will recognise current needs and realities, local and regional priorities, and what support can reaslistically be expected from elsewhere – from the industrialised countries and international organisations.

Rural poverty, and the need to create conditions for rural economic and social development, is the first need to be addressed. The necessary response is the installation of decentralised, locally appropriate and affordable renewable energy systems. This must be the **prime goal** in South Asia.

However energy is supplied, it should be conserved and used efficiently. Where energy is imported, the impacts on the balance of payments can be greatly reduced. The scope for improving energy efficiency is huge. Without efficiency improvements the supply of energy – whatever its source – is more difficult and costly. It is generally cheaper to invest in raising energy efficiency than in adding to capacity. For providing energy to large centres of population and large industrial complexes – which need to compete domestically and internationally — efficient and environmentally acceptable centralised systems of power generation and distribution are required. The **second goal** for energy policy in South Asia is to raise energy efficiency.

Pollution is a severe problem in South Asia. Energy provision and use is a major source of this pollution. In homes using fuelwood, agro wastes, and inefficient cooking stoves air quality is often foul and respiratory diseases and death an all too frequent result. The use of coal without adequate application of clean coal technology is another important source of pollution. The road transport sector is another, rapidly growing, source of pollution. The **third goal** is to address, much more vigorously than at present, the problems of local and regional pollution in South Asia by a combination of improved efficiency and better technology.

Many of the energy supply problems in South Asia – which inevitaby are reflected in demand side issues – flow from lack of infrastructure. The building up of the necessary infrastructure requires investment, which in turn means financing and with adequate rewards going to investors. This can only be achieved by a combination of mobilising domestic savings, supported by the necessary local institutions, and attracting inward foreign investment on acceptable terms. There are many competing opportunities for international investment. South Asia needs to create the conditions which make the Region a relatively attractive area to invest in. This means both adequate returns to foreign and domestic investors, and avoidance of undue procedural obstacles, and institutional barriers. The **fourth goal** for South Asia, therefore, is to strengthen the necessary infrastructure and attract the investment and its financing. This will require realism about economic decision-making and creating the conditions for attracting foreign investment. Improved public education and awareness will raise understanding of economic realities and receptivity to technological innovation.

Much of what has been stated in this Declaration means trying to kick-start sustainable energy development when, on the one hand, huge numbers of people in South Asia cannot afford commercial energy; while on the other hand, they will not obtain commercial energy unless they pay a price which covers the costs of provision. For as long as the electricity tariff in rural India is little more than 3% of total costs of supply, and little more than 25% of fuel costs alone, supplies will remain severely inadequate; the necessary infrastructural investments will not take place; and the technology and finance desired from other countries and international organisations will not be forthcoming. Innovative financing mechanisms will be required in many cases to help bridge the gap between the present situation and the introduction of full market prices. The **fifth goal** for South Asia, therefore, is to phase out energy subsidies and eventually to move to pricing energy at a level which covers all costs – including environmental costs.

Institutional changes are required to allow competition to work more effectively and efficiency to be increased. Liberalisation of policy and privatisation are required, with the retention of the necessary framework to ensure effective competition, conformation to legislative requirements, and the setting of tariffs which are mutually acceptable. There is a growing number of examples in South Asia of how industrial performance can be improved by new, commercial management – reflecting the manifold professional skills of the people of South Asia. These professional skills include engineering, financial, economic, and entrepreneurial skills as well as those of modern management. The availability of these skills is one of the necessary conditions for the future energy development of the Region. The **sixth goal** for the South Asia Region, therefore, is the liberalisation and privatisation of the whole energy supply sector in order that efficiency, availability, and investment can be improved.

Technological innovation and diffusion, together with supportive institutional changes, will have a powerful role to play in promoting sustainable energy provision and use. Underlying these developments will need to be effective Research, Development and Demonstration expenditures. The **seventh goal** for the South Asia region is, therefore, to raise the level and effectiveness of R&D expenditure.

Many of the challenges outlined above will be more sucessfully addressed within the context of International, including regional, co-operation. The **eighth goal** for South Asia is to reinforce the effectiveness of regional co-operation in particular.

Finally, the New Delhi Declaration notices that the industrialised countries have failed, and continue to fail, to fulfil the international commitments they have freely entered into to provide financing and technology transfers, and to meet the environmental goals they aim at and the wider responsibilities they acknowledge. The **ninth goal** for the South Asia Region is to call successfully upon the industrialised countries to meet their agreed obligations with respect to the Region.

NOTE: Asia Energy Vision - 2020 was held under the auspices of the Indian Member Committee of the World Energy Council (The Institution of Engineers — India). The World Energy Council is a multi-energy, non-governmental organisation with Member Committees in 98 countries.